Ampère's Electrodynamics

Analysis of the Meaning and Evolution of Ampère's Force between Current Elements, together with a Complete Translation of his Masterpiece:

Theory of Electrodynamic Phenomena, Uniquely Deduced from Experience

A. K. T. Assis
and
J. P. M. C. Chaib

Apeiron
Montreal

Published by C. Roy Keys Inc.
4405, rue St-Dominique
Montreal, Quebec H2W 2B2 Canada
http://redshift.vif.com

First Published 2015

National Library of Canada Cataloguing in Publication

Assis, André Koch Torres, 1962-
[Eletrodinâmica de Ampère. English]
Ampère's electrodynamics : analysis of the meaning and evolution of Ampère's force between current elements, together with a complete translation of his masterpiece : Theory of electrodynamic phenomena, uniquely deduced from experience / A.K.T. Assis and J.P.M.C. Chaib.

Translation of: Eletrodinâmica de Ampère.
Includes translation of Théorie des phénomènes électro-dynamiques, uniquement déduite de l'expérience.
Includes bibliographical references.
Issued in print and electronic formats.
ISBN 978-1-987980-03-5 (paperback).--ISBN 978-1-987980-04-2 (pdf)

1. Ampère, André-Marie, 1775-1836. 2. Electrodynamics--History. 3. Electric currents--History. I. Chaib, J. P. M. C., 1978-, author. II. Ampère, André Marie, 1775-1836. Théorie des phénomènes électro-dynamiques, uniquement déduite de l'expérience. English. III. Title. IV. Title: Eletrodinâmica de Ampère. English.

QC630.5.A8713 2015 537.609 C2015-906099-0
C2015-906100-8

AMPÈRE'S ELECTRODYNAMICS:
Analysis of the Meaning and Evolution of Ampère's Force between Current Elements, together with a Complete Translation of His Masterpiece, *Theory of Electrodynamic Phenomena, Uniquely Deduced from Experience*

Contents

Letter from Ampère to his son, from September 1820:[1]

> Depuis que j'ai entendu parler pour la première fois de la belle découverte de M. Oersted, professeur à Copenhague, sur l'action des courants galvaniques sur l'aiguille aimantée, j'y ai pensé continuellement, je n'ai fait qu'écrire une grande théorie sur ces phénomènes et tous ceux dejà connus de l'aimant, et tenter des expériences indiquées par cette théorie, qui toutes on réussi et m'ont fait connaitre autant de faits nouveaux.

Tricker:[2]

> At the beginning of the year 1820 nothing was known of the magnetic action of an electric current. By 1826 the theory for steady currents had been completely worked out. Since then, though newer methods may have made the handling of the mathematical apparatus simpler and more concise, nothing fundamental has been changed.
>
> [...]
>
> In the theory of gravitation, Newton was already provided with a knowledge of a range of the phenomena, mainly through the medium of Kepler's laws. Ampère had to discover the laws as well as provide the theory, and thus do the work of Tycho Brahe, Kepler and Newton rolled into one.

Maxwell:[3]

> The experimental investigation by which Ampère established the laws of the mechanical action between electric currents is one of the most brilliant achievements in science. The whole, theory and experiment, seems as if it had leaped, full grown and full armed, from the brain of the 'Newton of electricity.' It is perfect in form, and unassailable in accuracy, and it is summed up in a formula from which all the phenomena may be deduced, and which must always remain the cardinal formula of electro-dynamics.

Whittaker:[4]

> [Ampère] published his collected results in one of the most celebrated memoirs in the history of natural philosophy.

Williams[5] comparing Ampère's main work[6] with Newton's masterpiece of 1687, *Mathematical Principles of Natural Philosophy*:[7]

> Having established a noumenal foundation for electrodynamic phenomena, Ampère's next steps were to discover the relationship between the phenomena and to devise a theory from which these relationships could be mathematically deduced. This double task was undertaken in the years 1821-1825, and his success was reported in his greatest work, the *Mémoire sur la théorie mathématique des phénomènes électrodynamique, uniquement déduite de l'expérience* (1827). In this work, the *Principia* of electrodynamics, Ampère first described the laws of action of electric currents, which he had discovered from four extremely ingenious experiments.

[1] [Ampère, d] and [Launay (ed.), 1936a, pp. 562].
[2] [Tricker, 1965, pp. vii and 36].
[3] [Maxwell, 1954, vol. 2, article 528, p. 175].
[4] [Whittaker, 1973, p. 83].
[5] [Williams, 1981, p. 145]
[6] [Ampère, 1826f] and [Ampère, 1823c].
[7] [Newton, 1934], [Newton, 1990], [Newton, 1999], [Newton, 2008] and [Newton, 2010].

Acknowledgments

We thank Fapesp, Brazil, for the post-doctoral fellowship leading to the publication of the Portuguese version of this book by the University of Campinas Press (Editora da UNICAMP).[8] To the PRPG of the University of Campinas—UNICAMP, Brazil, for the PhD fellowship during which this work was initiated.[9] To the Physics Institute of the University of Campinas—UNICAMP and to the School of Physics of the Catholic University of Brasília—UCB, Brazil, which supplied the necessary conditions for the realization of this work.

We thank several people for their suggestions, support when we were performing some experiments, help in the preparation of some figures, references, ideas, etc.: Antonio Vidiella Barranco, Umberto Bartocci, Abdelmadjid Benseghir, Sérgio Luiz Bragatto Boss, Marcelo de A. Bueno, João José Caluzi, Hugo B. de Carvalho, Roberto Antônio Clemente (in Memoriam), Olivier Darrigol, David Dilworth, F. Doran, Samuel Doughty, C. Dulaney, Junichiro Fukai, Daniel Gardelli, Wolfgang Gasser, Daniel Gardelli, Albert Gerard Gluckman, J. Gottschalk, Neal Graneau, Peter Graneau (in Memoriam), Eduardo Greaves, J. Guala-Valverde (in Memoriam), Ivã Gurgel, Hermann Härtel, Howard Hayden, Mark A. Heald, Laurence Hecht, Peter Heering, John L. Heilbron, Júlio Akashi Hernandes, David de Hilster, James R. Hofmann, Dietmar Höttecke, Jan Olof Jonsson, Ricardo Avelar Sotomaior Karam, Wolfgang Lange, Ceno Pietro Magnaghi, Antônio Jamil Mania, Juan Manuel Montes Martos, P. G. Moyssides, Hector Munera, Marcos César Danhoni Neves, Edmundo Capelas de Oliveira, Itala M. L. D'Ottaviano, P. T. Pappas, Gerald Pellegrini, José Rafael Boesso Perez, Thomas E. Phipps Jr., Daniel Robson Pinto, John Plaice, Manfred Pohl, Fabio Miguel de Matos Ravanelli, Karin Reich, Varlei Rodrigues, Waldyr A. Rodrigues Jr., Germain Rousseaux, C. Roy Keys, T. Rütting, Abner de Siervo, Hector Torres Silva, Moacir Pereira de Souza Filho, Domina E. Spencer, Ralf Steiner, Friedrich Steinle, Sean M. Stewart, Martin Tajmar, Mário Noboru Tamashiro, J. Tennenbaum, Dario S. Thober, Christian Ucke, Greg Volk, James Paul Wesley (in Memoriam), Karl-Heinrich Wiederkehr (in Memoriam), Bertrand Wolff, Bernd Wolfram and Gudrun Wolfschmidt.

Some scientists read a first version of this book and helped us with their detailed comments, corrections, suggestions and references. We wish to thank in particular Alan Aversa, Michael D. Godfrey, Steve Hutcheon and Kenneth S. Mendelson.

Our special thanks go to Christine Blondel. Her book is at the basis of our work.[10] She received us in Paris and maintains a wonderful site on the history of electricity containing Ampère's papers, correspondence, manuscripts, etc.[11] The resources of this homepage enriched enormously our work.

A. K. T. Assis[12] and J. P. M. C. Chaib[13]

[8] [Assis and Chaib, 2011].

[9] [Chaib, 2009].

[10] [Blondel, 1982].

[11] [Blondel, 2005].

[12] Institute of Physics, University of Campinas—UNICAMP, 13083-859 Campinas, SP, Brazil, e-mail: assis@ifi.unicamp.br, homepage: `www.ifi.unicamp.br/~assis`.

[13] School of Physics, Catholic University of Brasília—UCB, 71966-700 Taguatinga, DF, Brazil, e-mail: jopachaib@gmail.com.

Foreword

A. M. Ampère was a key contributor to modern physics. This volume provides English translations of Ampère's first paper and of his main work, *Theory of Electrodynamic Phenomena Uniquely Derived from Experiments.* Detailed annotations are provided in order to clarify these ground-breaking publications. In addition, there is extensive discussion of Ampère's interaction with the scientific community in England, France and the rest of Europe. This provides important context for a better understanding of his scientific work and of the then current state of physics.

All of this material is of particular significance since much of Ampère's work has not been well-known. Part of the reason for this may be that it was highly controversial in his time. In addition, as a mathematician, he was outside the mainstream of physics. To physicists it seemed improbable that a mathematician could design, implement and make effective use of revolutionary experiments, many of which were complex, quite delicate and not previously contemplated. His genius as an experimentalist was not widely recognized.

Following Ampère, Maxwell further developed the subject into the General Theory of Electricity and Magnetism. With this theory in place, the experiments seemed more obvious. However, Maxwell made clear the significance of Ampère's work when he wrote:

> The experimental investigation by which Ampère established the law of the mechanical action between electric currents is one of the most brilliant achievements in science. The whole, theory and experiment, seems as if it had leaped, full grown and full armed, from the brain of the "Newton of Electricity". It is perfect in form, and unassailable in accuracy, and it is summed up in a formula from which all the phenomena may be deduced, and which must always remain the cardinal formula of electro-dynamics.
>
> [J. C. Maxwell, *A Treatise on Electricity and Magnetism,* Dover, New York, 1954, vol. 2, p. 175]

In conclusion, this treatise provides compeling reading for anyone with an interest in physics and its history.

Michael D. Godfrey
Statistics Department
Stanford University
email: michaeldgodfrey@gmail.com
web: `https://sites.google.com/site/michaeldgodfrey`
Archive: `https://archive.org/details/@m_d_godfrey`

General Remarks

This work is an English version of a book first published in 2011.[14]

All the words between square brackets [ ] in the quotations are ours. They were inserted to facilitate the understanding of some passages or to clarify the meaning of some terms.

When we define any physical concept in this book we utilize "≡" as a symbol of definition.

[14] [Assis and Chaib, 2011].

Part I

Ampère's Force between Current Elements and the Meaning of Its Terms

Chapter 1

Introduction

1.1 André-Marie Ampère

André-Marie Ampère, figure 1.1, was born in Lyon, France, on 20 January 1775. He died on 10 June 1836, during an inspection tour in Marseille, when he was 61 years old. He worked in many areas of knowledge including physics, mathematics, chemistry, language and philosophy.

Figure 1.1: Portrait of Ampère (1775-1836) near the time of his marriage in 1799 when he was 24 years old, [Hofmann, 1996, p. 12].

Some of the main aspects of Ampère's life and his electrodynamics have been discussed by different authors from several perspectives.[1] His correspondence has already been published.[2] His manuscripts are available in 40 boxes or *cartons* at the Archives of the Academy of Sciences of Paris.[3] There are several files or *chemises* in these boxes and they are quoted by the box and file numbers. They are available online at the excellent homepage on Ampère and the history of electricity.[4] Ampère's autobiography written in 1824 has been published recently[5] and the manuscript is kept at the Academy of Sciences of Paris.[6]

[1] [Launay, 1925], [Tricker, 1962], [Poudensan, 1964], [O'Rahilly, 1965], [Tricker, 1965], [Williams, 1981], [Blondel, 1982], [Hofmann, 1982], [Williams, 1983], [Williams, 1989a], [Williams, 1989b], [Assis, 1992], [Graneau and Graneau, 1993], [Assis, 1994], [Graneau, 1994], [Graneau and Graneau, 1996], [Hofmann, 1996], [Darrigol, 2000], [Bueno and Assis, 2001], [Steinle, 2003], [Steinle, 2005], [Bueno and Assis, 2015] and [Assis, 2015b].

[2] [Launay (ed.), 1936b], [Launay (ed.), 1936a] and [Launay (ed.), 1943].

[3] [Blondel, 1978].

[4] [Blondel, 2005].

[5] [Ampère, 1982].

[6] [Ampère, b, *carton* 22, *chemise* 314].

In this book we will concentrate on his work on electrodynamics which he developed between 1820 and 1826.

According to Williams:[7]

> By 1820 Ampère had achieved a certain reputation as both a mathematician and a somewhat heterodox chemist. Had he died before September of that year, he would be a minor figure in the history of science. It was the discovery of electromagnetism in the spring of 1820 which opened up a whole new world to Ampère and gave him the opportunity to show the full power of his method of discovery.

1.2 The Forces of Gravitation, Electrostatics and Magnetism

Until the beginning of the XIXth century there were some separated branches of physics like gravitation, electrostatics and magnetism. They were described by central forces which varied as the inverse square of the distance r between the interacting bodies.

In 1687 Isaac Newton (1642-1727) published his masterpiece, *Mathematical Principles of Natural Philosophy.*[8] In this work he presented his famous law of universal gravitation. The force of gravitation is proportional to the product of the masses m and m' of the two interacting bodies, being always attractive. It varies as the inverse square of the distance r between two point bodies. Mathematically the force is then proportional to:

$$\frac{mm'}{r^2} . \tag{1.1}$$

Augustin Coulomb (1738-1806) obtained in 1785 the law of force between two bodies electrified with charges q and q' separated by a distance r which was large compared with the diameters of the bodies. He presented his results in two papers of 1785, published in 1788.[9] He called these electrified bodies by different names, namely, "electrical masses," "electrified molecules," or "densities of electric fluids."[10]

In the case of bodies electrified with charges of the same sign, Coulomb expressed himself as follows:[11]

> *Fundamental Law of Electricity*
>
> *The repulsive force between two small spheres charged with the same sort of electricity is in the inverse ratio of the squares of the distances between the centers of the two spheres.*

For bodies electrified with charges of opposite signs, Coulomb concluded that:[12]

> We have thus come, by a method absolutely different from the first, to a similar result; we may therefore conclude that the mutual attraction of the electric fluid which is called positive on the electric fluid which is ordinarily called negative is in the inverse ratio of the square of the distances; just as we have found in our first memoir, that the mutual action of the electric fluid of the same sort is in the inverse ratio of the square of the distances.

Up to now Coulomb mentioned only how the electric force varied with the distance between the electrified bodies. It was only in the final section of his second memoir, when he recapitulated the major propositions that resulted from his researches, that he mentioned that this force was proportional to the product between the charges:[13]

> Recapitulation of the subjects contained in this Memoir
>
> From the foregoing researches, it follows that:
>
> 1. The electric action, whether repulsive or attractive, of the two electrified spheres, and therefore of two electrified molecules, is in the ratio compounded of the densities of the electric fluid of the two electrified molecules and inversely as the square of the distances; [...]

[7] [Williams, 1981, p. 143].
[8] [Newton, 1934], [Newton, 1990], [Newton, 1999], [Newton, 2008] and [Newton, 2010].
[9] [Coulomb, 1785a], [Coulomb, 1785b], [Potier, 1884] and [Coulomb, 1935a].
[10] [Gillmor, 1971b] and [Gillmor, 1971a, pp. 190-192].
[11] [Coulomb, 1785a, p. 572], [Potier, 1884, p. 110] and [Coulomb, 1935a].
[12] [Coulomb, 1785b, p. 572], [Potier, 1884, p. 123] and [Coulomb, 1935a].
[13] [Coulomb, 1785b, p. 611], [Potier, 1884, p. 146] and [Gillmor, 1971a, pp. 190-191].

Gillmor pointed out correctly that Coulomb did not specifically prove that the electric force law was proportional to the product of the charges.[14] Coulomb simply supposed this proportionality in qq', although he did not consider it important to demonstrate this result experimentally.

Let us suppose two electrified particles or point bodies at rest relative to one another, separated by a distance r and electrified with charges q and q'. This force will be attractive for charges of opposite signs and repulsive for charges of the same sign. Ampère used to consider an attractive force as positive and a repulsive force as negative. With this supposition, the electrostatic force between these electrified bodies is proportional to:

$$-\frac{qq'}{r^2} \ . \tag{1.2}$$

This force is very similar to Newton's law of gravitation, equation (1.1). Both force laws are directed along the straight line connecting the bodies, they follow the law of action and reaction, and vary as the inverse square of the distance between the bodies. Moreover, the electric force is proportional to the product of the two charges, while the gravitational force is proportional to the product of the two gravitational masses. It seems that Coulomb arrived at his force law more by analogy with Newton's law of gravitation than by his doubtful few measurements with the torsion balance.[15]

In order to describe the magnetic interaction between magnets, or the magnetic interaction between a magnet and the Earth, Coulomb proposed in 1785 an expression describing the force between magnetic poles considered as concentrated on particles or material points.[16] Coulomb called the intensities of these poles "magnetic densities of the fluids."[17] Nowadays these poles are called North pole of the magnet and South pole of the magnet, with the North pole being considered positive, by convention.

Coulomb expressed himself in the following words:[18]

> *The magnetic fluid acts by attraction or repulsion in a ratio compounded directly of the density of the fluid and inversely of the square of the distance of its molecules.*
>
> The first part of this proposition does not need to be proved; let us pass to the second. [...]

Let p and p' be the intensities of two magnetic poles (magnetic pole-strengths) separated by a distance r. The force will be attractive for poles of opposite type and repulsive for poles of the same type. We can represent the North pole as positive and the South pole as negative. We can also consider an attractive force as positive and a repulsive force as negative. We can then represent the magnetic force between two poles as being proportional to:

$$-\frac{pp'}{r^2} \ . \tag{1.3}$$

Gillmor also pointed out correctly that Coulomb did not deduce experimentally that the force between two magnetic poles was proportional to the product of the pole-strengths.[19] Coulomb only implied that this force was proportional to the product pp', although he did not perform experiments to test this statement. According with his words just quoted, he did not consider it necessary to prove experimentally this aspect of the law. This statement of Coulomb does not seem correct to us. It would be necessary to verify experimentally this essential aspect of the force between two magnetic poles, before one could reach the conclusion that this was a law of nature. The same happens with the electric force being proportional to the product of the two charges.

1.3 Ørsted's Experiment and Its Impact on Ampère

It was known for a long time that a horizontal magnetic needle, like a compass needle, which is free to rotate around a vertical axis connected to its center will orient itself relative to the ground. After being released from rest in an arbitrary orientation relative to the geographic North-South meridian of the Earth, it will normally acquire another orientation. This new orientation is called the local magnetic meridian, being

[14] [Gillmor, 1971b] and [Gillmor, 1971a, pp. 190-192].
[15] [Heering, 1992].
[16] [Coulomb, 1785b], [Potier, 1884] and [Coulomb, 1935b].
[17] [Gillmor, 1971b] and [Gillmor, 1971a, pp. 190-192].
[18] [Coulomb, 1785b, p. 593], [Potier, 1884, p. 130] and [Coulomb, 1935b, p. 417].
[19] [Gillmor, 1971b] and [Gillmor, 1971a, pp. 190-192].

an imaginary great circle line connecting the so-called magnetic South and North poles of the Earth. The magnetic South pole of the Earth is close to its geographic North pole, while the magnetic North pole of the Earth is close to its geographic South pole. In the XVIII and early XIXth centuries it was supposed that a magnetic needle was composed of a North pole and a South pole of equal magnitudes located at the extremities of a thin needle and separated by its length. In order to understand the orientation of a magnetic needle due to the magnetic influence of the Earth, it was usually supposed that the North pole of a magnetic needle was attracted by the magnetic South pole of the Earth, close to its geographic North pole, while the South pole of a magnetic needle was attracted by the magnetic North pole of the Earth, close to its South pole.

There was a great turning point for electric researches in 1800 when Alessandro Volta (1745-1827) published a work describing his invention of the electric pile or battery.[20] With this instrument and with the improved devices following Volta's discovery, scientists had at their disposal, for the first time, a reliable source of small voltage and constant electric current. Volta's invention created a revolution in technology and also in the experimental and theoretical study of electricity in motion.

Hans Christian Ørsted (1777-1851),[21] figure 1.2, was a Danish physicist and chemist who worked with the pile. In 1820 he observed the deflection of a magnetic needle from the magnetic meridian when there was a constant electric current flowing in a long wire which was close to the needle.[22] Ørsted's discovery marks the beginning of *electromagnetism*, that is, of the systematic study of the relation between electric and magnetic phenomena.

Figure 1.2: H. C. Ørsted.

Ørsted expressed some of his discoveries as follows:[23]

> The opposite ends of the galvanic battery were joined by a metallic wire, which, for shortness sake, we shall call the *uniting conductor*, or the *uniting wire*. To the effect which takes place in this conductor and in the surrounding space, we shall give the name of the *conflict of electricity*.
>
> Let the straight part of this wire be placed horizontally above the magnetic needle, properly suspended, and parallel to it. If necessary, the uniting wire is bent so as to assume a proper position for the experiment. Things being in this state, the needle will be moved, and the end of it next the negative side of the battery will go westward.
>
> If the distance of the uniting wire does not exceed three-quarters of an inch from the needle, the declination of the needle makes an angle of about 45°. If the distance is increased, the angle diminishes proportionally. The declination likewise varies with the power of the battery.
>
> [...]
>
> If the uniting wire be placed in a horizontal plane under the magnetic needle, all the effects are the same as when it is above the needle, only they are in an opposite direction; for the pole of the magnetic needle next the negative end of the battery declines to the east.

[20] [Volta, 1800a], [Volta, 1800b], [Volta, 1964] and [Magnaghi and Assis, 2008].

[21] The Danish name of Ørsted is Hans Christian Ørsted, [Ørsted, 1986, n. 2]. His Latinized name received several forms like Örsted, OErsted, Oersted or OErstedt. In this book we will utilize the Ørsted format, except when quoting original sources which utilized other formats of his name.

[22] [Oersted, 1820], [Oersted, 1965], [Franksen, 1981], [Ørsted, 1986], [Ørsted, 1998b], [Ørsted, 1998a] and [Wolff and Blondel, 2005]. Recently we reproduced Ørsted's original experiment with simple materials, [Chaib and Assis, 2007c].

[23] [Oersted, 1820, pp. 274-275], [Oersted, 1965, pp. [114-115] and [Ørsted, 1986, pp. 116-120].

> That these facts may be more easily retained, we may use this formula—the pole *above* which the *negative* electricity enters is turned to the *west*; *under* which, to the *east*.

This experiment is illustrated in figure 1.3.

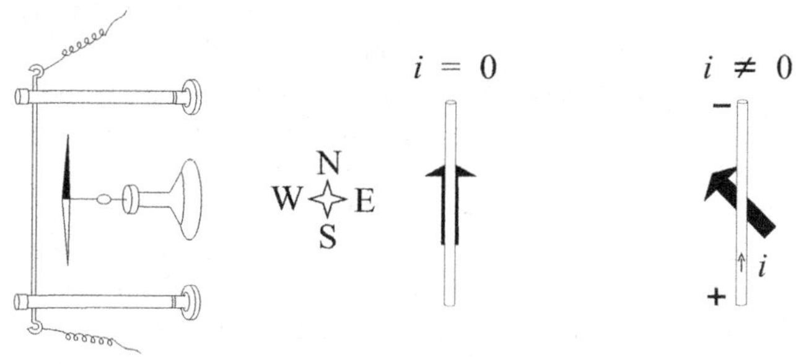

Figure 1.3: Representation of Ørsted's experiment with the horizontal wire above the magnetic needle. In (a) and (b) the needle points along the magnetic meridian while there is no electric current in the wire. In (c) there is a constant current flowing from the South towards the North. The needle is deviated from the magnetic meridian, with its North pole going westward.

When the horizontal wire is located below the needle, the opposite phenomenon takes place. In this case the North pole of the needle goes eastward, as represented in figure 1.4.

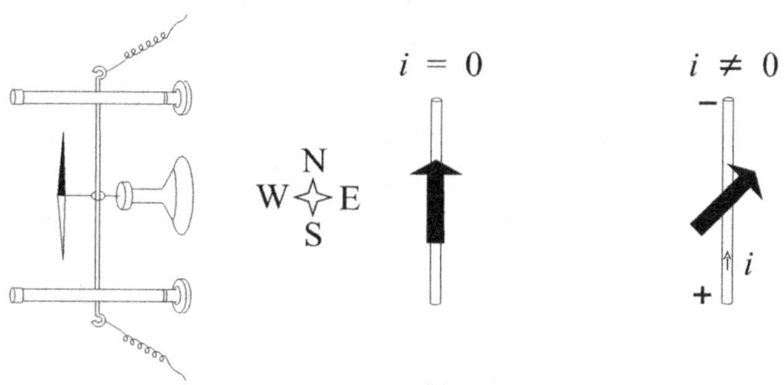

Figure 1.4: Representation of Ørsted's experiment with the horizontal wire below the magnetic needle. In (a) and (b) the needle points along the magnetic meridian while there is no electric current in the wire. In (c) there is a constant current flowing from the South towards the North. The needle is deviated from the magnetic meridian, with its North pole going eastward.

Ørsted did not publish his work in any scientific journal. He wrote it in Latin, with four pages, sending it as a brochure to several scientists on 21 July 1820. It caused a sensation, being translated and published in several scientific journals. Arago (1786-1853) described Ørsted's work to the Academy of Sciences in Paris on 4 September 1820. Due to the generalized disbelief, he repeated this experiment to the members of the Academy on 11 September 1820.

One of the reasons for this incredulity was due to the fact that Ørsted's experiment seemed to go against the ideas of symmetry of that time. Consider the situation of figure 1.3 (a) and (b) when there is no current in the wire. The horizontal wire and the magnetic needle define a vertical plane. There is nothing which seems to privilege one side of this vertical plane relative to the other side. However, Ørsted's experiment indicated that, when there was a constant electric current flowing in the wire, from the South towards the North, the North pole of the needle remained inclined westward relative to the vertical plane. That is, in the new equilibrium configuration of the needle its North pole pointed between the Earth's North and West directions. The angle of deviation of the axis of the needle relative to the magnetic meridian was shown to depend on the power of the battery and on the distance between the straight wire and the center of the needle. When this distance was 3/4 of an inch, Ørsted observed a deviation of 45°. There seemed to be a symmetry breaking in this experiment. It would be more natural to expect that the North pole of the needle were attracted or repelled by the current-carrying wire, remaining in the vertical plane. This deviation of the North pole of the needle towards one of the sides of the vertical plane was totally unexpected. This new effect attracted the attention of many scientists.

Ampère believed in the existence of the effect described by Oersted since he first heard of it, as we can conclude from the letter he sent to his son.[24] He also saw Arago's repetition of Ørsted's experiment made at the Academy of Sciences of Paris. He soon began to work intensely on this new subject. He interpreted Ørsted's experiment and all magnetic phenomena already known for a long time as being due to an interaction between current elements. To this end it was necessary to suppose the existence of electric currents inside the Earth and inside the normal magnets. According to Ampère, these electric currents would be responsible for the so-called magnetic properties of these bodies. All these phenomena would be then due to a single principle, namely, the force between current-carrying conductors. With this new hypothesis, Ampère expected to explain and unify not only the magnetic phenomena known for a long time, as the interaction between two magnets or the interaction between the Earth and a magnetic needle, but also the phenomenon discovered by Ørsted of a torque produced by a current-carrying wire and acting upon a magnetic needle. Moreover, from this hypothesis Ampère was able to predict a new phenomenon, not yet observed by anyone before him. This new phenomenon was the interaction between two current-carrying wires. He soon performed experiments showing the existence of this new interaction.

In 1822 Ampère arrived at his final mathematical expression describing the interaction between two current-carrying elements.[25] With this expression he could explain the magnetic phenomena, Ørsted's discovery and all of his own experiments describing the torque and force which he observed between current-carrying wires. In November 1826 he published his main work on this subject: *Theory of Electrodynamic Phenomena, Uniquely Deduced from Experience.*[26] It will be called here simply the *Théorie*. This work was also published in 1827 by the Academy of Sciences of Paris.[27] In all of our quotations of this work we will indicate the pages of the editions published in 1826 and 1827. The pagination of the 1827 edition coincides with that of the reprinted edition which took place in 1990.[28] English and Portuguese translations are available.[29] In Appendix VIII we include a complete and commented English translation of the *Théorie*.

In our book we will first present the meaning of the terms appearing in Ampère's force between current elements. We will then analyze in detail the path followed by Ampère in order to arrive at his mathematical expression. We will quote his papers published from 1820 onwards. We will also quote the *Collection of electrodynamic observations, containing several articles, notes, extracts from letters or from papers in scientific journals, relative to the mutual action between two electric currents, to the existing action between an electric current and a magnet or the terrestrial globe, and to the mutual action between two magnets.*[30] This work will be called here simply the *Recueil*. This Collection was published in 1822. Next year Ampère republished the work. All of our quotations will be taken from this work of 1823.[31] It includes a work of Savary which was read at the Academy of Sciences of Paris on April 18, 1823,[32] and a letter from Ampère to Faraday, dated 18 April 1823,[33] which does not appear in the Table of Contents printed at the end of this work. On the cover of the *Recueil* published in 1823 the publication date is erroneously given as 1822.

It should be mentioned here that some papers by Ampère were printed without the author's name. Some of these papers were written in the third person. However, it is known that these works were written by Ampère due to the fact that there are some of his manuscripts with the exact content of these works. Although written in the third person, the calligraphy is Ampère's. There are also some of his unpublished manuscripts written in the third person, like *chemise* 156 of *carton* 8.[34]

[24] [Ampère, d] and [Launay (ed.), 1936a, pp. 562].
[25] [Ampère, 1822o], [Ampère, 1822y] and [Ampère, 1885p].
[26] *Théorie des phénomènes électro-dynamiques, uniquement déduite de l'expérience*, [Ampère, 1826f].
[27] [Ampère, 1823c].
[28] [Ampère, 1990].
[29] [Ampère, 1965b], [Chaib, 2009], [Assis and Chaib, 2011] and [Ampère, 2012].
[30] *Recueil d'observations électro-dynamiques, contenant divers mémoires, notices, extraits de lettres ou d'ouvrages périodiques sur les sciences, relatifs à l'action mutuelle de deux courans électriques, à celle qui existe entre un courant électrique et un aimant ou le globe terrestre, et à celle de deux aimans l'un sur l'autre.*
[31] [Ampère, 1822w].
[32] [Savary, 1822].
[33] [Ampère, 1822l].
[34] [Ampère, b, *carton* 8, *chemise* 156].

1.4 The Introduction of the Words *Electromagnetism*, *Electromagnetic*, *Eletrodynamic* and *Electrostatics*

In order to characterize his discovery of an interaction between a current-carrying wire and a magnetic needle, Ørsted created two new words, namely, *electromagnetic* and *electromagnetism*. These expressions appeared for the first time in the titles of the articles which he published in 1820 and 1821, *New electro-magnetic experiments* and *Observations on electro-magnetism*.[35]

In 1820 Ampère discovered a new phenomenon, namely, the attractions and repulsions between current-carrying conductors. In order to distinguish this new set of phenomena from the electromagnetic phenomena discovered by Ørsted, Ampère created two new expressions, namely, *electrostatic* and *electrodynamic* phenomena. Electrostatics should include the attractions and repulsions between electrified bodies which were at rest relative to one another. Electrodynamics, on the other hand, should include the attractions and repulsions between current-carrying wires:[36]

> The word *electromagnetic*, given to the phenomena produced by the conducting wire of Volta's pile, could only describe these phenomena conveniently at the time in which there were known only those phenomena discovered by M. OErsted between an electric current and a magnet. I believe that I should utilize the denomination *electrodynamic*, in order to combine in a single word all those phenomena and, especially, to designate those phenomena which I discovered between two voltaic conductors. This name expresses the characteristic property of these phenomena, namely, to be produced by electricity in motion; while the attractions and repulsions known for a long time are the *electrostatic* phenomena produced by the unequal distribution of electricity at rest in the bodies in which these phenomena are observed.

In other publications Ampère made similar statements.[37] In particular, he mentioned that in electrodynamic phenomena the presence of a magnet is not necessary:[38]

> Ever since I discovered the mutual action between two voltaic conductors, which evidently has the same nature as the action of a conductor upon a magnetized bar, **and which acts without the assistance of any magnet**, the name of *electromagnetic action*, which I utilize here only to conform myself to the common use, would no longer be convenient to designate this kind of action. I think that it should be [known] under the name of *electrodynamic action*.

In this book we will utilize the following nomenclature:

- *Electrostatic phenomena:* Forces and torques between electrified bodies which are at rest relative to one another.
- *Magnetic phenomena:* Forces and torques between magnets, together with the torques exerted by the Earth on magnets (orientation of compass and dip needles).
- *Electromagnetic phenomena:* Forces and torques between a current-carrying conductor and a magnet, together with the forces and torques exerted by the Earth on current-carrying conductors.
- *Electrodynamic phenomena:* Forces and torques between current-carrying conductors.

[35] [Ørsted, 1998c, p. 421], [Ørsted, 1998f, p. 426], [Grattan-Guinness, 1990a, p. 920] and [Grattan-Guinness, 1991, p. 116].
[36] [Ampère, 1822d, p. 60].
[37] [Ampère, 1822i, p. 62], [Ampère, 1822j, p. 200], [Ampère, 1822e, p. 237], [Ampère, 1885e, p. 239], [Ampère, 1885d, p. 192], [Blondel, 1982, p. 78] and [Benseghir, 1989].
[38] [Ampère, 1822j, note on p. 200] and [Ampère, 1885e, note on p. 239], our emphasis in boldface.

Chapter 2

Ampère's Force and the Meaning of Its Terms

2.1 Ampère's Force between Current Elements

In 1822 Ampère obtained his final expression for the force acting between two current elements ids and $i'ds'$ separated by a distance r, namely:[1]

$$\frac{ii'dsds'}{r^n}(\sin\alpha\sin\beta\cos\gamma + k\cos\alpha\cos\beta) \ . \tag{2.1}$$

In the *Théorie* of 1826 this force appeared as follows:[2]

$$\frac{ii'dsds'}{r^n}(\sin\theta\sin\theta'\cos\omega + k\cos\theta\cos\theta') \ , \tag{2.2}$$

and

$$\frac{ii'dsds'}{r^n}(\cos\varepsilon + h\cos\theta\cos\theta') \ . \tag{2.3}$$

In these equations the letters n, k and h represented constants. Their values were obtained by Ampère and are given by, respectively:[3]

$$n = 2 \ , \tag{2.4}$$

$$k = -\frac{1}{2} \ , \tag{2.5}$$

and

$$h = k - 1 = -\frac{3}{2} \ . \tag{2.6}$$

Ampère's force between current elements has a much more complex structure than the gravitational, electric and magnetic forces expressed by equations (1.1), (1.2) and (1.3). In the following Sections we will discuss the meanings of the angles α, β and γ (or θ, θ' and ω, respectively), together with the meaning of the angle ε. In his works Ampère did not explain the reasons for utilizing different letters to represent the same angle, like the use of the letters α and θ representing the same angle.

[1] [Ampère, 1822o, pp. 412 and 420].

[2] [Ampère, 1826f, p. 32], [Ampère, 1823c, Ampère, 1990, p. 204] and [Ampère, 1965b, p. 176].

[3] [Ampère, 1822o], [Ampère, 1826f, pp. 32, 50, 55 and 202] and [Ampère, 1823c, Ampère, 1990, pp. 204, 232, 237 and 374].

2.1.1 Ampère's Force in Vector Notation and in the International System of Units

Let f represent the force between interacting bodies. It is possible to write equations (1.1) up to (1.3) in terms of equalities utilizing dimensionless constants of proportionality. We then obtain the gravitational, electric and magnetic forces as follows, respectively:

$$f = \frac{mm'}{r^2} , \tag{2.7}$$

$$f = -\frac{qq'}{r^2} , \tag{2.8}$$

and

$$f = -\frac{pp'}{r^2} . \tag{2.9}$$

In the International System of Units and in vector notation equations (2.7) up to (2.9) can be written as follows:

$$\vec{F}_{M' \text{ on } M} = -GMM'\frac{\hat{r}}{r^2} = -\vec{F}_{M \text{ on } M'} , \tag{2.10}$$

$$\vec{F}_{Q' \text{ on } Q} = \frac{QQ'}{4\pi\varepsilon_0}\frac{\hat{r}}{r^2} = -\vec{F}_{Q \text{ on } Q'} , \tag{2.11}$$

and

$$\vec{F}_{P' \text{ on } P} = \frac{\mu_0}{4\pi}PP'\frac{\hat{r}}{r^2} = -\vec{F}_{P \text{ on } P'} . \tag{2.12}$$

In these equations the forces $\vec{F}$ are expressed in newtons (N), the magnitudes M and M' represent the masses of the interacting bodies expressed in kilograms (kg), $G = 6.67 \times 10^{-11}\ Nm^2/kg^2$ is the constant of universal gravitation, the magnitudes Q and Q' represent the charges of the electrified bodies expressed in coulombs (C), $\varepsilon_0 = 8.85 \times 10^{-12}\ A^2s^2N^{-1}m^{-2}$ is an electric constant called vacuum permittivity or permittivity of free space, the magnitudes P and P' represent the magnetic poles of a magnet expressed in ampere-meters (Am), while $\mu_0 \equiv 4\pi \times 10^{-7}\ kgmA^{-2}s^{-2}$ is a magnetic constant called vacuum permeability or permeability of free space. The distance between the two point particles which are interacting with one another is represented by r, while $\hat{r}$ represents the unit vector pointing from M' to M, from Q' to Q, or from P' to P, respectively.

A comparison of equations (2.8) and (2.11) representing the force between two electrified bodies shows that the system of units with a dimensionless constant of proportionality can be expressed in the International System of Units by performing the following substitution:

$$qq' \Leftrightarrow \frac{QQ'}{4\pi\varepsilon_0} . \tag{2.13}$$

Likewise, a comparison of equations (2.9) and (2.12) representing the force between two magnetized bodies shows that the system of units with a dimensionless constant of proportionality can be expressed in the International System of Units by performing the following substitution:

$$pp' \Leftrightarrow \frac{\mu_0}{4\pi}PP' . \tag{2.14}$$

Let d^2f represent the force between two current elements. Equation (2.3) expressed in terms of an equality utilizing a dimensionless constant of proportionality can be written as follows:

$$d^2f = \frac{ii'dsds'}{r^n}(\cos\varepsilon + h\cos\theta\cos\theta') . \tag{2.15}$$

Likewise, in modern vector notation and in the International System of Units, Ampère's force $d^2\vec{F}^A_{I'ds' \text{ on } Ids}$, exerted by the current element $I'd\vec{s}\,'$ on the current element $Id\vec{s}$, is given by:[4]

$$d^2\vec{F}^A_{I'ds' \text{ on } Ids} = -\frac{\mu_0}{4\pi}II'\frac{\hat{r}}{r^2}\left[2(d\vec{s}\cdot d\vec{s}\,') - 3(\hat{r}\cdot d\vec{s})(\hat{r}\cdot d\vec{s}\,')\right] = -d^2\vec{F}^A_{Ids \text{ on } I'ds'} \ . \tag{2.16}$$

In this equation r is the distance between the centers of these two current elements, $I \geq 0$ and $I' \geq 0$ represent the current intensities in the International System of Units, that is, expressed in amperes (A), the magnitudes $d\vec{s}$ and $d\vec{s}\,'$ represent the infinitesimal lengths of the two current elements, pointing along the sense of the currents I and I' in each current element, while $\hat{r}$ is the vector of unit magnitude pointing from the center of $d\vec{s}\,'$ towards the center of $d\vec{s}$. Ampère's force between current elements is a central force, pointing along the straight line connecting both current elements. It satisfies Newton's action and reaction law in the strong form. That is, the force exerted by $I'd\vec{s}\,'$ on $Id\vec{s}$ is not only equal and opposite to the force exerted by $I'd\vec{s}\,'$ on $Id\vec{s}$, but is also along the straight line connecting these two current elements.

As will be discussed in this book, we have the following relations:

$$d\vec{s}\cdot d\vec{s}\,' = dsds'\cos\varepsilon \ , \tag{2.17}$$

$$\hat{r}\cdot d\vec{s} = ds\cos\theta \ , \tag{2.18}$$

and

$$\hat{r}\cdot d\vec{s}\,' = ds'\cos\theta' \ . \tag{2.19}$$

Applying equations (2.17) up to (2.19) into equation (2.16) yields:

$$d^2\vec{F}^A_{I'ds' \text{ on } Ids} = -\frac{\mu_0}{4\pi}\frac{II'dsds'}{r^2}(2\cos\varepsilon - 3\cos\theta\cos\theta')\,\hat{r} = -d^2\vec{F}^A_{Ids \text{ on } I'ds'} \ . \tag{2.20}$$

We can compare equation (2.20) with equation (2.15) utilizing equation (2.6). We conclude that all results obtained by Ampère for the force between current elements, for the force exerted by a closed circuit acting on a current element of another circuit, and also for the force between two closed circuits, can be expressed in the International System of Units utilizing the following substitution:

$$ii' \Leftrightarrow \frac{\mu_0}{2\pi}II' \ . \tag{2.21}$$

In this equation (2.21) the intensities of the currents i and i' are expressed in the system of units introduced by Ampère (called nowadays the electrodynamic system of units), while the intensities of the currents I and I' are expressed in the International System of Units, that is, in amperes (A).

Before showing the path followed by Ampère to arrive at his force between current elements, we will present the meaning of the main components of his force. This discussion will be helpful in understanding his approach and his explanation of the magnetic phenomena based only on the interaction between electric currents.

2.2 Ampère's Conception of an Electric Current

Ampère's excitement with Ørsted's new discovery and his full commitment to investigate this subject are extremely well described in a letter he wrote to his son between 19 and 25 September 1820:[5]

> [...] I regret for not sending this letter three days ago [...], but all my time has been taken up by an important circumstance in my life. Ever since I heard for the first time about the discovery by M. Oersted, professor at Copenhagen, of the action of galvanic currents on the magnetized needle, I have been thinking continuously on this subject, and the only thing I have been doing is to write a great theory about this phenomenon and about all those phenomena already known about the magnet, and to perform experiments suggested by this theory, all of which have been successful and made me know several new facts.

[4][Assis, 1992, Chapter 3], [Assis, 1994, Chapter 4], [Assis, 1998, Section 11.2], [Assis, 1999a, Section 11.2], [Assis, 1999b, p. 151], [Bueno and Assis, 2001, Section 5.1], [Assis and Hernandes, 2007, Section 1.4], [Assis and Hernandes, 2009, Section 1.4], [Assis and Hernandes, 2013, Section 1.4], [Assis, 2013, Section 2.8], [Assis, 2014, Section 2.8], [Bueno and Assis, 2015, Section 5.1] and [Assis, 2015b, Chapter 3].

[5][Ampère, d], [Launay (ed.), 1936a, pp. 561-562], [Blondel and Wolff, a] and [Blondel and Wolff, b].

This is an important letter in several aspects. In the first place, it shows Ampère's full commitment with this subject since he became aware of Ørsted's discovery. In the second place, it indicates his initial desire to present a new theory about the magnetic and electromagnetic phenomena. As will be seen in this book, this theory is based on the interaction between electric currents. In the third place, the letter indicates that the expression utilized initially by Ampère in order to describe what happens in a wire, by connecting it to the terminals of a voltaic battery, is "galvanic current", as was common at that time. He utilized this expression in other publications.[6] Ørsted had created a new name, "electric conflict".[7] A third expression which was utilized sometimes was "voltaic current". Ampère himself utilized the word "voltaic" in his first published paper about electrodynamics, when comparing the usual electrostatic attractions and repulsions with the new attractions and repulsions which he had discovered between conductors carrying steady currents:[8]

> Let us consider now to what [aspect] is due the difference of these two kinds of phenomena completely distinct, of which one consists in the tension and in the attractions or repulsions which have been known for a long time, and the other phenomenon consists in the decomposition of water and in a great number of other substances, in the changes of direction of the [magnetized] needle, and in a kind of attractions and repulsions totally different from the ordinary electric attractions and repulsions; which I believe has been first recognized by myself, and which I designated *voltaic attractions and repulsions*, in order to distinguish them from these last [ordinary electric attractions and repulsions].

In the errata appearing on page 223 of the *Annales de Chimie et de Physique* of 1820 we read that the word "voltaic" in the expression *attractions et répulsions voltaïques* should be replaced by the expression *des courans électriques*, "of the electric currents." The expression "voltaic current" still appeared in the *Théorie*.[9]

From 1820 onwards Ampère began to utilize systematically the expression "electric current" instead of "galvanic current" in almost all of his publications about electrodynamics:[10]

> [...] *the galvanic current*, a denomination which I believe should be changed to that of electric current, [...]

He was not the first to utilize the expression "electric current", as it already appeared, for instance, in Volta's paper of 1800 in which he described his invention of the electric pile:[11]

> All this shows that, if the contact of the metals with each other in some points only be sufficient (as they are excellent conductors) to give a free passage to a moderately strong *current of electricity*, the case is not the same with liquids, or bodies impregnated with moisture, which are conductors much less perfect; and which, consequently, have need of more ample contact with metallic conductors, and still more with each other, in order that the electric fluid may easily pass, and that it may not be too much retarded in its course; especially when it is moved with very little force, as in the present case.
>
> [...] One might be surprised that in this circle the *electric current* having a free passage through an uninterrupted mass of water, [...]

Probably Volta utilized the expression "electric current" in order to oppose the concept of "animal electricity" which was being advocated by Galvani (1737-1798). According to Volta, the phenomena observed in Galvani's experiments and with his own electric pile were similar to the discharge of Leyden jars (except for being permanent and almost constant in time, instead of being short lived). The Leyden flasks accumulated electricity generated by friction. In his *chemise* 156 Ampère admitted that the expression "electric current" had already been utilized by other researchers:[12]

> The pile itself acts in these experiments as any other part of the voltaic circuit [that is, it deflects a magnetic needle as in Ørsted's original experiment] with this single difference, namely, that the disposition[13] of electricity which happens in the conductor from the zinc pole [positive pole] to the copper

---

[6] [Ampère, 1820a, p. 238] and [Ampère, a, p. 1].

[7] [Oersted, 1820], [Oersted, 1965] and [Ørsted, 1986], [Martins, 1986], [Franksen, 1981, pp. 28-30: Why "Conflictus"], [Benseghir and Closset, 1993] and [Benseghir and Closset, 2001].

[8] [Ampère, 1820c, pp. 61-62] and [Chaib and Assis, 2007d, p. 93].

[9] [Ampère, 1826f, pp. 27 and 172], [Ampère, 1823c, Ampère, 1990, pp. 199 and 344] and [Ampère, 1965b, p. 172].

[10] [Ampère, 1820f, p. 197] and [Chaib and Assis, 2009b, p. 133].

[11] [Volta, 1800a, pp. 408-409 and 415], [Volta, 1800b, p. 293 and 299], [Volta, 1964, pp. 115 and 120] and [Magnaghi and Assis, 2008, pp. 124 and 129], our emphasis.

[12] [Ampère, b, *carton* 8, *chemise* 156].

[13] *Disposition* in French, which may also be understood as arrangement or motion in this case.

> pole [negative pole] exists on the contrary in the interior of the pile from the copper pole to the zinc pole; Ampère called this disposition the *electric current*, in agreement with the usage of other physicists, but without attempting to pronounce himself, by adopting this expression, on the mode of action of the electricity upon the voltaic circuit. The sense of the electric current being so defined [...]

In his first paper on electrodynamics, Ampère expressed his conception of an electric current.[14] He first distinguished the phenomena due to an "electric tension", from those due to an "electric current". In the first category, related to electric tensions, he included the usual phenomena of electrostatics, which take place when the positive and negative charges are separated by an insulator (like the electrified bodies attracting light bodies or the attraction between oppositely charged bodies, with these bodies separated by air or by another insulating medium). The discovery of insulators and conductors of electricity is due to Stephen Gray (1666-1736), being published in 1731.[15] The hypothesis about the existence of two kinds of electricity, nowadays called positive and negative charges, is due to Du Fay (1698-1739) in 1733.[16] A detailed discussion of the experimental and historical foundations of electricity, with many quotations from the works of Gray and Du Fay, and a reproduction of some of their main experiments with simple and cheap materials, can be found in our books published between 2010 and 2015.[17]

In the second category, related to electric currents, Ampère included the phenomena which take place when we connect bodies oppositely electrified through a conducting medium (chemical decomposition of substances, deflection of a magnetic needle from the magnetic meridian when the needle is close to a current-carrying wire, attractions and repulsions between current-carrying wires, etc.) He then said:[18]

> But when the two bodies, or two systems of bodies, between which the electromotive action takes place are in contact via conducting bodies between which the electromotive action is not equal and opposite to the first so as to maintain the state of electric equilibrium and hence the tensions, these tensions vanish, or at least become very small, and characteristic phenomena occur. Since the arrangement of the bodies between which the electromotive action takes place is otherwise the same, the action doubtless continues, and since the mutual attraction of the two electricities, as measured by the difference between the electric tensions which has become zero, or else is considerably diminished, can no longer balance this action, it is generally accepted that it continues to carry the two electricities in two senses as before; a double current thus results, the one positive electricity and the other negative electricity, moving in opposite senses from the points where the electromotive action takes place to meet again in the part of the circuit opposite these points.

In Ampère's model of electric current there would be two fluxes of electric charges at any point inside the wire, a flux of positive charges and a flux of negative charges, moving relative to the wire with opposite velocities, figure 2.1 (a).

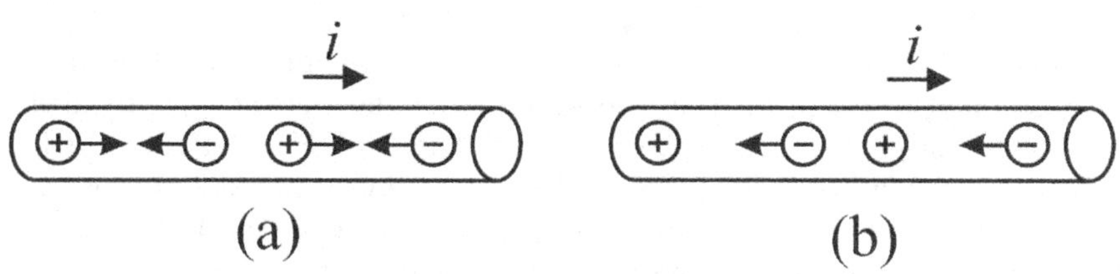

Figure 2.1: (a) According to Ampère, in a current-carrying wire there should be positive and negative charges moving in opposite directions relative to the wire. (b) Modern conception of current in a metal wire with the positive ions at rest relative to the wire, while only the negative electrons move relative to the wire.

[14] [Chaib and Assis, 2009a].

[15] [Gray,], with Portuguese translation in [Boss et al., 2012, Chapter 7].

[16] [Du Fay, 1733, pp. 464-465], [Du Fay, , pp. 263-264], [Roller and Roller, 1957, p. 586], [Heilbron, 1999, p. 257], [Blondel and Wolff, e] and [Blondel and Wolff, f].

[17] [Assis, 2010b], [Assis, 2010c], [Assis, 2011b], [Boss et al., 2012] and [Assis, 2015c].

[18] [Ampère, 1820c, p. 63], [Ampère, 1965a, p. 141] and [Chaib and Assis, 2007d, p. 94].

2.3 Relation between the Sense of the Current and the Motion of the Charges Inside the Wire

Ampère believed that in each point of a current-carrying wire there were positive and negative charges moving in opposite senses relative to the wire. But when referring to the *sense of the electric current* or to the *direction of the electric current*, he defined that he would be referring only to the motion of the positive charges:[19]

> For the sake of simplicity I shall call this state of the electricity in a series of electromotive and conducting bodies *electric current*; and since I shall continually have to speak of the two opposite senses in which the two electricities move, I shall invariably imply *positive electricity* by the words *sense of the electric current* to avoid unnecessary repetition; [...]

Ampère's meaning of the sense of the electric current is represented in figure 2.1 (a). At that time the electrons were not yet known. Nowadays we describe the current in metal wires as being due only to the motion of the conduction electrons relative to the wire, while the positive ions remain fixed in the metal lattice, 2.1 (b). In any event, even with this modification in our conception of the microscopic structure of the current, the convention for the direction of an electric current utilized for the first time by Ampère is still adopted nowadays. According to both conceptions related to the microscopic nature of an electric current, namely, Ampère's conception and the modern one, the sense of the electric current is assumed to be opposite to the direction of the motion of the negative charges relative to the wire, as represented in figures 2.1 (a) and (b).

Ampère established the following rule in order to find the sense or direction of the electric current:[20]

> The ordinary electrometer indicates the presence of tension and the intensity of this tension; there used to be no instrument for making known the presence of electric current in a battery or conductor and which would indicate its energy and direction. Such an instrument does exist today; it is sufficient to place the battery, or some portion of the conductor, roughly in the horizontal position in the direction of the magnetic meridian, and to place an apparatus similar to a compass (the only difference being the use to which it is put) on the battery or well above or below the portion of conductor: as long as the circuit is interrupted, the magnetized needle remains in its ordinary position; but it deviates away from it as soon as the current is established, and more so the greater its energy, and the direction can be told if the observer imagines himself to be placed in the direction of the current so that the current flows upwards from his feet to his head when facing the needle, for it is constantly to his left that the action of the current deflects the extremity which is pointing to the north, what I call the *austral pole of the magnetized needle* because it is the pole which is homologous to the south pole of the earth. This is what I express more concisely in saying that the austral pole of the magnet is carried to the left of the current acting on the needle. To distinguish this device from the ordinary electrometer, I think that it ought to be given the name *galvanometer* and it is appropriate to use it in all experiments on electric currents, as one habitually uses an electrometer with electric machines, so as to see at each instant if the current is there and find out its energy.

The word "galvanometer" was created by Ampère in this paper of 1820. But he did not create any instrument which would indicate quantitatively the intensity of the electric current utilizing the deflection of a magnetic needle. The first galvanometers were only built by Nobili (1784-1835) in 1825-1830, following Ampère's suggestion, and by Pouillet (1790-1868) in 1837, utilizing a tangent compass.[21] Ampère created also the expression "galvanoscope".[22] The instrument just described in the last quotation, which Ampère called galvanometer, should be called galvanoscope, as it indicates only qualitatively the direction and the intensity of the electric current, although it was not appropriate for a quantitative measurement of its intensity.

This imaginary observer mentioned by Ampère, placed along the wire and looking at the magnetic needle, with the current entering through his feet and leaving through his head, has been called *bonhomme d'Ampère*.[23] In this book it will be called "Ampère's observer", as Ampère himself utilized the word observer.[24]

[19] [Ampère, 1820c, p. 64], [Ampère, 1965a, p. 141] and [Chaib and Assis, 2007d, pp. 94-95].
[20] [Ampère, 1820c, pp. 66-67], [Ampère, 1965a, p. 143] and [Chaib and Assis, 2007d, p. 96].
[21] [Whipple, 1934], [Blondel, 1982, p. 73] and [Lauridsen and Abrahamsen, 1998].
[22] [Ampère, 1820a, p. 240], [Ampère, a, p. 3], [Blondel, 1982, pp. 73 and 184], [Steinle, 2003] and [Steinle, 2005].
[23] [Poudensan, 1964, p. 23], [Kastler, 1977, p. 144], [Blondel, 1982, p. 74] and [Blondel, 2004].
[24] [Ampère, 1820c, pp. 66-67], [Ampère, 1965a, p. 143] and [Chaib and Assis, 2007d, p. 96].

In order to understand Ampère's rule we should remember Ørsted's experiment. When the magnetic needle was under the wire, as in figure 1.3, the North pole of the needle moved westward when the current flowed from South to North. When the needle was over the wire, as in figure 1.4, the North pole of the needle moved eastward. Ørsted introduced the following rule in order to know the direction of the deflection of the needle: "The pole *above* which the *negative* electricity enters is turned to the *west*; *under* which, to the *east*".[25] Ampère, on the other hand, introduced his imaginary observer placed between the wire and the needle, looking at the needle, with the current coming from his feet towards his head. In figure 2.2 (a), we have this observer below the wire and above the needle, while in case (b) the observer is below the needle and above the wire.

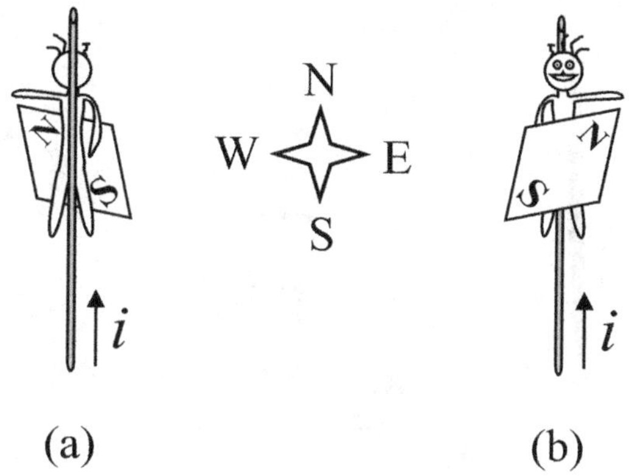

Figure 2.2: Representation of Ampère's observer in Ørsted's experiment. In both cases the observer is located between the current-carrying wire and the magnetic needle. In case (a) the needle is located below the wire, analogous to figure 1.3, while in case (b) the needle is above the wire, as in figure 1.4. When there is no current in the wire, the needle points in the North-South direction along the magnetic meridian. When the current flows in the wire from the South towards the North, then the North pole of the needle moves towards the left side of the observer in both cases.

When there is no current in the wire, the needle remains along the magnetic meridian due to the magnetic influence of the Earth. The North pole of the needle points along the magnetic South pole of the Earth, close to the geographic North pole of the Earth. When the current flows in the wire from South to North (or from the feet of Ampère's observer towards his head), the needle is deflected from its initial orientation. In both cases the North pole of the needle is deviated towards the left side of Ampère's observer. This is the mnemonic rule which he will adopt in all his works.

This observer appeared in *chemise* 156,[26] figure 2.3. Maybe this was the first time in which he utilized this representation.

Our figure 2.4 helps to understand Ampère's drawing. The horizontal magnetic needle of a compass can turn freely around a vertical axis passing through the center of the needle. When it is released at rest in an arbitrary orientation relative to the ground, the magnetic properties of the Earth orientate the needle. After reaching equilibrium, the North pole of the needle (or its austral pole, if we utilize Ampère's denomination) will point towards the magnetic South of the Earth, which is close to its geographic North pole, figure 2.4.

Ørsted's experiment indicated that the orientation of the magnetic needle is influenced by a nearby current-carrying wire. Beginning with this fact, Ampère supposed that the usual orientation of a compass due to the magnetic influence of the Earth might be due to electric currents flowing inside the Earth and around its surface. In order to explain this usual orientation of the needle, Ampère concluded that the supposed currents over the Earth's surface should be flowing from East to West along the terrestrial Equator. If this were the case, an observer lying on his back and looking at a compass needle above him would see, in equilibrium, the North pole of the needle pointing towards his left arm, that is, towards the geographic North pole of the Earth. In this position of the observer and the needle, the supposed terrestrial currents would be going from his feet towards his head, figure 2.4.

Ampère's rule is the forerunner of the right-hand rule and of the screw rule which are utilized nowadays in most textbooks of electromagnetism in order to indicate the direction of the magnetic field created by a current-carrying conductor. Ampère utilized his rule in order to establish the sense of the currents which he supposed to exist inside the Earth and inside permanent magnets.

[25][Oersted, 1820, p. 275], [Oersted, 1965, p. 115] and [Ørsted, 1986, p. 120].

[26][Ampère, b, *carton* 8, *chemise* 156].

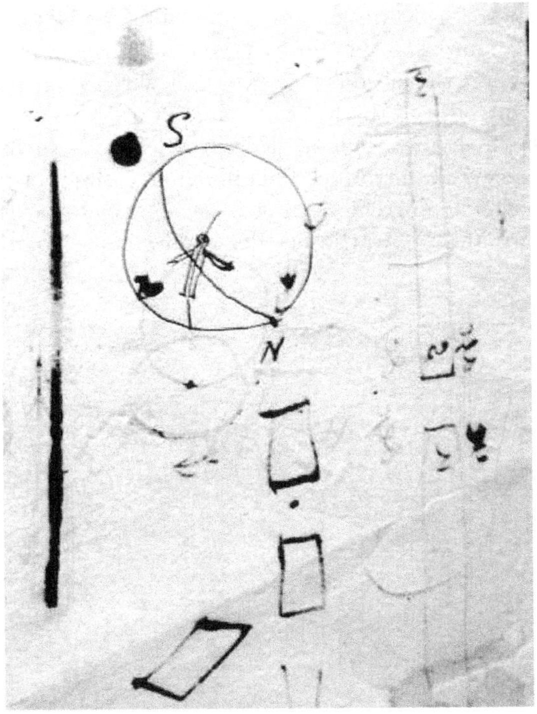

Figure 2.3: Ampère's representation of his observer.

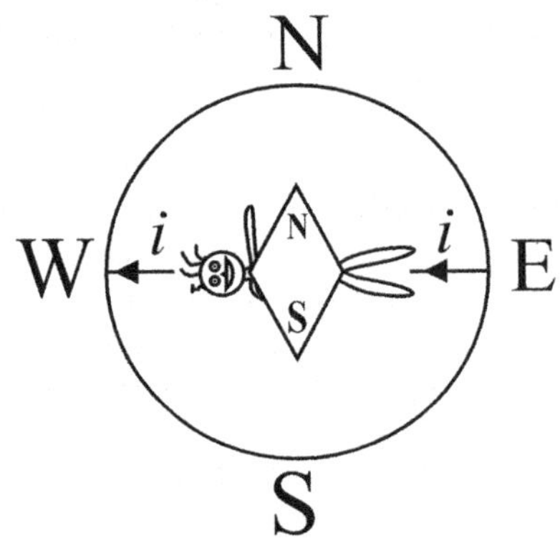

Figure 2.4: In equilibrium, the North pole of a compass will point approximately towards the geographic North pole of the Earth. Ampère's observer lying on his back and looking at the magnetic needle above it.

2.4 Different Meanings of the Expressions "Sense of the Current" or "Direction of the Current"

We saw in Section 2.3 that Ampère established a clear convention for the *sense* or *direction* of an electric current based on the sense of the motion of the positive charges which he supposed were flowing relative to the wire. Despite this convention, he utilized the expressions "sense of the current" or "direction of the current" with four distinct meanings, the same happening with other scientists of his age up to modern days. The adopted meaning depended on the context in which it was utilized. We can illustrate these four different meanings with figure 2.5.

There is a constant current flowing through a metal wire connected to the positive and negative terminals of a battery. Points A and I are inside the wire, while point J is inside the battery. The arrows indicate the sense of the current or the direction of the positive charges which Ampère supposed were moving relative to the conductors. That is, the back or origin of each arrow indicates the position of a supposed positive charge at a certain time t_0, while the head or tip of the arrow indicates the position of the same charge a little later, at a time $t_0 + \Delta t$. A figure like this, but without the external objects, appeared in a work of

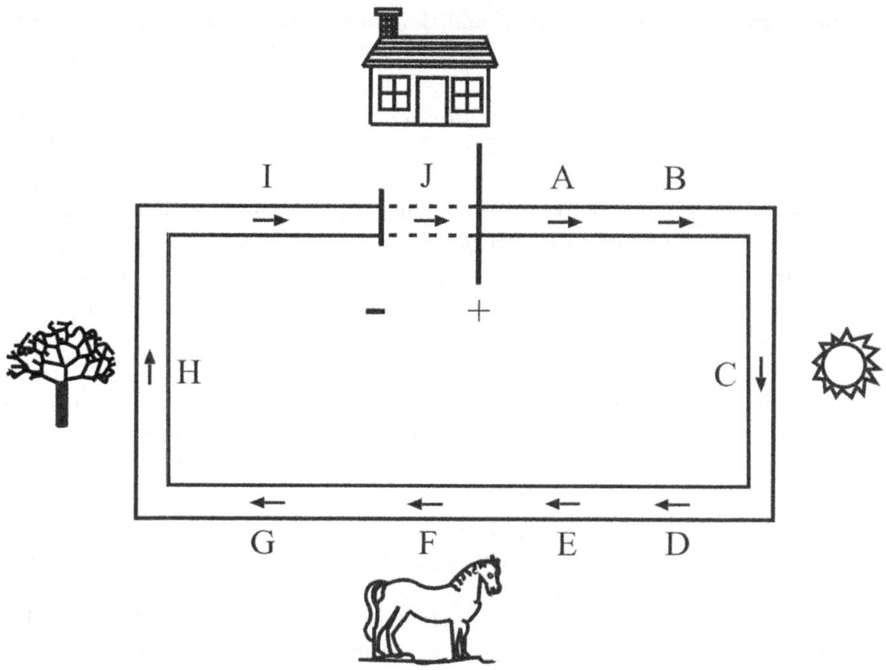

Figure 2.5: A constant current flowing through a resistive wire connected to a chemical battery.

Ampère and J. Babinet (1794-1872), with arrows only at points J and F, figure 2.6.[27] The positive pole of the battery is indicated in Ampère and Babinet's work by the symbol Z of zinc, while the negative pole is indicated by the symbol C of copper.

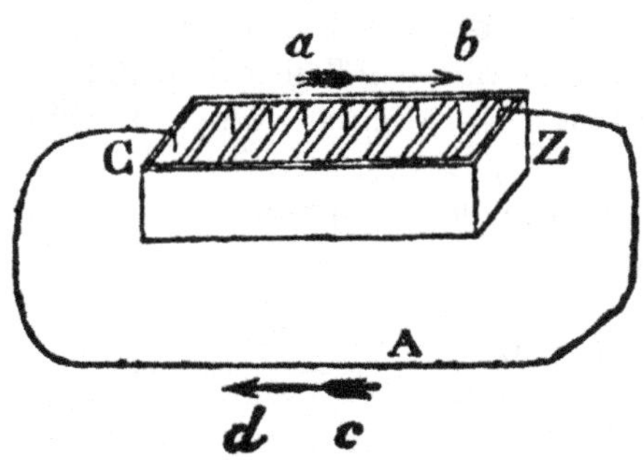

Figure 2.6: Experiment with which Ampère concluded that the current flows in closed circuits. That is, it flows not only along the metal wire, but also inside piles and batteries.

The most common meaning of the expression "sense of the electric current", or "direction of the motion of the positive electricity", utilized by Ampère, was referred implicitly to external bodies like the tree and the Sun of figure 2.5. We can say, for instance, that the currents in points I, J, A and B flow in the same sense, as they go from the tree towards the Sun, while the currents in points D, E, F and G flow in the opposite direction, from the Sun towards the tree. With this interpretation, the currents in points C and H would be orthogonal to the currents in points I, J, A, B, D, E, F and G. The current in point C would go from the house to the horse, while in point H it would be flowing in the opposite direction, that is, from the horse towards the house. Ampère utilized this meaning when he said, for instance, that a current element in point A repels a current element in point E due to the fact that these currents were flowing in opposite directions in the case of figure 2.5. An example of this utilization of the expression sense of a current can be seen in his first published paper:[28]

> I observed that by passing current through both parts at the same time, they were mutually attracted

[27][Ampère and Babinet, 1822a, p. 4, figure 1] and [Ampère and Babinet, 1822b, p. 169, figure 1].

[28][Ampère, 1820c, pp. 69-70], [Ampère, 1965a, pp. 144-145] and [Chaib and Assis, 2007d, p. 97].

> when both currents were in the same direction, and that they were repelled when the currents were in opposite directions.

Ampère also utilized another meaning for the expression "sense of the current". This new meaning was also connected to the motion of the supposed positive charges in a closed circuit. Ampère distinguished, in particular, two directions of the electric current, namely: (I) The positive charge going from the positive terminal towards the negative terminal of the battery, and (II) the positive charge going from the negative terminal towards the positive terminal of the battery.

One of the first important experiments of Ampère, performed in the beginning of September 1820, was to establish the sense of the current not only in the metal conductor connected to the terminals of a voltaic pile, but also inside the pile:[29]

> The first use to which I put this device was to check that the current which exists in the voltaic battery, from the negative extremity to the positive extremity, had the same influence on the magnetized needle as the current in a conductor which flows, on the contrary, from the positive extremity to the negative one.
>
> It is desirable to have for this two magnetized needles, one placed on the battery and the other above or below the conductor; it is seen that the austral pole of each needle is carried to the left of the current near to which it is placed. Thus, when the second [needle] is above the conductor, it is carried to the side opposite to that towards which the needle on the battery tends, since the currents have opposite directions in these two portions of the circuit. The two needles are, on the contrary, carried to the same side, remaining roughly parallel to each other, when one is above the battery and the other below the conductor.[30] As soon as the circuit is interrupted, they immediately revert, in both cases, to their ordinary orientation.

Ampère's experiment is illustrated in figure 2.7.[31] There are three compass needles (A, B and D) placed above the horizontal circuit, while needle C is below the circuit. Needle A is above a trough battery (indicated by the $+$ and $-$ symbols), needles B and D are above the wire connected to the terminals of the battery, while needle C is below the wire. When there is no current flowing in the circuit, the four needles point along the magnetic meridian. When there is a constant current i flowing in the circuit, all four needles are deviated from their equilibrium configurations. The North pole of needles A, B and C are deviated eastward, while the North pole of needle D is deviated westward. By supposing four observers of Ampère located between the circuit and the needles, looking at the needles with the current penetrating their feet and leaving through their heads, the North poles of all four needles will be deviated to the left sides of these observers, as pointed out by Ampère.

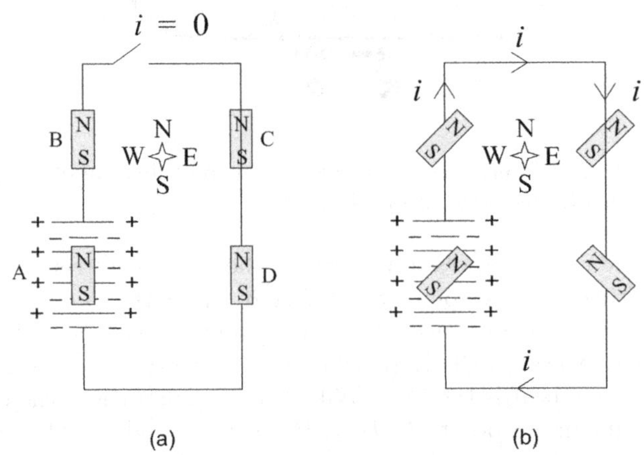

Figure 2.7: (a) Open circuit. The four compass needles point along the magnetic meridian, that is, along the North-South direction. (b) Constant current i flowing through the closed circuit. The compass needles are deviated not only when they are above or below the wire, but also when located above the battery.

[29] [Ampère, 1820c, pp. 67-68], [Ampère, 1965a, pp. 143-144] and [Chaib and Assis, 2007d, p. 96].

[30] [Note by Ampère:] For this experiment to leave no doubt as to the action of the current in the battery, it is convenient to use a trough battery with zinc and copper plates soldered together over the entire interface, and not just simply over a branch of metal which can rightly be regarded as a portion of conductor.

[31] See [Benseghir and Blondel, 2007].

In the case of figure 2.5, for instance, according to this new interpretation the currents in points A, B, ... and I would be flowing in the same direction, from the positive terminal towards the negative terminal of the battery, while the current in point J, inside the battery, would be flowing in the opposite direction, from the negative terminal towards the positive terminal.

This was a very important experiment performed by Ampère. In order to understand it, it should be kept in mind that he was utilizing a trough battery, *pile à auges*, presented in figure 2.8. These batteries were a development of Volta's pile, being due to William Cruickshank in July and September 1800.[32] Volta's original batteries were vertical and quickly dried, while Cruickshank's batteries were horizontal. These were powerful batteries in which the copper and zinc plates of the same size were soldered to one another. These bimetallic plates were placed vertically in a horizontal wooden vessel, covered with an insulating resin, with the spaces between the plates filled with a diluted solution of sulfuric acid.[33] The cross section of one of these batteries might have been, for instance, 20 cm by 20 cm, while the length might be 30 or 60 cm. It was then simple to place a small compass or magnetized needle above this battery in order to observe its deflection from the magnetic meridian when a constant current was flowing in the battery. With this experiment Ampère obtained the sense of the current not only in the external wire connecting the poles of the battery, but also inside the battery. This experiment showed that in the conductor the current flows from the positive terminal of the battery towards its negative terminal, while inside the battery it flows from the negative terminal towards the positive terminal.

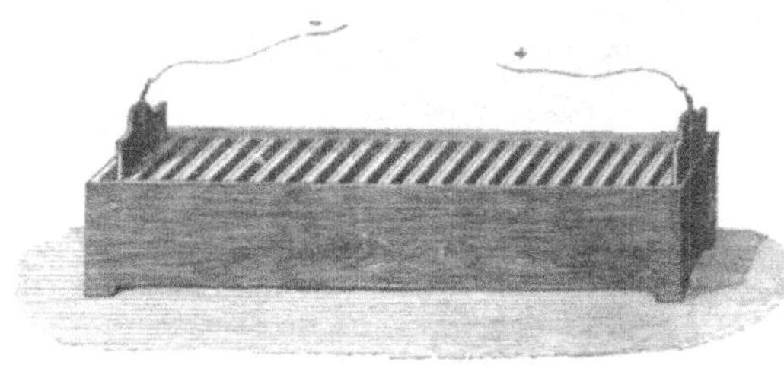

Figure 2.8: Trough battery.

Ampère's experiment showed for the first time that the current flows in a closed circuit in an ordinary circuit in which the terminals of a chemical battery are connected by a metal wire.

This was not a trivial conclusion. Suppose the battery were replaced by a charged capacitor composed of two parallel plates separated by a small distance d. We now connect the internal faces of these plates with a conductor A of large resistance, while the external faces are connected by another conductor B of large resistance. In this case the currents i and i' in conductors A and B, respectively, will flow from the positive plate towards the negative plate, as indicated in figure 2.9. The magnitudes of the currents i and i' may be equal or different from one another, depending on the values of the total resistances of conductors A and B. This means that in the internal conductor A of this example, the current flows in the opposite sense of the current in the interior of the chemical battery of Ampère's experiment, as can be seen comparing figures 2.6 and 2.9.

Ampère's experiments showed that in the interior of a battery the current flows from the negative pole towards the positive pole. From this fact it can then be concluded that there are forces of non electrostatic origin acting inside the battery.[34]

The third meaning utilized by Ampère to the words "sense" or "direction" was connected to the fact that the current flows in a closed circuit.[35] We can say, for instance, that all currents flowing clockwise are flowing in the same sense, while those anti-clockwise were flowing in the opposite sense. With this interpretation the current in all points of figure 2.5 flow in the same sense, that is, from the tree to the house, then to the Sun, to the horse and from there to the tree. Ampère utilized this meaning of the word when referring to the continuous circular motion of an extremity of a segment of a current-carrying wire which was moving around a pole of a magnet, or when referring to the continuous circular motion of the pole of a magnet around a current-carrying wire. These phenomena will be discussed in Section 7.1. They had been discovered by Michael Faraday (1791-1867), figure 2.10.

32[Neild, 1996].

33[Meyer, 1972, pp. 41-42, *Evolution of the battery and discoveries with electric currents*] and [Wolff and Blondel, 2005].

34[Varney and Fisher, 1980].

35[Ampère, 1822r, pp. 64-66], [Ampère, 1822j, pp. 203-205] or [Ampère, 1885e, pp. 241-243].

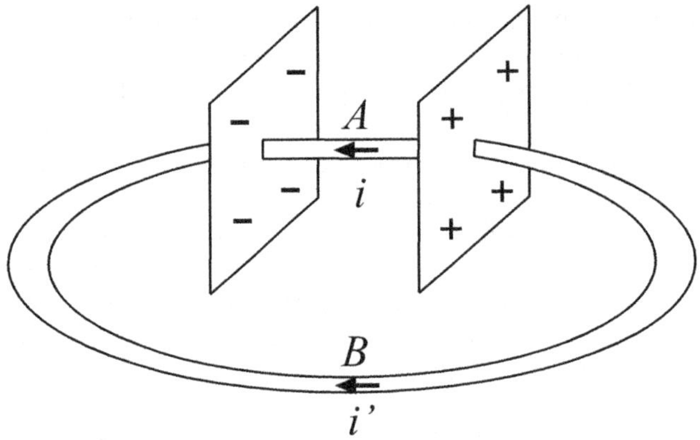

Figure 2.9: Currents i and i' flowing from the positive to the negative plates of a capacitor along the internal and external resistive conductors A and B.

Figure 2.10: M. Faraday.

Ampère, in particular, expressed himself as follows:[36]

> If one supposes the wire fixed and the magnet mobile, the same reasoning proves that the magnet should always rotate in the same sense around the wire, provided the electric current does not form a closed circuit of invariable shape.

He also utilized this meaning when referring to the possible circular motion of an arc of a circle around the axis of this circle,[37] or to the motion of a circular arc ABC around the axis of this circle:[38] "The arc is seen to turn in the sense CBA". As regards specifically the sense of the electric current, Ampère utilized this meaning of the word when discussing the attraction or repulsion of two circular current loops located in the same plane. The force would be attractive or repulsive depending on the direction of the currents:[39]

> As a matter of fact, it is easy to see, according to the laws of the mutual action between two electric currents, as the laws which I established, that the circular currents which turn in the same sense tend to repel one another and to change continuously their directions when they are in a single plane [...]

The fourth meaning of the expressions "sense of the current" and "direction of the current" is related to the fact that the current is approaching or moving away from a point, from a straight line or from a plane. Let us think of the point J of figure 2.5 as a fixed point inside the battery. We can say that the current in points A and B are in the same direction, moving away from J, while the current in point I would be in the opposite direction, moving towards J. In this example the current located exactly at the points J

[36] [Ampère, 1826c, pp. 434-435].
[37] [Ampère, 1826f, p. 24], [Ampère, 1823c, Ampère, 1990, p. 196] and [Ampère, 1965b, pp. 169-170].
[38] [Ampère, 1826f, p. 48] and [Ampère, 1823c, Ampère, 1990, p. 220].
[39] [Ampère, 1885o, p. 223].

and F would not be moving instantaneously towards point J nor away from this point. We can also think in a straight line connecting two fixed points like J and F, one inside the battery and the other inside the wire, or we can think of a vertical plane perpendicular to the circuit and passing through points J and F. In this case the currents in points A, B and G would be flowing in the same direction, moving away from the straight line JF, while the currents in points I, D and E would be flowing in opposite direction, that is, towards the straight line JF. In this case the currents in points J, C, F and H would not be moving towards the straight line JF nor away from this line. Ampère utilized this meaning of the expression "direction of a current" when referring to a principle of symmetry in order to establish if the force between two current elements in a specific configuration was a force of attraction, of repulsion or if there were no forces at all between them. Here is an example of how Ampère utilized this meaning for this expression:[40]

> Considering then two small portions of electric currents, one portion in the plane and the other portion directed perpendicularly to this plane, it was easy for me to realize that, in the first place, when this last portion was above or below the plane, the two senses in which this portion can be traveled by the electric current are different from one another by the following circumstance that, in one case, this current moves towards the plane and, in the other case, it moves away from the plane; [...]

2.5 The Direction of the Force and Its Algebraic Sign

In his entire electrodynamic researches Ampère always assumed that the force between two current elements was always along the straight line connecting their centers, following the principle of action and reaction. In a paper of 1820, for instance, he said the following:[41]

> M. Ampère's goal in this work is to show that all facts relative to the mutual action between two magnets, relative to the mutual action between a voltaic conductor and a magnet, discovered by M. OErstedt, and relative to the mutual action between two conductors which he was the first to observe, can be deduced from a single cause, namely, a force sometimes attractive and sometimes repulsive between the infinitely small portions of what he called electric currents, acting always along the straight line connecting their centers; this being the only direction in which the author thinks that an attractive or repulsive force can be exerted, no matter its nature. [...] But, as noted by M. Ampère in his work, his explanation of the action of voltaic conductors upon magnets has a double advantage, [...] To admit only attractive and repulsive forces between two points along the straight line connecting these two points.

A similar statement can be found in *chemise 162*, first published by Joubert in 1885.[42]

In Ampère's time the vector notation had not yet been developed. This notation appeared in complete form only in the middle of the XIXth century.[43] In order to characterize an attractive force, Ampère considered it as positive, while a repulsive force was considered as negative. His first attempts to obtain a mathematical expression describing the interaction between two current elements can be found on *chemise* 158 of box 8 of his manuscripts.[44] Blondel believes that this undated manuscript was probably written in the second half of October 1820. However, it was not published at this time and was not communicated to the members of the Academy of Sciences of Paris. Its first partial publication appeared only in Appendix III of Blondel's paper.[45] In this manuscript Ampère mentioned that "a repulsion should be considered as a negative attraction". Similar statements were published in his *Théorie*:[46]

> [...] we shall take the sign + when the two currents, flowing in the same direction, attract, and the sign − in the other case.

Similarly:[47]

> [...] this is what expresses the − sign found in front of the general expression [...] of this force, according to the common use in which the attractions are considered as positive forces and the repulsions are considered as negative forces.

[40] [Ampère, 1822p, p. 210] and [Ampère, 1885m, p. 247].
[41] [Hofmann, 1987, p. 326, n. 47], [Hofmann, 1996, pp. 253, 386 and 397] and [Ampère, 1820d, pp. 546 and 549].
[42] [Ampère, b, *carton* 8, *chemise* 162] and [Ampère, 1885j, pp. 134-135].
[43] [Crowe, 1985].
[44] [Ampère, b, *carton* 8, *chemise* 158].
[45] [Blondel, 1978].
[46] [Ampère, 1826f, p. 28], [Ampère, 1823c, Ampère, 1990, p. 200] and [Ampère, 1965b, p. 173].
[47] [Ampère, 1826f, pp. 130-131] and [Ampère, 1823c, Ampère, 1990, pp. 302-303].

2.6 The Current Intensity and the Size of the Current Element

As we have seen in equations (2.2) and (2.3), in the *Théorie* Ampère presented his force between two current elements as being proportional to the product between ids and $i'ds'$. These magnitudes were clearly defined as follows:[48]

> I will now explain how to deduce rigorously from these cases of equilibrium the formula by which I represent the mutual action of two elements of voltaic current, showing that it is the only force which, acting along the straight line joining their midpoints, can agree with the facts of the experiment. First of all, it is evident that the mutual action of two elements of electric current is proportional to their length; for, assuming them to be divided into infinitesimal equal parts along their lengths, all the attractions and repulsions of these parts can be regarded as directed along one and the same straight line, so that they necessarily add up. This action must also be proportional to the intensities of the two currents. To express the intensity of a current as a number, suppose that another arbitrary current is chosen for comparison, that two equal elements are taken from each current, and that the ratio is required of the actions which they exert at the same distance on a similar element of any other current if it is parallel to them, or if its direction is perpendicular to the straight lines which join its midpoint with the midpoints of two other elements. This ratio will be the measure of the intensity of one current, assuming that the other is unity.
>
> Let us put i and i' for the ratios of the intensities of two given currents to the intensity of the reference current taken as unity, and put ds and ds' for the lengths of the elements which are considered in each of them; their mutual action, when they are perpendicular to the line joining their midpoints, parallel to each other and situated a unit distance apart, is expressed by $ii'dsds'$; we shall take the sign $+$ when the two currents, flowing in the same direction, attract, and the sign $-$ in the other case.
>
> If it is desired to relate the action of the two elements to gravity, the weight of a unit volume of suitable matter could be taken for the unit of force. But then the current taken as unity would no longer be arbitrary; it would have to be such that the attraction between two of its elements ds and ds', situated as we have just said, could support a weight which would bear the same relation to the unit of weight as $dsds'$ bears to 1. Once this current was determined, the product $ii'dsds'$ would denote the ratio of the attraction of two elements of arbitrary intensity, still in the same situation, to the weight which would have been selected as the unit of force.

Ampère seems to have been the first scientist to express in words the idea that a current in a wire is proportional to the amount of electric charge passing through the cross section of the wire in unit time.[49] Nowadays it is common to express this relation by $i = dq/dt$, in which i represents the current intensity, while dq represents the infinitesimal amount of charge passing through the cross section of the wire in the infinitesimal time interval dt. Ampère also believed that the current intensity should be proportional to the product between the amount of electric charge and the velocity with which this charge flowed relative to the wire. Let ds represent an infinitesimal length along the direction of the current in a wire. Nowadays it is common to replace ids by qv where v represents the velocity of the charge q relative to the wire. In a manuscript related to a talk he gave to the Academy of Sciences of Paris on 4 December 1820, Ampère said the following:[50]

> g and h depend only on the [amount of] electricity flowing [through the cross section of the conductor] in equal times, the quantity compensating the velocity. It is [proportional to] the quantity of motion, provided the path is completely free, that is, provided the conductor is sufficiently thick.

In his first published paper he said the following:[51]

> The currents of which I am speaking accelerate until the electromotive force is balanced by the inertia of the electric fluids and the resistance they experience by the imperfections of even the best conductors, whereupon they continue indefinitely at a constant speed so long as this force conserves the same intensity; but they cease instantly whenever the circuit is interrupted.

Ampère was not always so clear in his statements. Initially he represented the current element ids by the symbol g and the current element $i'ds'$ by the symbol h. In particular, in his first manuscripts and verbal statements he made a misleading confusion of current element length and current intensity. For instance, in one of his early publications he expressed himself as follows:[52]

---

[48] [Ampère, 1826f, pp. 27-28], [Ampère, 1823c, Ampère, 1990, pp. 199-200] and [Ampère, 1965b, pp. 172-173].

[49] [Blondel, 1982, pp. 92 and 157] and [Benseghir, 1989, p. 459].

[50] [Ampère, b, *carton* 8, *chemise* 162].

[51] [Ampère, 1820c, p. 64], [Ampère, 1965a, p. 141] and [Chaib and Assis, 2007d, p. 94].

[52] [Ampère, 1820f, p. 174], [Hofmann, 1996, pp. 249-253] and [Chaib and Assis, 2009b, p. 122], our emphasis.

> In order to have a clear idea of this law [of force between current elements], one must imagine in space a line representing in magnitude and in direction the resultant of two forces which are similarly represented by two other lines, and suppose, in the directions of these three lines, three infinitely small portions of electric currents, *the intensities of which are proportional to their lengths*. The law at issue consists in the fact that the small portion of electric current directed along the resultant exerts, on another current or on a magnet, an attractive or repulsive action equal to what would result, in the same direction, from the combination of the two portions of current directed along the components.

It would be difficult for someone to understand what Ampère meant by "the intensities of electric currents proportional to their lengths". Gillet de Laumont (1747-1834) tried to clarify statements like this. In a paper published in 1820, in which he made a summary of Ampère's researches, he substituted this expression by the following:[53]

> When we wish to have a clear idea of the law expressed by M. Ampère in his work, one must imagine in space a line representing, in magnitude and in direction, the resultant of two forces which are similarly represented by two other lines, and suppose, in the directions of these three lines, three electric currents *of which the attractive or repulsive forces are proportional to their lengths*.

That is, Laumont clarified that the *forces* between two current elements are proportional to their lengths, instead of stating that the intensities of the electric currents were proportional to their lengths. In 1823 Ampère utilized a statement similar to Laumont's. Ampère replaced the expression "the intensities of electric currents proportional to their lengths" by the statement that the "attractive or repulsive forces are proportional to their lengths".[54]

The first substitution of the symbols g and h by ids and $i'ds'$, respectively, happened for the first time in 1822 in Ampère's fundamental paper in which he obtained the final value of his force between two current elements:[55]

> [...] the intensities of the action of two small portions of conductors which I called g and h in the note of the *Journal de Physique* will be represented here, as their lengths are ds and ds', by ids and $i'ds'$, [...].

In the *Théorie*,[56] Ampère had "put i and i' for the ratios of the intensities of two given currents to the intensity of the reference current taken as unity". We believe there are two possibilities for his choice of the symbol i to represent the intensity of the electric current. The first possibility is that this symbol is the first letter of the word "intensity". The second possibility is that in his first publications he called the intensities of two current-carrying wires by the symbols g and h, although still confusing the intensities of the currents in each element with their lengths. As the next letter of the alphabet after g and h is the letter i, Ampère may have chosen this letter for alphabetical reasons in order to represent the current intensity, while representing its infinitesimal length by ds. In any event, no matter his reasons for choosing the letter i to represent the intensity of a current, this convention is still adopted in all textbooks dealing with electromagnetism.

Moreover, it should be remarked that not only for Ampère, but also in modern textbooks, the magnitude i is always positive or zero, but never negative. This convention is not true for electric charges and for magnetic poles, which can assume positive or negative values. Ampère believed in a double current with positive and negative charges moving relative to the wire. As we said before, nowadays we assume that in metal wires only the negative conduction electrons move relative to the wire, while the positive ions remain fixed in the lattice. Despite this modern knowledge, we still adopt the convention that $i \geq 0$ and $i' \geq 0$.

2.7 The Distance between the Two Current Elements

In all his works Ampère represented the distance between two infinitesimal current elements ids and $i'ds'$ by r. However, these current elements are not point-like, having infinitesimal lengths ds and ds'. Therefore, it was necessary to specify what he meant by the distance between them. In the *Théorie* this distance was specified as follows:[57]

> Suppose we now consider two elements placed arbitrarily; their mutual action will depend on their lengths, on the intensities of the currents of which they are part, and on their relative position. This position can be determined by the length r of a straight line joining their midpoints, [...].

[53] [Laumont, 1820] and [Hofmann, 1996, p. 253].
[54] [Ampère, 1822p, pp. 212-213] and [Ampère, 1885m, pp. 248-249].
[55] [Ampère, 1822o, p. 412].
[56] [Ampère, 1826f, pp. 27-28], [Ampère, 1823c, Ampère, 1990, pp. 199-200] and [Ampère, 1965b, p. 173].
[57] [Ampère, 1826f, p. 28], [Ampère, 1823c, Ampère, 1990, p. 200] and [Ampère, 1965b, p. 173].

Analogous statements can be found in his earlier works.

2.8 The Angles Appearing in Ampère's Force

An important aspect to be emphasized here is that, for Ampère, the magnitudes i, i', ds, ds' and r which appeared in his force between current elements, equations (2.1) up to (2.3), were always positive. Therefore, only the angular components of this expression can transform a positive force into a negative force, that is, transform an attractive force into a repulsive force.

The main difference between Ampère's force and the other interactions known in his time (gravitational, electrostatic and magnetic forces) was exactly in the angular dependence of Ampère's expression. Ampère's force given by equations (2.1) up to (2.3) is much more complex than equations (1.1), (1.2) and (1.3). The distinction between Ampère's force and the other forces known in his time can be illustrated with a simple example in which two current elements ids and $i'ds'$ are interacting with one another separated by a fixed distance r. By changing only the spatial orientation of the two current elements it is possible to transform a positive force into a negative force. Nothing of this kind happens in the gravitational force between two bodies, in the electrostatic force between two electrified bodies, nor in the force between two magnetic poles.

In this Section we consider in detail the meaning of each one of the angles appearing in Ampère's force. Although this is a crucial portion of Ampère's work, it has received little attention in the literature. Moreover, although Ampère wrote a lot about these angles, we did not locate in his published papers any figure illustrating explicitly these angles.

2.8.1 The Angle between Two Current Elements

In the *Théorie* Ampère represented the angle between the two interacting current elements by ε:[58]

> This formula is simplified by introducing ε for the angle between the two elements in place of ω; [...]

Ampère presented for the first time a formula for the force between two current elements in a meeting of the Academy of Sciences of Paris on 4 December 1820. He wrote a paper describing what he had read at this meeting which was only published by Joubert in 1885. In this work Ampère considered the interaction between two small current-carrying line segments and called[59] "h the angle of the direction of the two lines". This last symbol h is different from the symbol h appearing in equation (2.3). He seldom utilized the letter h to represent the angle between two current elements. Usually he utilized the symbol ε to represent this angle. For this reason we will also utilize in this Section this symbol to represent this angle.

There are several possibilities to conceive the angle between two current elements. Let us consider two current elements of lengths ds and ds' directed along the senses of the two currents i and i' with their centers separated by a distance r. In order to understand the angle between these two elements we will superpose them in such a way that their centers coincide with one another. In this case they form a single plane which we will consider as being the plane of the paper in figure 2.11. There are two angles connecting the elements, ε_1 and $\varepsilon_2 = \pi - \varepsilon_1$, figure 2.11 (a). Ampère always considered the angles obtained from the directions of the two elements. In the *Théorie*, for instance, he specified the angles θ and θ' as follows:[60]

> Suppose we now consider two elements placed arbitrarily; their mutual action will depend on their lengths, on the intensities of the currents of which they are part, and on their relative position. This position can be determined by the length r of a straight line joining their midpoints, the angles θ and θ' between a continuation of this line and *the directions of the two elements in the same direction as their respective currents*, and finally by the angle ω between the planes drawn through each of these directions and the straight line joining the midpoints of the elements.

As he always considered the angles obtained from the directions of the two elements, we can be sure that what he understood by the angle ε between the two elements should be considered as the angle ε_1 of figure 2.11 (a) and not the angle ε_2.

But there are also two possibilities to consider the angle between two elements in the same directions as their respective currents, as represented in figure 2.11 (b). In this figure we have angles ε_1 and ε_3 between the two elements in the same directions as their respective currents, such that $\varepsilon_3 = 2\pi - \varepsilon_1$. Ampère's oldest

[58][Ampère, 1826f, p. 32], [Ampère, 1823c, Ampère, 1990, p. 204] and [Ampère, 1965b, p. 176].
[59][Ampère, 1885j, p. 134].
[60][Ampère, 1826f, p. 28], [Ampère, 1823c, Ampère, 1990, p. 200] and [Ampère, 1965b, p. 173], our emphasis.

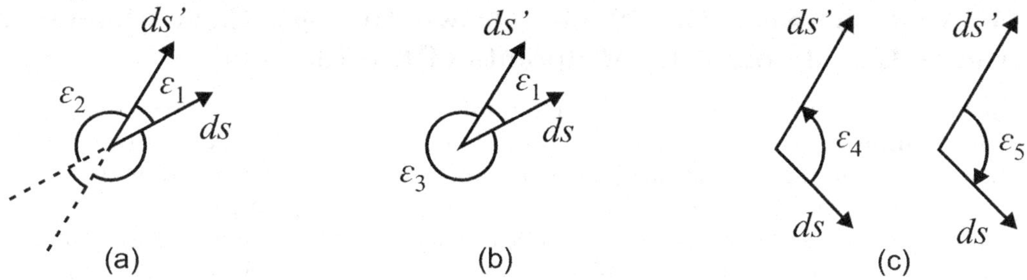

Figure 2.11: There are several ways of defining the angle between two current elements of directed lengths ds and ds'.

manuscript dealing with the force between current elements is from October 1820, being partially published by Blondel. In this paper he mentioned the following:[61]

> When the two currents are directed along the two lines which do not meet, instead of being directed along parallel lines, either in the same sense or in opposite senses, but in which their directions make an angle, there will be attraction when, by considering these lines in the same sense as the currents flowing in them, the angle which they form is acute; however, this attraction is always smaller than the attraction taking place when the currents are parallel and directed in the same sense, and this attraction always decreases as this angle increases, up to point in which this attraction goes to zero when the two currents flow in orthogonal directions. When the same angle is obtuse, the attraction is transformed into a repulsion; this repulsion [...] reaches its maximum value when this angle is equal to two right angles, that is, when the two currents are directed in opposite senses along two parallel lines.

Therefore, according to Ampère the angle between the directions of the two currents has a value between 0 rad and π rad. We can then conclude from this manuscript that he considered the angle between the two elements of figure 2.11 to be the angle ε_1 of figure 2.11 (b), and not the angle ε_3.

However, a third doubt still remains. Should we consider oriented angles? For instance, we might consider positive a clockwise angle and negative an anti-clockwise angle. In this case the angle ε_4 of figure 2.11 (c) going from ds to ds' would be negative, while the angle $\varepsilon_5 = -\varepsilon_4$ going from ds' to ds would be positive. But Ampère never mentioned oriented angles, nor negative angles. We can then conclude that according to Ampère's point of view the angle going from ds to ds' would be equal to the angle going from ds' to ds. For this reason we will not utilize oriented angles in our figures related to Ampère's electrodynamics, although Hofmann represented some of Ampère's angles as being oriented.[62] Hofmann did not justify his utilization of oriented angles. However, as he did not obtain any consequence or conclusion from this choice, it seems that choice of oriented angles was casual and of no relevance.

We can then finally conclude that the angle ε between two elements of lengths ds and ds', with these elements oriented along the directions of the currents flowing in them, is like the angle represented in figure 2.12, with this angle taking its value between the following limits: 0 rad $\leq \varepsilon \leq \pi$ rad. These general properties will also be valid for the other angles α, β and γ (or θ, θ' and ω). That is, each one of these angles has a value between 0 rad and π rad. Moreover, each one of these angles is not oriented, being always positive no matter if it is drawn clockwise or anti-clockwise.

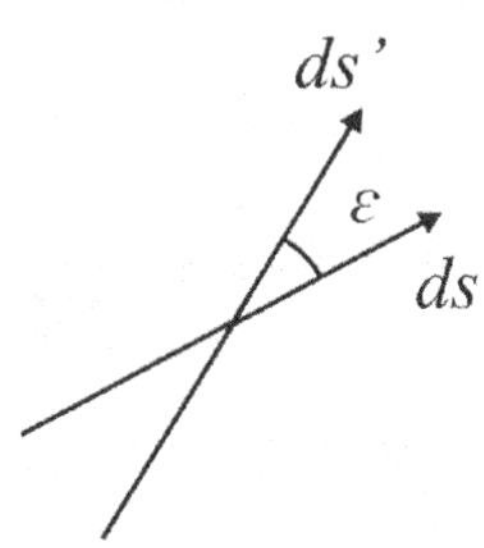

Figure 2.12: The angle ε between two oriented current elements ds and ds', as considered by Ampère.

[61] [Ampère, b, *carton* 8, *chemise* 158] and [Blondel, 1978].
[62] [Hofmann, 1987] and [Hofmann, 1996, p. 241, figure 5].

2.8.2 The Angle between the Planes Drawn through Each Element and the Straight Line Joining the Midpoints of the Elements

Although each current element of Ampère had an infinitesimal length, it was oriented along the direction of the current flowing through it and it was possible to think of the straight line passing through this element. Therefore, the straight line joining the midpoints of two elements forms a plane with the straight line passing through each element. In Ampère's force the angle between the planes drawn through each element and the straight line joining their midpoints also appears. In this case there is no ambiguity, provided we consider the smaller angle between two semi-planes. Each semi-plane was formed by the positive sense of the corresponding current element and the straight line connecting the two elements. The only aspect to take notice of here is the nomenclature utilized by Ampère. In his works of 1820 to 1822 this angle was represented by the symbol γ. However, in the *Théorie* he adopted the letter ω to represent the same angle. This angle is illustrated in figure 2.13. It is not oriented and has its value limited by 0 rad $\leq \gamma \leq \pi$ rad.

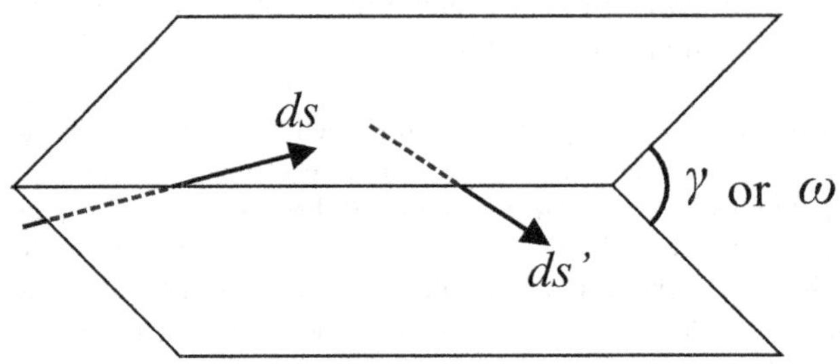

Figure 2.13: The angle between the planes drawn through each element and the straight line joining their midpoints.

2.8.3 The Angles between the Elements and the Straight Line Joining Their Midpoints

The angles requiring greater attention are those between each current element and the straight line connecting their midpoints.

In the *Théorie* Ampère introduced as follows the angles between the current elements and the straight line connecting their centers:[63]

> Suppose we now consider two elements placed arbitrarily; their mutual action will depend on their lengths, on the intensities of the currents of which they are part, and on their relative position. This position can be determined by the length r *of a straight line joining their midpoints, the angles* θ and θ' *between a continuation of this line and the directions of the two elements in the same direction as their respective currents*, and finally by the angle ω between the planes drawn through each of these directions and the straight line joining the midpoints of the elements.

According to this definition, there are two ways of representing the angles θ and θ', figure 2.14. That is, if we consider the continuation of the line going from ds' to ds, we have θ_1 and θ_1'. On the other hand, by considering the continuation of the line going from ds to ds', we have θ_2 and θ_2'.

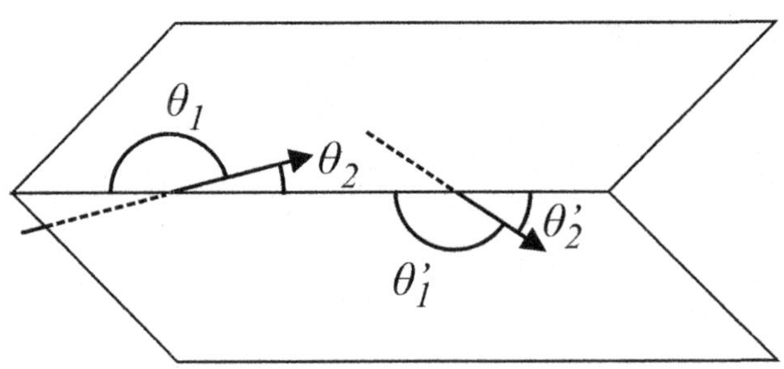

Figure 2.14: The angles between the current elements and the straight line connecting their midpoints.

From figure 2.14 we obtain that $\theta_2 = \pi - \theta_1$ and $\theta_2' = \pi - \theta_1'$. Therefore:

[63] [Ampère, 1826f, p. 28], [Ampère, 1823c, Ampère, 1990, p. 200] and [Ampère, 1965b, p. 173], our emphasis.

$$\cos\theta_2 = -\cos\theta_1 \ , \tag{2.22}$$

$$\sin\theta_2 = \sin\theta_1 \ , \tag{2.23}$$

$$\cos\theta_2' = -\cos\theta_1' \ , \tag{2.24}$$

and

$$\sin\theta_2' = \sin\theta_1' \ . \tag{2.25}$$

In Ampère's force, equation (2.2), we have the products $\sin\theta\sin\theta'$ and $\cos\theta\cos\theta'$. From equations (2.22) up to (2.25) we obtain that $\cos\theta_2\cos\theta_2' = \cos\theta_1\cos\theta_1'$ and $\sin\theta_2\sin\theta_2' = \sin\theta_1\sin\theta_1'$. Therefore, we can consider θ_1 and θ_1' as the angles between the elements and the straight line connecting them. These angles can also be considered as θ_2 and θ_2'. In both cases we will have the same force between the two current elements.

Ampère utilized the symbols θ and θ' in order to represent the angles between the two elements and the straight line connecting them only in the *Théorie*. In his earlier works he represented these angles by α and β.[64] In 1823, on the other hand, he represented[65] these angles by two symbols, namely, α and a curled beta.[66]

In conclusion, Ampère's three angles α, β and γ (or θ, θ' and ω) should be understood as the angles represented in figure 2.15. As happened with the angle ε between the two current elements, the following relations are also valid for these three angles: 0 rad $\leq \alpha \leq \pi$ rad, 0 rad $\leq \beta \leq \pi$ rad and 0 rad $\leq \gamma \leq \pi$ rad (or 0 rad $\leq \theta \leq \pi$ rad, 0 rad $\leq \theta' \leq \pi$ rad and 0 rad $\leq \omega \leq \pi$ rad).

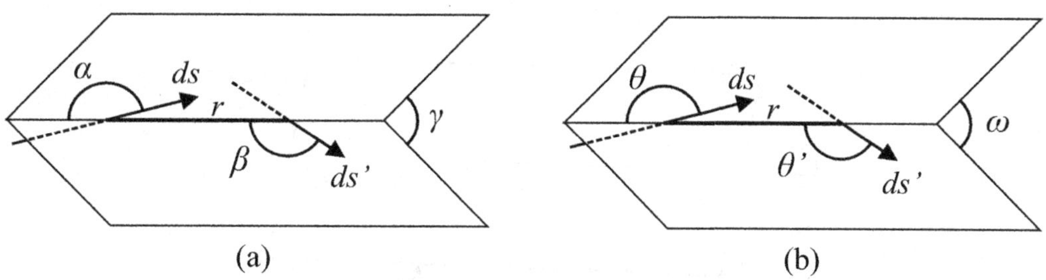

Figure 2.15: Let r be the distance between the midpoints of the current elements of lengths ds and ds'. (a) Representation of the angles α, β and γ according to Ampère's specifications. Analogous representation of the angles θ, θ' and ω.

A figure like this one appears in Hofmann's PhD dissertation,[67] and in Darrigol's book.[68] An illustration like figure 2.15 representing these three angles does not appear in some of the main works discussing Ampère's electrodynamics, namely, Maxwell,[69] Whittaker,[70] Tricker[71] and Williams.[72]

There are some works which represented these angles θ and θ' (or α and β) as being the angles between the current elements and the finite segment connecting their midpoints. As examples we can quote Tricker,[73] figure 2.16; Blondel,[74] figure 2.17; and Hofmann,[75] figure 2.18.

[64] [Ampère, b, *carton* 8, *chemise* 158], [Ampère, 1822o] and [Ampère, 1885j]. This last work was presented by Ampère to the Academy of Sciences of Paris in December 1820, being first published by Joubert in 1885.

[65] [Ampère, 1822p, p. 229] and [Ampère, 1885m, p. 262].

[66] That is:

ϐ

[67] [Hofmann, 1982, p. 264, figure 2].

[68] [Darrigol, 2000, p. 9, figure 1.3].

[69] [Maxwell, 1954, vol. 2, Chapter II: "Ampère's investigation of the mutual action of electric currents", articles 502-527, pp. 158-174].

[70] [Whittaker, 1973].

[71] [Tricker, 1965].

[72] [Williams, 1981], [Williams, 1983] and [Williams, 1989a].

[73] [Tricker, 1962].

[74] [Blondel, 1982, p. 84, figure 22].

[75] [Hofmann, 1987] and [Hofmann, 1996, p. 241, figure 5].

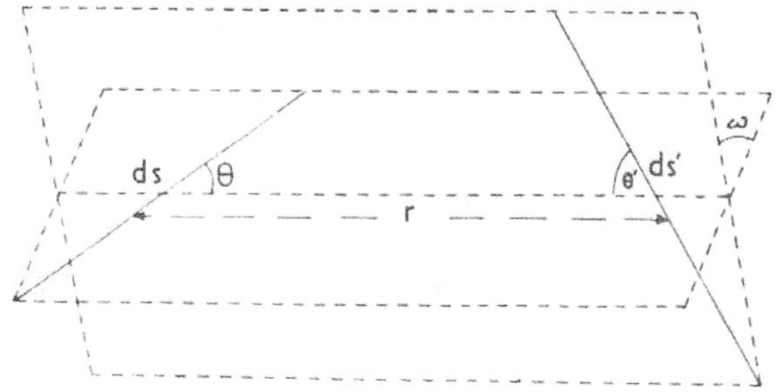

Figure 2.16: Representation of the angles θ, θ' and ω according to Tricker.

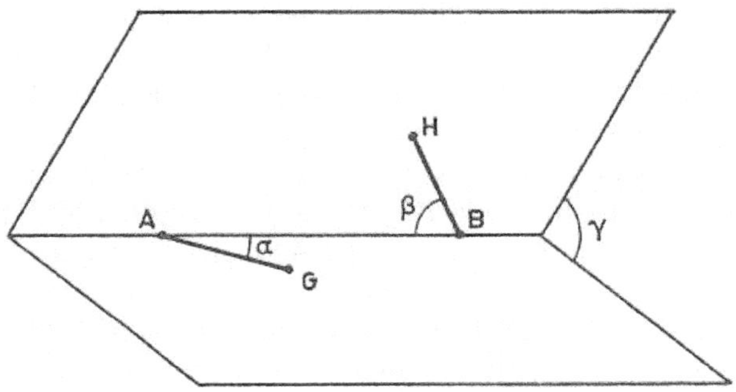

Figure 2.17: Representation of the angles α, β and γ according to Blondel.

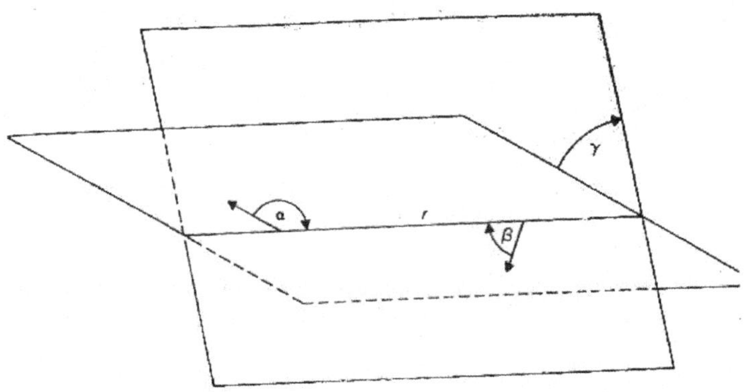

Figure 2.18: Representation of the angles α, β and γ according to Hofmann.

Ampère had specified that these angles θ and θ' should be considered between the directions of the two elements in the same direction as their respective currents and *a continuation of the line joining their midpoints.* These authors, on the other hand, considered these angles between the directions of the two elements and *the finite straight segment between their midpoints.* They represented, in particular, the angles α or θ as being the angle θ_2 of figure 2.14. Likewise, they represented the angles β or θ' as being the angle θ_1' of figure 2.14. With this representation, the angular portion of Ampère's force given by equation (2.2) would assume the following form (by utilizing equations (2.22) up to (2.25)):

$$\sin\theta\sin\theta'\cos\omega + k\cos\theta\cos\theta' = \sin\theta_2\sin\theta_1'\cos\omega + k\cos\theta_2\cos\theta_1'$$

$$= \sin\theta_1\sin\theta_1'\cos\omega \ - \ k\cos\theta_1\cos\theta_1' \neq \sin\theta_1\sin\theta_1'\cos\omega \ + \ k\cos\theta_1\cos\theta_1' \ . \qquad (2.26)$$

Equation (2.26) shows that the representations of the angles θ and θ' (or α and β) made in these specific works by Tricker, Blondel and Hofmann do not lead to Ampère's own force between current elements. Therefore, these representations do not agree with Ampère's own point of view.

Probably the reason why these authors arrived at this false representation of Ampère's angles is related to Ampère's own initial lack of clarity. In his first manuscripts of 1820, for instance, he said the following:[76]

> [...] I had also announced that this value [of the force between two current elements] should depend on the respective orientation of two small portions of current which are being considered. This orientation, in the general case, is determined by three angles, being the first two angles those which their directions [of the currents] make with the line connecting their midpoints, the third [angle] being the angle between the two planes passing through this last line and by the two small portions of electric current.
>
> Representing by r the distance between them [that is, between the midpoints of the two current elements], representing by g and h the intensities of their currents, representing by dx and dz their lengths, and representing finally by α and β the angles which they make with the line joining their centers, and representing by γ the angle between the two planes, [...]

In December of 1820 he expressed himself as follows, considering the length of each current element as being a "small line in space":[77]

> [...] I will only indicate here how I obtained the mathematical expression of the action between two currents or, rather, the way in which the angles determining, in general, the respective orientation of two small lines in space appear in this expression. [...] The angles referred here are, in general, three in number. Let α be the angle formed by one of these lines with that line joining the center [of the first element] to the center of the other [small line or current element], let β be the corresponding angle relative to this last [small line or current element], and let γ be the angle formed by the two planes passing through each of these small lines and by the line joining their centers.

By statements like these, one easily gets the impression that Ampère was referring the angles θ and θ' (or α and β) as being the angles of each current element with the line segment connecting their centers. It was only in 1822 that he clearly specified that these angles, called α and β in this paper, should be considered between each current element and *the same continuation of the straight line connecting their midpoints.* The following quotation should be considered in relation to figure 2.19:[78]

> [...] let ρ be a function of the three angles determining the relative orientation of two infinitely small portions of electric currents, being proportional to the force they exert on one another at a specific distance when this orientation is modified, and if these three angles are represented by α, β and γ; with α and β being the angles which the directions of these two small portions form with the straight line joining their centers, and γ being the mutual inclination of the planes of these two angles [...]. Let $Mm = ds$ and $M'm' = ds'$ represent two infinitely small portions of these conductors, and let their directions be determined by the two tangents MT and $M'T'$: if r represents the distance MM', [...] and if we consider the angles α and β in such a way as to have their openings towards the same side, as I supposed when calculating the value of ρ, the angle α being considered, for instance between the direction MT of Mm and the continuation MK of $M'M$, [then] the angle β should be considered between the direction $M'T'$ of $M'm'$ and the line $M'M$, [...].

Our figure 2.19 is a reproduction of figure 14 of Ampère's original paper showing the letters K, M, m, T, M', m' and T'.

Figure 2.20 presents a simplified version of figure 2.19. It includes the main elements of Ampère's original figure and also the angles α and β according to his specifications. From figure 2.20 we can see that Ampère's angles α and β coincide with the angles θ_1 and θ_1' of figure 2.14.

In 1823 Ampère presented similar specifications on how to consider the angle between each current element and the same continuation of the line connecting their midpoints.[79] In this paper these angles were represented by α and by a curled beta.[80] This work of 1823 will be discussed in Subsection 4.2.1, figures 4.11, 4.12, 4.13 and 4.14. These specifications were also presented by Ampère in his *Théorie*, this time representing these angles by θ and θ'.[81]

Beyond these specifications expressed in words, we were also able to find one of his drawings, unfortunately not yet published, in which he represented the angles θ and θ', figure 2.21.[82] Ampère's drawing is analogous to our figure 2.15 (b).

---

[76] [Ampère, b, *carton* 8, *chemise* 158].
[77] [Ampère, 1885j, pp. 133-134].
[78] [Ampère, 1822o, pp. 406 and 408], [Ampère, 1822y, pp. 303-305] and [Ampère, 1885p, pp. 277-279].
[79] [Ampère, 1822p, pp. 229-232] and [Ampère, 1885m, pp. 262-265].
[80] See the footnote 66 of our page 45.
[81] [Ampère, 1826f, p. 28], [Ampère, 1823c, Ampère, 1990, p. 200] and [Ampère, 1965b, p. 173].
[82] [Ampère, b, *carton* 9, *chemise* 173].

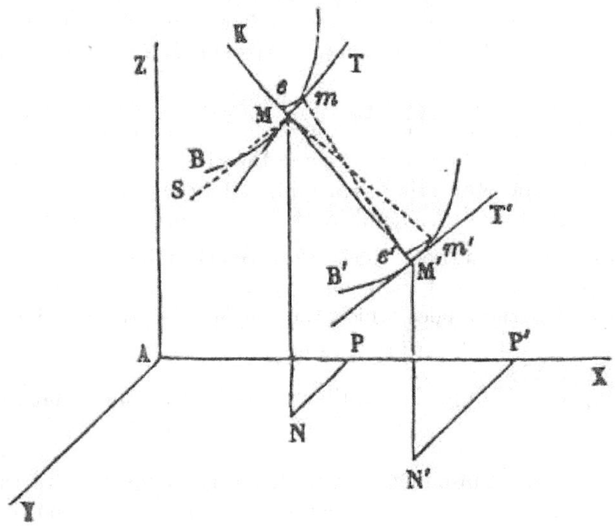

Figure 2.19: Ampère's original figure 14 of [Ampère, 1822o].

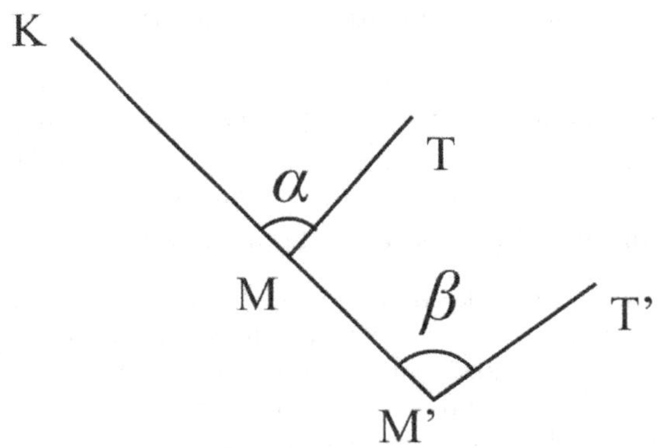

Figure 2.20: Representation of the angles α and β according to Ampère's own specifications.

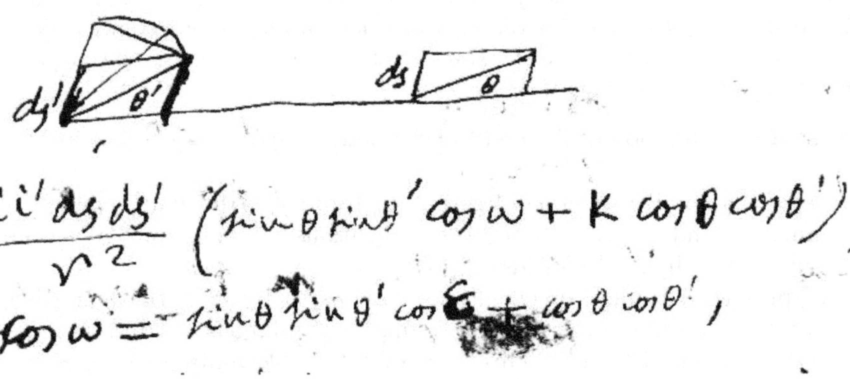

Figure 2.21: Ampère's drawing representing the angles θ and θ'.

Figure 2.22 presents Ampère's drawing representing the angles α and β.[83] This manuscript has not yet been published. Once more we can see that he understood these angles as having their openings to the same side.

In conclusion, figure 2.15 presents the correct representation of Ampère's angles α, β and γ (or θ, θ' and ω) according to Ampère's own specifications. These specifications appeared in his most important works from 1822 onwards.

[83] [Ampère, b, *carton* 11, *chemise* 206ter].

Figure 2.22: Ampère's drawing representing the angles α and β. This figure is analogous to our figure 2.15 (a).

It should be mentioned here that in Blondel and Wolff's recent works, they also utilized this representation of Ampère's angles.[84]

[84][Blondel and Wolff, d] and [Blondel and Wolff, c].

Part II

Origins and Evolution of Ampère's Force between Current Elements

We now discuss some of the main experiments, ideas and theoretical assumptions made by Ampère leading him to the final value of his force between current elements. The motivation for this analysis is a statement by James Clerk Maxwell (1831-1879), figure 2.23, one of the main scientists of the XIXth century.

Figure 2.23: J. C. Maxwell.

This statement refers to Ampère's deduction of his force law as presented in his masterpiece of 1826, the *Théorie*, based only on four cases of equilibrium:[1]

> The method of Ampère, however, though cast into an inductive form, does not allow us to trace the formation of the ideas which guided it. We can scarcely believe that Ampère really discovered the law of action by means of the experiments which he describes. We are led to suspect, what, indeed, he tells us himself[2], that he discovered the law by some process which he has not shewn us, and that when he had afterwards built up a perfect demonstration he removed all traces of the scaffolding by which he had raised it.

We will discuss the process followed by Ampère utilizing what he mentioned in his earlier papers, correspondence and manuscripts.

[1][Maxwell, 1954, vol. 2, article 528, pp. 175-176].

[2][Note by Maxwell:] *Théorie des phénomènes Électrodynamiques*, p. 9.

Chapter 3

Ampère's Initial Experiments

3.1 Ampère's Interpretation of Ørsted's Experiment

In Section 1.3 we described Ørsted's observations of the deflection of a magnetic needle from its original orientation along the magnetic meridian due to the influence of a nearby current-carrying wire. In order to interpret this experiment, Ampère had a very original, rich and fruitful idea which guided him to several new experiments and to a whole new set of phenomena which no one had observed before him. His main ideas were to suppose that all magnetic and electromagnetic phenomena were only due to interactions between current-carrying conductors, together with the assumption of the existence of electric currents in the Earth and also in magnets. This insight opened his mind to a whole set of new possibilities which might be tested and explored from several points of view, namely, experimental, conceptual and mathematical. He then looked for phenomena showing directly forces and torques between current-carrying conductors. He also began to look for a mathematical expression which might explain all these phenomena. In the end he was extremely successful in his endeavors.

Here are some of his words related to this topic:[1]

> This action that M. OErsted discovered led me to look for the interaction of two electric currents, the action of the Earth on a current, and how the electricity might produce all phenomena presented by the magnets, looking for a distribution [of electric currents inside each magnet] similar to that of a conductor of electric current, with closed curves perpendicular to the axis of each magnet. These points of view, most of which have only recently been confirmed by experiment, were communicated to the *Académie* in its session of 18 September 1820.

Another quotation of the same work:[2]

> I will only say that after having deduced only the first general result from the note of M. OErsted, I deduced from it the explanation of the magnetic phenomena, based upon the existence of electric currents in the globe of the Earth and in magnets.

The following statement is clearly revealing:[3]

> When first I wanted to find the causes of the new phenomena discovered by M. OErsted, I reflected that since the order in which two facts are discovered in no way affects any conclusions which can be drawn from the analogies they present, it might, before we knew that a magnetized needle points constantly from South to North, have first been known that a magnetized needle has the property of being influenced by an electric current into a position perpendicular to the current, in such a way that the austral pole of the magnet is carried to the left of the current, and it could then have subsequently been discovered that the extremity of the needle which is carried to the left of the current points constantly towards the North: would not the simplest idea, and the one which would immediately occur to anyone who wanted to explain the constant direction from South to North, be to postulate an electric current in the Earth in a direction such that the North would be to the left of a man who, lying on its surface facing the needle, received this current in the direction from his feet to his head, and to draw the conclusion that it takes place from East to West in a direction perpendicular to the magnetic meridian?

[1] [Ampère, 1820f, p. 196], [Ampère, 1965a, p. 152] and [Chaib and Assis, 2009b, p. 132].
[2] [Ampère, 1820f, pp. 200-201] and [Chaib and Assis, 2009b, p. 134].
[3] [Ampère, 1820f, pp. 202-203], [Ampère, 1965a, p. 152] and [Chaib and Assis, 2009b, pp. 135-136].

This was a radically new interpretation not only of Ørsted's experiment, but also of all magnetic phenomena related to the known attractions and repulsions between two magnets, or the orientation of the compass needle by the magnetic influence of the Earth. This insight led Ampère to a whole new research program. In particular, he predicted the interaction between current-carrying conductors which no one had observed before him. Ampère was also the first scientist to observe this phenomenon. Moreover, he predicted the interaction between the Earth and a current-carrying conductor. Once more he was also the first to observe this fact. Moreover, he was led to look for spatial distributions of electric currents which might mimic or reproduce the known interactions between magnets. That is, he tried to simulate the behavior of magnets utilizing only current-carrying conductors. He also tried to simulate terrestrial magnetism and the phenomena associated with it utilizing only electric currents flowing in wires. He tried to reproduce the interaction of a magnet with the Earth utilizing current-carrying conductors. He also tried to simulate Ørsted's experiment (orientation of a magnetic needle by a current-carrying wire) utilizing only current-carrying wires. This new conception led Ampère to look for a mathematical expression describing the interaction between current-carrying conductors with which he could not only explain quantitatively all these phenomena, but also predict new ones. All these aspects of his research did follow naturally from his new theoretical conception which was shown to be extremely fruitful and productive. Amazingly Ampère was able to accomplish all this immense research program in the time interval of six short years.

Although his interpretation may seem obvious, natural, simple or reasonable to us, it was rejected by the main scientists working on this subject in his times, like Ørsted, Biot, Savart and Faraday. His points of view were also rejected by later researchers, like Grassmann. Moreover, several key aspects of Ampère's interpretation are against the electromagnetic theory of the present time based on the concept of a magnetic field. All of these aspects will be discussed at length in Parts IV and V of the present book, called Controversies.

As mentioned by Bertrand in 1874:[4]

> The laws discovered by Ampère have remained in science as a solid and uncontested basis which support with confidence even those who tried to replace them with other [laws]. It is, in effect, by the complete agreement of the proposed principles with those of Ampère, in every case where verification is feasible, that it was thought possible to justify the new theories.

3.2 Orientation of a Magnetic Needle by a Current-Carrying Wire

Magnetism presents two general phenomena which Ampère called *attractive and repulsive action* and *directive action*.[5] The first phenomenon is the usual attraction and repulsion between two magnets. As an example, consider two magnetized bars aligned along their lengths. Let us suppose that their poles are aligned in the sequence N, S, N' and S', as in figure 3.1. When the two bars are released from rest relative to the ground in this orientation, they will attract one another, decreasing the distance between their opposite poles S and N'. If they were initially aligned with their poles in the sequence N, S, S' and N', then they would repel one another after being released from rest relative to the ground.

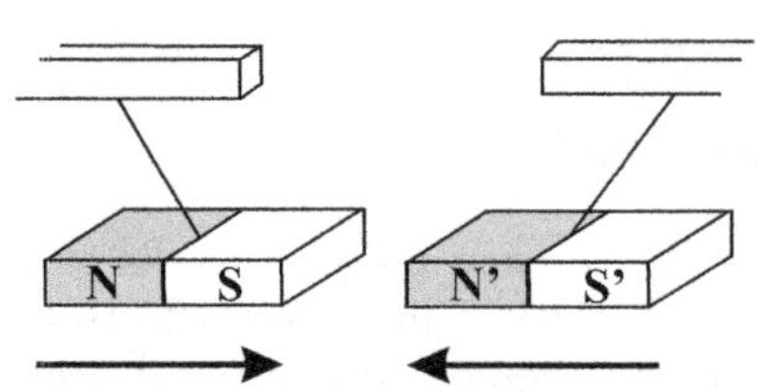

Figure 3.1: Attraction of two magnetized bars.

In the directive action, on the other hand, a magnetized needle suffers a torque due to the influence of terrestrial magnetism or due to the influence of other magnets. The magnetized needle will tend to turn around its center due to this directive action, acquiring a fixed orientation relative to the ground. Consider that we have a horizontal compass and release it in an arbitrary orientation relative to the ground, being free to turn around a vertical axis passing through its center. We observe that normally it will not remain in this initial orientation, but will stop at another very specific orientation due to the influence of

[4][Bertrand, 1874, p. 297].

[5][Ampère, 1820f, p. 197], [Chaib and Assis, 2009b, p. 132], [Ampère, 1820a, p. 238] and [Ampère, a, p. 1].

terrestrial magnetism. In this equilibrium orientation, the vertical plane passing through the needle will also pass through the terrestrial North and South magnetic poles. These poles are located along the terrestrial magnetic axis. This magnetic axis does not coincide with the axis of the diurnal rotation of the Earth relative to the frame of fixed stars. The intersection of this last axis with the surface of the Earth defines its geographic North and South poles. The meridian or line of longitude is an arc of a circle on the surface of the Earth passing through its geographic poles. The intersection of the magnetic axis of the Earth with its surface defines the magnetic poles of the Earth. The magnetic meridian is an arc of a circle on the surface of the Earth passing through its magnetic poles. The magnetic South pole of the Earth is located close to its geographic North pole, while the magnetic North pole of the Earth is located close to its geographic South pole. After reaching equilibrium, the magnetized needle will be parallel to the local magnetic meridian. The angle between this magnetic meridian and the geographic meridian is called magnetic declination.

Suppose that the first magnetic needle reaches equilibrium, aligned along the magnetic meridian. Now suppose that we place a second magnetic needle in the vertical plane passing through the first needle. Consider also that this second needle is free to turn around a horizontal axis passing through its center. After reaching equilibrium, it will acquire a specific orientation relative to the ground, being inclined relative to the horizontal plane. The angle between this horizontal plane and the direction of this second needle is called magnetic inclination. This second magnetic needle is usually called dip needle or inclination compass.

We now present Ampère's initial experiments. The several devices which will be quoted in this book were not constructed directly by Ampère. They were built by the instrument maker Hippolyte Pixii (1808-1835). He was a French engineer who died at the early age of 27. Ampère paid for these instruments from his own pocket. As he always had monetary problems, many times he was in debt with Pixii.[6]

Arago reported on Ørsted's discovery to the Academy of Sciences of Paris on 4 September 1820. One week later he repeated Ørsted's demonstration for the members of the Academy. In the meeting of 18th of September 1820, Ampère presented a very important new result. He had created a new instrument, which he called *astatic magnetic needle*.[7] The meaning of the word astatic, as utilized by Ampère, is that of neutral or indifferent equilibrium. That is, an astatic needle is a magnetized needle which remains in equilibrium, no matter its initial orientation relative to the ground. It is not affected by terrestrial magnetism. The astatic needle can be considered as a third compass (compared to the two compasses mentioned in the previous paragraphs) which can rotate freely around an axis passing through its center, with this rotation axis being parallel to a nearby dip needle. Suppose we consider a horizontal line passing through the center of this astatic needle. Suppose that initially this horizontal line makes an arbitrary angle ξ with the axis of the astatic needle. We will then observe that the astatic needle will remain at rest relative to the ground, no matter the value of this angle ξ.

Ampère's device can be seen in figure 3.2.[8]

His words describing this astatic needle:[9]

> By placing the [rotation] axis of the astatic needle parallel to the resultants of the actions of the [terrestrial] globe, the needle will only be able to move in the plane perpendicular to these resultants. In this way the action of the globe will be destroyed and the needle will remain indifferent in all its orientations, that is, it will be completely astatic.

Figure 3.3 (a) presents the graduated disc of this instrument. The compass 1 of figure 3.3 (b) represents a normal horizontal magnetic needle which, in equilibrium, points along the local North-South direction. The dip compass 2 in this figure is located in a vertical plane above needle 1. In equilibrium its axis indicates the local inclination angle of the Earth. Ampère's astatic compass is the needle AB of figure 3.3 (a). It can turn freely around the rotation axis CD of this figure. This rotation axis CD is parallel to a dip needle like that of figure 3.3 (b). It can be placed in any orientation along the arc LMN. Although it is free to turn, it will remain in equilibrium no matter its orientation, provided it is only under the magnetic influence of the Earth.

In modern terms it is possible to say that an astatic needle can turn freely around an axis which is parallel to the local magnetic field of the Earth. As it suffers no magnetic torque from the Earth, it remains in equilibrium no matter its initial orientation relative to the ground.

Ørsted's experiment showed that a current-carrying wire also has a directive action, just like the Earth, as it affects the orientation of a magnetic needle placed near it. In Section 1.3 it was discussed how Ørsted

[6][Blondel, 1982, pp. 103, 131, 145 and 165].

[7][Ampère, 1820f, p. 198], [Chaib and Assis, 2009b, p. 133], [Ampère, 1820a, p. 239] and [Ampère, a, p. 2].

[8][Ampère, 1820f, figure 8] and [Chaib and Assis, 2009b].

[9][Ampère, 1820a, p. 239] and [Ampère, a, p. 2].

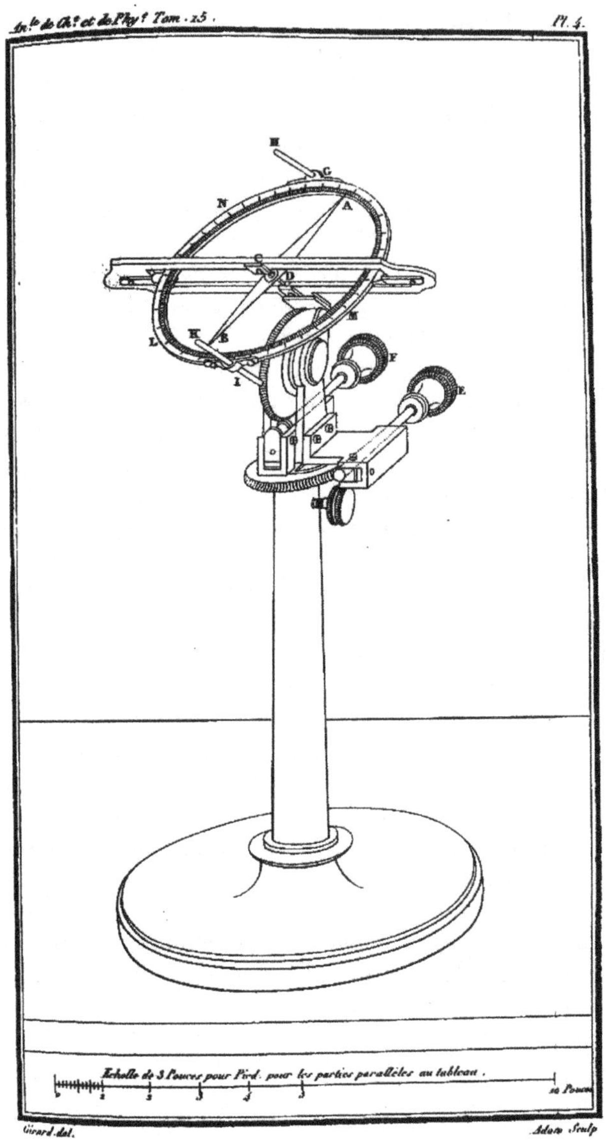

Figure 3.2: Ampère's astatic compass.

obtained a deflection of the needle relative to the magnetic meridian. The value of this deflection decreased with the increase of the distance between the needle's center and the wire. He obtained a deflection of 45° with a distance of 3/4th of an inch. Ørsted thought that this deflection was only due to the action of the current-carrying wire. Ampère, on the other hand, believed that this deflection angle was due to the combined influences of the Earth and current-carrying wire upon the needle. To avoid the directive influence due to terrestrial magnetism, he repeated Ørsted's experiment utilizing now his astatic needle. At the French Academy's meeting of September 18, 1820, he reported the results of his experiment as follows:[10]

> Then, when a galvanic current is close [to an astatic needle], its directive action will be the only one affecting the needle, and experiment shows that the needle always becomes exactly perpendicular to the direction of the current.

Ampère was the first scientist to show that a magnetized needle becomes perpendicular to a long and straight current-carrying wire, provided only the wire is exerting a torque upon the needle.

Ampère's rule to determine the deviation direction of a magnetic needle is to suppose an imaginary person along the wire, between the wire and the magnetized needle. The wire is at his back and the observer looks at the needle. Initially we can suppose the axis of the needle being parallel to this observer. When current

[10][Ampère, 1820a, p. 239] and [Ampère, a, p. 2].

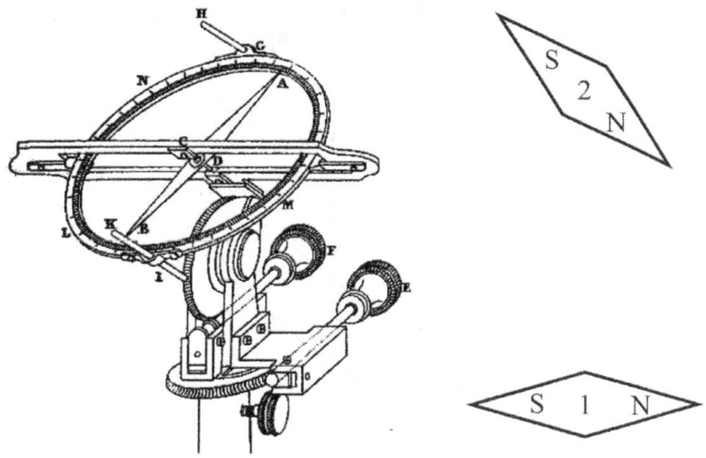

(a) (b)

Figure 3.3: (a) Ampère's astatic needle. (b) The magnetic needle 1 can turn freely around a vertical axis. In equilibrium it will point along the North-South direction. The magnetic needle 2 is vertically above needle 1 in equilibrium. Needle 2 can turn freely around a horizontal axis which is orthogonal to the local magnetic meridian. In equilibrium needle 2 will be inclined relative to the horizontal plane, indicating the local inclination or dip angle. The rotation axis CD of the astatic needle of figure 3.3 (a) is parallel to the dip needle 2.

flows in the wire, from the feet towards the head of this observer, he will observe the austral or North pole of the needle moving to his left side. The equilibrium configuration of figures 1.3 and 1.4 with the needle below or above the wire indicates these deflections, although in Ørsted's experiments there was also the magnetic influence of the Earth exerted on the needle.

3.3 Attraction and Repulsion between a Magnetic Needle and a Current-Carrying Wire

In the meeting of the Academy of Sciences of Paris which took place on 18 September 1820, Ampère showed that a current-carrying straight wire also has the attractive and repulsive action when interacting with a small magnetic needle. To this end he suspended the needle vertically by one of its poles, as if he was building a pendulum. He placed a horizontal straight wire perpendicular to this needle, in such a way that the wire and the center of the needle were in the same horizontal plane, with a small distance between them. He then observed an attraction of the needle towards the wire when there was a constant current flowing in one sense in the wire and a repulsion when the current flowed in the opposite sense. This attraction and repulsion was a *new fact*. Ørsted had not observed these net forces on the needle, as he only saw the wire deflecting the natural orientation of the needle. That is, Ørsted's observed in his experiment the wire exerting a torque on the magnetic needle. His experimental arrangement was not appropriate to indicate an attraction or repulsion of the needle. Ampère's original arrangement is presented in figure 3.4.[11]

Figure 3.5 (a) presents a simplified version of this experiment with the North pole of the needle suspended above its South pole, seen sideways. Figure 3.5 (b) presents this experiment seen from above. The letter F indicates the force exerted by the current-carrying wire on the needle. By reversing only the sense of the current, or the polarity of the needle, the force becomes attractive. By reversing simultaneously the polarity of the needle and the sense of the current, the force remains repulsive.

[11] [Ampère, 1820f, figure 9] and [Chaib and Assis, 2009b].

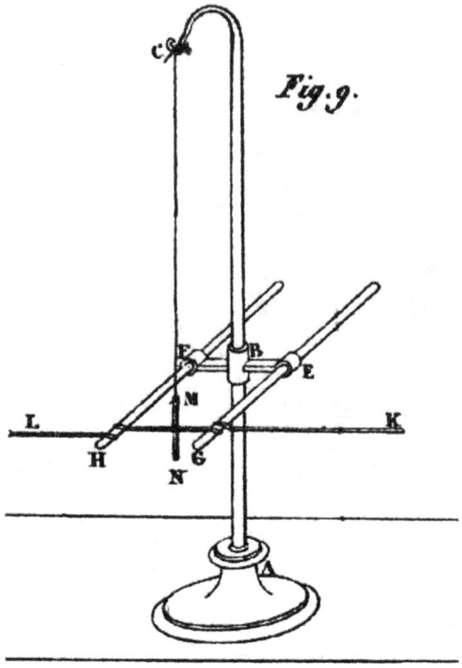

Figure 3.4: A vertical magnetic needle MN can be attracted or repelled by a long horizontal wire KL when there is a constant current in the wire.

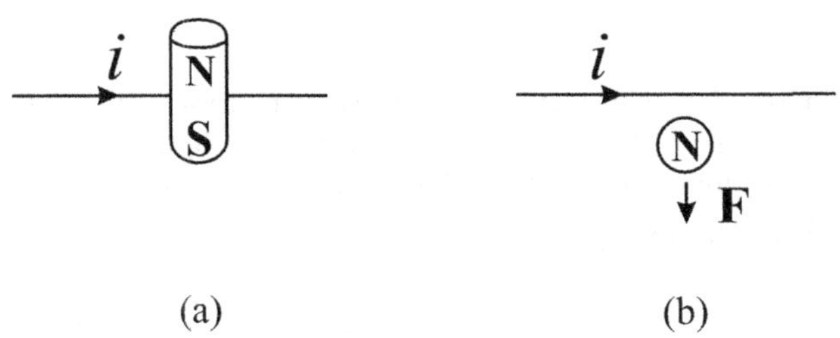

Figure 3.5: (a) Simplified representation of figure 3.4, seen sideways. (b) The experiment seen from above. The letter F represents the force exerted by the horizontal wire on the vertical needle.

3.4 Discovery of the Closed Currents

At the meetings of the French Academy which took place on 18 and 25 September 1820, Ampère described another very important experiment which we mentioned on Sections 2.3 and 2.4. He utilized his galvanoscope.[12] This experiment was also described in his first published paper dealing with electrodynamics.[13] With this experiment Ampère concluded that in Volta's original experiment in which the terminals of a battery were connected to a conducting wire, the electric current flowed in a closed circuit. That is, it flows not only along the metal wire, but also inside the battery. In order to reach this conclusion, he placed his galvanoscope not only above or below the wire, as Ørsted had done, but also above the trough battery. When a current was flowing in this circuit, he observed the deflection of the magnetized needle of his galvanoscope placed above the battery. This fact indicated that the current was also flowing inside the battery. He could also ascertain the sense of the current inside the battery by the direction of the deflection of the magnetic needle. He concluded that the current followed a closed path flowing always in the same sense (clockwise, for instance) along the whole circuit composed of the battery and connecting metal wire.[14]

Although this conclusion may seem trivial to us nowadays, this was a new experimental discovery at that time. Jean-Baptist Biot (1774-1862), for instance, one of the main researchers of electromagnetism in Ampère's time, had rejected explicitly in 1816 the possibility that an electric current might flow inside the

[12] See Section 2.3 on our page 32.
[13] [Ampère, 1820c], [Ampère, 1965a] and [Chaib and Assis, 2007d].
[14] [Blondel, 1982, pp. 72-75], [Steinle, 2000], [Steinle, 2005], [Wolff and Blondel, 2005] and [Benseghir and Blondel, 2007].

battery.[15] In his first paper on electrodynamics Ampère did not present a figure describing this experiment. However, such a representation was included in the work of Babinet and Ampère. In this representation we can see the trough battery, figure 2.6.[16]

We illustrated Ampère's experiment with figure 2.7.

3.5 Reproducing the Attraction and Repulsion between Two Magnets

In the letter to his son, written in September 1820, quoted in Section 2.2, Ampère mentioned that his initial motivation was to explain Ørsted's new discovery and the magnetic phenomena already known for a long time beginning from a single principle. In particular, he supposed the existence of electric currents inside the Earth and inside magnets. He then interpreted Ørsted's experiment as being due to a direct interaction between the current flowing in the conductor and the supposed currents existing inside the magnet. In order to test his hypothesis of the existence of electric currents in the Earth and in magnets, Ampère tried to reproduce the magnetic phenomena already known utilizing only electric circuits.

Consider two magnetic bars aligned along their parallel axes with their poles in the sequence $NSN'S'$. They were long known to attract one another, as indicated in figure 3.1. If they were aligned with their poles in the sequence $NSS'N'$ they would repel one another.

Ampère tried to replicate this behavior utilizing only current-carrying wires. To this end he simulated a bar magnet utilizing an electric circuit with the current describing a planar spiral, as in figure 3.6.

Figure 3.6: Wire coiled in a planar spiral.

The plane of this spiral should be considered orthogonal to the magnetic axis of the magnetized bar it was replacing. He then suspended this spiral in a vertical plane, like a pendulum, with its horizontal axis aligned with the North-South axis of another horizontal magnetized bar. When there was a constant current flowing through the spiral, he observed an attraction between the spiral and the bar magnet. By reversing only the sense of the current, or the pole of the magnet which was closest to the spiral, the attraction was changed into a repulsion. By reversing both, the sense of the current and the closest pole of the magnet, their attraction remained. He described this experiment as follows:[17]

> Now, if electric currents are the cause of the directive action of the Earth, then electric currents could also cause the action of one magnet on another magnet. It therefore follows that a magnet could be regarded as an assembly of electric currents in planes perpendicular to its axis, their direction being such that the austral pole of the magnet, pointing North, is to the right of these currents since it is always to the left of a current placed outside the magnet, and which faces it in a parallel direction, or rather that these currents establish themselves first in the magnet along the shortest closed curves, whether from left to right, or from right to left, and the line perpendicular to the planes of these currents then becomes the axis of the magnet and its extremities make the two poles. Thus, at each pole the electric currents of which the magnet is composed are directed along closed concentric curves. I simulated this arrangement as much as possible by bending a conducting wire in a spiral. This spiral was made from brass wire terminating in two straight portions enclosed in two glass tubes[18] so as to eliminate contact and attach them to the two extremities of the battery.

[15] [Grattan-Guinness, 1990a, p. 920].

[16] [Ampère and Babinet, 1822a, p. 4, figure 1] and [Ampère and Babinet, 1822b, p. 169, figure 1].

[17] [Ampère, 1820f, pp. 207-208], [Ampère, 1965a, pp. 152-153] and [Chaib and Assis, 2009b, pp. 137-138].

[18] [Note by Ampère:] I have since changed this arrangement as I shall show later.

> Depending on the direction of the current, such a spiral is greatly attracted or repelled by the pole of a magnet which is presented with its axis perpendicular to the plane of the spiral, according as the current of the spiral and of the pole of the magnet flow in the same or opposite directions.

Ampère's experiment can be illustrated by figure 3.7. Ørsted had observed the directive action of the current-carrying wire acting on the magnetic needle. Ampère, on the other hand, observed the attractive and repulsive actions of a bar magnetic acting on a conducting wire bent in a spiral.

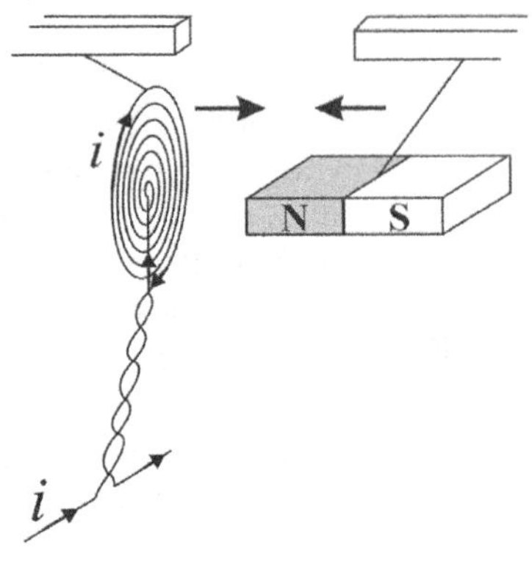

Figure 3.7: Attraction between a current-carrying wire bent in a spiral and a magnetized bar. The remainder of the closed circuit connecting the extremities of the wire to the battery is not shown. The wire connecting the spiral to the upper support is an insulator.

Ampère's important experiment with spirals can be easily repeated with simple materials.[19]

3.6 Interaction between Current-Carrying Wires

3.6.1 Interaction between Spirals

Ampère then performed a new experiment which opened a whole new area of research. This time, he replaced the magnetized bar of figure 3.7 by a second spiral. He had then two spirals in parallel vertical planes, with their centers at the same height above the ground and with both spirals sharing the same axis. When a constant electric current was flowing in both spirals, he observed their attractions or repulsions, depending on the senses of the currents in both spirals, as represented in figure 3.8.[20]

With this experiment he was able to reproduce the attraction and repulsion between two aligned magnetized bars, as in figure 3.1, utilizing two parallel spirals, as in figure 3.9.

Ampère described this observation as follows:[21]

> In replacing the magnet by another spiral with its current in the same direction, the same attractions and repulsions occur. It is in this way that I discovered that two electric currents attract each other when they flow in the same direction and repel each other in the other case.

This is one of the most important experiments ever performed in the history of physics. It showed for the first time the attraction and repulsion between current-carrying wires. This experiment created a whole new area of knowledge, namely, the interaction between electric currents. The magnetic properties of the Earth and other magnets were not relevant here. Later on Ampère called this new branch of science *electrodynamics*,[22] as described in Section 1.4. The experimental origin of this new science was this interaction between current-carrying spirals. Moreover, it was exactly this experiment which suggested to Ampère that two parallel electric currents should attract each other when they flowed in the same sense and should repel each other when they flowed in opposite senses. These spiral experiments were presented to the Academy of Sciences of Paris on 25 September 1820, being published in the same year.[23]

[19][Souza Filho et al., 2007] and [Assis et al., 2007].

[20][Ampère, 1820f, figure 11], [Ampère, 1965a, figure 47, p. 153] and [Chaib and Assis, 2009b, figure 11, p. 138].

[21][Ampère, 1820f, p. 208], [Ampère, 1965a, pp. 153-154] and [Chaib and Assis, 2009b, p. 138].

[22][Ampère, 1822d, p. 60], [Ampère, 1822j, note on p. 200], [Ampère, 1885e, note on p. 239], [Ampère, 1822e, note on p. 237] and [Ampère, 1885d, note on p. 192].

[23][Ampère, 1820f], [Ampère, 1965a, pp. 146-154] and [Chaib and Assis, 2009b].

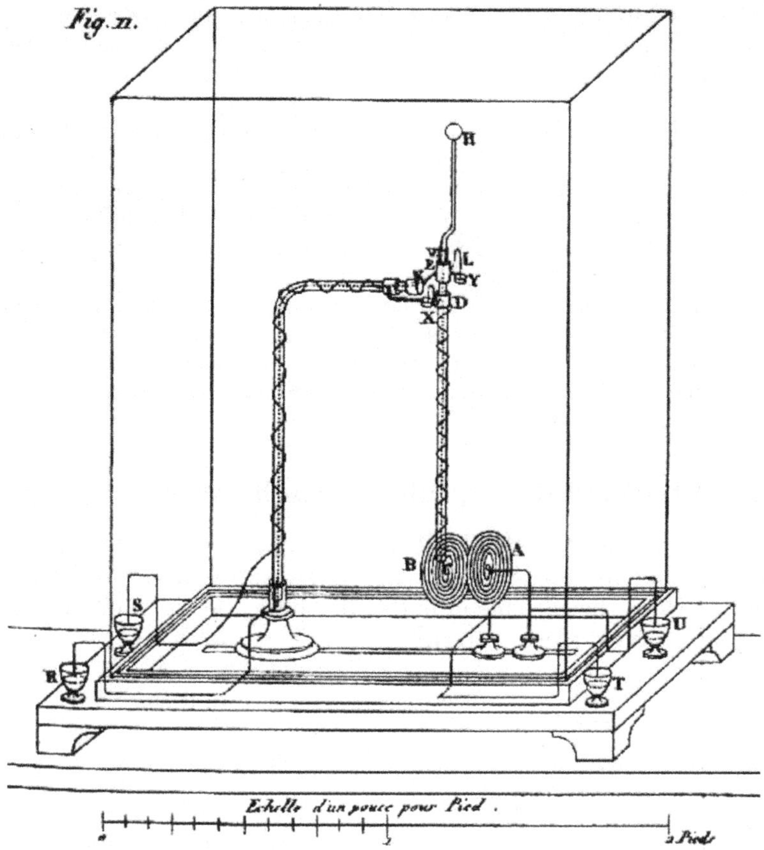

Figure 3.8: Ampère's first observation of attraction and repulsion between two current-carrying conductors. Spiral A remains fixed in the laboratory, while spiral B is mobile in such a way that it can move towards spiral A or away from it.

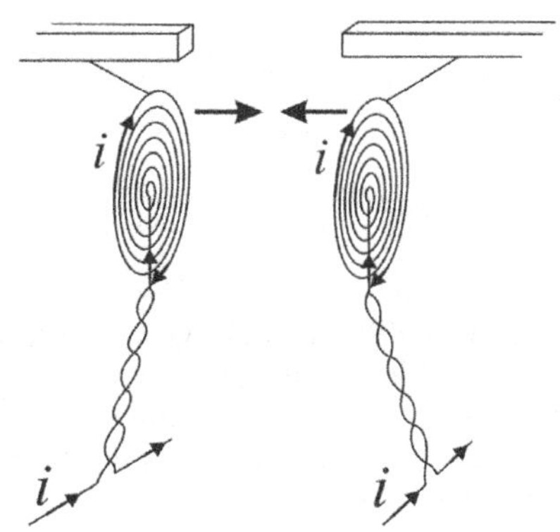

Figure 3.9: Attraction between two current-carrying spirals.

It might be thought that this important experimental discovery made by Ampère was a necessary consequence of Ørsted's experiment. However, Arago showed that this was not the case. After all, a magnet exerts forces upon a piece of soft iron but two pieces of soft iron are without effect upon each other.[24] Arago expressed himself as follows:[25]

> The infallible way of reducing to silence this passionate opposition, to undermine their objections, would be to quote an example of two bodies which, separately, act on a third body but which, on the other hand, do not exert any action on one another. A friend of Ampère [namely, Arago himself] observed that

[24][Whittaker, 1973, p. 84] and [Tricker, 1965, p. 23].

[25][Arago, 1854a, pp. 59-60].

> magnetism offers a phenomenon of this kind. He said the following to the complacent antagonists of the great geometer: "Here we have two soft iron keys. Each one of them attracts this compass. If you do not prove that, when presented to one another, these keys attract or repel one another, then the point of departure of all your objections will be false."

This counter example presented by Arago was expressed by Ampère in the following words:[26]

> When M. Oersted discovered the action which a conductor exerts on a magnet, it really ought to have been suspected that there could be interaction between two conductors; but this was in no way a necessary corollary of the discovery of this famous physicist. A bar of soft iron acts on a magnetized needle, but there is no interaction between two bars of soft iron. Inasmuch as it was only known that a conductor deflects a magnetic needle, could it have been concluded that electric current imparts to wire the property to be influenced by a needle in the same way as soft iron is so influenced without requiring interaction between two conductors when they are beyond the influence of a magnetized body? Only experiments could decide the question; I performed these in the month of September 1820, and the mutual action of voltaic conductors was demonstrated.

3.6.2 Interaction between Two Parallel Straight Wires

On 2 October 1820, Ampère presented to the Academy of Sciences of Paris his first published work on this subject.[27] On 9 October he showed the Academy an experiment in which two long parallel and straight wires attracted one another when the currents in both wires flowed in the same sense, repelling one another when they flowed in opposite senses, figure 3.10 (a).

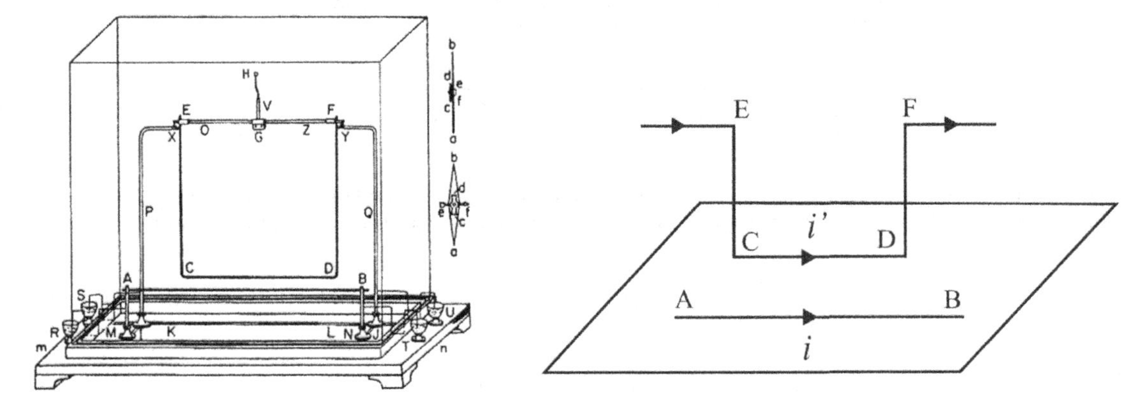

(a) (b)

Figure 3.10: (a) Ampère's demonstration that parallel straight wires carrying currents in the same sense attract one another, repelling when the currents flowed in opposite senses. Conductor AB is fixed in the laboratory, while the mobile conductor CD can turn around the horizontal axis EF, moving towards AB or away from it. The straight segments AB and CD are initially in the same horizontal plane, with the axis EF vertically above CD. (b) Our representation of this experiment, indicating the current senses.

Conductor AB is fixed in the laboratory, while the conductor $ECDF$ can turn around the horizontal axis EF, with the portion CD moving towards AB or away from it. Initially the straight segments AB and CD are located in the same horizontal plane. When the current flows from A to B and from C to D, the segment CD is attracted towards AB. By inverting the sense of the current in only one of these conductors, CD is then repelled by AB. On the other hand, by reversing the sense of the current in both conductors, the segment CD is once again attracted towards AB.

This experiment is one of the most famous demonstrations ever performed by Ampère, for several reasons. In the first place, this was the only figure appearing in his first published paper.[28] In the second place, it shows an interaction which Ampère considered fundamental, namely, the attraction and repulsion between rectilinear parallel current-carrying conductors. In the third place, this phenomenon is the basis of the

[26] [Ampère, 1826f, pp. 113-114], [Ampère, 1823c, Ampère, 1990, pp. 285-286] and [Ampère, 1965b, pp. 195-196].
[27] [Ampère, 1820c], [Ampère, 1965a, pp. 140-146] and [Chaib and Assis, 2007d].
[28] [Ampère, 1820c].

so-called *current balance* or *Ampère balance*. These balances, found in most didactic physics laboratories, are utilized to determine the intensities of electric currents. The current to be measured is passed in series through two pieces of wire, one of which is attached to one arm of a sensitive balance. The electrodynamic force between the two coils is measured by the amount of weight needed on the other arm of the balance to keep it in equilibrium.

Ampère's description of his experiment:[29]

> But the differences aforementioned are not the only ones which distinguish these two states of electricity. I discovered other more remarkable differences by disposing, in parallel directions, two straight parts of two conductors connected to the extremities of two voltaic batteries. One straight part was fixed, but the other, suspended from points and made highly mobile by a counter-weight, free to move parallel towards it or away from it.[30] I observed that by passing an electric current through both parts at the same time, they were mutually attracted when both currents were in the same sense, and that they repelled one another when the currents were in opposite directions.

3.7 Reproduction of the Directive Action of the Earth upon a Compass

As seen in Section 3.5, with two current-carrying spiral wires Ampère could reproduce the attractive and repulsive action between two aligned magnetized bars, figures 3.1 and 3.9.

Ampère tried now to reproduce the Earth's directive action upon a compass. It was known for centuries that a compass is aligned along the Earth's magnetic meridian. It was already known for a long time that a magnetized bar can also orient a magnetic needle. Ampère tried to reproduce the behavior of the compass in these experiments working only with a current-carrying wire. He tried to simulate the compass by a conducting helix coiled around a glass tube supported in the middle over a very thin rigid point, like the vertical needle over which a compass can turn freely in the horizontal plane.[31]

This current-carrying helix was oriented by a nearby bar magnet. In this way he succeed in reproducing the directive action between two bar magnets. This experiment happened at the end of September or beginning of October 1820. However, the Earth did not orient this current-carrying helix. He expected to see this helix oriented along the local magnetic meridian, but this effect did not take place. He suspected this lack of orientation might be due to the friction between the vertical pivot needle and the center of the glass tube at their point of contact. He then switched to a more mobile suspension shown in figure 3.11 (a).[32]

The helix was coiled around two hollow insulating glass tubes, ACD and BEF. A vertical wire HF entered the tube FEB in F, leaving at point B. It was then coiled into a helix around both glass tubes along $BECA$, entering the tube in A, leaving at point D, going down along DG. The helix coiled around both glass tubes could turn around the vertical axis HG. Ampère could also observe the orientation of this helix by approaching one of its extremities with a bar magnet. However, he was unsuccessful in orienting it by the Earth's magnetism, despite the extremely little friction of this apparatus. This anomaly was against his expectations and he was initially at loss to explain it.

On 17 October 1820, Ampère showed to Biot and Gay-Lussac (1778-1850) a simplified version of the instrument shown in figure 3.12.[33]

This instrument is similar to that presented in figure 3.10, but now with the two straight conductors making an adjustable angle with one another. Conductor SR is fixed in the laboratory. Its inclination relative to the vertical direction can be arbitrarily adjusted. Its distance to the conductor BC can also be adjusted at will. Conductor BC is mobile around the vertical axis passing through point D. Ampère described this instrument as follows:[34]

---

[29] [Ampère, 1820c, p. 69], [Ampère, 1965a, pp. 144-145] and [Chaib and Assis, 2007d, p. 97], our words in the footnote.

[30] In Ampère's figure 1, our figure 3.10 (a), the fixed straight part is represented by AB. The mobile wire is represented by $XCDY$. The counter-weight is represented by VH. The mobile portion can turn around the horizontal axis XY. With the counter-weight it is possible to adjust the distance between the rotation axis XY and the center of gravity of the mobile system (composed of the counter-weight plus the mobile portion $XCDY$). By decreasing this distance, the sensitivity of the balance is increased. This means that the mobile system will be deflected from the vertical plane by a greater angle when it is under the action of a horizontal force exerted by the fixed portion. This force can be attractive or repulsive.

[31] [Ampère, 1820f, p. 172] and [Chaib and Assis, 2009b, p. 121].

[32] [Ampère, 1820f, figure 3] and [Chaib and Assis, 2009b].

[33] [Ampère, 1820f, figure 6], [Ampère, 1965a, figure 45, p. 149] and [Chaib and Assis, 2009b, figure 6].

[34] [Ampère, 1820f, p. 182] and [Chaib and Assis, 2009b, p. 126].

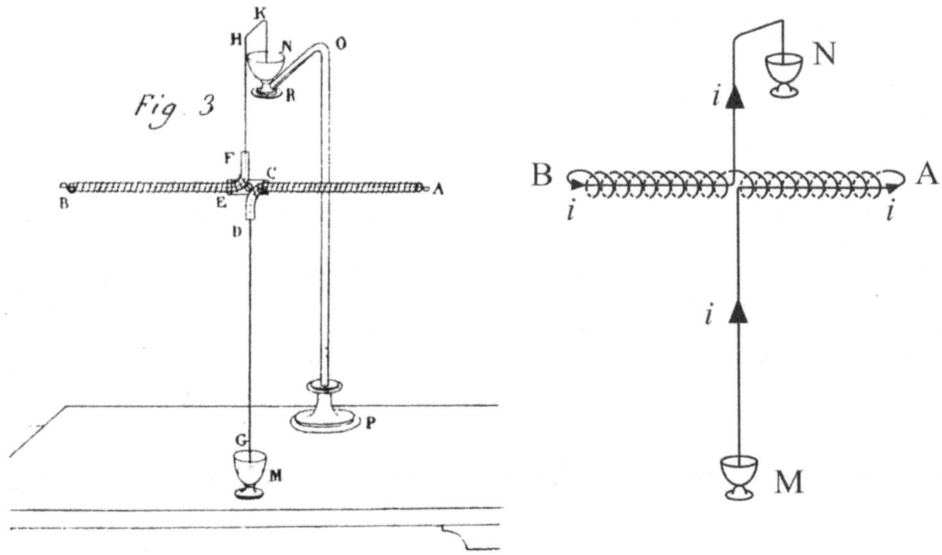

(a) (b)

Figure 3.11: (a) Ampère's helix. (b) Our reproduction of this figure indicating the current sense in this experiment.

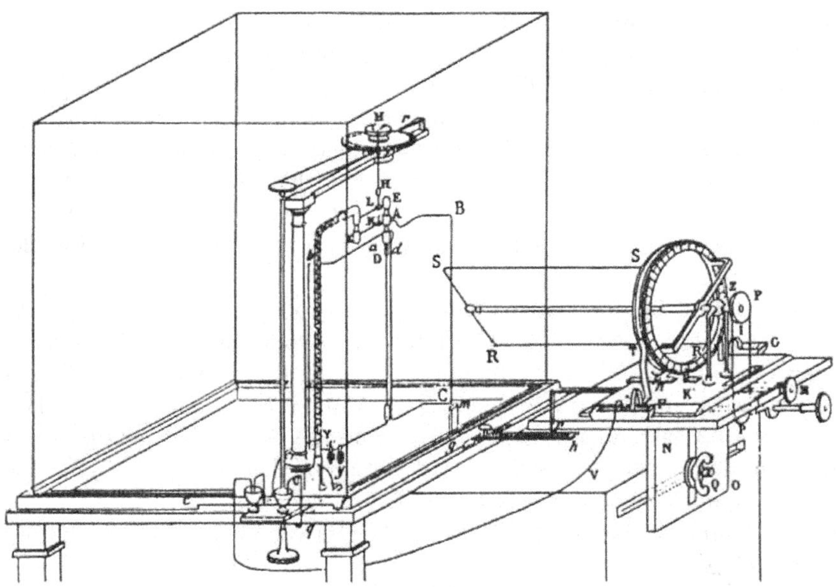

Figure 3.12: Instrument utilized to show the interaction between two conductors having variable orientations. Conductor SR is fixed in the laboratory, having an angular inclination which can be adjusted relative to the vertical direction. Conductor BC is mobile around the vertical axis passing through point D in such a way that it can move towards SR or away from it.

> I asked to be constructed, for these experiments, an instrument which I showed on 17 October [1820], to MM. Biot and Gay-Lussac, and which only differs from the apparatus represented in figure 1 [our figure 3.10] in the fact that the fixed conductor of this last instrument was replaced by a conductor connected to a circle which turned around a horizontal axis perpendicular to the direction of the mobile conductor, by means of a return pulley [*poulie de renvoi*], and graduated in such a way that we could see over the limb the angle formed by the directions of the two currents, in the different positions which could be done successively to the conductor carried by the graduated circle.

Ampère's intention to build such a complex instrument seems to have been to make a series of detailed measurements of the force and torque on the mobile conductor as a function of its distance and inclination to the fixed conductor. But apparently these measurements were never made. In any event, it was with this instrument that Ampère observed for the first time, completely by chance, without having anticipated this effect, the directive action of the Earth upon a mobile conductor carrying a steady current.[35]

What Ampère observed was the deflection of the mobile conductor BC around the vertical axis passing through point D when there was a current in this conductor. This deflection took place even when there was no current in the mobile conductor SR. He concluded that this torque on the mobile conductor was due to the Earth's electromagnetic action.

Ampère then realized that his failure in obtaining an orientation of the helix coil of figure 3.11 (a) by the Earth was not due to friction. This lack of motion was due to the small diameter of his coiled helices around the glass tube. He concluded that by increasing the diameter of the helices, he might increase the terrestrial torque acting on them, up to a point in which this effect might be visible and reproducible. He succeeded in showing this effect experimentally, proving that the torque upon a current-carrying loop did in fact increase with a larger diameter of the loop. Later on he built a large circular loop, almost closed, with 20 cm radius, which was freely suspended in a vertical plane by two vertical wires connected to the upper extremities of the loop. This circular loop could turn freely around a vertical axis passing through its center. When current flowed in this loop, it was oriented by terrestrial magnetism. In equilibrium the vertical plane of the loop remained orthogonal to the vertical plane passing through the local magnetic meridian, as represented in Ampère's figure 3.13.[36]

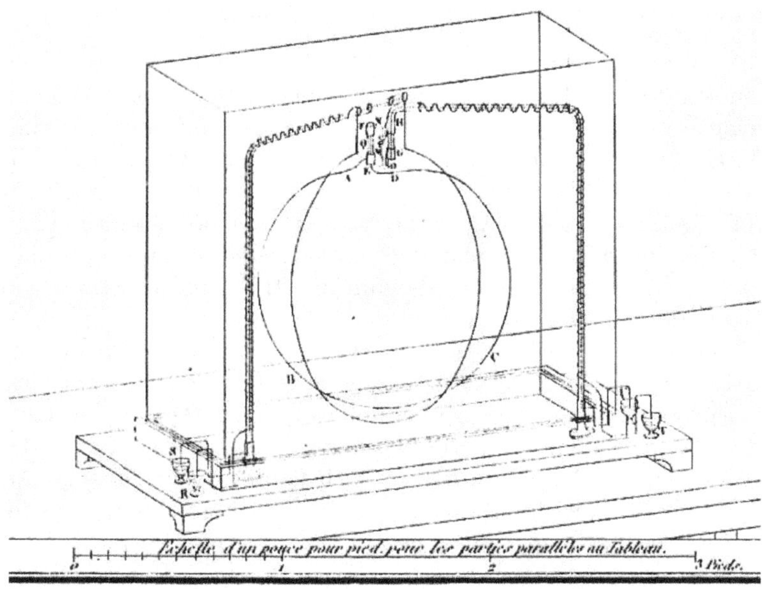

Figure 3.13: Current carrying circular loop which is oriented by terrestrial magnetism. The circular loop can turn freely around the vertical axis passing through its center.

That is, with the instrument of figure 3.13 Ampère succeeded in simulating with a current-carrying loop the behavior of an ordinary magnetic compass which is oriented along the local terrestrial magnetic meridian. In equilibrium the plane of Ampère's circular loop becomes orthogonal to the direction indicated by an ordinary horizontal compass needle.

Ampère then proceeded in order to simulate the behavior of a dip needle utilizing only current-carrying conductors. He was also successful in this endeavors. The instrument showing this effect appeared as figure 5 of the second part of his first published paper on electrodynamics,[37] reproduced here as our figure 3.14 (a).[38]

There is a rectangular loop of sides $BC = 30$ cm and $CD = 60$ cm. This rectangular loop can turn freely around the horizontal axis GH or QS. The battery should be connected to the cups U and T. Supposing

[35] [Ampère, 1820f, pp. 172-173 and 182-183] and [Chaib and Assis, 2009b, pp. 121-122 and 126].

[36] [Ampère, 1820f, figure 7], [Ampère, 1965a, figure 46, p. 151] and [Chaib and Assis, 2009b, figure 7].

[37] [Ampère, 1820f] and [Chaib and Assis, 2009b].

[38] [Ampère, 1820f, figure 5] and [Chaib and Assis, 2009b, figure 5].

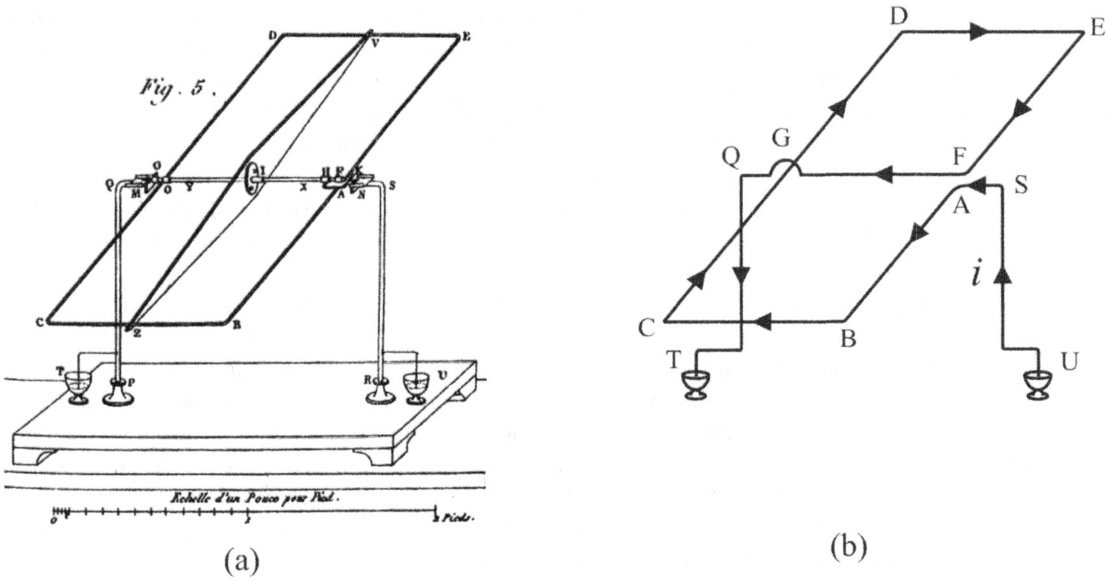

Figure 3.14: (a) Rectangular current-carrying loop which is oriented by terrestrial magnetism, in analogy with a dip needle. (b) Our figure indicating the sense of the electric current in this experiment.

that the positive terminal of the battery is connected at cup U, the electric current will follow the path $USABCDEFGQ$, leaving at cup T which is connected to the negative terminal of the battery. Suppose the rectangular loop $ABCDEF$ is initially at rest in a horizontal plane. By passing a constant current in this loop, it becomes inclined relative to the horizontal plane due to its interaction with terrestrial magnetism. In its new orientation of equilibrium, the plane of the rectangular loop becomes orthogonal to the axis of a dip magnetic needle. That is, the $ABCDEF$ plane becomes parallel to the equatorial plane of a dip needle. With this experiment Ampère reproduced electrodynamically the behavior of a dip needle utilizing only a current-carrying rectangular loop of large area.

We could not locate any modern author discussing this important experiment made by Ampère. The reason for this lack of attention of modern authors may be related to the fact that although this experiment is discussed in Ampère's paper, this specific figure is not mentioned explicitly in his article. Poudensan,[39] for instance, published in 1964 a summary of the two parts of Ampère's paper of 1820.[40] He reproduced Ampère's first four figures, skipped Ampère's original fifth figure, replacing it by Ampère's original sixth figure, but renumbering is as figure 5, without warning his readers about this fact.[41] The same also happened with a reproduction of Ampère's first paper published in 1921.[42] Ampère and Babinet's book is one of the rare works discussing this experiment.[43]

Ampère presented these results at the Academy of Sciences of Paris on 30 October 1820. With these electromagnetic experiments he could reproduce the main magnetic phenomena already known for a long time, replacing now the magnetic needle by appropriate current-carrying conductors.

3.8 Ørsted's Electrodynamic Experiment

Ampère was also successful in reproducing Ørsted's experiment utilizing only current-carrying conductors. That is, he replaced the magnetic needle of Ørsted's original experiment by a current-carrying wire which was oriented by another long straight wire carrying a steady current. This experiment performed by Ampère will be called here *Ørsted's electrodynamic experiment*. It was presented in the second part of his first published paper, figure 3.15.[44]

[39] [Ampère, 1964].

[40] [Ampère, 1820c] and [Ampère, 1965a, pp. 140-146], [Ampère, 1820f] and [Ampère, 1965a, pp. 146-154], [Chaib and Assis, 2007d] and [Chaib and Assis, 2009b].

[41] [Ampère, 1964, p. 39].

[42] [Ampère, 1921].

[43] [Ampère and Babinet, 1822b, p. 188] and [Ampère and Babinet, 1822a, p. 23].

[44] [Ampère, 1820f, figure 2] and [Chaib and Assis, 2009b, figure 2].

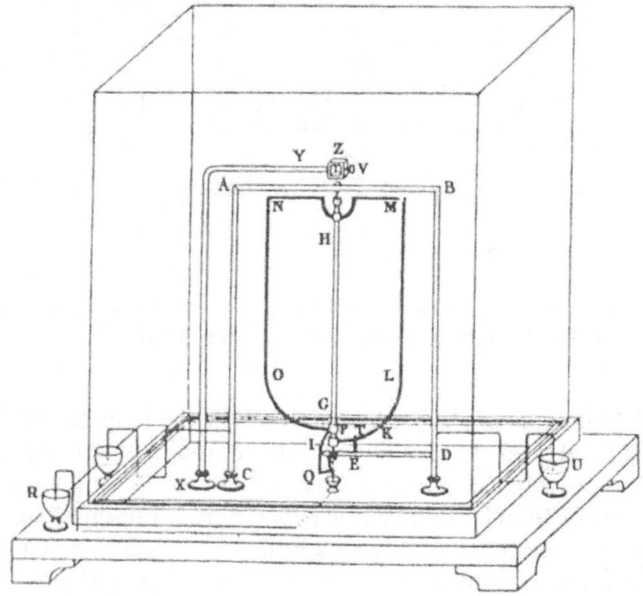

Figure 3.15: Ampère reproduced Ørsted's original experiment utilizing now only current-carrying conductors. In Ampère's original figure 2 of his paper, [Ampère, 1820f], the letters G and H along the central vertical axis do not appear. As these letters were mentioned by Ampère in his paper, they were included here in order to facilitate the comprehension of this experiment. A figure with these letters as they are shown here appeared in [Ampère, 1964, p. 36]. Conductor AB is fixed in the laboratory, while conductor MN is mobile. It can turn freely around the vertical axis $ZHGP$.

Ampère's own description of his experiment:[45]

> The electric current, arriving in this instrument by the support CA (figure 2 [our figure 3.15]), follows initially the conductor AB, returning by the support BDE. From this support, by the small steel goblet F, where I placed a globule of mercury, and inside which turned the steel pivot of the glass axis GH, the current communicated itself to the copper joint I and to the conductor $KLMNOPQ$, the extremity Q of which plunged into the mercury which was placed in communication with the other extremity of the battery. Everything being so arranged, it is clear that, in the situation in which this conductor is represented and in which I initially placed it, supporting it against the appendix T of the first conductor, the current of the part MN flowed in opposite sense as regards the current of AB, whereas when one made [the current] describe a semi-circumference $KLMNOPQ$, the two currents flowed in the same sense.
>
> Then the effect I expected was produced. At the moment in which the circuit was closed, the mobile portion of the instrument turned by the mutual action between this portion and the fixed conductor AB, until the situation in which the currents, which were initially in contrary senses, remained in such a configuration in which they were parallel and [flowed] along the same sense. The acquired speed made the circuit go beyond this last position; but it returned to it, went a little beyond it, and finished at rest after a few oscillations.

The mobile conductor $KLMNOPQ$ could turn around the vertical axis GH. The horizontal fixed conductor AB and the horizontal upper segment MN of the mobile conductor were always orthogonal to the vertical axis connecting the centers of these two conductors. The angle between the upper conductor MN and the fixed conductor AB could be changed by the torque exerted on the mobile conductor. We will represent by the symbol δ the angle between the vertical planes passing through each one of these conductors, namely, AB and MN, such that 0 rad $\leq \delta \leq \pi$ rad, figure 3.16.

When both currents flow in the same sense as, for instance, to the right in figure 3.16, with AB superimposed over the segment MN, we have $\delta = 0$ rad. When MN is orthogonal to AB we have $\delta = \pi/2$ rad. Moreover, $\delta = \pi$ rad when both currents flow in opposite senses as, for instance, with the current flowing to the right in AB and to the left in MN, with AB superimposed over the segment MN. With this electrodynamic experiment Ampère observed that the final orientation of equilibrium is the position in which

[45][Ampère, 1820f, p. 171] and [Chaib and Assis, 2009b, p. 121].

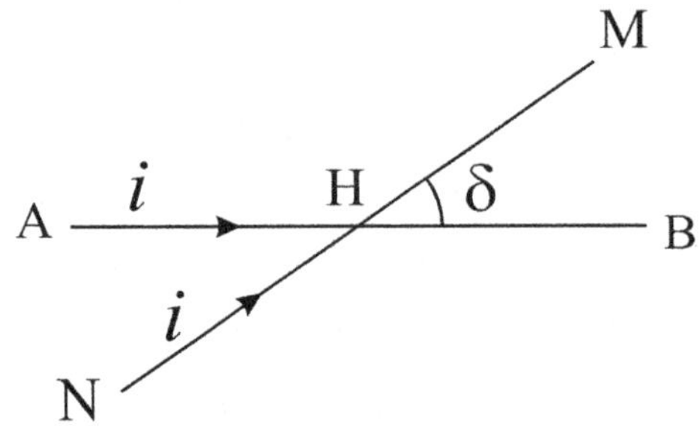

Figure 3.16: This drawing presents figure 3.15 as seen from above. The horizontal conductor AB is fixed in the laboratory, while the mobile horizontal conductor MN can turn around the vertical axis passing through point H.

$\delta = 0$ rad, no matter the initial value of the angle δ with which the loop $KLMNOPQ$ is released from rest. That is, the torque exerted by AB on the mobile loop always tends to decrease the angle δ.

Figure 3.17 presents qualitatively the torque acting on the mobile conductor as a function of the angle δ taking into account the results of this electrodynamic experiment represented in figure 3.15. This experiment is somehow analogous to that experiment represented in figure 3.12. The difference between them is that in the experiment of figure 3.15 Ampère was observing the torque between the interacting circuits, while in the experiment of figure 3.12 he was observing the net force acting on them.

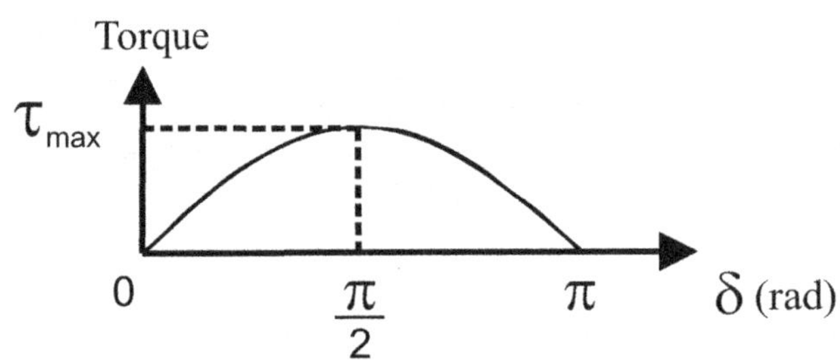

Figure 3.17: Qualitative behavior of the torque acting on the mobile circuit MN as a function of the angle δ in the experiment of figure 3.15.

According to figure 3.17, there is zero torque for $\delta = 0$ rad and for $\delta = \pi$ rad. The maximum torque happens at the angle $\delta = \pi/2$ rad. When $\delta = 0$ rad, there is stable equilibrium. On the other hand, when $\delta = \pi$ rad, there is unstable equilibrium. In this figure we are utilizing the convention that a positive torque tends to decrease the angle δ between the conductors, in the same way that a positive force tends to decrease the distance r between the current-carrying conductors.

This experiment had a great importance in the formulation of the mathematical expression describing Ampère's force between two current elements. Let us consider two current elements in the same plane with both currents flowing towards a certain point V, as figure 3.18 (a), or away from this point, as in figure 3.18 (b). In order to explain the origin of the torque on the mobile conductor of Ørsted's electrodynamic experiment, Ampère concluded that the current elements of situations (a) and (b) of figure 3.18 should attract one another. With this assumption he could explain how the torque on the mobile conductor in this experiment might make the angle δ go to zero. Ampère's explanation was based on an attraction between segments NH and AH of figure 3.16, together with another attraction between segments HB and HM.

Let us now consider the situation in which one of the currents was flowing towards point V while the other current was flowing away from V, as in cases (a) and (b) of figure 3.19. In order to explain Ørsted's electrodynamic experiment Ampère concluded that these current elements ids and $i'ds'$ should repel one another in both cases. With this assumption there would be a repulsion between segments NH and BH of figure 3.16, together with another repulsion between segments AH and HM. He could then explain the torque on the mobile conductor in such a way as to make the angle δ go to zero.

3.9 Ørsted's Inverse Experiment

In his original experiment, which he sent to several scientists in July 1820, Ørsted had modified the natural orientation of a magnetic needle due to the influence of a nearby long straight wire carrying a steady

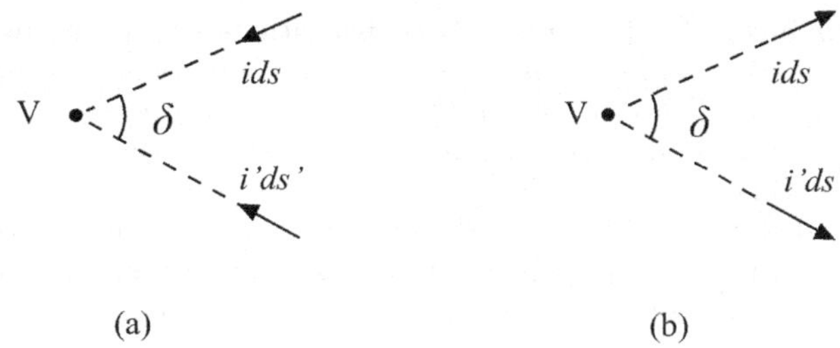

Figure 3.18: (a) Current elements ids and $i'ds'$ in the same plane directed towards a certain point V. (b) Current elements in the same plane directed away from point V. According to Ampère, the current elements ids and $i'ds'$ in cases (a) and (b) were attracting one another.

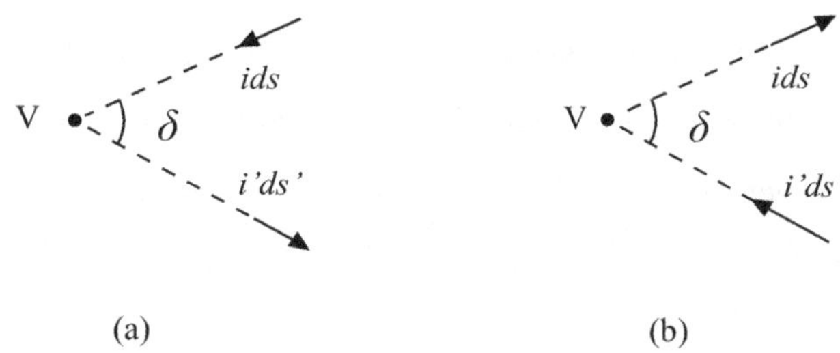

Figure 3.19: Electric currents in the same plane. One of them flows towards a certain point V, while the other flows away from it. According to Ampère, a repulsion should take place between the current elements ids and $i'ds'$ in both situations, (a) and (b).

current.[46] Two months later, he published another work in which he described his observations of the inverse phenomenon. He supposed Newton's law of action and reaction to be valid for the interaction between a magnet and a current-carrying wire. In this specific case, if a current-carrying wire generates a torque on a magnetized needle, then the magnet should generate an opposite torque on the wire. He then predicted that a magnet fixed in the laboratory might produce a rotation, relative to the laboratory, of a loop carrying a steady current. He performed the experiment and observed the expected effect.[47] A battery was connected to the current-carrying wire. This battery was called a "galvanic element" by Ørsted. The system composed of the battery and current-carrying wire was suspended by a vertical string in such a way that it could turn around this string. A powerful magnet was placed close to the system. His prediction of what should take place has been expressed as follows:[48]

> As a body cannot put another in motion without being moved itself, provided it is moveable, it is easy to foresee that the galvanic element must acquire some motion from the magnet.

His conclusions after performing this experiment:[49]

> If, instead of a moveable needle, one of the poles of a powerful magnet is presented to one of the extremities ff [of a brass wire connected to the galvanic element in such a way that a constant current flowed in this wire], the galvanic apparatus will be put in motion and will rotate around the extended axis of the wire ab [that is, around the vertical axis along which it was suspended] according to the nature of the pole.

That is, if the North pole of the magnet was presented to a specific region of the closed circuit composed of the battery and current-carrying wire, the system would turn clockwise around the vertical axis. If the South pole of the magnet was presented to the same region of the closed circuit, an anti-clockwise rotation would take place.

Ampère also did succeed in obtaining this inverse effect. As described in the second part of his first paper on electrodynamics of 1820, he utilized the instrument illustrated in figure 3.15:[50]

[46][Oersted, 1820], [Oersted, 1965] and [Ørsted, 1986].
[47][Ørsted, 1998c, p. 423].
[48][Ørsted, 1998c, p. 423].
[49][Ørsted, 1998c, p. 423].
[50][Ampère, 1820f, pp. 216-217] and [Chaib and Assis, 2009b, p. 141].

By replacing the fixed conductor AB by a magnetized bar situated horizontally in a direction perpendicular to that of this conductor, and in such a way that the currents of this magnet flow in the same sense as the electric current established at first in the fixed conductor, one then passes the current only through the mobile conductor, and one observes that this one turns by the action of the magnet precisely as it would turn in the experiment in which the current had been established in the two conductors, and in which there was no magnetized bar. It was with the goal to attach this bar, that I connected to this apparatus the support XY, terminated in Y by the box Z open in both ends where one fixes the magnet in the position which I have just explained by means of the pressure screw V.

3.10 Summary of Ampère's Initial Experiments

Ampère succeeded in obtaining several new phenomena which no one had observed before him:

1. Forces of attraction and repulsion between a rectilinear wire carrying a steady current and a magnet.
2. Attraction and repulsion between planar spirals carrying steady currents.
3. Attraction and repulsion between straight conductors carrying steady currents.
4. Torque exerted by a current-carrying loop and acting on another current-carrying loop.
5. Torque exerted by the Earth on a current-carrying loop.
6. He discovered that the electric current also flows inside the battery. He also obtained the sense of this current. These were important results in order to establish the concept of a closed circuit or the concept of a current following a closed path.
7. He reproduced electrodynamically the torques and forces between interacting magnets. To this end he replaced the magnets by planar spirals or by helices coiled around glass tubes.
8. He reproduced the directive action of the Earth, that is, the orientation of a magnetic needle along the magnetic meridian due to the magnetic influence of the Earth. To this end he replaced the compass needle by a large circular loop carrying a steady current.
9. He reproduced Ørsted's original experiment electrodynamically. To this end he replaced Ørsted's magnetic needle by a mobile loop carrying a steady current.

Items (2), (3) and (4), in particular, created a whole new research area, namely, the interaction between conductors carrying steady currents. Since the beginnings of his researches, Ampère had two clear goals in mind. The first goal was to obtain a mathematical expression with which he might explain quantitatively the interaction between current-carrying conductors. The second goal was to explain quantitatively all magnetic phenomena already known for a long time (orientation of a compass by the Earth, attractions and repulsions between magnets), together with the new electromagnetic phenomena discovered by Ørsted and himself (torques and forces between a magnet and a current-carrying wire, orientation of a current loop by the Earth). In order to reach this second goal he assumed the existence of electric currents inside the Earth and inside permanent magnets. He would then need to obtain how these supposed currents were flowing inside the Earth and in magnets. In this book we will follow the path he followed successfully.

Chapter 4

Initial Formulations of the Force between Current Elements

4.1 First Trial

Since the beginning of his researches on the force between two conductors carrying steady currents, Ampère assumed some principles or hypotheses, namely: (A) The force between two finite conductors could be obtained by the integration over each circuit of a supposed infinitesimal force between two current elements. (B) This infinitesimal force was assumed to act along the straight line connecting the centers of these current elements. (C) This force should comply with Newton's action and reaction law. (D) He also initially assumed, in analogy with the other known forces of gravitation, electrostatics and magnetostatics, that this force should vary with the inverse square of the distance between the current elements. (E) He supposed that this force should be proportional to the product of the intensities of both current elements and also to the product of their infinitesimal lengths, although initially he was not yet clear about the distinction between current intensity and the length of the current element.

In Section 3.6 it was seen that with the instrument presented in figure 3.10 (a) Ampère showed, in September 1820, that two rectilinear and parallel conductors attracted one another when the current was flowing in the same sense in both wires. When the currents flowed in opposite senses the wires repelled one another. He utilized this experimental fact as another extremely important principle in order to arrive at the force between two current elements.

In October 1820 the instrument of figure 3.12 was built. With this apparatus he could analyze the interaction between straight wires which were inclined relative to one another. There is a fixed conductor SR with an adjustable angle relative to the vertical. The vertical conductor BC can turn around a vertical axis passing through point D, moving towards SR or away from it. Conductors SR and BC were always orthogonal to the straight line connecting the centers of these two conductors. The angle between the conductor SR and the vertical could be changed at will. Let δ be the angle between the plane formed by each one of these conductors, SR and BC, with the straight line connecting their centers, such that $0 \text{ rad} \leq \delta \leq \pi$ rad, figure 4.1.

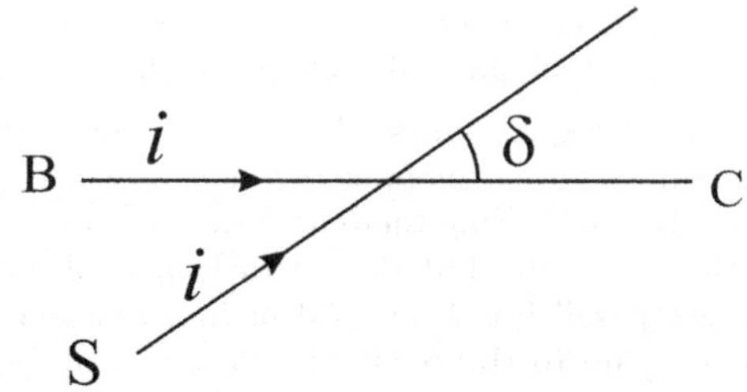

Figure 4.1: This illustration presents the experiment of figure 3.12 as seen along the straight line connecting the centers of conductors SR and BC. Conductor SR is fixed in the laboratory, while conductor BC is mobile, so that it can move towards SR or away from it.

The configuration of this experiment is analogous to figures 3.15 and 3.16. This time, however, an attraction or repulsion is observed between the conductors, instead of their mutual torques. The fixed conductor SR of figures 3.12 and 4.1 plays the same role as the fixed conductor AB of figures 3.15 and 3.16, while the mobile conductor BC plays the same role as the mobile conductor MN. In the situation of figures 3.12 and 4.1 the angle δ coincides with the angle ε between conductors SR and BC. When the currents in both conductors flow in the same sense as, for instance, vertically upwards, we have $\delta = 0$ rad. When SR is horizontal and BC vertical we have $\delta = \pi/2$ rad. On the other hand, $\delta = \pi$ rad when both currents flow in opposite senses as, for instance, vertically upwards in BC and vertically downwards in SR.

With a qualitative experiment in which he collected no numerical data, Ampère observed that conductor BC was attracted with maximal intensity when $\delta = 0$ rad, there was smaller attraction when $0 \text{ rad} < \delta < \pi/2$ rad, null force when $\delta = \pi/2$ rad, repulsion when $\pi/2 \text{ rad} < \delta < \pi$ rad and a repulsion with maximal intensity when $\delta = \pi$ rad. Figure 4.2 presents the qualitative behavior of this force acting on the mobile conductor as a function of the angle δ. We are here assuming that a positive force is attractive, while a negative force is repulsive.

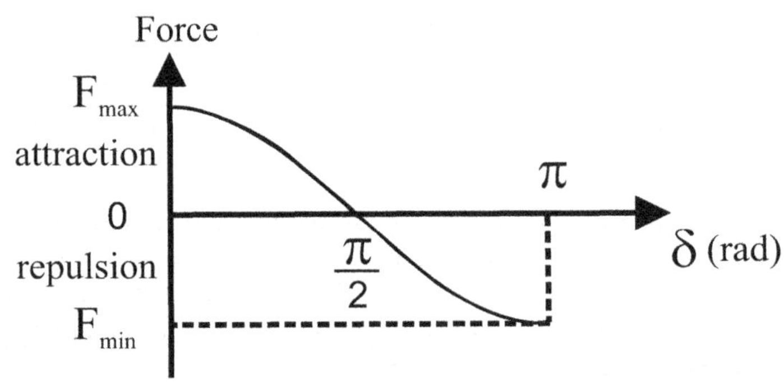

Figure 4.2: Qualitative behavior of the force acting on the mobile conductor BC as a function of the angle δ in the experiment of figure 3.12.

The experiments of figures 3.12 and 3.15 showed qualitative results illustrated in figures 4.2 and 3.17. However, these qualitative results are not conclusive for several reasons. In the first place, Ampère did not present precise quantitative measurements of the torque and force intensities, only their qualitative behaviors. Moreover, the actions observed on the mobile conductor BC were due not only to the fixed conductor SR, but also to the remaining portions of both circuits, and also due to the magnetic action of the Earth. Another difficulty in these experiments is that, at the time of Ampère, it was still difficult to keep a battery maintaining an electric current with constant intensity, as the energy of the batteries did decrease quickly with the passage of time. The inevitable sources of friction always present in these experiments should also be mentioned. Ampère did not estimate the order of magnitude of these frictional forces. He did not compare as well the magnitude of the forces of friction with the magnitude of the electrodynamic forces between the conductors.

In any event, in *chemise* 158, written around the second week of October 1820, he began to outline an expression for the force between two current elements utilizing these qualitative results. Figure 4.3 illustrates one page of this manuscript.[1]

This *chemise* 158 is the oldest document in which Ampère discussed mathematical details of his force between current elements. It is worthwhile to quote some of the main portions of this manuscript:[2]

> Before occupying myself with the mutual action between an electric current and a magnet, I will add the following observations relative to what I already said related to the action between two currents.
>
> When the two currents are directed along two lines which do not meet, instead of being directed along parallel lines [and flowing] in the same sense or in opposite senses, in such a way that their directions form an angle, there is attraction when, by considering these lines directed in the senses of the currents flowing in them, the angle formed by them is acute; but this attraction is always smaller than the attraction taking place when the currents are parallel and directed in the same sense, and this attraction always decreases when this angle increases, up to the point in which this attraction goes to zero when both currents flow through rectangular directions [that is, perpendicular to one another and perpendicular to the straight line connecting their centers].

[1] [Ampère, b, *carton* 8, *chemise* 158].

[2] [Ampère, b, *carton* 8, *chemise* 158], [Blondel, 1978] and [Hofmann, 1996, pp. 240-244].

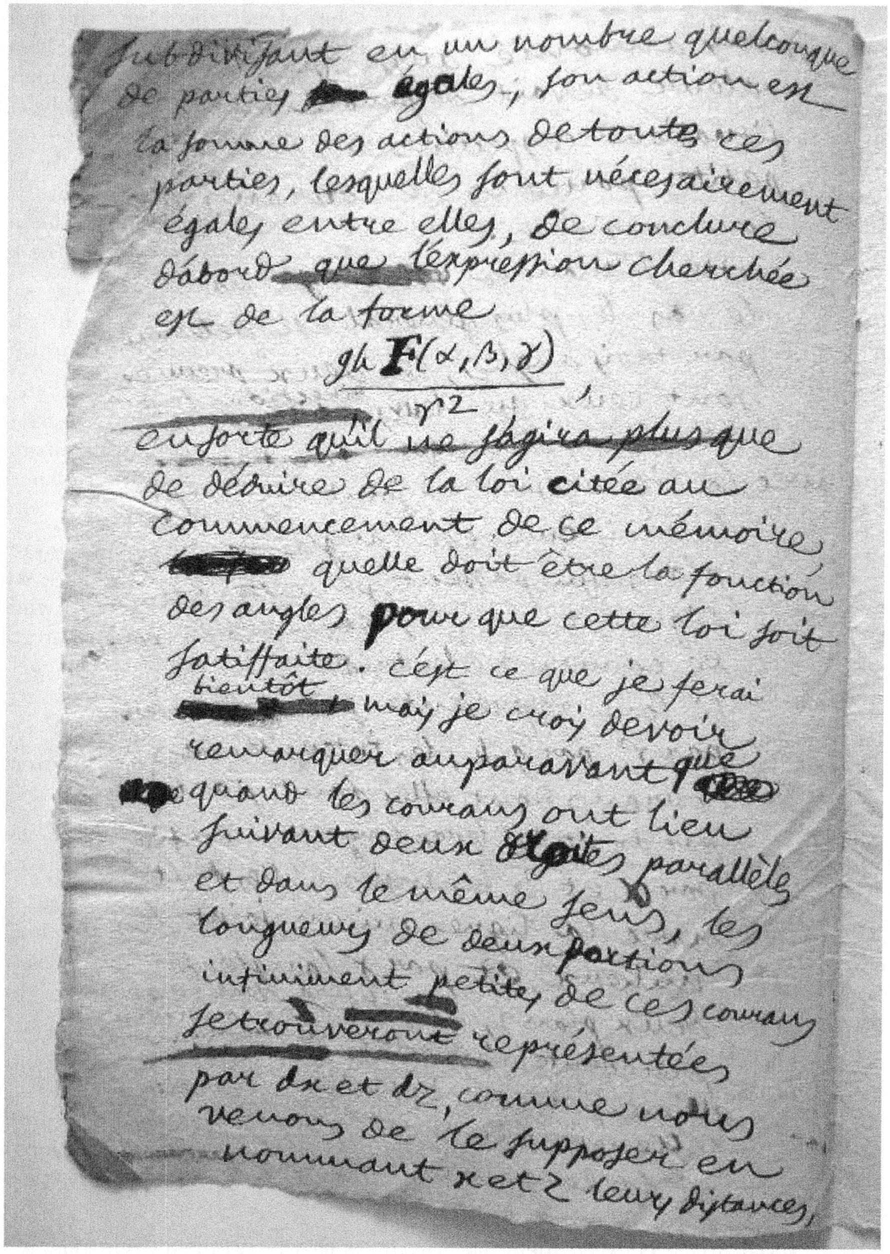
subdivisant en un nombre quelconque
de parties égales, son action est
la somme des actions de toutes ces
parties, lesquelles sont nécessairement
égales entre elles, de conclure
d'abord que l'expression cherchée
est de la forme
$\frac{gh\,F(\alpha,\beta,\gamma)}{r^2}$,
en sorte qu'il ne s'agira plus que
de déduire de la loi citée au
commencement de ce mémoire,
quelle doit être la fonction
des angles pour que cette loi soit
satisfaite. c'est ce que je ferai
bientôt mais je crois devoir
remarquer auparavant que
quand les courans ont lieu
suivant deux droites parallèles
et dans le même sens, les
longueurs de deux portions
infiniment petites de ces courans
se trouveront représentées
par dx et dz, comme nous
venons de le supposer en
nommant x et z leurs distances,

Figure 4.3: Ampère's *chemise* 158 showing his first algebraic formulation of the force between two current elements.

When this angle is obtuse, the attraction is transformed into a repulsion; this repulsion [...] [reaches] its maximum value when this angle is equal to two right angles, that is, when both currents are directed in contrary senses along two parallel straight lines.

As the repulsion should be considered a negative attraction, this experiment shows that the attraction diminishes with the cosine of the angle between the two currents, going then to zero [when this angle is a right angle] and changes sign together with it. Therefore, when the intensity and length of both currents remain constant, just like the shortest distance between their directions [remains constant], this action should be expressed as a function of odd degree in the cosine of this angle.

I suppose here that this shortest distance passes always by the midpoints of the two currents, in such a way that only the angle between the directions of these currents may change, considering [constant] all other magnitudes on which their mutual action may depend.

Furthermore, this function of the cosine of the angle between the directions of the two electric currents can have a simple form only when one considers infinitely small portions of these currents. It is probable that in that case it reduces to the first power of this cosine; at least this is the first supposition that is suitable to test in the comparison of a hypothesis about the law of attractions and repulsions with the results of experience. It is necessary to include in what has already been said the decrease of the attractions and repulsions when the distance increases, in the inverse square ratio of this distance, in conformity to what is observed for all genres of action more or less analogous to this one. Moreover, when the line connecting the center of the two portions of current, supposed infinitely small, is not perpendicular to their directions, the mutual action between them should still depend on the angles which these directions form with the straight line of which we have been talking about. In this last case, moreover, it seems that the cosine of the angle formed by the directions of the two currents when the line connecting their centers is perpendicular to them should be replaced by the cosine of the angle formed by the two planes formed by this line and by the directions of the two currents. When the two currents are in the same plane, this cosine is always equal to one, as this angle is always zero.

It was upon these general considerations that I had constructed an expression for the attraction of two infinitely small currents which was, in truth, only a hypothesis, but the simplest one that could be adopted and, consequently, the one that should be tried first.

I attempted to conclude from it the effects that would have to result, both for linear electric currents of finite extension, and for circular currents like those I have shown to exist in cylindrical magnets, and for the currents which take place in copper wires bent into helices, because of the various experiments that I had performed on the latter kind of currents.

I proposed to compare the results of these calculations with experiments in which one could measure the intensity of the action of two linear conductors of finite length, the angle between which could be varied at will, because it is impossible to experiment with infinitely small currents, and also when a magnet is employed, [in this case,] no matter how small is the magnet, the action is even more complicated, as it acts on currents forming closed curves determined by the cross sections of the magnet considered perpendicular to the line connecting its poles.

For these measurements, I had an apparatus constructed which I showed to Biot and Gay-Lussac last October 17th [1820]; I procured myself another one to observe the action between two currents bent into helices.

The experiments I tried with these two instruments caused me to discover two new facts[3] which complicated the results of them and consequently forced me to suspend the verifications of the results of my calculations that I had proposed to make with the aid of these apparatuses.

It was with this instrument that I observed for the first time the action exerted by the terrestrial globe on electric currents, [an action] which I had tried in vain to produce with instruments less appropriate to the success of this experiment, due to the fact that the net force of this action only tried to move the mobile portion of the instrument by a very short arm of the lever.

[...] I had also announced that this value [of the force between two current elements] should depend on the respective orientation of the two small portions of current which were being taken into account. This orientation, in the most general case, is determined by three angles; the first two angles are those which the directions [of the currents] make with the line connecting their midpoints, and the third angle is the angle formed by the two planes passing through this last line and by the two small portions of electric current.

Representing by r the distance between them [that is, between the two small portions of current], by g and h the intensities of their currents, by dx and by dz their lengths and, finally, by α and β the angles they make with the line connecting their centers, and by γ the angle between the two planes, it will be easy, after the preceding considerations and observing that an infinitely small portion exerts an action necessarily proportional to the length — for, by dividing it in an arbitrary number of equal parts, its action is the sum of the actions of all these parts, which are necessarily equal to one another —, to conclude then that the expression we are looking for has the form

$$\frac{ghF(\alpha,\ \beta,\ \gamma)}{r^2}\ , \tag{4.1}$$

so that it is only necessary to deduce from the law quoted in the beginning of this memoir, what should be the function of the angles for which this law should be satisfied.

Several important aspects should be highlighted in this text. In the first place, equation (4.1) represents Ampère's oldest mathematical expression describing the force between two current elements. Initially Ampère considered two currents flowing through finite straight wires which did not meet and for which the straight line connecting their centers was perpendicular to their directions. Based on his experiments and utilizing a principle of mathematical simplicity, he concluded that the force between two currents of finite lengths

[3]These "new facts" will be discussed in Subsection 4.2.1.

should be proportional to the cosine of the angle between the straight lines in which they were flowing. This cosine can be represented by $\cos\varepsilon$. In this example the angle ε is equal to the angle γ between two planes, each one formed by the line connecting the midpoints of the two finite straight lines with the directions of the currents. When the straight line segment connecting the centers of the two current elements was no longer perpendicular to their directions, Ampère mentioned that the expression $\cos\varepsilon$ should be replaced by $\cos\gamma$, although he did not justify this conclusion. He expressed the infinitesimal lengths of the two current elements by dx and dz. This notation reveals his intention to integrate the force between current elements in order to obtain the force between two segments of finite length.

By considering rigorously this situation, any sum of terms of the form $\cos^m\gamma$, in which m was an odd number, would be in agreement with the findings of this qualitative experiment. He utilized an argument of simplicity in order to keep only the first power of $\cos\gamma$ in his force expression. This argument may be questioned, although it seems reasonable to begin with the simplest hypothesis.

Although this *chemise* 158 was not published in Ampère's time, in another published paper of 1820 he said the following:[4]

> [...] moreover, in the expression [describing the force between two rectilinear currents] an odd power of the cosine of the angle between their directions should appear, because the attraction is transformed into a repulsion when this cosine changes sign. It is probable that this function is only the first power of this cosine, this being at least the simplest function, that function which should be tried in the first place.

As described in Section 2.5, Ampère always assumed, since the beginnings of his researches, that the attractive and repulsive forces between two current elements should be along the straight line connecting their midpoints. Moreover, here we can observe that he also assumed that this force should vary as the inverse square of the distance between the elements. From the experiments he performed with the instrument of figure 3.12 and supposing the simplest law satisfying the results of these experiments, he was then led to assume that when the straight conductors were orthogonal to the line connecting their centers the force between them should be proportional to:

$$\frac{gh\cos\gamma}{r^2}\,, \tag{4.2}$$

in which gh indicated both the current intensities and the sizes of the current elements.

It was still necessary to determine the dependence of the force on the angles α and β when the current elements were not orthogonal to one another. These are the angles between the two elements and the straight line connecting them, as represented in figure 2.15 (a). Blondel, Hofmann and Darrigol[5] suggested that Ampère around this time arrived at the following expression:

$$\frac{\sin\alpha\sin\beta\cos\gamma}{r^2}\,. \tag{4.3}$$

Figure 4.4 illustrates how he concluded that the force should be proportional to $\cos\gamma$. We are considering here current elements, instead of working with finite straight segments. The element ds is located in the plane of the paper, being orthogonal to the straight line connecting its center to the center of element ds', with $\alpha = \pi/2$ rad. Element ds' is also orthogonal to the straight line connecting its center to the midpoint of element ds, with $\beta = \pi/2$ rad. The plane formed by ds with the straight line connecting the elements makes an angle γ with the plane formed by ds' with the straight line connecting the elements.

Ampère initially considered two parallel current elements attracting one another with currents flowing in the same sense, with the force having its greatest intensity, with the direction of both current elements being orthogonal to the line connecting their centers, that is, with $\gamma = 0$ rad. In his mind he increased this angle γ. When $0 \text{ rad} < \gamma < \pi/2$ rad, both current elements still attracted one another, but with an intensity smaller than the maximal intensity which took place for $\gamma = 0$ rad. When $\gamma = \pi/2$ rad their mutual force vanished. In his mind he increased the angle γ even more. When $\pi/2 \text{ rad} < \gamma < \pi$ rad, the elements repelled one another. Ampère increased the angle γ even more, reaching the situation in which both elements were once again parallel to one another, but now with currents flowing in opposite senses, with $\gamma = \pi$ rad. In this configuration their repulsion had maximal intensity. Ampère then supposed that the simplest function which should be adopted in order to represent this attractive and repulsive behavior should be proportional to $\cos\gamma$.

---

[4] [Ampère, 1820a, p. 248] and [Ampère, a, p. 11].

[5] [Blondel, 1982, p. 84], [Hofmann, 1982, p. 266], [Hofmann, 1996, pp. 240-241] and [Darrigol, 2000, p. 9].

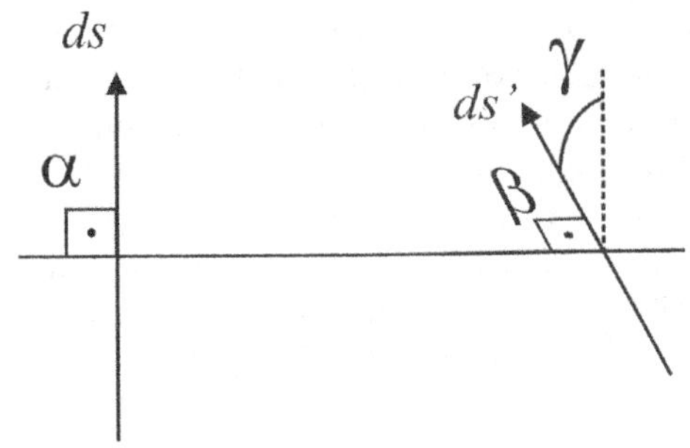

Figure 4.4: The element ds is located in the plane of the paper, being orthogonal to the straight line connecting its midpoint to the center of element ds', with $\alpha = \pi/2$ rad. Element ds' is also orthogonal to the straight line connecting its midpoint to the center of element ds, making an angle $\beta = \pi/2$ rad with this line connecting both elements. The angle γ is the angle between the planes formed by the direction of each element with the line connecting their midpoints.

Figure 4.5 illustrates the reasoning presented by Blondel, Hofmann and Darrigol in order to conclude that the force between two current elements should be proportional to $\sin\alpha\sin\beta$, as represented by equation (4.3). We are considering the angles α, β and γ as represented by figure 2.15 (a).

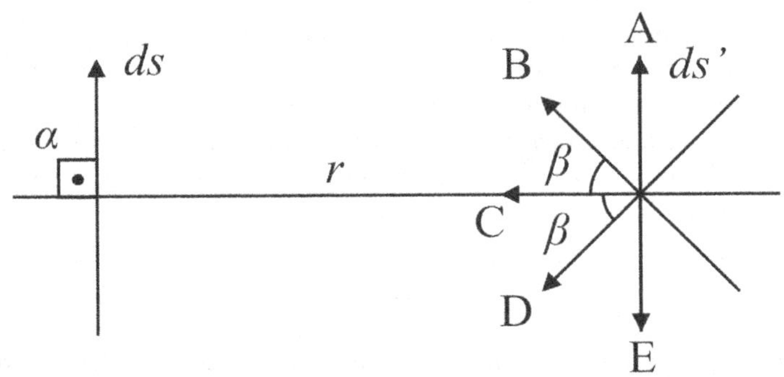

Figure 4.5: Interaction between two current elements located in the same plane. In configurations A and B we have $\gamma = 0$ rad, in C the angle γ is undefined, while in configurations D and E we have $\gamma = \pi$ rad.

We have two current elements of lengths ds and ds' located in the same plane. They make angles α and β with the same continuation of the straight line connecting their centers, r being the constant distance between their midpoints. Once more we will begin with the same initial state of figure 4.4, namely, with $\alpha = \beta = \pi/2$ rad and $\gamma = 0$ rad. In this configuration both elements are parallel to one another, they are orthogonal to the line connecting their centers, and both currents flow in the same sense. They attract one another with maximal intensity. The final configuration will also be the same as that of figure 4.4, namely with $\alpha = \beta = \pi/2$ rad and $\gamma = \pi$ rad. In this final configuration they are once more parallel to one another, they are orthogonal to the line connecting their center, but now the currents flow in opposite directions. They repel one another with maximal intensity. As we go from the initial configuration to the final configuration we will keep $\alpha = \pi/2$ rad $=$ constant, with both elements being constantly in the same plane. From configurations A to C of figure 4.5 the angle β will go from $\beta = \pi/2$ rad to $\beta = 0$ rad. From C to E the angle β goes from $\beta = 0$ rad to $\beta = \pi/2$ rad. In configurations A and B we have $\gamma = 0$ rad, in C the angle γ is undefined, while in configurations D and E we have $\gamma = \pi$ rad.

Elements ds and ds' attract one another with maximal intensity in configuration A of figure 4.5, with $\alpha = \beta = \pi/2$ rad and $\gamma = 0$ rad. They repel one another with maximal intensity in configuration E, with $\alpha = \beta = \pi/2$ rad and $\gamma = \pi$ rad. It would be natural to expect attraction in configuration B, although with a smaller intensity than the attraction of configuration A. In B we have $\alpha = \pi/2$ rad, $0 \text{ rad} < \beta < \pi/2$ rad and $\gamma = 0$ rad. In configuration C there should be no attraction nor repulsion. In this case we have $\alpha = \pi/2$ rad, $\beta = 0$ rad and γ does not exist or is undefined. In configuration D a repulsion should take place, although with a smaller intensity than the repulsion of configuration E. In configuration D we have $\alpha = \pi/2$ rad, $0 \text{ rad} < \beta < \pi/2$ rad and $\gamma = \pi$ rad. In configuration E we have $\alpha = \beta = \pi/2$ rad and $\gamma = \pi$ rad.

As we have seen, Ampère utilized an argument of mathematical simplicity in order to conclude that the force between two current elements should be proportional to $\cos\gamma$. Utilizing the same argument it might then be concluded that the simplest mathematical expression representing the force in this thought experiment should be proportional to $\sin\beta$. The same conclusion might be reached by rotating ds' clockwise around an axis passing through its center and orthogonal to the plane of the paper, from configuration A up to configuration E.

Supposing now element ds' fixed in the laboratory, while element ds rotates clockwise or anti-clockwise around an axis passing through its center and orthogonal to the plane of the paper, we would then conclude that the force between the two current elements should be also proportional to $\sin\alpha$. By following this procedure it might then be possible to arrive at equation (4.3) indicating that the force between two current elements should be proportional to $\sin\alpha\sin\beta\cos\gamma$, as discussed by Blondel, Hofmann and Darrigol.

We were unable to locate any manuscript by Ampère in which he presented a reasoning like this one. However, two facts indicate that this possible route outlined by Blondel, Hofmann and Darrigol is very reasonable. As will be seen later on in Section 4.2, Ampère's first published expression of his force law between two current elements was represented as follows:[6]

$$\frac{gh}{r^2}\left(\sin\alpha\sin\beta\cos\gamma+\frac{n}{m}\cos\alpha\cos\beta\right) . \tag{4.4}$$

This equation is analogous to equation (2.1). In his paper of 1820 he mentioned that the second term in equation (4.4) should be much smaller than the first term, he even suggested that the second term might be exactly null. He maintained this idea until 1822, when he finally concluded that $n/m = k = -1/2$, concluding that both terms of equations (2.1) and (4.4) should have the same order of magnitude. This fact indicates that his initial belief was that the force between two current elements should really be given by an expression like equation (4.3).

There is also a second fact suggesting the feasibility of the path indicated by Blondel, Hofmann and Darrigol. There is a Section of a joint work by Ampère and his friend Babinet, edited in 1822, containing a similar reasoning. This section was written by Babinet, who had written the major part of this work in July 1821.[7] Babinet did not present the mathematical reasoning leading to the force between two current elements being proportional to $\sin\alpha\sin\beta$. He also did not mention that the function $\cos\gamma$ was responsible for making the attraction in configurations A and B of figure 4.5 be transformed into the repulsion of configurations D and E of this figure. In any event, Babinet presented a figure similar to our figure 4.5. In discussing qualitatively his figure, he mentioned that the attraction between the two current elements should decrease when the angle between element ds' with the line connecting it to element ds was smaller or bigger than $\pi/2$ rad. He also mentioned that this attraction should go to zero when the element ds' was along the straight line connecting the elements ds and ds'. Moreover, he concluded that the attraction should be transformed into a repulsion when the currents in both elements were in opposite senses as regards the straight line connecting their centers.

4.2 Ampère's First Publication with an Expression for the Force between Two Current Elements

4.2.1 The Addition Law

In *chemise* 158 quoted in Section 4.1, Ampère mentioned that he gave up trying a quantitative verification of his force law between current elements utilizing the instrument of figure 3.12. He abandoned this route due to two "new facts" which complicated the results he expected to achieve. As mentioned in Section 3.7, he had not obtained the expected orientation of the helix of figure 3.11 (a) due to the influence of terrestrial magnetism. The first time he obtained this terrestrial orientation of a current-carrying wire took place, quite unexpectedly, with the apparatus of figure 3.12.

This observed orientation of a current-carrying wire due to the action of the Earth was the first "new fact" mentioned in *chemise* 158. Ampère then understood the reason why he had failed to observe this orientation with the helix of figure 3.11 (a). It was due to the fact that he was utilizing helices with very small cross sections:[8]

[6] [Ampère, 1820e].

[7] [Ampère and Babinet, 1822a, Sections 14 to 17, pp. 17-21] and [Ampère and Babinet, 1822b, Sections 14 to 17, pp. 182-186].

[8] [Ampère, 1820f, pp. 172-173] and [Chaib and Assis, 2009b, pp. 121-122].

> [...] But as regards the directive force of the Earth, the apparatus [of our figure 3.11 (a)] was not mobile enough, and this force acted by a very short lever arm to produce the desired effect.

By increasing the arm of the lever,[9] or the area of the loop, he was able to obtain in a reproducible way the orientation of a magnet due to terrestrial magnetism. This time he replaced the magnetic compass by a current-carrying circular loop with a large radius, of the order of 20 cm, figure 3.13.

The second "new fact" which he discovered had a crucial importance in all his researches leading to the final expression of his force between current elements. Two bar magnets, aligned along their axes, attract and repel one another according to the orientation of their poles, as indicated in figure 3.1. Ampère was able to reproduce this phenomenon electrodynamically by utilizing two planar spirals placed in front of one another in parallel vertical planes, as in figures 3.8 and 3.9. There is another situation in which two bar magnets are easily seen to attract and repel one another, namely, when they are placed side by side. If the poles of the same kind point towards the same side, the bars repel one another, as in figure 4.6 in which both North poles are at the right extremities. If the poles of the same kind point to opposite sides, the bars will attract one another.

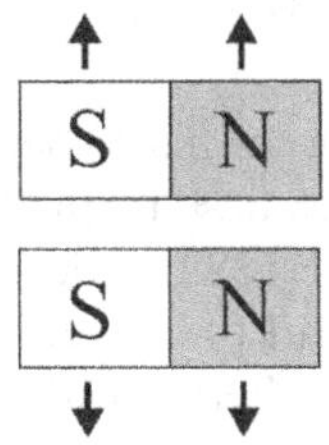

Figure 4.6: Two bar magnets repelling one another.

Ampère tried to replicate this situation electrodynamically. To this end he coiled two helices with a metal wire. He then put these two helices side by side in an instrument like that of figure 3.10 (a). We were unable to locate any figure or drawing in which Ampère described this instrument. In any event, by his descriptions we can represent it, following Blondel, by figure 4.7.[10]

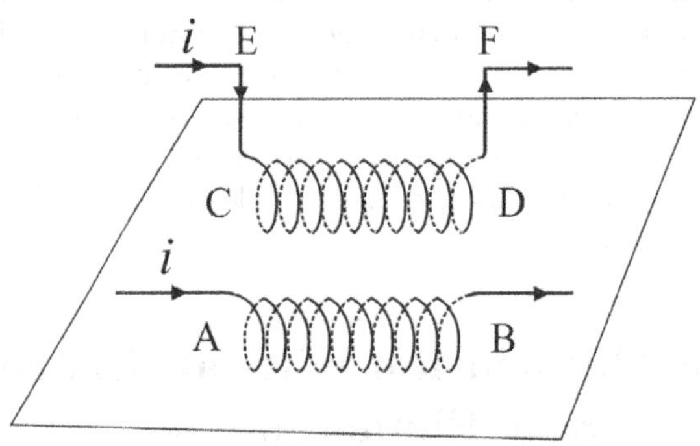

Figure 4.7: Interaction between two helices.

Helices AB and CD are located in the same horizontal plane. Helix AB is fixed in the laboratory, while the circuit $ECDF$ containing the helix CD can turn around the horizontal axis EF placed vertically above the helix CD, in such a way that CD can move towards AB or away from it. With this configuration of the instrument and with the currents flowing along the directions indicated in figure 4.7, Ampère expected to reproduce electrodynamically the repulsion between two bar magnets as indicated in figure 4.6. However, by performing the experiment in October 1820, he observed the helix CD being attracted by the helix AB, instead of being repelled by it!

This was the second "new fact" mentioned previously in Section 4.1. Ampère was totally surprised, as the result of this experiment was totally against his expectations.

The helix of figure 3.11 (a) did behave in his experiments like an ordinary bar magnet. On the other hand, the helices AB and CD of figure 4.7 did not behave like ordinary bar magnets. Ampère tried to figure out

[9] For a detailed discussion of the law of the lever and about the center of gravity of bodies, see [Assis, 2008], [Assis, 2010a] and [Assis, 2011a].

[10] [Blondel, 1982, p. 87].

the reason for their different behavior. He concluded that this anomaly was related to the different way in which these helices were coiled. The helix of figure 3.11 (a) and (b) was axially compensated. There was no such axial compensation in the helices AB and CD of figure 4.7. Initially Ampère had erroneously neglected the longitudinal contributions of these helices AB and CD. In his initial helices, like the helix of figure 3.11 (a) and (b), the current flowed not only in the helix coiled around the glass tube (for instance, from left to right), but also returned along the axis of the tube (for instance, from right to left). The longitudinal action of the current flowing in the helix around the glass tube was compensated by the return current flowing through the axis of the tube. There was no such longitudinal compensation in the helix AB and CD of figure 4.7. His first helices had been coiled totally by chance like those of figure 3.11 (a) and (b), only to facilitate their rotation around the vertical axis, without knowing about the longitudinal compensation of their currents.

We can understand his initial expectation related to the helices of figure 4.7 by considering figure 4.8 (a).

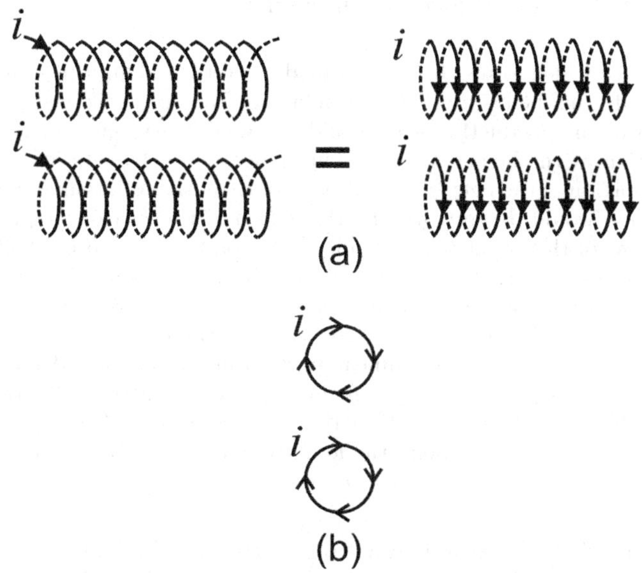

Figure 4.8: (a) Ampère expected initially that each one of these helices were equivalent to a cylinder in which the currents were flowing only along the azimuthal direction φ. (b) Cross sections of two side by side cylinders carrying azimuthal currents. The currents which are closer to one another flow in opposite senses. According to Ampère's expectations, these two helices should repel one another.

The currents flow in the same sense in two parallel helices. Looking at the helices along their axes, there are two clockwise or anti-clockwise currents, as in figure 4.8 (b). Parallel wires carrying currents in the same sense attract one another, while they will repel one another if the currents were flowing in opposite senses. As these forces decrease their intensities by increasing the distance between the wires, Ampère expected a repulsion between the helices of figure 4.8.

After observing that these helices unexpectedly attracted one another, he concluded that he could no longer neglect the longitudinal component of the current which was flowing in each helix. Each helix of figure 4.7 should in reality be considered as the sum of a rectilinear current flowing through the axis of the helix, together with a purely circular current like the azimuthal current flowing over the surface of a cylinder, as represented in figure 4.9.

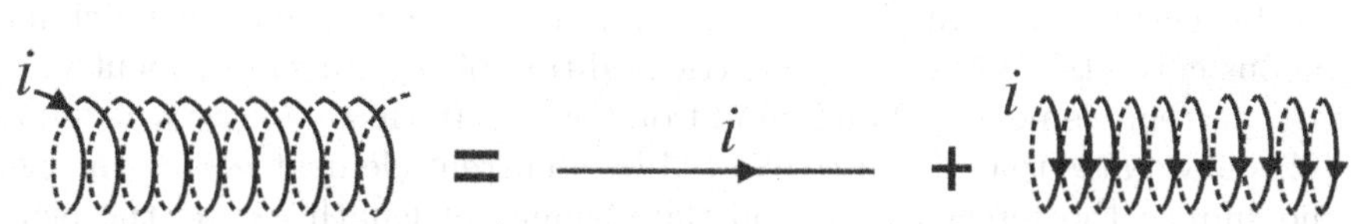

Figure 4.9: A helix which is not axially compensated is equivalent to a rectilinear current along the axis of the helix, together with a purely azimuthal current flowing circularly over the surface of a cylindrical shell along the φ direction.

The interaction of the helices AB and CD of figure 4.7 should be then considered as a fourfold interaction, namely: (I) The rectilinear component of AB interacting with the rectilinear component of CD; (II) the rectilinear component of AB interacting with the azimuthal current of CD; (III) the azimuthal current of AB interacting with the rectilinear current of CD; and (IV) the azimuthal current of AB interacting with the azimuthal current of CD. Straight wires carrying currents in the same sense attract one another. Likewise, two circular coils in the same plane carrying anti-clockwise currents like those of figure 4.8 (b) repel one another. A repulsion will also take place between two parallel finite cylinders carrying anti-clockwise currents over their surfaces. The magnitude of each one of these fourfold interactions will depend on the distance between the rectilinear components, on the radii of the cylinders and on the distance between their centers. In Ampère's specific experiment described in figure 4.7, the attraction between the longitudinal components of the currents flowing in his helices was greater than the repulsion between the azimuthal components of the currents flowing over the surfaces of the cylindrical shells. The net attraction between his two helices was due to their small radii. This attraction was totally against his initial expectations.

He expressed his ideas at this time as follows, our emphasis:[11]

> When the conducting wire which forms the helix is made to return along this axis, by enclosing it inside a glass tube placed inside this helix in order to insulate it from the coils composing it, the current of this rectilinear portion of the conducting wire would be in contrary sense to the current which would correspond to the portion of the helix's action taking place parallel to its axis, it will repel what this [equivalent rectilinear portion of the helix's current] would attract, and it will attract what it would repel. Therefore, this last action would be destroyed by the action of the rectilinear portion of the conductor, and from the joint actions of this rectilinear portion of the conductor with the helix, there would remain only the action of the transverse circular currents, [an action] perfectly similar to the action of a magnet. This assembly did take place in the instrument represented in figure 3 [our figure 3.11 (a)], *although I had not planned its advantages*, and for this reason this instrument presented exactly the effects of a magnet, while the helices for which the rectilinear portion of conductor did not return inside the axis presented, on the contrary, the effects of a rectilinear conductor equal to the axis of these helices. As the radius of the cylindrical surfaces over which the helix was located were very small in the helices which I utilized, it was the very effects along the longitudinal sense which were more relevant, *and this phenomenon astonished me before I could discover its cause.*

Therefore, in order to cancel the longitudinal action exerted by the helices of figure 4.7, it is necessary to compensate axially these helices, with the current returning along the axis of the helices after being coiled around the cylindrical surface, as was the case with the axially compensated helix of figure 3.11 (a) and (b). Figure 4.10 indicates the equivalence between the helix of figure 3.11 and a cylinder with the current flowing only azimuthally around its surface, without axial component.

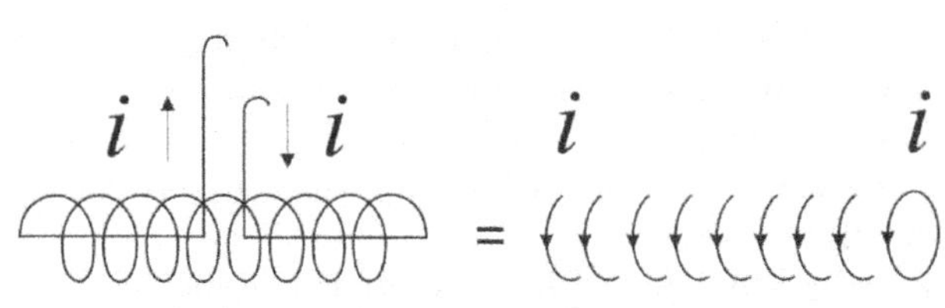

Figure 4.10: An axially compensated helix is equivalent to a purely circular or azimuthal current flowing over the surface of a cylinder, without a longitudinal or axial component.

The discovery of this new fact led Ampère to a very important law or principle which has received several names: "Law of sinuous currents",[12] "law of geometric addition of the current elements",[13] "principle of the vector addition of the current elements",[14] and "addition law".[15] In this book we will call this new discovery the *addition law*. This law states that the force exerted by a current element ids on another current element $i'ds'$ is equal to the sum of the forces exerted on the element of length ds' by the element of length ds decomposed in two or three directions, provided there is the same current intensity i flowing in each one of

[11] [Ampère, 1820f, pp. 175-176] and [Chaib and Assis, 2009b, p. 123].
[12] [Joubert, 1885, p. 134, n. 1].
[13] [Blondel, 1982, p. 86].
[14] [Blondel, 1982, p. 88].
[15] [Blondel, 1982, pp. 88-96] and [Hofmann, 1996, pp. 246-250].

these components. Ampère referred to this fact as a "law" or "principle". He presented it to the Academy of Sciences of Paris on 6 November 1820:[16]

> This law consists in the following, if one considers an infinitely [small] portion of electric current, and if one considers in the same point in space other infinitely small portions of electric currents which are, relative to the first portion, in intensity and in direction, the same as are the components of a force in magnitude and in direction relative to this force, then the combination of these portions of electric currents, corresponding to the component forces, will exert, in every case, precisely the same action corresponding to the resultant of these portions.

The addition law was also presented as follows:[17]

> Afterwards I required an instrument to be built similar to that of figure 1,[18] in which the fixed conductor and the mobile conductor were replaced by brass helices surrounding glass tubes, but whose prolongations, instead of returning inside these tubes, were placed in communication with the two extremities of the battery, as the straight conductors of figure 1.[19] It was by utilizing this instrument that I discovered a new fact which initially did not seem to agree with the other phenomena which I had until then observed in the mutual action between two electric currents, or between a current and a magnet. Later on I recognized that this new fact did not contradict the set of these phenomena. But in order to explain it, it is necessary to admit as a general law of the mutual action between electric currents, a principle which I verified up to now only as regards the currents of metal wires bent into helix, but which I believe to be generally valid, as regards infinitely small portions of electric current, which one conceives as composing every current of finite length, whether [the electric current] follows a straight line or a curve.
>
> In order to have a clear idea of this law, it is necessary to conceive in space a line representing in magnitude and in direction the resultant of two forces, which are similarly represented by two other lines, and suppose, in the directions of these three lines, three infinitely small portions of electric currents, the intensities of which are proportional to their lengths. The law at issue consists in the fact that the small portion of electric current, directed along the resultant, exerts, in any direction, upon another current or upon a magnet, an attractive or repulsive action equal to what would result, in the same direction, from the combination of the two portions of current directed along the components.

At this time Ampère was still unclear, stating that the current intensities were proportional to their lengths. Laumont and Ampère himself later on replaced this statement by saying that the attractive and repulsive forces were proportional to the lengths of the elements, as we discussed in Section 2.6. Leaving aside this aspect of the paper, Ampère's presentation of the addition law in this work is reasonably clear. We can understand what he had in mind by saying that each current element should be decomposed in two other lines, by observing a figure he presented in 1823 in the *Recueil*,[20] our figure 4.11. At the upper right corner of the original figure there appeared by a typographical error the letter F, while the correct letter should be D. We included this letter D in our figure. In this figure the plane $ABDC$ is orthogonal to the plane $ABFE$.

In figure 4.12 we present the angles $\alpha = \pi/2$ rad, $\beta = \pi/2$ rad and γ superimposed to figure 4.11. In order to illustrate the angle γ in figure 4.12 (b), we joined the centers of the two current elements and placed these elements in the plane of the paper. The configuration of figures 4.11 and 4.12 is analogous to that of figure 4.4.

In figures 4.11 and 4.12, the points A and B represent the centers of the two current elements. Ampère illustrated only half of each element. The segments AG and BH are orthogonal to the straight line segment AB connecting the centers of both elements. Initially Ampère considered two planes, $ACDB$ and $AEFB$, orthogonal to one another. He decomposed the segment AG into two portions, AM and MG. Likewise, the segment BH was decomposed into two portions, BP and PH. Ampère had a figure like this in his head when he mentioned that "it is necessary to conceive in space a line [AG] representing in magnitude and in direction the resultant of two forces, which are similarly represented by two other lines [AM and MG]". He utilized the addition law in order to replace the interaction between AG and BH by the sum of four other interactions, namely, between AM and BP, between AM and PH, between MG and BP, together with the interaction between MG and PH.

The configuration shown in figures 4.11 and 4.12 represents only the particular case in which $\alpha = \beta = \pi/2$ rad. Ampère considered also the general case of figure 4.13.[21]

[16] [Ampère, 1820e].

[17] [Ampère, 1820f, pp. 173-174] and [Chaib and Assis, 2009b, p. 122].

[18] [N. T.] See our figure 3.10 (a).

[19] [N. T.] Ampère's helices arranged in an instrument similar to that of his figure 1 may be represented by our figure 4.7.

[20] [Ampère, 1822p, figure 21] and [Ampère, 1885m, p. 263, figure 7].

[21] [Ampère, 1822p, figure 22] and [Ampère, 1885m, p. 265, figure 8].

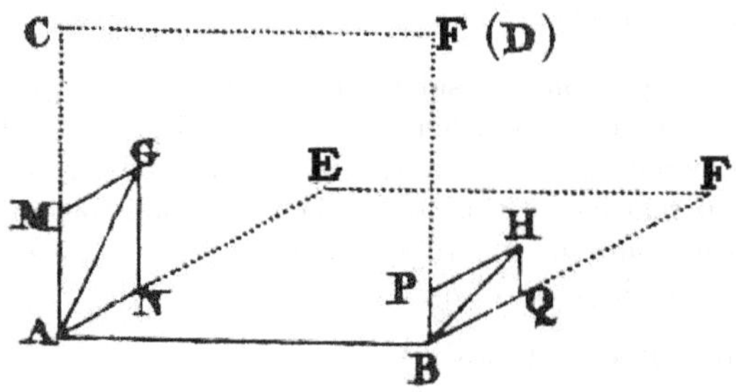

Figure 4.11: Ampère's figure with which it is possible to illustrate the addition law in the particular case of the interaction between the current elements AG and BH which are orthogonal to the straight line AB connecting their midpoints. The plane $BACF(D)$ is orthogonal to the plane $BAEF$.

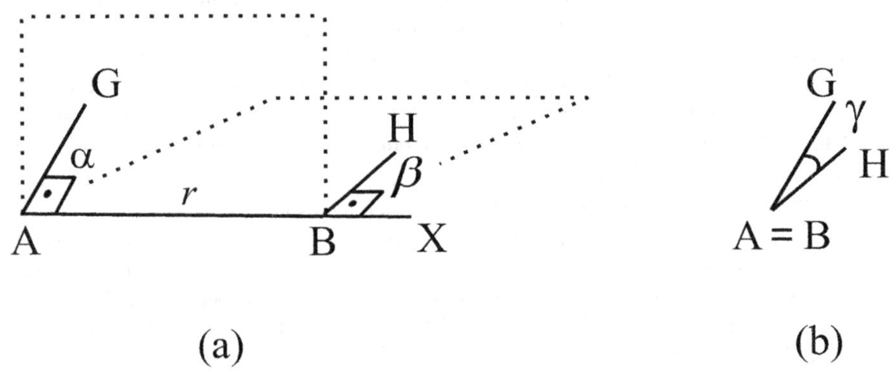

Figure 4.12: Here we included the angles α, β and γ in the configuration of figure 4.11. (a) We have $GAB = \alpha = \pi/2$ rad and $HBX = \beta = \pi/2$ rad. The point X is along the continuation of the straight line AB. (b) Superposition of AG and BH, with point A coinciding with point B, indicating the angle $GAH = \gamma$ with AG and BH in the plane of the paper.

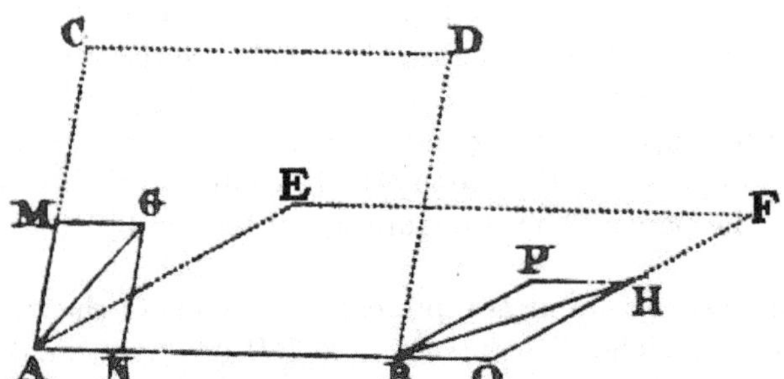

Figure 4.13: Ampère's figure illustrating the addition law in the general case. The planes $ABDC$ and $AQFE$ make an angle γ between them.

In figure 4.14 we present the angles α, β and γ superimposed on figure 4.13, following Ampère's specifications.

Figures 4.13 and 4.14 present the general case in which the elements AG and BH are no longer perpendicular to the straight line AB connecting the centers of the two current elements. In this case Ampère considered two planes $ACDB$ and $AEFQ$, each one formed by the direction of one current element with the straight line connecting the centers of both elements. These two planes make an angle γ with one another. Ampère decomposed AG into two portions, AM and MG. The segment BH, on the other hand, was decomposed into BP and PH. He utilized once more the addition law in order to replace the interaction between AG and BH by the sum of four interactions between the decompositions of each one of these current

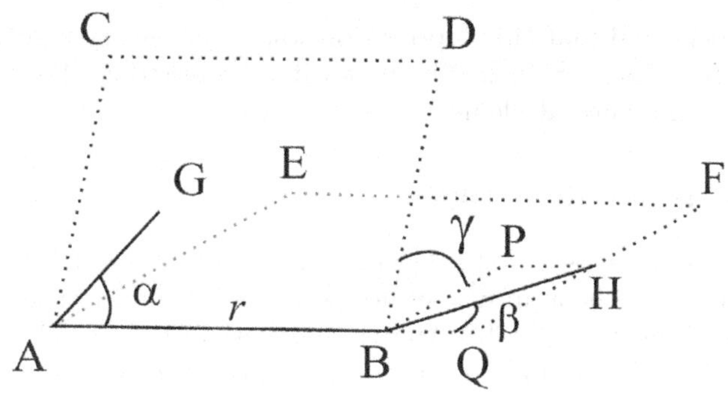

Figure 4.14: Angles α, β and γ superimposed on figure 4.13, with $GAB = \alpha$, $HBQ = \beta$ and $DBP = \gamma$.

elements.

4.2.2 Theorem of the Nonexistence of Interaction between Orthogonal Current Elements

In order to arrive at his first published version of the interaction between two current elements, Ampère utilized also a new principle or theorem, namely, he supposed there were no attractions or repulsions between two current elements which were orthogonal to one another.[22] In particular, Ampère considered that there were no interactions between the current elements of lengths ds and ds' when they were orientated as in figure 4.15 (a), (b) and (c). In this figure ds' is orthogonal to the straight line connecting the centers of the two elements, such that $\beta = \pi/2$ rad. Moreover, the midpoint of ds' is located in the plane formed by ds with the straight line connecting the centers of the two elements. The plane in which ds is located has been considered to be the plane of the paper in this drawing. In figure 4.15 (a), we have $\alpha = \pi/2$ rad, $\gamma = \pi/2$ rad, the segment ds is orthogonal to ds', with ds being also orthogonal to the straight line connecting the midpoint of both current elements. In figure 4.15 (b) we have $\alpha = \pi$ rad, the angle γ is undefined or does not exist, while ds is directed along the straight line connecting the centers of both elements. In figure 4.15 (c), on the other hand, we have 0 rad $< \alpha < \pi$ rad and $\gamma = \pi/2$ rad.

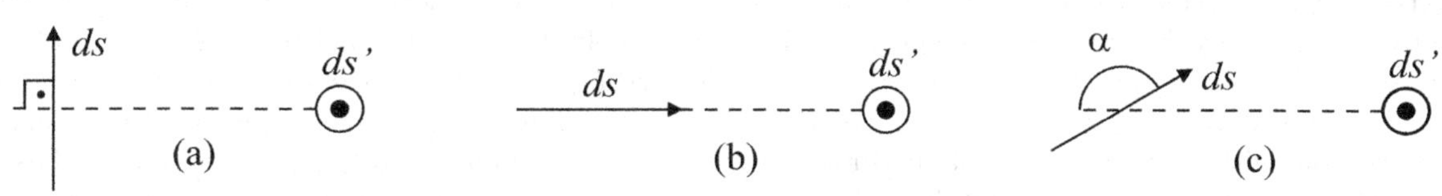

Figure 4.15: Ampère considered that the electrodynamic force between the current elements of lengths ds and ds' vanishes in these three configurations. The element ds' is orthogonal to the plane of the paper, with its center in this plane. The element ds, on the other hand, is located in the plane of the paper.

In *chemise* 162 and in the published summary of his presentation to the Academy of Sciences of Paris which took place on 4 December 1820, he expressed himself as follows:[23]

> Beyond the previously announced law [that is, beyond the addition law], M. Ampère admits as a necessary result of all the circumstances which present the effects of the action [between two current elements of lengths ds and ds'] that one wishes to express analytically, that [this action] vanishes in all cases in which one of the two small portions of electric current [as ds of our figure 4.15] is located in the plane passing through the center of the other perpendicularly to its direction [that is, when the plane in which ds is located passes through the center of ds' and cuts it at a right angle]. For the rest, this [lack of action] will no longer take place, as results from the obtained expression itself, if the length of this last portion of electric current is no longer infinitely small.

In this first work Ampère did not justify this theorem of the nonexistence of interaction between orthogonal current elements. However, in Subsection 4.2.4, we will see how he justified it in another work.

[22] [Ampère, 1820e], [Ampère, b, *carton* 8, *chemise* 162], [Blondel, 1982, pp. 92-94] and [Hofmann, 1996, p. 251].
[23] [Ampère, b, *carton* 8, *chemise* 162] and [Ampère, 1820e, pp. 226-227].

In this work of 1820 Ampère said that this supposition was a "necessary result".[24] In the *Théorie* he said that this supposition was a "theorem".[25] In this book it will be referred as the "theorem of the nonexistence of interaction between orthogonal current elements".

4.2.3 The Article of December 1820

In this Subsection we present how Ampère first presented the law of interaction between two current elements utilizing the addition law and the theorem of the nonexistence of interaction between orthogonal current elements.[26] The magnitudes g and h were called the "intensities" of the two electric currents. It was not yet clear if he was referring to what later would be represented as the current intensities i and i', or the lengths ds and ds' of the two elements, or their products ids and $i'ds'$. Initially we consider the configuration in which the two current elements are orthogonal to the straight line connecting them, as in figures 4.11 and 4.12. The angle $\widehat{GAE}$ was called ζ by Ampère, while the angle $\widehat{HBF}$ was called η. Utilizing the addition law, Ampère represented the two components of AG, namely, AM and MG, as $g \sin \zeta$ and $g \cos \zeta$. Likewise, the two components of BH, namely, BP and PH, were represented as $h \sin \eta$ and $h \cos \eta$. He moved MG to AN, while PH was moved to BQ, so that all components of the elements had the same center. He justified this decision due to the small values of these displacements compared with the distance between the elements. In this way he needed to consider four interactions between the components. There were two interactions between orthogonal elements, namely, between AM and BQ, and also between AN and BP. Utilizing the theorem of the nonexistence of interaction between orthogonal elements discussed in Subsection 4.2.2, he concluded that these two interactions vanished, figure 4.15 (a). There were also two interactions between parallel elements, namely, between AM and BP, and also between AN and BQ. The sum of these two last interactions was represented as follows:

$$\frac{gh\,(\cos\zeta\cos\eta + \sin\zeta\sin\eta)}{r^2} = \frac{gh\cos(\zeta-\eta)}{r^2} = \frac{gh\cos\gamma}{r^2} . \tag{4.5}$$

In this equation $\gamma = \zeta - \eta$ represents the angle between the planes formed by each element with the straight line connecting them.

This result is extremely interesting. In Section 4.1 we saw that Ampère had concluded that the force between two elements should be proportional to $\cos\gamma$ by analyzing the experiment of figure 3.12 between rectilinear conductors of finite length and utilizing, moreover, a principle of mathematical simplicity which was extremely subjective. He was now obtaining the same result by considering directly the addition law for current elements, together with a very specific principle, namely, the non interaction between two elements which were orthogonal to one another and which were also orthogonal to the line connecting them. Figure 4.2 represents qualitatively the force observed in the experiment of figure 3.12. This experimental observation was compatible with any odd power of the function $\cos\gamma$, being also compatible with any sum of these odd powers. However, by utilizing his addition law, Ampère was now obtaining only the first power of $\cos\gamma$. Therefore, he no longer needed to utilize the principle of mathematical simplicity in order to eliminate higher powers of $\cos\gamma$ in his force law. It was now only necessary to utilize the addition law. He arrived at the addition law by considering the different behaviors shown by his two kinds of helices, namely, the axially compensated helix of figure 3.11 (a), and the helix of figure 4.7 which had no such axial compensation. Ampère emphasized this fact in his paper:[27]

> In the first case, the value of the action sought is attractive when $\gamma < \pi/2$; it vanishes with the cosine of γ when $\gamma = \pi/2$, as it should, according to what was stated before; and it changes to a repulsion when $\gamma > \pi/2$, as $\cos\gamma$ is negative, in agreement with the experiment. At last, when $\gamma = \pi$, the exerted action is expressed by $-gh/r^2$, that is, a repulsion with the same magnitude as the attraction which takes place when $\gamma = 0$. Every odd power of $\cos\gamma$, if it were utilized instead of $\cos\gamma$, would also lead to these last results, but would not be in agreement with the [addition] law presented in the memoir read to the *Académie des Sciences* in the meeting of 6 November [1820], while the previous formula [which depends only on the first power of $\cos\gamma$], is a necessary consequence of this law.

In the next step Ampère considered the general case represented by figures 4.13 and 4.14. The angle GAN was called α, the angle HBQ was called β, while the angle between the planes $ACDB$ and $AEFQ$

[24] [Ampère, 1820e, pp. 226-228].
[25] [Ampère, 1826f, p. 30], [Ampère, 1823c, Ampère, 1990, p. 202] and [Ampère, 1965b, p. 174].
[26] [Ampère, b, *carton* 8, *chemise* 162] and [Ampère, 1820e].
[27] [Ampère, b, *carton* 8, *chemise* 162] and [Ampère, 1820e].

was called γ. In this general case he replaced each current element by two components, one component along the straight line connecting the midpoints of both elements, and the other component orthogonal to this straight line. Therefore the element AG was decomposed into $AM = AG \sin\alpha$ and $MG = AG \cos\alpha$. The component MG was displaced to AN. The element BH was likewise decomposed into $BP = BH \sin\beta$ and $PH = BH \cos\beta$. The component PH was displaced to BQ. The current intensities were represented by g and h. Therefore, these four components were expressed by Ampère as $g \sin\alpha$, $g \cos\alpha$, $h \sin\beta$ and $h \cos\beta$, respectively. By the theorem of the nonexistence of interaction between orthogonal elements presented in Subsection 4.2.2, figure 4.15 (b), Ampère concluded that there was no interaction between AN and BP. Likewise, there was no interaction between AM and BQ. The interaction between AM and BP had already been considered by Ampère. Let γ be the angle between the planes $ACDB$ and $AEFQ$. The interaction between AM and BP can then be written as:

$$\frac{(g \sin\alpha)(h \sin\beta) \cos\gamma}{r^2} = \frac{gh \sin\alpha \sin\beta \cos\gamma}{r^2} . \tag{4.6}$$

The interaction between the collinear elements AN and BQ should still be taken into account. Ampère said that the action between these collinear elements might be expressed as:[28]

> a part represented by the n/m fraction of the part which the same small portions of current would exert in the most favorable situation for their mutual action, that is, when they are parallel to one another and perpendicular to the line connecting their centers.

Therefore, the interaction between the collinear components AN and BQ could be written as:

$$\frac{n}{m} \frac{(g \cos\alpha)(h \cos\beta)}{r^2} . \tag{4.7}$$

The total force between elements AG and BH was then written as:

$$\frac{gh}{r^2} \left(\sin\alpha \sin\beta \cos\gamma + \frac{n}{m} \cos\alpha \cos\beta \right) . \tag{4.8}$$

Initially Ampère supposed that there should be no action between two collinear elements, without explaining the reasons for this belief:[29]

> I will also suppose that the attractive or repulsive action between two current portions infinitely small vanishes in another circumstance, namely, when they are along the same straight line.

This configuration is represented, for instance, by the current elements A and B of figure 2.5. Ampère did not present any justification for this conclusion. In any event, he kept his belief that the value of n/m of equation (4.8) should be exactly zero, or that $k = 0$ in equations (2.1) and (2.2), up to the middle of the year 1822, when he first concluded that $n/m = k = -1/2$, as will be fully discussed in Section 7.5.

Ampère had also assumed the lack of interaction when the elements were orthogonal to one another and also orthogonal to the straight line connecting them, as discussed in Subsection 4.2.2. In Subsection 4.2.4 we will see that he justified this lack of interaction between orthogonal elements by the principle of symmetry.

When he published this law, he did not explicitly write $n/m = 0$, leaving the value of n/m undefined as in equation (4.8).[30] He said that this was the most general law compatible with the addition law. However, he also expressed his beliefs as regards the magnitude of this component of his force law:[31]

> If one wants to assume that $n = m$, then it will be found that the factor gh/r^2 will be equal to the cosine of the angle formed by the directions of the two small portions of current, in such a way that the action would be null even for two currents of finite length when their directions formed a right angle, a result against the experiment; which shows evidently that n is much smaller than m, and it seems unlikely that n is not absolutely null. [...] However, M. Ampère thinks that, without inconvenience, this formula might be reduced to
>
> $$\frac{gh \sin\alpha \sin\beta \cos\gamma}{r^2} .$$
>
> He had presented this formula with this format to several scholars before reading his memoir to the *Académie*.

[28] [Ampère, b, *carton* 8, *chemise* 162] and [Ampère, 1820e].
[29] [Ampère, b, *carton* 8, *chemise* 162].
[30] [Ampère, 1820e, p. 229].
[31] [Ampère, b, *carton* 8, *chemise* 162] and [Ampère, 1820e, p. 229].

From this statement we can see that he firmly believed at this time that $n/m = 0$, or at least $n/m \ll 1$, which implied a lack of interaction between collinear elements. What made him hesitate in stating clearly that $n/m = 0$ was an experiment performed by Gay-Lussac (1778-1850) and Thénard (1777-1857). They had a powerful battery with a wire connected to the positive wire and another wire connected to the negative terminal. By bringing the free extremities of these two wires together, an electric spark was produced, allowing a current to flow in this closed circuit. They observed that, at the moment of this spark, just before the contact between the extremities of both wires, there was a strong attraction between them. Apparently Gay-Lussac and Thénard never published the result of this experiment, although it was quoted many times by Ampère.[32] For two collinear elements with currents flowing in the same sense, we have $\alpha = \beta = 0$ rad (or $\alpha = \beta = \pi$ rad), while the angle γ is not defined. If the attraction observed by Gay-Lussac and Thénard were due to a direct interaction between current elements, not being due to other causes (like an electrostatic attraction between the opposite charges located at the extremities of the wires, for instance), then equation (4.8) would indicate that $n/m > 0$. Ampère found an alternative explanation for this experiment, attributing the observed attraction to the vacuum created between the free extremities of the wires due to the current flow.[33] In *chemise* 189, Ampère stated:[34]

> As I had not [yet] found this explanation [based on the vacuum effect] for Gay-Lussac and Thénard's experment, I postponed for 15 days the publication of my first paper in which I utilized only the term $\sin\alpha\sin\beta\cos\gamma$, and stated hesitantly that the other term [for the action between current elements] were null at the 4th December meeting of the *Académie.*

Ampère published his reflections about this experiment in the summary of his presentation to the Academy of Sciences of Paris which took place on 4 December 1820. He asked himself if there was any action between collinear elements, answering as follows:[35]

> M. Ampère does not believe [in this possible action between two collinear elements], due to a series of phenomena, especially after he realized that an observation by MM. Gay-Lussac and Thénard, which seemed to indicate an action of this kind between the extremities of two conducting wires of the great battery of the *École Polytechnique*, when the current seems to have been established before there was a real contact between them, can be explained by the atmospheric pressure, due to a vacuum formed by the current from one [extremity of the wire] to the other [extremity of the other wire], [...]

In conclusion, Ampère presented publicly his force in the form of equation (4.8). But he also simultaneously expressed his points of view according to which the second term of this equation should be exactly null, or at least should have a much smaller magnitude than the first term. Therefore, at this time he believed the force law between two current elements should take the form of equation (4.6). He began to work with this last equation from then onwards, integrating it in order to explain known experiments and in order to make new predictions. He utilized this simplified formula in a series of memoirs read to the Academy of Sciences of Paris on 11 and 26 December 1820, as well as in 8 and 15 January 1821.[36] Although he did not publish the contents of these memoirs, the manuscripts of what he read are still extant in the archives of the Academy of Sciences of Paris.

4.2.4 The Principle of Symmetry

We now discuss how Ampère justified the lack of interaction between the elements in the configurations of figures 4.15 (a), (b) and (c) of Subsection 4.2.2. He utilized a powerful principle of symmetry, expressed as follows in the *Recueil* of 1823:[37]

> The first general fact which I deduced from my experiments on the mutual action between two voltaic conductors was that, if two portions of voltaic conductors [each one of them having an arbitrary orientation in space] act on one another, and if the sense of the electric current in one of them be reversed, without making any changes in their respective orientations, the action which existed between them, if attractive, was transformed into an equal repulsive action and, if the action was repulsive, [was transformed] into an attraction of equal intensity. I applied this result, confirmed by all experiments which I

[32] [Ampère, b, *carton* 8, *chemise* 162], [Ampère, b, *carton* 10, *chemise* 189] and [Ampère, 1820e].
[33] [Blondel, 1982, p. 94], [Hofmann, 1982, pp. 294-295, 300 and 361] and [Hofmann, 1996, pp. 237, 261-263 and 303].
[34] [Ampère, b, *carton* 10, *chemise* 189].
[35] [Ampère, 1820e, p. 229].
[36] [Ampère, 1821b].
[37] [Ampère, 1822p, pp. 208-211] and [Ampère, 1885m, pp. 245-247].

> had made with electric currents of finite length, to two infinitely small portions of these currents, which I could not submit immediately to the experiment. From this application I obtained this conclusion, that the mutual action between two infinitely small portions of conducting wires vanishes necessarily whenever there is not, in their respective orientations, any circumstance which might distinguish from one another the two senses according to which the electric current might flow in one of these two small portions; because in this case, by reversing the sense of this current, the action, if it existed, should remain the same; attractive if it was attractive, and repulsive if it was repulsive, while, on the contrary, according to the general fact aforementioned, the attraction is transformed into a repulsion, and the repulsion [is transformed] into attraction. Therefore, only when the respective orientations of two small portions of conducting wires present circumstances allowing the distinction, in each one of these portions, between the two senses according to which the electric current may flow in them alternatively, will it be possible that an attractive or repulsive action between them be exerted according to the sense, determined by the circumstances, which is given to the electric current.
>
> Considering then two small portions of electric current, with one portion in one plane and the other portion directed perpendicularly to this plane, it was easy for me to realize in the first place that, when this last portion is above or below the plane, the two senses according to which it can be traversed by the electric current are distinct from one another by this circumstance, namely, that in one case this current is moving towards the plane, and in the other case is moving away from it. Consequently, there is nothing against the existence of an action, be it attractive or repulsive, between the two small portions which are being considered, provided the sense of the current which is located on the plane can also be determined by circumstances which depend on the respective configuration of these two small portions, as I will soon consider. In the second place [it was easy for me to realize] that if, on the contrary, the center of the infinitely small portion perpendicular to the plane were located on this plane, everything being equal on both sides of this plane, there is no longer any difference between the two senses according to which the electric current can traverse this portion which might depend of its situation as regards the other [portion] which is on the plane, and the two portions of conducting wires can no longer exert any action on one another; a result which can be enunciated generally as follows: The attractive or repulsive action between two infinitely small portions of electric currents vanishes necessarily when their relative configuration in space is such that, by passing through the center of one of these portions a plane perpendicular to its direction, the straight line representing the direction of the other portion is totally in this plane.

In this quotation Ampère utilized the expression "sense or direction of the electric current" with the fourth meaning described in Section 2.4. With this principle of symmetry he concluded that there is no force between the two current elements of lengths ds and ds' shown in figure 4.15 not only in cases (a) and (b), but also in case (c).

This principle of symmetry, together with the addition law, justifies the deduction presented in Subsection 4.2.3 of the interaction between two current elements in the format of equation (4.8).

4.3 Cases of Equilibrium

4.3.1 Methods to Obtain the Force between Infinitesimal Elements

One of Ampère's main contributions to experimental science was his creation of what he called "cases of equilibrium" and all the theoretical consequences he obtained from these experiments.[38] These cases of equilibrium were also called "null method"[39] and "null experiments".[40] The instruments which he utilized in these experiments have been called "equilibrium apparatus".[41]

Ampère's goal was to determine the force between two current elements. In 1822 he mentioned the existence of two methods in order to arrive at a mathematical expression of this force.[42] The first method is based on a direct measurement of the force and torque exerted between current-carrying conductors of finite length, as measured at different distances and with different orientations of the circuits. After these experiments, an initial hypothesis or guess is made specifying how the force should be between current elements. This infinitesimal interaction is then integrated along both finite circuits in order to see if it agrees with the values obtained experimentally for the force and torque between finite conductors. If there is an agreement, fine. When there is disagreement, the initial hypothesis should be modified until there is

[38] [Ampère, 1826f, pp. 13-27], [Ampère, 1823c, Ampère, 1990, pp. 185-199] and [Ampère, 1965b, pp. 160-173].
[39] [Maxwell, 1954, vol. 2, article 503, p. 159], [Blondel, 1982, pp. 101 and 127], [Darrigol, 2000, p. 11] and [Blondel, 2004].
[40] [Tricker, 1965, index on p. 217].
[41] [Hofmann, 1996, pp. 252-261 and 293-302].
[42] [Ampère, 1822o].

agreement between the integrated calculation and the experimental data. This is one of the usual procedures utilized in order to obtain an expression for the force in different physical conditions and in different branches of physics (electrostatics, magnetism, etc.). It was utilized, for instance, by Biot in his electromagnetic researches.

He then presented the second method which might be utilized in order to obtain the infinitesimal force between two current elements:[43]

> However, there is another way of obtaining the same goal more directly. It was the procedure I have utilized since then and which conducted me to the desired results. It consists in observing, experimentally, that the mobile portions of conductors remain, in certain cases, exactly in equilibrium between equal forces, or between equal torques, whatever, moreover, the shape of the mobile part, and looking for, directly, with the help of the calculus, what should be the value of the mutual action between two infinitely small portions [of current-carrying conductors], in order to obtain an equilibrium which is effectively independent of the shape of the mobile part.

In the first method there is a direct measurement of forces on a mobile circuit at different distances to other circuits, or a direct measurement of torques on a mobile circuit at several orientations relative to other circuits. In the second method, on the other hand, forces and torques are not measured. Instead of these measurements, the mobile circuit is placed between two or more circuits which are fixed in the laboratory, exerting forces or torques on the mobile circuit. Appropriate symmetrical conditions are then found in which the mobile circuit remains in equilibrium, at rest relative to the laboratory, due to the opposite actions exerted by the fixed circuits. Although the mobile circuit is free to move relative to the laboratory, it remains in equilibrium due to opposite forces and opposite torques acting on it. Once more an initial hypothesis suggesting a specific formula for the force between two current elements is necessary. This force is then integrated along the fixed and mobile conductors. One then tries to find if the integrated expression yields zero net force and zero net torque acting on the mobile conductor under the specific configurations in which it was observed to remain stationary.

This method can only work if the central conductor is observed to move when it is under a small electrodynamic force different from zero. That is, the lack of motion in the equilibrium configuration should not be due to friction. The mobility of the circuit can be guaranteed observing its motion or rotation when under the action of only one fixed circuit acting upon it, in which case there is no equilibrium configuration. Another way in which the mobility of the test circuit can be guaranteed is to observe its motion or rotation when there is no longer a symmetrical configuration (for instance, by deforming one of the fixed circuits, or by moving one of the fixed circuits away from its symmetrical configuration).

Ampère utilized two of these cases of equilibrium in order to present a more general justification for the addition law, for his supposition about the nonexistence of interaction between orthogonal elements and for his principle of symmetry which had been discussed in Subsections 4.2.1, 4.2.2 and 4.2.4, respectively. With other cases of equilibrium he obtained finally the value $k = -1/2$ in his force law, equations (2.1) and (2.2). He utilized in three of his most famous cases of equilibrium a special kind of rectangular current loop which was indifferent for terrestrial magnetism. This so-called *astatic coil* did not respond to the actions of the Earth and was not orientated by terrestrial magnetism. This coil is presented in the next Subsection.

4.3.2 Astatic Coils

The astatic coils are mobile current-carrying loops which are immune to terrestrial magnetism. Ampère first described them in 1820.[44] As discussed in Section 3.7, it was with the simplified version of the instrument of figure 3.12 that Ampère first observed unexpectedly the Earth affecting a current-carrying conductor. He then asked to be made a vertical circular current loop of 20 cm diameter which might rotate freely around a vertical axis passing through its center, figure 3.13. This circular loop was orientated by terrestrial magnetism. In particular, he observed that, after being released in an arbitrary orientation relative to the ground:[45]

> [...] *the plane of the conductor rotates and remains at rest in a vertical plane perpendicular to the magnetic meridian*, in such a way that the electric current, in the lower portion of the conductor, is directed from East to West, or from the right to the left of an observer looking at the magnetic North.

[43] [Ampère, 1822o, p. 400].

[44] [Ampère, 1820f], [Ampère, 1965a, pp. 146-154] and [Chaib and Assis, 2009b].

[45] [Ampère, 1821c, p. 91], emphasis by Ampère.

He then predicted that he might join two of these coils side by side, with the same size and area, one coil with a current flowing clockwise and the other coil with a current of the same intensity flowing anti-clockwise, in such a way that the whole system would not rotate due to terrestrial magnetism, although it was free to rotate around a common vertical axis passing through their centers. He confirmed this prediction utilizing the rectangular coil shown in figure 4.16 (a).[46]

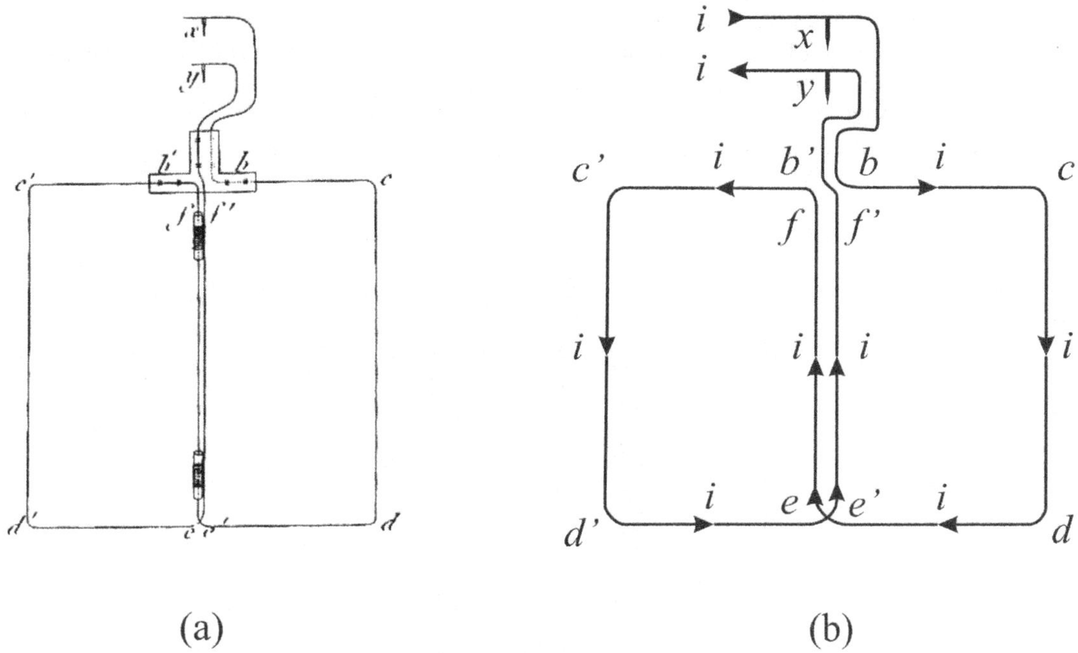

Figure 4.16: (a) Astatic coil indifferent for terrestrial magnetism. This vertical coil can rotate around the vertical symmetry axis passing through x and y. (b) Our representation of the sense of the current along this astatic coil. It flows clockwise in $bcdef$ and anti-clockwise in $b'c'd'e'f'$.

This vertical coil can rotate around the vertical axis xy. The current flows, for instance, in the sense $xbcdefb'c'd'e'f'y$, that is, clockwise in $bcdef$ and anti-clockwise in $b'c'd'e'f'$. When there was a constant current flowing through this coil, Ampère observed that it remained motionless, no matter the initial orientation of the vertical plane of the coil relative to the magnetic meridian. His astatic needle discussed in Section 3.2 was a compass which remained at rest no matter its initial inclination relative to the vertical or horizontal directions. As the coil of figure 4.16 was immune to terrestrial magnetism, Ampère said that it became, analogously, *astatic*.[47] That is, the Earth exerts no net magnetic torque on this coil as a whole, no matter the orientation of its vertical plane relative to the magnetic meridian. In this book we will call them *astatic coils*. In the paper in which he first presented a mobile astatic coil, he said explicitly that its possible motion had no influence from the Earth:[48]

> Therefore this last mobile conductor [i.e., the astatic coil] turns by the action of the fixed conductor without the influence of the Earth, as there is a compensation between the actions exerted by the Earth upon the two equal and opposite halves of the mobile conductor.

4.3.3 Case of Equilibrium of the Sinuous Wire

Ampère utilized a case of equilibrium in order to justify his addition law with more solid and convincing experimental basis. Until December 1820 the addition law was based only on the behavior of axially compensated and non-compensated helices. On 26 December 1820, he described to the Academy of Sciences of Paris an instrument in which a small magnetized needle was suspended horizontally by a vertical thread passing through its center.[49] This vertical thread was at equal distances to two vertical conductors of the

[46] [Ampère, 1820f, figure 10], [Chaib and Assis, 2009b, figure 10] and [Ampère, 1821c, figure 9].
[47] [Ampère, 1822w, pp. 89-90].
[48] [Ampère, 1820f, p. 188] and [Chaib and Assis, 2009b, p. 128].
[49] [Ampère, b, *carton* 8, *chemise* 164], [Ampère, 1821b] and [Hofmann, 1996, pp. 254-256].

same length. A constant current flowed upwards in one conductor, while in the other conductor a current of the same intensity flowed downwards. In order to prove his addition law, Ampère said that the current might flow downwards along a straight line, or along a wire bent into many small sinuosities, without altering the zero net torque on the magnetic needle. Instead of two conductors equally distant to the magnetic needle, he might utilize a single hollow cylindrical support with a current flowing vertically upwards by the center of the support and another current of the same intensity flowing downwards in a crooked path with small sinuosities around the first current. He said that also in this case there would be no net torque acting on the magnetized needle due to this up and down current.

This was his first example of a case of equilibrium. At that moment he did not attach a great importance to this apparatus. Although this instrument was mentioned in some of his publications, it had not yet been built on 26 December 1820.[50] Another analogous apparatus was built later, although the exact date is not known. The most famous representation of this instrument is that presented by Ampère in his *Recueil* of 1823,[51] our figure 4.17.

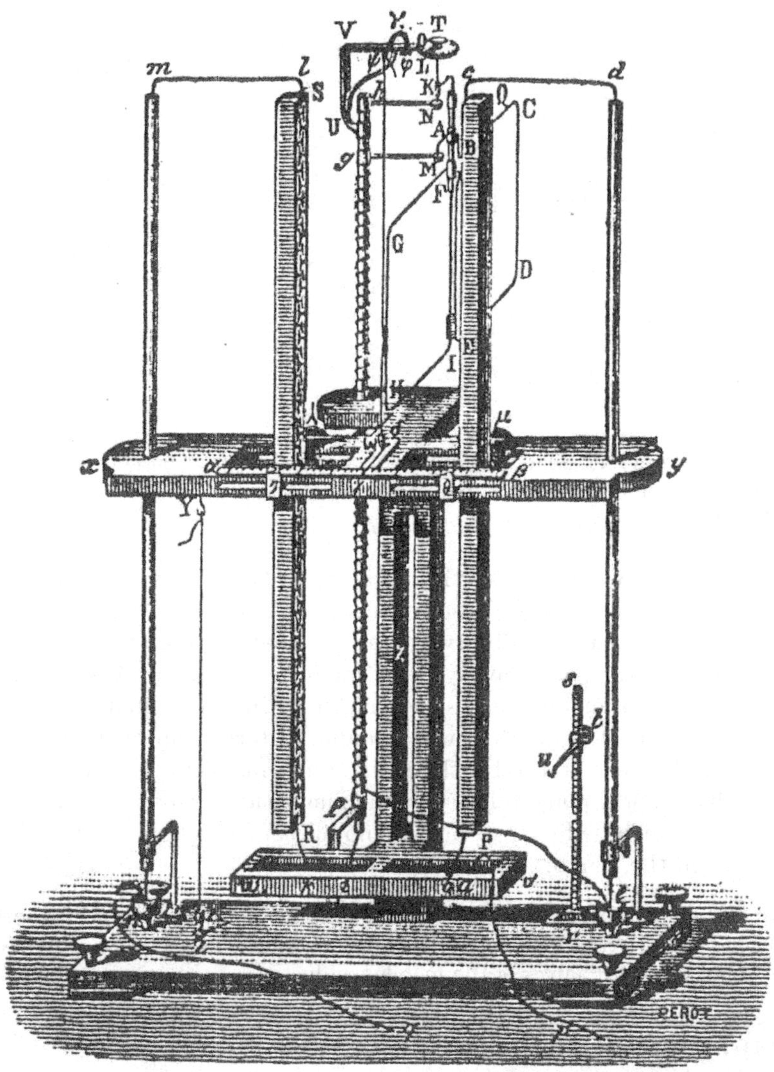

Figure 4.17: Case of equilibrium of the sinuous wire. The fixed sinuous wire is SR, while the fixed vertical wire is QP. The astatic coil $ABCDEFGHIM$ can turn freely around a vertical axis passing through FI, in such a way that the mobile vertical portion GH can move towards SR or away from it.

This experiment will be called the *case of equilibrium of the sinuous wire*. In 1823 Ampère was totally

[50] [Ampère, 1820b] and [Ampère, 1821b].

[51] [Ampère, 1822w, pp. 89 and 216].

aware of the importance of these cases of equilibrium. In this instrument the same constant current flows upwards by two vertical conductors fixed in the laboratory, a straight conductor QP and a sinuous or zigzag conductor SR. These two wires are located at equal distances to the vertical portion GH of an astatic coil $GCDHG$. The magnetized needle described in the previous paragraphs was replaced in this apparatus by an astatic coil analogous to that of figure 4.16. This astatic coil is indifferent for terrestrial magnetism. However, this coil is free to rotate around the vertical axis FI due to the joint influences of the vertical conductors QP and SR. The main portion of this astatic coil is the vertical portion GH, in which the same current which had flown upwards in QP and in SR flows downwards. This portion GH is then repelled by SR and by QP. Figure 4.18 (a) presents a schematic representation of this experiment indicating the senses of the currents.

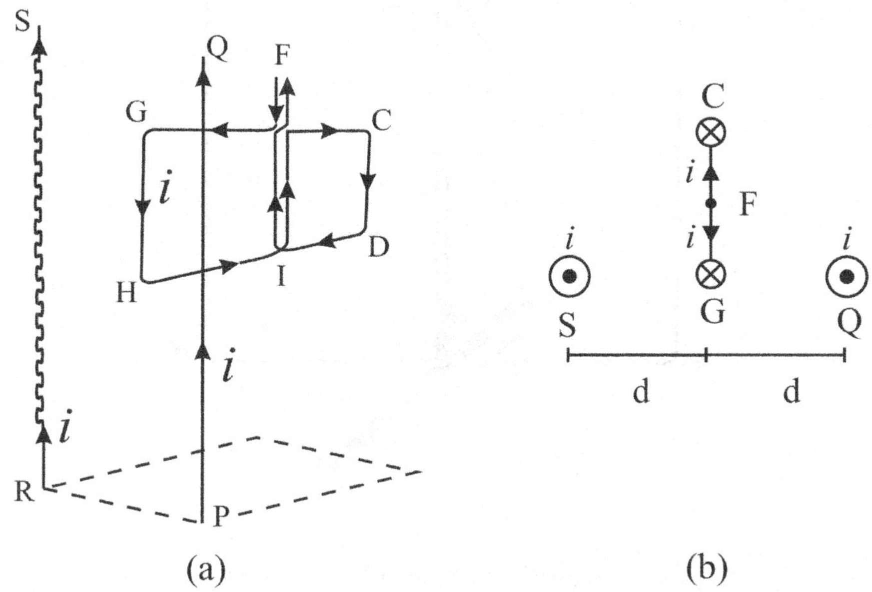

Figure 4.18: (a) Schematic representation of the direction of the current in the experiment of figure 4.17. (b) Case of equilibrium of the sinuous wire seen from above.

Figure 4.18 (b) presents this instrument as seen from above, with the directions of the currents and the most important letters. The astatic coil $GCDHG$ can turn around the vertical axis passing through point F. The current flows sinuously upwards by conductor S, while flowing vertically upwards by conductor Q.

Ampère observed zero torque acting on the rectangular astatic coil $GCDHG$ when portion GH was at equal distances to conductors SR and QP, with a current of the same intensity flowing in all these conductors. This was the experimental observation which justified more completely his addition law. His words in the *Recueil*:[52]

> [Electrical] connections are established in such a way that the current flowing in the two halves of the circuit [SR and QP] exert actions on the mobile conductor which tend to destroy one another. As these two halves of the circuit differ only in the fact that a rectilinear part of one half corresponds to a part of the other half, curved and surrounded as we have said, it is evident that, in the case in which the sinuosities and contours of this last half made its action bigger or smaller than the action of the rectilinear part of the other half of the circuit, the mobile conductor would be deflected by a force equal to the difference of these two actions. However, if the law stated above [that is, the addition law] is exact, then this [mobile] conductor remains in the position in which it was placed before establishing the [electrical] connections, [namely,] in equilibrium between two equal forces. By observing that this [equilibrium] takes place, the experiment demonstrates the exactitude of this law.

Chronologically this was the first case of equilibrium obtained by Ampère utilizing current-carrying conductors.[53] In the *Théorie*, on the other hand, this case of equilibrium appeared in the second place.[54]

[52] [Ampère, 1822w, p. 90].

[53] [Ampère, 1820b, p. 555].

[54] [Ampère, 1826f, p. 16 and figure 2], [Ampère, 1823c, Ampère, 1990, p. 188 and figure 2] and [Ampère, 1965b, p. 164 and figure 49].

4.3.4 Case of Equilibrium of Anti-Parallel Currents

We now discuss the first case of equilibrium discussed in the *Théorie*.[55] It was first presented by Ampère in the *Recueil*[56], our figure 4.19. It will be called *case of equilibrium of anti-parallel currents*.

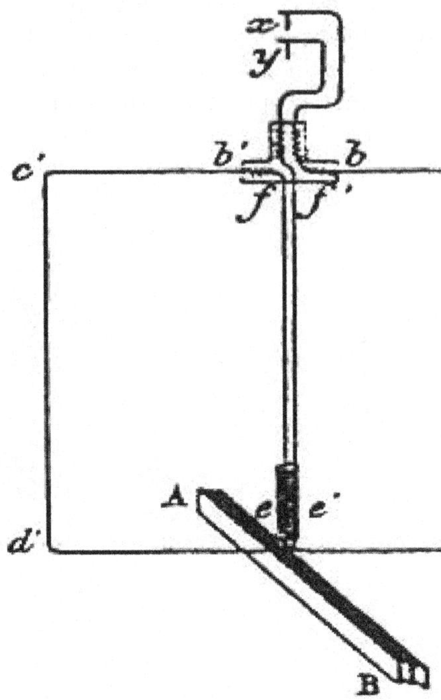

Figure 4.19: Case of equilibrium of anti-parallel currents.

As was seen in Subsections 4.2.1 up to 4.2.4, in order to arrive at his first published formula for the interaction between two current elements, equation (4.8), Ampère utilized not only his addition law, but also the principle of symmetry in order to eliminate the interactions between orthogonal current elements. He based this principle of symmetry on something even more fundamental, namely, the supposition that the force between two current elements changes from attractive to repulsive, or vice-versa, having the same intensity, when the sense of the current is reversed in only one of both elements. He discussed this topic in the *Recueil* of 1823:[57]

> The first general fact which I deduced from my experiments on the mutual action between two voltaic conductors is that, whenever two portions of voltaic currents [with each one of them arbitrarily orientated in space] act on one another, and if the sense of the electric current in one of them is reversed, without making any change in their respective orientations, then the action which was being exerted between them, if it was attractive, is transformed into an equal repulsive action and, if the action was repulsive, [is transformed] into an attraction of equal intensity. I applied this result, which was confirmed by all experiments which I had made with electric currents of finite length, [...]

Despite this generic statement, it was only in the *Recueil* itself that Ampère presented a specific case of equilibrium in order to justify this fundamental supposition. Once more he utilized the astatic vertical coil represented in figure 4.16. A rectilinear horizontal conductor AB was placed below this astatic coil, figure 4.19. In principle this conductor AB might exert a torque on the astatic coil. However, Ampère did not observe any net torque on the astatic coil, no matter the angle between the conductor AB and the vertical plane of the astatic needle. This lack of a net torque on the astatic needle was his experimental proof of the principle of symmetry discussed in Subsection 4.2.4. Ampère's words:[58]

---

[55] [Ampère, 1826f, p. 14 and figure 1], [Ampère, 1823c, Ampère, 1990, p. 186 and figure 1] and [Ampère, 1965b, p. 162 and figure 48].

[56] [Ampère, 1822y, p. 301] and [Ampère, 1885p, p. 275].

[57] [Ampère, 1822p, pp. 208-211] and [Ampère, 1885m, pp. 245-247].

[58] [Ampère, 1822y, pp. 300-302] and [Ampère, 1885p, pp. 275-276].

> Therefore, it is by the observation of cases of equilibrium which are independent on the form of the conductors that it is convenient to determine the value of the sought force. I already recognized three of these [cases of equilibrium]. The first one consists in the equality of the absolute values of the attraction and repulsion which is produced when the same current flows alternately, in two opposite senses, in a fixed conductor, the orientation and the distance to the body on which it acts remaining constants. This equality results from the simple observation that two equal portions of one and the same conductor wire which is covered in silk to prevent communication, [that is, to prevent electrical contact,] well twisted together to form round each other two equal helices, in which the same electric current flows, one current in one sense and the other current in the opposite sense, exert no action on either a moving conductor or a magnet. This [equality of the absolute values] can also be established by the help of the mobile conductor which is illustrated in plate I, figure 9 of the *Annales de Chimie et de Physique*, volume XVIII,[59] [see figure 4.16,] relating to the description of the electrodynamic apparatus of mine which is represented here (figure 1 [our figure 4.19]).
>
> To this end, a horizontal straight conductor AB, doubled several times over, is placed slightly below the lower part $dee'd'$ of this [mobile] conductor, in an arbitrary orientation, such that its midpoint in length and in thickness is in the vertical line through the point x about which the mobile conductor turns freely. It is seen that this [mobile] conductor stays in the orientation where it is placed, which proves that there is equilibrium between actions exerted [that is, between the torques exerted] by the fixed conductor on the two equal and opposite portions of the voltaic circuit $bcde$ and $b'c'd'e'$, which differ only in that the current flows towards the fixed conductor AB in the one [portion], and away from it in the other [portion], whatever the angle between the fixed conductor [AB] and the [vertical] plane of the mobile conductor. Now, considering first the two actions exerted between each portion of the voltaic circuit and the half of the conductor AB which is the nearest, and then the two actions between each of the two portions and the half of the conductor which is the furthest away, it will be seen without difficulty:
>
> 1. That the equilibrium under consideration cannot occur at all angles except insofar as there is equilibrium separately between the first two actions and the last two.
> 2. That if one of the first two [actions] is attractive because current flows in the same direction along the sides of the acute angle formed by the portions of the conductors, the other [action] will be repulsive because the same current flows in opposite senses along the two sides of the equal and opposite angle at the vertex.
>
> So that, initially, for there to be equilibrium between these actions, this attraction and this repulsion, which tend to make the mobile conductor turn, the one in one sense, and the other in the opposite sense, must be equal to one another; then, the last two actions, the one attractive and the other repulsive, between the sides of the two obtuse and opposite angles at the vertex and the supplements of those about which we have just been speaking, must also be equal to each other. Needless to say, these actions [that is, these torques] are really sums of products of forces which act on each infinitesimal portion of the mobile conductor, multiplied by their distance to the vertical about which this conductor is free to turn; however, the corresponding infinitesimal portions of the two arms $bcde$ and $b'c'd'e'$ always being equal at equal distances from the vertical about which they turn, the equality of the torques makes it necessary for the forces to be equal.

It is worthwhile to discuss in detail this experiment, not only because it was the first experiment discussed in Ampère's *Théorie*, but also because it belongs to the four cases of equilibrium appearing in his masterpiece and which were utilized in order to deduce the complete expression of his force between two current elements.

Ampère did not explain what he meant by "a horizontal straight conductor AB, doubled several times over" (*un conducteur rectiligne horizontal plusieurs fois redoublé AB*). The meaning of this expression can be found in the first didactic textbook on electrodynamics published in 1823 by Demonferrand (1795-1844), a disciple of Ampère. When discussing a similar experiment, Demonferrand said that:[60]

> [...] the current dj is doubled several times over to increase its intensity.

As Ampère's force is proportional to the current intensity flowing in the wire, by making a coil with N turns, the torque and force exerted by this coil on another conductor will be increased approximately N times, provided the internal resistance of this coil can be neglected in comparison with the total resistance of the electrical circuit to which it is connected. Therefore, the horizontal straight conductor AB of figure 4.19, doubled several times over, can be represented as the upper portion of the rectangular coil of figure 4.20.

---

[59] [Ampère, 1821c, plate I, figure 9, pp. 88-89].

[60] "[...] le courant dj est redoublé plusieurs fois pour augmenter son intensité, [Demonferrand, 1823, p. 15].

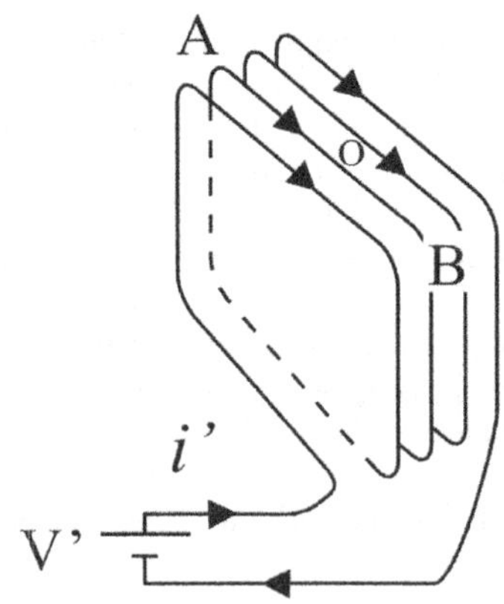

Figure 4.20: A rectangular coil with N turns, with a current intensity i' flowing in each turn. Point O is the midpoint of the upper portion AB.

The case of equilibrium of anti-parallel currents can then be represented as in figure 4.21. In this case the astatic coil with current i is located in a vertical plane, being free to turn around its vertical axis passing through the points x and y. The projection of the vertical axis meets the upper horizontal portion AB of the rectangular coil at its midpoint O. In this figure, separate batteries feed independent currents i' and i in the lower rectangular coil and in the upper astatic coil, respectively, in order to represent the most generic configuration. However, in most of his experiments, Ampère utilized a single battery in order to generate the same current flowing through all his conductors.

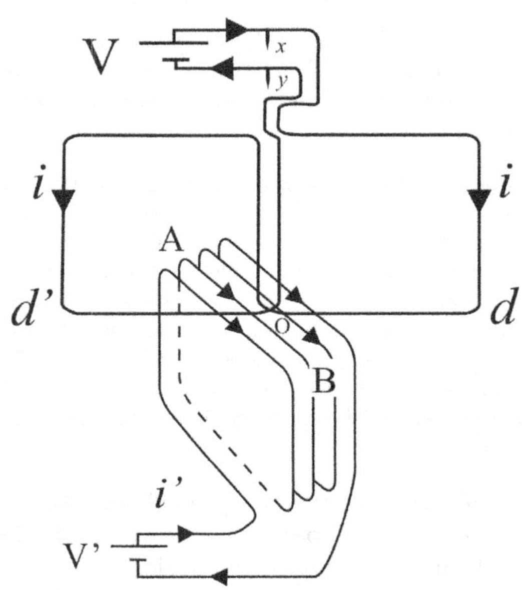

Figure 4.21: Rectangular coil below an astatic coil. The rectangular coil is fixed in the laboratory, while the mobile astatic coil can turn freely around its vertical axis of symmetry.

Figure 4.22 presents only the horizontal components which are closer to one another, namely, portions de and $d'e'$ of the astatic coil with current i, together with the portion AB of the rectangular coil which is fixed in the laboratory. A fixed coil with N turns, with a current i' flowing in each turn, is equivalent to a single turn coil AB with center O carrying a current Ni'.

Ampère's experiment showed that the mobile astatic coil remained in equilibrium, at rest relative to the ground, no matter the angle δ between the fixed horizontal conductor AB and the vertical plane of the astatic coil $d'e'ed$, figure 4.22.

Johann Salomo Christoph Schweigger (1779-1857) in 1820 utilized this idea of increasing the effect generated by a single current-carrying wire when acting on a magnetic needle by making N turns of the conduc-

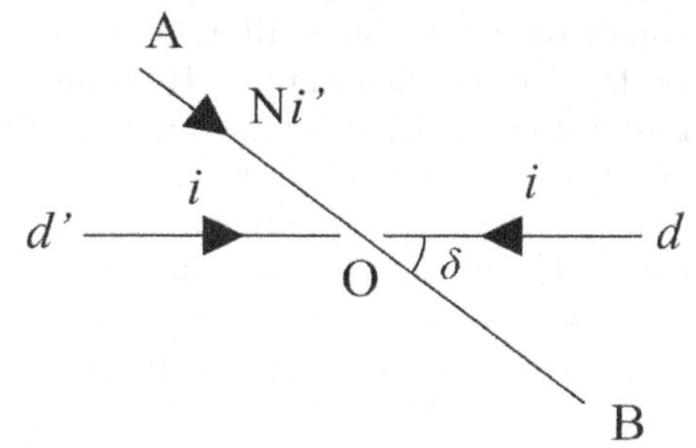

Figure 4.22: Upper view of figure 4.21 indicating only the lower portions $d'e'$ and de of the astatic coil, together with the upper horizontal portion AB of the rectangular coil below the astatic coil.

tor.[61] This instrument has been called "multiplier". Ampère arrived independently at the same idea, as he mentioned in a letter to G. de la Rive from March 25, 1821. He created an instrument in order "to increase the force of the currents", in which "the same conducting wire passes 120 times between two stands always in the same direction, with the goal to exert a stronger action in this proportion".[62] Figure 4.23 presents one of these instruments created by Ampère.[63]

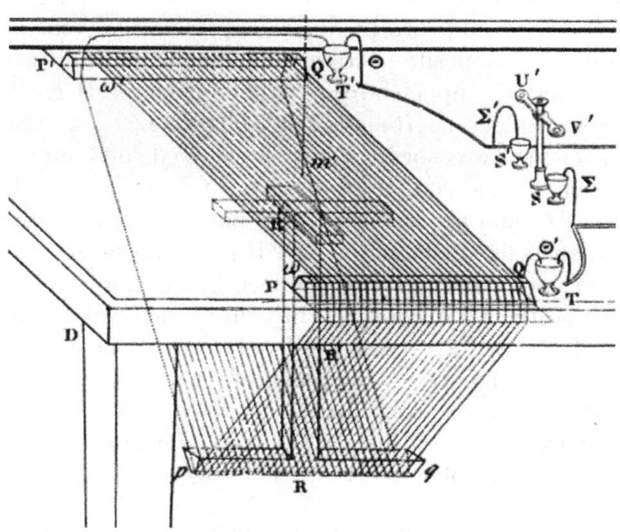

Figure 4.23: Ampère's multiplier.

In this instrument a metal wire makes 30 turns around three wooden stands PQ, $P'Q'$ and pq. The current beginning, for instance, at cup Q', makes several turns in the three stands $P'pP$, getting out at cup Q. Ampère's statements when describing this instrument:[64]

> When M. Ampère utilized this conductor, he could not have been aware of M. Schweigger's ingenious instruments; he [Ampère] had an idea similar to that of this skilled physicist, although he was not able to obtain the several applications made by M. Schweigger about the properties which he recognized on a conducting wire returning in this way around itself.

We now reproduce the main portions of Ampère's long text quoted after figure 4.19. We include new figures in order to illustrate what he wanted to say. We also include our comments. This text has been reproduced almost verbatim in his *Théorie* published in 1826. The only thing he could observe in this experiment of the case of equilibrium of anti-parallel currents was the possible rotation of the astatic coil

[61] [Schweigger, 1821], [Joubert, 1885, p. 13] and [Blondel, 1982, pp. 50 and 104].
[62] [Launay (ed.), 1936a, pp. 567-569].
[63] [Ampère, 1821c, plate 2, figure 12].
[64] [Ampère, 1821c, note on pp. 95-96].

around its vertical axis passing through points x and y after being released from rest, due to the torque exerted by the lower rectangular coil fixed in the laboratory. He observed no torque acting on the astatic coil, although it was free to turn around its vertical axis of symmetry. This experiment was not adapted in order to observe a net force acting on the astatic coil, as only a net torque might be observed. Therefore, whenever Ampère mentioned the "action" exerted by the fixed conductor AB acting on the astatic coil, he was referring to the "torque" acting on it. However, sometimes he utilized the expressions "attractive action" and "repulsive action", referring to the possible attractive and repulsive "forces" acting on the portions of the astatic coil. When discussing Ampère's text, we will clearly distinguish *force* and *torque* in order to clarify his reasonings.

Here are the relevant portions of Ampère's text,[65] together with our clarifications:

> To this end, a horizontal straight conductor AB, doubled several times over, [that is, the upper portion of a rectangular vertical N turn coil,] is placed slightly below the [horizontal] lower part $dee'd'$ of this [mobile] conductor, in an arbitrary orientation, such that its midpoint [O] in length and in thickness is in the vertical line through the point x [and y] about which the mobile [astatic] conductor turns freely. It is seen that this [mobile astatic] conductor stays in the orientation where it is placed, which proves that there is equilibrium between actions exerted [that is, between the torques exerted] by the fixed conductor [AB] on the two equal and opposite portions of the voltaic circuit $bcde$ and $b'c'd'e'$, which differ only in that the current flows towards the fixed conductor AB in the one [portion], and away from it in the other [portion], whatever the angle [δ] between the fixed conductor [AB] and the [vertical] plane of the mobile [astatic] conductor.

Ampère considered the astatic coil, composed of the circuits $bcde$ and $b'c'd'e'$ of figure 4.19, as being a single rigid conductor $c'cdd'c'$ which could turn freely around its vertical axis of symmetry passing through points x and y. The "two equal and opposite portions of the voltaic circuit $bcde$ and $b'c'd'e'$" to which he was referring in this quotation were the upper portion, composed of parts bc and $b'c'$, in which the current was flowing away from the fixed conductor AB according to figure 4.21, and the lower portion, composed of parts de and $d'e'$, in which the current was flowing towards the fixed conductor AB. The astatic coil has two other portions, namely, the central axis, composed of parts ef and $e'f'$, in which the current was flowing away from the fixed conductor AB, and the vertical lateral portion, composed of parts cd and $c'd'$, in which the current was flowing towards the fixed conductor AB. It is not necessary to take into account the action on the central axis because, no matter the net force exerted by AB on this central axis, this supposed net force will not exert any torque no the astatic coil, as the arm of the lever of this torque has zero length.

In the sequence of his analysis, Ampère considered only one of the portions of the astatic coil. The considered portion might be the upper portion, composed of parts bc and $b'c'$, the lower portion, composed of parts de and $d'e'$, or the lateral portion, composed of parts cd and $c'd'$. We will illustrate his reasoning considering only the lower portion, although his statements could also be applied to the upper or lateral portions. Here are the relevant portions of Ampère's text,[66] together with our clarifications:

> Now, considering first the two actions [torques] exerted between each [lower] portion of the voltaic circuit and the half of the [fixed] conductor AB which is the nearest, and then the two actions [torques] between each of the two [lower] portions and the half of the [fixed] conductor [AB] which is the furthest away, it will be seen without difficulty:
>
> 1. That the equilibrium under consideration cannot occur at all angles except insofar as there is equilibrium separately between the first two actions [torques] and the last two [torques].
> 2. That if one of the first two [forces] is attractive because current flows in the same direction along the sides of the acute angle formed by the portions of the conductors, [that is, with the two currents flowing towards the center O of the fixed conductor AB, or flowing away from this center,] the other [force] will be repulsive because the same current flows in opposite senses along the two sides of the equal and opposite angle at the vertex, [that is, with one of the currents flowing towards the center O of the fixed conductor AB, while the other current is flowing away from it].
>
> So that, initially, for there to be equilibrium between these actions [torques], this attraction and this repulsion, which tend to make the mobile conductor turn, the one in one sense, and the other in the opposite sense, [that is, the attractive force tending to make the astatic coil turn in one direction and the repulsive force tending to make the astatic coil turn in the opposite direction,] must be equal to one another; then, the last two actions [forces], the one attractive and the other repulsive, between the

[65] [Ampère, 1822y, pp. 301-302] and [Ampère, 1885p, pp. 275-276].
[66] [Ampère, 1822y, pp. 301-302] and [Ampère, 1885p, pp. 275-276].

> sides of the two obtuse and opposite angles at the vertex and the supplements of those about which we have just been speaking,[67] must also be equal to each other. Needless to say, these actions [that is, these torques] are really sums of products of forces which act on each infinitesimal portion of the mobile conductor, multiplied by their distance to the vertical about which this conductor is free to turn; however, the corresponding infinitesimal portions of the two arms *bcde* and *b'c'd'e'* always being equal at equal distances from the vertical about which they turn, the equality of the torques makes it necessary for the forces to be equal.

Figures 4.24 and 4.25 illustrate what Ampère wanted to say. The attractive force exerted by the half part AO of the fixed conductor, acting on the lower part $d'e'$ of the astatic coil which is closer to AO, is represented by $\vec{F}_1$. The repulsive force exerted by OB on de is represented by $\vec{F}_2$. Force $\vec{F}_1$ is attractive due to the fact that both currents flow towards point O. Force $\vec{F}_2$, on the other hand, is repulsive, because one of the currents flows towards O, while the other current flows away from O, as can be seen in figures 3.18 and 3.19. Ampère concluded that these two forces should be equal in magnitude and direction, namely, $\vec{F}_1 = \vec{F}_2$. According to Ampère, only with this equality could they generate equal and opposite torques on the lower portion $d'e'ed$ of the astatic coil. That is, the torque generated by the force $\vec{F}_1$ tends to turn the astatic coil clockwise around the vertical passing through point O, while the torque generated by the force $\vec{F}_2$ tends to turn the astatic coil anti-clockwise.

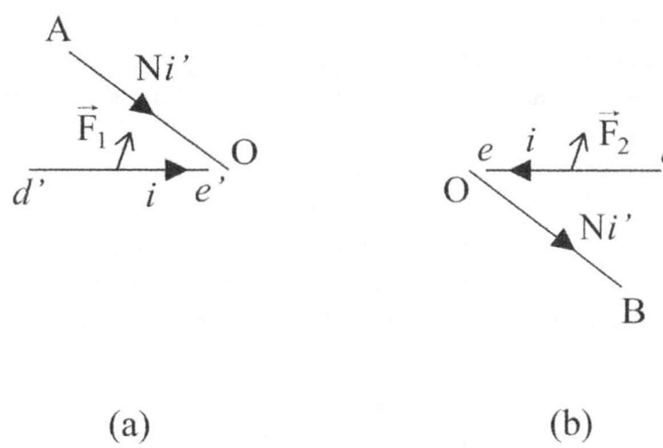

Figure 4.24: (a) $\vec{F}_1$ represents the attractive force exerted by the half AO of the fixed conductor acting on the lower portion $d'e'$ of the astatic coil which is closer to AO. (b) $\vec{F}_2$ represents the repulsive force exerted by OB acting on de.

Figure 4.25 represents the remaining forces exerted by the fixed conductor AB acting on the lower portion $d'e'ed$ of the astatic coil. The repulsive force exerted by the half OB of the fixed conductor acting on the farthest portion $d'e'$ of the astatic coil is represented by $\vec{F}_3$, while the attractive force exerted by AO and acting on the portion de is represented by $\vec{F}_4$. Force $\vec{F}_3$ is repulsive because one of the currents flows towards point O while the other current flows away from this point. Force $\vec{F}_4$, on the other hand, is attractive because both currents flow towards point O, see figures 3.18 and 3.19. Ampère concluded that these two forces should have the same magnitude and direction, that is, $\vec{F}_3 = \vec{F}_4$. Only with this force equality could they generate equal and opposite torques on the lower portion $d'e'ed$ of the astatic coil. That is, the torque generated by force $\vec{F}_3$ tends to turn the astatic coil clockwise around the vertical passing through point O, while the torque generated by force force $\vec{F}_4$ tends to turn the astatic coil anti-clockwise.

A similar analysis can be made as regards the forces and torques exerted by the fixed conductor AB acting on the upper portion $c'b'bc$ of the astatic coil. Analogously, a similar analysis can be made as regards the forces and torques exerted by the fixed conductor AB acting on the lateral portions cd and $c'd'$ of the astatic coil.

The experimental result is that the astatic coil remains in equilibrium no matter the angle δ between the vertical plane passing through the fixed conductor AB and the vertical plane of the astatic coil. This lack of a net torque acting on the mobile astatic coil in this configuration can only be explained by the fact that the force between two current elements changes from attractive to repulsive, or from repulsive to attractive, maintaining the same intensity, whenever only one of these currents reverses its sense. For this reason we have called this experiment the *case of equilibrium of anti-parallel currents*. The anti-parallel currents in the case of figure 4.24, for instance, would be $d'e'$ and de. That is, parallel and collinear currents flowing in

[67]Two angles are supplementary if they add up to 180° or π rad.

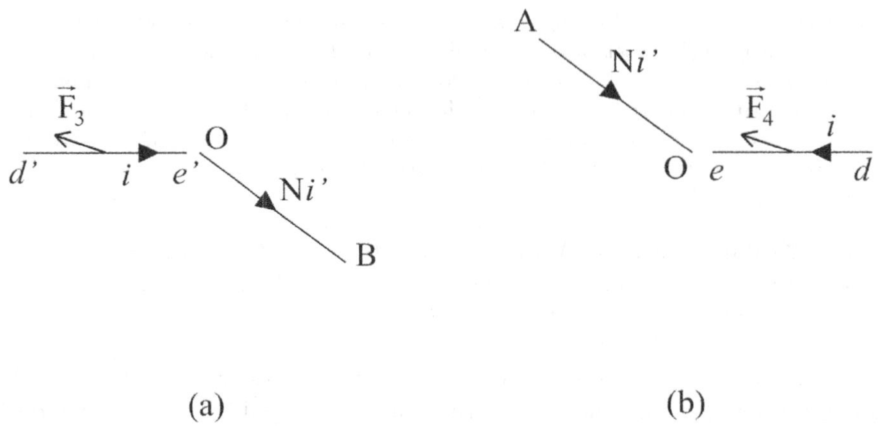

Figure 4.25: (a) $\vec{F}_3$ represents the repulsive force exerted by the half OB of the fixed conductor acting on the farthest lower portion $d'e'$ of the astatic coil. (b) $\vec{F}_4$ represents the attractive force exerted by AO acting on de.

opposite senses, one current flowing from left to right towards point O, while the other current flows from right to left towards point O. The anti-parallel currents in this figure might also be AO and OB. That is, parallel and collinear currents, one of them flowing towards the rotation axis, while the other current flows away from this axis.

By turning figure 4.24 (a) clockwise by an angle of π rad around the vertical axis passing through point O, we arrive at the configuration of figure 4.26. This configuration is analogous to the configuration 4.24 (b) with OA instead of OB and with $e'd'$ instead of ed. However, in figure 4.26 the current in the fixed conductor flows from point A towards point O. There is now an attractive force $\vec{F}_1$ on the mobile portion $e'd'$ of the astatic coil pointing towards OA, instead of the repulsive force $\vec{F}_2$ of figure 4.24 (b).

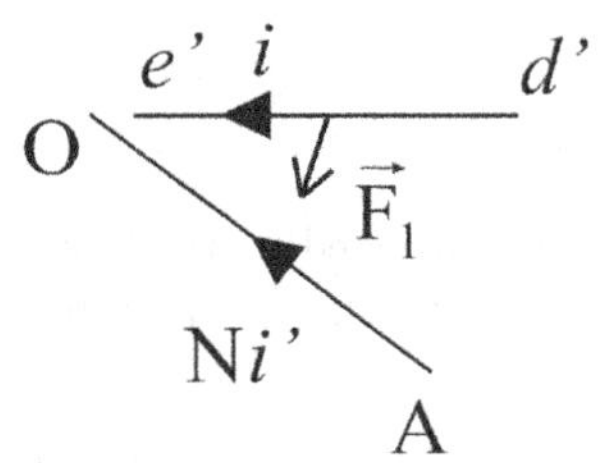

Figure 4.26: Configuration of figure 4.24 (a), after rotating the whole system clockwise by π rad around the vertical axis passing through point O.

By changing the sense of the current only in the fixed conductor OA of figure 4.26, so that it now flows from O to A instead of flowing from A to O, the attractive force $\vec{F}_1$ which was acting on the mobile portion ed of the astatic coil is transformed into a repulsive force having the same intensity as force $\vec{F}_1$. We then arrive exactly at the configuration of figure 4.24 (b).

Ampère could then justify his addition law and his principle of symmetry utilizing the case of equilibrium of the sinuous current together with the case of equilibrium of anti-parallel currents. In this way he could conclude that equation (4.8) was the only expression compatible with the experimental behavior of nature. From this time onwards he needed only to determine experimentally the value of the constant which he represented by n/m or, equivalently, by k.

Chapter 5

Ampère's Conception of Magnetism

5.1 Magnetism being due to Macroscopic Electric Currents Flowing in Magnets and in the Earth

The nomenclature utilized in this book has been presented in Section 1.4:

- *Electrostatic phenomena:* Forces and torques between electrified bodies which are at rest relative to one another.
- *Magnetic phenomena:* Forces and torques between magnets, together with the torques exerted by the Earth on magnets (orientation of compass and dip needles).
- *Electromagnetic phenomena:* Forces and torques between a current-carrying conductor and a magnet, together with the forces and torques exerted by the Earth on current-carrying conductors.
- *Electrodynamic phenomena:* Forces and torques between current-carrying conductors.

The words *magnet*, *magnetic* and *magnetism* come from a region called Magnesia, where the ancient Greeks found the naturally occurring magnetic mineral magnetite, an iron oxide which had the property of attracting small pieces of iron.[1] The word *electromagnetism* was introduced by Ørsted, while the expressions *electrostatic* and *electrodynamic* were introduced by Ampère, as discussed in Section 1.4.

In Section 2.3 and in Chapter 3 we presented Ampère's initial reactions to Ørsted's experiment. He supposed that the magnetic and electromagnetic effects might be due purely to interactions between current-carrying conductors. That is, Ampère believed these phenomena resulted from electrodynamic interactions. In order to test experimentally this hypothesis, he tried to reproduce electrodynamically the magnetic and electromagnetic phenomena, that is, utilizing current-carrying conductors replacing the magnets and the Earth. To this end he had to find the distribution of the supposed electric circuits inside the Earth and also inside magnets.[2] In order to test theoretically this hypothesis, he tried to find a mathematical expression for the force between two current elements which, after being integrated, might reproduce the known formulas describing the magnetic and electromagnetic interactions.

Initially he supposed macroscopic currents flowing in the Earth, as discussed in Section 2.3, figures 2.3 and 2.4. Although he was not very specific about these hypothetic terrestrial electric currents, it is possible that he had in mind something as represented in figure 5.1. That is, these electric currents should describe circles centered along the North-South magnetic axis of the Earth, with all of these currents flowing in planes orthogonal to this axis.

As regards the distribution of the supposed electric currents inside a magnet, Ampère was very specific, as discussed in Section 3.5. It is relevant to reproduce once more his words of 1820 illustrating what current distribution he had in mind:[3]

> Now, if electric currents are the cause of the directive action of the Earth, then electric currents could also cause the action of one magnet on another magnet. *It therefore follows that a magnet could be regarded*

[1] [Assis, 2010b], [Assis, 2010c], [Assis, 2011b] and [Assis, 2015c].

[2] [Chaib and Assis, 2007a].

[3] [Ampère, 1820f, pp. 207-208], [Ampère, 1965a, pp. 152-153] and [Chaib and Assis, 2009b, pp. 137-138], our emphasis.

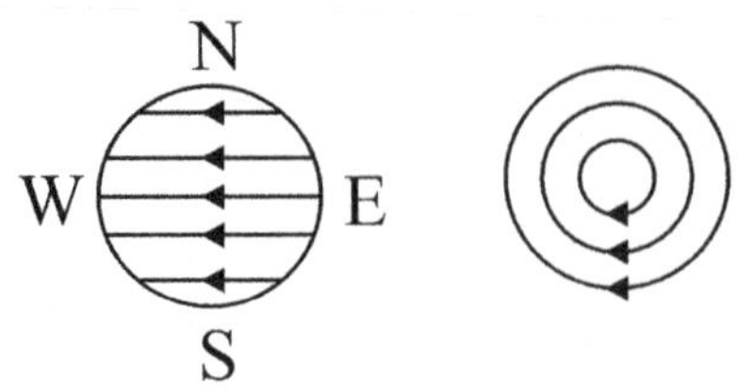

(a) (b)

Figure 5.1: Possible representation of the supposed terrestrial electric currents according to Ampère. (a) Currents over the surface of the Earth. (b) Concentric currents inside the Earth along its magnetic equatorial plane, with the North pole above the paper and the South pole below the paper.

> *as an assembly of electric currents in planes perpendicular to its axis*, their direction being such that the austral pole of the magnet, pointing North, is to the right of these currents since it is always to the left of a current placed outside the magnet, and which faces it in a parallel direction, or rather that these currents establish themselves first in the magnet along the shortest closed curves, whether from left to right, or from right to left, and the line perpendicular to the planes of these currents then becomes the axis of the magnet and its extremities make the two poles. Thus, at each pole the electric currents of which the magnet is composed are directed along closed concentric curves.

Figure 5.2 presents a representation of the currents for a cylindrical magnet, following Ampère's prescription. In figure 5.2 (a) Ampère's observer is lying on his back with the currents flowing from his feet towards his head. The North pole of this magnet is at the right of this observer. Figure 5.2 (b) presents a cross section of this magnet, where the North pole is above the paper and the South pole is below it. According to Ampère's initial conception, the electric currents should form macroscopic circles centered along the North-South axis of the magnet.

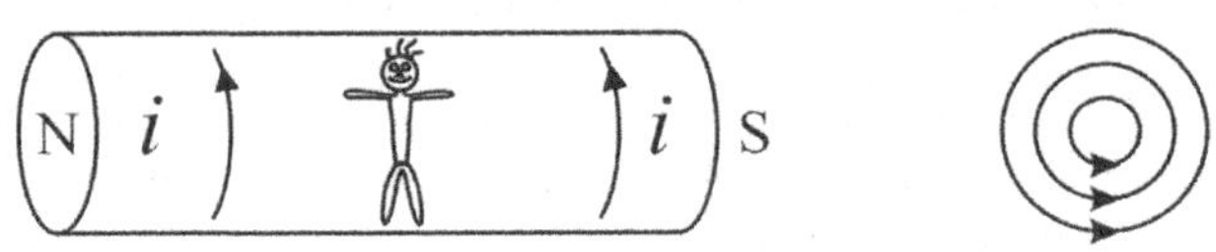

(a) (b)

Figure 5.2: Electric currents flowing on the surface and inside a cylindrical magnet, according to Ampère's initial conception of magnetism.

The fact that these supposed electric currents should flow in circles around the axis of cylindrical magnets has been explicitly mentioned by Ampère in another portion of this paper:[4]

> From M. Biot's splendid experiment, currents which are in one and the same plane perpendicular to the axis of a magnet, must be regarded as having the same intensity, since it results from the experiment where he compared the effects produced by the action of the Earth on two similarly magnetized bars of the same size and shape, of which one was hollow and the other solid, that the motive force is proportional to the mass and that in consequence the causes to which it is due act with the same intensity on all particles of one and the same cross-section perpendicular to the axis, the intensity varying from section to section according as these sections are close to or far from the poles. When the magnet is a solid of rotation about the line joining its two poles, all the currents of one and the same section must be circles; [...]

Figure 5.3 presents Ampère's initial conceptions about the macroscopic currents in a rectangular magnet. Figure 5.3 (a) shows Ampère's original figure.[5] Figure 5.3 (b) is our redrawing of this figure.

[4] [Ampère, 1820f, p. 179], [Ampère, 1965a, p. 147] and [Chaib and Assis, 2009b, pp. 124-125].

[5] [Ampère and Babinet, 1822a, p. 32, figure 15] and [Ampère and Babinet, 1822b, p. 197, figure 15].

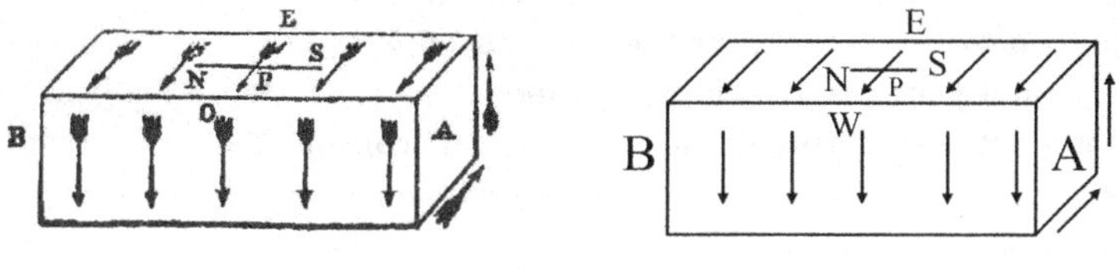

(a) (b)

Figure 5.3: Electric currents flowing in a rectangular magnet according to Ampère's initial conception.

5.2 Fresnel's Contributions

As seen in Section 5.1, initially Ampère conceived macroscopic currents flowing in magnets and in the Earth. In January 1821 he changed his conception and began to adopt the idea that magnets and also the Earth were composed of assemblies of particles. Electric currents would be circulating perpetually round these particles. These microscopic electric currents should flow especially around magnetized particles or molecules of iron or steel. He attributed this conception to Augustin-Jean Fresnel (1788-1827), figure 5.4.

Figure 5.4: A.-J. Fresnel.

Fresnel began to study at the *École Polytechnique* in 1804, the same year in which Ampère began to teach at this institution. They began to have close contact around 1814 when Fresnel began his interests in optics. In order to improve his financial situation, Ampère used to rent a room is his Parisian home to some friends. Fresnel lived there from 1822 until his early death in 1827, when he was 39 years old.[6]

Ampère initially accepted the corpuscular theory of light, which he also called the "system of emission". Fresnel, on the other hand, defended the wave theory of light. Ampère changed his mind in 1816 due to Fresnel's influence.[7] Fresnel initially believed that the luminous waves were due to longitudinal vibrations in the ether, in analogy with sound waves in air. Ampère suggested to Fresnel to consider luminous transverse waves instead of longitudinal ones, in order to explain satisfactorily several phenomena, including light polarization. Fresnel himself acknowledged that this fundamental idea of transverse vibrations orthogonal to the direction of propagation of the luminous wave was due to Ampère.[8]

As regards the electrodynamic conception of magnetism, it was Fresnel who supplied a fundamental contribution to Ampère's thinking. Initially Ampère believed in macroscopic electric currents flowing inside magnets and in the Earth in order to explain their magnetic properties. Fresnel suggested to him the conception of microscopic currents flowing around the iron or steel particles of a magnet, instead of macroscopic currents flowing around the axis of the magnet.[9] Fresnel presented a comparison between these two electrodynamic conceptions of magnetism in two papers.[10] The second work, dated 5 July 1821, probably reflects his mature thinking on this subject due to his exchange of ideas with Ampère during the previous months.

[6][Hofmann, 1996, pp. 135-136 and 223].

[7][Rosmorduc, 1977, p. 162] and [Hofmann, 1996, pp. 216-217 and 221-222].

[8][Rosmorduc, 1977, pp. 165-167] and [Hofmann, 1996, p. 222].

[9][Blondel, 1982, p. 98] and [Hofmann, 1996, p. 283].

[10][Fresnel, 1885a] and [Fresnel, 1885b].

These two papers were not published in 1821. They were found with Ampère's manuscripts, being published by Joubert in 1885 in his *Collection of memoirs related to physics.*[11]

In these two works Fresnel presented several arguments against the conception of macroscopic currents and in favor of microscopic currents around the particles of iron or steel composing a magnet. Title of his first paper:[12] *Comparison of the supposition of currents around the axis [of a magnet] with currents around each molecule.*[13] Title of his second work:[14] *Second note on the hypothesis of currents belonging to the particles.*[15]

In his first paper Fresnel argued as follows:[16]

> With the supposition of [macroscopic] currents around the axis [of a magnet], when we cut longitudinally a hollow magnetized steel cylinder, it should lose immediately its magnetic property. [...]
>
> With the supposition of currents around each particle [of the magnet], a hollow magnet can be cut longitudinally, without destroying its magnetism.

This statement is a strong argument in favor of microscopic currents. After all, when performing this experiment, cutting longitudinally a hollow magnet, its magnetism is not destroyed.

Fresnel presented other arguments against the conception of macroscopic currents. Since the early experiments with Volta's battery around 1800 it was known that a resistive wire heats when a current flows through it. A permanent magnet, on the other hand, usually has the same temperature of the surrounding bodies. A bar of soft iron can be magnetized by passing a current through a helix coiled around it. The bar is not heated by this procedure. How could these phenomena be explained based on Ampère's conception that magnetism was due to electric currents inside the magnet? Fresnel utilized these phenomena to argue in favor of molecular currents:[17]

> I point out, moreover, the peculiarity that a magnet is not hot, although it should be [hot] with the hypothesis of [macroscopic] currents around the axis [of the magnet], is not difficult to explain with the hypothesis of currents around the particles [of the magnet]; for, if a [macroscopic] electric current, by flowing through a mass of particles of a conducting body, heats it, then it is not necessary that the [microscopic] currents around the particles of a similar mass should also heat it [the body], because the circumstances are not the same. The cause of the heat produced by an [macroscopic] electric current is not well known, and we have only very incomplete ideas about the constitution of bodies in order to know if, in this case [of molecular currents], the electricity should produce heat.

5.3 Ampère and the Molecular Currents

In an unfinished memoir on the theory of magnetism, Ampère attributed to Fresnel the suggestion of the hypothesis of currents flowing around the particles or molecules composing a magnet. There is a manuscript of this work, written with his letter, mentioning this fact. The following fragment of this memoir has been published by Joubert:[18]

> This hypothesis (the hypothesis of currents around the particles) has been communicated to me by M. Fresnel, who found several advantages in considering in this way the electric currents in a magnet.

In 26 December 1820 and in 15 January 1821, Ampère presented at meetings of the Academy of Sciences of Paris the hypothesis according to which the supposed currents in magnets do not need necessarily to be coaxial.[19] His words:[20]

[11] *Collection de Mémoires relatives a la Physique.*

[12] [Fresnel, 1885a].

[13] French original title: *Comparaison de la supposition des courants autour de l'axe avec celle des courants autour de chaque molécule.*

[14] [Fresnel, 1885b].

[15] French title: *Deuxième note sur l'hypothèse des courants particulaires.* The adjective *particulaire* can be translated as *particulate.* It is related to the particles composing the bodies (the magnet, in his example), [Littré, 1877]. This word should not be confused with the adjectives *particulier* and *particulière,* having the meaning of something "specific" or "particular".

[16] [Fresnel, 1885a, p. 143].

[17] [Fresnel, 1885b, p. 144].

[18] [Ampère, 1885a, p. 140].

[19] [Blondel, 1982, p. 98].

[20] [Ampère, 1820b, pp. 557-558], [Ampère, 1821b, p. 163], [OErsted et al., 1822, pp. 91-92] and [Ampère, 1885a, pp. 139-140].

> In a last memoir, read to the *Académie des Sciences* in 8 and 15 January 1821, M. Ampère presented some approximate calculations, relative to the mutual action between a [current-carrying] connecting wire and a magnet, utilizing the formulas from which he can deduce, from the law aforementioned, all the circumstances of this action. He finished examining a question which does not seem to him susceptible of being exactly solved, before these calculations have been advanced and their results compared with those of experiments, a situation for which no precise observation has already been made. This question is to know if the closed curves, where the electric currents providing the magnetized steel the properties characterizing it flow, are situated concentrically around the line connecting the two poles of the magnet, or if these currents are distributed over all its mass around each one of its particles, always in planes orthogonal to this line. Several considerations which the author has not yet developed seem to favor this last proposal of conceiving the existence of electric currents in the magnets. However, as all phenomena known up to now can be equally well explained with the first [hypothesis], he considered it better to leave this question open for the time being, until new calculations and new experiments offer all the necessary elements for a solution.

Figure 5.5 illustrates Ampère's initial conception of molecular currents for the cylindrical magnet.

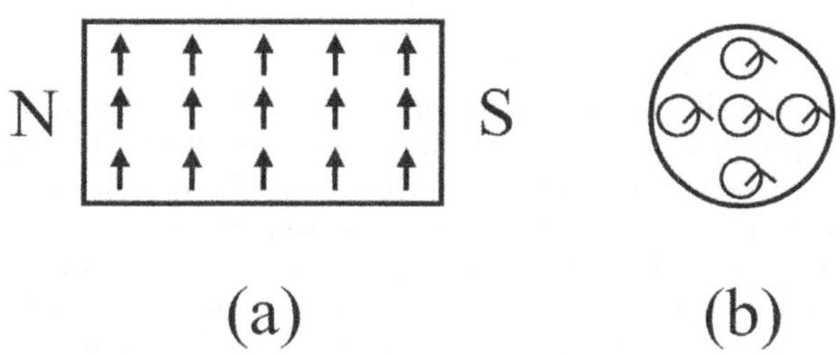

Figure 5.5: Ampère's initial conception of molecular currents for a cylindrical magnet. (a) Directions of the molecular currents at the surface of a cylindrical magnet NS, as seen in profile. (b) Cross section of this magnet indicating the directions of the currents. The North pole is above the paper and the South pole below it.

According to this quotation, Ampère initially considered even molecular currents as being located in planes orthogonal to the straight line connecting the magnetic poles of a magnet. Figure 5.5 (a) shows a longitudinal cut of a cylindrical magnet with its North pole at the left of this figure. The small arrows pointing upwards indicate the senses of the microscopic currents at the upper parts of the particles composing the magnet, that is, the parts of these particles which are closer to the eyes of someone looking at this figure. Figure 5.5 (b) shows a cross section of this magnet, with the North pole above the paper and the South pole below it. The small circular loops represent the currents flowing around the particles of the magnet.

Between 1821 and 1822, Gaspard de la Rive (1770-1834), Albert van Beek (1787-1856) and Faraday performed some experiments showing that the poles of a cylindrical magnet are not located exactly at the extremities of the magnet. These experiments forced Ampère to modify his conception of microscopic currents.[21] In a letter addressed to Gaspard de la Rive, dated 12 June 1822, Ampère included figure 5.6 (a).[22] It presents a longitudinal cut of a cylindrical magnet with its North pole at point A and its South pole at point B. This figure presents the equilibrium configuration of the microscopic currents around the particles of the magnet, due to the interaction of all microscopic currents. That is, due to the collective interactions between the small current-carrying loops, the planes of these molecular currents should no longer remain orthogonal to its magnetic axis.[23] Figure 5.6 (b) presents the same image indicating the North and South poles by the letters N and S, respectively.

This final conception of molecular currents presented by Ampère, with their planes inclined relative to the axis of an uniformly magnetized bar, is accepted in its essence up to the present time.

5.4 Names Given to the Molecular Currents

As discussed in Sections 5.2 and 5.3, Fresnel suggested to Ampère the hypothesis of electric currents flowing around the particles or molecules of a magnet. Fresnel mentioned currents around each *molecule* or *particle*

[21] [Blondel, 1982, pp. 98 and 123-125] and [Hofmann, 1996, pp. 282-290].
[22] [Ampère, 1822m, p. 257 and plate 6, figure 25], [Ampère, 1885i, p. 155] and [Blondel, 1982, p. 124].
[23] [Hofmann, 1996, pp. 288-289].

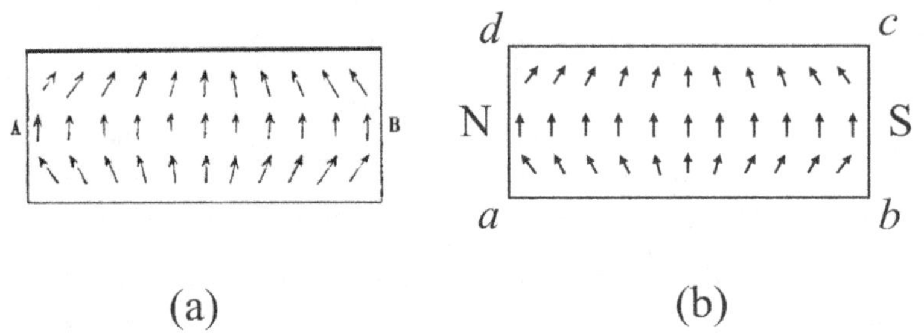

Figure 5.6: Ampère's improved conception of the molecular currents for a cylindrical magnet.

of a magnet, calling them *courants particulaires*, that is, *particulate currents* or currents around the particles or molecules composing the magnet.

Ampère initially adopted this nomenclature. For instance, in a letter dated 10 July 1822, addressed to Faraday, he said:[24]

> This objection [to Wollaston's theory] does not appear in my explanation of the phenomena, because I initially admit that the preexisting currents around each particle of iron or steel are directed in all directions before the magnetization, in such a way that it can be concluded that the total action they exert on an external point is necessarily null. I consider as something probable, [the existence of] similar currents around the particles of all bodies, but I suppose that they remain rigidly in this configuration in which they cannot act externally, whenever these bodies are not susceptible of being magnetized. In iron, nickel and cobalt, on the other hand, this configuration can be modified, and this modification takes place by the action of a conducting wire, of the Earth, or of a magnet orienting these particulate currents [*courants particulaires*], exactly as this action orientates, in my experiments, a conductor bent such that it forms an almost closed circle; [...]

In the original manuscript and in Joubert's publication the expression *courants particulaires* is explicitly mentioned.[25] In Ampère's published correspondence published by Launay, on the other hand, this expression has been wrongly transcribed as *courants particuliers*.[26] In the *Théorie* Ampère also utilized the expression *courants particulaires*.[27]

The expression *molecular currents*, *courants moléculaires*, has also been utilized by Ampère.[28] He utilized this expression, for instance, in letters of 1833 addressed to Auguste de la Rive (1801-1873)[29] and to Faraday.[30] In this last example, for instance, Ampère expressed himself as follows:

> But since the experiment I performed with M. Auguste de la Rive obliged me to retract this and admit the production of currents by influence, I thought that the big question of the pre-existence or otherwise of molecular currents of metals susceptible to magnetization, could no longer be determined in this way, that it should remain undecided until it could be resolved by other means, and I put no more importance on these experiments than that I had been wrong not to study more thoroughly.

Nowadays the most common nomenclature is that of *molecular currents*.[31] For instance, an editorial note published in *Nature* in 1914 describing the discovery of superconductors by H. K. Onnes had the following title: "Experimental demonstration of an Ampère molecular current in a nearly perfect conductor".[32] This paper includes the following sentence:

> As there was practically no resistance, there was practically no dissipation of energy, and the system behaved like the imagined molecular currents of Ampère, and realised the conception of Maxwell as to a conductor without resistance.

[24] [Ampère, e], [Ampère, 1885g, p. 293], [Launay (ed.), 1936a, pp. 586-592] and [Blondel, 1982, p. 110, n. 2].
[25] [Ampère, e] and [Ampère, 1885g, p. 293].
[26] [Launay (ed.), 1936a, p. 587].
[27] [Ampère, 1826f, note on p. 197] and [Ampère, 1823c, Ampère, 1990, note on p. 369].
[28] [Blondel, 1982, pp. 120 and 122].
[29] [Launay (ed.), 1936a, p. 763].
[30] [Launay (ed.), 1936a, pp. 765-766] and [James, 1993, Letter 654].
[31] [Maxwell, 1954, vol. 2, Chapter XXII, "Ferromagnetism and diamagnetism explained by molecular currents"], [Tricker, 1965, pp. 29-31 and 39-41] and [Hofmann, 1996, pp. 282-290].
[32] [Editor-News, 1914].

In his lecture when he received the Nobel prize, Onnes stated the following:[33]

> Now that we are able to use these metals, which are easy to work, all types of electrical experiments with resistance-free apparatus have become possible. To take one example: a self-contained coil, cooled in the magnetic field, should, when the field is removed, be able to simulate for some time an Ampère molecular current.

[33] [Onnes, 1967, pp. 333-334].

Chapter 6

The Contributions of Biot and Savart

6.1 The Experiment with the Straight Wire

Jean-Baptist Biot (1774-1862), figure 6.1 (a), also began to research the torques generated by a current-carrying wire acting on a magnetized needle after Ørsted's 1820 experiment. He worked together with Félix Savart (1791-1841), figure 6.1 (b), with whom he collaborated since 1819. In October 30, 1820, they presented their first work on the action of a straight wire acting on a magnetic needle to the Academy of Sciences of Paris. In 18 December 1820, they presented another work on the action of a bent wire. The reports read at these occasions were never published.[1]

Figure 6.1: (a) J.-B. Biot. (b) F. Savart.

In 1820 they published their first paper on this subject.[2] Another work presented by Biot to the Academy of Sciences of Paris in 2 April 1821, has been published in 1821.[3] Biot also published a summary of these researches, without describing the experiments in detail, in the second edition of his book *Précis élémentaire de Physique expérimentale* of 1821.[4] The detailed exposition of his electromagnetic researches appeared only in the third edition of this book published in 1824.[5]

We now describe the first work of Biot and Savart presented to the Academy of Sciences of Paris in 30 October 1820. They suspended a small magnetized needle AB by a vertical silk thread, with the projection of this thread passing through the center of gravity of the needle. In order to cancel the magnetic action of the Earth acting on the magnetic needle, they utilized another large and powerful magnet $A'B'$ placed

[1] [Biot and Savart, 1885, p. 80], [Hofmann, 1982, p. 230] and [Hofmann, 1996, p. 377, n. 9].

[2] [Biot and Savart, 1820]. This work has already been translated to English and to Portuguese, [Biot and Savart, 1965b] and [Assis and Chaib, 2006], respectively.

[3] [Biot, 1821b].

[4] [Biot, 1821a].

[5] [Biot and Savart, 1824], [Farrar, 1826], [Biot and Savart, 1885] and [Biot and Savart, 1965a].

horizontally at the same height as the small needle. The large magnet was placed in an appropriate orientation and distance to the needle. With this procedure, the needle could remain at rest in any orientation, so that its axis AB might make, for instance, an arbitrary angle with the East-West direction. They then placed a long vertical wire CMZ close to the needle. In their first published paper they did not supply any figure describing their experimental apparatus. However, in his book of 1824 Biot presented the image appearing in our figure 6.2.[6]

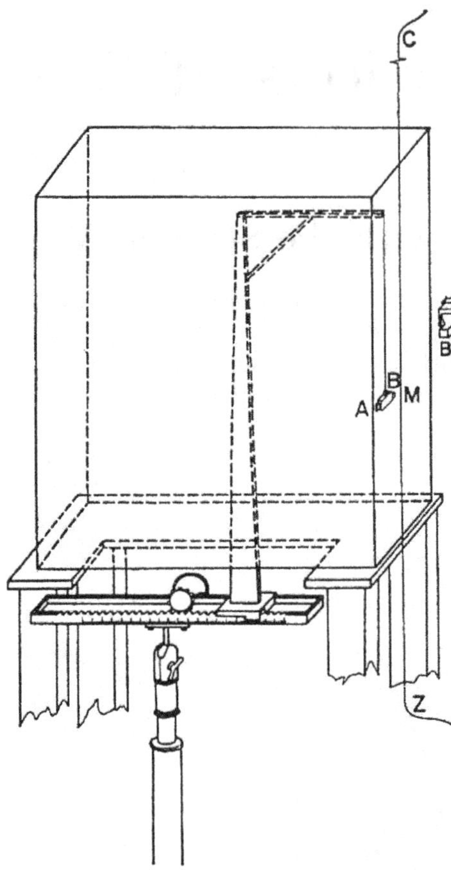

Figure 6.2: Biot and Savart's experiment with the straight wire. The electric current flows from Z to C. The small magnetized needle AB can oscillate in a horizontal plane around the vertical silk thread.

Their first conclusion when a constant current was flowing through the vertical wire was the following:[7]

> By these procedures MM. Biot and Savart arrived at the following result which rigorously represents the action experienced by a molecule of austral or boreal magnetism when placed at some distance from a fine and indefinite cylindrical wire which is made magnetic by voltaic current. Drawing a perpendicular to the axis of the wire from the point where the magnetic molecule resides, the force influencing the molecule is perpendicular to this line and to the axis of the wire.

What they called a "magnetic molecule" should be understood as a supposed particle of austral or boreal fluid, that is, a particle containing a North pole or a South pole, respectively.

However, as a matter of fact, they did not observe a "force influencing the magnetic molecule". What they observed was a torque exerted by the current-carrying wire acting on the magnetized needle. When the horizontal magnetic needle was in equilibrium under the action of this torque, at rest relative to the ground, with the auxiliary magnet $A'B'$ canceling terrestrial magnetism, its orientation was orthogonal not only to the vertical current-carrying wire, but also to the shortest straight line connecting the center of the needle to the vertical wire. If the needle were removed from this specific orientation and released from rest, it would

[6] [Biot and Savart, 1824, plate II, figure 7, p. 707], [Farrar, 1826, p. 308, figure 134] and [Tricker, 1965, p. 120, figure 41].
[7] [Biot and Savart, 1820], [Biot and Savart, 1965b, p. 118] and [Assis and Chaib, 2006].

suffer a torque tending to return it to this orientation. This orthogonal orientation of the needle had already been emphasized previously by Ampère in the September 18th meeting of the Academy of Sciences of Paris utilizing his astatic magnetic needle, as discussed in Section 3.2, figure 3.2.

Biot and Savart interpreted their experiment based on supposed forces acting on the magnetic poles of the needle. They concluded that the direction of this force acting on a point magnetic fluid was orthogonal to the plane formed by the vertical straight wire and by the pole. Moreover, the direction of this force pointed in opposite senses if it was acting on an austral or on a boreal fluid (that is, if it was acting on a North pole or on a South pole of the needle). Their conclusion can be illustrated by means of figure 6.3.[8] The straight wire carrying a current i and the particle of austral or boreal magnetism m are located in the plane of the paper. When the current flows as indicated by the arrow, the force acting on the particle will be orthogonal to the paper. When it is an austral magnetic fluid (that is, a North pole), the force will point downwards. When it is a boreal magnetic fluid (that is, a South pole), the force will point upwards.

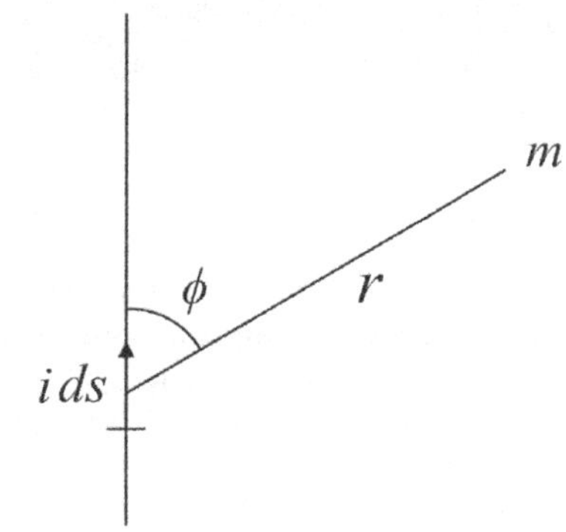

Figure 6.3: A particle of austral or boreal magnetism is represented by the letter m.

They described their conclusion utilizing an observer analogous to that of Ampère, who had already published his rule previously, as discussed in Section 2.3. Biot and Savart utilized a battery in which the current flowed along the vertical wire from the terminal connected to zinc to the terminal connected to copper. Their observer, as described in 1824, was described as follows:[9]

> If one conceives an observer placed on this wire, with his head at the copper extremity, the feet at the zinc extremity and facing the [magnetized] needle, the force emanating from the wire will act on the elements of austral magnetism from right to left of this observer, and [will act on] the elements of boreal magnetism from his left to the right, orthogonally to the shortest distance from these elements to the wire.

In the situation of figure 6.3 the zinc terminal would be in the lower extremity of the wire, the copper terminal in the upper extremity, with the current flowing along the arrow direction. An observer lying on his back in the wire and facing a particle m of austral magnetism would notice this particle being forced from his right to his left, that is, with the force pointing downwards in the plane of the paper. The force would point in the opposite sense for a particle of boreal magnetism. Therefore, a magnetized needle would remain in equilibrium with its horizontal axis orthogonal to the plane passing through the vertical wire and by the center of the needle.

Biot and Savart also allowed this needle to perform small oscillations around this equilibrium orientation, measuring the periods of these oscillations at several distances between the center of the needle and the current-carrying vertical wire. From these measurements they concluded that the intensity of the supposed force acting on a magnetic pole of the needle "is inversely proportional to the distance".[10]

In 1824 Biot mentioned in the third edition of his book that, from the results of these experiments, Pierre-Simon Laplace (1749-1827), figure 6.4, concluded by a theoretical calculation that the intensity of the elementary force exerted by each current element of the wire acting on a magnetic molecule should vary as the inverse square of their distance.

Biot's words:[11]

[8] [Blondel, 1982, p. 55, figure 10].

[9] [Biot and Savart, 1824, p. 725] and [Biot and Savart, 1885, p. 102]. See also: [Farrar, 1826, p. 345].

[10] [Biot and Savart, 1820], [Biot and Savart, 1965b, pp. 118-119] and [Assis and Chaib, 2006].

[11] [Biot and Savart, 1824, p. 740], [Farrar, 1826, p. 334], [Biot and Savart, 1885, pp. 113-114] and [Biot and Savart, 1965a, pp. 133-134].

Figure 6.4: P. S. Laplace.

> The action of an indefinite and rectilinear connecting wire on a magnetic element, such as that obtained in the foregoing experiments, is still nothing but a composite result; for by imagining the length of wire to be divided into infinitely many fine sections of very low height, it is seen that each section must act on the [magnetized] needle with a different energy according to its distance and direction [to the center of the needle] and the direction in which its action is exerted. Now these elementary forces are just the simple result which it is specially important to know, for the total force exerted by the wire is nothing other than the arithmetic sum of these effects. However, calculation is sufficient to rise from the resultant to the simple action. This is what Laplace did. From our observations he deduced mathematically the law of the force exerted individually by each section of the wire on each magnetic molecule presented to it. This force, like the total action, is perpendicular to the plane drawn through the longitudinal element of wire in the shortest distance between this element and the magnetic molecule which is influenced. The intensity of the force, as in other magnetic actions, is reciprocal to the square of the distance.

Biot did not present the details of these calculations performed by Laplace. To our knowledge Laplace also never published them. It should be kept in mind that in 4 December 1820, Ampère had already published the first version of his force between current elements which also varied as the inverse square of their distance, depending also on the angles between the elements and the straight line connecting them, equation (4.8).

This conclusion of Laplace and Biot was wrong. It is not possible to *deduce* an elementary force exerted by a portion of the wire beginning only with the integrated force exerted by the whole wire. The only thing which can be shown is that a specific elementary force, after being integrated, is compatible or not compatible with the experimental result due to the influence of the whole wire. However, there can be two or more different elementary forces which, after integration, yield the same macroscopic result. This fact can be shown with a single counterexample. Hofmann, for instance, supposed a wire of infinite length along the x axis as in figure 6.5.[12]

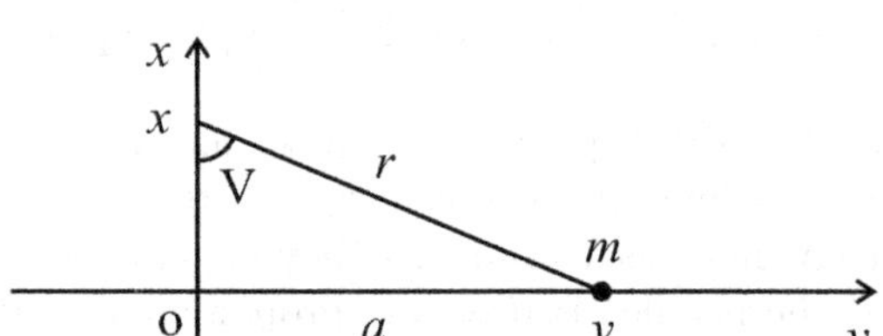

Figure 6.5: A current-carrying wire along the x axis with a magnetic particle m at a distance a to the wire.

He supposed a particle of magnetic fluid located along the y axis, at a distance a from the center of the wire located at $x = 0$. An infinitesimal element of the wire located at position x has a length dx. He called V the angle between the wire and the straight line connecting the current element to the magnetic pole, with r being their distance. We then have $x = r\cos V$ and $a = r\sin V$. He then supposed a force between

[12][Hofmann, 1982, pp. 240-241].

the current element and the magnetic pole proportional to:[13]

$$\begin{cases} e^{ax}dx = e^{r^2 \cos V \sin V} dx \,, & \text{for} \quad -\infty < x < 0 \,, \\ e^{-ax}dx = e^{-r^2 \cos V \sin V} dx \,, & \text{for} \quad 0 < x < \infty \,. \end{cases} \tag{6.1}$$

This supposed force pointed along the z axis, that is, orthogonal to the xy plane, just like Biot's supposed force.

After integrating this force with x going from $-\infty$ to ∞, Hofmann obtained a force proportional to $1/a$. That is, inversely proportional to the distance between the magnetic pole and the wire. The same proportionality had been obtained experimentally by Biot and Savart for the action of a long straight wire acting on a magnetic needle.

Although the integrated result of Hofmann's force was proportional to $1/a$, equation (6.1) is not proportional to $1/r^2$. This counterexample proves that it is not possible to *deduce* that the elementary force is inversely proportional to the square of the distance between the current element and the magnetic pole, beginning with the experimental result that the integrated force exerted by an infinite wire acting on a magnetic pole is inversely proportional to the distance between this pole and the wire.

6.2 The Experiment of the Bent Wire

In their work presented to the Academy of Sciences of Paris on 18 December 1820, Biot and Savart described a new experiment in which they measured the periods of oscillation of a horizontal magnetized needle around its equilibrium orientation, this time with the needle being influenced by a bent wire. Figure 6.6, which appeared in Biot's book of 1824, presents their apparatus.[14]

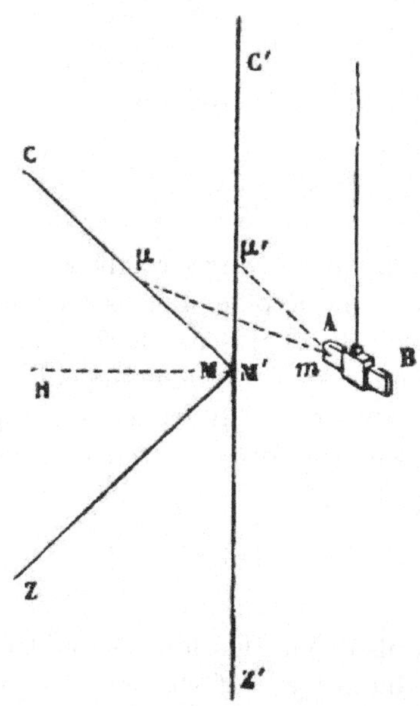

Figure 6.6: Biot and Savart's experiment with the bent wire. The current flows from the zinc Z towards the copper C, that is, in the direction $Z'M'C'$ along the vertical wire, and in the direction ZMC along the bent wire located in this vertical plane. Points M and M' are electrically insulated from one another. The small magnetized needle AB can oscillate in a horizontal plane passing through HMM' around the vertical thread holding it. The letter m represents a magnetic pole of the needle, while μ and μ' represent small elements of the current-carrying wires.

The small magnetized needle AB is once more suspended horizontally by a vertical thread whose projection passes through the center of the needle. The vertical wire $C'M'Z'$ has the same role as the vertical wire CMZ of figure 6.2. Letter C indicates that this extremity of the wire is connected to the copper terminal of the battery, while letter Z indicates that this extremity is connected to the zinc terminal of the battery.

[13] [Hofmann, 1982, pp. 240-241].

[14] [Biot and Savart, 1824, pp. 716-717], [Farrar, 1826, p. 334, figure 136], [Biot and Savart, 1885, p. 93] and [Biot and Savart, 1965a, p. 135, figure 43].

Letters M and M' indicate the midpoints of the wires. These points are at the same height from the ground as the center of the needle. The horizontal segment HM is at the same height as the needle. Letters μ and μ' indicate small lengths of the current-carrying wires. Letter m indicates a molecule of austral or boreal magnetism, that is, a North or a South pole of the needle. In their new experiment, Biot and Savart utilized a bent wire CMZ located in a vertical plane. They called the angle ZMH as the inclination i of the wire to the horizontal. This letter i should not be confused with Ampère's current intensity i.

Initially Biot and Savart maintained a constant angle $ZMH = i$ and changed the distance c between the center of the wire and the center of the needle. Later on they kept a constant distance c and varied the angle $ZMH = i$. They measured the period for small oscillations of the needle relative to its equilibrium orientation for several values of the distance c and for several values of the angle $ZMH = i$. From these measurements, they arrived at the following conclusion, as expressed by Biot in 1821, at the second edition of his book *Précis élémentaire de Physique expérimentale*:[15]

> I thus found that, for both the bent wire and the straight wire, the action was inversely proportional to the distance to the points M and M'; but the absolute intensity was weaker for the bent wire than for the straight wire, in the proportion of the angle ZMH to unity. This result, analyzed by calculation, appeared to indicate to me that the action of each element μ of the oblique wire on each molecule m of austral or boreal magnetism is inversely proportional to the square of its distance μm to this molecule, and proportional to the sine of the angle $m\mu M$ formed by the distance μM with the direction of the wire.

That is, the supposed force exerted by the bent wire on the magnetic pole located at a distance c from the vertex of the bent wire should be proportional to:

$$\frac{\text{angle(ZMH)}}{c} \ . \tag{6.2}$$

On the other hand, the force exerted by the current element μ on the magnetic pole m, located at a distance μm from the current element, should be proportional to:

$$\frac{\sin(m\mu M)}{(\mu m)^2} \ . \tag{6.3}$$

Let us consider figure 6.3. There is an infinitely long straight wire, with the arrow indicating the direction of the current. The letter m indicates a magnetic pole. Let us represent by ds the length of an infinitesimal element of the wire located at a distance r from this magnetic pole. The angle between the wire and the straight line connecting the current element to the magnetic pole is represented by ϕ. This notation is more similar to the notation utilized by Ampère, facilitating the comparison of both theories. Utilizing this notation, then according to Biot and Savart the force exerted by the current element on the magnetic pole should be proportional to:

$$\frac{\sin\phi}{r^2} \ . \tag{6.4}$$

Implicitly Biot and Savart assumed also that this force should be proportional to the intensity of the magnetic pole m, proportional to the intensity i of the current flowing through the element, and also proportional to the infinitesimal length ds of the element. They needed this force exerted by a current element in order to obtain, by integration, the force exerted by a long straight or bent wire acting on a magnetic pole. With this notation, the result obtained by Biot and Savart can be expressed mathematically by saying that their force exerted by a current element on a magnetic pole should be proportional to:

$$\frac{mids\sin\phi}{r^2} \ . \tag{6.5}$$

As will be seen in Section 9.7, Biot obtained erroneously, from his experimental data, that the force exerted by a bent wire acting on a magnetic pole should be given by an expression like equation (6.2). Later on he needed to correct this expression, as will be discussed in Section 9.7. He made also a theoretical mistake in assuming that it would be possible to deduce an infinitesimal force exerted by a current element acting on a magnetic pole beginning only with the integrated result of the macroscopic force exerted by a bent wire. After all, as shown by Hofmann's counterexample, it is possible to have two or more elementary

[15][Biot, 1821a, p. 123], [Biot, 1885a] and [Hofmann, 1982, p. 242].

forces which are different from one another, although both of them yield the same integrated result for a specific configuration. For the time being we will not discuss these aspects of Biot's work. They will be discussed in detail in Section 9.7 and in Chapter 17.

6.3 An Unexpected Result for Ampère: The Case of Equilibrium of Orthogonal Currents

In Sections 6.1 and 6.2 we discussed Biot and Savart's experiments presented to the Academy of Sciences of Paris with the straight and bent wires at the meetings of 30 October and 18 December 1820, respectively. Consider a long straight wire and a magnetic particle located in the same vertical plane. According to their results, the force exerted by the wire on the magnetic particle should always be in a horizontal plane. When the distance c between the particle and the wire is constant, this force will always have the same intensity, no matter the inclination between the wire and a horizontal straight line belonging to this vertical plane. Let b represent this angle between the wire and the horizontal, figure 6.7.[16]

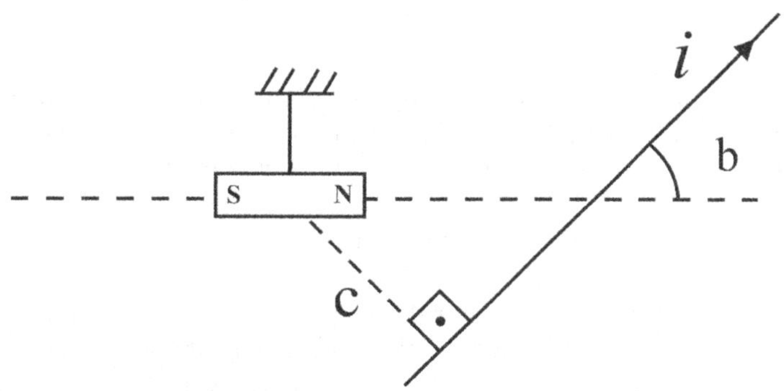

Figure 6.7: A long straight wire located in the vertical plane of the paper, making an angle b with the horizontal. A small horizontal magnetized needle has its center at a distance c to the wire.

Suppose now there is a small horizontal magnetized needle pointing along the local magnetic meridian, suspended by a vertical thread passing above its center of gravity. This needle is located in the same vertical plane of the wire. If the distance c between the wire and the center of the needle is great compared with the length of the needle, then the intensity of the torque exerted by the current-carrying wire and acting on the needle will not depend on the angle b, according to Biot and Savart's conceptions. That is, with a constant distance c this force will be independent of the angle b. Likewise, the period of small oscillations of a small magnet in the horizontal plane suspended by a vertical thread will not depend on the angle b, considering a constant distance c between this center of the magnet and the wire.

Ampère was suspicious of this conclusion based on Biot and Savart's conceptions. At this time, between December 1820 and January 1821, he considered that $n/m = 0$ in his equation (4.8), or that $k = 0$ in equations (2.1) and (2.2). Ampère did not work with magnetic poles, considering all magnetic effects of the magnets as being due to electric currents flowing through them. Initially he assumed that these supposed equivalent electric current loops followed concentric circles which were orthogonal to the axis of this small cylindrical magnet. He then supposed $n/m = k = 0$ in his force between two current elements and calculated the torque exerted by the wire of figure 6.7 acting on the ensemble of these equivalent electric circuits. He obtained that the intensity of this torque was proportional to:[17]

$$\frac{1 + \sin^2 b}{c} . \tag{6.6}$$

That is, according to Ampère this torque should depend on the angle b. The torque, in particular, should double its intensity when we go from $b = 0$ rad to $b = \pi/2$ rad. Ampère's torque should depend also on the sense of the current, reversing its sign by reversing the current's sense.

He then imagined an experiment with which he could distinguish his own theory from that of Biot and Savart. In this experiment two long straight wires are placed in a vertical plane, one of them horizontal and

[16] [Hofmann, 1982, pp. 317-328] and [Hofmann, 1996, pp. 271-274 and 322-323].

[17] [Ampère, b, *carton* 8, *chemise* 166], [Hofmann, 1982, p. 323] and [Hofmann, 1996, p. 272].

the other vertical, as in figure 6.8.[18]

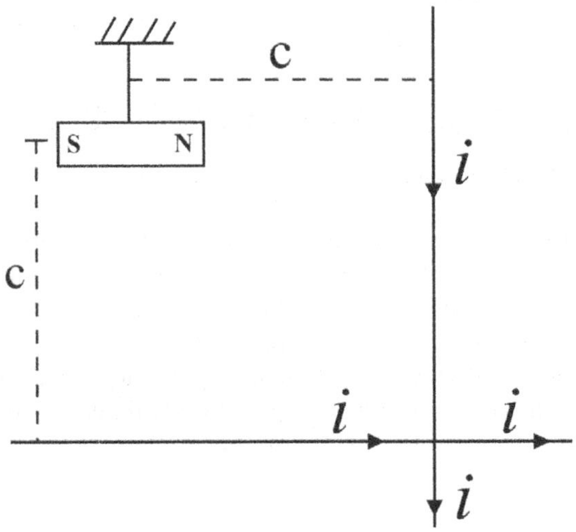

Figure 6.8: A vertical wire and a horizontal wire carry currents of the same intensity, being at equal distance c from the center of a small cylindrical magnet which has its center in the same vertical plane formed by these wires. The magnet can turn in the horizontal plane around the vertical thread holding it.

Ampère supposed that the vertical plane formed by both wires coincided with the local magnetic meridian. He also supposed a small horizontal magnet suspended by a vertical thread, with the axis of this magnet initially oriented along the terrestrial magnetic meridian. The center of the magnet was located at the same distance c from both wires. He modeled this magnet electrodynamically, that is, as an ensemble of concentric circles carrying steady currents. These equivalent electric circuits were located in planes orthogonal to the axis of this small cylindrical magnet. Moreover, he supposed this distance c much larger than the length ℓ of the small magnet, $c \gg \ell$.

Ampère then predicted what should happen with the magnet when a current of the same intensity flowed in both wires, with both of them flowing towards the junction of both wires or away from this junction. According to Biot and Savart's conclusions, the torques exerted by these wires should have the same intensity and opposite senses. Therefore, the small magnet should remain stationary along the magnetic meridian. According to equation (6.6) obtained by Ampère, on the other hand, the magnet should no longer remain in its equilibrium orientation along the magnetic meridian. He made this prediction on 15 January 1821. It seems that he intended to publish his calculations, as there is a copy of his *chemise* 166 containing these calculations in a format ready for publication.[19]

He performed the experiment together with César-Mansuète Despretz (1798-1863) five days later, on his forty-sixth birthday, 20 January 1821.[20] The magnet did not move! We will call this experiment the *case of equilibrium of orthogonal currents.*

The result of this experiment was a confirmation of Biot and Savart's theory and a refutation of Ampère's prediction based on equation (6.6). This wrong prediction was based on Ampère's belief at this time that the coefficient n/m in his equation (4.8), should be null, $n/m = 0$, or that equivalently $k = 0$ in equations (2.1) and (2.2). At this time Ampère was totally convinced that $n/m = k = 0$ in his force between current elements. Despite the negative result of this experiment, he did not doubt that the constants n/m or k should have a non-zero value.

On the other hand, if he had performed the same calculations utilizing his complete equation (2.1) or (2.2), he would had obtained a torque intensity exerted by a single long straight wire acting on a small magnet as given by:[21]

$$\frac{(1-k)+(1+2k)\sin^2 b}{c} . \tag{6.7}$$

[18][Hofmann, 1982, pp. 317-328] and [Hofmann, 1996, pp. 271-274 and 322-323].

[19][Ampère, b, *carton* 8, *chemise* 166].

[20][Ampère, b, *carton* 8, *chemise* 166], [Ampère, 1823a, p. 21], [Ampère, 1822h, p. 343], [Hofmann, 1982, p. 325] and [Hofmann, 1996, pp. 272-273].

[21]These detailed calculations can be found in: [Hofmann, 1982, Appendix C, pp. 454-462].

Therefore, the torque intensity exerted by the horizontal wire ($b = 0$ rad) located below the magnetized needle of figure 6.8 would be then proportional to $(1 - k)/c$. The torque exerted by the vertical wire of this figure with $b = \pi/2$ rad, on the other hand, would point in the opposite sense of the previous torque, being proportional to $(2 + k)/c$. The net torque due to both wires would be then proportional to $(1 + 2k)/c$. Utilizing this theoretical prediction and the negative result obtained in this experiment performed with Despretz showing a zero net torque, Ampère might have concluded that $k = -1/2$. He might had arrived at this conclusion as early as January 1821, provided he had performed these calculations.

He might also have concluded that $k = -1/2$ by performing experiments with a single wire and varying the angle b of inclination of the wire to the horizontal, observing that the torque intensity experienced by the magnetized needle was independent of b. According to equation (6.7) he would then conclude that $k = -1/2$. As a matter of fact, however, he only arrived at this correct value of k one year later, utilizing a completely different experiment, as will be discussed in Section 7.5.

However in 1821 Ampère did not perform any of these generic calculations leading to equation (6.7). This experiment had a profound impact on him. He gave up publishing this new memoir and remained in silence for two years. He only mentioned this experiment in 1823,[22] after obtaining the final value of his force between current elements with $k = -1/2$. Only then could he explain quantitatively the null result of his experiment performed with Despretz utilizing his own force law, as will be fully discussed in Section 9.8.

[22] [Ampère, 1823a, pp. 20-21], [Ampère, 1822h, pp. 342-343] and [Hofmann, 1982, p. 325].

Chapter 7

How Ampère Obtained the Final Expression of His Force between Current Elements

7.1 Faraday's Experiment of Continuous Rotation

The first phase of Ampère's electrodynamic researches began in September 1820 with the announcement of Ørsted's discovery, and continuing to January 1821. He then interrupted his researches due to illness and fatigue.[1] Another reason for stopping his studies at this moment should certainly be related to his prediction of 15 January 1821, followed by the disappointing refutation obtained with his own experiment of 20 January 1821, as discussed in Section 6.3.

In September 1821 Faraday announced a discovery which made Ampère return to his researches. He found how to produce a continuous rotation of the extremity of a current-carrying wire around a fixed magnet and also how to produce the rotation of the extremity of a magnet around a fixed current-carrying wire.[2] Figure 7.1 presents Faraday's illustration of 1822 describing this experiment.[3]

At the upper portion of this figure there are two glass cups filled with mercury. In the left side, the upper extremity of the magnet rotates around the vertical current-carrying wire fixed in the ground. In the right side, the lower extremity of the current-carrying wire rotates around the upper pole of the fixed vertical magnet. In the left lower portion of this figure there is a portable apparatus which Faraday sent to some scientists, including Ampère. In this instrument a wire can turn around a magnetized bar of soft iron when the lower portion of the wire is in contact with mercury and a constant current is flowing in the wire.[4] The sense of rotation of these instruments is reversed when only the polarity of the magnet is reversed, or when only the sense of the current is reversed. By reversing simultaneously the sense of the current and the polarity of the magnet, the sense of rotation of the mobile portion of this instrument remains the same.

Faraday informed Ampère of his discovery on 18 October 1821, sending him his portable instrument of figure 7.1. This event marked the beginning of Ampère's second phase of researches, culminating with his obtaining the final value $k = -1/2$ of the constant appearing in his force law given by equation (2.1). In a letter to Faraday dated 10 July 1822, Ampère mentioned that:[5]

> His discoveries enriching physics with new facts are the main cause of what I could add to what I had done for two years regarding electrodynamic phenomena.

Figure 7.2 presents the situation in which the lower extremity of the current-carrying wire rotates around the fixed magnet. It indicates the sense of the current, the poles of the magnet and the sense of rotation of the wire. In this configuration the force acting on the wire is leaving the paper. According to many authors, this instrument is considered the first electric motor, being called Faraday's motor.[6]

---

[1] [Blondel, 1982, p. 103].

[2] [Faraday, 1822e], [Faraday, 1822d], [Faraday, 1952d] and [Faraday, 1952b].

[3] [Faraday, 1822a] and [Faraday, 1952a].

[4] Recently a historic and didactic reproduction of this small instrument has been published, [Hottecke, 2000].

[5] [Ampère, e], [Ampère, 1885g, p. 293], [Launay (ed.), 1936a, pp. 586-592] and [Blondel, 1982, p. 110, n. 2].

[6] [Silva and Laburú, 2009].

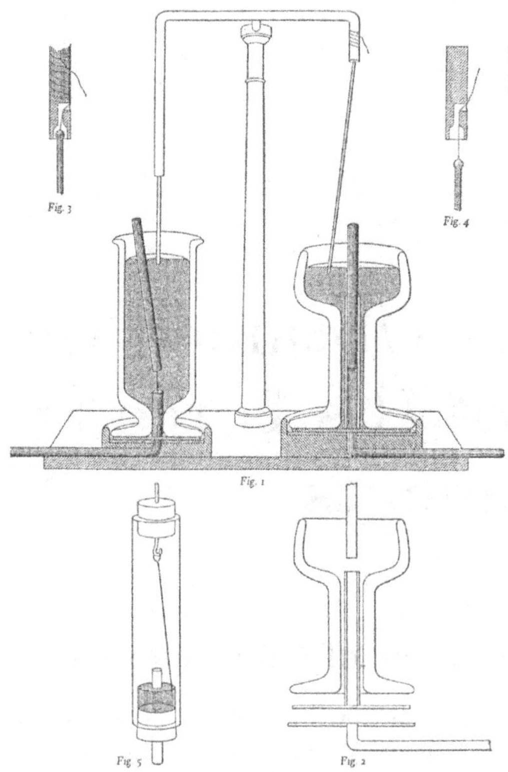

Figure 7.1: Faraday's instruments showing continuous rotation of the extremity of a current-carrying wire around a fixed magnet or the rotation of the extremity of a magnet around a fixed current-carrying wire.

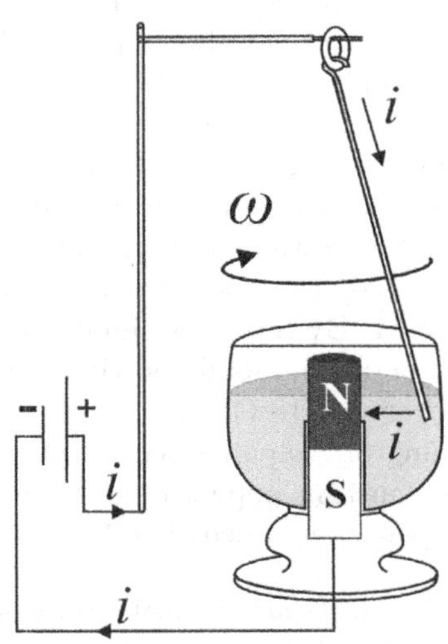

Figure 7.2: Illustration of the main elements of Faraday's instrument in which the extremity of a wire carrying a current i rotates with angular velocity ω around a magnet NS.

7.2 Ampère's Initial Experiments on Continuous Rotation

7.2.1 Reproduction of Faraday's Experiments

Ampère became fascinated by Faraday's discovery, as it presented an effect which he had not predicted. After reproducing these experiments with the portable instrument which Faraday had sent him, Ampère began to develop new results. Initially he replaced the mobile conductor by another conductor in the shape of a horseshoe, like the inverted letter U. The lower portions were connected to a crown or cylinder. He replaced the mercury by acidulated water, increasing the mobility of the system and increasing its rotation rate, presenting his results to the Academy of Sciences of Paris on 3 December 1821.

One of his experiments is shown in figure 7.3.[7] The mobile circuit is the horseshoe conductor LOM. The current flowed downward along the branches OL and OM. There was a metal ring soldered to the lower portion of this mobile conductor. The whole system floated in acidulated water. The current passed through the acid going then along the SN wire. A magnet might be placed below the system in positions AB, AB' or AB''. In all these cases the mobile circuit LOM turned around the vertical axis AKO coinciding with its axis of symmetry.

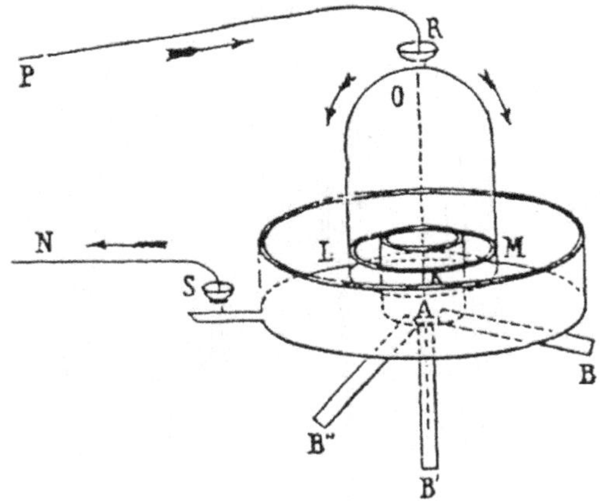

Figure 7.3: The horseshoe conductor turns around its axis relative to the ground when a constant current flows through it with a magnet placed below the system. The arrows indicate the directions of the current.

Ampère succeeded also in eliminating the external battery by working with conductors made of different metals, as shown in figure 7.4.[8]

The mobile portion of this instrument is the conductor LOM of figure 7.5, made of copper.[9] It was placed in the container $ABCD$ of figure 7.4. A system of vertical magnets is placed below the center of the container $ABCD$, with their homologous poles pointing towards the same side (that is, with all North poles above the South poles, or with all South poles above the North poles). The container $ABCD$ and the horseshoe conductor EF were made of zinc, the same material of the lower cylinder G, soldered to the horseshoe EF. When acidulated water was placed in the container $ABCD$, a constant current flowed through the closed circuit. The current flowed upwards by the arms LO and MO of the mobile conductor, flowing downwards along the vertical axis $OIHF$. When this happened, the mobile conductor LOM rotated continuously around its central vertical axis OI. Its sense of rotation, clockwise or anti-clockwise, depended on the orientation of the magnets placed below the container $ABCD$.

This apparatus is extremely interesting, as an external source of power is not necessary in order to obtain a continuous rotation. The instrument itself is a battery. It is composed of different metals, copper and zinc, connected by a liquid conductor. Therefore, connecting wires are not necessary as well.

7.2.2 Obtaining Continuous Rotation Only with Terrestrial Magnetism

Ampère was the first scientist to obtain continuous rotation of a conductor utilizing only terrestrial magnetism, without the presence of any other magnet. In this case his mobile conductor was made of two horseshoes orthogonal to one another, soldered to a circular ring, as shown in figure 7.6. This ring had a large diameter so that the Earth might exert a great electromagnetic torque even with small currents flowing through it, as the arm of the lever was large. Moreover, Ampère utilized acidulated water which offered less mechanical resistance than Faraday's mercury. In this way Ampère succeeded in obtaining a continuous

[7][Ampère, 1823b], [Ampère, 1822l] and [Ampère, 1885h].
[8][Ampère, 1821e].
[9][Ampère, 1821e, figure 21].

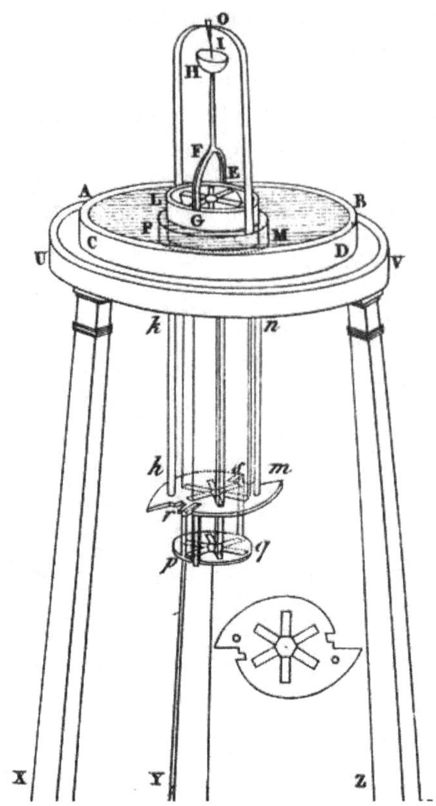

Figure 7.4: Another instrument with which Ampère obtained continous rotation.

Fig. 21.

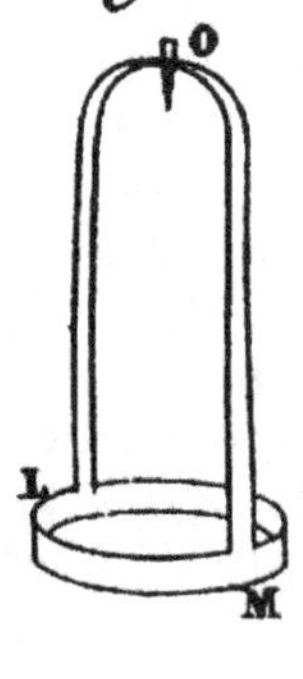

Figure 7.5: Mobile conductor of the instrument of figure 7.4.

rotation of his conductor around the central vertical axis of the system relative to the ground when the current-carrying conductors were under the action of terrestrial magnetism.[10]

Ampère presented his results to the Academy of Sciences of Paris at the meeting of 10 December 1821, with the following words:[11]

> By removing the magnets and replacing the mobile conductor by another conductor represented in figure 23 [our figure 7.6], M. Ampère observed this conductor, under the action of the Earth, turn slowly and constantly, from East to West through the South. When the austral [North] pole of a magnet was placed below the conductor and close to it, the conductor is made to turn in the opposite direction; returning to its initial motion by removing the magnet.

[10][Ampère, 1821e, figure 23].
[11][Ampère, 1821e].

Fig. 23.

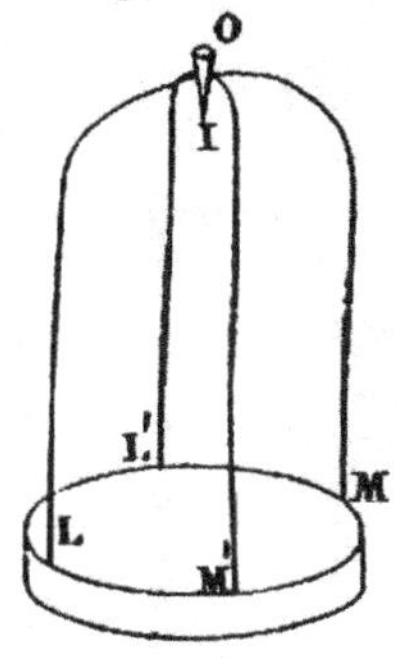

Figure 7.6: With this instrument Ampère obtained a continuous rotation utilizing only the influence of terrestrial magnetism.

7.2.3 Rotation of a Magnet around Its Axis

In his original article describing the discovery of a continuous rotation of one extremity of a current-carrying wire around a magnet, as well as the rotation of one extremity of a magnet around a current-carrying wire, Faraday stated the following:[12]

> Having succeeded thus far, I endeavoured to make a wire and a magnet revolve on their own axis by preventing the rotation in a circle round them, but have not been able to get the slightest indications that such can be the case; nor does it, on consideration, appear probable.

Ampère, on the other hand, considered that this new kind of motion might be produced in the laboratory. He was also the first to obtain it experimentally. He communicated his discovery to the Academy of Sciences of Paris in 7 January 1822.[13] The essence of this experiment can be visualized by means of figure 7.7.[14]

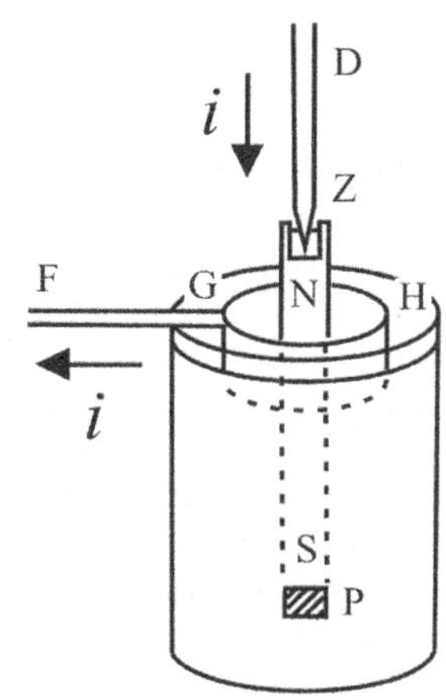

Figure 7.7: The magnet NS floats in mercury with the help of the counterweight P in its lower extremity. At the upper extremity of the magnet there is a cavity Z filled with mercury. The electric current i flows down along DZ, leaving laterally along GF. There is a metal ring GH connected to the mercury. In this configuration the magnet turns around its axis relative to the ground.

In order to obtain continuous rotation of a magnet around its axis, Ampère initially floated it in mercury by the help of a counterweight in its lower extremity. By closing the circuit, a constant current i flowed vertically downwards through the upper extremity of the magnet, leaving laterally along its lower portion and going through the mercury. When this constant current was flowing through the magnet, it rotated around its axis relative to the ground. Ampère's original illustration of this device appears in figure 7.8.[15]

[12] [Faraday, 1822e, p. 79] and [Faraday, 1952d, p. 798].
[13] [Ampère, 1822d], [Ampère, 1822e], [Ampère, 1885d] and [Blondel, 1982, pp. 114-115].
[14] [Blondel, 1982, p. 115].
[15] [Ampère, 1822d, Ampère, 1822e, Ampère, 1885d, figures 7 and 8].

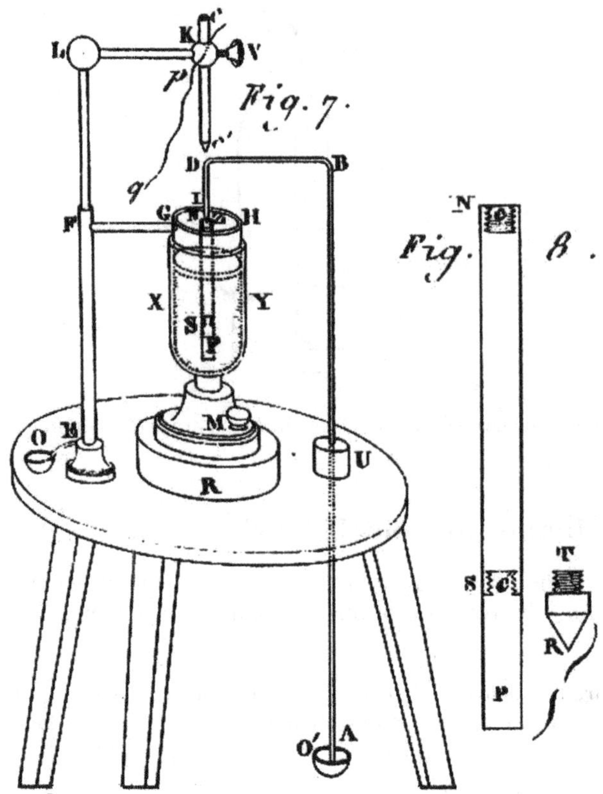

Figure 7.8: With this instrument Ampère was the first scientist to obtain a continuous rotation of a magnet around its axis, a phenomenon Faraday considered improbable to take place.

Ampère's description of his experiment:[16]

> According to what came before [Ampère had reproduced Faraday's experiment of continuous rotation], the motion of translation of the magnet takes place whenever it is only traversed by currents entering it from one side and leaving from the other [side]; however, by making all [currents] penetrating the magnet, or making all of them leaving from it, only a motion of rotation of the magnet around itself will result. In order to produce this motion, a motion which I was the first to obtain, mercury is placed on the upper cavity of the cylindrical bar cc' (figure 8 [our figure 7.8; this cylindrical bar is magnetized, with its North pole in the upper extremity and its South pole in the lower extremity, the magnet kept floating vertically with the help of the counterweight P]), and the wire Z is immersed in it [that is, the wire Z of Ampère's original figure, our figure 7.8, is immersed in the mercury filled upper cavity of the magnet]. In this configuration all currents diverge from the axis of the magnet towards the copper ring [that is, towards the copper ring HIG of Ampère's original figure 7, our figure 7.8].

In the same paper Ampère described how he obtained the rotation around its axis of a current-carrying wire.

Nowadays it is easy to obtain continuous rotation of a magnet around its axis. All that is necessary is a common battery, a screw and a small powerful magnet, as in figure 7.9.

This figure should be considered in the vertical plane. When the screw made of iron or steel is placed closed to the strong magnet, it is attracted towards it. It remains attached to the magnet and also becomes magnetized. The screw-magnet system behaves then as Ampère's single magnet. The screw-magnet system is put close to the vertical battery. This system is attracted towards the metallic portion of the battery. The battery can be suspended by the hand while the screw-magnet system will be connected to the negative terminal of the battery through the tip of the screw. The circuit is then closed by connecting the extremity of a copper wire to the upper positive terminal of the battery, while the free extremity of the wire is connected through a sliding contact to the side of the magnet. A constant current i then flows through the whole circuit. At this moment the screw-magnet system begins to rotate around the vertical axis of the battery,

[16] [Ampère, 1822d, pp. 70-71], [Ampère, 1822e, pp. 247-248] and [Ampère, 1885d, pp. 201-202].

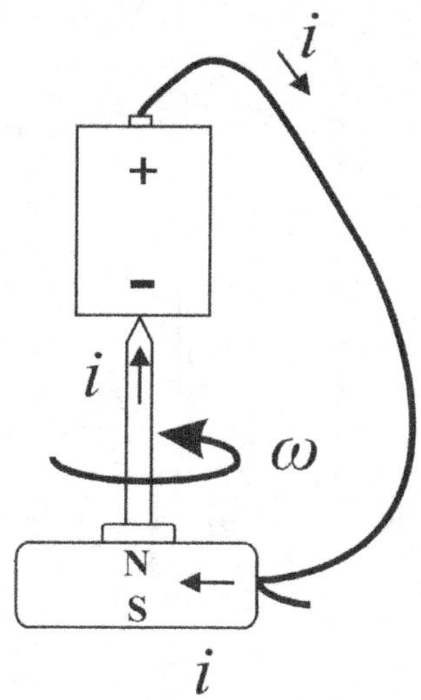

Figure 7.9: A magnetized screw rotates with angular velocity ω, together with the magnet NS attached to it, around the axis of the battery, when there is a constant current i flowing in the circuit. There is a sliding contact between the wire and the magnet.

relative to the ground, with a great angular velocity ω. By fixing the battery and the wire in our hands, only the screw-magnet system will rotate relative to the ground. There is a very low friction between the tip of the screw and the lower extremity of the battery. The screw-magnet system soon attains a fast and constant rate of rotation relative to the ground.

In Section 21.3 we will discuss different mechanisms trying to explain this rotation of a magnet around its axis. Ampère was the first to predict this phenomenon and also the first scientist to realize it experimentally. Therefore, we will call it *Ampère's motor*.[17] This nomenclature seems more appropriate than other names suggested in the literature, like "homopolar motor" or "unipolar motor".[18] Nowadays it is so simple to construct this apparatus that it also received the name of "the world's simplest motor".[19]

7.2.4 Obtaining Continuous Rotation Utilizing Only Current-Carrying Wires

One of Ampère's initial projects was to reproduce all magnetic and electromagnetic phenomena electrodynamically, that is, utilizing only current-carrying wires. By following this goal, he succeeded in obtaining a continuous rotation without utilizing permanent magnets. Continuous rotation was also obtained without the influence of terrestrial magnetism. According to a letter addressed to Van Beek,[20] Ampère performed this experiment on 27 March 1822. In this case he obtained continuous rotation of a current-carrying conductor due to the action of another current-carrying conductor. This second current-carrying conductor replaced the vertical magnet which was located below Faraday's original instrument. That is, Faraday's magnet was replaced electrodynamically by concentric horizontal loops, figure 7.10. He published his experiments in 1822.[21]

In this experiment the current flows upwards along the central axis SS', flows radially outwards along the horizontal radial conductors $S'E$ and $S'F$, and flows downwards along the vertical lateral axes ED and FG of the mobile conductor $DEFGHD$ floating in acidulated water filling the container ABC, so that the mobile conductor can rotate around its central axis relative to the ground. The electric current leaves the acid through conductors connected to cups S'' and S'''. The circuit is closed by connecting these cups S'' and S''' to one of the terminals of the battery, with its other terminal connected to cup S. Another electric current flows in the direction ABC through fixed horizontal concentric circular loops located at the base of the mobile conductor. Ampère made 10 or 12 turns in his loops, forming planar spirals superimposed on one another, with the wires having silk insulation to prevent electrical contact between them. The instrument of

[17] [Assis and Chaib, 2012] and [Chaib and Assis, 2013].

[18] [Guala-Valverde and Mazzoni, 2001], [Guala-Valverde et al., 2002], [Bednarek, 2002], [Kagan, 2005], [Sugimoto and Kawada, 2006], [Stewart, 2007], [Munoz, 2007], [Wong, 2009a] and [Wong, 2009b].

[19] [Chiaverina, 2004], [Schlichting and Ucke, 2004a], [Schlichting and Ucke, 2004b] and [Schlichting and Ucke, 2004c].

[20] [Ampère, 1821d, p. 467], [Ampère, 1822x, p. 198] and [Ampère, 1885o, p. 236].

[21] [Ampère, 1822d], [Ampère, 1822e] and [Ampère, 1885d].

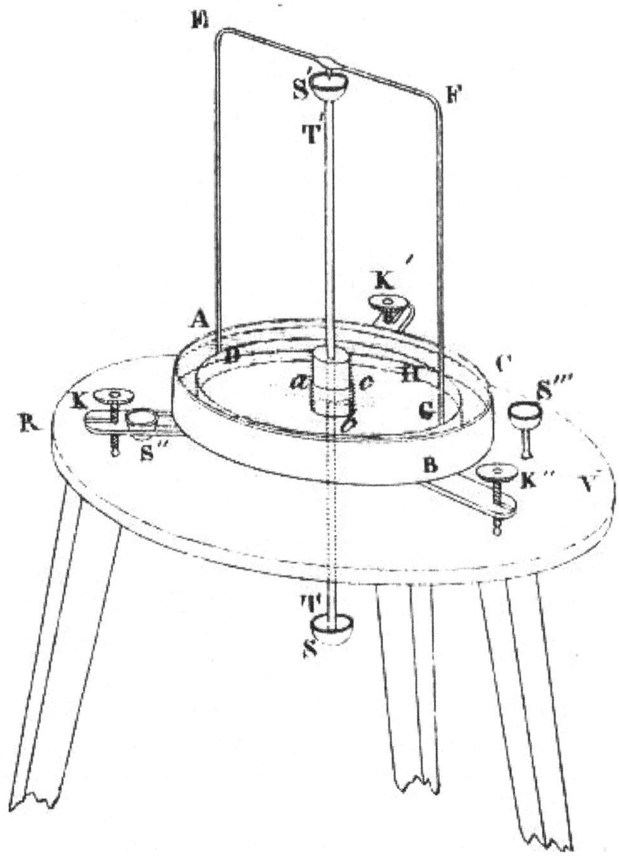

Figure 7.10: Ampère's instrument to obtain a continuous rotation utilizing only current-carrying conductors. He did not employ magnets or terrestrial magnetism.

figure 7.10 is similar to the instrument of figure 7.4. In figure 7.10 there is an ensemble of circular loops with the current flowing around ABC. In figure 7.4, on the other hand, there is an ensemble of vertical magnets placed below the central portion of the container $ABCD$.

In order to explain continuous rotation utilizing his force between current elements, Ampère utilized the following consequence which he had drawn from his initial electrodynamic researches:[22]

> [...] there exists attraction between two voltaic conductors forming a right angle when the electric current passing through them flows in both [conductors] towards or away from the common perpendicular which measures the shortest distance between them, and a repulsion takes place when one of the current flows away from this perpendicular and the other [current] flows towards it, [...]

We can understand this quotation utilizing figure 7.11. The straight segments AC and DF are orthogonal to one another. They are also orthogonal to the segment BE connecting their midpoints. There is a current i flowing in AC and a current i' flowing in DF.

Consider first the situation of figure 7.11 (a). According to Ampère, there is attraction between AB and DE as both currents flow towards BE. There is also attraction between BC and EF because the currents in these segments flow away from BE. Segments AB and EF, on the other hand, repel one another because the current in one of these segments flows towards BE, while the other current flows away from BE. For the same reason, there is repulsion between BC and DE.

Consider now the situation of figure 7.11 (b). According to Ampère, there is repulsion between AB and DE and another repulsion between BC and EF. There is attraction between AB and EF and another attraction between BC and DE.

Ampère then presented figure 7.12, in which D and G represent the projections of the vertical currents coming down from the mobile lateral conductors DE and FG of figure 7.10. The segments lKL and $l'K'L'$, on the other hand, represent two portions of a single horizontal circular loop placed below the base of the

[22][Ampère, 1822d, p. 62], [Ampère, 1822e, p. 239] and [Ampère, 1885d, pp. 194-195].

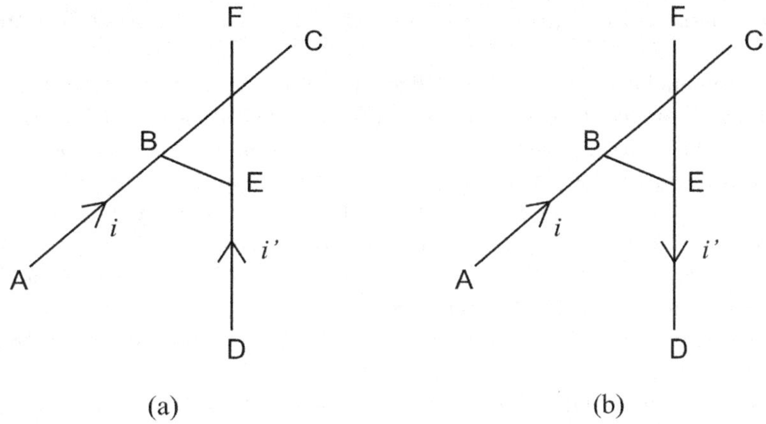

Figure 7.11: (a) Attraction between current-carrying conductors AB and DE. (b) Repulsion between current-carrying conductors AB and DE.

mobile conductor. This circular loop is fixed in the laboratory and there is a constant current flowing through it following the arrow directions.

Figure 7.12: Ampère's explanation of continuous rotation.

According to Ampère, the conductor FG with the current flowing downwards is attracted by lK and repelled by KL. The net effect of these two actions is a force acting on FG, directed from L towards l. The segment DE with the current flowing downwards is attracted by $l'K'$ and repelled by $K'L'$. The net effect of these two actions is a force on DE, directed from L' towards l'. Therefore, when the current in the horizontal circular loop flows clockwise in the direction $lKLl'K'L'$, the mobile conductor will be under the action of a torque making it turn anti-clockwise around the vertical axis passing through the center of the circular loop. Ampère described the sense of rotation of the mobile conductor with the following words:[23]

> Then these two forces combine into a single force tending to turn the [vertical] wire in the opposite sense of the current in the [horizontal] spiral. This new force combines with the previous one, and similar actions are renewed in each position of the [vertical] wires, the whole system of the mobile conductor turns continuously, in opposite sense of the current in the spiral, during the time interval in which the communication [with the external battery] is maintained.

In just a few months Ampère not only reproduced Faraday's experiment of continuous rotation, but also produced many new phenomena of continuous rotation.

7.2.5 Distinction between Continuous Rotation and Continuous Revolution

Ampère distinguished Faraday's original experiments, described in Section 7.1, from the experiment described in Subsection 7.2.3 in which a magnet rotated around its own axis. He made a conceptual distinction between these two kinds of experiments. This distinction affected his explanations of these different motions described by the magnet. In Faraday's original experiment one of the poles of the magnet was along the vertical axis of rotation, while the other pole described a circular motion around this axis. In Ampère's own experiment,

[23][Ampère, 1822d, p. 62], [Ampère, 1822e, p. 239] and [Ampère, 1885d, p. 195].

on the other hand, the magnet rotated around its own axis, so that both poles of the magnet were along the axis of rotation.

The most common nomenclature which he utilized in order to describe Faraday's original experiment was "continuous revolution" or "continuous translation". The expression which he utilized to describe the rotation of a magnet around its axis, on the other hand, was "continuous rotation". He utilized the words "revolution" and "rotation" in analogy with the motion of the Earth in the solar system, relative to the background of fixed stars. That is, the Earth rotates daily around its axis, while it has a translational motion, or annual revolution around the Sun, with a period of 365 days. In some of his works Ampère did not make this distinction between the motions of the magnet, calling both phenomena "continuous rotation".

Ampère explained the conceptual distinction between these two kinds of motion of the magnet in two letters addressed to Gherardi (1802-1879), one of them from 1825 and the other from 1826.[24] This distinction was also discussed in the *Théorie*.[25]

We will not discuss this conceptual difference in this book, calling both phenomena simply "continuous rotations". In addition we will not discuss Ampère's explanation for continuous revolution, which was different from his explanation for continuous rotation. Our only goal in this Subsection is to call attention to the different nomenclatures Ampère utilized.

7.3 Ampère's Crucial Experiment

7.3.1 Ampère's Wrong Prediction

In the sequence of his experiments, Ampère imagined that he could obtain a continuous rotation utilizing astatic coils like those described in Subsection 4.3.2. This prediction appeared in a manuscript which has never been completely published, namely, *chemise* 206 bis, figure 7.13.[26]

Ampère did not specify his motivations to perform this crucial experiment.[27]

He was probably only refining his experiments described in Section 7.2. He had already reproduced Faraday's experiment of continuous rotation. He had also produced continuous rotations in several new configurations, namely: continuous rotation due only to the influence of the Earth; continuous rotation of a magnet around its axis; rotation of a current-carrying wire around its axis, a subject which we do not discuss in this book; and rotation of a conductor having the shape of the inverted letter U, due to a circular current loop located under the base of the mobile conductor. In this last experiment, in particular, he had produced continuous rotation of a current-carrying conductor around another current-carrying conductor, without utilizing magnets or the magnetic influence of the Earth, as described in Subsection 7.2.4. In all these configurations he utilized a liquid conductor which might be mercury or acidulated water. Maybe he thought of his crucial experiment as a possibility of producing continuous rotation utilizing only current-carrying conductors, without the presence of any conducting liquid. Alternatively, he may have intended to replace the mobile conductor of figure 7.10, analogous to the mobile conductor of figure 7.6, by an astatic coil.

As discussed before, with the instrument of figure 7.10 Ampère obtained, in March 1822, continuous rotation of a mobile conductor due to the action of a circular current loop. The current flowed from one extremity of the battery, upwards along the central axis SS' of the mobile conductor, went laterally along the horizontal radial conductors $S'E$ and $S'F$, flowing downwards along the two vertical conductors ED and FG, which were floating in the acidulated water placed in the vessel ABC. The voltaic circuit was closed by connecting this acid to the other extremity of the battery. Maybe Ampère was thinking of performing an experiment similar to this one utilizing now an astatic coil as that of figure 4.16 in order to avoid the presence of the conducting liquid, working only with rigid conductors.

At this moment Ampère made a prediction of what to expect from this experiment, a prediction which turned out to be refuted by his later experiment. This prediction did not appear in any of his publications, being stated only in a manuscript which has never been completely published, figure 7.13.

[24] [Ampère, 1825b], [Ampère, 1887b] and [Ampère, 1825a], [Ampère, 1827c] and [Ampère, 1887a].

[25] Continuous revolution or continuous translation: [Ampère, 1826f, pp. 4-5, 118-121, 167, 172 and 197] and [Ampère, 1823c, Ampère, 1990, pp. 176-177, 290-293, 339, 344 and 369] and [Ampère, 1965b, pp. 155-156]. Continuous rotation: [Ampère, 1826f, pp. 16, 22, 37, 49, 59, 88, 101, 108, 111, 113, 119-121, 124-125, 153, 167, 172-177, 181 and 187] and [Ampère, 1823c, Ampère, 1990, pp. 188, 194, 209, 221, 231, 260, 273, 280, 283, 285, 291-293, 296-297, 325, 339, 344-349, 353 and 359] and [Ampère, 1965b, pp. 163-164, 167-168, 180-181, 187-188 and 194-196].

[26] [Ampère, b, *carton* 11, *chemise* 206 bis].

[27] We are utilizing the word *crucial* in this Section with the same meaning adopted by Hofmann in his PhD thesis of 1982, *The Great Turning Point in André-Marie Ampère's Research in Electrodynamics: A Truly "Crucial" Experiment*, [Hofmann, 1982].

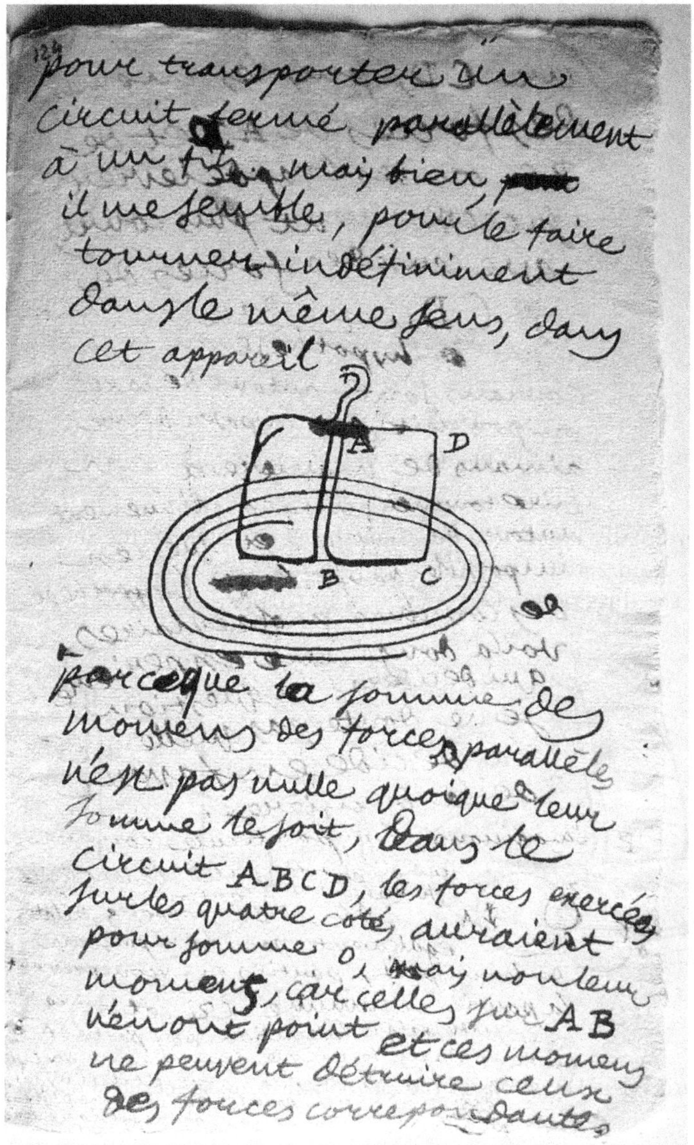

pour transporter un circuit fermé parallèlement à un fil mais bien il me semble, pour le faire tourner indéfiniment dans le même sens, dans cet appareil

A D B C

parceque la somme de momens des forces parallèles n'est pas nulle quoique leur somme le soit, dans le circuit ABCD, les forces exercées sur les quatre côtés auraient pour somme 0, mais non leurs momens, car celles sur AB n'en ont point et ces momens ne peuvent détruire ceux des forces correspondantes

Figure 7.13: Ampère's Crucial Experiment.

Ampère's prediction as contained in this manuscript:[28]

> A closed circuit can never be transported parallel to a wire. However, it seems to me that it can be made to rotate indefinitely in the same direction in this apparatus: [here Ampère presented our figure 7.14]:

Soon after this figure, Ampère justified his prediction of a continuous rotation due to a supposed net torque acting on the mobile circuit, namely:[29]

> This is because the sum of the moments of the parallel forces is not zero, even though their sum is zero.[30] In the circuit $ABCD$, the forces exerted on the four sides will sum to zero, but their moments will not; this is because those on AB will vanish, and these moments will not be able to cancel those of the corresponding forces on CD, since the moments of the forces on AD and BC have respective lever arms shorter than those of the forces on CD.

Ampère therefore predicted a continuous rotation with this apparatus. When he wrote this manuscript, he still assumed that $k = 0$ in equations (2.1) and (2.2), and that $n/m = 0$ in equation (4.8). If these

[28][Ampère, b, *carton* 11, *chemise* 206 bis] and [Hofmann, 1996, p. 300].

[29][Ampère, b, *carton* 11, *chemise* 206 bis] and [Hofmann, 1996, p. 300].

[30]That is, although the net force acting on the mobile circuit is zero, the net torque is different from zero.

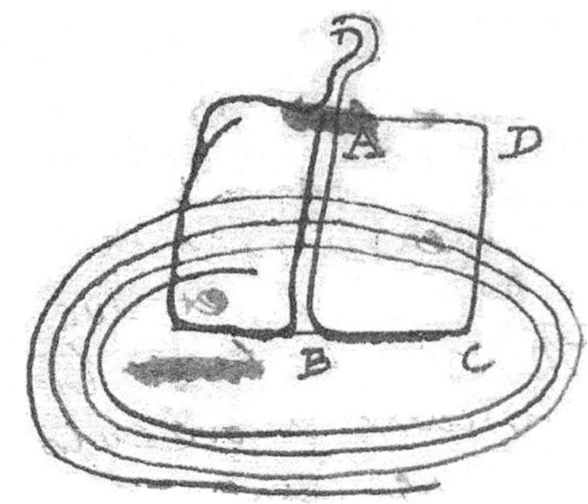

Figure 7.14: Ampère's prediction of continuous rotation of an astatic coil.

constants k and n/m were really equal to zero, then the net torque acting on the astatic coil of figure 7.14 would in fact be different from zero.

Figure 7.15 indicates the directions of the currents in this experiment.

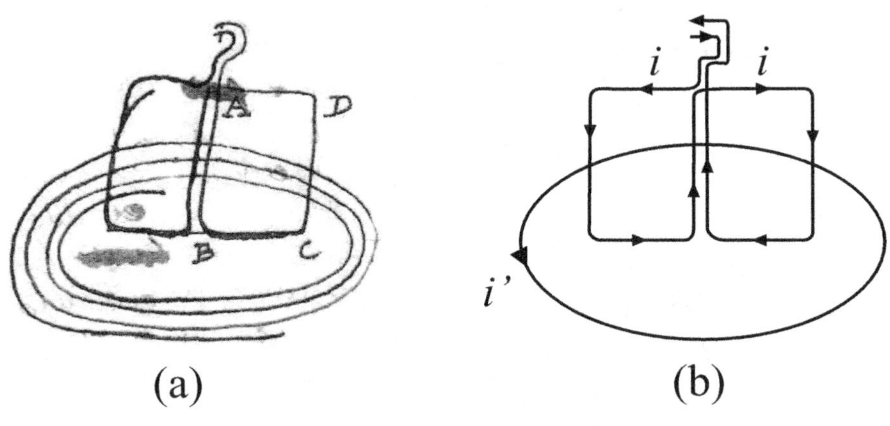

Figure 7.15: Directions of the currents in Ampère's crucial experiment.

7.3.2 An Experimental Anomaly: The Case of Equilibrium of the Nonexistence of Continuous Rotation

Ampère performed the experiment described in Subsection 7.3.1 in March 1822.[31] He did not observe the expected rotation of the astatic coil! This experiment was first presented to the Academy of Sciences of Paris in June 1822 in the form of figure 7.16 (a).[32] In this publication Ampère did not mention his wrong prediction. He only informed that he did not obtain a continuous rotation with this apparatus.

The horizontal circular loop, placed at the bottom of the astatic coil of figure 7.16, appears in figure 7.17.[33] This circular loop is fixed in the laboratory, with its axis of symmetry coinciding with the vertical axis of symmetry of the astatic coil.

Ampère described this crucial experiment as follows:[34]

> [...], at the rod TT' (figure 16 [our figure 7.16 (a)]) is adapted an annular cup O which is insulated from the rod by a glass tube Mm, and which is in communication with the cup S'' by the copper miter NnS''.[35]
>
> The spiral represented in figure 2 [our figure 7.17, showing the circular coils placed below the astatic coil], with which is produced the continuous motion in the apparatus of figure 1 [our figure 7.10], plunges into

[31] [Ampère, b, *carton* 9, *chemise* 173].

[32] [Ampère, 1822o].

[33] [Ampère, 1822o].

[34] [Ampère, 1822o, pp. 414-415].

[35] Original French text: équerre en cuivre NnS''.

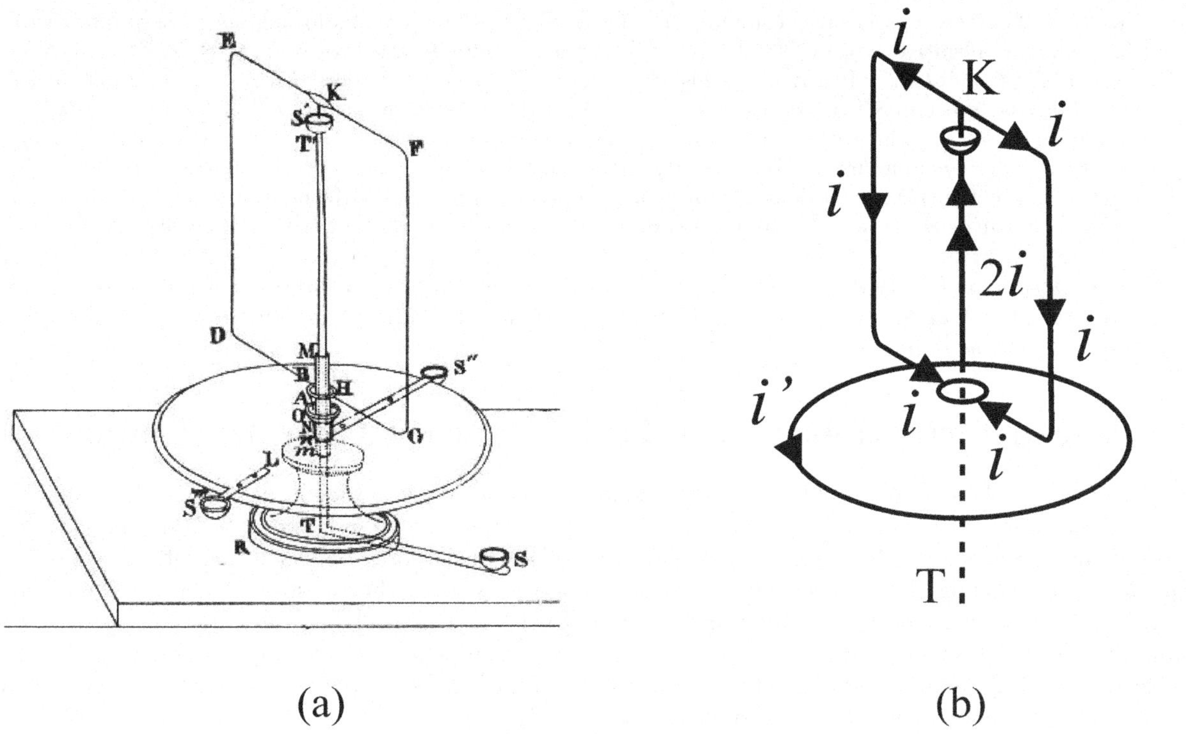

Figure 7.16: (a) Case of equilibrium of the nonexistence of continuous rotation. (b) Directions of the currents in Ampère's crucial experiment.

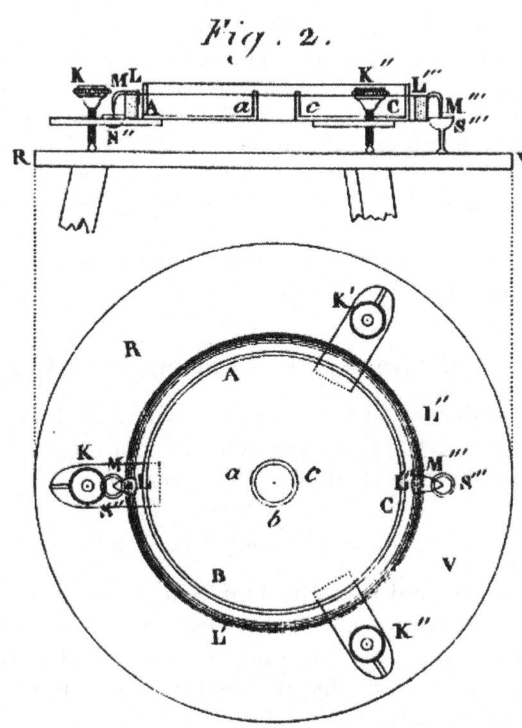

Figure 7.17: Circular current-carrying coil located at the base of the astatic coil of figure 7.16.

cups S'' and S''' by its two extremities. The mobile conductor [that is, the astatic coil $KEDBHGFK$] supported by point K at cup S' is composed of two parts, $KFGH$ and $KEDB$, equal and similar to one another, in such a way that the Earth does not act on this conductor. They [these two parts] are gathered by a ring BH concentric to the rod TT'. This ring is connected by a point A which plunges into the

> mercury of cup O. The [electric] communication is established by plunging, for instance, the positive wire in S and the negative wire in S'''. The current is then divided into two directions, $STKEDBAONS''$ and $STKFGHAONS''$. It arrives in this way at cup S'', follows the spiral $LL'L''$ (figure 2 [our figure 7.17]), arriving at cup S''' (figure 16 [our figure 7.16 (a)]), where the appendix $L'''M'''$ is plunged (figure 2 [our figure 7.17]), which is put into communication with the negative extremity of the battery by the conducting wire coming from this extremity also plunging there. When everything is so arranged, the mobile conductor $BDEFGH$ does not turn continuously, as happened with the conductor of figure 1 [our figure 7.10], it does not acquire any motion or then it oscillates around a position of stable equilibrium.

This experiment, contradicting Ampère's expectations, had a crucial importance in his researches. He began to consider it as a new case of equilibrium, which will be called here the *case of equilibrium of the nonexistence of continuous rotation.*

7.4 Transformations of the Force between Current Elements

Ampère utilized the case of equilibrium of nonexistence of continuous rotation in order to finally obtain the correct value of the constant k of equations (2.1) and (2.2). Between March and June 1822, he analyzed this experiment theoretically. He utilized the general expression of his force between current elements which he had obtained utilizing the addition law, with an arbitrary value of the constant k. He presented his final results to the Academy of Sciences of Paris on June 10, 1822. His fundamental work was published in the same year.[36] A detailed summary of Ampère's calculations can be found in the work of Grattan-Guinness.[37]

In this work Ampère considered the general expression of the force between current elements, equation (2.1), which was compatible with his addition law, namely:

$$\frac{ii'dsds'}{r^n}(\sin\alpha\sin\beta\cos\gamma+k\cos\alpha\cos\beta)\ . \tag{7.1}$$

For the sake of generality, Ampère worked with a force inversely proportional to an arbitrary power n of the distance r between both elements, although he still believed that $n = 2$ by analogy with the gravitational, electrostatic and magnetic forces. As he said in another work:[38]

> [...] the consideration of the several attractions observed in nature makes me believe that the attraction [between current elements], whose expression I am looking for, also acts in the inverse square of the distance; [however,] I will suppose it, for the sake of generality, in the inverse ratio of the nth power of this distance, with n being a constant whose value is to be determined.

It should be remarked that the power n of the distance r appearing in equation (7.1) has no relation with the constant n appearing in equation (4.8).

7.4.1 Force as a Function of the Angle between the Current Elements

Before obtaining the final value of the constant k appearing in equation (7.1), Ampère presented two transformations of this force. One of these transformations related the four angles discussed in Section 2.8, namely, α, β, γ and ε. This relation is expressed by the following expression:

$$\cos\varepsilon=\sin\alpha\sin\beta\cos\gamma+\cos\alpha\cos\beta\ . \tag{7.2}$$

In the article published in 1822 he said that this equation "is evident by the fundamental principle of spherical trigonometry".[39] In the *Théorie* he mentioned that this expression was obtained "by considering the spherical triangle with sides θ, θ' and ε".[40] The angle θ is equivalent to the angle α, while the angle θ' is equivalent to the angle β. He did not present a figure illustrating how to obtain this relation. In any event, equation (7.2) can be understood utilizing our figure 7.18.

In this spherical triangle, ABC represents a portion of the surface of a sphere centered in O, with radius $OA = OB = OC = R$. The projection of point B onto the plane AOC is the point P. The point M is chosen along OA in such a way that BM is orthogonal to OA. Likewise, point N is chosen along OC in such

[36] [Ampère, 1822o].
[37] [Grattan-Guinness, 1990a, pp. 930-933].
[38] [Ampère, 1822p, p. 229] and [Ampère, 1885m, p. 263].
[39] [Ampère, 1822o, p. 411].
[40] [Ampère, 1826f, p. 32], [Ampère, 1823c, Ampère, 1990, p. 204] and [Ampère, 1965b, p. 176].

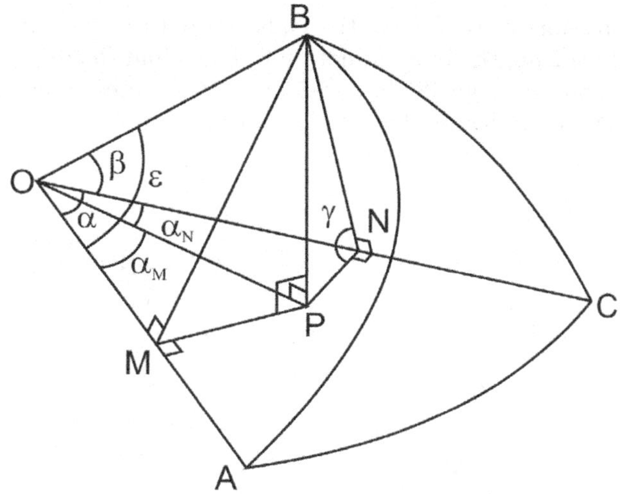

Figure 7.18: Spherical triangle.

a way that BN is orthogonal to OC. The infinitesimal element ds is chosen along OA, the element ds' is chosen along OB and the straight line r connecting these elements is chosen along OC. With these choices the angle α is given by COA, the angle β is given by COB, the angle ε is given by AOB and the angle γ is given by BNP. Moreover, the angle α_M is defined by POM and the angle α_N is defined by NOP, such that $\alpha = \alpha_M + \alpha_N$.

From the triangle ONB one obtains that $\cos\beta = ON/R$ and $\sin\beta = BN/R$. From the triangle OMB one obtains that $\cos\varepsilon = OM/R$ and $\sin\varepsilon = BM/R$. From the triangle OMP one obtains that $\cos\alpha_M = OM/OP$ and $\sin\alpha_M = PM/OP$. And from the triangle ONP one obtains that $\cos\alpha_N = ON/OP$ and $\sin\alpha_N = PN/OP$. From these relations it follows that:

$$OM = R\cos\varepsilon = OP\cos\alpha_M = OP\cos(\alpha - \alpha_N)$$

$$= OP(\cos\alpha\cos\alpha_N + \sin\alpha\sin\alpha_N) \ . \tag{7.3}$$

Utilizing the relations of the triangle OMB together with equation (7.3) one obtains:

$$R\cos\varepsilon = OM = (OP\cos\alpha_N)\cos\alpha + (OP\sin\alpha_N)\sin\alpha \ . \tag{7.4}$$

Utilizing the relations of the triangle ONP in equation (7.4) yields:

$$R\cos\varepsilon = ON\cos\alpha + PN\sin\alpha \ . \tag{7.5}$$

Utilizing the relations of the triangle ONB together with equation (7.5) yields:

$$R\cos\varepsilon = R\cos\alpha\cos\beta + PN\sin\alpha \ . \tag{7.6}$$

From the triangle BPN one obtains $PN = BN\cos\gamma$, while from the triangle ONB one obtains $BN = R\sin\beta$. Applying these two last results into equation (7.6) one obtains finally equation (7.2). Utilizing equation (7.2) into equation (7.1) yields:

$$\frac{ii'dsds'}{r^n}[\cos\varepsilon + (k-1)\cos\alpha\cos\beta] = \frac{ii'dsds'}{r^n}(\cos\varepsilon + h\cos\alpha\cos\beta) \ . \tag{7.7}$$

Equation (7.7) is equivalent to equation (2.3) with α and β replacing θ and θ', respectively, and with $h = k - 1$.

7.4.2 Force Expressed in Terms of Partial Derivatives

In this very important paper of 1822 Ampère presented for the first time another fundamental transformation of his force between current elements. He now expressed this force in terms of partial derivatives of the distance r between the two current elements, relative to the location of the infinitesimal lengths ds and ds'.

Although he utilized the expression "partial derivatives", he wrote these partial derivatives with the symbols dr/ds and dr/ds', instead of utilizing the modern notation $\partial r/\partial s$ and $\partial r/\partial s'$.

He considered figure 2.19, on our page 48. In our figure 7.19 we present the main elements of Ampère's original figure in order to understand his mathematical transformation.

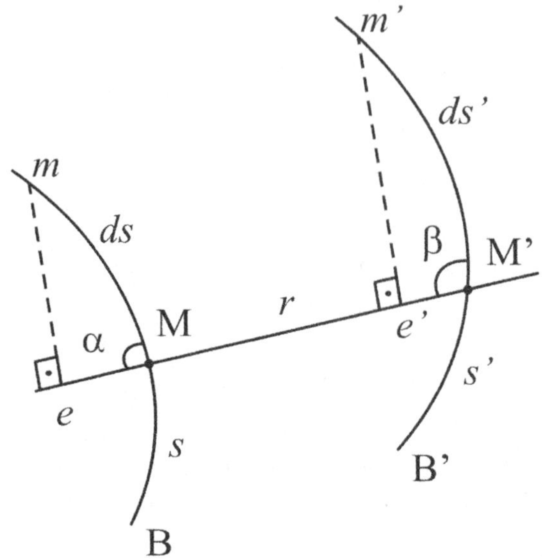

Figure 7.19: The infinitesimal elements ds and ds' are arbitrarily orientated in three-dimensional space.

The lengths of the curved current-carrying wires are represented by $BM = s$ and $B'M' = s'$. The infinitesimal increments of these distances are given by $Mm = ds$ and $M'm' = ds'$. In figure 7.19 these elements are purposely represented with a very large size. These elements are located in space, with arbitrary orientations. Therefore, they do not need to be located in a single plane. The distance between these elements is given by $MM' = r$. This distance is a function of two independent variables, s and s', such that, $r = r(s,\ s')$. The perpendiculars me and $m'e'$ are drawn along the straight line passing through MM'. According to Ampère's specification, the angle α is given by mMe, while the angle β is given by $m'M'e'$. The distance between M and M' is given by $MM' = r$, while the distance between points m and m' is given by mm'. According to figure 7.19, and taking into account that we are dealing with infinitesimal distance, the variation dr is then given by:

$$dr = eM - e'M' = \cos\alpha ds - \cos\beta ds' . \tag{7.8}$$

This equation leads to the following conclusion:[41]

$$\cos\alpha = \frac{dr}{ds} , \tag{7.9}$$

and

$$\cos\beta = -\frac{dr}{ds'} . \tag{7.10}$$

Ampère took the partial derivative of equation (7.10) with respect to s and obtained:

$$\frac{d\beta}{ds}\sin\beta = \frac{d^2r}{dsds'} . \tag{7.11}$$

---

[41] In modern notation, we would have:

$$dr = \left(\frac{\partial r}{\partial s}\right) ds + \left(\frac{\partial r}{\partial s'}\right) ds' ,$$

$$\cos\alpha = \frac{\partial r}{\partial s} ,$$

and

$$\cos\beta = -\frac{\partial r}{\partial s'} .$$

He then observed that by going from point M to point m, the curved length s is transformed into $s+ds$, while the angle β decreases its value. This reduction in the value of β is given by the projection of the angle $MM'm$ onto the plane $MM'm'$, that is, $d\beta = -MM'm\cos\gamma$. Here γ represents the angle between planes $M'Mm$ and $MM'm'$. However, the angle $MM'm$ is approximately given by:

$$MM'm \approx \frac{me}{eM'} \approx \frac{me}{M'M} = \frac{ds\sin\alpha}{r} \ . \tag{7.12}$$

Combining this equation with the previous value of $d\beta$ one obtains:

$$\frac{d\beta}{ds} = -\frac{\sin\alpha\cos\gamma}{r} \ . \tag{7.13}$$

The combination of equations (7.11) and (7.13) yields:

$$\sin\alpha\sin\beta\cos\gamma = -r\frac{d^2r}{dsds'} \ . \tag{7.14}$$

Plugging equations (7.9), (7.10) and (7.14) into equation (7.2) yields:

$$\cos\varepsilon = -r\frac{d^2r}{dsds'} - \frac{dr}{ds}\frac{dr}{ds'} \ . \tag{7.15}$$

Equations (7.2), (7.9), (7.10), (7.14) and (7.15) applied into equation (7.1) yields finally:

$$\begin{aligned}
&\frac{ii'dsds'}{r^n}(\sin\alpha\sin\beta\cos\gamma + k\cos\alpha\cos\beta) \\
&= \frac{ii'dsds'}{r^n}[\cos\varepsilon + (k-1)\cos\alpha\cos\beta] \\
&= -\frac{ii'dsds'}{r^n}\left(r\frac{d^2r}{dsds'} + k\frac{dr}{ds}\frac{dr}{ds'}\right) \\
&= -ii'dsds'r^{1-n-k}\left(r^k\frac{d^2r}{dsds'} + kr^{k-1}\frac{dr}{ds}\frac{dr}{ds'}\right) \\
&= -ii'dsds'r^{1-n-k}\frac{d\left(r^k\frac{dr}{ds'}\right)}{ds} = -ii'dsds'\frac{r^{1-n-k}}{1+k}\frac{d^2\left(r^{1+k}\right)}{dsds'} \ .
\end{aligned} \tag{7.16}$$

Equation (7.16) presents the various force expressions which Ampère included in his paper of 1822 representing the interaction between two current elements.

This mathematical transformation is valid no matter the value of the constant k. In the particular case in which $k = 1$ one obtains:

$$\sin\alpha\sin\beta\cos\gamma + \cos\alpha\cos\beta = -\frac{d^2(r^2/2)}{dsds'} \ . \tag{7.17}$$

In the paper of 1822, Ampère also utilized rectangular coordinates in order to represent the location of point M by $(x,\ y,\ z)$, while the location of point M' was represented by $(x',\ y',\ z')$. The distance MM' can then be written as:

$$r = \sqrt{(x'-x)^2 + (y'-y)^2 + (z'-z)^2} \ . \tag{7.18}$$

By taking the partial derivative of $r^2/2$ with respect to s', he obtained:

$$\frac{d(r^2/2)}{ds'} = (x'-x)\frac{dx'}{ds'} + (y'-y)\frac{dy'}{ds'} + (z'-z)\frac{dz'}{ds'} \ . \tag{7.19}$$

Taking the partial derivative of this equation with respect to ds, he obtained:

$$\frac{d^2(r^2/2)}{dsds'} = -\frac{dx}{ds}\frac{dx'}{ds'} - \frac{dy}{ds}\frac{dy'}{ds'} - \frac{dz}{ds}\frac{dz'}{ds'} \ . \tag{7.20}$$

Plugging equation (7.20) into equations (7.2) and (7.17) yields

$$\cos\varepsilon = \sin\alpha\sin\beta\cos\gamma + \cos\alpha\cos\beta = \frac{dx}{ds}\frac{dx'}{ds'} + \frac{dy}{ds}\frac{dy'}{ds'} + \frac{dz}{ds}\frac{dz'}{ds'} \ . \tag{7.21}$$

7.5 Obtaining the Value $k = -1/2$

In his paper of 1822, Ampère said that the case of equilibrium of the nonexistence of continuous rotation was a "new fact". He explained this new fact as follows:[42]

> This fact can be stated as follows: A closed circular circuit can never produce a continuous motion taking place always in the same sense, when acting on a mobile conductor of arbitrary form beginning from a point located along the perpendicular to the plane of this circuit passing through the center of the circle relative to which the circuit forms a circumference and ending at another point along the same axis, while the mobile conductor can only move by rotating around this axis.

In order to justify this new fact, he presented publicly for the first time his case of equilibrium of the nonexistence of continuous rotation. This experiment was discussed in Subsection 7.3.2.

We now analyze how Ampère obtained the final value of his constant k. He first presented figure 7.20.[43] The coordinate system utilized by Ampère was left or levogyrous, that is, with the x and y axes inverted relative to the common orientation utilized nowadays which follows the right-hand rule.

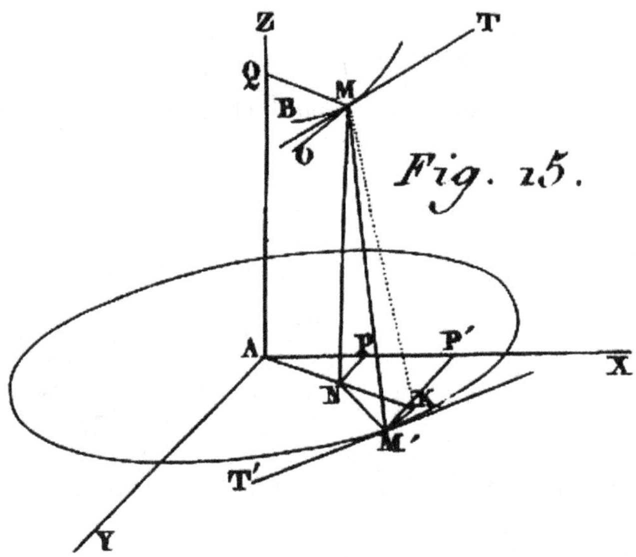

Figure 7.20: Ampère's original figure.

There is a circle of radius a located in the xy plane and centered at the origin of the coordinate system. A constant current i' flows along this circle. This current-carrying circuit exerts a torque upon a current element ids located at point M. This current element ids belongs to another circuit which can turn around the z axis. Ampère's goal was to calculate the net torque exerted by the circular loop acting on this current element ids. He would then impose that this net torque must vanish, as shown by his experiment of the case of equilibrium of the nonexistence of continuous rotation. The desired value of the constant k would be obtained by imposing a zero net torque acting on this current element.

He was very succinct in his paper of 1822 as regards the deduction of the value of the constant k. In this Section we detail the mathematical passages of his short deduction. Moreover, in order to simplify the comprehension of his mathematical arguments, we utilize a coordinate system oriented according to the right-hand rule, as usual, instead of utilizing Ampère's levogyrous system. We also utilize the symbol ∂ in order to express partial derivatives. He utilized a cylindrical coordinate system with variable $u = \sqrt{x^2 + y^2}$ representing the distance of an arbitrary point $(x,\ y)$ to the z axis of rotation and with variable t representing the azimuthal angle. Instead of Ampère's symbols, we utilize the more usual letter $\rho = \sqrt{x^2 + y^2}$ to represent the distance of the z axis, while φ will represent the azimuthal angle. In figure 7.21 we included the main elements which will interest us here, as taken from Ampère's original image.

There is a circle of radius a located in the xy plane, centered at the origin O of a Cartesian coordinate system xyz. A constant current i' flows in this circular circuit. We utilize cylindrical coordinates $(\rho,\ \varphi,\ z)$.

[42] [Ampère, 1822o, p. 414].

[43] [Ampère, 1822o, figure 15].

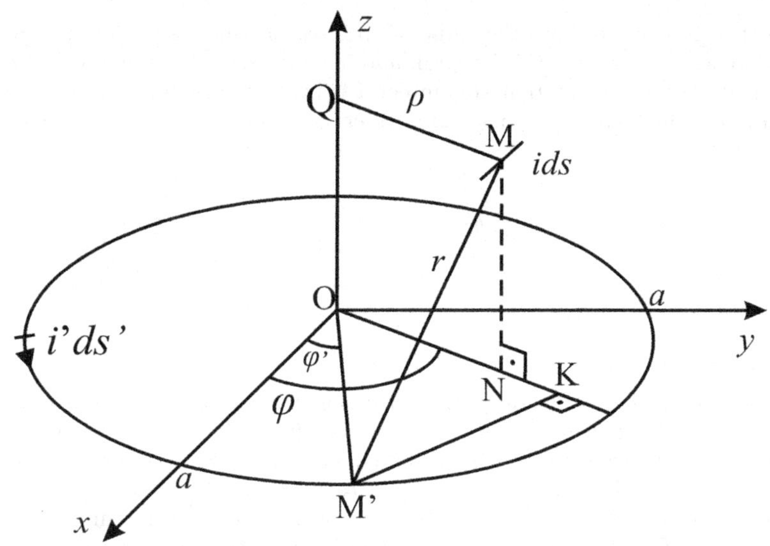

Figure 7.21: The main elements of Ampère's figure which were utilized to obtain the correct value of the constant k.

A current element $i'ds' = i'ad\varphi'$ is located at point $M' = (a,\ \varphi',\ 0)$. It exerts a torque on another current element ids located at point $M = (\rho,\ \varphi,\ z)$. The vertical projection of point M onto the xy plane is the point N. The straight line passing through ON is cut at point K orthogonally by another straight line passing through point M'. The angle between the x axis and the segment OM' is given by φ', while φ represents the angle between the x axis and the segment ON. Therefore, $\sin(\varphi - \varphi') = M'K/a$. The distance between the two current elements is given by $M'M = r = \sqrt{\rho^2 + a^2 - 2\rho a \cos(\varphi - \varphi') + z^2}$. With these definitions one obtains:

$$\frac{\partial r}{\partial s'} = \frac{1}{a}\frac{\partial r}{\partial \varphi'} = -\frac{\rho}{r}\sin(\varphi - \varphi') \ . \tag{7.22}$$

According to equations (7.16) and (7.22), Ampère's force exerted by current element $i'ds'$ and acting on the current element ids can be written as:

$$-ii'dsds'r^{1-n-k}\frac{\partial}{\partial s}\left(r^k\frac{\partial r}{\partial s'}\right) = ii'dsad\varphi' r^{1-n-k}\frac{\partial}{\partial s}\left[r^{k-1}\rho\sin(\varphi - \varphi')\right] \ . \tag{7.23}$$

This force is directed along the straight line $M'M = r$ connecting both current elements. In this experiment only the torque acting on the circuit containing the element ids matters in order to make it turn around the z axis. Therefore, the only component of the force given by equation (7.23) which interests us here is the force component perpendicular to the plane ONM. In order to obtain this component, we only need to multiply the total force by the factor $M'K/M'M = a\sin(\varphi - \varphi')/r$, yielding:

$$ii'dsa^2d\varphi' r^{-n-k}\sin(\varphi - \varphi')\frac{\partial}{\partial s}\left[r^{k-1}\rho\sin(\varphi - \varphi')\right] \ . \tag{7.24}$$

In order to obtain the torque around the z axis, one only needs to multiply equation (7.24) by the arm ρ of the lever. Integrating, over the whole circuit S, this torque acting on the element $i'ds'$ yields:

$$ii'a^2d\varphi'\int_S \rho r^{-n-k}\sin(\varphi - \varphi')\frac{\partial}{\partial s}\left[r^{k-1}\rho\sin(\varphi - \varphi')\right] ds \ . \tag{7.25}$$

According to Ampère and his case of equilibrium of the nonexistence of continuous rotation, this integral must vanish whenever two points of the circuit S pass through the z axis, no matter the shape of the circuit S. This integral can only vanish for all circuits S of arbitrary form when the integrand of equation (7.25) is an exact differential relative to the three integration variables, namely, r, ρ and φ. At this point Ampère concluded succinctly his analysis with the following words:[44]

[44] [Ampère, 1822o, p. 418].

> Now, it is known that, in order to have the value of an integral which is in this way independent of the relations of the variables appearing in this integral, and having always the same [value] between the same limits [of integration], it is necessary that this integral is presented in the form of an exact differential between these variables which are considered independent from one another, and this condition will only be fulfilled provided that:
>
> $$k - 1 = -n - k ,$$
>
> or
>
> $$k = \frac{1-n}{2} .$$
>
> The experiment [that is, the case of equilibrium of the nonexistence of continuous rotation,] shows that this is the relation between k and n. When $n = 2$ one has $k = -1/2$. However, no matter the force of the analogies which make us think that n is indeed equal to 2, there is no proof [of this fact] deduced directly from experiment, as all experiments made on this subject were performed by making a voltaic conductor act on a magnet and, consequently, they can only be applied, by an extension which cannot be considered as a complete proof, to the mutual action between two infinitely small portions of electric currents.

In order to understand how Ampère obtained this mathematical relation between k and n, it is necessary to continue the calculations beginning once more from equation (7.25). Calculating the partial derivative on the square brackets yields the following torque:

$$ii'a^2d\varphi' \int_S \rho r^{-n-k} \sin(\varphi - \varphi') \left[(k-1)r^{k-2}\frac{\partial r}{\partial s}\rho \sin(\varphi - \varphi') \right.$$

$$\left. + r^{k-1}\frac{\partial \rho}{\partial s} \sin(\varphi - \varphi') + r^{k-1}\rho\cos(\varphi - \varphi')\frac{\partial \varphi}{\partial s} \right] ds$$

$$= ii'a^2d\varphi' \int_S [(k-1)r^{-n-2}\rho^2\sin^2(\varphi - \varphi')dr$$

$$+ r^{-n-1}\rho\sin^2(\varphi - \varphi')d\rho + r^{-n-1}\rho^2\sin(\varphi - \varphi')\cos(\varphi - \varphi')d\varphi] . \tag{7.26}$$

We have then a line integral expressed in terms of three variables, namely, r, ρ and φ. This line integral will be independent of the path of integration only if there is a function $\phi(r,\ \rho,\ \varphi)$ in such a way that the line integral can be expressed as:

$$\int \left(\frac{\partial \phi}{\partial r}dr + \frac{\partial \phi}{\partial \rho}d\rho + \frac{\partial \phi}{\partial \varphi}d\varphi \right) . \tag{7.27}$$

That is, equation (7.26) will be independent from the path of integration when there is a function ϕ satisfying the following relations:

$$(k-1)r^{-n-2}\rho^2\sin(\varphi - \varphi') = \frac{\partial \phi}{\partial r} , \tag{7.28}$$

$$r^{-n-1}\rho\sin^2(\varphi - \varphi') = \frac{\partial \phi}{\partial \rho} , \tag{7.29}$$

and

$$r^{-n-1}\rho^2\sin(\varphi - \varphi')\cos(\varphi - \varphi') = \frac{\partial \phi}{\partial \varphi} . \tag{7.30}$$

Integrating these three equations one obtains three functions, namely:

$$\phi_1 = -\frac{k-1}{n+1}r^{-n-1}\rho^2\sin^2(\varphi - \varphi') + f_1(\rho,\ \varphi) , \tag{7.31}$$

$$\phi_2 = r^{-n-1}\frac{\rho^2}{2}\sin^2(\varphi - \varphi') + f_2(r,\ \varphi) , \tag{7.32}$$

and

$$\phi_3 = r^{-n-1}\rho^2\frac{\sin^2(\varphi-\varphi')}{2} + f_3(r,\ \rho)\ . \tag{7.33}$$

We can only have $\phi_1 = \phi_2 = \phi_3$ for any $(r,\ \rho,\ \varphi)$ when $f_1(\rho,\ \varphi) = f_2(r,\ \varphi) = f_3(r,\ \rho) =$ constant and when the following relation is also satisfied:

$$-\frac{k-1}{n+1} = \frac{1}{2}\ . \tag{7.34}$$

That is:

$$k = \frac{1-n}{2}\ . \tag{7.35}$$

Equation (7.35) represents the equation presented by Ampère in his fundamental paper of 1822.

Combining equation (7.35) with his supposition that $n = 2$, Ampère finally concluded that $k = -1/2$. These conclusions represent the main portion of the work which he presented to the Academy of Sciences of Paris on 10 June 1822.[45] He then arrived at the final value of his force between two current elements. Following equation (7.16), his force can then be written as:

$$\begin{aligned} &\frac{ii'dsds'}{r^2}\left(\sin\alpha\sin\beta\cos\gamma - \frac{1}{2}\cos\alpha\cos\beta\right) \\ &= \frac{ii'dsds'}{r^2}\left(\cos\varepsilon - \frac{3}{2}\cos\alpha\cos\beta\right) \\ &= -\frac{ii'dsds'}{r^2}\left(r\frac{d^2r}{dsds'} - \frac{1}{2}\frac{dr}{ds}\frac{dr}{ds'}\right) \\ &= -ii'dsds'r^{-1/2}\left(r^{-1/2}\frac{d^2r}{dsds'} - \frac{1}{2}r^{-3/2}\frac{dr}{ds}\frac{dr}{ds'}\right) \\ &= -ii'dsds'r^{-1/2}\frac{d\left(r^{-1/2}\frac{dr}{ds'}\right)}{ds} = -2ii'dsds'r^{-1/2}\frac{d^2\left(r^{1/2}\right)}{dsds'}\ . \end{aligned} \tag{7.36}$$

This equation with $k = -1/2$ represents the final expression of Ampère's force between current elements as expressed in different formats.

7.6 Two Remarkable Results Obtained by Ampère

On 24 June 1822, Ampère presented to the Academy of Sciences of Paris two remarkable results (his words) which he obtained utilizing the final value of his force between current elements. These results were published in the same paper in which he presented this final expression of his force with the value $k = -1/2$, equation (7.36).

Ampère integrated this expression in order to obtain the net force acting on a test current element and being due to a closed circuit of arbitrary form. He obtained a null component of this net force acting along the direction of the test current element. This was his first "remarkable result":[46]

> [...] therefore, the integral [of the net force component acting along the direction of the current element and being due to a closed circuit of arbitrary form] will vanish, which means that the net effect of all actions exerted by a closed circuit on a small portion of a conductor is always perpendicular to the direction of this small portion. I observe that the same [result] should take place for any set of closed circuits and, consequently, for a magnet, when it is considered as such [that is, when the magnet is supposed to consist of an ensemble of closed current-carrying circuits], in agreement with my opinion as regards the causes of the magnetic phenomena, and it is, in effect, what results from various experiments due to several physicists.

[45] [Ampère, 1822o].

[46] [Ampère, 1822o, pp. 419-420].

This consequence will be discussed in Sections 10.1 and 10.4.

The second remarkable result which he obtained from his force with the value $k = -1/2$ was presented by Ampère as follows:[47]

> [...] two small portions [of current-carrying conductors] should repel one another when they are located in the same straight line and when they are directed towards the same point in space. [...] The repulsion, in this case, was so unexpected, that it was necessary to verify it; later on I performed the experiment with M. Auguste de la Rive, and it was completely successful.

This remarkable result will be discussed in Section 8.2.

[47] [Ampère, 1822o, p. 420].

Part III

The Last Period of Ampère's Electrodynamic Researches

Chapter 8

Ampère's New Experiments

8.1 The Case of Equilibrium of the Currents in a Semicircle

As Ampère obtained the final and correct value of the constant k utilizing an experiment describing a case of equilibrium, he began to give even more importance to situations of null force and null torque. In his original paper of 1822,[1] he presented only the case of equilibrium of the nonexistence of continuous rotation. When this article was reprinted in the *Recueil* in 1823, Ampère included, in the new version, two other cases of equilibrium, namely, the experiment of the sinuous wire, figure 4.17, and the experiment of antiparallel currents, figure 4.19.[2] It is possible that these two new experiments were first performed after September 1822, when Ampère became totally convinced of the importance of the cases of equilibrium.

Ampère visited Auguste de la Rive, in Geneva, in September 1822, performing a new experiment with him. He replaced the horizontal circular conductor utilized in the case of equilibrium of the nonexistence of continuous rotation with a conducting semicircle. We will call this experiment the *case of equilibrium of the currents in a semicircle*. Ampère's motivation to perform this experiment may have been related to possible criticisms which might be raised against the experiment of the case of equilibrium of the nonexistence of continuous rotation. It might be argued, in particular, that in this last situation the lack of motion might be due to equal and opposite torques acting on the astatic coil, with these actions which were canceling one another being due to opposite portions of the fixed circular conductor. When describing the new case of equilibrium of the currents in a semicircle, Ampère himself mentioned that:[3]

> Therefore, in this [new] experiment it is not possible to attribute the immobility of the conductor, suspended in such a way that it can turn freely around a vertical axis, to the compensation of two equal and contrary actions produced by two portions of the fixed semicircular conductor, as it [that is, this immobility] happens in all orientations in which the mobile conductor is placed.

The case of equilibrium of the currents in a semicircle is shown in figure 8.1.[4]

In this experiment there is a horizontal circuit $CDEFG$ doubled several times over (*plusieurs fois redoublé*) and containing the semicircle DEF.[5] As seen in Subsection 4.3.4, "doubled several times over" means that there were many turns of the wire composing the circuit in order to increase the effect of the current acting upon an external body. Figure 8.2 (a) represents the circuit $CDEFG$ as seen from above, with the battery closing the circuit, and with four turns of the wire along the semicircular portion DEF. A system with N turns in the semicircle, with a current of intensity i' in each turn, is equivalent to another circuit of the same size, with a single turn of the wire carrying a current of intensity Ni', as represented in figure 8.2 (b). We are here supposing that the resistance of the remainder of the circuit is much greater than the resistance of the wire coiled in semicircle, so that we can neglect this last resistance compared with the resistance of the remainder of the circuit. The center O of the straight segment DF is located along the vertical projection of the axis of symmetry of the astatic coil of figure 8.1. In figure 8.2 (b) we represent the two lower horizontal portions of the astatic coil, $d'O$ and Od, together with the horizontal circuit in semicircle, $DEFD$, located below the astatic coil.

[1] [Ampère, 1822o].

[2] [Ampère, 1822y, p. 300] and [Ampère, 1885p, p. 275].

[3] [Ampère, 1885n, p. 332].

[4] [Ampère, 1822k, pp. 145-146], [Ampère, 1822b, pp. 320-321], [Ampère, 1822y, pp. 301 and 312-313] and [Ampère, 1885p, p. 275].

[5] [Ampère, 1822y, p. 312] and [Ampère, 1885p, p. 284].

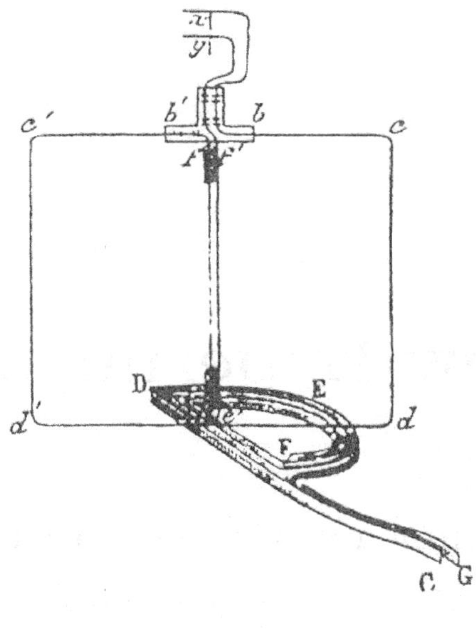

Figure 8.1: Case of equilibrium of the currents in a semicircle.

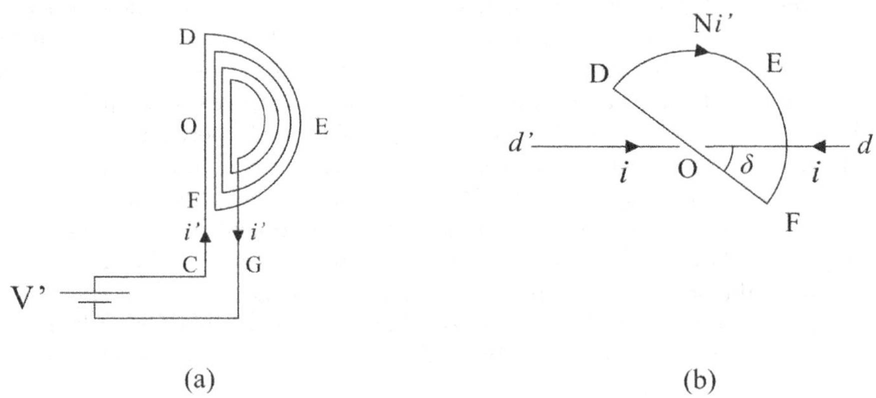

Figure 8.2: (a) Horizontal circuit in semicircle DEF doubled several times over. (b) Angle δ between the lower horizontal portions of the astatic coil and the straight segment DF of the circuit in semicircle located just below the astatic coil.

Once more Ampère did not obtain a continuous rotation of the astatic coil in this experiment, no matter the angle δ between the straight segment DF of the circuit in semicircle and the lower horizontal portions of the astatic coil, figure 8.2 (b). There was no continuous rotation, despite the large current intensity flowing through the lower circuit in semicircle and the negligible friction acting on the astatic coil.

The work which Ampère presented to the Academy of Sciences of Paris on 16 September 1822 describing this experiment performed with August de la Rive was initially published only as an abstract.[6] The complete work was only published by Joubert in 1885 from Ampère's manuscripts.[7] In any event, Ampère presented in 1823, in the *Recueil*, the main aspects of this experiment, together with figure 8.1.[8] In the paper of 1822, Ampère mentioned this experiment only in a footnote, without any accompanying figure.[9]

8.2 Ampère's Bridge Experiment

Two weeks after presenting to the Academy of Sciences of Paris his final formula with the value $k = -1/2$, Ampère obtained two remarkable consequences from this expression, as discussed in Section 7.6. We first

[6] [Ampère, 1822k].
[7] [Ampère, 1885n].
[8] [Ampère, 1822y, p. 301] and [Ampère, 1885p, p. 275].
[9] [Ampère, 1822o, pp. 415-416].

discuss Ampère's second consequence. This second "remarkable result" was expressed as follows:[10]

> [...] two small portions [of current-carrying conductors] should repel one another when they are located in the same straight line and when they are directed towards the same point in space. [...] The repulsion, in this case, was so unexpected, that it was necessary to verify it; later on I performed the experiment with M. Auguste de la Rive, and it was completely successful.

Until the middle of 1822 Ampère believed that $n/m = 0$ in equation (4.8) or, equivalently, $k = 0$ in equations (2.1) and (2.2), although he was not completely sure of the values of these constants. If $n/m = k = 0$, then two parallel current elements aligned along the same straight line should not act on one another. In any event, if the constant k were different from zero, then Ampère expected that it should be positive, as suggested by the experiment of Gay-Lussac and Thénard discussed in Subsection 4.2.3. The consequence of a positive constant k would be the attraction between two parallel current elements aligned along the same straight line with currents flowing in the same sense. When Ampère finally concluded that $k = -1/2$, he was skeptical about this unexpected theoretical conclusion that two collinear and parallel current elements should repel one another when both currents flowed in the same direction towards the same point in space. He then performed with Auguste de la Rive, in September 1822, in Geneva, an experiment to test this prediction, as shown in figure 8.3 (a). This figure appears in the paper of Auguste de la Rive,[11] as well as in Ampère's works.[12] This experiment has received several names in the literature: "Ampère's floating wire experiment",[13] "Ampère's hairpin experiment"[14] and "Ampère's bridge experiment".[15] We will call it *Ampère's bridge experiment.*

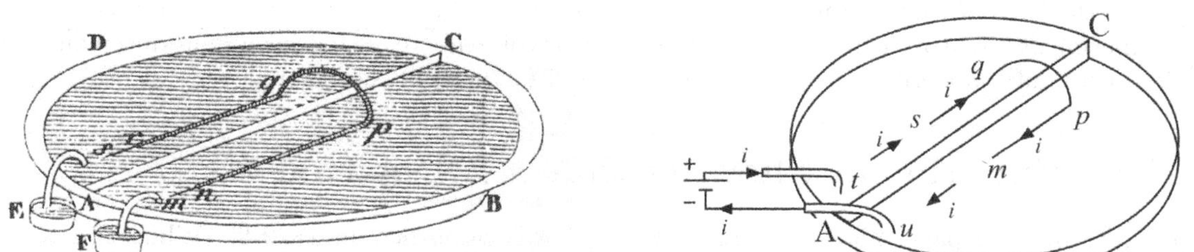

(a) (b)

Figure 8.3: (a) Ampère's bridge experiment. (b) Our illustration indicating the directions of the current. The bridge is represented by the segment *sqpm*.

Ampère described his prediction as follows:[16]

> Two very interesting electro-magnetic experiments have lately been made by M. Ampere, in the laboratory of M. de la Rive at Geneva. M. Ampere had been induced, from his mathematical investigations, to expect a repulsion between two portions of an electrical current passing in the same direction, and in the same right line, or that every part of an electrical current would repel the other parts, a result which may be comprehended by conceiving an endeavour in the current to elongate itself. The experiment which M. Ampere has contrived to illustrate this action of the current, and which our readers may compare with one described, Vol. XII., page 420 of this Journal, consisted of dividing a dish into two parts by a division across the middle, and filling each division with mercury, a piece of wire was then bent into the form of the letter U, but the curved part was bent to one side, so that the two limbs of the wire might lie on the mercury one on each cell, and the bent part pass over the division without touching it. The wire was covered with silk, except a small portion at each extremity, by which the communication was established with the mercury. The poles of a voltaic apparatus were then connected with the cells of mercury near to the division, so as to be in the lines of the limbs, in which case the wire moved parallel to the division, so as to elongate the current through which the electricity was passing.

[10] [Ampère, 1822o, p. 420].
[11] [de la Rive, 1822a], [de la Rive, 1822b], [de la Rive, 1822c] and [de la Rive, 1885].
[12] [Ampère, 1885n], [Ampère, 1826f, p. 39 and figure 8] and [Ampère, 1823c, Ampère, 1990, p. 211 and figure 8].
[13] [Hofmann, 1996, p. 317].
[14] [Graneau and Graneau, 1996, p. 61].
[15] [Wesley, 1991, pp. 225-232, Section 6.4: "Determination of the force on Ampere's bridge"] and [Wesley, 2002, pp. 305-321].
[16] [Ampère, 1823d].

Auguste de la Rive presented a very clear prediction of this experiment:[17]

> During his stay in Geneva, M. Ampère, after performing some new experiments, wished that I presented two main and important ones in the sequence of this memoir.
>
> The first experiment is a confirmation of M. Ampère's theoretical points of view, which, utilizing his formula, concluded that two portions of current directed in the same sense along the same straight line should repel one another, and that all portions of the same current should repel one another.
>
> As a matter of fact, in a dish $ABCD$, [figure 8.3 (a),] separated in two equal compartments by the division AC, with each compartment filled with mercury, a brass wire covered with silk is placed, so that the portions qr and pn may float in the mercury parallel to the division AC. The uncovered extremities rs and mn plunge into mercury. By connecting the poles [of the battery] into cups E and F, two currents are established which are independent from one another, and in which each one of them has as a conductor a portion of the mercury and a solid portion. No matter the direction of the current, it is always observed the two wires rq and pn moving parallel to the division AC going towards the opposite side from where it [the current] arrives, and this fact indicates a repulsion for each wire between the current established in the mercury and its continuation in the wire itself.

We made figure 8.3 (b) in order to facilitate the comprehension of this important experiment. Ampère's bridge is represented by the segment $sqpm$. When a constant current flows in this circuit, clockwise or anti-clockwise, the bridge moves from A to C. According to Ampère, this motion was due mainly to the repulsion between the current i flowing in the segment ts of the mercury and the current i flowing in the segment sq of the bridge, together with a repulsion between the current i flowing in the segment um of the mercury and the current i flowing in the segment mp of the bridge.

This experiment is one of the most important discoveries of Ampère. It has been reproduced and discussed in the literature until modern days.[18] It will be discussed further in Section 21.2.

8.3 The Experiment Showing that $n > 1$ or that $k < 0$

Ampère's fundamental paper showing that $k = -1/2$ was published in 1822.[19] When this article was reprinted in the *Recueil*, with some modifications, Ampère introduced a very interesting note which is worthwhile to reproduce here:[20]

> Due to equation $k = \frac{1-n}{2}$, the value k can only be negative when n is greater than 1. This is the reason why, before I had verified by the described experiment [namely, Ampère's bridge experiment, Section 8.2], that this value [of the constant k] is in fact negative, I assured myself that the value of n is greater than 1. To this end, after obtaining by a simple calculation that, by assuming $n = 1$, a fixed [plane] conductor, of arbitrary shape, cannot exert any action on a circular conductor placed in the same plane, and that the action between the circular conductor and a rectilinear conductor should be attractive or repulsive when these conductors have the same relative position, depending if n is greater or smaller than 1, I asked this experiment to be made in May 1822 and observed that this action is not null, and it results, from the sense it happens [that is, from the observed sense of motion of the astatic coil after being removed from the equilibrium configuration and released from rest], that n is greater than 1 so that, consequently, k is negative, by utilizing the mobile conductor represented in $xabcdefghiky$, [see our figure 8.4,] upon which the vertical conductor AB acted. The figure presented here seems to me sufficient in order to have a complete idea [of the experiment] so that a detailed description is unnecessary.

Despite Ampère's words, it seems relevant to us to discuss what he may have observed in this experiment. There is once more an astatic coil in a vertical plane, which can turn around the vertical axis xy. Although Ampère mentioned attractive or repulsive actions, what he observed was the torque acting on this astatic coil, or its rotation around the vertical axis xy, due to the influence of the rectilinear vertical conductor AB. Figure 8.5 presents Ampère's experiment as seen from above.

Figure 8.5 represents a horizontal cross section of figure 8.4, passing through the center of the vertical conductor AB. Crosses and circles with points represent the senses of the currents, namely, penetrating the paper and leaving out of the paper, respectively. The five points L, M, N, O and P are located at the same

[17] [de la Rive, 1822a, pp. 47-48], [de la Rive, 1822b, pp. 46-47], [de la Rive, 1822c, pp. 284-285] and [de la Rive, 1885, pp. 327-328].

[18] For modern references consult [Assis and Bueno, 1996], [Bueno and Assis, 2001] and [Bueno and Assis, 2015].

[19] [Ampère, 1822o].

[20] [Ampère, 1822y, p. 317].

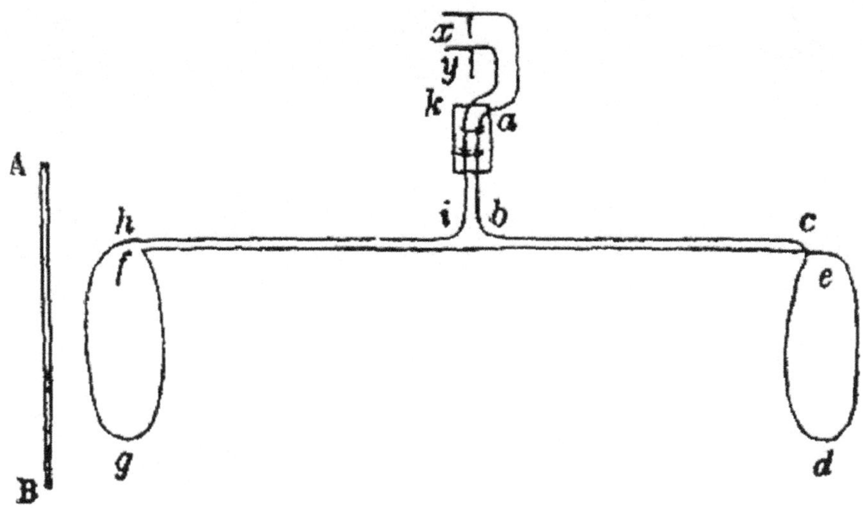

Figure 8.4: Experiment showing that $n > 1$.

Figure 8.5: Illustration of figure 8.4 as seen from above. (a) Astatic coil in equilibrium, aligned with the rectilinear vertical conductor AB. (b) Astatic coil removed from the equilibrium configuration and torque τ acting on it, forcing the astatic coil to return to the equilibrium configuration.

height in a single horizontal plane. Point L is at the center of the vertical conductor AB; point M is a point in the astatic coil closer to AB; etc. The electric currents in L, M and P leave the paper vertically, while the currents in N and O penetrate the paper. When these points L, M, N, O and P are located along the same straight line, with the currents flowing as indicated in figure 8.5 (a), the astatic coil remains in equilibrium. When the astatic coil is rotated around a vertical axis passing through is axis of symmetry xy, clockwise or anti-clockwise, it suffers a torque making it return to the previous configuration of equilibrium, as indicated in figure 8.5 (b). Ampère probably observed this torque or the rotation of the astatic coil after being released from rest from this last configuration. From this experiment he concluded that $n > 1$.

Combining this conclusion that $n > 1$ with the previous relation between the constants n and k, namely, $k = (1 - n)/2$, he concluded that $k < 0$.

Chapter 9

The Contributions of Savary

The first stage of Ampère's electrodynamic researches began in September 1820 with the announcement of Ørsted's discovery and finished in January 1821 with Ampère's prediction about the case of equilibrium of orthogonal currents and his experiment refuting his earlier prediction. The second stage began in September 1821 with Faraday's discovery of continuous rotation and finished in September 1822 when Ampère obtained the value $k = -1/2$ in his force law between current elements, together with his new predictions obtained from the final form of the force law. This time his new predictions were completely corroborated by the experiments. Ampère then abandoned for several months his experimental and theoretical researches on electrodynamics due to many bureaucratic and didactic activities he was required to accomplish in this period. The last great stage of his electrodynamic researches began in February 1823, with the work of his student Félix Savary (1797-1841) and finished in August 1826, with the publication of his main work, the *Théorie*. Savary was his student at the *École Polytechnique*.[1]

On 3 February 1823, Félix Savary presented an important work to the Academy of Sciences of Paris related to Ampère's force between current elements. This work was published in the same year.[2] In 28 July 1823, Savary presented to the Academy a complement of his earlier work which was also published in the same year.[3] We present here some of his results.

9.1 Obtaining a New Relation between the Constants n and k

The first relevant result obtained by Savary was a new relation between the constants n and k appearing in Ampère's force between current elements, equation (7.16).

In 1820 Gay-Lussac and J. J. Welter (1763-1852) had performed a very interesting experiment. Initially they utilized an unmagnetized steel ring which did not interact with a compass needle. If this ring was broken into pieces, its pieces also had no influence upon the magnetized needle. They then coiled a toroidal helix around this ring and a constant current flowed through it. The current was then turned off and the helix was removed out of the ring. The ring did not interact with a compass needle placed nearby. However, when the ring was broken into pieces, each piece did now interact with the magnetized needle. Each piece behaved now as a small magnet. That is, each small piece of the ring was magnetically polarized with a North and a South pole, so that it became magnetized. We were unable to locate any paper published by Gay-Lussac and Welter describing this experiment. In any event, Ampère and Savary described several times the outcome of this experiment.[4] This experiment is illustrated in figure 9.1.

The scientists Davy, Paul Erman and De la Borne generalized this experiment of Gay-Lussac and Welter, this time utilizing flexible cables which were then longitudinally magnetized.[5] De la Borne, for instance, coiled a flexible iron wire around a hollow rectilinear glass tube in such a way that the iron wire was coiled like a helix or spring. A straight wire was placed along the axis of the glass tube. He then discharged a Leyden jar through the straight wire, observing that the iron helix became magnetized along its length. In

[1] Savary should not be confused with Biot's collaborator, Félix Savart (1791-1841).

[2] [Savary, 1823a], [Savary, 1822], [Savary, 1823b], [Savary, 1823c] and [Savary, 1885b].

[3] [Anonymous, 1823] and [Savary, 1885a].

[4] [Ampère, 1823a, pp. 9 and 20], [Ampère, 1822h, pp. 331 and 342], [Savary, 1823a, p. 93], [Savary, 1822, p. 346], [Savary, 1823c, p. 12], [Savary, 1823b, p. 2], [Anonymous, 1823, p. 414], [Launay (ed.), 1936a, p. 624], [Ampère, 1826b, p. 37], [Ampère, 1826f, p. 185], [Ampère, 1823c, p. 357], [Savary, 1885b, pp. 340, 349 and 351] and [Savary, 1885a, p. 370].

[5] [De la Borne, 1821], [Ampère, 1823b], [Ampère, 1822l, note on p. 374], [Ampère, 1826f, p. 185], [Ampère, 1823c, Ampère, 1990, p. 357], [Savary, 1885a, p. 374] and [Ampère, 1885h, note on p. 391].

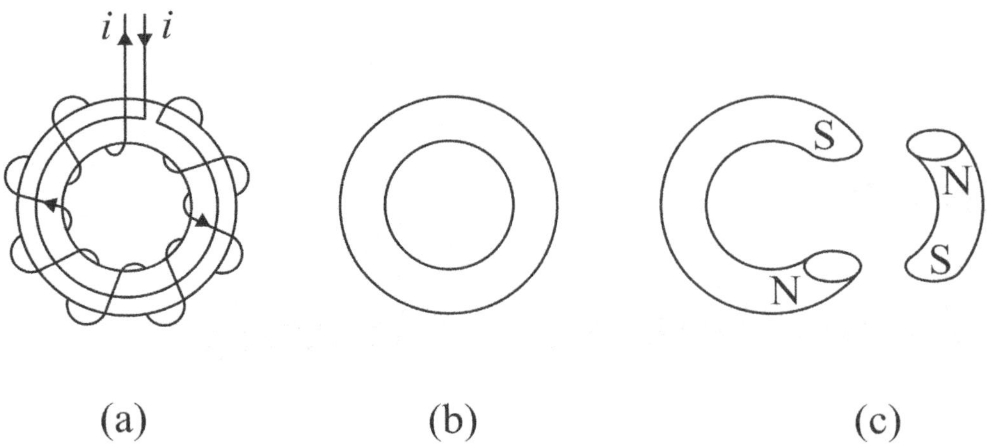

Figure 9.1: (a) A constant current flows through a toroidal wire coiled around an unmagnetized steel ring. (b) The wire is removed and the ring does not interact with a compass needle. (c) The ring is broken into pieces and each piece behaves like a small magnet, interacting with a compass needle.

this way the flexible iron helix had a North pole in one extremity and a South pole in the other extremity. These polarities were ascertained by approaching each extremity of the iron spring with a magnet. By connecting the two extremities of this iron helix he observed that it did not act any longer upon a compass needle placed nearby:[6]

> Such a helix then presents the singular configuration of a flexible and elastic magnet, which can be coiled, stretched, shortened and that, according to the theory normally accepted, should no longer act as a magnet upon a compass needle if, by connecting its two extremities, a ring is formed. This, indeed, is what is observed, at least perceptibly.

Savary decided to analyze this experiment theoretically utilizing Ampère's force between current elements. To this end he imagined the electrodynamic equivalent of a cylindrical magnet of radius r and length ℓ according to Ampère's conceptions. As mentioned by Savary, the electrodynamic analog of this magnet had been called an *electrodynamic cylinder* by Ampère. We quote Savary's description of an electrodynamic cylinder and also of an *electrodynamic ring*:[7]

> The main goal of the following calculations is to study the analogy between magnets and the ensembles of circular electric currents having planes parallel to one another and orthogonal to the line connecting the centers of the circles on which the currents flow. M. Ampère called this ensemble an *electrodynamic cylinder* when this line is straight and an *electrodynamic ring* when this line forms the circumference of a circle. We built [such an ensemble] utilizing a conductor wire coiled like a helix, returning then along the axis of this helix, in such a way to neutralize the longitudinal effect of the loops, at least at great distances relative to the radius of these loops, and this radius will always be considered very small.

An electrodynamic cylinder is composed of a system of circular electric currents having the same radius r, placed in parallel planes and orthogonal to the straight line connecting their centers which are equally spaced, figure 9.2 (a). The distance between the left and right centers of the circular coils has the same value ℓ of the length of the cylindrical magnet. Ampère's axially compensated helix, figure 4.10, represents another electrodynamic equivalent of a cylindrical magnet, figure 9.2 (b). Another electrodynamic equivalent of a cylindrical magnet is a cylindrical shell over which a current of surface density K flows azimuthally, figure 9.2 (c).

The electrodynamic cylinders of figure 9.2 are equivalent to the cylindrical magnet of figure 9.3 (a). That is, the electrodynamic cylinders of figure 9.2 act upon a compass needle just like the magnet of figure 9.3 (a), provided the magnet and the electrodynamic cylinders have the same length and radius. The intensity of the effect produced upon the compass needle will depend upon the current intensity of the electrodynamic cylinder and also upon the intensity of the magnetization of the magnet. It is possible to find the appropriate current intensity flowing through the electrodynamic cylinder which will produce the same effects (that is,

[6] [De la Borne, 1821].

[7] [Savary, 1823b, Savary, 1823c, pp. 1-2] and [Savary, 1885b, pp. 338-339].

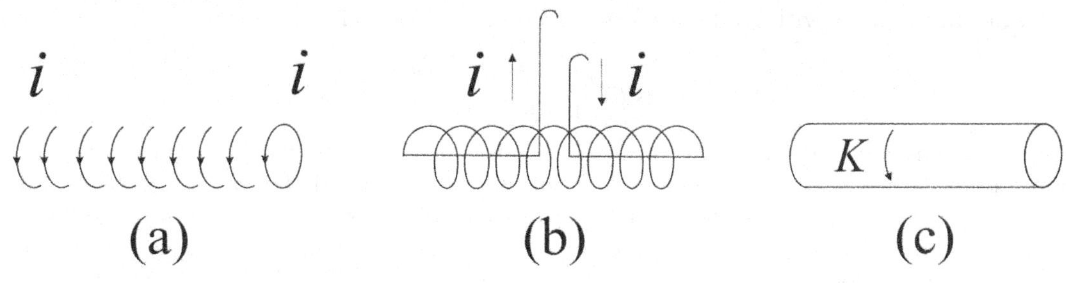

Figure 9.2: Electrodynamic cylinders which produce the same effects of a cylindrical magnet when acting upon a compass needle.

the same force and torque) on the compass needle as the effects produced on the same compass needle by a given magnet replacing the electrodynamic cylinder.

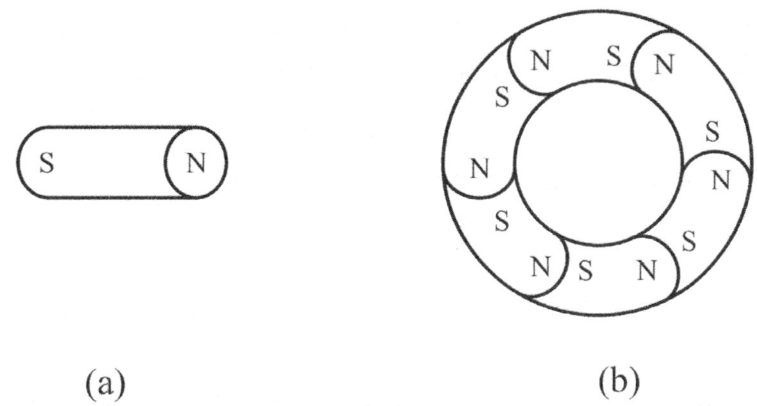

Figure 9.3: (a) Cylindrical magnet. (b) Toroidal magnet.

Figure 9.3 (b) presents a toroidal magnet of major radius R and minor radius r which is analogous to the magnetized ring of Gay-Lussac and Welter's experiment. This toroidal magnet is similar to a ring uniformly magnetized along its perimeter of length $2\pi R$.

Figure 9.4 presents three electrodynamic analogs of the toroidal magnet. That is, ensembles of current-carrying conductors which behave like the toroidal magnet of Gay-Lussac and Welter. In particular, these three current-carrying systems do not act on a nearby magnetic compass. Figure 9.4 (a) presents Ampère's *electrodynamic ring*. That is, a system of circular current-carrying electric circuits having the same radius r. The centers of these circular loops are equally spaced along a circle of radius $R > r$. In figure 9.4 (b) there is a toroidal coil compensated along its hollow circular interior. In figure 9.4 (c) there is a toroidal shell with a density of surface current K flowing along the poloidal direction.

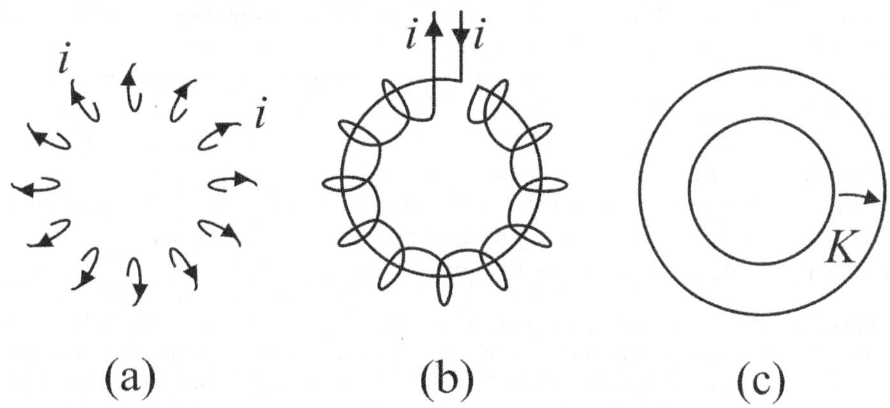

Figure 9.4: Electrodynamic rings.

Savary began his analysis with Ampère's force given by equation (7.16), namely:

$$-ii'dsds'\frac{r^{1-n-k}}{1+k}\frac{d^2\left(r^{1+k}\right)}{dsds'} \ . \tag{9.1}$$

He supposed a circular current loop of radius r located in the xy plane, centered at the origin of the coordinate system, carrying a constant current i. Initially he calculated the force exerted by this circular loop acting on a current element $i'ds'$ located in an arbitrary point in space. He then supposed several loops having the same radius r, with a current i flowing in each one of them, equally spaced in space in such a way that the ensemble formed an electrodynamic ring of radius R. He integrated the force exerted by all these circular loops forming the ring of radius R when acting upon the external current element $i'ds'$. In this calculation he assumed that the distance between the current element $i'ds'$ and the center of each one of these circular loops was much larger than the radius r of each loop. He obtained that this net force was proportional to $kn+1$. By utilizing the analogous result of Gay-Lussac and Welter's experiment which showed no net force between a toroidal magnet and a compass needle, Savary concluded that $kn+1$ should also be null:

$$kn+1=0 \ . \tag{9.2}$$

He utilized also equation (7.35) which had been obtained by Ampère, expressing it as:

$$n-1+2k=0 \ . \tag{9.3}$$

The combination of equations (9.2) and (9.3) yields:

$$2k^2-k-1=0 \ , \tag{9.4}$$

and

$$n^2-n-2=0 \ . \tag{9.5}$$

Equation (9.4) has two solutions, namely, $k=1$ and $k=-1/2$. Combining these two values of the constant k with the corresponding values of the constant n given by equation (9.2), one obtains two possible sets of values for these magnitudes, namely:

$$k=1 \quad \text{and} \quad n=-1 \ , \tag{9.6}$$

or

$$k=-\frac{1}{2} \quad \text{and} \quad n=2 \ . \tag{9.7}$$

Ampère's bridge experiment had shown a repulsion between colinear elements carrying currents in the same sense, which meant that $k<0$, as discussed in Section 8.2. Ampère had also shown experimentally that $n>1$, as discussed in Section 8.3. Savary then concluded that the only solution of equation (9.4) compatible with Ampère's experimental results was given by equation (9.7).

As a matter of fact, Gay-Lussac and Welter's experiment showed only the lack of interaction between a toroidal magnet and a small magnetized compass needle. Savary's calculations, on the other hand, refer to the lack of interaction between an electrodynamic ring and an external current element. Therefore, there is no complete analogy between these two situations; they are not really equivalent. We can only compare these two situations with the assumption that a magnet is equivalent to an ensemble of microscopic electric currents. However, Ampère himself performed an experiment which was electrodynamically equivalent to Gay-Lussac and Welter's experiment. He utilized only current-carrying conductors without the presence of any magnet. That is, Ampère's experiment was the electrodynamic analog of Gay-Lussac and Welter's magnetic experiment. We will call Ampère's experiment the *case of equilibrium of the toroidal ring*. Savary described Ampère's experiment with the following words:[8]

> In order to make the determination of the constants which appear in his formula independent of all analogy between magnets and electric currents, M. Ampère imitated the steel ring by coiling a portion of a silk covered conducting wire in the shape of a helix around another portion of the same wire, in such a way that the electric current of this last portion might destroy the effect of the longitudinal projections

[8][Savary, 1823b, Savary, 1823c, p. 12] and [Savary, 1885b, p. 352].

> of the loops of the first [portion], making with this helix a circular ring composed of several turns of the same helix, taking care that the remaining portions of the conducting wire, which served to put it into communication with the extremities of the battery, were coiled together until a certain distance of the ring, in such a way that their actions were completely neutralized. M. Ampère assured himself that this apparatus does not exert any action upon a mobile portion of [another] conductor which had an arbitrary shape. This result is independent of the radius of the [larger] ring, but it supposes the [smaller] radius of the loops to be extremely small in comparison with the distance to the mobile conductor.

That is, Ampère built an electrodynamic ring like that of figure 9.4 (b) and observed that it did not act upon an external current-carrying conductor of arbitrary shape.

In an anonymous work of 1823, written in the third person, although evidently written by Ampère,[9] this experiment was described as follows:[10]

> The results obtained by M. Savary when calculating, according to this formula, an observation of MM. Gay-Lussac and Welther made in 1820, have made M. Ampère try to observe the action, exerted on a mobile conductor, exerted by a system of circular electric currents having planes, extremely close to one another, perpendicular to a circumference passing through the centers of the circles described by these electric currents. He obtained, as it was easy to predict according to the aforementioned experiment [of Gay-Lussac and Welter], that this system did not exert any action upon the mobile conductor no matter the position in which it was presented [to the electrodynamic ring].

Savary did not present any illustration of Ampère's experiment in his article. We could not find it as well in Ampère's papers. However, there is a representation of this situation in a book by Verdet (1824-1866),[11] as shown in our figure 9.5. The left figure shows an open electrodynamic ring. It exerts a net force upon an external current element. When the electrodynamic ring is closed, as in the right figure, the net force vanishes.

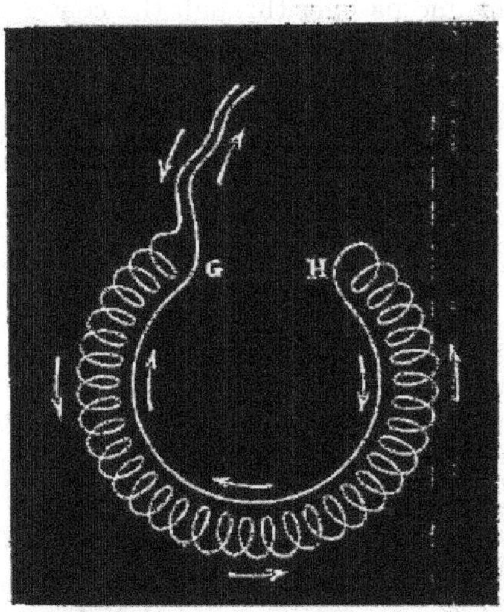

Figure 9.5: (a) When the electrodynamic ring is open, it exerts a net force upon an external current element. (b) When the electrodynamic ring is closed, it exerts no net force upon an external current element.

This experiment of Ampère, together with Savary's calculations, leads then to equation (9.2). This equation, combined with the previous results of Ampère, led to equation (9.7). Savary utilized these values of the constants n and k in the remainder of his paper.

9.2 The Electrodynamic Analog of a Magnetic Pole

Savary also considered the case of a semi-infinite cylinder carrying a constant current in the azimuthal direction. By integrating Ampère's expression he calculated the force exerted by this cylinder acting on

[9] [Joubert, 1885, p. 338] and [Hofmann, 1996, p. 387, reference Ampère (1822g)].
[10] [Ampère, 1823a, p. 9] and [Ampère, 1822h, p. 331].
[11] [Verdet, 1868, p. 326, figures 185 and 186].

an external current element idz. In all his calculations he considered the radius of the cylinder negligible compared with the distance between the current element idz and any circular loop of the electrodynamic cylinder. He concluded that this force[12]

> [...] is perpendicular to the plane formed by the current element and by corresponding extremity of the cylinder, is proportional to the sine of the angle between this element and the straight line connecting it to this extremity, and [varies] in the inverse ratio of the distance [between the current element and the corresponding extremity of the electrodynamic cylinder].

He expressed mathematically this proportionality by the following expression:

$$\frac{dz \sin V}{r^2} . \tag{9.8}$$

Here r is the distance between the current element and the extremity of the semi-infinite electrodynamic cylinder, while V is the angle between this distance r and the current element of length dz, as represented in figure 9.6 (a). This force is orthogonal to the plane formed by dz and by the straight line connecting the midpoint of the element to the center of the extremity of the cylinder carrying azimuthal current. In our figure we considered a particular example in which dz, the distance r and the axis of the electrodynamic cylinder are all located in the plane of the paper. In this situation the force acting on idz is orthogonal to the paper, pointing vertically upwards. It acts on the center of this current element. Its value does not depend on the orientation of the axis of the cylinder on the plane of the paper. The force reverses its sense when only the sense of the current in the current element is reversed, or when only the sense of the azimuthal current in cylinder is reversed. When the senses of the currents in the element and in the cylinder are simultaneously reversed, the force remains pointing vertically upwards.

By action and reaction, the force exerted by the current element idz acting on the semi-infinite electrodynamic cylinder is equal and opposite to the force exerted by the cylinder on the element. As regards Ampère's force, this reaction force also acts along a straight line passing through the center of the current element idz.

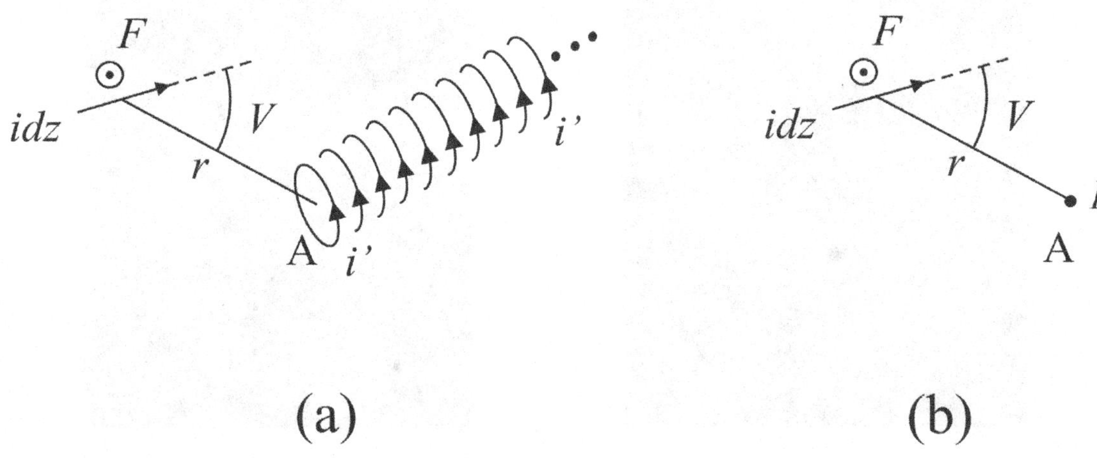

Figure 9.6: (a) Force F exerted on a current element by a semi-infinite cylinder. By action and reaction, the current element exerts an equal and opposite force on the cylinder. (b) This electrodynamic cylinder behaves like the supposed magnetic pole of Biot and Savart located at the extremity A of the cylinder, as the current element idz exerts the same force on the electrodynamic cylinder and on the magnetic pole.

Equation (9.8) represents the electrodynamic analog of Biot and Savart's result expressing the force exerted by a current element acting on a supposed magnetic pole. We expressed Biot-Savart's result in the form of equations (6.4) and (6.5). Figure 9.6 (b) represents the equivalence between Savary's semi-infinite cylinder and Biot-Savart's supposed magnetic pole. That is, if a magnetic pole p is located at point A coinciding with the center of the extremity of the semi-infinite cylinder, then Biot-Savart's force exerted by the current element on the pole will coincide with the force exerted by this element acting on Savary's semi-infinite current-carrying cylinder, as both forces are proportional to the expression given in equation (9.8).

[12][Savary, 1823b, Savary, 1823c, p. 14] and [Savary, 1885b, p. 354].

Moreover, both for Ampère and for Biot-Savart, the magnitude of the reaction force exerted by the current element is equal to the magnitude of the force acting on the current element, while the force acting on the current element passes through its center. But there is a distinction between Ampère and Biot-Savart's conceptions. Ampère compared these two conceptions in his main work, the *Théorie*. According to Biot and Savart, the reaction force exerted by the current element idz acting on the magnetic pole p located at point A acts on the pole. According to Ampère, on the other hand, the net reaction force exerted by the current element idz acting on the semi-infinite cylinder with azimuthal current, with the center of its extremity located at point A, acts also on the center of the current element idz and not at point A.

That is, according to Ampère, the action and reaction between the current element and the semi-infinite cylinder are parallel to one another and have the same magnitude, while acting along the same straight line passing through the midpoint of the current element, in such a way that they do not generate a net torque acting on the system. According to Biot and Savart, on the other hand, the action and reaction between the current element and the magnetic pole are parallel to one another and have the same magnitude, but they do not act along the same straight line. According to Biot and Savart, the force exerted by the magnetic pole acts on the midpoint of the current element, while the reaction force exerted by the current element acts on the magnetic pole. As they act along parallel straight lines, this action and reaction pair of Biot-Savart generates a primitive couple on the system composed of the current element and the magnetic pole. That is, this pair of internal action and reaction forces would generate a primitive torque on the system. Ampère argued strongly against this primitive couple, as will be discussed in Section 20.2.

The magnitude of the force given by equation (9.8) depends on the distance between the element and the extremity of the semi-infinite cylinder, depending also on the angle between this straight line and the element. It does not depend on the orientation of the cylinder in space. Therefore, a semi-infinite cylinder with azimuthal current is the electrodynamic analog of a magnetic pole.

In his following work presented at the Academy of Sciences of Paris on July 28, 1823, Savary generalized this result. Instead of utilizing a semi-infinite cylinder, he considered the configuration in which the centers of the circular loops were located along an arbitrary curve in space, as in figure 9.7. Ampère called this system an *electrodynamic solenoid*.[13] Figure 9.7 (a) represents what Ampère called a *definite solenoid*.[14] That is, a system of N circular current-carrying loops between points A and B, which are the extremities of the solenoid. Point A should be considered at the center of the left extremity of the solenoid, while point B should be considered at the center of its right extremity. The planes of these circular loops are orthogonal to a curve of arbitrary shape passing through the centers of these loops, with the centers of the loops equally spaced along this curve. There is a constant current i flowing in each one of these loops, always in the same sense. Figures 9.7 (b), (c) and (d) represent, respectively, what Ampère called a *closed solenoid*, an *indefinite solenoid in both directions* and a *simply indefinite solenoid*.[15] This last solenoid has an extremity A, while the other extremity is located at infinity. Sometimes Ampère called it an *indefinite solenoid*.

Savary calculated the interaction between an indefinite solenoid and a current element, as in figure 9.8. He showed that when one of the extremities of this system went to infinity or were indefinitely far away from the current element, with the other extremity located at the previous point A, then the force exerted by this system acting on the current element idz was also given by equation (9.8). That is, this force depends only on the distance between the current element and the extremity of the system which is close to the element, depending also on the angle V between this distance and the direction of the current element idz.

As was seen in Section 9.1, Savary had obtained that an electrodynamic ring exerted no net force on an external current element. With the result of the previous paragraph he could generalize this result. That is, he now considered the configuration in which the centers of the circular current-carrying loops of radii r were located along a curve of arbitrary form (in the case of an electrodynamic ring this curve was a circle of radius $R > r$). When this curve of arbitrary shape was closed, forming a closed solenoid as represented in figure 9.9, then Savary showed that the system with azimuthal current exerted no net force when acting on an external current element.

His words:[16]

> It follows that, when the two extremities of the axis are united [in such a way that the centers of the system with circular azimuthal current loops are located along a *closed* curve], no matter its shape [that is, the closed curve can have an arbitrary shape], the action vanishes.

---

[13] See our Section 10.2 and [Ampère, 1824c, p. 135].

[14] [Ampère, 1824c, p. 156] and [Ampère, 1885f, p. 399].

[15] [Ampère, 1824c, pp. 154 and 156] and [Ampère, 1885f, p. 399].

[16] [Savary, 1885a, p. 374].

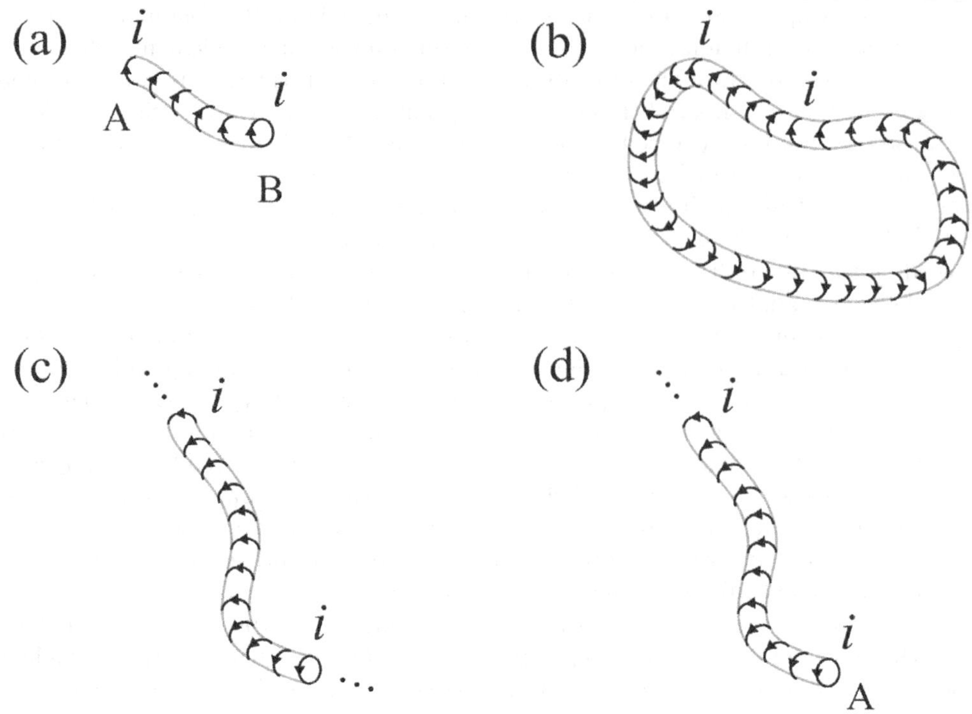

Figure 9.7: Ampère's electrodynamic solenoids: (a) Definite, (b) closed, (c) indefinite in both directions, and (d) simply indefinite.

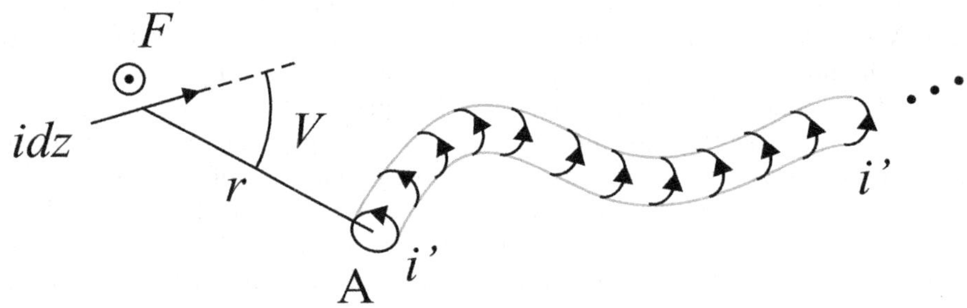

Figure 9.8: Interaction between a current element and an indefinite solenoid.

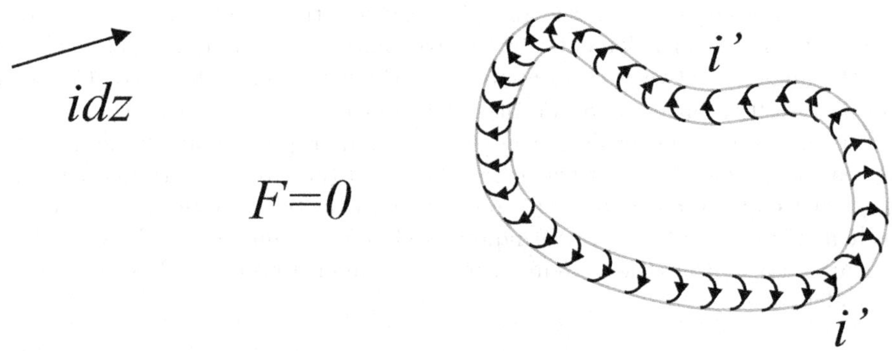

Figure 9.9: There is no interaction between a closed solenoid and an external current element.

We now return to our discussion of Savary's work presented to the Academy of Sciences of Paris on 3 February 1823. Savary integrated the force exerted by an infinite straight wire acting on a semi-infinite cylinder carrying azimuthal current, obtaining that it varies as the inverse of the distance g between the extremity of the cylinder and the straight wire. Moreover, this force was perpendicular to the plane formed by this distance g and the straight wire. In the case of figure 9.10, this force F penetrates the plane of the paper.

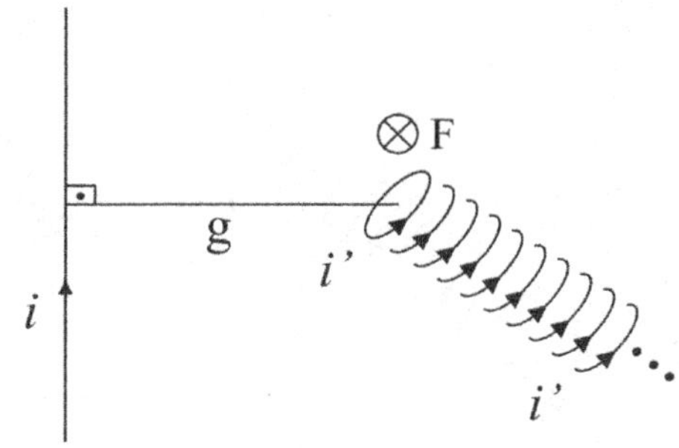

Figure 9.10: Force F exerted by a straight wire on a semi-infinite cylinder.

Savary's result was the electrodynamic analog of what Biot and Savart had obtained as regards the interaction between a current-carrying straight wire and a magnetic pole, as discussed in Section 6.1. That is, both forces are inversely proportional to a distance and were orthogonal to the plane formed by this distance and the straight wire. Once more the force calculated by Savary was shown to be independent of the orientation of the axis of the semi-cylinder in space.

In the complement of the first paper presented in July 1823, Savary generalized this result by considering an indefinite solenoid, as in figure 9.11. Once more this force was shown to be inversely proportional to the distance g between with straight wire and the extremity of the indefinite solenoid, being orthogonal to the plane formed by this distance and the straight wire.

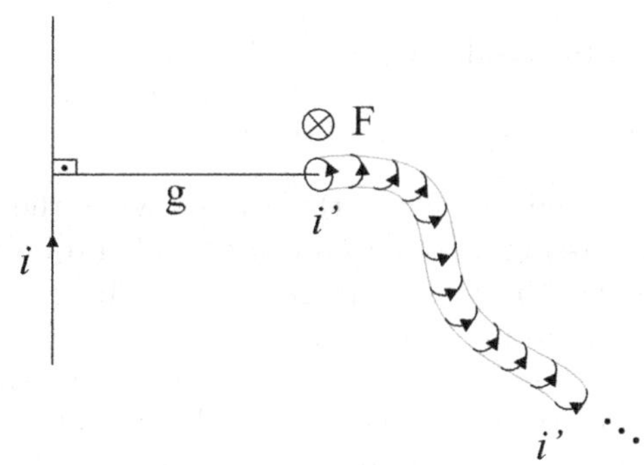

Figure 9.11: Force F exerted by a straight wire on an indefinite solenoid.

9.3 Torque Exerted by a Current-Carrying Straight Wire Acting on an Electrodynamic Cylinder

Savary also calculated the force and torque between a current element and an electrodynamic cylinder of finite length. He then integrated this last result in order to obtain the force and torque between an infinite current-carrying straight wire and this finite electrodynamic cylinder. A finite electrodynamic cylinder AB of radius r can be considered as a superposition of two semi-infinite cylinders of the same radius r having their axes of symmetry along AB, being indefinitely extended along the same side. One of these semi-infinite cylinders has its extremity located at point A, while the other has its extremity located at point B. Moreover, the currents in these two semi-infinite cylinders flow in opposite directions, figure 9.12.

The net force exerted by the infinite straight wire carrying a constant current and acting upon the finite cylinder is composed of two forces acting on the extremities of this cylinder. Each one of these forces is inversely proportional to the distance between the straight wire and the corresponding extremity of the cylinder, being perpendicular to the plane formed by this distance and by the straight wire. Savary then said the following:[17]

[17][Savary, 1823b, Savary, 1823c, p. 17] and [Savary, 1885b, p. 357].

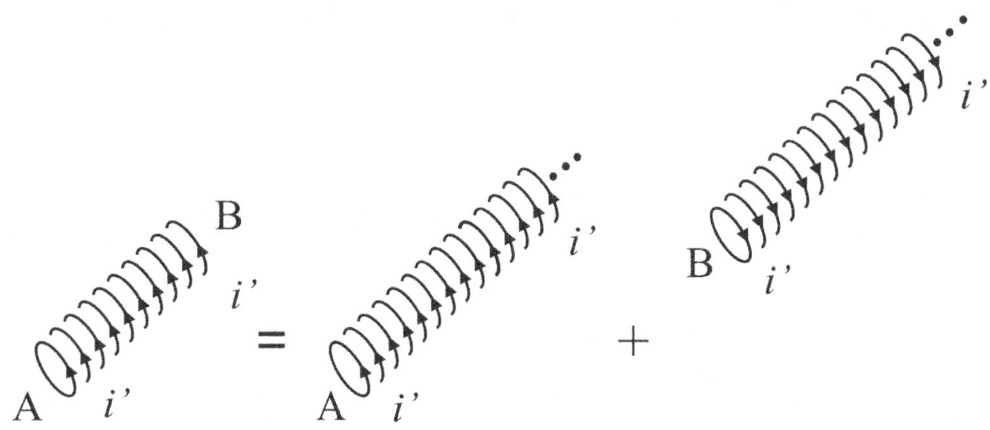

Figure 9.12: A finite current-carrying cylinder AB can be considered as a superposition of two semi-infinite cylinders, one finishing at point A and the other at point B. The current in the cylinder finishing at A flows in the same sense as the current in the finite cylinder AB, while the current in the cylinder finishing at B flows in the opposite sense.

> This result, applied to an infinitely short cylinder, is the law which had been proposed by M. Biot (*Annales de Chimie et de Physique*, volume XV, pp. 222 and 223),[18] to represent the action exerted by a magnetic molecule on an indefinite wire.

Savary concluded this Section of his paper with the following words, referring to the torque exerted by the cylinder on the wire, which can be thought as being due to two forces exerted by each extremity of the cylinder and acting on the straight wire:[19]

> When the cylinder is parallel to the conducting wire, the two forces are equal and act along parallel straight lines in opposite senses. Therefore, there only remains an action tending to turn the wire in order to leave it in a plane orthogonal to the axis of the cylinder.

Suppose now the straight wire is fixed relative to the ground while the cylinder is free to turn around an axis perpendicular to the cylinder, passing through its center and through the infinite wire. By action and reaction, the current-carrying straight wire will exert an opposite torque on the cylinder. This torque will turn the cylinder. After reaching equilibrium, remaining at rest relative to the ground, the axis of symmetry of the cylinder will be orthogonal to the plane containing the straight wire and the center of the cylinder. Moreover, this torque is inversely proportional to the distance between the straight wire and the center of the cylinder. This torque supplies a quantitative explanation of Ørsted's electrodynamic experiment described in Section 3.8.

9.4 Explanation of the Case of Equilibrium of Orthogonal Currents

Savary utilized his results in order to explain the case of equilibrium of orthogonal currents described in Section 6.3. To this end he performed calculations appropriate for the configuration of figure 6.7. Ampère had previously performed these calculations in 1820 supposing $k = 0$. Savary, on the other hand, made the same calculations but utilizing now Ampère's force between current elements in its final form, equation (7.36). That is, he utilized not only $n = 2$, but mainly $k = -1/2$.

He supposed[20] "a very short electrodynamic circuit, mobile around a vertical axis passing through its center and under the action of a conducting [straight] wire located in the vertical plane passing through the axis of the cylinder". He calculated the torque exerted by this infinite straight wire acting upon an electrodynamic cylinder of length 2λ which was considered much smaller than the distance g between the center of the cylinder and the wire, that is, $2\lambda \ll g$. He showed that this torque was proportional to $2\lambda/g$. Savary then concluded his analysis as follows:[21]

[18] [Biot and Savart, 1820], [Biot and Savart, 1965b] and [Assis and Chaib, 2006].
[19] [Savary, 1823b, Savary, 1823c, p. 17] and [Savary, 1885b, p. 357].
[20] [Savary, 1823b, Savary, 1823c, p. 17] and [Savary, 1885b, p. 357].
[21] [Savary, 1823b, Savary, 1823c, p. 18] and [Savary, 1885b, p. 358].

> Therefore, [this torque varies] in inverse ratio of the distance g, and does not depend at all on the orientation of the [straight] conductor, which can be vertical, horizontal, or inclined [to the horizon].
>
> This [result] is what M. Ampère observed with a very short magnet suspended in the corner and in the plane of two very long conducting wires, one horizontal and the other vertical [as in the configuration of our figure 6.8]. When the currents [of the same intensity] in these conductors flow simultaneously away from the vertex of the angle [formed by the junction of the two straight wires], or when they flow simultaneously towards this vertex, and when the distances of the center of the magnet to both wires are equal to one another, the magnet remains in equilibrium. Therefore, the actions which it suffers [from both current-carrying wires] are equal and of opposite signs, as it should happen according to the previous calculation.

Therefore, Savary finally obtained an explanation for the experiment performed by Ampère and Despretz utilizing Ampère's final force between current elements with $k = -1/2$. Ampère himself had never published anything related to this case of equilibrium of orthogonal currents up to this time. Moreover, Ampère had made a prediction about the outcome of this experiment, utilizing $k = 0$, which was refuted by his own experiment. Savary had been a student of Ampère. Ampère himself probably informed him about this experiment. Certainly Ampère was thrilled by this theoretical explanation Savary gave to this embarrassing experiment, especially because it gave another independent verification that the constant k of his force between current elements should have the value $-1/2$.

9.5 Mutual Action between Two Electrodynamic Cylinders

Another extremely important result Savary obtained referred to the interaction between two finite electrodynamic cylinders. We will represent the first electrodynamic cylinder by AB, with points A and B representing the centers of its extremities. The second electrodynamic cylinder will be represented by CD, with points C and D representing the centers of its extremities. They have arbitrary orientations in space. Savary integrated the force exerted by AB acting on CD and showed that it can be expressed as a sum of four terms. Each term acted along the straight line connecting one extremity of AB with one extremity of CD, varying inversely as the square of the distance between these extremities. This force might be attractive or repulsive, depending on the direction of the currents in both cylinders. The first term was directed along the straight segment AC, varying inversely as the square of this distance. The second term was directed along the straight segment AD, varying inversely as the square of this distance. The third term was directed along the straight segment BC, varying inversely as the square of this distance. And the fourth term was directed along the straight segment BD, varying inversely as the square of this distance.

Figure 9.13 is a representation of his calculations.

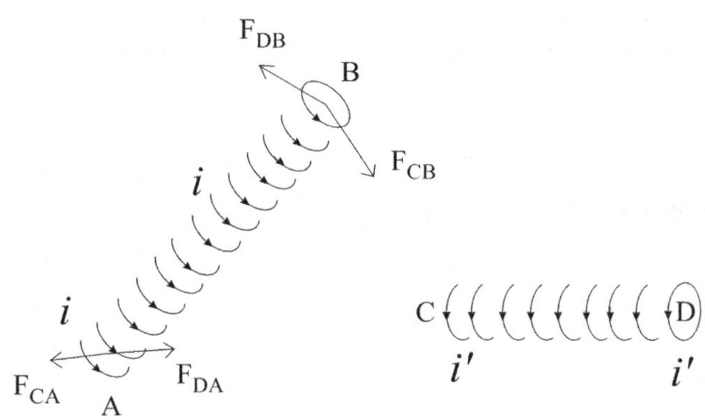

Figure 9.13: Electrodynamic cylinders AB and CD.

The electrodynamic cylinder AB carries a current i in each circular loop. There is a current i' in each circular loop of the electrodynamic cylinder CD. The force exerted by CD on AB can be considered as a sum of four terms, namely, the force F_{CA} directed along CA, the force F_{CB} directed along CB, the force F_{DA} directed along DA and the force F_{DB} directed along DB. Each one of these forces is inversely proportional to the square of the distance between the corresponding extremities of both cylinders.

Savary's result represents the electrodynamic analog of the observed interaction between two cylindrical magnets AB and CD, as represented by Coulomb's law between magnetic poles. Suppose magnet AB has

North and South poles represented by N and S, respectively, while the magnet CD has North and South poles represented by N' and S', figure 9.14. That is, the force and torque between these two magnets can be obtained considering the sum of the forces and torques exerted between each magnetic pole of AB with each magnetic pole of CD. The elementary force here is that given by Coulomb. That is, it takes place along the straight line connecting each pair of poles and varies inversely with the square of the distance between these poles, as given by equation (1.3).

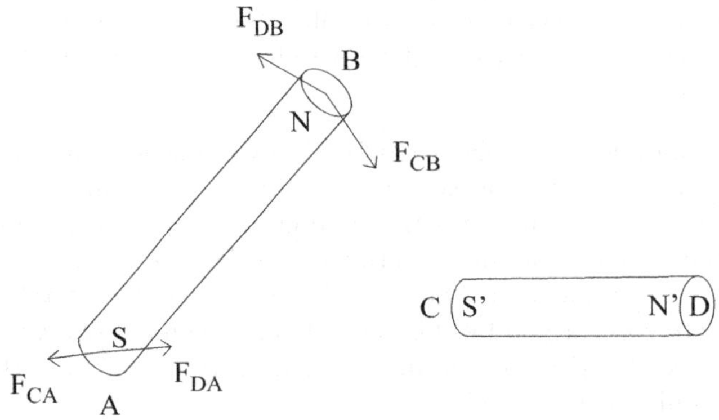

Figure 9.14: Forces exerted by the cylindrical magnet CD acting on the cylindrical magnet AB.

Savary expressed this analogy in the following terms:[22]

> Therefore, the action between two [electrodynamic] cylinders is reduced to four forces, of which two are attractive and two are repulsive, directed along the straight lines connecting the pairs of their extremities; as if these points exerted an action on one another, in order to attract or to repel one another, with this action varying in the inverse ratio of the square of the distance. It is easy to see, following the order of the calculation, that these forces are attractive between two extremities, on which one extremity is at the right and the other at the left of the currents of the cylinders to which they belong, and [these forces are] repulsive between two extremities, one extremity in each cylinder, located at the same side of these currents.
>
> By replacing at the extremities of the [electrodynamic] cylinders the poles of two magnets, the previous result turns out to be the law with which Coulomb represented his experiments about the orientation followed by a mobile magnetized needle under the action of a long magnetized bar, at least at great distances of the poles.

In order to know if the force between the extremities of the electrodynamic cylinders was attractive or repulsive, Savary considered two observers of Ampère, one observer in each cylinder, as represented in figure 9.15 (a).

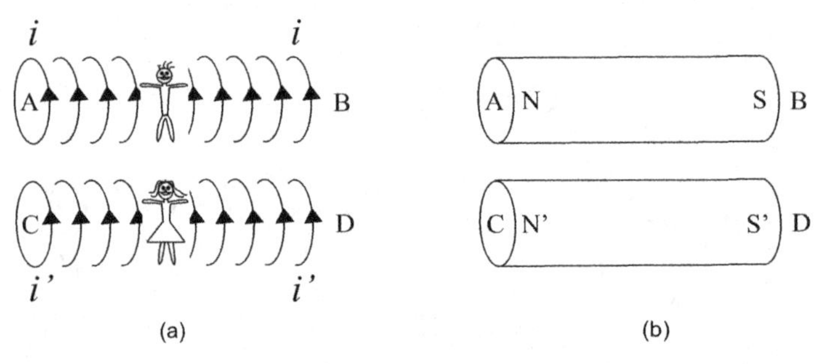

Figure 9.15: (a) Ampère's observers located on two electrodynamic cylinders. (b) The current-carrying cylinders are the electrodynamic analogs of two magnets.

In this particular example the axes of the electrodynamic cylinders AB and CD are parallel to one another. The arrows indicate the directions of the currents in each cylinder. Ampère's observers are lying on their backs on each cylinder, with the currents going from their feet towards their heads. In the case of this figure, extremity A is at the right of the observer located on the cylinder AB, while extremity D is at the

[22] [Savary, 1823b, Savary, 1823c, p. 24] and [Savary, 1885b, p. 367].

left of the observer located on the second cylinder CD. According to Savary's conclusion, this force between extremities AD is attractive, acting along the line AD. This force represents the electrodynamic analog of two cylindrical magnets, AB and CD, with their North poles located at A and C, while their South poles are located at B and D, respectively, as represented in figure 9.15 (b).

9.6 The Electrodynamic Analog of the Experiment of the Bent Wire

In the same paper Savary also considered theoretically the electrodynamic analog of Biot and Savart's experiment of the bent wire discussed in Section 6.2. To this end he considered a horizontal electrodynamic cylinder of length 2λ which might turn around a vertical axis passing through its center O. A bent wire DCD' carrying a current i acts on this cylinder. The bent wire is located in a vertical plane with its midpoint C located in the same horizontal plane of the center O of the cylinder. Savary represented the distance between points C and O by $CO = c$. Figure 9.16 presents this experiment as seen in a vertical plane. Point E is located along the prolongation of the horizontal segment OC. Savary indicated the angle DCE by a curled beta.[23] In this book this angle curled beta utilized by Savary will be replaced by ϱ.

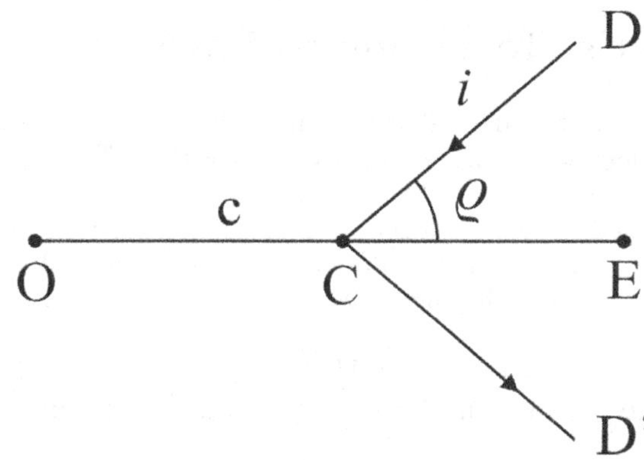

Figure 9.16: Representation in a vertical plane of the electrodynamic analog of the experiment of the bent wire.

Figure 9.17 presents this case as seen in a horizontal plane. Savary calculated the force exerted by the bent wire acting on the short electrodynamic cylinder. He showed[24] that "the [horizontal] cylinder remains in equilibrium when its axis is orthogonal to the [vertical] plane CDE".

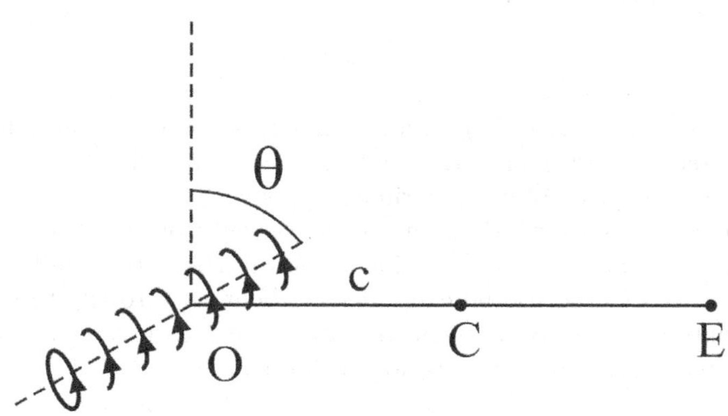

Figure 9.17: Representation in a horizontal plane of the electrodynamic analog of the experiment of the bent wire.

He also calculated the torque exerted by the bent wire acting on the cylinder when the axis of the cylinder is inclined by an angle θ relative to the direction of equilibrium. According to Savary's calculations, this torque is proportional to:

[23] See the footnote 66 of our page 45.

[24] [Savary, 1823b, Savary, 1823c, p. 21] and [Savary, 1885b, p. 363].

$$\frac{2\lambda\theta}{c}\tan\frac{\varrho}{2} \ . \tag{9.9}$$

He then concluded:[25]

> Therefore, [the torque] varies in inverse ratio of the simple distance [between the vertex of the bent wire and the center of the cylinder] and proportionally to the tangent of half of the inclination of the conductor relative to the horizontal plane, which differs little from the result obtained experimentally by M. Biot.

We have seen in Section 6.2, equation (6.2), that Biot obtained experimentally that this torque was inversely proportional to the distance c between the vertex of the bent wire and the center of the small magnet. Moreover, Biot obtained that this torque was also proportional to the angle which is being represented here by the symbol ϱ. When $\varrho \ll 1$ rad we have $\tan(\varrho/2) \approx \varrho/2$. Therefore, Biot's experimental result is close to Savary's theoretical calculation, when Biot's magnet is replaced by Savary's electrodynamic cylinder.

The results presented in this Section represent some of the main conclusions obtained by Savary in this extremely important paper. In Section 9.7 we present the reactions of Biot and Savart to this work, while in Section 9.8 we present Ampère's reactions.

9.7 Biot and Savart's Reactions to Savary's Work

In Section 6.2 we saw that Biot and Savart concluded the following with their experiment of the bent wire presented to the Academy of Sciences of Paris on 18 December 1820:[26]

> I thus found that, for both the bent wire and the straight wire, the action was inversely proportional to the distance to the points M and M'; but the absolute intensity was weaker for the bent wire than for the straight wire, in the proportion of the angle ZMH to unity.

Representing the distance to the points M and M' by c and the angle ZMH by the symbol ϱ, in order to compare their result with the calculation of Savary, the torque obtained by Biot and Savart was then proportional to:

$$\frac{\varrho}{c} \ . \tag{9.10}$$

Savary, on the other hand, obtained theoretically that the torque exerted by the bent wire and acting on the electrodynamic cylinder was proportional to the expression given in equation (9.9). That is, the torque was proportional to the tangent of $\varrho/2$ instead of being proportional to ϱ. For angles ϱ which are much smaller than 1 radian we can apply the following approximation:

$$\tan\frac{\varrho}{2} \approx \frac{\varrho}{2} \ . \tag{9.11}$$

Combination of equations (9.9) and (9.11) indicate that in this approximation the torque obtained by Savary can also be considered proportional to the angle ϱ. Therefore, when $\varrho \ll 1$ rad, Savary's theoretical result may be considered equivalent to Biot's experimental result.

However, according to Savary's calculation, the torque is only proportional to ϱ for small angles. A rectilinear wire is equivalent to a bent wire with an inclination to the horizontal being given by $\varrho = \pi/2$ rad. We can then compare the ratio of the torque exerted by a bent wire to the torque exerted by a straight wire according to the results obtained by Biot-Savart and Savary, as given by equations (9.10) and (9.9), respectively. According to Biot and Savart this ratio is given by:

$$\frac{\varrho}{\pi/2} = \frac{2\varrho}{\pi} \ . \tag{9.12}$$

According to Savary, on the other hand, this ratio is given by:

$$\frac{\tan(\varrho/2)}{\tan(\pi/4)} = \tan\frac{\varrho}{2} \ . \tag{9.13}$$

[25] [Savary, 1823b, Savary, 1823c, p. 22] and [Savary, 1885b, p. 364].
[26] [Biot, 1821a, p. 123] and [Biot, 1885a].

Equations (9.12) and (9.13) are different from one another. In principle it might be possible to distinguish between the predictions of Biot-Savart and Savary by means of a precise experiment.

Biot and Savart performed once again their experiment, this time more carefully. They now concluded that the correct result expressing the torque exerted by a bent wire and acting on a small magnet is proportional to the tangent of half of the inclination of the conductor relative to the horizontal plane, in agreement with Savary's calculation, instead of the earlier results which they had presented in 1821 by saying that the torque was proportional to the angle of inclination. Their new result was presented in Biot's *Précis élémentaire de Physique expérimentale*.[27] In the third edition of this book, published in 1824, the previous quotation of 1820, repeated in the second edition of Biot's book,[28] was replaced by the following statement:[29]

> The early tests soon made me see that the action of the oblique wire diminishes with decreasing angle between the two arms, *and this seemed to be proportional*; such a law was quite compatible with the limits of the phenomenon, for the action must obviously be zero if the angle is zero when the two halves of the wire are bent together with the current flowing in opposite directions; and this action and that of the straight wire must become equal when each arm is at 90° to the horizontal, since they then both form a vertical straight line. But because of imperfections in the experiments, other laws were equally admissible and, for example, the inclination i to the horizontal[30] could have been replaced by the tangent to half of the inclination, that is to say, $\tan\frac{1}{2}i$;[31] then, by putting F for the observed action on the needle of the vertical wire at a certain distance, $F\tan\frac{1}{2}i$ would be the action of the oblique wire animated by the same current;[32] formerly, it would have been represented by $\frac{Fi}{90^\circ}$,[33] a value which can only ever differ from the foregoing in hundredths.

Biot then described his new precise experiments in order to distinguish if the torque was proportional to ϱ or to $\tan\frac{\varrho}{2}$, concluding that this last expression was the one which, indeed, was compatible with his new experimental results:[34]

> Therefore, it is not possible to doubt that this expression[35] does not represent in general the total action of a bent wire folded in two branches forming an angle i with one another.[36]

That is, with his new precise experiments Biot confirmed Savary's theoretical calculations given by equation (9.9). This means that the torque generated by a bent wire is indeed proportional to the tangent of half of its opening angle, instead of being proportional to this angle. Biot therefore corrected the interpretation of his own previous experiment!

Let us now see what Biot said related to the supposed elementary force exerted by each current-carrying element of the bent wire acting on a magnetic particle:[37]

> Now, considering an infinitely fine section of a similar [bent] wire situated at μ (*fig.* 9),[38] where μm or R is the distance from the wire to the particle m of boreal or austral magnetism, we know from our previous experiments that the action of this section on the particle is reciprocal to the square of the distance μm multiplied by an unknown function of the angle $m\mu M$ for which we put ω. It therefore only remains to find a form for this function such that a resultant proportional to $\frac{\tan\frac{1}{2}i}{R}$ if formed by the total sum of the actions of all the wire sections exerted on m perpendicular to the plane CMZ.[39] This condition is satisfied by taking $\sin\omega$ for the required function; this makes the elementary action of a section proportional to $\frac{\sin\omega}{R^2}$; using this experimentally determined expression, knowing the absolute direction of the force to be perpendicular to a plane through each distance in the direction of each

[27] [Biot, 1821a] and [Biot, 1885a].

[28] [Biot, 1821a, p. 123] and [Biot, 1885a].

[29] [Biot and Savart, 1824], [Farrar, 1826, p. 336], [Biot, 1885b, p. 116], [Biot and Savart, 1885, p. 116] and [Biot and Savart, 1965a, pp. 136-137], our emphasis and our words in the footnotes.

[30] Biot represented the inclination of the bent wire to the horizontal by the symbol i. This inclination is being represented in this Section by the symbol ϱ to avoid confusion with the current intensity.

[31] This expression is being represented in this Section by $\tan\frac{\varrho}{2}$.

[32] That is, the torque would be proportional to $F\tan\frac{\varrho}{2}$.

[33] Or equivalently by $\frac{F\varrho}{90^\circ}$.

[34] [Biot, 1885b, p. 119].

[35] That is, the expression written by Biot as $F\tan\frac{1}{2}i$, which is being represented in this book by $F\tan\frac{\varrho}{2}$.

[36] This angle i of Biot is equivalent to our angle ϱ.

[37] [Biot and Savart, 1824], [Biot, 1885b, pp. 119-120], [Biot and Savart, 1885, p. 119] and [Biot and Savart, 1965a, p. 139]. See also [Farrar, 1826, p. 339].

[38] Our figure 6.6.

[39] That is, proportional to our function $\frac{\tan\frac{\varrho}{2}}{R}$.

> longitudinal element of wire, the total resultant of the action exerted by a wire, or some straight or curved, limited or unlimited portion of wire, can be assigned by calculation.[40]

Biot's conclusion is extremely curious. Biot and Savart had initially concluded, utilizing that the force exerted by the bent current-carrying wire acting on a supposed magnetic pole was proportional to the angle ϱ, that the elementary force between a current element and a magnetic pole should be proportional to $\sin\phi$ or to $\sin\omega$. After Savary's theoretical calculations, Biot and Savart performed an improved series of experiments with the bent wire, concluding now that the action of the bent wire was proportional to $\tan\frac{\varrho}{2}$. They also concluded that the elementary function which would supply, after being integrated along the whole bent wire, this action of the bent wire proportional to $\tan\frac{\varrho}{2}$, was still proportional to $\sin\phi$ or to $\sin\omega$! However, it is impossible that the same elementary action will yield simultaneously, after being integrated in the same configuration, a force proportional to ϱ and a force proportional to $\tan\frac{\varrho}{2}$. Therefore Biot and Savart had made at least two mistakes in their initial works related to the bent wire. The first mistake was the wrong conclusion of what was the behavior, observed experimentally, of the force acting on a supposed magnetic pole as a function of the opening angle of the bent wire. Instead of being proportional to the angle ϱ, the observed force was indeed proportional to $\tan\frac{\varrho}{2}$. The second mistake was to suppose that an elementary force proportional to $\sin\phi$ or to $\sin\omega$ should lead to an integrated force of the bent wire proportional to ϱ. As a matter of fact, the integration of an elementary force proportional to $\sin\phi$ or to $\sin\omega$ leads to a net integrated force of the bent wire proportional to $\tan\frac{\varrho}{2}$.

However, by an incredible coincidence, these two mistakes canceled one another. That is, the elementary action presented by Biot and Savart in 1821, given by equation (6.4), happened to be valid according to their more precise experimental results of 1824!

9.8 Ampère's Reactions to Savary's Work

Ampère became very motivated with Savary's work. From his manuscripts it is possible to conclude that he himself wrote a great portion of the papers which were published having only Savary as their author.[41] He also paid from his own pocket for the publication of 500 copies of Savary's paper.[42]

Ampère's reaction can be easily understood. He realized that Savary's work represented an important advance related to his own research program. Ampère had obtained the final value of his force law with $k = -1/2$ in June 1822, presenting his results to the Academy of Sciences of Paris. With this final expression he had explained quantitatively all his electrodynamic experiments related to the interaction between current-carrying wires. With Savary's work he realized that it was possible, beginning with a single force law, to explain quantitatively all known electromagnetic experiments, like the experiments of Ørsted and those of Biot and Savart with the straight and bent wires. To this end, it was only necessary to suppose the existence of microscopic currents inside magnets and inside the Earth, as Ampère had always assumed. Moreover, Ampère now realized that it was also possible to explain, quantitatively utilizing his force between current elements, all known magnetic phenomena. The basis of this later accomplishment was Savary's deduction of the electrodynamic analog of the force between magnetic poles given by equation (1.3). After all, the interaction between two magnets and also the interaction between a magnet and the Earth can be modeled based on the interaction between magnetic poles. Ampère now realized that he could explain all purely magnetic phenomena from an electrodynamic point of view, namely, utilizing his force between current elements and supposing the existence of microscopic electric currents in the Earth and in magnets. Savary's work opened the door to a mathematical unification of three areas of knowledge, namely: magnetism (interaction between two magnets and interaction between a magnet and the Earth), electromagnetism (interaction between a current-carrying wire and a magnet, as in the experiments of Ørsted, Biot and Faraday), and electrodynamics (interaction between two current-carrying wires, as in most experiments performed by Ampère).

Savary's work contained some additional features. In the first place, by obtaining a new relation between the constants k and n appearing in Ampère's force between current elements, Savary was able to deduce separately that $k = -1/2$ and $n = 2$. Previously Ampère had only been able to obtain $k = -1/2$ by assuming that $n = 2$, in analogy with the other known forces of nature. In the second place, Savary was able to explain

[40] Note by R. A. R. Tricker on page 139 of his book *Early Electrodynamics: The First Law of Circulation*, [Tricker, 1965, p. 139]: This account confirms Ampère's criticisms. The proportionality of the force to $\tan\frac{i}{2}$ was not suggested by Biot's experiments. It was only confirmed by them after it had been pointed out that the Biot-Savart law did not follow from the previous result on which it had been based, namely that the force was proportional to i.

[41] [Ampère, b, *carton* 9, *chemise* 172].

[42] [Grattan-Guinness, 1990a, p. 935] and [Savary, 1823c].

quantitatively Ampère and Despretz's experiment, namely, the case of equilibrium of orthogonal currents discussed in Section 6.3. The unexpected outcome of this experiment bothered Ampère since 1821 and he had not published anything related to this experimental anomaly. From now on Ampère always mentioned this experiment as another proof of his theory!

In a letter to Auguste de la Rive, written the day after Savary's presentation at the Academy of Sciences of Paris, Ampère said the following:[43]

> Yesterday's meeting of the *Académie des Sciences* marks a new epoch in the history of dynamic electricity [...]. It follows from the whole [that is, from the presentations of Savary and Demonferrand] that all the facts which have not yet been completely explained, observed by MM. Gay-Lussac and Welter, Pouillet, Biot and Savart, Coulomb's experiments with magnets, your own experiments, those of MM. Faraday and Barlow, the known law of the inclination of the magnetic needle, etc., are necessary consequences of my formula.

All these aspects were mentioned by Ampère in a paper he published in the *Recueil* soon after the reprint of Savary's paper.[44] In his *Exposé méthodique* of 1823, Ampère mentioned publicly for the first time the experiment he had performed with Despretz in 1821, the case of equilibrium of orthogonal currents discussed in Section 6.3:[45]

> An indefinitely long linear conductor has the same rotational action on a circular current located in a plane perpendicular to the plane which passes through the conductor and the center of the circle described by this current, about the common intersection of these two planes, provided that, the distance from the center of the circular current to the conductor remaining the same, the conductor is placed in various orientations with respect to this intersection. This agrees with an experiment done the 20th of January, 1821 by Ampère and Despretz.

Ampère's theory became essentially complete with this work of Savary. From now on Ampère began to elaborate and complete his electrodynamic theory, incorporating also the electromagnetic and magnetic phenomena, which were now explained quantitatively, culminating in the *Théorie* of 1826, his masterpiece.

---

[43] [Launay (ed.), 1936a, p. 624] and [Blondel, 1982, p. 142].
[44] [Ampère, 1822u] and [Ampère, 1885l].
[45] [Ampère, 1823a, pp. 20-21], [Ampère, 1822h, pp. 342-343] and [Hofmann, 1982, p. 325].

Chapter 10

Some Later Developments

10.1 The Directrix, the Directing Plane and the Force Exerted by a Closed Circuit of Arbitrary Form Acting on an External Current Element

As seen in Section 7.5, on 10 June 1822 Ampère presented to the Academy of Sciences of Paris his force between current elements in its final form, that is, with $k = -1/2$. It was expressed in several formats, as given by equation (7.36).

On June 24, 1822, Ampère presented to the Academy two remarkable results which he obtained with this final value of his force between current elements, as discussed in Section 7.6. One of these remarkable results was discussed in Section 8.2. The other remarkable was the following:[1]

> [...] the net effect of all actions exerted by a closed circuit on a small portion of a conductor is always perpendicular to the direction of this small portion.

In Section 10.4 we will discuss Ampère's experiment confirming this statement, namely, the case of equilibrium of the nonexistence of tangential forces.

In memoirs read at the Academy of Sciences of Paris on 22 and 29 December 1823 and in another one read on 5 January 1824, Ampère extended even more this last result. This development was based on Savary's works discussed in Chapter 9.[2] Ampère initially worked with his force between current elements in the form of equation (7.16), utilizing also that

$$2k = 1 - n \; . \tag{10.1}$$

He then calculated the force acting on a current element $i'ds'$ being exerted by "an arbitrary system of currents forming closed circuits, or being extended indefinitely in both directions".[3] Figure 10.1 (a) presents a closed circuit S of arbitrary shape carrying a current i interacting with an external current element $i'ds'$ located at an arbitrary point in space. Figure 10.1 (b) presents a circuit S of arbitrary shape, carrying a current i, extending indefinitely in both directions. In this last configuration the extremities of the circuit are located at infinite distances from ds'. This is the case, for instance, of a straight and filiform circuit of infinite length acting on $i'ds'$.

Ampère calculated the net force acting on a current element $i'ds'$ which did not belong to the circuit S of arbitrary form represented by figures 10.1 (a) and (b). He obtained once more that this net force was always orthogonal to the direction of the element $i'ds'$. Moreover, he now concluded that this net force belonged to a certain plane which he called the *directing plane.*[4] The straight line orthogonal to this directing plane and passing through the midpoint of the current element $i'ds'$ was initially called *normal to the directing plane.*[5] In the *Théorie* Ampère called it *directrix*, a term which will be adopted in this book.[6]

[1] [Ampère, 1822o, pp. 419-420].
[2] [Ampère, 1824c], [Ampère, 1824f], [Ampère, 1824e] and [Ampère, 1885f].
[3] [Ampère, 1824c, p. 136] and [Ampère, 1824e, pp. 12, 19, 28 and 44].
[4] [Ampère, 1824c, p. 142], [Ampère, 1824e, p. 11] and [Ampère, 1885f, p. 398].
[5] [Ampère, 1824e, p. 47], [Ampère, 1826b, p. 41], [Ampère, 1885f, p. 398], [Ampère, 1887c, p. 195] and [Ampère, 1887f, p. 210].
[6] [Ampère, 1826f, p. 43] and [Ampère, 1823c, Ampère, 1990, p. 215].

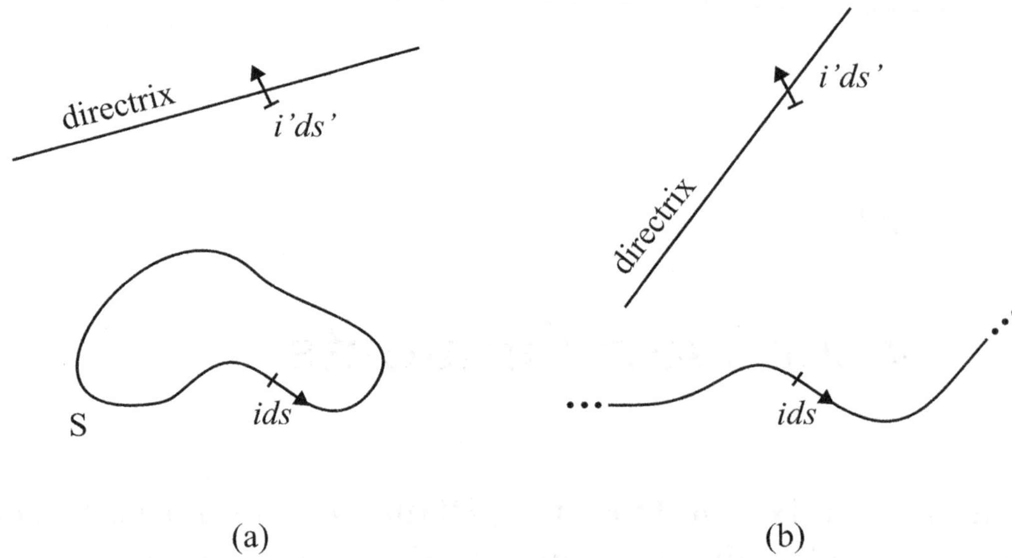

Figure 10.1: (a) Directrix due to a closed circuit of arbitrary shape passing through the center of the current element $i'ds'$ located at an arbitrary point in space. (b) Directrix passing through the center of $i'ds'$, due to a circuit of arbitrary shape extending indefinitely in both directions.

Figure 10.1 (a) presents the directrix due to the closed circuit S passing through the center of the current element $i'ds'$ located at an arbitrary point in space. Figure 10.1 (b) presents the directrix passing through the center of the current element $i'ds'$, with this directrix being due to the circuit of arbitrary shape extending indefinitely in both directions.

Ampère called ε the angle between the direction of the element $i'ds'$ and the directrix. This angle ε should not be confused with the angle ε between two arbitrary current elements ids and $i'ds'$. Figure 10.2 presents the directing plane, the directrix D, the infinitesimal length ds' of the current element $i'ds'$ with center A and the angle ε between this element and the directrix.[7]

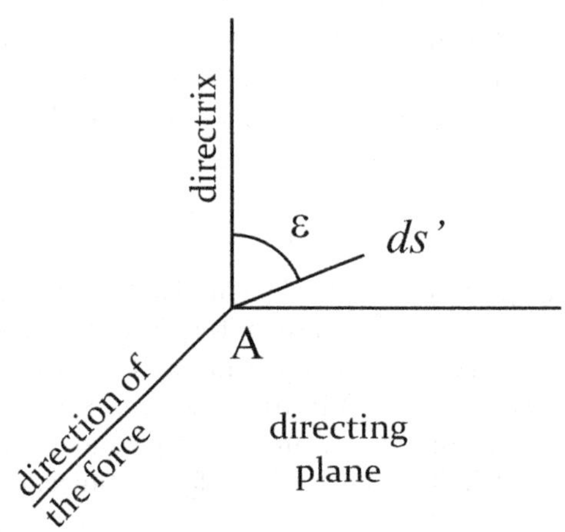

Figure 10.2: The directing plane and the directrix orthogonal to this plane, with the angle ε between the directrix and the direction of the current element $i'ds'$.

Let us give a specific example. Consider a constant current i flowing along the positive z direction through an infinite and straight conductor. The center of an external current element $i'ds'$ is located at point A. In this configuration the directing plane containing point A passes through the infinite current-carrying wire along the z axis, figure 10.3 (a). The directrix due to this wire at the location of the external current element is orthogonal to the directing plane, passing through point A. The force exerted by the infinite wire on the current element $i'ds'$ is always in the directing plane, no matter the orientation of $i'ds'$ relative to the infinite wire. It is orthogonal to the directrix and also orthogonal to $i'ds'$. Figure 10.3 (b) presents this configuration

[7] [Blondel, 1982, pp. 143-145].

with the z axis orthogonal to the plane of the paper.

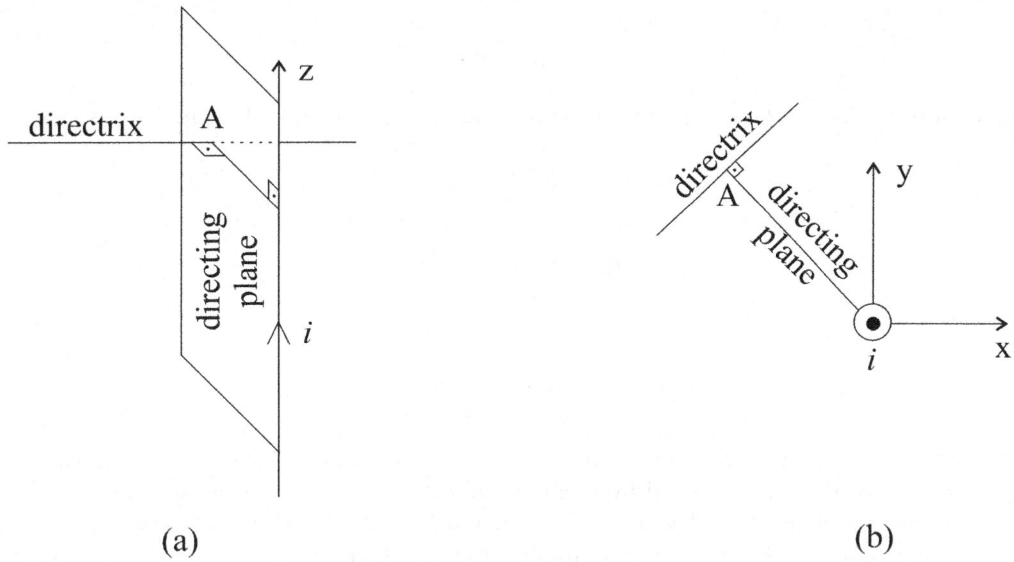

Figure 10.3: (a) Directing plane and directrix due to an infinite straight wire acting on a current element whose midpoint is at point A. (b) Configuration seen in the xy plane.

Ampère supposed the midpoint of the element ds' at the origin of a Cartesian coordinate system. He considered an element ds of the closed circuit of arbitrary form as located at point $(x,\ y,\ z)$, with the decompositions of the element ds along the three orthogonal axes being represented by dx, dy and dz. The distance between the centers of ds and ds' was represented by r. By utilizing equation (7.16) he obtained that the magnitude of the net force acting on $i'ds'$ being exerted by a closed current of arbitrary form and intensity i was given by:[8]

$$\frac{Dii'ds'}{2}\sin\varepsilon\ . \tag{10.2}$$

In this equation the magnitude D was defined by Ampère as being given by:

$$D \equiv \sqrt{A^2 + B^2 + C^2}\ , \tag{10.3}$$

where the magnitudes A, B and C were defined by, respectively:

$$A \equiv \int \frac{ydz - zdy}{r^{n+1}}\ , \tag{10.4}$$

$$B \equiv \int \frac{zdx - xdy}{r^{n+1}}\ , \tag{10.5}$$

and

$$C \equiv \int \frac{xdy - ydx}{r^{n+1}}\ . \tag{10.6}$$

In these definitions the integrals are performed along the complete closed circuit of arbitrary form which is carrying a current i and acting on the current element $i'ds'$.

Utilizing equation (10.2) it can be seen that the force vanishes when $\varepsilon = 0$ rad or when $\varepsilon = \pi$ rad, that is, when ds' is directed along the directrix. There will be a force of maximal intensity when ds' is along the directing plane, that is, when $\varepsilon = \pi/2$ rad. The force will always have this maximal value, no matter the orientation of ds' in the directing plane. The magnitude D depends only on the geometrical form of the closed circuit. It does not depend on the intensities of the currents i and i'. It also does not depend on the orientation of the element ds' relative to the directrix.

[8] [Ampère, 1824c, pp. 143 and 148].

After working with undefined values of the constants n and k in this paper, Ampère began to utilize the values he had already obtained, namely:

$$n = 2 \quad \text{and} \quad k = -\frac{1}{2} \ . \tag{10.7}$$

With these values of n and k the magnitudes A, B and C assume the following values, respectively:

$$A \equiv \int \frac{ydz - zdy}{r^3} \ , \tag{10.8}$$

$$B \equiv \int \frac{zdx - xdy}{r^3} \ , \tag{10.9}$$

and

$$C \equiv \int \frac{xdy - ydx}{r^3} \ . \tag{10.10}$$

In note A at the end of his work of 1824, the *Précis*, Ampère explored these results further. Once more his goal was to calculate the force exerted by a closed circuit of arbitrary form carrying a constant current i and acting on a current element $i'ds'$ which did not belong to the closed circuit. He represented the point $(x,\ y,\ z)$ by the Cartesian coordinates of the midpoint of an arbitrary current element of length ds belonging to the closed circuit. Analogously the Cartesian coordinates of the midpoint of element ds' were represented by $(x',\ y',\ z')$. The distance r between these two elements was then given by:

$$r = \sqrt{(x - x')^2 + (y - y')^2 + (z - z')^2} \ . \tag{10.11}$$

He represented the x, y and z components of the force exerted by this closed circuit and acting on the current element $i'ds'$ by Xds', Yds' and Zds', respectively. He then showed that these components might be written as:[9]

$$Xds' = \frac{1}{2}ii'(Cdy' - Bdz') \ , \tag{10.12}$$

$$Yds' = \frac{1}{2}ii'(Adz' - Cdx') \ , \tag{10.13}$$

and

$$Zds' = \frac{1}{2}ii'(Bdx' - Ady') \ . \tag{10.14}$$

In these equations the magnitudes dx', dy' and dz' represent the projections of the infinitesimal length ds' along the orthogonal axes x, y and z, respectively. The magnitudes A, B and C, on the other hand, were expressed in terms of a line integral along the whole closed circuit which was generating the force, namely:

$$A \equiv \int \frac{(y - y')dz - (z - z')dy}{r^{n+1}} \ , \tag{10.15}$$

$$B \equiv \int \frac{(z - z')dx - (x - x')dy}{r^{n+1}} \ , \tag{10.16}$$

and

$$C \equiv \int \frac{(x - x')dy - (y - y')dx}{r^{n+1}} \ . \tag{10.17}$$

Utilizing these equations Ampère obtained two other mathematical expressions which he wrote as:[10]

$$Xds'dx' + Yds'dy' + Zds'dz' = 0 \ , \tag{10.18}$$

[9] [Ampère, 1824e, p. 46].

[10] Due to a typographical misprint, equation (10.18) was originally published in 1824 as:

$$Xds'dx + Yds'dy + Zds'dz = 0 \ .$$

and

$$AXds' + BYds' + CZds' = 0 . \tag{10.19}$$

The meanings of equations (10.18) and (10.19) were expressed as follows by Ampère:[11]

> [...] the direction of the net action exerted by the whole system on the element ds', which forms with the three axes the angles having cosines which are proportional to, respectively, Xds', Yds' and Zds', is perpendicular to the direction of this element whose angles with the axes have their cosines proportional to dx', dy' and dz', and is perpendicular to the straight line drawn through the center of element ds' in such a way that the cosines of the angles it forms with the same axes are proportional to A, B and C.

That is, Ampère showed that the force exerted by a closed circuit of arbitrary form acting on a current element not belonging to this closed circuit is orthogonal not only to the direction of this element, but also orthogonal to the directrix passing through the midpoint of ds'. This straight line represented by the directrix was due to the closed circuit. Section 10.4 will present Ampère's experimental proof of this statement.

10.1.1 The Directrix Expressed in Vector Notation

The equations presented in Section 10.1 can be expressed in modern vector notation. In this modern format it may be easier to understand the results obtained by Ampère and their extremely important meanings.

We will here consider the particular case in which $n = 2$, as this was also the final result obtained by Ampère. The position of the center of the current element relative to the origin O of a Cartesian coordinate system can be represented by $\vec{r}_1 \equiv x\hat{\imath} + y\hat{\jmath} + z\hat{k}$, where $\hat{\imath}$, $\hat{\jmath}$ and $\hat{k}$ are the unit vectors pointing along the x, y and z axes, respectively. Analogously, the position of the midpoint of the current element $i'd\vec{s}\,'$ can be expressed by $\vec{r}_2 \equiv x'\hat{\imath} + y'\hat{\jmath} + z'\hat{k}$. Vector $\vec{r}$ pointing from ds to ds' is then given by:

$$\vec{r} \equiv \vec{r}_2 - \vec{r}_1 \equiv (x' - x)\hat{\imath} + (y' - y)\hat{\jmath} + (z' - z)\hat{k} . \tag{10.20}$$

The distance r between these two current elements is given by:

$$r \equiv |\vec{r}| \equiv |\vec{r}_2 - \vec{r}_1| = \sqrt{(x - x')^2 + (y - y')^2 + (z - z')^2} . \tag{10.21}$$

The infinitesimal vector $d\vec{s}$, of length ds and pointing in each point of the closed circuit S along the direction of the current i can be written as $d\vec{s} = dx\hat{\imath} + dy\hat{\jmath} + dz\hat{k}$. Analogously, the infinitesimal vector $d\vec{s}\,'$, of length ds' and pointing along the direction of i', can be written as $d\vec{s}\,' = dx'\hat{\imath} + dy'\hat{\jmath} + dz'\hat{k}$.

It is possible to define a "directrix vector", represented by $\vec{D}$, by the following expression:

$$\vec{D} \equiv A\hat{\imath} + B\hat{\jmath} + C\hat{k} , \tag{10.22}$$

in which its components A, B and C are defined by equations (10.15) up to (10.17). This directrix vector does not depend on the current intensities i and i'. It does not depend as well on the direction of the current element $i'd\vec{s}\,'$ which suffers the force exerted by the closed circuit S. The vector $\vec{D}$ depends only on the size and shape of the closed circuit S. The direction of the directrix vector $\vec{D}$ in each point of space coincides with Ampère's directrix, although Ampère himself never defined this concept of a directrix vector.

Supposing $n = 2$, the directrix vector $\vec{D}$ due to a closed circuit S of arbitrary form can be expressed as follows by means of a vector product:

$$\vec{D} = \oint_S \frac{d\vec{s} \times \vec{r}}{r^3} . \tag{10.23}$$

The force $d\vec{F}$ exerted by the closed circuit S and acting on a current element $i'd\vec{s}\,'$, given by equations (10.12) up to (10.14), can then be written as:

$$d\vec{F} = \frac{ii'}{2} d\vec{s}\,' \times \vec{D} = \frac{ii'}{2}(dx'\hat{\imath} + dy'\hat{\jmath} + dz'\hat{k}) \times (A\hat{\imath} + B\hat{\jmath} + C\hat{k}) = i'd\vec{s}\,' \times \left(\frac{i}{2}\oint_S \frac{d\vec{s} \times \vec{r}}{r^3}\right) . \tag{10.24}$$

The representation of equations (10.18) and (10.19) in vector notation is then given by, respectively:

[11][Ampère, 1824e, pp. 46-47].

$$d\vec{F} \cdot d\vec{s}\,' = 0 \ , \tag{10.25}$$

and

$$d\vec{F} \cdot \vec{D} = 0 \ . \tag{10.26}$$

Therefore the force $d\vec{F}$ exerted by a current-carrying closed circuit of arbitrary form and acting on a current element $i'd\vec{s}\,'$ is then orthogonal not only to this current element, but also to the directrix vector $\vec{D}$.

Equation (2.21) relates the current in the electrodynamic system of units utilized by Ampère with the current expressed in the International System of Units, namely:

$$ii' \Leftrightarrow \frac{\mu_0}{2\pi} II' \ . \tag{10.27}$$

Equation (10.24) can then be expressed in the International System of Units by means of equation (10.27), yielding:

$$d\vec{F} = I'd\vec{s}\,' \times \left(\frac{\mu_0}{4\pi} \oint_S \frac{Id\vec{s} \times \vec{r}}{r^3} \right) \ . \tag{10.28}$$

In this equation $\vec{r}$ represents the vector pointing from the current element $Id\vec{s}$ towards the current element $I'd\vec{s}\,'$.

Equation (10.28) is only obtained from Ampère's force between current elements after integrating it along a *closed* circuit S of arbitrary form, as represented in figure 10.1 (a), or when integrating it along a circuit S of arbitrary form extending indefinitely in both directions, that is, when both extremities of the circuit S are located at infinite distances to the element $I'd\vec{s}\,'$, as represented in figure 10.1 (b).

Combining equation (10.28) with the directrix vector $\vec{D}$ given by equation (10.23), Ampère's force exerted by a closed circuit S of arbitrary form acting on a current element $I'd\vec{s}\,'$ which does not belong to S can then be written as follows in the International System of Units:

$$d\vec{F} = I'd\vec{s}\,' \times \left(\frac{\mu_0}{4\pi} I\vec{D} \right) \ . \tag{10.29}$$

Defining by ε the angle between $I'd\vec{s}\,'$ and the directrix vector $\vec{D}$, the magnitude $dF \equiv |d\vec{F}|$ of the force acting on this current element can then be written in terms of the magnitude $D \equiv |\vec{D}|$ of the directrix as follows:

$$dF = I'ds' \left(\frac{\mu_0}{4\pi} ID \right) \sin \varepsilon \ . \tag{10.30}$$

10.1.2 Relating the Directrix with the Magnetic Field

Ampère never worked with the concept of a "magnetic field". However, nowadays all textbooks on electromagnetism utilize this concept. It is possible to express Ampère's integrated results given by the equations of Subsection 10.1.1 in terms of this magnetic field. In this way it may be easier to understand the origin of the formulas appearing in the textbooks.

In modern textbooks the magnetic field $d\vec{B}$ at a point $\vec{r}_2 \equiv x'\hat{\imath} + y'\hat{\jmath} + z'\hat{k}$ due to a current element $Id\vec{s}$ located at point $\vec{r}_1 \equiv x\hat{\imath} + y\hat{\jmath} + z\hat{k}$ is defined by the following expression:[12]

$$d\vec{B} \equiv \frac{\mu_0}{4\pi} \frac{Id\vec{s} \times \vec{r}}{r^3} \ . \tag{10.31}$$

In this equation the vector $\vec{r}$ given by equation (10.20) points from $\vec{r}_1$ to $\vec{r}_2$, that is, from the midpoint of element 1 to the midpoint of element 2. Moreover, $r = |\vec{r}|$ represents the distance between these two points. Some authors call this equation the law of Biot-Savart,[13] although Biot and Savart never mentioned in any of their works a "magnetic field" generated by a current element.

The magnetic field $\vec{B}$ due to a closed circuit S of arbitrary form is obtained by the integration of equation (10.31) along S, namely:

[12] [Tricker, 1962, pp. 453-455], [Tricker, 1965, p. 25], [Whittaker, 1973, pp. 82-83], [Purcell, 1980, p. 187, equation (38)], [Reitz et al., 1982, p. 167, equation (8-27)] and [Griffiths, 1989, p. 222].

[13] For references see [Chaib and Assis, 2007b].

$$\vec{B} \equiv \frac{\mu_0}{4\pi} \oint_S \frac{Id\vec{s} \times \vec{r}}{r^3} \ . \tag{10.32}$$

Combining equations (10.23) and (10.32) one then obtains:

$$\vec{B} = \frac{\mu_0 I}{4\pi} \vec{D} \ . \tag{10.33}$$

Let $D \equiv |\vec{D}|$ be the magnitude of the directrix vector $\vec{D}$. In this case the factor $\mu_0 ID/4\pi$ will be equal to the magnitude of the magnetic field $B \equiv |\vec{B}|$ generated by the closed circuit S and acting at the location of the current element $I'ds'$, namely:

$$B = \frac{\mu_0 ID}{4\pi} \ . \tag{10.34}$$

Therefore, the direction of the straight line represented by Ampère's directrix coincides not only with the directrix vector given by equation (10.23), but also with the direction of the magnetic field of the modern textbooks which is due to a closed circuit S when this magnetic field is calculated at the location of the element ds', as can be seen by equation (10.33).

The combination of equations (10.28) and (10.32) yields Ampère's force exerted by a closed current-carrying wire S of arbitrary form acting on a current element $I'd\vec{s}\,'$ located at point $\vec{r}_2 \equiv x'\hat{\imath} + y'\hat{\jmath} + z'\hat{k}$ as expressed in terms of a magnetic field, namely:

$$d\vec{F} = I'd\vec{s}\,' \times \vec{B} \ . \tag{10.35}$$

Let ε represent the angle between vectors $d\vec{s}\,'$ and $\vec{B}$. The magnitude of the force, $dF = |d\vec{F}|$, can then be expressed by the magnitude of the magnetic field, $B = |\vec{B}|$, by the following expression:

$$dF = I'ds'B \sin \varepsilon \ . \tag{10.36}$$

Equations (10.35) and (10.36) represent the modern expressions of the supposed force exerted by a magnetic field on a current element.

Ampère's original results of 1824, on the other hand, can be expressed as equations (10.12) up to (10.14). They can also be expressed in vector notation as equation (10.24). It should be emphasized that according to Ampère's original conceptions, this force resulted from a direct action between the closed conductor S carrying a current I and the current element $I'ds'$. That is, according to Ampère this direct net force was not mediated through a "magnetic field". According to Ampère, the net force was the result of a direct action at a distance between current elements.

As we have seen, given the location of a closed circuit carrying a steady current and a current element of another circuit located at an external point, Ampère's directrix will be a certain straight line passing through the center of the current element. The direction of this directrix coincides with the direction of the directrix vector passing through this point and also with the direction of the magnetic field vector passing through this point due to the closed current-carrying circuit. Despite the similarities of these concepts, another important difference between them should be emphasized, beyond that difference mentioned in the previous paragraph. Ampère's original directrix represents a simple straight line. It is not a directed straight line. The directrix vector and the magnetic field vector, on the other hand, represent directed magnitudes. They are obtained by the right-hand rule, as indicated by equations (10.23) and (10.32). Consider, for instance, an infinite straight conductor carrying a constant current i along the positive z direction and an external point A, as in figure 10.3. The directing plane containing point A passes through the z axis. The directrix due to the infinite wire will be a straight line orthogonal to the directing plane, figure 10.4 (a). The magnetic field vector, on the other hand, will be an orientated line segment directed according to the right-hand rule, figure 10.4 (b).

10.2 The Introduction of the Electrodynamic Solenoid

In a work presented to the Academy of Sciences of Paris on 22 December 1822, Ampère introduced the word "solenoid" and the concept of an electrodynamic solenoid. In particular,[14] he deduced the value of the force

[14][Ampère, 1824c, p. 135]. See also [Ampère, 1885f, p. 399].

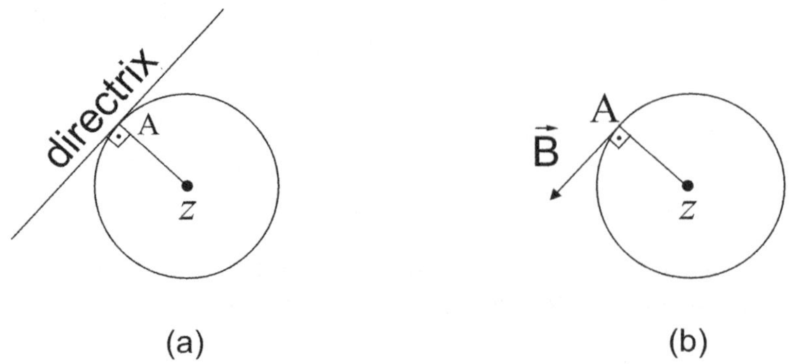

Figure 10.4: (a) Directrix due to an infinite straight wire carrying a current i along the positive z direction, orthogonal to the plane of the paper. (b) Magnetic field $\vec{B}$ due to this current-carrying wire.

> [...] between an element [that is, between a current element] and a system of circular currents having a very small diameter [compared with the distance between the midpoint of the current element and the center of this circular loop], whose planes were everywhere perpendicular to a line straight or curved passing through the centers of the circumferences followed by the currents. I believe I should call by the name *solenoid* this kind of system, whose shape is that of a surface usually called a *surface channel*, obtained from the Greek word σωληνοειδής, derived from σωλήν, channel, meaning precisely that it has the shape of a channel.

Figure 10.5 (a) represents what Ampère called a *definite solenoid*.[15] That is, a system of N circular current-carrying loops between points A and B, which are the extremities of the solenoid. Point A should be considered at the center of the left extremity of the solenoid, while point B should be considered at the center of its right extremity. The planes of these circular loops are orthogonal to a curve of arbitrary shape passing through the centers of these loops, with the centers of the loops equally spaced along this curve. There is a constant current i flowing in each one of these loops, always in the same sense. Figures 10.5 (b), (c) and (d) represent, respectively, what Ampère called a *closed solenoid*, an *indefinite solenoid in both directions* and a *simply indefinite solenoid*.[16] This last solenoid has an extremity A, while the other extremity is located at infinity. Sometimes Ampère called it an *indefinite solenoid*.

Ampère also defined the *directrix of a solenoid* as "a line straight or curved passing through the centers of all circular currents composing it".[17] This *directrix* should not be confused with the *normal to the directing plane* discussed in Section 10.1, which was also called *directrix* in the *Théorie*. The *directrix of the electrodynamic cylinders* represented in figure 9.2, for instance, is the straight line passing through the axes of the cylinders. Let us consider now an electrodynamic ring, like those represented in figure 9.4. If each small circular loop of these solenoids has a radius r and the centers of these circular loops follow a circle of radius $R > r$, then the *directrix of these electrodynamic rings* will be the circumference of length $2\pi R$. In the case of figure 10.5 the directrices of the solenoids always pass through the center of the corresponding small circular current-carrying loops and are given by: (a) an open and curved line of finite length connecting points A and B, (b) a closed and curved line, (c) a curved line extending to infinity in both directions, and (d) a curve line beginning in point A and extending to infinity at the other extremity.

In 1824 Ampère considered all small current-carrying loops of a solenoid to be circular with the same radius m and area πm^2. In the *Théorie* he generalized this result and considered the loops forming a solenoid as being small circuits of arbitrary form having an area λ.[18]

10.2.1 Interaction between a Solenoid and a Current Element

In his works of 1824, Ampère integrated his force law in order to obtain the force on a current element exerted by a simply indefinite solenoid having its extremity at point L'. The midpoint of the current element was placed at point A. He showed that the directrix at point A, due to this simply indefinite solenoid, is a straight line connecting this point A to the extremity L' of the solenoid:[19]

[15] [Ampère, 1824c, p. 156] and [Ampère, 1885f, p. 399].
[16] [Ampère, 1824c, pp. 154 and 156] and [Ampère, 1885f, p. 399].
[17] [Ampère, 1824c, pp. 151-152].
[18] [Ampère, 1826f, pp. 94-99], [Ampère, 1823c, Ampère, 1990, pp. 266-271] and [Ampère, 1965b, pp. 181-186].
[19] [Ampère, 1824f, p. 246] and [Ampère, 1824e, p. 31].

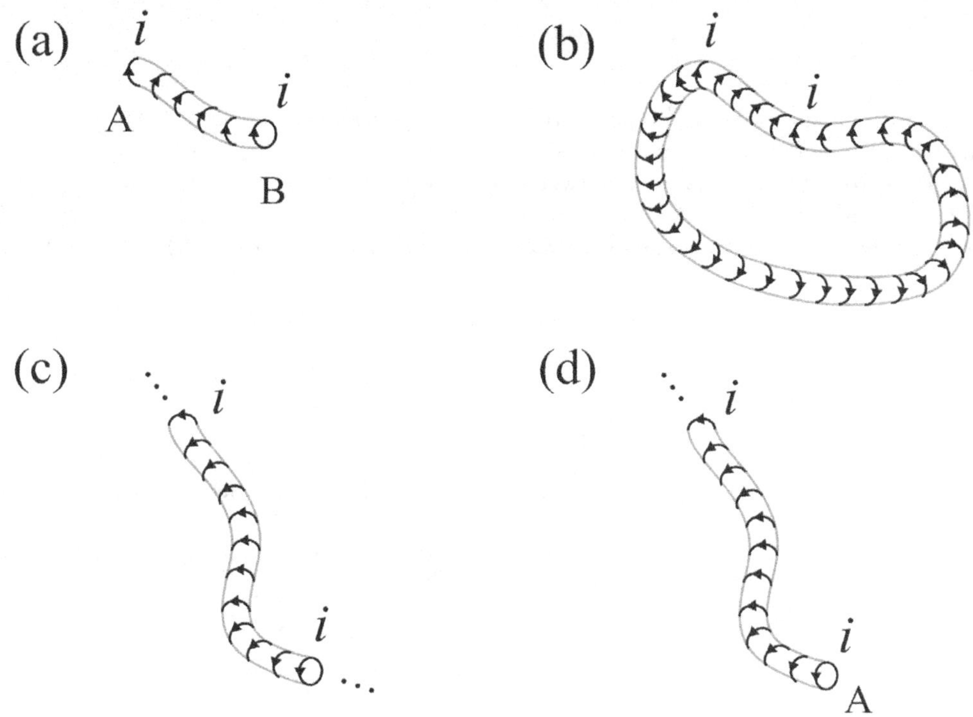

Figure 10.5: Ampère's electrodynamic solenoids: (a) Definite, (b) closed, (c) indefinite in both directions, and (d) simply indefinite.

> When the system of closed circuits just considered is itself an indefinite solenoid, the normal to the directing plane passing through point A is, as we have just seen, the straight line connecting this point A to the extremity of the solenoid.

This directrix can be seen in figure 10.6.

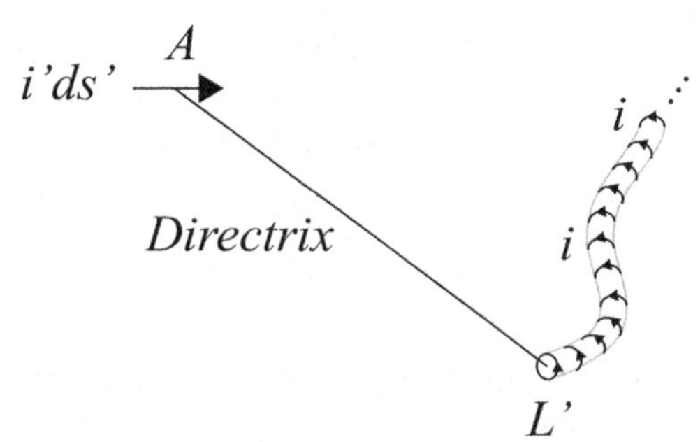

Figure 10.6: Directrix due to an indefinite solenoid calculated at the midpoint A of a current element $i'ds'$.

He then calculated the direction of the force exerted by this indefinite solenoid acting on the current element $i'ds'$ located at point A and obtained the following result:[20]

> [...] the action [force] of an indefinite solenoid on a [current] element is perpendicular to the straight line connecting the center of this element to the extremity of the solenoid and, as it should also be perpendicular to the element, it will be [perpendicular] to the plane formed by this element and by [the straight line connecting its center to] the extremity of the solenoid.

He found the magnitude of this force to be given by the following expression:

[20][Ampère, 1824c, p. 155] and [Ampère, 1824e, p. 24].

$$\frac{\pi m^2 ii' ds' \sin\varepsilon'}{2g\ell'^2} . \tag{10.37}$$

In this equation the magnitude πm^2 represented the area of each circular loop of the solenoid of radius m carrying a current i, the magnitude g represented the distance between the centers of two consecutive loops of the solenoid, ℓ' represented the distance between the midpoint of the current element and the extremity of the solenoid, while ε' represented the angle between ds' and the normal to the directing plane passing through it (that is, the angle between the current element $i'ds'$ and the directrix), as shown in figure 10.7.

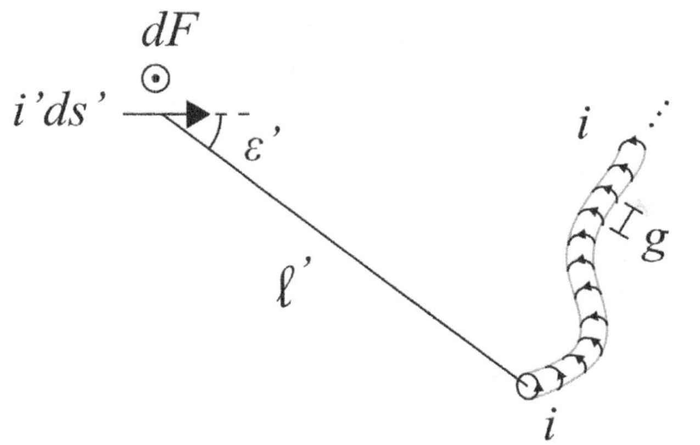

Figure 10.7: Force dF acting on a current element $i'ds'$ and being due to an indefinite solenoid.

With point A representing the midpoint of the current element ab which was under the action of the indefinite solenoid, Ampère expressed himself as follows:[21]

> It is then seen that the action of an indefinite solenoid with extremity located at L' exerted on element ab is normal at A to the plane bAL',[22] is proportional to the sine of the angle bAL' and in inverse ratio of the square of the distance AL', and [it is seen] that it remains always the same no matter the form and direction of the indefinite curve $L'L''O$[23] on which the centers of all circular currents composing the indefinite solenoid are located.

As we saw in Section 6.2, Biot and Savart had expressed the force exerted by a current element acting on a supposed magnetic pole by equations (6.4) and (6.5). Representing the intensity of the magnetic pole by p, the current element by $i'ds'$, their distance by ℓ', and the angle between the element and the straight line connecting it to the magnetic pole by symbol ε', then the force proposed by Biot and Savart is found to be proportional to the following magnitude:

$$\frac{pi'ds' \sin\varepsilon'}{\ell'^2} . \tag{10.38}$$

There will be an equivalence between equations (10.37) and (10.38) when there is a proportionality between the intensity p of the magnetic pole and the magnitude $\pi m^2 i/(2g)$ of the simply indefinite solenoid. Let us represent this proportionality by the symbol $\propto$:

$$\frac{\pi m^2 i}{2g} \propto p . \tag{10.39}$$

If the solenoid has an area of cross section λ of arbitrary form, instead of being a circle of area πm^2, then this proportionality will be represented by:

$$\frac{\lambda i}{2g} \propto p . \tag{10.40}$$

---

[21] [Ampère, 1824c, pp. 155-156] and [Ampère, 1824e, pp. 24-25], our words in the footnote.

[22] That is, the force is orthogonal to the plane formed by the element and by the directrix connecting its midpoint A to the extremity L' of the solenoid

[23] The curve of arbitrary shape, beginning at point L', goes to infinity.

Ampère also calculated the interaction between a definite solenoid of finite length and an external current element, expressing his conclusions as follows:[24]

> When it is desired to go from this case[25] to the case of a definite solenoid having its two extremities located at points L' and L'', then it will be enough to suppose a second indefinite solenoid beginning at point L'' of the first [solenoid], coinciding with it from this point up to infinity, having currents of the same intensity, but directed in opposite senses, [so that] the action of this last [solenoid] will have an opposite sign of the action of the first indefinite solenoid beginning at point L', destroying it over the whole portion $L''O$ on which they are superimposed. Therefore, the action [force] of the solenoid $L'L''$ will be the same action which would be exerted by the combination of these two indefinite solenoids and will be composed, consequently, by the force just calculated[26] and by another force acting in opposite sense, passing by the same point A perpendicularly to the plane bAL'', expressed by:
>
> $$\frac{\pi m^2 ii' ds' \sin \varepsilon''}{2g\ell''^2} , \tag{10.41}$$
>
> where ε'' is the angle bAL'' while ℓ'' is the distance AL''.

Finally, Ampère showed that a closed solenoid of arbitrary form, like that of figure 10.5 (b), exerts no net force on an external current element. He also showed that there is no net force on a current element exerted by an indefinite solenoid in both directions of arbitrary form, like that of figure 10.5 (c).[27]

With these works Ampère succeeded in explaining electrodynamically the experiments described in Section 9.1 of Gay-Lussac, Welter, Davy, Erman and De la Borne.

10.2.2 Interaction between a Solenoid and a Closed Circuit of Arbitrary Form

By considering the action and reaction of the calculation presented in Subsection 10.2.1, Ampère also obtained the force exerted by the current element acting on the indefinite solenoid. He then integrated expression (10.37) and obtained the force that a closed circuit of arbitrary form exerted on an indefinite solenoid located at an arbitrary point in space, as in figure 10.8 (a). He also obtained the force that a circuit of arbitrary form, extending in both directions to infinity, exerts on an indefinite solenoid, as in figure 10.8 (b). In this last configuration the extremities of the circuit of arbitrary form extending in both directions are at infinite distances to the extremity of the indefinite solenoid.

In both cases he showed that these forces might be considered as being applied at the extremity A of the indefinite solenoid, pointing along the directrix at this point.[28] This force might be due to the closed circuit, as in figure 10.8 (a). This force might also be due to the circuit extending to infinity in both directions, as in figure 10.8 (b).

Figure 10.8 (a) presents the directrix due to the closed circuit S. It passes through the extremity of the indefinite solenoid located at an arbitrary point in space. The force exerted by the closed circuit S on the indefinite solenoid is directed along this directrix. The sense of this force along the directrix depends on the sense of the current on the closed circuit and also on the sense of the current flowing on the surface of the solenoid. Figure 10.8 (b) presents the directrix due to the circuit of arbitrary shape extending indefinitely in both directions. The force exerted by the circuit of arbitrary shape extending indefinitely in both directions acting on the indefinite solenoid is directed along this directrix.

Ampère showed that the intensity of this force in both cases was given by:

$$\frac{\pi m^2 ii' D}{2g} . \tag{10.42}$$

Here $D = \sqrt{A^2 + B^2 + C^2}$ represents the intensity of the directrix due to the closed circuit (or due to the circuit extending to infinity in both directions) and acting on the extremity of the solenoid, with the constants A, B and C being given by equations (10.8) to (10.10) integrated along the closed circuit.

This force changes sign when only the sense of the current in the closed circuit is reversed, or when only the sense of the current in all loops of the solenoid is reversed. When the sense of the current in the closed

[24] [Ampère, 1824c, p. 156] and [Ampère, 1824e, p. 25], our words in the footnotes.

[25] That is, from the force exerted by a simply indefinite solenoid on a current element.

[26] The intensity of this force is given by equation (10.37).

[27] [Ampère, 1824c, pp. 153-154], [Ampère, 1824e, pp. 22-23], [Ampère, 1826f, p. 97], [Ampère, 1823c, Ampère, 1990, p. 269] and [Ampère, 1965b, pp. 183-184].

[28] [Ampère, 1824c, p. 161] and [Ampère, 1824e, p. 30].

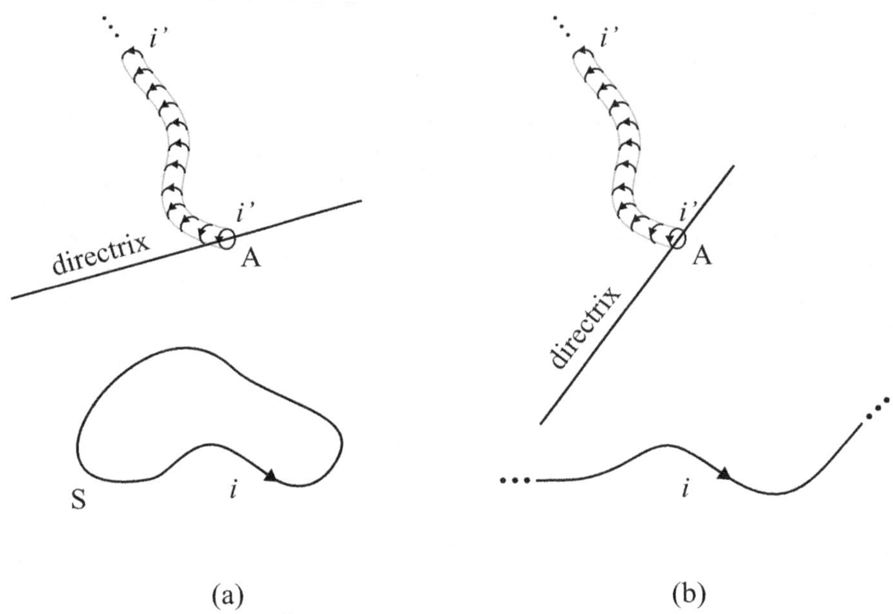

Figure 10.8: (a) Directrix due to a closed circuit of arbitrary shape acting passing through the extremity of an indefinite solenoid located at an arbitrary point in space. (b) Directrix due to a circuit of arbitrary shape extending indefinitely in both directions passing through the extremity of an indefinite solenoid.

circuit is reversed simultaneously with the sense of the current in all loops of the solenoid, then the net force between them does not change its sign.

This result obtained by Ampère can be expressed nowadays in terms of the magnetic field $\vec{B}(\vec{r})$ due to a closed circuit and acting at a magnetic pole of intensity P located at point $\vec{r}$. This magnetic pole might be considered as being equivalent to Ampère's simply indefinite solenoid, that is, as if the magnetic pole were located at the extremity of the solenoid. If P represents a North pole, the force acting on it will be along the direction of the magnetic field $\vec{B}$. If, on the other hand, P represents a South pole, the force acting on it will be opposite to the direction of $\vec{B}$. Mathematically the force $\vec{F}$ acting on the pole would be then given by:

$$\vec{F} = P\vec{B} \ . \tag{10.43}$$

Here P represents the intensity of the magnetic pole and $\vec{B}$ represents the magnitude and direction of the magnetic field generated by the closed circuit at the location of the magnetic pole. A North pole is considered positive and a South pole negative.

10.2.3 Interaction between Two Simply Indefinite Solenoids

In 1824 Ampère considered also the interaction between two simply indefinite solenoids.[29] Let i be the current intensity in each loop of the first solenoid, g the distance between the centers of two consecutive loops, each one of area λ. In the particular case in which the solenoids have a circular cross section of radius m, then the area λ is given by $\lambda = \pi m^2$. In this Subsection we will only represent the generic expressions valid for solenoids with cross sections of arbitrary forms, as this was the final treatment given by Ampère in the *Théorie*. The equivalent magnitudes of the second simply indefinite solenoid are represented by i', g' and λ', respectively. Let L_1 and L' be the centers of the extremities of these solenoids, as in figure 10.9 (a).

Ampère showed that the force between these two solenoids acted along the straight line connecting their extremities. After all, the directrix due to the first solenoid at an arbitrary point A is along the straight line connecting this point to the extremity L_1 of this indefinite solenoid. The force exerted by a closed circuit, or by a system of closed circuits, acting on a second solenoid is along the directrix of this system, passing through the extremity L' of this second indefinite solenoid. Therefore, the force exerted by the first indefinite solenoid acting on the second indefinite solenoid will be along the straight line L_1L' connecting their extremities, as expressed by Ampère in the following words:[30]

[29] [Ampère, 1824f] and [Ampère, 1824e].
[30] [Ampère, 1824f, p. 246] and [Ampère, 1824e, p. 31].

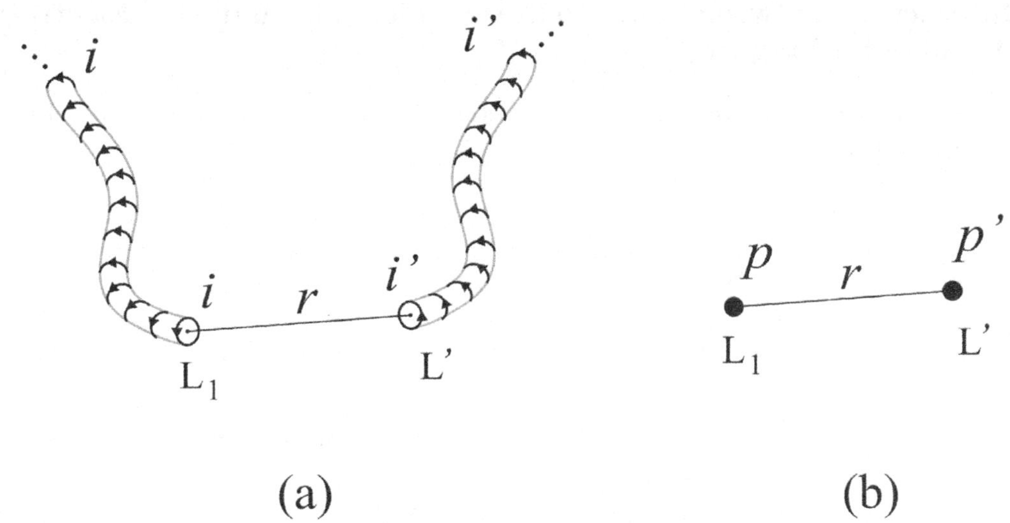

Figure 10.9: (a) Two simply indefinite solenoids with the centers of the extremities located at points L_1 and L', separated by a distance r. (b) Magnetic poles p and p', separated by a distance r and located at points L_1 and L'.

> The mutual action between two [simply] indefinite solenoids happens along the straight line connecting the extremity of one solenoid with the extremity of the other solenoid.

Ampère found the intensity of this force as given by:[31]

$$-\frac{\lambda\lambda' ii'}{2gg'r^2} \, . \tag{10.44}$$

Here r is the distance between the centers of the extremities L_1 and L' of both solenoids. This force acts along the straight line connecting the extremities L_1 and L'. That is, it behaves as a central force acting between the extremities of these simply indefinite solenoids.

This expression is analogous to the interaction between two magnetic poles obtained by Coulomb and represented by equation (1.3). In figure 10.9 (b) we replaced both indefinite solenoids by two magnetic poles of intensities p and p', located at points L_1 and L', and separated by a distance r. The force between these two poles is proportional to the product of the intensities p and p', varying as the inverse square of their distance r, pointing along the straight line connecting them, namely:

$$-\frac{pp'}{r^2} \, . \tag{10.45}$$

Comparing equations (10.44) and (10.45), it might be said that a simply indefinite electrodynamic solenoid is equivalent to a magnetic pole. The force between two indefinite solenoids is equivalent to the force between two magnetic poles provided one makes the following substitution:

$$\frac{\lambda\lambda' ii'}{2gg'} \Leftrightarrow pp' \, . \tag{10.46}$$

Utilizing equations (2.14) and (2.21), one observes that equation (10.46) can be written as follows in the International System of Units:

$$\frac{\lambda\lambda' II'}{gg'} \Leftrightarrow PP' \, . \tag{10.47}$$

[31] [Ampère, 1824c, pp. 135 and 162], [Ampère, 1824f, p. 246], [Ampère, 1824e, pp. 4, 31-32], [Ampère, 1826f, p. 105], [Ampère, 1823c, Ampère, 1990, p. 277] and [Ampère, 1965b, p. 191].

10.2.4 Interaction between Two Definite Solenoids and the Electrodynamic Analog of a Magnet

In his work of 1824 Ampère also considered[32] the interaction between two definite solenoids with extremities L_1, L_2, L' and L'', as represented in figure 10.10.

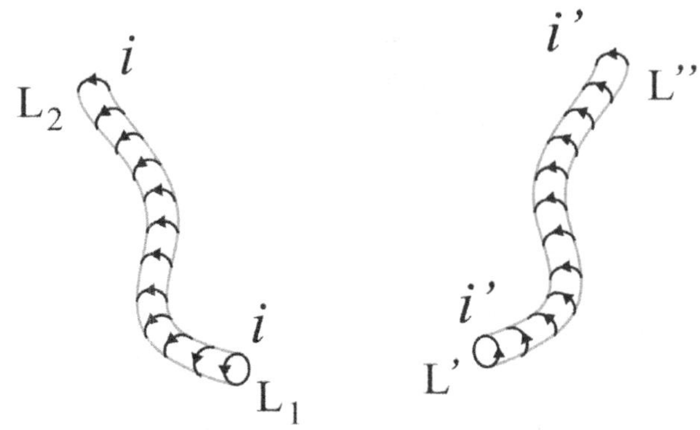

Figure 10.10: Two definite solenoids with the centers of their extremities located at points L_1, L_2, L' and L''.

Ampère showed that the interaction between solenoid L_1L_2 and solenoid $L'L''$ can be expressed as a sum of four components directed along the straight lines L_1L', L_2L', L_1L'' and L_2L''. Moreover, each one of these components varied as the inverse square of the distance between the corresponding extremities in agreement with equation (10.44).

Consider now two long, thin and curved magnets like those of figure 10.11. When they are uniformly magnetized along their axes, their magnetic poles can be considered as being located at their extremities.

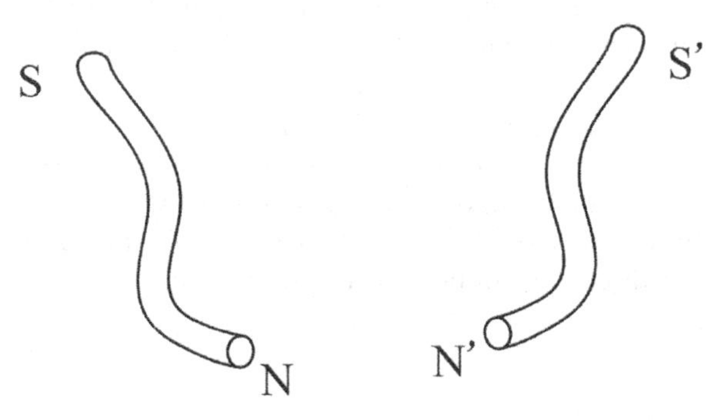

Figure 10.11: Two long, thin and curved magnets. Their magnetic poles are N, S, N' and S'.

The force between these two magnets can also be expressed as a sum of four components, namely, the interaction between the pole N of one magnet with the poles N' and S' of the other magnet, together with the interaction between the pole S of a magnet with the poles N' and S' of the other magnet. Each one of these interactions is given by equation (10.45).

Ampère then concluded that a definite electrodynamic solenoid is equivalent to a long, thin and curved magnet, provided the magnet and solenoid have the same shape, with the North and South poles of the magnet located at the extremities of the solenoid. The force between two definite solenoids is equivalent to the force between two thin, curved magnets, uniformly magnetized along their axes, provided the magnets have the same sizes and shapes as the corresponding definite solenoids, being positioned at the same locations. This equivalence can be obtained mathematically by the substitution given by equation (10.46).

[32] [Ampère, 1824f] and [Ampère, 1824e].

10.3 The Contributions of Poisson

Siméon Denis Poisson (1781-1840), figure 10.12, presented a mathematical formulation for magnetism at the meetings of the Academy of Sciences of Paris of 2 February and 27 December 1824. These works were only published in 1826.[33]

Figure 10.12: S. D. Poisson.

He supposed the existence of two magnetic fluids, austral and boreal. Nowadays they are called North and South fluids, or North and South poles, respectively. He assumed also the validity of Coulomb's force law between these fluids. That is, the force, pointing centrally along the direction connecting the fluids and following the principle of action and reaction, was proportional to the product of the intensities of these fluids and varied as the inverse square of their distance, as given by equation (1.3). He also supposed that in magnetized substances these fluids remained confined in the interior of small "magnetic elements" by some unknown force. Each one of these magnetic elements would be equivalent to a small magnetic dipole with the North and South poles of the same intensity separated by a small distance. In 1825 Ampère also utilized the expression "magnetic element" to refer to a system which is called nowadays a magnetic dipole.[34]

Poisson calculated the force exerted by one of these magnetic dipoles on a particle of magnetic fluid (that is, the force exerted on another pole, be it a North pole or a South pole). He supposed this external particle located at a point M with Cartesian coordinates $(x,\ y,\ z)$. An arbitrary point C, inside the magnetic element, had coordinates $(x',\ y',\ z')$. He represented by h^3 the volume of the dipole, ε being the distance between the austral and boreal fluids of the element. The distance between points M and C was represented by ρ. He also defined a magnitude δ, proportional to the distance between the centers of mass of the two magnetic fluids inside the element. Nowadays this magnitude is called the magnetic moment per unit volume. The direction or "axis" of δ is the straight line connecting the centers of mass of the dipole. He also represented the angles between δ and the Cartesian axes x, y and z by the letters a, b and c, while l, l' and l'' represented the angles between CM and the axes x, y and z, respectively. Supposing $\rho \gg \varepsilon$, he obtained the following result expressing the three components $(\lambda,\ \lambda',\ \lambda'')$ of the force per unit of magnetic fluid exerted by the dipole on the external magnetic particle:[35]

$$\lambda = \frac{h^3\delta}{\rho^3}(\cos a - 3\cos i \cos l) \ , \tag{10.48}$$

$$\lambda' = \frac{h^3\delta}{\rho^3}(\cos b - 3\cos i \cos l') \ , \tag{10.49}$$

and

$$\lambda'' = \frac{h^3\delta}{\rho^3}(\cos c - 3\cos i \cos l'') \ . \tag{10.50}$$

Utilizing vector notation, these equations take the following form in the International System of Units:

[33] [Poisson, 1822a] and [Poisson, 1822b].

[34] [Ampère, 1826b, p. 42] and [Ampère, 1887f, p. 211].

[35] [Poisson, 1822a, p. 268].

$$-\frac{\mu_0}{4\pi}\frac{1}{\rho^3}[\vec{m} - 3(\hat{\rho}\cdot\vec{m})\hat{\rho}] \ . \tag{10.51}$$

Here $\hat{\rho}$ is the unit vector connecting the center of the dipole to the external magnetic particle, the distance between the center of the dipole and the magnetic particle is ρ, while $\vec{m}$ represents the magnetic moment of the element. When the dipole is composed of a North pole P and a South pole $-P$ separated by a small distance ℓ, then vector $\vec{m}$ points from the South to the North pole, with its magnitude given by $m = |\vec{m}| = P\ell$.

In his paper, Poisson considered a magnetic shell or magnetic surface, that is, a magnetic dipole layer, figure 10.13. It was a sheet containing a uniform distribution of austral poles on one side and a uniform distribution of boreal poles of the same intensity on the other side, like a sandwich. The North-South axis of each dipole was considered orthogonal to the surface in each point. Some unknown force kept these two fluids separated from one another by a small distance. He then integrated the force this magnetic dipole layer exerted on a magnetic particle external to the sheet. He also calculated the interaction between two magnetic dipole layers.

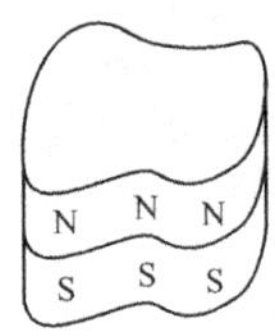

Figure 10.13: A magnetic dipole layer.

As will be seen in Section 10.7, Ampère obtained results analogous to the formulas obtained by Poisson at the end of 1825, although Ampère considered only the interaction between closed circuits carrying steady currents.

10.4 The Case of Equilibrium of the Nonexistence of Tangential Force

We now consider the first remarkable result obtained by Ampère in June 1822, by utilizing his force law between current elements in its final form with $k = -1/2$, equation (7.36), as discussed in Section 7.6. The first "remarkable result" is that, after integrating this expression in order to obtain the net force exerted by a closed circuit of arbitrary form acting on an external current element, he obtained a null component of this force acting along the direction of the test element:[36]

> [...] the integral[37] will be, necessarily, zero, from which it follows that the resultant of all the actions exerted by a closed circuit on a small portion of a conductor is always perpendicular to the direction of this small portion. As regards this aspect I note that the same[38] should take place for any ensemble of closed circuits and, consequently, for a magnet, when it is considered as such,[39] in conformity with my opinion as regards the causes of the magnetic phenomena and, as a matter of fact, this is what results from several experiments due to several physicists.

This first remarkable result that Ampère predicted from his final formula between current elements suggested to him a new case of equilibrium. It will be called here the *case of equilibrium of the nonexistence of tangential force*. With this new experiment Ampère obtained the same relation between his constants n and k which he had already obtained, as given by equations (7.34) and (7.35). Moreover, with this new case of equilibrium he confirmed experimentally his theoretical prediction about the nonexistence of a tangential force acting along the direction of a current element and being due to a closed circuit of arbitrary form. He made this prediction in June 1822, being mentioned in his crucial paper published in the same year.[40] Despite this paper, the real experiment with which he confirmed this prediction was first mentioned in a letter to

[36][Ampère, 1822o, pp. 419-420], our words in the footnotes.

[37]Integral of the component of the force acting along the direction of the current element and being due to a closed circuit of arbitrary form.

[38]That is, the same null value for the net force along the direction of the element.

[39]That is, when a magnet is considered as being composed of an ensemble of small closed circuits.

[40][Ampère, 1822o].

Gherardi, dated 16 August 1825.[41] The detailed description of the case of equilibrium of the nonexistence of tangential force and figure 10.14 describing the experiment appeared in two papers published in 1825.[42]

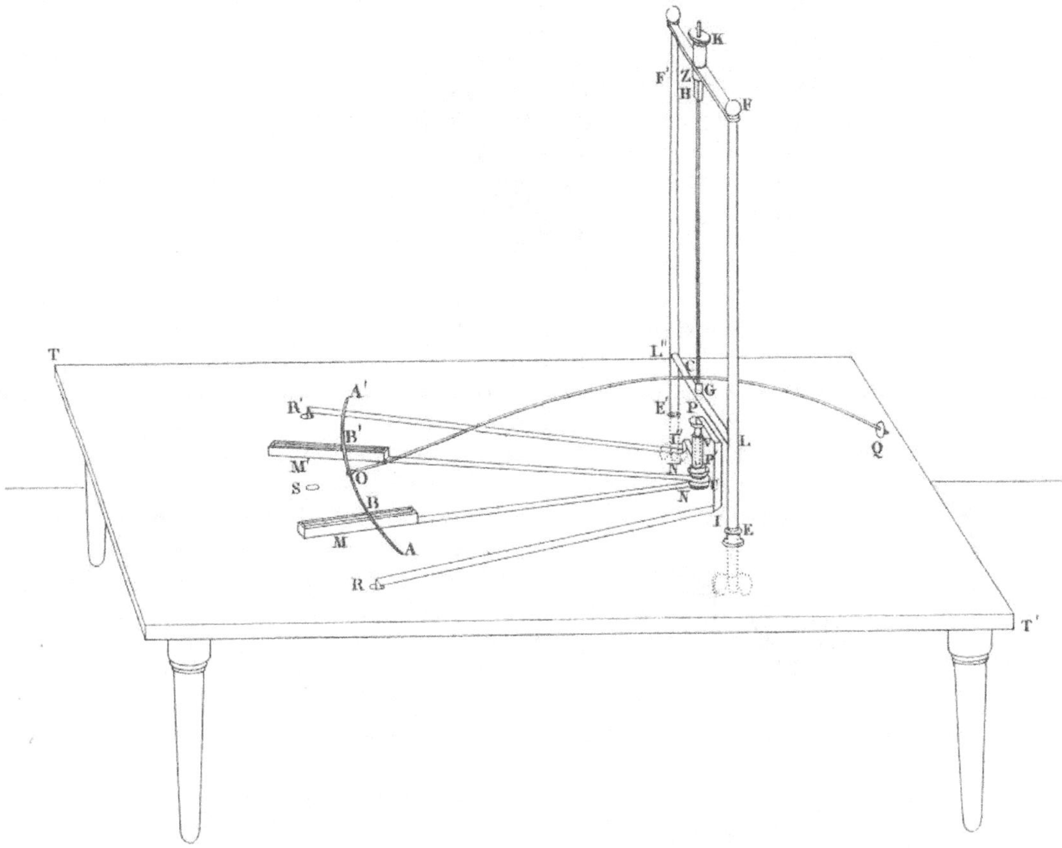

Figure 10.14: Case of equilibrium of the nonexistence of tangential force.

An arc of a horizontal circle AA' floats at points B and B' on two troughs M and M' full of mercury, balanced by a counterpoise Q connected to a support OGQ. The arc AA' is free to turn around the vertical axis GH. There is a hinge at O with which it is possible to adjust the angle between the arc AA' and the support OGQ at the point O.

Figure 10.15 shows our representation of this experiment as seen from above. Current i from a battery V flows along the following path: $VRGBOB'GR'SV$. Circuit $VRGR'SV$ has an arbitrary shape. When the center of the arc BOB' is located at point C, which is vertically below the vertical axis of rotation GH, the arc remains in equilibrium, at rest relative to the laboratory, no matter the shape of the circuit $VRGR'SV$ and no matter the values of the opening angles ω and ω' which can be independently adjusted.

Suppose now that by means of the hinge O the arc AA' is positioned such that its center lies at point C_2 outside the axis GH, so that it will float in the mercury filling troughs M and M' at points B_2 and B_2', as in figure 10.16 (a). After being released from rest the arc will move and slide on the mercury of the troughs owing to the action of the closed curvilinear current flowing through $VRGR'SV$ and also due to the electromagnetic action of the Earth. During this motion the angle between the arc AA' and the straight line OQ, at their point O of junction, will remain constant. The arc will, for instance, pass through the configuration shown in figure 10.16 (b) in which the current is flowing through the path $VRGB_3OB_3'SV$, with the center of the arc AA' being at this moment located at point C_3. The angle B_2OG is equal the angle B_3OG at the hinge O.

Ampère's description of this experiment and the conclusions he obtained from it:[43]

[41][Ampère, 1825b], [Ampère, 1887b] and [Ampère, 1825a]; [Ampère, 1826e] and [Blondel, 1982, pp. 146-148].

[42][Ampère, 1825d, figure 3], [Ampère, 1825c, figure 3], [Ampère, 1825g] and [Ampère, 1826e].

[43][Ampère, 1825d, p. 384], [Ampère, 1825e, pp. 5-6] and [Ampère, 1825c, p. 375], our emphasis in italics and our words in the footnote.

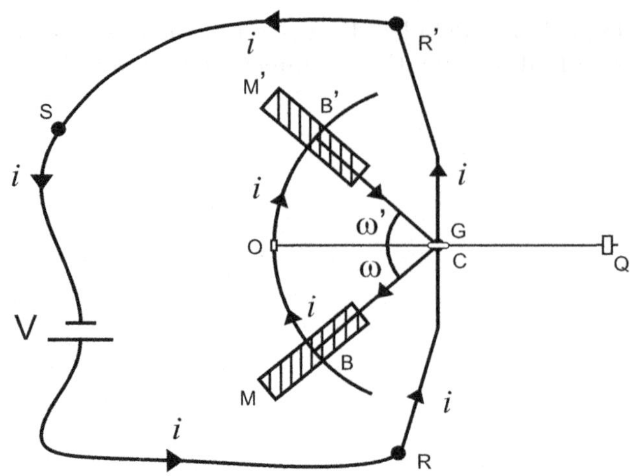

Figure 10.15: Arc BB' with its center C along the vertical rotation axis GH remains in equilibrium.

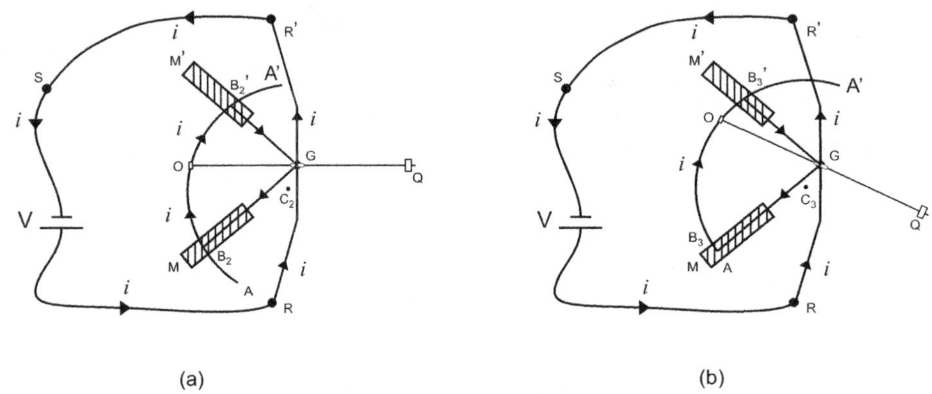

Figure 10.16: (a) Arc B_2B_2' with its center C_2 lying outside the vertical rotation axis GH. After being released from rest it does not remain in equilibrium, rotating around the vertical axis. (b) One of the configurations of the arc during its sliding on the mercury of the troughs. The angle B_2OG is equal the angle B_3OG at the hinge O.

> When, by means of the joint O, the arc AA′ is placed in such a position that its centre is outside the axis GH, this arc begins to move, and slides on the mercury of the little troughs MM′ by the force of the action of the complete curved current, which runs from R′ into S. If on the contrary its centre is in the axis, it remains immoveable: the complete circuit has therefore no action to make it turn round the axis, and that whatever be the size of the part BB′ determined by the opening of the angle of the conductors MN, M′N′. If, therefore, we take two arcs BB′ differing little from each other, as the torque[44] is null for either of them, it will be null for their little difference, and therefore for every element of the circumference, the centre of which is in the axis; *whence it follows that the direction of action which the complete circuit exercises on the element, passes through this axis, and is thus perpendicular to the element.*
>
> When the arc AA′ is situated so that its centre is in the axis, the portions of the conductors MN, M′N′ exert on the arc BB′ equal and opposite repulsive actions, in such a manner that no effect can result from it; and since there is no motion, we are sure there is no torque produced by the complete circuit.
>
> When the arc AA′ moves in the other situation in which we supposed it first [that is, with its center outside the axis GH], the actions of the conductors MN and M′N′ are no longer equal. One might be led to believe that the motion is owing only to this difference; but in proportion as we approach or remove the curved conductor running from R′ to S, the movement is increased or diminished; which circumstance leaves no room for doubt that the complete circuit bears a great share in the effect noticed.

Ampère integrated his equation (2.1) or (2.2) in order to obtain the tangential component of the force acting on a current element due to a closed circuit of arbitrary form. He could only obtain a null value

[44]Original French expression: *Moment de rotation*. Its meaning is torque or moment. It was translated as *momentum of rotation* in the Philosophical Magazine of 1825.

for this component of the net force acting along the direction of the current flowing in the element when equation (7.34) was valid. Assuming the validity of this last equation, as corroborated by his experiment of the case of equilibrium of the nonexistence of tangential force, he was then the first scientist to predict theoretically a null component of the net force acting on a current element along the direction of its current when this net force was due to a closed circuit of arbitrary shape. His case of equilibrium of the nonexistence of tangential force was also the first experimental proof of this remarkable fact.

As was seen in Section 10.1, this result of Ampère has been incorporated in the modern expression of the force $d\vec{F}$ acting on a current element $I'd\vec{s}\,'$ due to a magnetic field $\vec{B}$, namely:

$$d\vec{F} = I'd\vec{s}\,' \times \vec{B} \ . \tag{10.52}$$

That is, no matter the direction of the magnetic field $\vec{B}$, the net force $d\vec{F}$ acting on a current element will be located along a plane which is orthogonal to the direction $d\vec{s}\,'$ of the current element. There is no component of the force along the direction of the current element.

The case of equilibrium of the nonexistence of tangential force was so important to Ampère that he included it as the third case of equilibrium discussed in the *Théorie* in order to deduce his force between current elements.[45]

10.5 The Case of Equilibrium of the Law of Similarity

As discussed in Chapter 9, the theoretical results obtained by Savary were very important. They showed that it was possible to deduce the value $n = 2$ in Ampère's force, equations (2.1) up to (2.3), without assuming any analogy with the known gravitational, electric, or magnetic forces. All configurations and calculations performed by Savary were confirmed by Ampère in 1824.[46]

In 1825 Ampère obtained a new case of equilibrium which gave directly the expected value $n = 2$ for the power law in the distance between two current elements appearing in his force between current elements. This experiment was presented to the Academy of Sciences of Paris on November 21, 1825.[47] This experiment can be called the *case of equilibrium of the law of similarity* or the *case of equilibrium of the law of similitude.*

In order to understand this experiment it is better to consider initially figure 10.17 (a). There are two current elements ids and $i'ds'$ separated by a distance r, making angles α and β with the same continuation of the straight line connecting their midpoints, while γ is the angle between the planes drawn through each of these directions and the straight line joining the elements. Let F be the force exerted by $i'ds'$ on ids, while $F' = -F$ is the reaction force of ids on $i'ds'$. These two forces act along the direction connecting the midpoints of the elements.

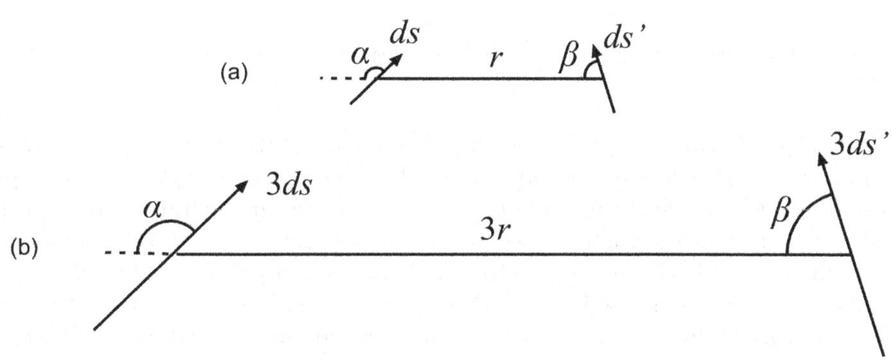

Figure 10.17: (a) Current elements of lengths ds and ds' separated by a distance r. (b) The lengths of both current elements and also the distance between them was multiplied by a single factor (like 3 in this example). When the mutual force between the current elements does not change its intensity, it is said that it follows the law of similarity. In this situation it can be concluded that $n = 2$.

Suppose now the lengths of both current elements and also their distance is multiplied by a single factor Q, while the values of i, i', α, β and γ remain unaltered, as in figure 10.17 (b) with $Q = 3$. Then, according to

[45] [Ampère, 1826f, p. 22 and figure 3], [Ampère, 1823c, Ampère, 1990, p. 194 and figure 3] and [Ampère, 1965b, p. 168 and figure 50].

[46] [Ampère, 1824c], [Ampère, 1824f] and [Ampère, 1885f].

[47] [Ampère, 1826b], [Ampère, 1887c] and [Ampère, 1887f].

equation (2.1), the forces F and F' are transformed into Q^2F/Q^n and Q^2F'/Q^n, respectively. Therefore, if $n = 2$, then the magnitudes of these forces remain constant, no matter the value of Q. When a mathematical formula remains the same when all geometric factors appearing in it are multiplied by the same factor Q, it is said that this formula satisfies the law of similarity, law of similitude, or principle of similarity.[48]

Let us imagine now three current elements which are parallel to one another with a current of the same intensity flowing in all of them along the same direction, as in figure 10.18. The lengths of these current elements are supposed to follow a continuous geometric proportion of ratio $Q = 3$. This is also the supposed ratio of the distance between these elements, that is:

$$\frac{ds}{ds'} = \frac{ds'}{ds''} = \frac{OO'}{O'O''} = Q = 3 \ . \tag{10.53}$$

The central current element ds' is attracted by the left element ds'' and also by the right element ds. However, according to Ampère's expression, equation (2.1), there will be no net force acting on the central element when the law of similarity is satisfied, provided their dimensions follow equation (10.53). That is, there will be no net force acting on the central element ds' in this configuration only if $n = 2$. When $n = 2$ then the net force acting on the middle current element will vanish no matter the value of the ratio Q appearing in equation (10.53).

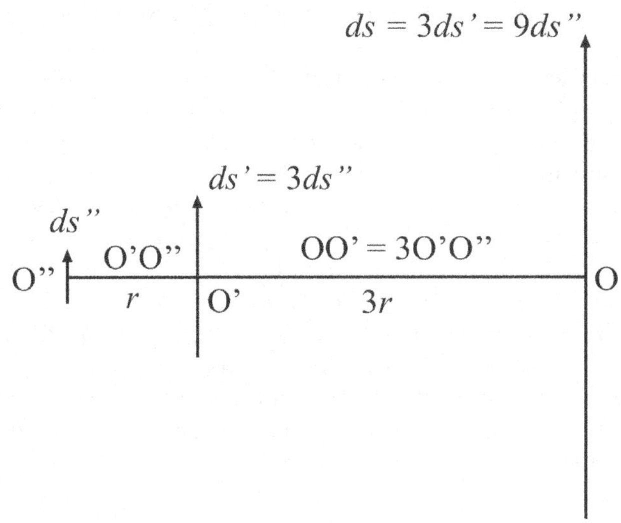

Figure 10.18: Three parallel current elements with their dimensions and distances satisfying the relation $ds/ds' = ds'/ds'' = OO'/O'O'' = 3$.

It is possible to verify experimentally if a specific phenomenon satisfies or not the law of similarity. As regards electrodynamics, Ampère devised the experiment depicted in figure 10.19.[49] This experiment is being called here the *case of equilibrium of the law of similarity.* There are three circular conductors located in the same horizontal plane. The right conductor CDE has center O and radius R, the middle conductor MNP has center O' and radius R', while the left conductor VXY has center O'' and radius R''. A constant current of the same intensity flows anti-clockwise in these three conductors. The left and right conductors are fixed in the laboratory, while the middle conductor can rotate around the vertical axis GH, moving towards or away from any one of the lateral conductors. The circular currents in the left and right conductors repel the central circular current.

When discussing this experiment, Kastler said that the horizontal circuits had elliptical shapes.[50] He was probably led to this mistake by Ampère's own figure in which the circuits seem to have elliptical shapes. However, this was only an effect of perspective in Ampère's drawing. After all, Ampère mentioned explicitly that the three circuits were circular.[51] Graneau and Graneau, on the other hand, mentioned that the circular

[48] [Kastler, 1977].

[49] [Ampère, 1826b, p. 38, figure A], [Ampère, 1887f, p. 206, figure 1], [Ampère, 1826f, p. 25, figure 4], [Ampère, 1823c, Ampère, 1990, p. 197, figure 4] and [Ampère, 1965b, p. 171, figure 51].

[50] [Kastler, 1977, p. 154].

[51] [Ampère, 1826b, p. 38], [Ampère, 1887f, pp. 206-207], [Ampère, 1826f, pp. 25-26], [Ampère, 1823c, Ampère, 1990, pp. 197-198] and [Ampère, 1965b, p. 171].

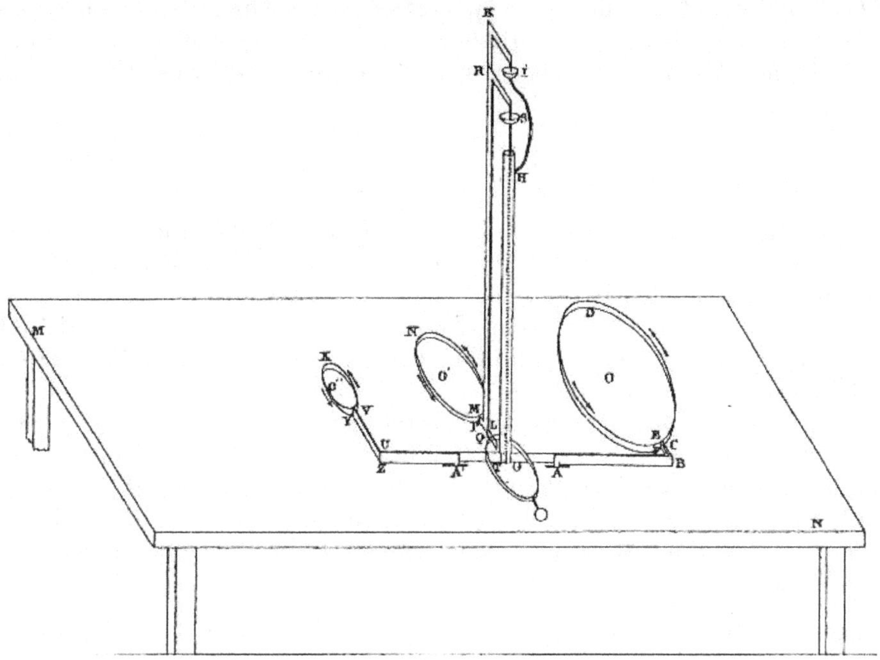

Figure 10.19: Case of equilibrium of the law of similarity. The centers O″, O′ and O of the left, center and right circles, respectively, do not appear in figure 4 of Ampère's *Théorie* published in 1826.

circuits were located in vertical planes.[52] Once more it should be emphasized that Ampère said explicitly that the circular circuits were located in the same horizontal plane. One example of a quotation by Ampère related to figure 10.19 saying not only that the mobile central circuit is circular, but also that it is located in a horizontal plane:[53]

> The moving conductor $IKLMNPQRS$, of which the part MNP is circular, starts from the cup I; the parts MLK and PQR are insulated by a silk covering. The conductor is held horizontal by the counterweight a fixed on the circumference of a circle formed around the tube GH by the continuation bcg of the sheet constituting the moving conductor.

Figure 10.20 presents the experiment of figure 10.19 as seen from above.

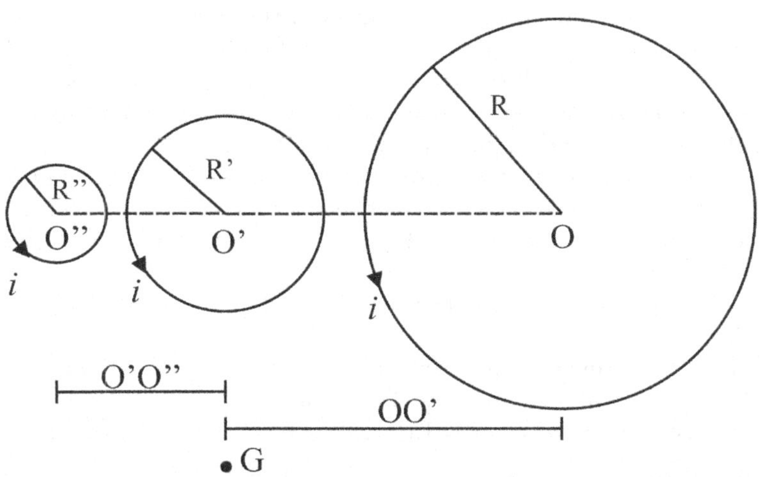

Figure 10.20: Case of equilibrium of the law of similarity seen from above.

[52] [Graneau and Graneau, 1996, p. 6].

[53] [Ampère, 1826b, p. 38], [Ampère, 1887f, pp. 206-207], [Ampère, 1826f, p. 26], [Ampère, 1823c, Ampère, 1990, pp. 197-198] and [Ampère, 1965b, p. 171].

The centers O, O' and O'' of the three circular portions are located along a single straight line. The circles are made in such a way that their radii follow a continuous geometric proportion, while the ratio between distances OO' and $O'O''$ has the same value as the consecutive terms of this geometric proportion, namely:

$$\frac{R}{R'} = \frac{R'}{R''} = \frac{OO'}{O'O''} \ . \tag{10.54}$$

Let us suppose that the middle conductor is initially at rest in the laboratory and no current is flowing through the three circuits. The middle circle is free to turn around the vertical axis GH, moving away or towards any one of the lateral circuits. If $n = 2$ then the middle circuit will remain at rest in the laboratory even when the same constant current i is flowing through the three circular circuits. This experimental configuration represents the new case of equilibrium that Ampère designed. When the central circle is moved to the right by turning it around the vertical axis GH, being released from rest, it is then repelled from the right circle by a force of higher intensity than the repulsion due to the left circle. In this case the central circuit will oscillate around the central position of equilibrium, until it stops due to the presence of friction. This experiment shows that the equilibrium of the central circuit is stable.

Although Ampère presented a detailed description of this experiment in 1825 and 1826, it was not apparently performed at that time, as he himself admitted. Probably he never performed it in its final format. At the end of the *Théorie* he said the following:[54]

> In finishing this Treatise I think that I should observe that I have not had time to build the instruments shown in figure 4 of the first plate and figure 20 of the second plate.[55] Therefore, the experiments for which they are intended have not yet been done,[56] but since these experiments are only designed to verify results obtained by other means and that, on the other hand, it would be useful to perform them as a counterproof of those that provided these results, I have not thought it necessary to remove the description.

We did not locate any other work or manuscript by Ampère in which he performed this experiment after 1826. In any event, variations of this experiment were performed in 1878 by Albert von Ettingshausen[57] (1850-1932) and in 1882 by Riccardo Felici[58] (1819-1902). They had three circular circuits with their dimensions and distances following the same geometric ratio, as in equation (10.54). They placed the three similar circles in three vertical planes which were parallel to one another, with the centers of the three circles located along the same horizontal straight line, as in figures 10.21 and 10.22. The central conductor B was mobile around a vertical axis passing through point r of figure 10.21 or through point T of figure 10.22. The plane of the central conductor was chosen so that it was orthogonal to the local magnetic meridian, in such a way that it was not affected by terrestrial magnetism. It is much easier to calculate the net force acting on the central circle and being due to any one of the lateral circles in this configuration than in Ampère's original configuration in which the three circular circuits were located in the same horizontal plane.

In Ettingshausen's experiment,[59] for instance, a current of the same intensity flowed in all three circuits of figure 10.21. The ratio of the radii of circuits A and B was 2:1, like the ratio of the radii of circuits B and C. In the first situation the currents in all three circuits flowed in the same sense (for instance, all of them clockwise). In this case he observed that circuit B remained stationary, after being released from rest, when the distances between the centers of the circuits was in the ratio $AB : BC = 2 : 1$. In the second situation the current in circuit B flowed in opposite sense from the currents in circuits A and C. In this case circuits A and B repelled one another; the same happened between circuits B and C. Then, no matter the initial position of circuit B, it stopped at rest when the distance of its center to the centers of the other circuits was in the ratio $AB : BC = 2 : 1$.

These experimental observations allowed Ettingshausen and Felici to conclude, like Ampère predicted, that $n = 2$ in his force law.

Recently Tricker has also performed a similar experiment which led to the same conclusion that $n = 2$.[60] Instead of working with similar circular conductors, Tricker worked with similar solenoids. In any event, the final conclusion of the experiments performed by Ettingshausen, Felici and Tricker is that the only value of

[54] [Ampère, 1826f, p. 201] and [Ampère, 1823c, Ampère, 1990, p. 373].
[55] [N. T.] These figures appear in our pages 355 and 383, respectively.
[56] [N. T.] See our discussion on this topic in Section 10.5.
[57] [Ettingshausen, 1878a], [Ettingshausen, 1878b] and [Ettingshausen, 1879].
[58] [Felici, 1882] and [Bouty, 1883].
[59] [Ettingshausen, 1878b, pp. 133-134].
[60] [Tricker, 1965, pp. 46-48].

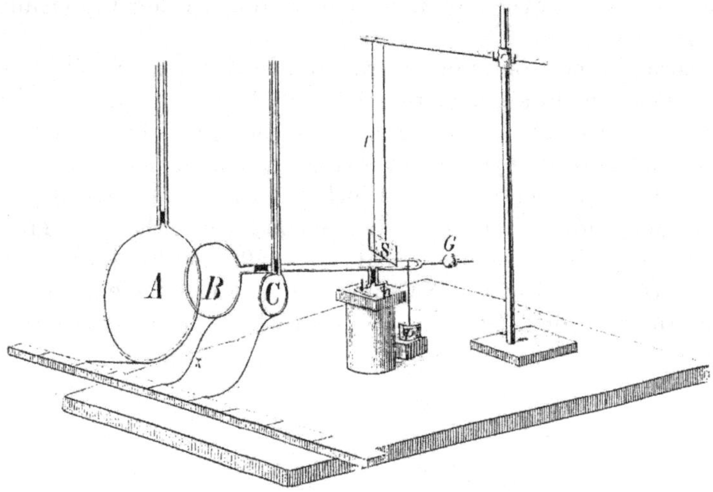

Figure 10.21: Ettingshausen's experiment.

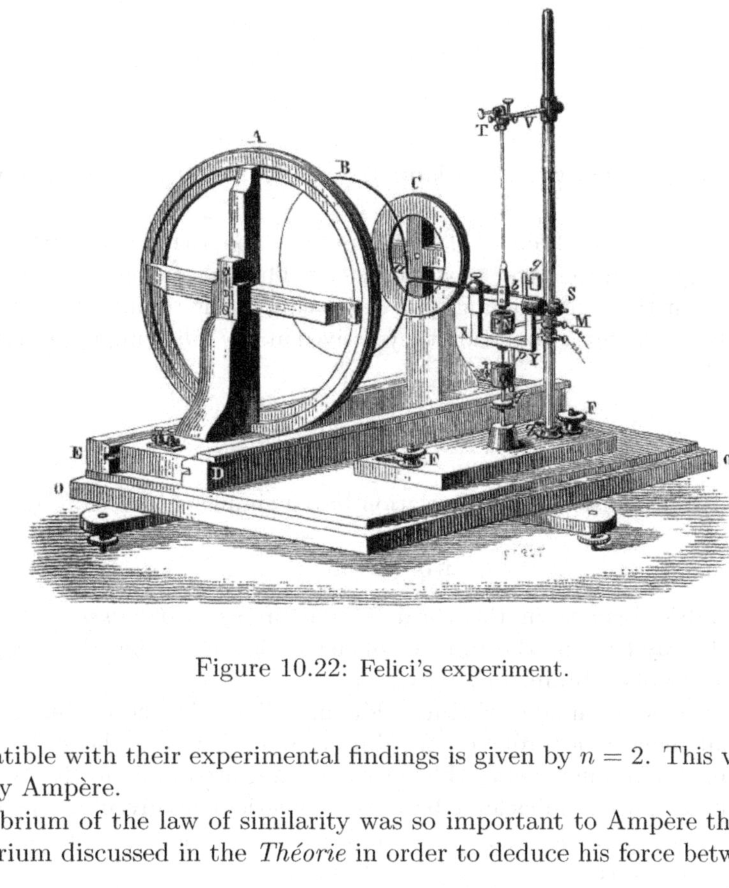

Figure 10.22: Felici's experiment.

the constant n compatible with their experimental findings is given by $n = 2$. This value of n coincides with the value predicted by Ampère.

The case of equilibrium of the law of similarity was so important to Ampère that he included it as the fourth case of equilibrium discussed in the *Théorie* in order to deduce his force between current elements.[61]

10.6 Mapping Terrestrial Magnetism

Suppose that a magnetic compass is free to turn around all directions relative to a point which is fixed relative to the ground. After being released from rest in an arbitrary orientation, it will oscillate around a specific orientation. The amplitudes of these oscillation decrease as a function of time due to the inevitable presence of friction, until the compass remains at rest pointing along its equilibrium orientation. This direction of

[61][Ampère, 1826f, p. 25 and figure 4], [Ampère, 1823c, Ampère, 1990, p. 197 and figure 4] and [Ampère, 1965b, p. 170 and figure 51].

equilibrium depends on its location over the Earth, being defined by the local value and direction of terrestrial magnetism.

A compass which can turn in a horizontal plane around a vertical axis passing through its center reaches an equilibrium orientation along the local magnetic meridian. The geographic meridian, on the other hand, is connected with the diurnal rotation of the Earth relative to the frame of the fixed stars. The angle between the magnetic meridian and the geographic meridian is called magnetic declination.

A dip needle or dip circle is a magnetic needle which can turn in a vertical plane around a horizontal axis passing through the center of gravity of the needle. In equilibrium the needle remains normally inclined relative to the horizon. The angle between the axis of the needle and the horizon is called dip angle. This phenomenon was known since the end of the XVIth century. Measurements made during the XVII and XVIIIth centuries showed that the dip angle was related with the terrestrial latitude, as shown in figure 10.23.

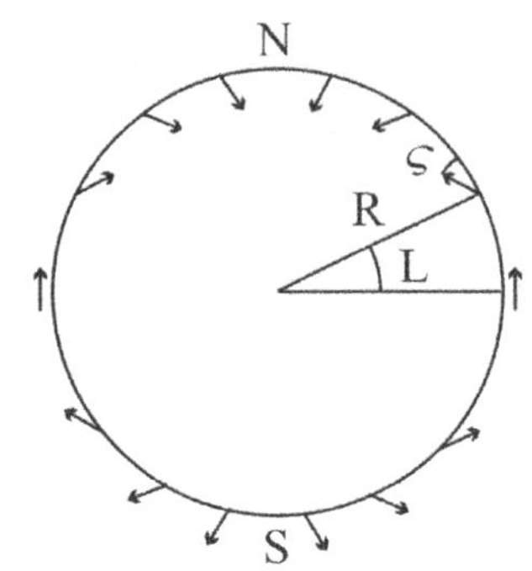

Figure 10.23: Dip angle ζ as a function of the latitude angle L. The letter R indicates the Earth's radius.

In 1804 Alexander von Humboldt (1769-1859) and Biot were able to explain mathematically the observed relation between the dip angle and the terrestrial latitude.[62] This mathematical relation represents a direct consequence of assuming that the torque acting on a dip needle is due to an infinitely small magnet located at the center of the Earth. With this assumption they arrived at the following relation between the dip angle ζ and the latitude angle L:

$$\tan(\zeta + L) = \frac{\sin 2L}{\cos 2L - 1/3} \ . \tag{10.55}$$

In 1809 the scientist Kraft transformed this relation into a simpler expression, namely:

$$\tan \zeta = 2 \tan L \ . \tag{10.56}$$

Biot incorporated Kraft's formula in the third edition of his book *Précis élémentaire de Physique expérimentale*, of 1824, by saying that the tangent of depression is double the tangent of magnetic latitude.[63] Equation (10.56) was also obtained by Bowditch.

Figure 10.24 (a) illustrates the model of Humboldt and Biot. At the center of the Earth there is a magnetic dipole. It would generate a torque on the dip needles located at the surface of the Earth, which would then, in equilibrium, remain inclined to the horizon as indicated in figure 10.23. It is possible to show mathematically, beginning with Coulomb's force between magnetic poles given by equation (1.3), that in this case the relation between the dip angle ζ and the latitude angle L where the dip needle is located satisfies equation (10.56), which agrees with the experimental data.

In 1823 Savary considered this problem utilizing Ampère's force between current elements.[64] He considered a small electrodynamic cylinder located at the center of the Earth and interacting with another small electrodynamic cylinder located at the surface of the Earth. The first cylinder replaced Humboldt and Biot's magnetic dipole, while the last cylinder would be the electrodynamic equivalent of a dip needle. He supposed that the distance between these cylinders was much larger than their lengths or radii. Utilizing Ampère's force between current elements he calculated the equilibrium orientation of the second cylinder

[62] [Humboldt and Biot, 1804] and [Savary, 1885b, Note of Joubert on p. 369].

[63] [Biot, 1824, p. 85] and [Farrar, 1826, p. 266].

[64] [Savary, 1823a, p. 97], [Savary, 1823b, pp. 24-25], [Savary, 1823c, pp. 24-26], [Savary, 1885b, pp. 343 and 367-370] and [Blondel, 1982, p. 141].

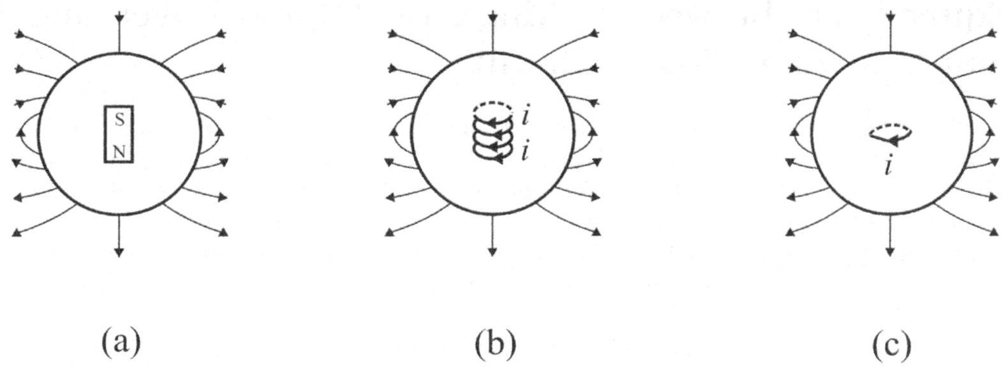

Figure 10.24: (a) In the model of Humboldt and Biot a small magnetic dipole located at the center of the Earth inclines the dip needles located at the surface of the Earth. (b) Savary obtained the same effect replacing the magnetic dipole by an electrodynamic cylinder. (c) Ampère obtained the same effect replacing the magnetic dipole by a small planar loop of arbitrary shape.

due to the influence of the first cylinder. He obtained that this orientation was exactly the orientation given by equation (10.56). Figure 10.24 (b) represents Savary's model.

In his works of November 21, 1825, Ampère generalized Savary's calculations related to terrestrial magnetism.[65] He replaced the electrodynamic cylinder by a small planar loop of arbitrary form located at the center of the Earth with its plane located along the terrestrial magnetic equator. He calculated the directrix (or the normal to the directing plane) due to this planar loop at a point located far away from the loop. He arrived at the following conclusions when considering the infinitesimal loop centered at the origin of a coordinate system with its plane located along the xy plane:[66]

> 1st: That the straight line which I called the normal to the directing plane of the electrodynamic action in relation to the point under consideration is located in the plane drawn from this point perpendicularly over the plane of the small circuit.
>
> 2nd: That this straight line is orientated, in relation to this last plane, in the same way as a magnetized dip needle is orientated in relation to the magnetic equator of our globe. That is, it forms with the straight line drawn from the point under consideration up to the origin, an angle the tangent of which is half of the tangent of that angle which the same straight line forms with the z axis.

What Ampère expressed with these words is equivalent to our equation (10.56). Ampère's model is illustrated in figure 10.24 (c). The fourth Note at the end of the *Théorie* published in 1826, corresponding to the third Note of the *Théorie* published in 1827, presents a mathematical deduction of this relation; see our Sections 30.4 and 31.3.[67]

As seen at Subsection 10.2.2, the extremity of an electrodynamic solenoid suffers a force due to a closed circuit acting along the directrix due to this circuit at the location of the extremity of the solenoid. When the solenoid is definite and very small, one of its extremities will suffer a force acting at a certain sense along this directrix, while its other extremity will suffer a force of the same magnitude in the opposite sense along the same directrix. Therefore it will normally suffer a torque, unless it is orientated along this directrix. In this particular orientation it will suffer no net torque from the closed circuit, remaining in equilibrium at rest relative to the ground. It is then possible to map the direction of this directrix at any point in space utilizing small definite solenoids.

The magnetic analog of this situation is the orientation of a compass by terrestrial magnetism. The magnetic action of the Earth can be mapped utilizing the equilibrium orientations of small compasses spread over the surface of the Earth.

Nowadays most physicists work with the concept of a magnetic field. Ampère's result about the orientation of the directrix is expressed nowadays by saying that the magnetic field at the Earth's surface has the same direction as the magnetic field due to a small magnetic dipole placed at the center of the Earth, with its axis orientated along the terrestrial magnetic axis.

[65] [Ampère, 1826b, p. 41], [Ampère, 1887c, p. 195] and [Ampère, 1887f, p. 210].
[66] [Ampère, 1887f, p. 210].
[67] [Ampère, 1826f, pp. 214-215] and [Ampère, 1823c, Ampère, 1990, pp. 379-381].

10.7 Equivalence between a Magnetic Dipole Layer and a Current-Carrying Closed Circuit

As seen in Section 10.3, Poisson developed a magnetic theory based on Coulomb's force between magnetic poles, presenting his results in February and December 1824. He calculated, in particular, the force acting on a magnetic particle (or magnetic pole) due to a magnetic surface composed of two sheets separated by a small distance, one sheet composed only of austral fluids and the other sheet composed only of boreal fluids. These two fluids had opposite signs and the same intensity. It was like a magnetic dipole layer. The three components of the force exerted by a magnetic dipole acting on a magnetic pole were expressed by Poisson through equations (10.48) up to (10.50).

In 1825 Ampère obtained analogous results, although expressing himself in terms of electric currents.[68] He presented his calculations at the meetings of 12 September, 21 November and 28 November 1825 of the Academy of Sciences of Paris.[69]

Ampère obtained, in particular, results analogous to equations (10.48) up to (10.50). Poisson considered the force exerted by a magnetic dipole layer acting on a magnetic particle. Ampère, on the other hand, instead of a magnetic dipole layer, considered a closed circuit having the same size and shape as Poisson's magnetic dipole layer, supposing that a constant current flowed along this closed circuit. Ampère calculated the force exerted by this current-carrying closed circuit acting on a simply indefinite solenoid. He supposed the extremity of this indefinite solenoid to be located at the same position of Poisson's magnetic particle. He then obtained[70] equations analogous to those obtained earlier by Poisson, as regards the angles appearing in these equations. The only difference between Poisson and Ampère's results appeared in the constant coefficients multiplying the angular portions of these equations. In analogy with equation (10.46), Ampère obtained an electrodynamic equivalence for Poisson's purely magnetic results.

Ampère expressed this equivalence by means of two theorems, namely:[71]

> Consider an arbitrary surface terminated by a closed circuit with given size and position in space. Suppose that very small magnets are placed in points infinitely close to one another along this whole surface, all magnets having the same length, this length being of the same order of magnitude as their mutual distances, with all magnets of the same magnitude and equally spaced, in such a way that the poles of the same name are all located on the same side of the surface, and that the straight lines connecting the poles of each magnet have their central points along this surface, being directed according to the normals [that is, magnets orthogonal to the surface in each point]. From the calculations which I have the honor to present to the *Académie*, it then follows:
>
> 1st: That the action exerted by this ensemble of magnets on an austral or boreal pole of another magnet, arbitrarily placed in relation to the boundary of this surface, does not depend on the shape of this surface, depending only on its contour.
>
> 2nd: That this action is exactly the same action which results from the formula by means of which I expressed the mutual action between two elements of voltaic conductors, [from the interaction] between an electric current that traversed the contour of the surface and the extremity of an electrodynamic solenoid located at the point where the pole being acted upon by all infinitely small magnets which are orthogonal to the surface is supposed to be located.

Ampère then considered a small magnet interacting with a magnetic pole. He represented by μ the infinitely small intensity of the magnetic force of one of these small magnets.[72] Let us suppose that this magnet has a North pole, p_1, and a South pole, $-p_1$. This magnet would be interacting with another magnetic pole p_2. The force exerted by each one of the poles of the magnet acting on the other pole would be proportional to the product p_1p_2, according to equation (1.3). The magnitude μ, which Ampère called the infinitely small intensity of the magnetic force of the magnet, would be then given by p_1, which would be proportional to the product p_1p_2.

Initially Ampère obtained expressions analogous to those obtained earlier by Poisson, namely, equations (10.48) up to (10.50). Later on he showed that the x, y and z components of this force by unit magnetic pole could be written in terms of his integrals A, B and C given by equations (10.8) up to (10.10).

[68] [Blondel, 1982, pp. 150-153], [Blondel, 1989] and [Hofmann, 1996, pp. 324-336].

[69] [Ampère, 1825d] and [Ampère, 1825c], [Ampère, 1825e], [Ampère, 1825g] and [Ampère, 1826e], [Ampère, 1826b], [Ampère, 1827a], [Ampère, 1828], [Ampère, 1887c], [Ampère, 1887f] and [Ampère, 1887e].

[70] [Ampère, 1887c, p. 202].

[71] [Ampère, 1887c, pp. 198-199].

[72] [Ampère, 1887c, pp. 199 and 202].

We represent this equivalence by means of figure 10.25. The magnetic dipole layer of figure 10.25 (a), composed of many NS dipoles orientated orthogonally to the surface, acts on an external magnetic pole N' located at point A. In figure 10.25 (b) there is a closed circuit following the contour of the magnetic dipole layer. There is a constant current i flowing through this closed circuit. It acts on a simply indefinite solenoid composed of many loops carrying a constant current i'. The extremity of this indefinite solenoid is located at the same point A. Ampère showed that the force acting on the configurations of figures 10.25 (a) and (b) have the same behavior, as regards the angles and distances involved in these problems. These two forces differ only as regards the constant coefficient multiplying each one of these interactions.

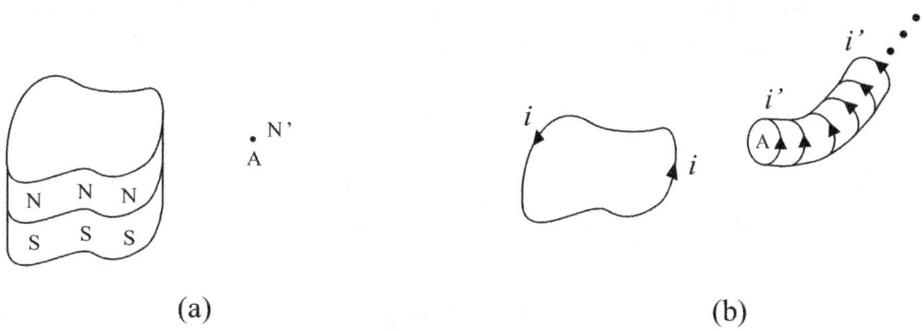

Figure 10.25: (a) Magnetic surface composed of a magnetic dipole layer interacting with a magnetic pole N' located at point A. (b) Current carrying closed circuit following the contour of the magnetic surface and interacting with a simply indefinite solenoid with its extremity located at point A. The forces acting on these two configurations, (a) and (b), have the same behavior as regards the angles and distances involved in these problems.

In another work of 1825 Ampère deepened his comparison of these two configurations.[73] He now considered the force between two magnetic dipoles of lengths $\delta\rho$ and $\delta\rho'$ separated by a distance r which was much greater than their lengths, $r \gg \delta\rho$ and $r \gg \delta\rho'$. He called the intensities of their magnetic poles by μ and μ', respectively. Figure 10.26 represents a dipole 1 of length $\delta\rho$, North pole μ and South pole $-\mu$. Dipole 2 has a length $\delta\rho'$, North pole μ' and South pole $-\mu'$. They are separated by a distance r.

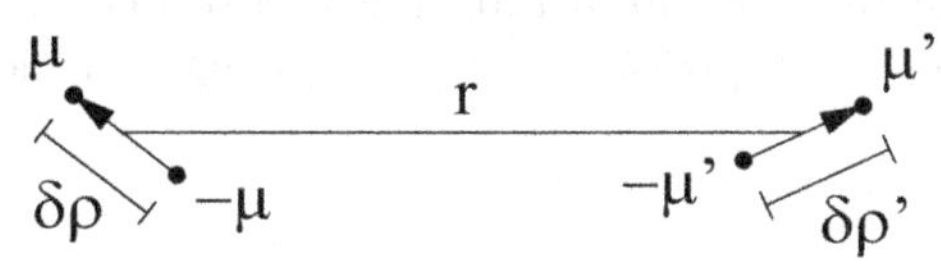

Figure 10.26: Two magnetic dipoles separated by a distance r.

Ampère also considered the force between two current-carrying loops separated by a distance r which was much larger than their radii or sizes. Figure 10.27 represents a circular loop 1 of radius r_1 and area λ carrying a constant current i with versor $\hat{n}_1$ orthogonal to its surface, together with a circular loop 2 of radius r_2 and area λ' carrying a constant current i' with versor $\hat{n}_2$ orthogonal to its surface. The centers of these two loops are separated by a distance r much larger than their radii, such that $r \gg r_1$ and $r \gg r_2$.

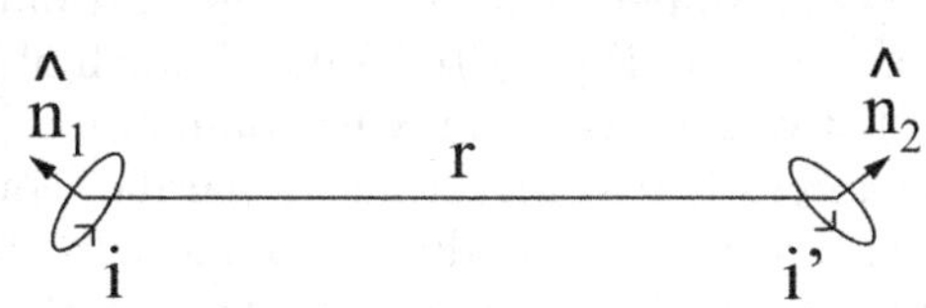

Figure 10.27: A closed loop of area λ, current intensity i and unit normal vector $\hat{n}_1$ interacting with another closed loop of area λ', current intensity i' and unit normal vector $\hat{n}_2$.

Ampère concluded his comparison of the situations represented by figures 10.26 and 10.27 as follows:[74]

> [...] supposing the axes of the two magnetic elements[75] normal to the planes of the two circuits, and that the centers of the axes [are located] at the determined points of the areas circumscribing the same

[73] [Ampère, 1826b] and [Ampère, 1887f].

[74] [Ampère, 1826b, pp. 42-43] and [Ampère, 1887f, pp. 211-212], our words in the footnotes.

[75] That is, the axes of the two magnetic dipoles.

> circuits, I obtained in these two configurations values[76] which differ only in the constant coefficient appearing in these values. This constant coefficient is $ii'\lambda\lambda'/2$ when dealing with the mutual action between two circuits, with i and i' being the current intensities, while λ and λ' represent the areas of the circuits. When one considers the values of the forces produced by the mutual action between two magnetic elements,[77] the same factor is given by $\mu\mu'\delta\rho\delta\rho'$, where μ and μ' represent the intensities of the attractive and repulsive forces of the poles of these elements, while $\delta\rho$ and $\delta\rho'$ represent the lengths of their axes

That is, when the normal to each current-carrying circuit is parallel to the magnetic axis of the corresponding dipole, then Ampère found the force between the dipoles of figure 10.26 to be equivalent to the force between the current-carrying circuits of figure 10.27, provided there was the following equivalence:

$$\frac{\lambda\lambda' ii'}{2} \Leftrightarrow \mu\mu'\delta\rho\delta\rho' \ . \tag{10.57}$$

This equation is analogous to the result given by equation (10.46).

Utilizing equations (2.14) and (2.21), then equation (10.57) can be expressed in the International System of Units as follows:

$$\lambda\lambda' II' \Leftrightarrow PP'\delta\rho\delta\rho' \ . \tag{10.58}$$

Here P and P' represent the intensities of the magnetic poles of the two dipoles in the International System of Units.

In Section 29.19 of the *Théorie*, Ampère further developed this identity between a magnetic dipole and a current-carrying loop.[78]

This theorem first developed by Ampère is usually represented nowadays in vector notation in the International System of Units. Consider a magnetic dipole composed of a North pole P and a South pole $-P$ separated by a small distance ℓ. Let $\vec{\ell}$ be the vector pointing from $-P$ to P, with magnitude $|\vec{\ell}| = \ell$ being the distance between the two poles. The magnetic moment $\vec{m}$ of this dipole is then given by $\vec{m} \equiv P\vec{\ell}$. Consider now a loop of area A carrying a constant current of intensity I. Its magnetic moment is given by $\vec{m} = IA\hat{u}$, where $\hat{u}$ is a unit vector orthogonal to the area of the loop. The direction of the versor $\hat{u}$ is given by the right-hand rule. Let us assume that $\vec{\ell}$ and $\hat{u}$ are parallel to one another, pointing in the same sense. The equivalence obtained by Ampère is expressed by the following equation:

$$P\ell = IA \ . \tag{10.59}$$

When equation (10.59) is satisfied it is said that the current-carrying loop and the magnetic dipole have the same magnetic moment. They exert the same force and the same torque on another magnetic dipole with magnetic moment $P'\ell'$. They also exert the same force and the same torque on another current-carrying loop with magnetic moment $I'A'$.

10.8 Final Synthesis

After six years working on this subject, Ampère realized that he had obtained the goals he had imagined when first hearing about Ørsted's experiment. He created a new branch of physics, namely, electrodynamics, dealing with the forces and torques between current-carrying conductors. He obtained the main phenomena of this area of knowledge, namely: forces between rectilinear parallel conductors, forces between current-carrying flat spirals, torques between a rectilinear conductor and a helix, force exerted by a closed circuit of arbitrary form acting on a current element of another circuit, force acting on a mobile portion of a circuit due to the remainder of this closed circuit (as in Ampère's bridge experiment), etc. Moreover, he obtained an algebraic expression for the force between two current elements. With this expression he could not only explain quantitatively the phenomena he had observed, but he could also predict new facts which were later on corroborated experimentally. In particular, he showed theoretically and experimentally that the force exerted by a closed circuit of arbitrary shape acting on an external current element is always orthogonal to this element. This force is located in the directing plane, being normal to the directrix at the location of

[76] That is, a value for the forces between two magnetic dipoles and a value for the force between two current-carrying closed loops.

[77] That is, between two magnetic dipoles.

[78] [Ampère, 1826f, pp. 131-151] and [Ampère, 1823c, Ampère, 1990, pp. 303-323].

the element, with this directrix being due to the closed circuit. The force exerted by this closed circuit of arbitrary form acting on a simply indefinite solenoid acts at the extremity of this solenoid, pointing along the directrix due to the closed circuit. He also showed that the force between two definite solenoids is composed of four components, pointing along the straight lines connecting the extremities of these solenoids and varying as the inverse square of these distances.

Moreover, he presented a new conception of magnetism. It was no longer necessary to suppose the existence of austral and boreal fluids (our North and South fluids, or North and South poles, respectively). Instead of these fluids, Ampère postulated the existence of microscopic electric currents flowing in the molecules of the magnetized substances. He showed that the interaction between two simply indefinite solenoids is electrodynamically analogous to Coulomb's force between two magnetic poles. Ampère also showed that the interaction between two magnetic dipoles is equivalent to the interaction between two definite solenoids carrying constant currents. He explained quantitatively the experiments of Ørsted, Biot, Savart and Faraday relative to the interaction between a current-carrying conductor and a magnet. To this end he utilized only his force between current elements. He could also reproduce these experiments replacing the magnets (or the Earth) by appropriate current-carrying conductors. He could also reproduce the usual behavior of a compass or the behavior of a dip needle utilizing appropriate current-carrying conductors instead of the compass or the dip needle.

In this way he unified three branches of physics, namely: magnetism (interaction between two magnets or the interaction of the Earth with a magnet), electromagnetism (interaction of a magnet with a current-carrying conductor or the interaction of the Earth with a current-carrying conductor) and electrodynamics (interaction between two current-carrying conductors). According to Ampère these three branches of knowledge were only due to forces and torques acting between current-carrying conductors. Moreover, the experimental phenomena of these three branches were explained theoretically utilizing the integration of his force law between current elements.

In 1826 he wrote his masterpiece, the *Theory of Electrodynamic Phenomena, Uniquely Deduced from Experience*.[79] This work presented a final synthesis of his experimental and theoretical works.

Although he published a few other works on electrodynamics later on, he soon lost interest in this subject. He dedicated most of his time to the philosophy of science, writing a text presenting a classification of all sciences known in his time. The first volume was published in 1834,[80] while the second volume was published posthumously in 1843,[81] seven years after his death in 1836.

[79] [Ampère, 1826f].
[80] [Ampère, 1834].
[81] [Ampère, 1843].

Part IV

Controversies, Part 1: Most Scientists Against Ampère

We now present some controversies which began in Ampère's time, some of which have not yet been solved. These electromagnetic controversies help to illuminate Ampère's own masterpiece, the *Théorie*, and the criticisms he incorporated in this work. Tricker, in the beginning of a chapter in which he presented some criticisms made to Ampère's work after his death, stated the following:[1]

> The theory of the electrodynamics of steady currents developed by Ampère was accepted by his contemporaries immediately.

However, this statement is far from the truth. Ampère's work was criticized by the main contemporary scientists working on this subject, namely, Ørsted, Biot, Savart and Faraday, among others. In the next Chapters we present some of the main criticisms presented by these authors.

[1] [Tricker, 1965, p. 98].

Chapter 11

Ørsted *Versus* Ampère

11.1 Ørsted's Interpretation of His Own Experiment

As was seen in Section 1.3, Ørsted presented on July 21, 1820, his fundamental discovery of the deflection of a magnetized needle due to a long and straight current-carrying conductor. His most important experiment is represented in figures 1.3 and 1.4. Figure 11.1 reproduces the first of these images.

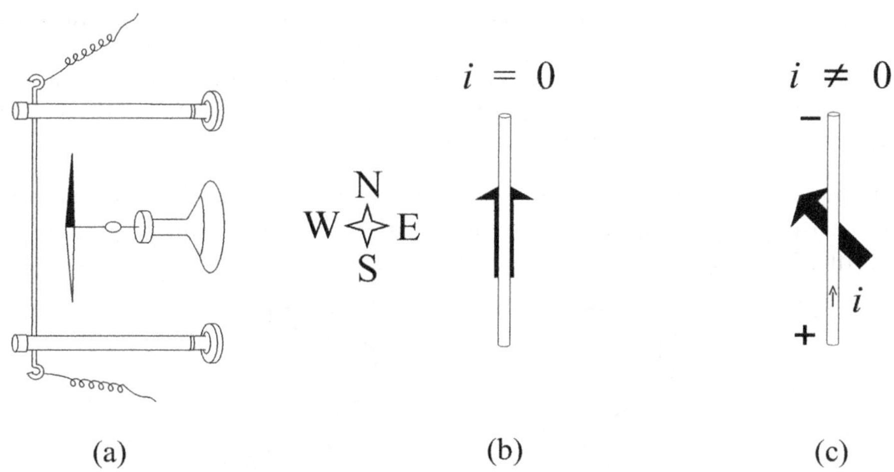

Figure 11.1: Representation of Ørsted's experiment with the horizontal wire above the magnetic needle. In (a) and (b) the needle points along the magnetic meridian while there is no electric current in the wire. In (c) there is a constant current flowing from the South towards the North. The needle is deviated from the magnetic meridian, with its North pole going westward.

Ørsted interpreted his crucial observation as follows:[1]

> We may now make a few observations towards explaining these phenomena.
>
> The electric conflict acts only on the magnetic particles of matter. All non-magnetic bodies appear penetrable by the electric conflict, while magnetic bodies, or rather their magnetic particles, resist the passage of this conflict. Hence they can be moved by the impetus of the contending powers.
>
> *It is sufficiently evident from the preceding facts that the electric conflict is not confined to the conductor, but dispersed pretty widely in the circumjacent space.*
>
> From the preceding facts we may likewise collect[2] that this conflict performs circles; for without this condition, it seems impossible that the one part of the uniting wire, when placed below the magnetic pole, should drive it towards the east, and when placed above it towards the west; for it is the nature of a circle that the motions in opposite parts should have an opposite direction. Besides, a motion in circles,

[1][Oersted, 1820, p. 276], [Oersted, 1965, pp. 116-117] and [Ørsted, 1986, pp. 121-122], our emphasis in italics.

[2]In the original English translation of Oersted's paper we read, [Oersted, 1820, p. 276] and [Ørsted, 1998a, p. 419]: "From the preceding facts we may likewise collect..." In Tricker's book we read, [Oersted, 1965, p. 116]: "From the preceding facts we may likewise infer..."

joined with a progressive motion, according to the length of the conductor, ought to form a conchoidal or spiral line; but this, unless I am mistaken, contributes nothing to explain the phenomena hitherto observed.

All the effects on the north pole[3] *above-mentioned are easily understood by supposing that negative electricity moves in a spiral line bent towards the right, and propels the north pole, but does not act on the south pole.* The effects on the south pole are explained in a similar manner, if we ascribe to positive electricity a contrary motion and power of acting on the south pole, but not upon the north. The agreement of this law with nature will be better seen by a repetition of the experiment than by a long explanation. The mode of judging of the experiments will be much facilitated if the course of the electricities in the uniting wire be pointed out by marks or figures.

I shall merely add to the above that I have demonstrated in a book published five years ago that heat and light consist of the conflict of the electricities. From the observations now stated, we may conclude that a circular motion likewise occurs in these effects. This I think will contribute very much to illustrate the phenomena to which the appellation of polarization of light has been given.

Copenhagen, July 21, 1820. John Christian Oersted.

Ørsted did not present figures to illustrate his explanation. Figure 11.2 illustrates what he may have imagined according to his own description. In figure 11.2 (a) we have positive charges flowing along the magnetic meridian, from the South towards the North, inside the horizontal current-carrying wire, while the negative charges flow in the opposite sense. The North pole of the magnetized needle placed below the wire, which originally pointed along the NS direction when there was no current in the wire, is displaced westwards when a constant current flows in the wire. For instance, in this article he observed that:[4]

> If the distance of the uniting wire does not exceed three-quarters of an inch from the needle, the declination of the needle makes an angle of about 45°. If the distance is increased, the angle diminishes proportionally.

Figure 11.2 (b) presents the electric conflict flowing helically outside the wire, according to Ørsted's conception. According to his original interpretation, negative electricity would "propel the north pole" of the magnetized needle.

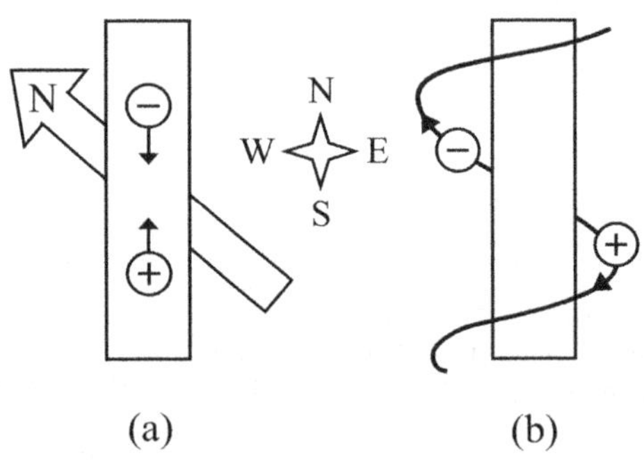

Figure 11.2: (a) Ørsted's experiment. (b) His interpretation of this observation in which he supposed positive and negative electricities describing helical paths around the wire and propelling the magnetic poles of the magnet. The arrows indicate the directions of motion of the supposed positive and negative charges moving (a) inside and (b) outside the wire.

It is curious to observe that the longitudinal component, parallel to the wire, of the motion of the supposed negative charges outside the wire point in the opposite sense of the motion of the negative charges inside the wire. The same behavior happens with the supposed motions of the positive charges inside and outside the wire.

Ørsted initially considered that the deflection of the needle was due only to the action of the current-carrying wire. Ampère, on the other hand, realized that this deflection was due to the joint action of the wire and the Earth acting on the needle. Ampère was the first to show, with his astatic needle, that, by

[3][*Note by R. A. R. Tricker, [Oersted, 1965, p. 117]:*] Oersted's expressions are "Omnes in polum septentrionalem" and "Effectus in polum meridionalem". If by "a spiral line bent towards the right" he means a right-handed screw, then he must be using septentrionalem in the same sense that boreal was used at the time, namely to indicate a south-seeking pole. The term north pole in the translation would therefore mean a south-seeking pole—i.e. one homologous with the earth's north pole.

[4][Oersted, 1820, p. 274], [Oersted, 1965, p. 114] and [Ørsted, 1986, p. 119].

eliminating the magnetic action of the Earth, the needle is orientated orthogonally to a current-carrying wire, as discussed in Section 3.2. Ampère presented this result at the Academy of Sciences of Paris in September 1820, publishing his results at the end of this year.

When Ørsted became aware of Ampère's experiments, he changed his interpretation of his own experiment. In an article of 1821, for instance, he said the following:[5]

> I shall here state, rather more in detail than I have done in my first publication, the rule by which I think all electro-magnetic effects are governed. It is this: *When opposite electrical powers*[6] *meet under circumstances which offer resistance, they are subjected to a new form of action, and in this state they act upon the magnetic needle in such a manner that positive electricity repels the south, and attracts the north pole of the compass: and negative electricity repels the north, and attracts the south pole;*[7] *but the direction followed by the electrical powers in this state is not a right line, but a spiral one, turning from left hand to the right.*

In his original paper of 1820, Ørsted mentioned the *negative electricity*. He now infers *negative electrical power*. These expressions would represent the analogous to our *negative electric charge*, or *negatively electrified particle*, the same being valid for the positive sign.

In his first interpretation of 1820, Ørsted believed that negative electricity *propelled* the North pole of a magnet, while the positive electricity *propelled* the South pole. He was certainly thinking of forces exerted by contact or collision. Similar examples are the orientation of a windsock by air currents in an airport or a flowing river propelling a water wheel or watermill. In the case of Ørsted's experiment, he imagined a material flow of charged particles pushing the poles of the magnet. These magnetic poles would not be penetrable by the electric charges, resisting their passage. As he said, the magnetic particles of matter would be moved by the impetus of the electric charges. In this paper of 1821, on the other hand, he mentioned the *attractions and repulsions* exerted by the electricities flowing helically outside the wire and acting on the poles of the magnetic needle. In 1820 he suggested actions transmitted through mechanical contact, charges propelling magnetic poles, while in 1821 he suggested actions at a distance (attractions and repulsions). In any event, it is not clear to us how these supposed attractions and repulsions acting between the charges of the electric conflict and the poles of the magnet might orientate the needle, as shown by his experiment.

We present here another quotation of Ørsted from 1821 in which he mentioned these attractions and repulsions acting between the electric charges and the poles of a magnet:[8]

> Given all this, the north pole of a magnetic needle is repelled by the negative electricity and attracted by the positive. Naturally, the south pole of the magnetic needle has the same relation with the positive electricity.

As can be seen from these quotations, from 1820 to 1821 he changed his conceptions of how the electricities flowing outside the current-carrying wire interacted with the magnetic poles of the needle. Moreover, later on he also began to talk about circles around a current-carrying wire, instead of considering helices or spirals around it. An example can be seen in a paper he published on 1830 related to thermoelectricity. What he called austral magnetism is the North pole of a magnetized needle, that is, the pole pointing approximately towards the terrestrial geographic North. His words:[9]

> If we now suppose that the electricity of the current enters the conductor at the right hand of the observer,[10] the austral magnetism (the same which predominates in the north-end of the needle,) will, upon the superior surface of the conductor go off from the observer; on the side most distant from the observer, the austral magnetism goes downwards; on the inferior surface it goes towards the observer; on the side nearest the observer it goes upwards. This is represented in figure 1 [figure 11.3], where BA is the conductor in which the direction of the current is AB, the circle *cedf* [sic][11] represents a plane perpendicular to the conductor, in which the magnetical circulation takes place. This plane is here and

[5][Ørsted, 1998f, p. 432], italics in the original.

[6][Note by Ørsted:] I here repeat what I have already stated in other works, that by *electrical forces*, I mean only the unknown cause of electrical phenomena, whether it belong to imperceptible matter or independent motion.

[7][Note by Ørsted:] In my first memoir, I grounded all explanations upon the repulsions only which are exerted by electrical and magnetic forces; but I soon discovered, that from the fear of assuming more than the phenomenon required, I drew an unjust inference; for if magnetic forces are the same as electrical under another form of action, it follows, that opposite forces ought to attract each other reciprocally, and forces of the same kind to repel each other.

[8][Ørsted, 1998e, p. 426].

[9][Ørsted, 1998g, p. 549], our explanations in the footnotes.

[10]That is, for an observer looking at figure 11.3 at this instant.

[11]These letters are not indicated in Ørsted's figure.

in the other figures represented as if it were material and opaque. The little arrows show the direction of the austral magnetism.[12]

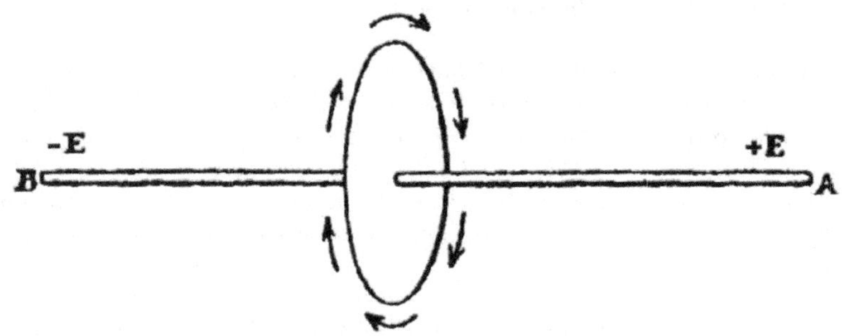

Figure 11.3: Ørsted's figure representing the magnetic action of a straight current-carrying conductor.

As can be seen from this quotation, after becoming aware of Ampère's experiment with his astatic needle, Ørsted was convinced that when a magnetic compass is only under the influence of a long, straight current-carrying conductor, without being influenced by terrestrial magnetism, the needle will become orientated in equilibrium orthogonally to the wire.

11.2 Ørsted Against Ampère

Ørsted never completely accepted Ampère's interpretations as regards the electromagnetic and electrodynamic phenomena. He considered his own theory correct as opposed to that of Ampère. He even performed an experiment, in 1830, with which he believed to have refuted Ampère's theory.[13] Ørsted mentioned several times that his theory was different from that of Ampère. For instance:[14]

> [...] and if I adopt a theory of magnetism differing from his [that is, differing from Ampère's theory], I shall never cease to acknowledge the great merit of his labours.

Some aspects which Ørsted criticized in Ampère's interpretations:

1. The mathematical complication of Ampère's theory.

2. The supposition of a direct interaction between two current-carrying conductors, without the intermediation of a flux of charges circulating around the wire.

3. The hypothesis of the existence of microscopic or molecular electric currents flowing inside a magnet. Ørsted, on the other hand, always considered magnetism as being due to a distribution of real magnetic particles distributed inside a magnet and inside the Earth.

We now present some of his quotations expressing these criticisms.

11.2.1 The Mathematical Complication of Ampère's Theory

We first quote a paper by Ørsted in which he expressed not only that his theory opposed that of Ampère, but also mentioned the mathematical difficulties found in Ampère's theory:[15]

---

[12]That is, the direction of motion of the North pole of a magnet, due to the influence of the straight current-carrying conductor.

[13][Ørsted, 1998d].

[14][Ørsted, 1998f, p. 437].

[15][Ørsted, 1998d, p. 539], our emphasis in italics.

> Councillor of State[16] and Professor Ørsted has informed the Society of a new electromagnetic experiment which he believes to be inconsistent with Ampère's theory.[17] It is a familiar experience in the history of science that *opposing theories* about a natural phenomenon are able to persist for a long time even though there may be arguments which should decide the issue. In such a case an attempt must be made to devise an experiment which cannot possibly be explained in two ways. If one stopped at a crossroads where one did not know which direction to take, such an *experimentum crucis*, as Baco[18] called it, would show the right way. *The controversy between the explanation of the electromagnetic effects given by Ampère and the one given by the discoverer [that is, by Ørsted himself] may be said more or less to stand at such a point.* Admittedly, Ampère's theory has not retained many defenders outside France, and even there opinions are divided, but *the profusion of mathematical expositions which makes it difficult to assess this theory* has also prevented many physicists from deciding in favour of one view.

Another quotation showing that Ørsted considered Ampère's theory very complicated:[19]

> By these suppositions, and a considerable exertion of mathematical skill, he [Ampère] is enabled to make this view represent well enough the phenomena, *though his theory is very complicated.*

11.2.2 Direct Action between current-carrying conductors, Without being Mediated by a Flux of Electric Charges Circulating around the Wire

Ørsted's main criticism against Ampère was that Ampère did not accept the rotary action of the electric current, that is, Ampère did not accept Ørsted's conception according to which there were electric charges circulating externally around a current-carrying wire. This supposed circulation of electric charges around the wire was the basis of Ørsted's explanation of his own experiment of the deflection of a compass needle. Ampère, on the other hand, explained the torque exerted by the current-carrying wire in this experiment as being due to a direct interaction between the macroscopic current in the wire and the microscopic currents he supposed to exist inside the magnetic needle. Moreover, according to Ampère, this macroscopic torque was due to an integration of microscopic forces acting between a current element in the wire and a current element of a microscopic current inside the magnet, with these forces acting along the line connecting these two current elements and following the principle of action and reaction. Ørsted accepted the existence of real magnetic particles of austral and boreal magnetism. Ampère, on the other hand, explained the magnetic and electromagnetic interactions by supposing the existence of microscopic electric currents not only inside magnets, but also inside the Earth. Moreover, according to Ampère the so-called magnetic properties of magnets and the Earth were in fact due to electrodynamic interactions between these microscopic electric currents. Ampère's suppositions were contrary to those of Ørsted, who considered them unacceptable, as he expressed several times. Ørsted never abandoned the idea of the real existence of magnetic particles or poles. He also never abandoned his initial supposition of the existence of a helical or circular flow of electric charges moving externally around a current-carrying wire and being responsible for the magnetic properties of this wire. We quote here some statements by Ørsted related to these topics as expressed in 1830:[20]

> M. Ampère, to whom we are indebted for the discovery of the mutual attractions and repulsions of the electrical currents, considers the law of this action as a fundamental one, at least so far as our present knowledge extends. *He thus admits no rotative action in the electrical current, but he transports it to the magnet, in which he supposes electrical currents, revolving in planes perpendicular or nearly perpendicular to the axis of the magnet.*

Ampère wished that Ørsted would accept his theory. This never happened, although they met one another in Paris and had the opportunity to discuss magnetism. In a letter written to his wife on 25 April 1822, Ørsted described Ampère's discomfort when Ørsted maintained his own theory:[21]

> Ampère, *who has worked so much with my discovery and has founded a very elaborate theory on it, was greatly annoyed that I still keep to mine which is extremely simple. In order to have a conversation with me about this in the company of several scholars, he invited me to a dinner-party where* Fourier, Dulong, Chevreul, Friedrich Cuvier, Savary *and* Montferrand *were present too. The latter two are young disciples*

[16][Between 1827 and 1840 Ørsted refers to himself in these reports as *Etatsraad*, an honorary title meaning Councillor of State.]

[17][Summary in Oken's *Isis*, Vol. 22, Col. 260-62, Jena 1829.]

[18][Bacon.]

[19][Ørsted, 1998g, p. 568], our emphasis.

[20][Ørsted, 1998g, p. 568], our emphasis.

[21][Franksen, 1981, p. 32].

of Ampère. *After the meal the conversation began and lasted for nearly three hours. I quite succeeded in proving that my theory accounts for all the phenomena, and what was most remarkable, I had to prove to* Fourier *that my theory was older than* Ampère's *which was, however, easy, seeing that I have already given it in my first publication. Even* Ampère's *two disciples declared that my theory was able to explain all the phenomena. They declare that so will* Ampère's, *and as his theory is nothing but the reverse of mine, he having removed the circuits of forces, discovered by me, from the conductor to the magnet, it will no doubt be difficult to find any entirely decisive objection to his theory, but I do not care for that either.*

Chapter 12

Biot and Savart *Versus* Ampère

12.1 Biot and Savart's Interpretation of Ørsted's Experiment

Biot and Savart presented an interpretation of Ørsted's experiment which was different from that of Ørsted, Section 11.1, and also different from that of Ampère, Sections 1.3 and 3.1.

Biot and Savart supposed that the wire became magnetized by the current flowing through it. There would be then an interaction between the magnetic poles of this wire and the magnetic poles of the compass needle. This interpretation can be gathered not only from the title of their paper, "Note on the magnetism of Volta's battery", but also from their own words when describing the experiment which they performed, namely:[1]

> By these procedures MM. Biot and Savart arrived at the following result which rigorously represents the action experienced by a molecule of austral or boreal magnetism when placed at some distance from a fine and indefinite cylindrical wire *which is made magnetic by voltaic current.* Drawing a perpendicular to the axis of the wire from the point where the magnetic molecule resides, the force influencing the molecule is perpendicular to this line and to the axis of the wire.

Similar statements can be found in a paper by Biot published in 1821, "On the magnetization of metals by electricity in motion":[2]

> However, what nobody had supposed, before M. Oersted made the observation, is that the electric current has another power. When it flows along metal bodies of any nature, *it gives them momentarily a magnetic virtue*; they are then able to attract soft and non magnetized iron. When one presents to them a steel magnetized needle, they attract one of its poles and repel the other, but only according to the parts of their surfaces which it faces. At last, what completes the characteristic of a magnetic action, they do not act at all on silver or copper needles, but act only on substances which are capable of magnetization.

Analogous statements were made in 1824.[3] One example:[4]

> To obtain, therefore, the abstract law of the forces, which must be the first principle and the determining cause of all the effects produced by the electro-magnetic bodies of whatever figure, it remains to be learned how each infinitely small particle of the uniting wire contributes to the total action of the lamina of which it is a part. This determination is in fact the sole means of ascertaining with certainty the nature of the modification produced in the particles of the metal by the voltaic current, in virtue of which the electromagnetic effects take place. It is the only means of knowing, for example, whether these effects result from a proper action immediately exerted by the electric current upon the particles of magnetism presented to it, *or whether, as all analogy seems to indicate, they are only the secondary consequence of a true magnetism impressed by the voltaic current upon the metallic conductors*, differing not in principle, but in its distribution merely, from the longitudinal magnetism which we have as yet been able to produce, in certain metals, by friction only.

Biot and Savart also mentioned the rotational or gyratory character of the force exerted by a current-carrying wire on a magnetized needle. For instance:[5]

---

[1] [Biot and Savart, 1820, p. 223], [Biot and Savart, 1965b, p. 118] and [Assis and Chaib, 2006, p. 308], our emphasis.
[2] [Biot, 1821b, p. 225] and [Biot, 1821a, p. 117], our emphasis.
[3] [Biot and Savart, 1824, pp. 704 and 768], [Farrar, 1826, pp. 305 and 359] and [Biot and Savart, 1885, pp. 80 and 121].
[4] [Biot and Savart, 1824, p. 768], [Farrar, 1826, p. 359] and [Biot and Savart, 1885, p. 121], our emphasis.
[5] [Biot, 1821b, pp. 226-227].

> [...] it should necessarily be concluded from these effects that the connecting wire moves the needle by a force emanating from the wire itself, which is directed transversely to the length of the wire, revolving around its axis, being always parallel to the portion of its circular contour indicated by the needle. This was also the consequence drawn by M. Oersted from his observations. Now, this rotational character of the force, and rotational following a definite sense, [...]
>
> *The first fact of the momentary magnetization impressed to the connecting wire by the voltaic current* might appear to a common observer; I don't know if some aspects of this property had not been perceived and indicated before. In any event, to recognize this particular character of the force [that is, its rotational character around the current-carrying wire], and to have assigned it according to the phenomena, without hesitations, without uncertainties, this belongs really to M. Oersted, and this is what really constitutes a totally new condition in the motion of electricity.

Similar statements appear in another work of Biot.[6] Biot has stated this idea of a "revolving force" as follows in 1824:[7]

> [...] we infer as a necessary consequence, that the uniting wire deranges the needle by a force proceeding from itself, directed transversely with respect to the length of the wire, and revolving about its axis, and acting always parallel to the part of its circular outline presented to the needle. This is also the conclusion drawn by M. Oersted from his first observation. Now, the circumstance of the force revolving, and revolving in a determinate direction, in a medium which, like copper or silver, or any other metal, seems perfectly identical in all its parts, is a very remarkable phenomenon, of which only one example was before known, namely, that relating to the theory of light, which consists, as will be shown hereafter, in the deviations which certain liquids cause in the planes of polarization of the luminous rays.

12.2 Biot and Savart Against Ampère

Biot and Savart never accepted Ampère's interpretation of Ørsted's experiment. They also rejected completely Ampère's research program. The main criticisms they made of Ampère's work were presented as follows:

1. According to Biot and Savart, the interaction between current-carrying wires was a secondary effect arising from something more basic, namely, the magnetization of these wires due to the passage of the electric current. Accordingly, the interaction between two current-carrying wires would be due to the supposed magnetic poles spread over the cross section of one wire interacting with the magnetic poles spread over the cross section of the second wire. According to Biot and Savart, the fundamental interaction took place between the molecules of austral and boreal fluids (that is, the North and South poles of each magnet belonging to the first current-carrying wire interacting with the North and South poles of each magnet belonging to the second current-carrying wire). According to Ampère, on the other hand, the fundamental interaction took place between current-carrying elements.

2. Biot and Savart rejected not only Ampère's conception of a force between current elements depending on the spatial orientations of these elements, but also the more basic supposition of the existence of current elements which were spatially orientated. According to Biot and Savart there was nothing similar to this hypothesis in the known interactions (gravitational, electrostatic and magnetostatic). These last three interactions are central ones; they act between point-like objects and they depend only on the distance between the interacting particles.

3. Biot and Savart considered Ampère's supposition of molecular currents an unnecessary complication.

All these points were emphasized by Biot in the third edtion of his book *Précis élémentaire de Physique expérimentale* of 1824:[8]

> M. Ampère proposed to make these phenomena the fundamental principle of the whole theory of electromagnetism, by considering them, not as compound results in the way we have done, but as simple effects resulting from an attraction or repulsion, which the electric currents would exert upon each other immediately, without sensible tension, according as they are transmitted through the metallic conductors in the same direction or in opposite directions. This hypothesis, which attributes to fluid currents an

[6] [Biot, 1821a, p. 119].
[7] [Biot and Savart, 1824, p. 706], [Farrar, 1826, p. 307] and [Biot and Savart, 1885, p. 82].
[8] [Biot and Savart, 1824, pp. 771-772], [Farrar, 1826, pp. 362-363] and [Biot and Savart, 1885, pp. 124-125].

> attractive property, depending on their different or similar directions, is in the first place completely opposed in itself to all the analogy observed in the other laws of attraction. It would, moreover, be necessary to modify it by another entirely arbitrary circumstance, in order to deduce from it the variation of intensity which is observed in the transverse action of the elementary laminae of the uniting wire, according to the obliquity of their direction to the lines which separate them from the magnetic particles subjected to the action; whereas this particular may be considered as only a compound result of the unknown distribution of the elementary magnetism, when we attribute the magnetic action of the wires to such an action. Finally, upon the latter supposition, the influence of the uniting wires upon magnets is referred to the general analogy of the action of magnetic bodies upon each other; while, in order to explain this influence according to the hypothesis in question, M. Ampère is obliged to make a multitude of other still more complicated suppositions; for he is under the necessity of considering all the mutual actions of magnetic bodies in general as produced by voltaic currents circulating about the metallic particles which compose them, in a manner greatly resembling the vortices of Descartes. Hence arises a complication of arrangements and suppositions very difficult to be explained; while, on the other hand, these phenomena, although not yet rendered capable of being calculated in their compound character, considered as depending upon an elementary magnetism imparted by the voltaic current, offer nothing in themselves which may not be easily conceived. For this reason, I have thought proper to give the observations of M. Ampère, without adopting his explanation, presenting them merely as compound results derived from the more simple phenomena, instead of recognising in them a simple primitive principle from which all the results are to be deduced.

Biot's statement according to which Ampère was "obliged to make a multitude of other still more complicated suppositions", or that it would be necessary to modify Ampère's hypothesis "by another entirely arbitrary circumstance" seems incorrect to us. It also seems to us inappropriate to compare Ampère's molecular currents with the vortices of Descartes (1596-1650).

As regards the first two statements, it will be seen in Part V of this book that Ampère considered the word "complicated" differently from Biot. Biot believed that the complication appeared in the consequences of Ampère's postulates. Biot considered these consequences of Ampère's theory abnormal. According to Ampère, on the other hand, complicated would be to create postulates or additional entities in order to keep the consequences of theories more in agreement with intuition. Moreover, as will be seen in Chapter 21, instead of having entirely arbitrary circumstances, Ampère's theory explained qualitatively and quantitatively all phenomena known in his time.

We now discuss Biot's third statement, according to which Ampère's molecular currents resembled the vortices of Descartes. Descartes proposed vortices of a subtle matter rotating around the Sun in order to offer a mechanical explanation for the planetary system. These vortices would push the planets around the Sun, exerting in them a tangential force parallel to their motion. It would be analogous to a mechanical wind pushing a windmill and making it turn around its axis. Ampère, on the other hand, supposed that the magnetic properties of magnets were due to electric currents flowing around the particles composing the magnet. According to Ampère, the interactions between two current elements would obey Newton's third law of motion, that is, would act along the straight line connecting their midpoints. He could describe the interactions between closed current-carrying loops by means of the integration of a direct interaction between two current elements. According to Ampère it was not necessary to suppose a mechanical connection between two electric currents. No vortex or wind was necessary to explain the interaction between two current-carrying conductors. There is no correspondence between Ampère's molecular currents and the vortices of Descartes.

In Section 16.2 we will discuss Ampère's criticisms against Biot and Savart's conceptions.

Chapter 13

Faraday *Versus* Ampère

13.1 Faraday's Interpretation of Ørsted's Experiment

The main electromagnetic researches of Faraday began after Ørsted's announcement of his discovery in July 1820. Starting in 1813 Faraday had been an assistant of Humphry Davy (1778-1829) at the Royal Institution in London.[1] Faraday initially reproduced some of the main experiments performed by Ørsted, Ampère, and a few other scientists and then began his own researches. Between 1821 and 1822 he published a paper, in three parts, presenting a historical sketch of electromagnetism.[2] Although published anonymously, he later on assumed its authorship.[3]

In 1821 Faraday performed some experiments analyzing the torque acting on a horizontal magnetic needle close to a vertical wire carrying a constant current. He interpreted his observations in terms of the forces exerted by the current-carrying wire and acting on the magnetic poles of the needle. From these experiments he concluded that these poles were not located exactly at the extremities of the needle. Moreover, the forces exerted by the wire on the magnetic pole did not point towards the wire. They were orthogonal to the wire and to the straight line connecting the pole to the wire. These forces should cause the rotation or revolution of the pole around the wire. These rotational, rotary, or revolutive forces were not attractive nor repulsive. Although in these experiments Faraday did not observe the motion of the current-carrying wire due to the forces exerted by the magnet, he believed these opposite forces should be present. Probably he was thinking in terms of Newton's action and reaction law. Faraday described his experiments, somewhat analogous to those of Ørsted, as follows:[4]

> It is evident from this that the centre of the active portion of either limb of the needle, or the true pole, as it may be called, is not at the extremity of the needle, but may be represented by a point generally in the axis of the needle, at some little distance from the end. It was evident, also, that this point had a tendency to revolve round the wire, and necessarily, therefore, the wire round this point; and as the same effects in the opposite direction took place with the other pole, it was evident that each pole had the power of acting on the wire by itself, and not as any part of the needle, or as connected with the opposite pole.
>
> [...]
>
> Several important conclusions flow from these facts; such as that there is no attraction between the wire and either pole of a magnet; that the wire ought to revolve round a magnetic pole, and a magnetic pole round the wire; that both attraction and repulsion of connecting wires, and probably magnets, are compound actions; [...]
>
> The revolution of the wire and the pole round each other being the first important thing required to prove the nature of the force mutually exerted by them, various means were tried to succeed in producing it. [...]

In September 1821, Faraday succeeded with these experiments of continuous rotation of the extremity of a current-carrying wire around a magnet, and also with the experiments of continuous rotation of the extremity of a magnet around a current-carrying wire, as discussed in Section 7.1. Once more he interpreted

[1] [Davy, 1821, p. 18].
[2] [Faraday, 1821a], [Faraday, 1821b] and [Faraday, 1822b].
[3] [Williams, 1989b, p. 85, n. 14].
[4] [Faraday, 1822e, p. 76] and [Faraday, 1952d, p. 797].

these experiments based on the existence of magnetic poles, together with the forces between these poles and the current-carrying wire. According to Faraday, these forces followed the principle of action and reaction. However, they were not directed along the shortest distance connecting each pole with the long straight wire. According to Faraday, these forces were orthogonal to this shortest line, being also orthogonal to the wire. They caused the rotation of the pole around the wire, together with the opposite rotation of the wire around the pole. These opposite forces causing a mutual rotation are represented in figure 13.1. Each one of these forces might be reversed by reversing the direction of the current or the type of magnetic pole. By reversing simultaneously the direction of the current and the type of magnetic pole, the forces would remain pointing as indicated in this figure.

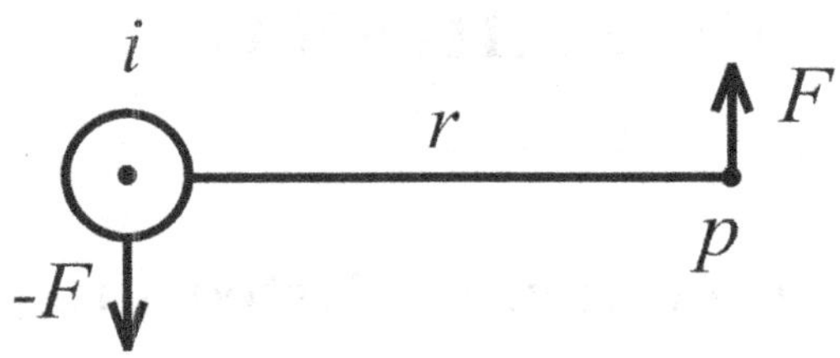

Figure 13.1: A long straight wire normal to the plane of the paper, with a current i coming out of the paper. The arrows indicate the forces exerted between a North pole p of the magnetized needle and the current-carrying wire, according to Faraday's conceptions.

13.2 Faraday Against Ampère

As discussed in Section 7.1, Ampère abandoned most of his electrodynamic researches between January and September 1821. Faraday's discovery of continuous rotation made Ampère resume his researches. Ampère recognized the importance of Faraday's discovery as regards his own motivation. However, he emphasized that Faraday's conceptions used to explain these phenomena were contrary to his own interpretations. In a letter addressed to C. J. Bredin, dated December 3, 1821, Ampère mentioned this controversy:[5]

> When arriving here [in Paris,] metaphysics occupied my thoughts; however, after Faraday's work, I think only of electric currents. This memoir contains very singular electromagnetic facts, which perfectly confirm my theory, although the author tries to fight against it by replacing it with a [theory] of his creation.

We list here some objections Faraday presented against Ampère's conceptions and explanations:

1. Faraday was always skeptical about the idea that an electric current is due to the motion of electric charges.

2. Faraday doubted Ampère's magnetic conception, according to which the magnetic properties of the Earth and magnets were due to electric currents flowing in the Earth and in magnets.

3. According to Faraday the simplest or basic cases to be considered were the circular motion of a magnetic pole around a current-carrying wire and the opposite circular motion of a current-carrying wire around a magnetic pole. He believed that the attraction and repulsion between two current-carrying wires should be considered a complex phenomenon, which might be explained in terms of simpler configurations not involving the direct interaction between current-carrying conductors.

In his historical sketch of electromagnetism, published between 1821 and 1822, Faraday expressed his skepticism relative to the usual conception of an electric current as being due to the flow of electric charges:[6]

> Those who consider electricity as a fluid, or as two fluids, conceive that a current or currents of electricity are passing through the wire during the whole time it forms the connection between the poles of an active [voltaic] apparatus. There are many arguments in favour of the materiality of electricity, and but few

[5] [Launay (ed.), 1936a, pp. 576-577] and [Blondel, 1982, pp. 109-110].
[6] [Faraday, 1821a, p. 196] and [Blondel, 1982, p. 52].

> against it; but still it is only a supposition; and it will be as well to remember, while pursuing the subject of electro-magnetism, that we have no proof of the materiality of electricity, or of the existence of any current through the wire.

Faraday's doubts mentioned in item 2 were expressed as follows in a letter to De la Rive of 12 September 1821:[7]

> But yet I am by no means decided that there are currents of electricity in the common magnet. I have no doubt that electricity puts the circles of the helix into the same state as those circles are in that may be conceived in the bar magnet, but I am not certain that this state is directly dependent on the electricity, or that it cannot be produced by other agencies; and therefore, until the presence of electrical currents be proved in the magnet by other than magnetical effects, I shall remain in doubt about Ampère's theory.

Only after his own discovery of electromagnetic induction in 1831 did Faraday begin to have a more positive attitude towards Ampère's magnetic theory. He mentioned the following in a letter to Phillips, dated 29 November 1831:[8]

> Then I found that magnets would induce just like voltaic currents, and by bringing helices and wires and jackets up to the poles of magnets, electrical currents were produced in them; these currents being able to deflect the galvanometer, or to make, by means of the helix, magnetic needles, or in one case even to give a spark. Hence the evolution of *electricity from magnetism*. The currents were not permanent. They ceased the moment the wires ceased to approach the magnet, because the new and apparently quiescent state was assumed, just as in the case of the induction of currents. But when the magnet was removed, and its induction therefore ceased, the return currents appeared as before. These two kinds of induction I have distinguished by the terms *volta-electric* and *magneto-electric* induction. Their identity of action and results is, I think, a very powerful proof of M. Ampère's theory of magnetism.

In his fundamental article of 1831 describing his discovery of the induction of electric currents, Faraday expressed himself as follows:[9]

> The similarity of action, almost amounting to identity, between common magnets and either electro-magnets or volta-electric currents, is strikingly in accordance with and confirmatory of M. Ampère's theory, and furnishes powerful reasons for believing that the action is the same in both cases; but, as a distinction in language is still necessary, I propose to call the agency thus exerted by ordinary magnets, *magneto-electric* or *magnelectric* induction.

We now present the criticism of item 3 made by Faraday against Ampère's ideas. In his paper of 1821 describing the discovery of continuous rotation, Faraday mentioned the following:[10]

> A simple case which may be taken of magnetic motion, is the circle described by the wire or the pole round each other.

On the other hand, as regards the phenomenon Ampère discovered of the attraction and repulsion between parallel current-carrying wires, Faraday made the following evaluation:[11]

> [...] the attractions and repulsions of M. Ampère's wires are not simple, but complicated results.

Faraday believed that it would be possible to explain the interactions between current-carrying conductors supposing only the interactions between a current-carrying wire and the supposed magnetic poles existing in the other current-carrying wire.

In his letter to De la Rive, dated 12 September 1821, Faraday emphasized this point of view as follows:[12]

> I find all the usual attractions and repulsions of the magnetic needle by the conjunctive wire are deceptions, the motions being not attractions or repulsions, nor the result of any attractive or repulsive forces, but the result of a force in the wire, which, instead of bringing the pole of the needle nearer to or further from the wire, endeavours to make it move round it in a never-ending circle and motion whilst the battery remains in action. I have succeeded not only in showing the existence of this motion theoretically, but experimentally, and have been able to make the wire revolve round a magnetic pole, or a magnetic pole round the wire, at pleasure. The law of revolution, and to which all the other motions of the needle and wire are reducible, is simple and beautiful.

---

[7] [Jones, 1870a, p. 317].
[8] [Jones, 1870b, p. 8].
[9] [Faraday, 1952c, §58, p. 273].
[10] [Faraday, 1822e, p. 79] and [Faraday, 1952d, p. 799].
[11] [Faraday, 1822e, p. 79] and [Faraday, 1952d, p. 799].
[12] [Jones, 1870a, p. 316] and [Gross, 2009].

Chapter 14

Grassmann *Versus* Ampère

14.1 Grassmann's Force between Current Elements

Hermann Günther Grassmann (1809-1877) was a German linguist and mathematician, figure 14.1. He studied theology at Berlin University. Apparently he did not follow the mathematics nor the physics courses of this university. After graduation he returned to his native town, Stettin, remaining there until the end of his life, working as a high school teacher of mathematics. He never taught at a university and apparently never performed experiments on electromagnetism. He developed a new algebra and was a member of the Göttingen Academy of Sciences. He married in 1849 and had 11 children. He translated the *Rig Veda* from Sanscrit to German.

Figure 14.1: H. G. Grassmann.

In 1845 he published a paper on a new theory of electrodynamics.[1] In 1877 he published another work on this subject showing that Clausius's expression for the force between current elements, also published in 1877, was identical to his own expression published in 1845.[2]

In his paper of 1845 he presented a force between current elements which was different from Ampère's expression. Grassmann accepted Ampère's conception of current elements orientated and infinitely small, denominating them a and b. According to Grassmann these magnitudes represent the following:[3]

> [...] where a and b are the circuit elements, that is, the infinitely small linear portions multiplied by the current intensity, in which the currents move; [...]

In this work he also utilized,[4] in particular, $a = ids$. We can then similarly write: $b = i'ds'$. Grassmann presented a new mathematical expression describing the force exerted by a on b. In order to understand his formula we first present figure 14.2. In figure 14.2 (a), there is an xy plane containing the length ds of the current element $a = ids$, the straight segment r connecting the midpoint of a to the midpoint of the

[1] [Grassmann, 1845] and [Grassmann, 1965].
[2] [Grassmann, 1877] and [Clausius, 1877].
[3] [Grassmann, 1845, p. 3] and [Grassmann, 1965, p. 203].
[4] [Grassmann, 1845, pp. 6 and 9] and [Grassmann, 1965, pp. 205 and 208].

current element $b = i'ds'$. In this plane the projection of the length ds' of the current element b on the xy plane, namely, b_l, is also represented. The current element a makes an angle α with the segment r, with $0 \leq \alpha \leq \pi$ rad. Figure 14.2 (b) presents a plane orthogonal to the xy plane containing the z axis passing through the midpoint of the current element $b = i'ds'$, containing the length ds' of the current element b and containing also the projection b_l of the element b on the xy plane.

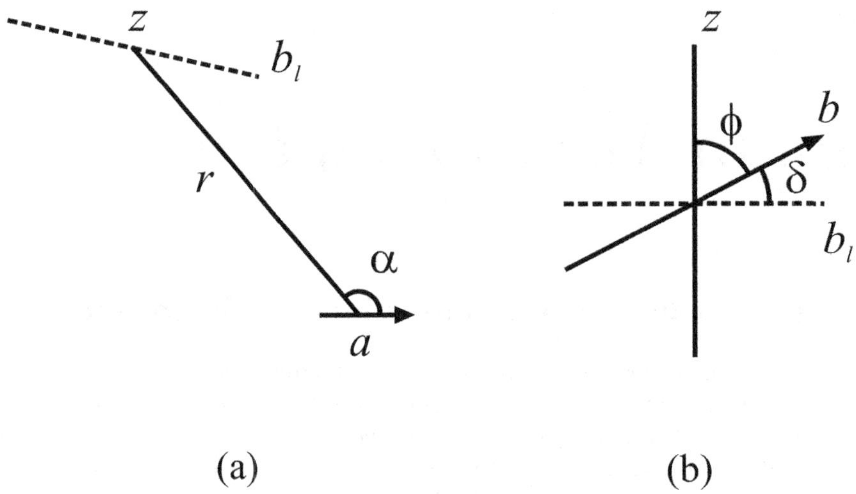

Figure 14.2: (a) Plane xy formed by the infinitesimal length ds of the current element $a = ids$ and by the segment r. (b) Plane formed by the z axis, by the length ds' of the current element $b = i'ds'$ and by the projection b_l of the current element b on the xy plane.

The current element $b = i'ds'$ makes an angle δ with the xy plane. It also makes an angle $\phi = \pi/2 - \delta$ with the normal to the xy plane. That is, the current element $b = i'ds'$ makes an angle ϕ with the z axis. The segment b_l is a projection of $b = i'ds'$ on the xy plane. We can then write the following relation:

$$b_l = b \cos \delta = b \sin \phi \ . \tag{14.1}$$

It is possible to utilize trigonometric relations in order to express equation (14.1) in terms of Ampère's angles, although this was not made by Grassmann. Consider Ampère's angles β and γ as defined in Section 2.8. That is, β is the angle which the current element $b = i'ds'$ makes with the segment r, while γ is the angle between the planes formed by each element with the straight line connecting their midpoints. Equation (14.1) can then be written as follows:

$$b_l = b \cos \delta = b \sin \phi = b\sqrt{1 - \sin^2 \beta \sin^2 \gamma} \ . \tag{14.2}$$

Figure 14.3 presents all relevant magnitudes in perspective.

Grassmann's force exerted by a on b was presented as follows:[5]

$$\frac{ab_l}{r^2} \sin \alpha \ . \tag{14.3}$$

Equation (14.3) was described by the following words:[6]

> We then obtain $(ab_l/r^2) \, . \sin \alpha$ as the expression for the effect exerted by a current element a on another b, distant r from it, the vertical projection of the second element on the plane through a and r being equal to b_l, while α represents the angle formed between a and the line drawn to b. The movement [of b] then occurs perpendicularly to b (or b_l) in the plane through a and r, [...]

That is, the force acting on b is always orthogonal to this element. The intensity or magnitude of Grassmann's force is given by equation (14.3).

It is important to point out that, according to Grassmann's expression, there is no component of the force exerted by $a = ids$ on $b = i'ds'$ which acts along the direction of this current element b. Therefore,

[5] [Grassmann, 1845, p. 9] and [Grassmann, 1965, p. 208].
[6] [Grassmann, 1845, p. 9] and [Grassmann, 1965, p. 208].

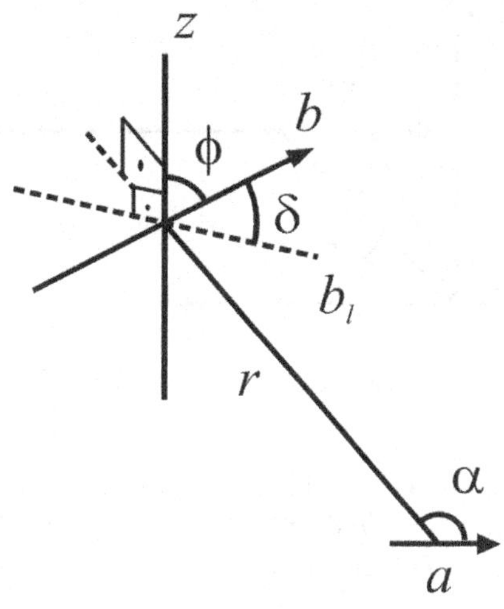

Figure 14.3: The current element $a = ids$ exerts a force on the current element $b = i'ds'$.

when we have two current elements pointing along the straight line connecting their midpoints, they exert no net force on one another according to Grassmann's expression. In this case we have $\alpha = 0$ rad or $\alpha = \pi$ rad, while $\beta = 0$ rad or $\beta = \pi$ rad. According to Ampère's expression, on the other hand, these two aligned and parallel current elements should attract (when the parallel currents flow toward each other) or repel (when both parallel currents flow towards the same side) one another.

In his paper of 1877, Grassmann utilized a kind of vector formulation. He considered the *external product* between two segments a and b as being the area of the parallelogram formed by these segments, representing this external product by $[a \cdot b]$. Figure 14.4 presents segments ds and ds' with the angle ε between them. The area of this parallelogram is given by $dsds' \sin\varepsilon$.

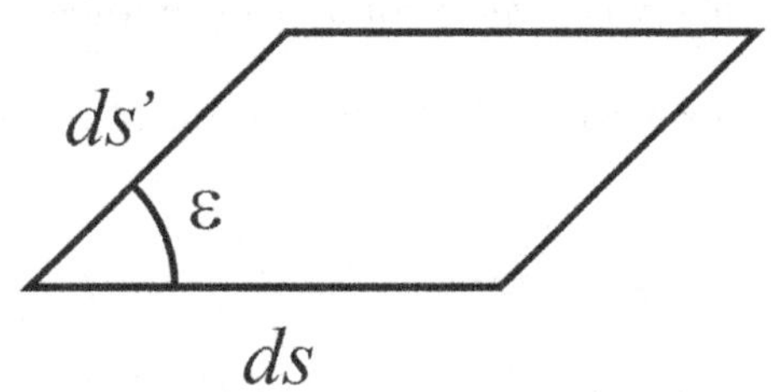

Figure 14.4: Angle ε formed by segments ds and ds'.

Therefore, Grassmann's external product is similar to the magnitude of the modern vector product. That is, although Grassmann utilized a point in his external product, his product does not represent the modern scalar product, which is also represented with a point. Suppose we have two orientated segments $d\vec{s}$ and $d\vec{s}\,'$ with an angle ε between them. Grassmann's external product, $[ds \cdot ds']$, would be represented nowadays by $|d\vec{s} \times d\vec{s}\,'| = dsds' \sin\varepsilon$, as represented in equation (14.4).

$$[ds \cdot ds']_{\text{Grassmann}} = |d\vec{s} \times d\vec{s}\,'|_{\text{modern}} = dsds' \sin\varepsilon \; . \tag{14.4}$$

In this article of 1877 Grassmann also defined the *internal product* between an area λF with a segment b as being given by the product between λF and the projection of b on this area. He represented this internal product by $[F|b]$. Figure 14.5 shows the element b making an angle δ with the area λF. The z axis is orthogonal to this area, making an angle $\phi = \pi/2 - \delta$ with b.

The internal product between b and λF is given by $b\cos\delta\lambda F = b\sin\phi\lambda F$. Suppose we have three orientated segments, namely, $\vec{a}_1$, $\vec{a}_2$ and $\vec{b}$, with $\vec{a}_1$ and $\vec{a}_2$ composing the area λF. In modern vector notation we have $\lambda F = |\vec{a}_1 \times \vec{a}_2| = a_1 a_2 \sin\varepsilon$, where ε represents the angle between $\vec{a}_1$ and $\vec{a}_2$. The magnitude represented by Grassmann as $[F|b]$ would be then represented nowadays as $|(\vec{a}_1 \times \vec{a}_2) \times \vec{b}|$, namely:

$$[F|b]_{\text{Grassmann}} = |(\vec{a}_1 \times \vec{a}_2) \times \vec{b}|_{\text{modern}} = (a_1 a_2 \sin\varepsilon) b \sin\phi \; . \tag{14.5}$$

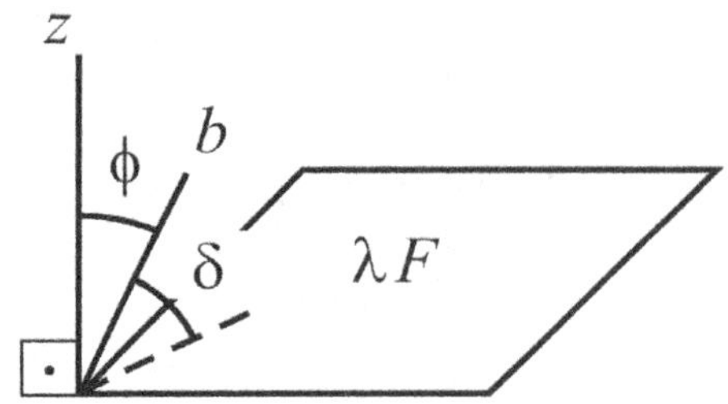

Figure 14.5: The z axis is orthogonal to the area λF, while the b segment makes an angle δ with this area.

In his paper of 1877 Grassmann inverted the denominations he had utilized in 1845 for the current elements. In 1845 he had $a = ids$ and $b = i'ds'$, while in 1877 he utilized $a = i'ds'$ and $b = ids$. In 1877 he expressed the force P exerted by $\underline{a} = i'\underline{ds'}$ and acting on $\underline{b} = i\underline{ds}$ as being given by:[7]

$$P = \frac{k}{r^3}\left[\underline{r}\cdot\underline{a}|\underline{b}\right] . \tag{14.6}$$

In this equation the magnitude k represents a numerical constant, the value of which was not specified by Grassmann. In the paper of 1845 Grassmann had adopted $k = 1$.

14.1.1 Grassmann's Force in Modern Vector Notation

Utilizing modern vector notation, Grassmann's force $d^2\vec{F}^G_{i'ds' \text{ on } ids}$ exerted by the current element $i'd\vec{s}\,'$ and acting on the current element $id\vec{s}$ is given by:[8]

$$d^2\vec{F}^G_{i'ds' \text{ on } ids} = id\vec{s} \times \frac{ki'd\vec{s}\,' \times \hat{r}}{r^2} . \tag{14.7}$$

In this equation r represents the distance between the midpoints of both current elements, k is a constant depending on the system of units, while $\hat{r}$ is a versor of unit magnitude pointing from the center of $i'd\vec{s}\,'$ to the center of $id\vec{s}$.

Grassmann's force $d^2\vec{F}^G_{ids \text{ on } i'ds'}$ exerted by the current element $id\vec{s}$ and acting on the current element $i'd\vec{s}\,'$ is given by:

$$d^2\vec{F}^G_{ids \text{ on } i'ds'} = -i'd\vec{s}\,' \times \frac{kid\vec{s} \times \hat{r}}{r^2} \neq -d^2\vec{F}^G_{i'ds' \text{ on } ids} . \tag{14.8}$$

In the International System of Units, Grassmann's force $d^2\vec{F}^G_{I'ds' \text{ on } Ids}$ exerted by the current element $I'd\vec{s}\,'$ and acting on the current element $Id\vec{s}$ is given by:[9]

$$d^2\vec{F}^G_{I'ds' \text{ on } Ids} = Id\vec{s} \times \left[\frac{\mu_0}{4\pi}\frac{I'd\vec{s}\,' \times \hat{r}}{r^2}\right] = -\frac{\mu_0}{4\pi}\frac{II'}{r^2}\left[(d\vec{s}\cdot d\vec{s}\,')\hat{r} - (d\vec{s}\cdot\hat{r})d\vec{s}\,'\right] . \tag{14.9}$$

In this equation r represents the distance between the midpoints of the two current elements, while $\hat{r}$ is the versor of unit magnitude pointing from the center of $d\vec{s}\,'$ towards the center of $d\vec{s}$. The currents I and I' are in amperes (A).

Grassmann's force $d^2\vec{F}^G_{Ids \text{ on } I'ds'}$ exerted by the current element $Id\vec{s}$ and acting on the current element $I'd\vec{s}\,'$ is given by:

$$d^2\vec{F}^G_{Ids \text{ on } I'ds'} = I'd\vec{s}\,' \times \left[\frac{\mu_0}{4\pi}\frac{Id\vec{s} \times (-\hat{r})}{r^2}\right] = \frac{\mu_0}{4\pi}\frac{II'}{r^2}\left[(d\vec{s}\cdot d\vec{s}\,')\hat{r} - (d\vec{s}\,'\cdot\hat{r})d\vec{s}\right] \neq d^2\vec{F}^G_{I'ds' \text{ on } Ids} . \tag{14.10}$$

Grassmann's force, in general, does not act along the straight line connecting the midpoints of the interacting current elements. Despite this fact, in some particular configurations it will follow the principle

[7] [Grassmann, 1877, pp. 58 and 64].

[8] [Assis, 1992, Chapter 3], [Assis, 1994, Chapter 4], [Bueno and Assis, 2001, Section 5.2], [Bueno and Assis, 2015, Section 5.2] and [Assis, 2015b, Chapter 3].

[9] [Assis, 1992, Chapter 3], [Assis, 1994, Chapter 4], [Bueno and Assis, 2001, Section 5.2], [Bueno and Assis, 2015, Section 5.2] and [Assis, 2015b, Chapter 3].

of action and reaction in the weak form. That is, the force exerted by $Id\vec{s}$ on $I'd\vec{s}\,'$ will be equal and opposite to the force exerted by $I'd\vec{s}\,'$ on $Id\vec{s}$, although these two forces of action and reaction will not point along the straight line connecting the midpoints of both elements. However, in other configurations Grassmann's force will not satisfy the principle of action and reaction, not even in the weak form. That is, there are situations in which the force exerted by $Id\vec{s}$ on $I'd\vec{s}\,'$ will not be equal and opposite to the force exerted by $I'd\vec{s}\,'$ on $Id\vec{s}$, so that these two opposite forces will either have different magnitudes, or will not be parallel to one another, as will be discussed in Section 20.3.

14.2 Grassmann Against Ampère

Grassmann's force given by equations (14.7) up to (14.10) is different from Ampère's force given by equations (2.1) up to (2.20).

In his paper of 1845 Grassmann made three criticisms against Ampère's force, namely:

1. Ampère's force between current elements is complicated.

2. There is no *a priori* reason justifying why the force between two current elements should always act along the straight line connecting their midpoints.

3. A specific aspect of Ampère's force seemed improbable to Grassmann. He considered a special case in which two current elements, $a = i'ds'$ and $b = ids$, were parallel to one another, as in figure 14.6. In this case we have $\alpha = \beta$ and $\varepsilon = 0$. Ampère's force between two parallel current elements changes from attraction to repulsion at a critical angle between the direction of the elements and the straight line connecting their centers given by $\arccos\sqrt{2/3} = 35.26^o$. That is, when $\alpha = \beta$ is smaller than 35.26^o they will repel one another, while when $\alpha = \beta$ is greater than 35.26^o they will attract one another. Grassmann did not accept this property of Ampère's force.

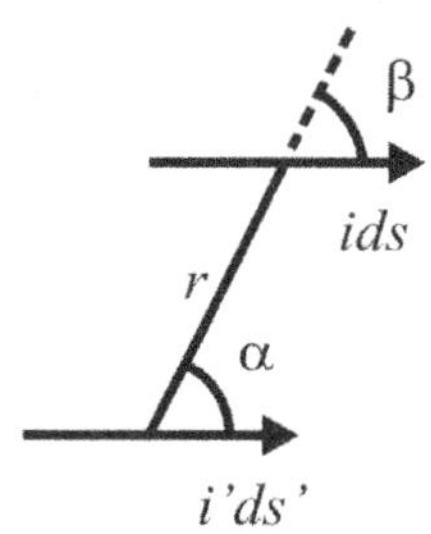

Figure 14.6: Two parallel current elements.

Grassmann's statements presenting his criticisms:[10]

> When I submitted the explanation offered by Ampère for the interaction of two infinitely small current-sections on one another to a more exacting analysis, *this explanation seemed to me a highly improbable one*; [...]
>
> (2) *Ampère was obliged, therefore, in order to obtain his formula, to use an arbitrary assumption* together with the experimental results. The assumption used for this purpose is, at first glance, very simple and natural, consisting in the supposition that two infinitely small circuit elements exert forces on each other along the straight line connecting their midpoints, either of attraction or repulsion. By means of this assumption Ampère is able to proceed from the experimental results directly to his basic formula, according to which the force exerted by one infinitely small circuit element a on another such element b is proportional to the expression:

$$(2\cos\varepsilon - 3\cos\alpha\cos\beta)\frac{ab}{r^2}\ , \tag{14.11}$$

[10][Grassmann, 1845, pp. 1 and 3-4] and [Grassmann, 1965, pp. 201-203], our emphasis.

> where a and b are the circuit elements, that is, the infinitely small linear portions multiplied by the current intensity, in which the currents move; r is the distance of the midpoints of these elements from each other; ε is the angle between the two circuit portions; and α and β are the angles formed by the elements a and b respectively with the line drawn between the two midpoints.
>
> (3) *The complicated form of this formula arouses suspicion*, and the suspicion is heightened when an attempt is made to apply it. If, for example, the simplest case is considered, in which the circuit elements are parallel, so that $\varepsilon = 0$ and $\alpha = \beta$, the Ampère expression becomes
>
> $$\left(2 - 3\cos^2\alpha\right)\frac{ab}{r^2} , \quad (14.12)$$
>
> from which it appears that, when $\cos^2\alpha$ is equal to 2/3 or, which comes to the same thing, $\cos 2\alpha$ is equal to 1/3, that is, if the position of the midpoint of the attracted element lies on the surface of a cone whose apex is at the attracting element, and whose apex angle is arccos 1/3, there is no interaction; while for smaller angles there is repulsion, and for larger ones attraction. *This is such an unlikely result, that the principle from which it is derived must come under the gravest suspicion and with it the supposition that the force in question must show an analogy with all other forces.* It must be concluded that there is little reason to apply this analogy to our present field. Since in the case of all other forces it is originally point elements, without any definite direction, which interact with each other, so that the mutual interaction must *a priori* be regarded as necessarily operating along the line connecting them, it is hard to see any justification for transferring this analogy to an entirely foreign field in which the elements are arranged in definite directions. The formula itself, which in no way resembles that for gravitational attraction, also indicates that there is no real analogy.

Grassmann then proposed equation (14.3) in order to circumvent these problems which he saw in Ampère's force between current elements.

Chapter 15

The Field Concept *Versus* Ampère's Conception

15.1 Multiple Definitions of the Magnetic Field

Nowadays Ampère's force between current elements usually does not appear in the textbooks. In order to explain magnetic, electromagnetic and electrodynamic interactions; the textbooks usually utilize the concept of a "magnetic field". In this Section we discuss this concept.[1]

Faraday utilized the word "field" for the first time on 7 November 1845 in his diary.[2] However, long before this time he already utilized expressions with analogous meanings, as "magnetic curves" or "lines of magnetic force".[3] He began to mention the "magnetic field" in his publications presented to the Royal Society in 1845, published in 1846.[4]

However, it was only in a paper of 1851 that he clearly defined this concept (the numbers in parenthesis refer to the paragraphs of Faraday's work):[5]

> 2806. I will now endeavour to consider what the influence is which paramagnetic and diamagnetic bodies, viewed as conductors (2797), exert upon the lines of force in a magnetic field. *Any portion of space traversed by lines of magnetic power, may be taken as such a field*, and there is probably no space without them. The condition of the field may vary in intensity of power, from place to place, either along the lines or across them; but it will be better to assume for the present consideration a field of equal force throughout, and I have formerly described how this may, for a certain limited space, be produced (2465).

That is, according to Faraday a magnetic field may be considered as any portion of space traversed by lines of magnetic power (or by lines of force, another expression which he sometimes used instead of lines of magnetic power). According to Faraday, these lines of magnetic power might be visualized by means of iron fillings.[6]

Maxwell adopted Faraday's conceptions and expressed them mathematically. In a paper of 1864 on the theory of the electromagnetic field he presented the following definition:[7]

> (3) The theory I propose may therefore be called a theory of the *electromagnetic field*, because it has to do with the space in the neighbourhood of the electric and magnetic bodies, and it may be called a *dynamical* theory, because it assumes that in that space there is matter in motion, by which the observed electromagnetic phenomena are produced.

---

[1] [O'Rahilly, 1965, Vol. 2, Chapter 13, Section 4: The 'Field', pp. 645-661], [Larson, 1982], [Gardelli, 2004], [Krapas and Silva, 2005], [Anderton, 2007], [Silva and Krapas, 2007], [Krapas and Silva, 2008], [Ribeiro, 2008], [Ribeiro et al., 2008], [Assis et al., 2009], [Assis and Chaib, 2011], [Assis, 2013] and [Assis, 2014, Section 3.1: Multiple definitions of the field concept].

[2] [Nersessian, 1989, Note 17].

[3] [Faraday, 1952c, p. 281, §114], [Faraday, 1952c, p. 595, §2149] and [Faraday, 1952c, p. 758, §3071].

[4] [Faraday, 1952c, p. 608, §2247] and [Faraday, 1952c, p. 608, §2252].

[5] [Faraday, 1952c, p. 690, §2806], our emphasis.

[6] [Faraday, 1952c, p. 281, §114] and [Faraday, 1952c, p. 758, §3071].

[7] [Maxwell, 1865, p. 460] and [Maxwell, 1965, p. 527]. Digits (3) and (4) refer to the numeration of Maxwell's paper. Emphasis given by Maxwell.

> (4) The electromagnetic field is that part of space which contains and surrounds bodies in electric and magnetic conditions.

In his book *A Treatise on Electricity and Magnetism*, published in 1873, we find the same definition of "field" as a region of space around electrified and magnetized bodies:[8]

> 44.] The electric field is the portion of space in the neighbourhood of electrified bodies, considered with reference to electric phenomena.

Maxwell utilized a similar definition for the "magnetic field" generated by a current-carrying wire (he was referring to Ørsted's experiment):[9]

> 476.] It appears therefore that in the space surrounding a wire transmitting an electric current a magnet is acted on by forces depending on the position of the wire and on the strength of the current. *The space in which these forces act may therefore be considered as a magnetic field*, and we may study it in the same way as we have already studied the field in the neighbourhood of ordinary magnets, by tracing the course of the lines of magnetic force, and measuring the intensity of the force at every point.

Faraday and Maxwell's definitions of the field concept coincide with one another. According to these authors, the electric and magnetic fields are regions of space around electrified particles, magnets and current-carrying wires. These definitions were also accepted by other authors of the past and are still adopted nowadays. Joseph John Thomson (1856-1940), for instance, after stating the basic triboelectric phenomena stated the following:[10]

> The sealing-wax and the flannel are said to be *electrified*, or to be in a state of *electrification*, or to be charged with *electricity*; and the region in which the attractions and repulsions are observed is called the *electric field*.

Jeans (1877-1946) presented the same definition:[11]

> 30. The space in the neighbourhood of charges of electricity, considered with reference to the electric phenomena occurring in this space, is spoken of as the electric field.

Heilbron resumed these definitions as follows:[12]

> *Field* in general signifies a region of space considered in respect to the potential behaviour of test bodies moved about in it; the electricians of 1780 lacked the word but not the concept, which they called 'sphere of influence', *sphaera activitatis*, or *Wirkungskreis*. [...]

Faraday, Maxwell and Hendrik Antoon Lorentz (1853-1928) believed that the space between electrified particles, between magnets, or between current-carrying wires, was filled with a material medium. They believed that this material medium was responsible for the interaction between charges, magnets, or current-carrying wires. This material medium was called "ether". The existence of this medium was at the basis of the mathematical formulations of Maxwell and Lorentz.

In his paper of 1905 introducing the special theory of relativity, Albert Einstein (1879-1955) made the ether superfluous and considered that light and the electromagnetic waves propagate in empty space, our emphasis:[13]

> Examples of this sort, together with the unsuccessful attempts to discover any motion of the Earth relatively to the "light medium," suggest that the phenomena of electrodynamics as well as of mechanics possess no properties corresponding to the idea of absolute rest. They suggest rather that, as has already been shown to the first order of small quantities, the same laws of electrodynamics and optics will be valid for all frames of reference for which the equations of mechanics hold good. We will raise this conjecture (the purport of which will hereafter be called the "Principle of Relativity") to the status of a postulate, and also introduce another postulate, which is only apparently irreconcilable with the former, namely, that *light is always propagated in empty space with a definite velocity* c which is independent of

[8][Maxwell, 1954, vol. 1, §44, p. 47].

[9][Maxwell, 1954, vol. 2, §476, p. 139], our emphasis.

[10][Thomson, 1921, p. 1].

[11][Jeans, 1941, p. 24].

[12][Heilbron, 1981, p. 187].

[13][Einstein, 1952, pp. 37-38] and [Einstein, 1978, p. 48].

> the state of motion of the emitting body. These two postulates suffice for the attainment of a simple and consistent theory of electrodynamics of moving bodies based on Maxwell's theory for stationary bodies. The introduction of a "luminiferous ether" will prove to be superfluous inasmuch as the view here to be developed will not require an "absolutely stationary space" provided with special properties, nor assign a velocity-vector to a point of the empty space in which electromagnetic processes take place.

Later on Einstein and Infeld expressed themselves as follows:[14]

> Our only way out seems to be to take for granted the fact that space has the physical property of transmitting electromagnetic waves, and not to bother too much about the meaning of this statement.

Since then this material medium, the ether, was considered superfluous and was abandoned in physics.

In any event it should be emphasized that Faraday, Maxwell and the modern textbooks present many other alternative definitions and properties associated with the field concept:[15]

- The field is a region of space around gravitational masses, around electric charges, around magnetic poles, around magnets, and around current-carrying wires.
- A field is a real physical entity filling the space.
- The field is a vector quantity (with magnitude and direction).
- The electromagnetic field propagates in a material medium according to Maxwell.
- The electromagnetic field propagates in empty space according to Einstein.
- The field stores energy, linear momentum and angular momentum.
- The field mediates the action between gravitational masses, between electric charges, between magnetic poles, between magnets, between current-carrying wires, mediating also the interaction between a magnet and a current-carrying wire.
- Field is a magnitude with dimensions.
- The field as the lines of force taken together.
- The field as a state of the space.
- The field is generated or produced by source bodies like gravitational masses, electric charges, magnetic poles, magnets and electric currents.
- The field due to source bodies generates or produces a force on other test bodies like gravitational masses, electric charges, magnetic poles, magnets and electric currents.
- A field can be transformed into another field.
- A field changing in time can produce or induce another field.
- Condensations of the electromagnetic field are the elementary particles of matter.
- Etc.

Many of these definitions or properties are contradictory to one another. They also contradict Faraday and Maxwell's initial definitions. For instance, how is it possible for a region of space to propagate in space? How is it possible for a region of space to have magnitude and direction? How can a region of space distort space? How can a region of space, something immaterial, interact with a material body like an electrified particle or a current-carrying wire? As we have discussed these contradictions in other publications, we will not deal with this topic here.[16]

---

[14] [Einstein and Infeld, 1938, p. 159].

[15] For the original sources and references presenting these alternative formulations see, for instance, [Gardelli, 2004], [Krapas and Silva, 2005], [Silva and Krapas, 2007], [Krapas and Silva, 2008], [Ribeiro, 2008], [Ribeiro et al., 2008], [Assis et al., 2009], [Assis and Chaib, 2011], [Assis, 2013] and [Assis, 2014, Section 3.1: Multiple definitions of the field concept].

[16] [Assis, 2013, Section 2.9] and [Assis, 2014, Section 3.2: The Different Field Definitions Contradict One Another].

15.2 The Sources of the Magnetic Field

The textbooks normally consider four magnitudes as possible sources of a magnetic field: the magnetic poles of a magnet, the current elements belonging to a conducting wire, electrified particles moving in space, and the temporal variations of an electric field. As in this textbook we are only considering constant currents, we will not discuss this last source of a magnetic field.

Maxwell, in his *Treatise*, presented the following definition:[17]

> The ends of a long thin magnet are commonly called its Poles.

In the same section he presented Coulomb's law for magnetism as follows:[18]

> The repulsion between two like magnetic poles is in the straight line joining them, and is numerically equal to the product of the strengths of the poles divided by the square of the distance between them.

In §383 of his *Treatise*, Maxwell mentioned the *magnetic potential* due to a pole of a magnet. In §389 he discussed the potential energy of the poles of a magnet placed in a magnetic field generated by other magnets. The magnetic field generated by a magnetic pole was analogous to the electrostatic field due to a charge at rest, being also analogous to the gravitational field of a stationary mass. That is, the field at a certain point of space, due to a magnetic pole located at another point of space, would be central, pointing along the straight line connecting the first point to the pole, and varying inversely as the square of this distance. It would be also directly proportional to the magnitude of the pole generating the field.

In vector notation and in the International System of Units, the magnetic field $\vec{B}(\vec{r})$ at a certain point $\vec{r}$ of an inertial coordinate system, generated by a magnetic pole P' located at the origin of this coordinate system is given by:

$$\vec{B}(\vec{r}) = \frac{\mu_0}{4\pi}\frac{P'}{r^2}\hat{r} \ . \tag{15.1}$$

In this equation $r \equiv |\vec{r}|$ represents the distance between the pole and the point where the field is being calculated, while $\hat{r} \equiv \vec{r}/r$ represents the unit vector pointing from the origin towards the point under consideration.

Nowadays it is not so common to talk about magnetic poles, as usually only electric currents are considered as sources of a magnetic field. For instance, when dealing with electromagnetism in his *Treatise*, Maxwell initially mentioned the "lines of magnetic force" produced by a long current-carrying wire.[19]

Let us suppose a current element $I'd\vec{s}\,'$ with its center located at the origin of a Cartesian coordinate system pointing along the direction of the current I'. The magnetic field $d\vec{B}(\vec{r})$ at point $\vec{r}$ generated by this current element is given by:[20]

$$d\vec{B}(\vec{r}) = \frac{\mu_0}{4\pi}\frac{I'd\vec{s}\,' \times \hat{r}}{r^2} \ . \tag{15.2}$$

Sometimes textbooks present the formula expressing the magnetic field $\vec{B}(\vec{r})$ at a point $\vec{r}$ of an inertial coordinate system, with the field generated by a point charge q' which is passing through the origin while moving with velocity $\vec{v}\,'$. This field is given by:[21]

$$\vec{B}(\vec{r}) = \frac{\mu_0}{4\pi}\frac{q'\vec{v}\,' \times \hat{r}}{r^2} \ . \tag{15.3}$$

The meaning of this velocity $\vec{v}\,'$ of the charge q' has changed over the years. Nowadays it is interpreted as being the velocity of the charge q' *relative to an inertial frame of reference.*[22]

---

[17] [Maxwell, 1954, vol. 2, §373, p. 3].

[18] [Maxwell, 1954, vol. 2, §373, p. 3].

[19] [Maxwell, 1954, vol. 2, §477, p. 139].

[20] [Tricker, 1962, pp. 453-455], [Tricker, 1965, p. 25], [Whittaker, 1973, pp. 82-83], [Purcell, 1980, p. 187, equation (38)], [Reitz et al., 1982, p. 167, equation (8-27)] and [Griffiths, 1989, p. 222].

[21] [Reitz et al., 1982, p. 162, equation (8-4)].

[22] [Feynman et al., 1964, p. 13-6], [Assis and Peixoto, 1992], [Assis, 1992, Appendix A, pp. 114-117], [Assis, 1994, Appendix A: "The origins and meanings of the magnetic force $\vec{F} = q\vec{v} \times \vec{B}$", pp. 238-241], [Assis, 2013, Section 14.5], [Assis, 2014, Section 15.5: Origins and meanings of the velocity $\vec{v}$ which appears in the magnetic force $q\vec{v} \times \vec{B}$, pp. 273-283] and [Assis, 2015b, Appendix B, pp. 161-170].

15.3 The Force Exerted by a Magnetic Field

In Section 15.2 it was seen that the magnetic field may have different sources. Likewise it is usually supposed that it can act on different entities. When it is considered that it acts on a magnetic pole P located at the position vector $\vec{r}$ of a Cartesian coordinate system, the magnetic force $\vec{F}$ is given by:

$$\vec{F} = P\vec{B}(\vec{r}) \ . \tag{15.4}$$

A North pole is usually considered positive, while a South pole is negative. Therefore, the force exerted by a magnetic field on a North pole points along the direction of $\vec{B}$, while the force acting on a South pole acts against the magnetic field.

When it is considered that the magnetic force acts on a current element $Id\vec{s}$ pointing along the direction of the current I with its center located at the position vector $\vec{r}$, the magnetic force $d\vec{F}$ is given by:[23]

$$d\vec{F} = Id\vec{s} \times \vec{B}(\vec{r}) \ . \tag{15.5}$$

It should be observed that, by replacing the magnetic field $\vec{B}$ of equation (15.5) by the magnetic field $d\vec{B}$ generated by another current element $I'd\vec{s}\,'$, as given by equation (15.2), we obtain Grassmann's force between current elements discussed in Section 14.1.

Consider now a point electrical charge q located at the position vector $\vec{r}$ and moving with velocity $\vec{v}$ relative to an inertial frame of reference. According to classical electromagnetism the electromagnetic force acting on q is given by:[24]

$$\vec{F} = q\vec{E}(\vec{r}) + q\vec{v} \times \vec{B}(\vec{r}) \ . \tag{15.6}$$

In this equation $\vec{E}(\vec{r})$ and $\vec{B}(\vec{r})$ represent the electric and magnetic fields, respectively, at the location $\vec{r}$ of the test charge q. These electric and magnetic fields are due to other source charges. As stated before, $\vec{v}$ represents the velocity of the test charge q *relative to an inertial frame of reference.*[25]

Equation (15.6) was first obtained by J. C. Maxwell between 1861 and 1873, and by H. A. Lorentz in 1895.[26] For this reason it will be called in this book the Maxwell-Lorentz force.

The magnetic component of the Maxwell-Lorentz force acting on the test charge q is given by:[27]

$$\vec{F} = q\vec{v} \times \vec{B}(\vec{r}) \ . \tag{15.7}$$

Once more the velocity $\vec{v}$ represents the velocity of the charge q relative to an inertial frame of reference, while $\vec{B}(\vec{r})$ represents the magnetic field, due to other charges, acting at the position vector $\vec{r}$ of the test charge q.

15.4 The Field Concept Against Ampère's Conception

According to Ampère, the interaction between two magnets or the interaction between a magnet and a current-carrying conductor were due only to interactions between electric currents. He supposed the existence of electric currents flowing inside magnets and also inside the Earth. He could then reduce the magnetic and electromagnetic interactions to electrodynamic interactions. He also speculated that maybe these electrodynamic interactions might take place with the intermediation of a material medium, the ether. But he never developed this last idea quantitatively.

[23][Purcell, 1980, p. 175, equation (6)], [Reitz et al., 1982, p. 164, equation (8-8)] and [Heald and Marion, 1995, p. 21, equation (1.54)].

[24][Lorentz, 1895], [Feynman et al., 1964, p. 1-2, equation (1-1) and p. 13-1, equation (13.1)], [Purcell, 1980, p. 141, equation (1)], [Pais, 1982, p. 125], [Reitz et al., 1982, p. 162, equation (8-5)], [Pais, 1986, p. 76], [Griffiths, 1989, p. 198, equation (5.2)] and [Heald and Marion, 1995, p. 21, equation (1.53)].

[25][Feynman et al., 1964, p. 13-6], [Assis and Peixoto, 1992], [Assis, 1992, Appendix A, pp. 114-117], [Assis, 1994, Appendix A: "The origins and meanings of the magnetic force $\vec{F} = q\vec{v} \times \vec{B}$", pp. 238-241], [Assis, 2013, Section 14.5], [Assis, 2014, Section 15.5: Origins and meanings of the velocity $\vec{v}$ which appears in the magnetic force $q\vec{v} \times \vec{B}$, pp. 273-283] and [Assis, 2015b, Appendix B, pp. 161-170].

[26][Marinov, 1990, p. 31], [Ribeiro, 2008], [Curé, 2009, Section 4.6: On the paternity of Lorentz's force, pp. 122-128], [Huray, 2010, p. 22], [Tombe, 2012a], [Tombe, 2012b], [Assis, 2013, Section 14.5] and [Assis, 2014, Section 15.5: Origins and meanings of the velocity $\vec{v}$ which appears in the magnetic force $q\vec{v} \times \vec{B}$, pp. 273-283].

[27][Reitz et al., 1982, p. 162, equation (8-3)] and [Heald and Marion, 1995, p. 21, equation (1.52)].

In any event it should be emphasized that even when Ampère speculated that the interaction between two current elements might take place with the intermediation of a material medium like the ether, he believed that Newton's third law of motion should remain valid in its strong form. That is, the forces of action and reaction should take place along the straight line connecting the interacting bodies.[28]

Ampère's electrodynamics was based essentially on two suppositions: (a) his force law between current elements, equations (2.1) up to (2.20); and (b) the existence of electric currents flowing inside magnets and inside the Earth.

We have then essentially direct interactions at a distance between current-carrying conductors. Ampère's force between current elements is central and satisfies the principle of action and reaction in the strong form, acting along the straight line joining the midpoints of both current elements.

With the field concept, on the other hand, the interaction is conceived differently. It is said, for instance, that a current element ids generates a magnetic field. This magnetic field propagates in space, typically at light velocity. When this magnetic field reaches the position of a second current element $i'ds'$, the magnetic field exerts a force on this second current element. Analogously, this second current element $i'ds'$ would produce another magnetic field which would also propagate in space, acting on the first current element ids after reaching it. However, in the textbooks it is not explained how the magnetic field interacts with the current element on which it is exerting a force. That is, no mechanism is presented to explain how an immaterial entity like the magnetic field might act on matter, that is, on a current-carrying wire.

Maxwell, for instance, expressed clearly the opposition between his points of view (following those of Faraday and based on the field concept) and the suppositions of other scientists based on action at a distance:[29]

> For instance, Faraday, in his mind's eye, saw lines of force traversing all space where the mathematicians saw centres of force attracting at a distance: Faraday saw a medium where they saw nothing but distance: Faraday sought the seat of the phenomena in real actions going on in the medium, they were satisfied that they had found it in a power of action at a distance impressed on the electric fluids.

As regards specifically the interactions between two current-carrying conductors, Maxwell expressed himself as follows:[30]

> 645.] In explaining the electromagnetic force by means of a state of stress in a medium, we are only following out the conception of Faraday[31], that the lines of magnetic force tend to shorten themselves, and that they repel each other when placed side by side. All that we have done is to express the value of the tension along the lines, and the pressure at right angles to them, in mathematical language, and to prove that the state of stress thus assumed to exist in the medium will actually produce the observed forces on the conductors which carry electric currents.
>
> We have asserted nothing as yet with respect to the mode in which this state of stress is originated and maintained in the medium. We have merely shewn that it is possible to conceive the mutual action of electric currents to depend on a particular kind of stress in the surrounding medium, instead of being a direct and immediate action at a distance.

That is, the main criticisms of Faraday and Maxwell against Ampère's conception was related to the concept of action at a distance, which Maxwell did not accept. This concept appears in Newton's law of gravitation, in the force between stationary electric charges, in the force between magnetic poles and also in Ampère's force between current elements. Instead of action at a distance, Faraday and Maxwell suggested the field concept as the intermediary agent responsible for the interaction between current-carrying conductors.

Despite this different epistemological point of view, Maxwell always expressed a great admiration for Ampère's theoretical and experimental works. Instead of trying to replace Ampère's force between current elements by something different, Maxwell said[32] that Ampère's formula should always remain the cardinal formula of electrodynamics, as mentioned in Section 2.1.

---

[28] [Ampère, 1826f, pp. 112-113] and [Ampère, 1823c, Ampère, 1990, pp. 284-285].

[29] [Maxwell, 1954, vol. 1, preface to the first edition, p. ix].

[30] [Maxwell, 1954, vol. 2, §645, p. 282].

[31] [Note by Maxwell: *Exp. Res.*, 3266, 3267, 3268.] Maxwell was quoting Faraday's work: [Faraday, 1952c, §§3266, 3267 and 3268].

[32] [Maxwell, 1954, vol. 2, article 528, p. 175].

Part V

Controversies, Part II: Ampère Against Most Scientists

Chapter 16

Ampère Against His Main Opponents

Ampère's main experimental and theoretical works on electrodynamics ranged from 1820 to 1827. In this period he developed his extremely original conception that magnetism is due to electric currents flowing in magnets and in the Earth. Moreover, he obtained his force between current elements and several extremely relevant results from the integration of this expression. At this phase he received the support of some colleagues and students like Arago, Gaspard and Auguste de la Rive, Fresnel, Savary, Liouville and Demonferrand. Despite this support, the main scientists working on electromagnetism at this time were totally against several aspects of his conceptions. Among his main opponents we can quote Ørsted, Biot, Savart, Faraday and Grassmann, although this last scientist published his criticisms after Ampère's death.

In this part of the book we discuss how Ampère answered the main criticisms to his work and what his main objections were against the theories of these scientists. Initially we list Ampère's arguments against his main opponents, presenting some specific arguments against some of them. We then discuss some generic arguments which can be applied to the conceptions of one or more of these authors.

Before discussing specific topics, it is worthwhile presenting one of the main points of controversy between Ampère and his opponents, in particular Ørsted, Biot, Savart and Faraday.[1] This aspect of the disagreement is reflected in the interpretation of Ørsted's original experiment of 1820. All of these authors, with the exception of Ampère, interpreted this experiment based on gyratory forces or rotatory actions exerted by the current-carrying wire on the supposed magnetic poles of the magnetized needle placed close to the wire. Let i be the constant electric current flowing in a long and straight wire. The magnetic poles of the needle had a real existence according to all opponents of Ampère. The forces exerted by the wire on the poles of the magnet would act in planes orthogonal to the straight wire, being also orthogonal to the straight line connecting the midpoint of the magnet to the wire. These rotatory forces would act in opposite senses on the North and South poles of the magnet which were initially aligned with the wire, as indicated in figure 16.1 (a). Figure 16.1 (b) presents a cross section of figure 16.1 (a).

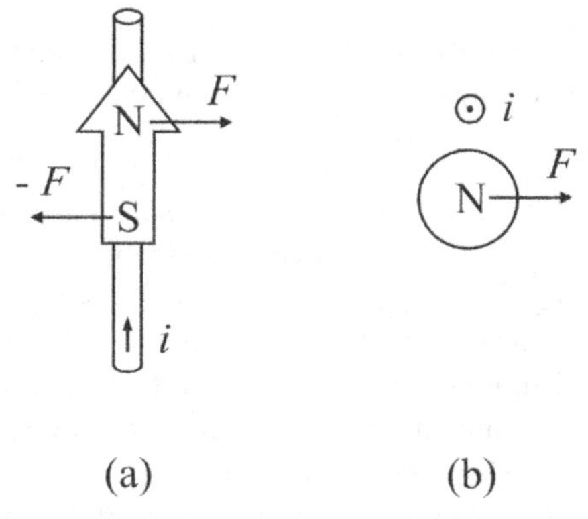

Figure 16.1: (a) Forces exerted by a long and straight current-carrying wire and acting on the supposed magnetic poles of a compass needle NS aligned with the wire, according to the conceptions of Ørsted, Biot, Savart and Faraday. (b) A cross section of the previous situation. The current i of the wire is leaving the paper in this representation and we see only the upper North pole of the magnet.

[1] [Heilbron, 1981].

A similar interpretation is obtained nowadays in terms of the magnetic field produced by the current-carrying wire, as represented in figure 16.2 (a). Figure 16.2 (b) represents the forces acting on the North and South poles of a compass needle aligned with the wire, according to equation (15.4). Figure 16.2 (c) represents a cross section of this configuration, indicating only the force acting on the North pole of the magnet.

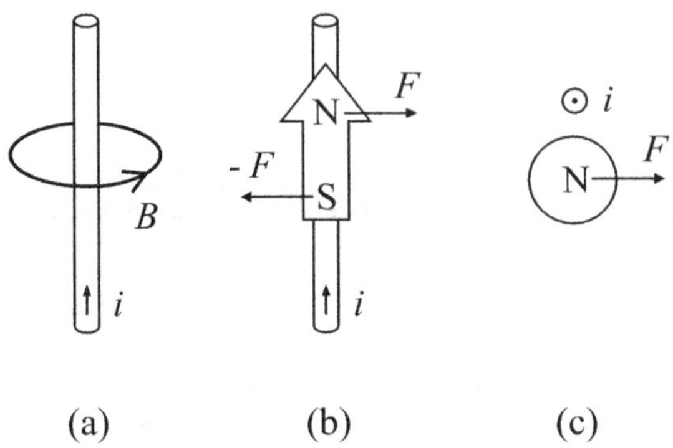

Figure 16.2: (a) Magnetic field produced by a long and straight current-carrying wire. (b) Forces exerted by the magnetic field on the North and South poles of a compass needle aligned with the magnet. (c) Cross section of this configuration showing only the force acting on the North pole of the magnet.

Ampère's interpretation of what happened in Ørsted's experiment was completely different. Instead of supposing the real existence of magnetic poles on the compass needle, Ampère, after Fresnel's suggestion, supposed there were microscopic currents flowing around the particles of the needle. The magnetic properties of magnets would be due to these molecular currents, as represented in figure 16.3 (a) and (b). The magnitude of each one of these molecular currents was represented by i'. These molecular currents would cancel one another inside the magnet, as they would be flowing in opposite directions in each internal point, there remaining only an effective electric current i' on the surface of the magnet, figure 16.3 (c).

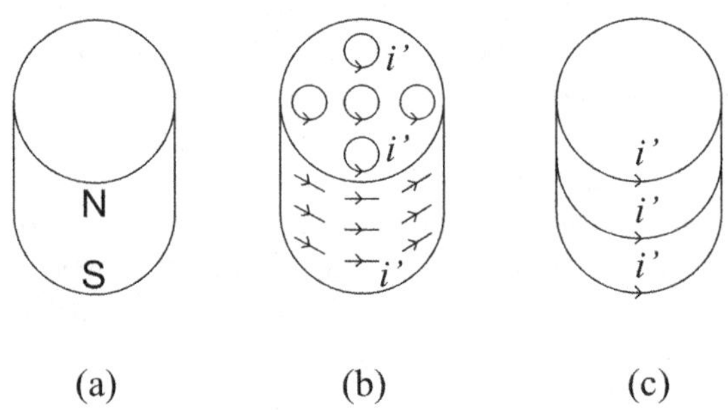

Figure 16.3: (a) Cylindrical magnet. (b) Ampère's molecular currents of intensity i'. (c) Effective electric currents flowing on the surface of the magnet due to the cancellation of the microscopic currents in any internal point of the magnet.

The torque exerted by a long straight wire carrying a constant current i acting on the magnet would be due to forces acting between each current element ids of the wire interacting with each current element $i'ds'$ of the microscopic currents of the magnet. Figure 16.4 presents a simplified view indicating the essence of Ampère's interpretation of Ørsted's experiment. In configuration (a) we have a long straight wire carrying a constant current i and a cylindrical magnet NS which is initially aligned with the wire. Figure 16.4 (b) presents a simplified view of Ampère's conception of a magnet, as indicated in figure 16.3. The cylindrical NS magnet has been replaced by only three circular current loops flowing on the surface of the magnet, each one with the same intensity i' of a microscopic current. Figure 16.4 (b) also indicates the net forces F and $-F$ exerted by the straight wire and acting on the surface currents of the magnet. These forces are parallel to the straight wire carrying a current i. Figure 16.4 (c) presents a cross section of this situation indicating only the current i of the wire leaving the plane of the paper and a single circular current loop i' in the plane of the paper representing the upper current i' flowing through the surface of the magnet. The force F acting on the left side of the circular loop leaves the paper, while the force $-F$ acting on the right side of the circular loop penetrates the paper. These opposite forces exert a torque on the magnet.

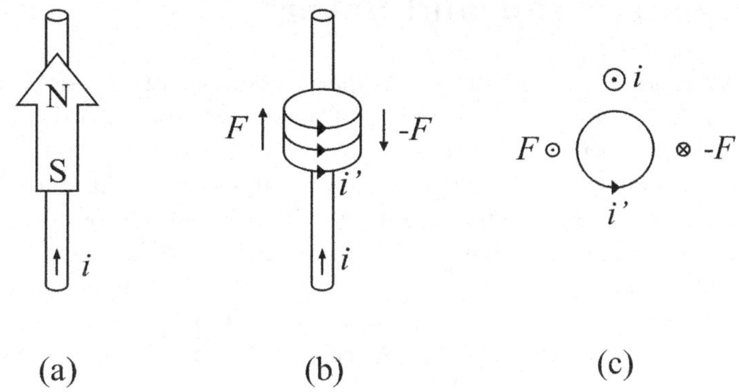

Figure 16.4: (a) Cylindrical magnet NS which is initially parallel to a long straight wire carrying a constant current i. (b) According to Ampère, the magnet would be equivalent to a series of current loops of intensity i' flowing through its surface. The forces F and $-F$ exerted by the wire on the current loops of the magnet are parallel to the wire. (c) Cross section view of this configuration showing a single current loop flowing through the surface of the magnet and the opposite forces F and $-F$ exerted on its left and right sides.

There are then two opposite interpretations of Ørsted's experiment. The one based on the existence of magnetic poles requires the existence of opposite gyratory forces acting tangentially like circles around the straight wire, as represented in figures 16.1 and 16.2. Ampère's explanation, on the other hand, is based only on the interaction between current elements and leads to opposite forces parallel to the wire, as represented in figure 16.4. In both cases the same torque is predicted acting on the compass needle, forcing it to point orthogonally to the wire. However, the mechanism responsible for these forces is completely different. These two opposing paradigms are discussed at length in this book.

16.1 Ampère Against Ørsted

As seen in Sections 1.3 and 3.1, Ampère's interpretation of Ørsted's experiment was totally different from Ørsted's own conception discussed in Section 11.1.

Ørsted was against Ampère's conception. Likewise, Ampère rejected Ørsted's interpretation. His main criticisms as regards Ørsted's theory were as follows:

1. Ampère did not believe in the existence of magnetic poles. He interpreted all magnetic and electromagnetic phenomena based on the interaction between electric currents, as will be discussed in Section 19.1.

2. Ampère was against Ørsted's idea of a vortex of electric charges flowing around a current-carrying wire, as discussed in Section 20.1.

3. Ampère pointed out that Ørsted's conception did not establish the identity between a magnetic fluid and a galvanic fluid, as discussed in Section 19.2.

4. Ampère identified two problems in Ørsted's idea about the existence of magnetic poles and in the supposed interaction of a magnetic pole with an electric charge. Ørsted had assumed this interaction in his own interpretation of his experiment, according to which electric charges flowing around the wire would propel the magnetic poles of the magnet. The two problems pointed out by Ampère were: (a) The interaction between distinct magnitudes, that is, the interaction of a magnetic pole with an electric charge. (b) Ørsted's conception did not lead to a real unification of the electromagnetic phenomena. In order to overcome these problems, Ampère proposed the interaction between current elements and the hypothesis of electric currents flowing in magnets and in the Earth, as discussed in Section 19.3.

5. According to Ampère's point of view, Ørsted never succeeded in explaining convincingly the interaction between two current-carrying wires, a phenomenon first observed and explained by Ampère. This topic is discussed in Subsection 21.1.2.

16.2 Ampère Against Biot and Savart

Biot and Savart criticized Ampère's suppositions. Ampère, likewise, attacked head on the hypotheses of these authors. In January 1821, Ampère even thought of an experiment in order to distinguish his theory from that of Biot, as discussed in Section 6.3. At this time Ampère utilized an incomplete form of his force law between current elements, believing that the constant k appearing in his force law given by equations (2.1) and (2.2) should vanish, that is, believing at that time that $k = 0$. He performed this experiment with Despretz and obtained a result against his own theory in its incomplete form, with $k = 0$. The outcome of this experiment was favorable to Biot's theory. It was only in 1823 that Savary was able to show, utilizing Ampère's complete force with $k = -1/2$, that this theory led to the same result for this specific experiment as the result given by Biot's theory, as discussed in Sections 6.3 and 9.8. Ampère always remained totally against Biot and Savart's conceptions.

Ampère's main criticisms against Biot and Savart's theory were as follows:

1. Ampère was against Biot and Savart's hypothesis about the existence of magnetic fluids (called austral and boreal fluids, or North and South poles, respectively), as discussed in Section 19.1.

2. Ampère pointed out mistakes made by Biot and Savart when they "deduced" the supposed force exerted by a current element acting on a magnetic pole, as discussed in Chapter 17.

3. According to Ampère, the primitive fact yielding the main elements capable of explaining all known phenomena was the interaction between current-carrying elements (or the interaction between current-carrying wires). Biot and Savart, on the other hand, believed that the interaction of a current-carrying wire with a magnet was the basic or primitive fact. This topic will be discussed in Section 19.3.

4. According to Ampère, the explanation Biot and Savart gave to account for Ørsted's experiment, based on the hypothesis of the magnetization of the wire by the passage of a current though it, led to contradictions related to the observed phenomena, as discussed in Section 18.1.

5. According to Ampère, moreover, Biot and Savart's hypothesis of the magnetization of the wire did not explain quantitatively nor qualitatively the interaction between two current-carrying conductors, as discussed in Subsection 21.1.3.

6. Biot and Savart's magnetization hypothesis was also against the phenomena of continuous rotation observed in Faraday and Ampère's experiments, as discussed in Section 18.3.

7. Ampère discussed in the *Théorie* the wrong explanation given by Biot as regards the experiment of continuous revolution of a magnet around a current-carrying wire.[2]

8. Ampère was also against Biot's explanation of the experiment of the rotation of a magnet around its own axis, as discussed in Section 21.3.

9. According to Ampère, the force proposed by Biot and Savart acting between a current element and a magnetic pole led to a primitive couple which violated Newton's third law in its strong form. That is, Biot and Savart's force was not a central force. It satisfied the principle of action and reaction. However, the action and reaction were not directed along the same straight line, being directed along parallel straight lines. Biot and Savart's action and reaction pair would generate a primitive couple in the system composed of a current element and a magnetic pole. Ampère's force between current elements, on the other hand, was central and always complied with Newton's third law of motion in its strong form, as discussed in Section 20.2.

10. Ampère showed that Biot and Savart's fundamental assumption in order to explain Ørsted's experiment, based only on the interaction between magnetic poles, did not lead to the unification of magnetic, electromagnetic and electrodynamic phenomena, as discussed in Section 22.1.

[2] [Ampère, 1826f, pp. 166-167] and [Ampère, 1823c, Ampère, 1990, pp. 338-339].

16.3 Ampère Against Faraday

The main criticisms Ampère raised against Faraday's conceptions were as follows:

1. Ampère was against Faraday's belief in the real existence of magnetic poles, as discussed in Section 19.1.

2. Faraday had argued in favor of a law of revolution, as discussed in Section 13.2. Ampère always fought against this rotational, rotary, gyratory, or revolutive action, as discussed in Section 20.1.2.

3. Ampère always considered that a primitive, fundamental, or primordial action must take place between magnitudes of the same type. Faraday, on the other hand, considered as a simple action the force he supposed to be acting between a current-carrying wire and a magnetic pole. This topic will be discussed in Section 19.3.

4. Ampère never accepted Faraday's explanation for the attractions and repulsions between current-carrying wires, a phenomenon first observed and explained by Ampère. This topic is dicussed in Subsection 21.1.4.

5. Ampère showed experimentally that Faraday was wrong when he did not believe it would be possible to produce a rotation of a magnet around its own axis of symmetry utilizing only electric currents, as discussed in Subsection 7.2.3.

6. When Faraday heard of Ampère's experiment producing the rotation of a magnet around its own axis, he offered an explanation based on a supposed torque exerted by the magnetic pole of the magnet acting on electric currents flowing inside the magnet. Ampère rejected Faraday's explanation as it violated Newton's third law, as discussed in detail in Section 21.3.

16.4 "Ampère" Against Grassmann

Ampère died in 1836. Therefore, he did not know Grassmann's criticisms against his force law, published in 1845, and could not answer them. This is the reason for the quotation marks in the title of this Section. In any event, from all that we know from his works and correspondence, it is easy to imagine some arguments he might utilize to defend against Grassmann's criticisms presented in Section 14.2. Some of these arguments might be the following:

1. Grassmann said that Ampère's force between current elements was complicated. Ampère might answer saying that his force law was the simplest of all formulas which was in agreement with the experimental results.

2. Grassmann said that Ampère was obliged to utilize an arbitrary supposition, namely, that the force between two current elements had to act along the straight line connecting them. Ampère might have replied that is was possible to describe all experimental results related to the interaction between current-carrying wires utilizing this supposition. Therefore, it would be preferable to assume this hypothesis than any other one. After all, the other forces of nature also behave like that. Examples: gravitational, electrostatic and magnetic forces. It does not make sense to complicate a theory when this is not necessary.

3. As discussed in Section 14.2, Grassmann considered it improbable that a force between two parallel current elements should change from attraction to repulsion at a critical angle between the direction of the elements and the straight line connecting them. Ampère might had replied that no experiment had been shown contrary to this prediction.

It is also possible to imagine some arguments which he might utilize against Grassmann's force. These arguments arise from his personality, correspondence and papers. They can also be obtained from the criticisms he presented against the works of Ørsted, Biot, Savart and Faraday:

1. Grassmann never presented experimental results which were in disagreement with Ampère's force between current elements. All criticisms he presented against Ampère's force were subjective, like

considering some of its aspects mathematically complicated or improbable. Ampère might utilize the same arguments against Grassmann's force.[3]

2. In some situations Grassmann's expression presents a rotational, rotary, or revolutive force acting between two current elements. Ampère criticized the rotational actions appearing in the conceptions of Biot, Savart and Faraday in the following terms:[4]

   > One is not obliged to admit as a simple and primitive fact a rotational action, an action which nature has never supplied any other example.

   Ampère might then utilize an analogous argument against Grassmann's force, by saying something like:

   > One is not obliged to admit as *a simple and primitive fact an action which does not comply with Newton's third law of motion, as is the case of Grassmann's force.* After all, nature has never supplied any other example of a force behaving like this.

   This topic is discussed in Section 20.2.

3. Ampère might have criticized Grassmann's force due to the fact that it does not predict a repulsion between two aligned and parallel current elements when both currents flow in the same sense along the straight line connecting their midpoints. After all, Ampère considered that he had observed this repulsion with his bridge experiment. Ampère's force predicts such a repulsion. He proposed the bridge experiment in order to verify the existence of this repulsion and the experiment confirmed his prediction, as discussed in Section 8.2. The alternative explanations of this experiment utilizing Ampère and Grassmann's forces are discussed in Section 21.2.

4. Ampère would certainly criticize the primitive couple appearing in Grassmann's force. As discussed in Section 20.2, there are situations in which Grassmann's forces between two current elements do not act along the straight line connecting their midpoints, giving rise to a primitive internal torque acting on this system.

5. Ampère would certainly be against Grassmann's force between current elements as in general it does not comply with Newton's action and reaction law. As discussed in Section 20.3, there are situations in which Grassmann's force exerted by a current element 1 on another current element 2 is not equal and opposite to the force exerted by 2 on 1. Therefore, as always emphasized by Ampère, Grassmann's force should be rejected by every scientist following the principles of Newtonian mechanics.

16.5 "Ampère" Against the Field Concept

The field concept was introduced in physics after Ampère's death, so that he could not react against this concept. The criticisms presented here are our adaptations of some similar arguments he presented against some analogous concepts presented by his contemporaries. This is the reason for the quotation marks in the title of this Section.

As seen in Chapter 15, Faraday and Maxwell defined the magnetic field as a region of space in the neighborhood of magnets and current-carrying wires. Moreover, they argued in favor of the existence of a material medium, the ether, filling the space between magnets and between current-carrying conductors. Faraday and Maxwell did not believe in a direct interaction between these bodies. The interaction between two magnets or between two current-carrying conductors should be mediated by stresses happening in the intervening medium.

As also discussed in Chapter 15, since the works of Einstein the concept of a material ether has been abandoned in physics as being superfluous. The modern electric and magnetic fields do not have a material support any longer.

Some of the main criticisms that Ampère could make to the field concept might be as follows:

1. Ampère would be against a revolutive magnetic field around a long straight wire carrying a constant current, as discussed in Subsection 20.1.3.

[3] [Tricker, 1965, p. 100].

[4] [Savary and Ampère, 1821, p. 374], [Savary and Ampère, 1822, p. 162] and [Savary and Ampère, 1885, p. 187].

2. According to Ampère, every elementary or fundamental force must act between magnitudes of the same type. Therefore, he would be against the idea of a "field" acting on an electric charge. He would also be against the idea of a "field" acting on a current element or on a current-carrying wire. This topic is discussed in Section 19.3.

3. Ampère was the first to succeed in making a magnet spin around its own axis due to the action of electric currents. This Ampère motor was discussed in Subsection 7.2.3. The modern explanation of this motor is based on the magnetic field concept, as discussed in Section 21.3. The modern explanation assumes that the magnetic field due to the magnet acts on currents flowing inside and/or around the surface of the magnet, generating an internal torque which would make the magnet spin around its axis even in the presence of friction. Ampère would certainly be against this explanation as it blatantly violates Newton's third law of motion.

4. According to the actual classical electromagnetic theory, a current element generates a magnetic field which can act on another current element. The expression satisfied by this magnetic field due to a current element is usually called Biot-Savart's law, although Biot and Savart did not work with the field concept. The modern force present in the textbooks is that given by Grassmann. Ampère would certainly be against three specific consequences of this interaction: (a) The absence of a net force between two collinear and parallel current elements, as discussed in Section 21.2; the primitive couple due to this magnetic field generated by a current element, as discussed in Section 20.2; and (c) the inequality between action and reaction in some specific situations, as discussed in Section 20.3.

Chapter 17

Errors Made by Biot and Savart in the "Deduction" of a Supposed Force Exerted by a Current Element and Acting on a Magnetic Pole (the so-called Biot-Savart's Law)

As discussed in Sections 6.1 and 6.2, Biot and Savart said that they deduced a supposed force exerted by a current element acting on a magnetic pole from the experiments which they performed with straight and bent wires acting on a compass needle. Ampère presented three main criticisms as regards this aspect of Biot and Savart's works, namely:

1. It is not possible to "deduce" an elementary force exerted by a current element utilizing a supposed integrated result valid for a macroscopic current-carrying wire.

2. Moreover, even neglecting the previous remark, Biot and Savart made a miscalculation when they tried to obtain the elementary force from the integrated result. They began with an integrated result A (force exerted by a bent wire acting on a compass) and obtained an elementary force B (force exerted by a current element acting on a magnetic pole). Biot and Savart believed that the elementary force B was compatible with A, although they did not check their equivalence. Savary and Ampère showed later on, in particular, that after integrating B they arrived at an integrated result C which was different from A. Even more remarkable, this integrated result C, the force exerted by a bent wire acting on a compass needle, did not correspond in general to the experimental value obtained by Biot himself in 1820, except in some particular situations.

3. Ampère pointed out that the initial experimental results of Biot and Savart related to the bent wire, published in 18 December 1820, were wrong.

17.1 First Error

On 30 October 1820, Biot and Savart described to the Academy of Sciences of Paris an experiment in which a small magnetized needle, suspended horizontally, was only under the action of a long vertical wire. They utilized an auxiliary magnet in order to eliminate the magnetic action of the Earth on the needle. They concluded that the equilibrium orientation of the needle was obtained when its North-South axis became orthogonal to the shortest straight line connecting its center to the wire and also orthogonal to the direction of the wire. They also measured the period of small oscillations of the needle around the vertical suspension thread, after being released with small deviations from the equilibrium orientation, at different distances from the wire. They concluded that the magnitude of the supposed force acting on each magnetic pole of the needle was inversely proportional to the distance between the pole and the wire. They said that, from this global result, Laplace had "mathematically deduced" that the force exerted by each current element of the wire acting on the magnetic pole would be inversely proportional to the square of the distance between the pole and the current element.

Laplace never published this "deduction". Biot and Savart accepted this deduction and quoted it without any criticism. However, as discussed in Section 6.1, it is not possible to deduce a force exerted by a current element utilizing only the integrated result of the force exerted by a wire of finite length. After all, there are other possible forces between a current element and a magnetic pole which do not vary as the inverse square of their separation but which, nevertheless, after being integrated lead to the result that the force exerted by a long straight wire acting on a magnetic pole varies as the inverse of the distance between the pole and the wire.

In conclusion, it is not possible to "deduce" an elementary force from an integrated result. What can happen, on the other hand, is that there are some elementary forces different from one another which, after integration in a specific configuration, lead to the same global effect. There are also other elementary forces, once more different from one another, which do not lead to this global effect after integration. It is then possible to say that the first group of elementary forces is compatible with this specific global integrated result, while the second group is not compatible with it. The possible existence of this first group was not considered by Biot.

Ampère criticized this aspect of Biot and Savart's work in his *Théorie*. His criticism was not only directed to the configuration of the straight wire, but also to the configuration of the bent wire. He said many things, from which we quote only an ironic statement:[1]

> [...] [this calculation] would have as a goal to determine the value of a differential beginning with the value of the definite integral obtained between given limits, but it does not seem to me that, up to now, any mathematician had considered [this calculation] something possible to be done.

17.2 Second Error

On 18 December 1820, Biot and Savart presented to the Academy of Sciences of Paris the experiment of the bent wire discussed in Section 6.2. Biot published a description of this experiment in 1821, in the second edition of his book *Précis élémentaire de Physique expérimentale*. According to their experimental results, the action of the wire on the magnetic pole would be proportional to half the opening angle of the bent wire, or proportional to the inclination of the bent wire with the horizon:[2]

> I thus found that, for both the bent wire and the straight wire, the action was inversely proportional to the distance to the points M and M'; but the absolute intensity was weaker for the bent wire than for the straight wire, in the proportion of the angle ZMH to unity.

From the analysis of the empirical result giving the action of the bent wire on the magnetized needle, they concluded that the supposed force of each current element acting on a magnetic pole would be proportional to the sine of the angle formed by the straight line connecting the pole to the element and by the straight line parallel to the wire at the position of this current element:[3]

> This result, analyzed by calculation, appeared to indicate to me that the action of each element μ of the oblique wire on each molecule m of austral or boreal magnetism is inversely proportional to the square of its distance μm to this molecule, and proportional to the sine of the angle $m\mu M$ formed by the distance μM with the direction of the wire.

They certainly made a mistake in their calculation. Unfortunately they did not present the "analysis by calculation" of how they arrived at the force produced by a current element beginning with the action produced by a bent wire of finite length. It is easy to confirm the existence of an error in this calculation by performing the opposite calculus. That is, one supposes initially that the force exerted by a current element on a magnetic pole is proportional to the sine of the angle formed by the direction of the current at the position of the element and by the line connecting the element to the pole. After integrating this force, one obtains that the force exerted by a bent wire on a magnetic pole is proportional to the *tangent of half the inclination of the bent wire to the horizon*. Therefore, this integrated force *is not proportional to the angle of inclination of the bent wire to the horizon*, although this last result was the departure point of Biot

[1] [Ampère, 1826f, p. 217], [Ampère, 1887d, p. 186] and [Ampère, 1823c, Ampère, 1990, pp. 382-383].

[2] [Biot, 1821a, p. 123] and [Biot, 1885a].

[3] [Biot, 1821a, p. 123], [Biot, 1885a] and [Hofmann, 1982, p. 242].

and Savart's calculation. These aspects were shown by Savary in 1823, as discussed in Sections 9.2 and 9.6. Ampère discussed in detail these calculations in the fifth note he included at the end of his *Théorie* of 1826.[4]

Ampère pointed out this error of Biot and Savart as follows:[5]

> It is very remarkable that this law[6] [...] was initially obtained by an error of calculation.

Another even more emphatic statement of Ampère in the *Théorie*:[7]

> This force, in inverse ratio of BM $= a$, is then, for the same value of a, proportional to the tangent of half of the angle CMH, and is not proportional to this very angle, although it has been claimed that the following value,
>
> $$\frac{\rho \sin \theta ds}{r^2},$$
>
> expressing the force exerted by the [current] element ds on the [magnetic] pole B, had been obtained *analyzing by calculation* the supposition that the force produced by the conducting wire CMZ was proportional to the angle CMH. It is not possible to doubt that there was some error in this calculation. But it would be equally curious to know it,[8] a calculation intended to determine the value of a differential beginning with the value of the definite integral obtained between given limits, but it does not seem to me that, up to now, any mathematician had considered [this calculation] something possible to be done.

17.3 Third Error

As discussed in Sections 9.6 and 9.7, Savary obtained in 1823, beginning with Ampère's force between current elements in its final form with $k = -1/2$, the integrated action of a bent wire acting on an electrodynamic cylinder. That is, the force and torque exerted by the bent wire acting on a cylindrical shell of finite length carrying an azimuthal current (a current around the surface of the cylinder, directed at each point of the surface along the angle φ of cylindrical coordinates). This current-carrying cylinder would be the electrodynamic equivalent of a small cylindrical magnet having the same length as the cylinder. Savary obtained that the action of the bent wire acting on the electrodynamic cylinder was proportional to the tangent of half of the inclination of the bent wire. This theoretical result is different from the experimental result described earlier by Biot and Savart, of 18 December 1820, according to which the action of the bent wire acting on the small magnet was proportional to the inclination of the bent wire.

If Biot and Savart experimental results had been really precise, this calculation of Savary would represent a serious blow against Ampère's theory. But what happened was exactly the opposite. Utilizing Savary's theoretical result as a guide, Biot and Savart once more performed their experiment in 1824, but now with much greater precision. Instead of maintaining their earlier result of 1820, they now concluded that the experimental result was proportional to the tangent of half of the inclination of the bent wire, exactly as predicted theoretically by Savary! Biot's words describing their new findings were described in Section 9.7.

Biot and Savart had then made a mistake in 1820 when they claimed that the result of their own experiments was proportional to the angle of inclination of the bent wire with the horizon. They recognized this mistake in 1824, after Savary's theoretical results were obtained utilizing Ampère's force between current elements. In the *Théorie* Ampère emphasized that the new experimental results of Biot and Savart of 1824 were in agreement with Savary's calculations of 1823 and not with the earlier experimental results of Biot and Savart of 1820:[9]

> It seems, moreover, that M. Biot recognized this error,[10] because in the third edition of the same work which has just been published,[11] he gives, as a matter of fact without quoting the memoir where it was

[4][Ampère, 1826f, pp. 216-217] and [Ampère, 1887d, pp. 185-186].

[5][Ampère, 1826f, p. 110], [Ampère, 1823c, Ampère, 1990, p. 282], [Ampère, 1887d, p. 99] and [Blondel, 1982, p. 141], our words in the footnote.

[6]That is, the law of Biot and Savart.

[7][Ampère, 1826f, p. 217], [Ampère, 1887d, p. 186] and [Ampère, 1823c, Ampère, 1990, pp. 382-383], our words in the footnote.

[8]That is, to know this wrong calculation.

[9][Ampère, 1826f, p. 110], [Ampère, 1823c, Ampère, 1990, p. 282] and [Ampère, 1887d, p. 99], our words in the footnotes.

[10]In the calculation of the elementary force exerted by a current element acting on a magnetic pole, beginning with the integrated result of the force exerted by a long current-carrying wire acting on a magnetic needle.

[11]That is, the *Précis élémentaire de Physique Expérimentale* of 1824, [Biot, 1824].

corrected,[12] new experiments in which the intensity of the total force is, in agreement with M. Savary's calculations, proportional to the tangent of half the angle ZHM, and concludes once again, with more reason than he had done in his first experiments, that the force which he calls elementary is, at equal distances, proportional to the sine of the angle included between the direction of the element of the conducting wire and the direction of the straight line connecting its midpoint to the magnetic molecule. (*Précis élémentaire de physique expérimentale*, third edition, volume II, pp. 740-745.)

[12]That is, without quoting Savary's paper.

Chapter 18

Criticism Against the Hypothesis of a Magnetization of the Wire Due to the Flow of an Electric Current

Some authors like Biot and Savart tried to explain Ørsted's experiment and the interaction between current-carrying wires by assuming a supposed magnetization of the wires. This magnetization would be due to the flow of the electric current through the wire. We can only work with this hypothesis by specifying the location and orientation of the small magnets which were supposed to be created by the passage of a current through the wire.

Ampère was against this hypothesis of the magnetization of the wire, as it did not lead to an explanation of Ørsted's experiment nor to an explanation of his own experiments showing the attractions and repulsions between current-carrying wires.

18.1 The Hypothesis of the Magnetization of the Wire Did Not Explain Ørsted's Experiment

There are two main criticisms against Biot and Savart's explanation of Ørsted's experiment supposing a magnetization of the wire due to the flow of an electric current, namely:

1. This hypothesis does not explain qualitatively the equilibrium orientation of a magnetized needle in all points around a current-carrying wire.

2. Even at the specific points in which this qualitative explanation of the orientation of the needle is correct, this hypothesis leads to a wrong quantitative law.

18.1.1 Qualitative Error

Biot presented in 1824 figure 18.1 in order to explain Ørsted's experiment with his hypothesis of the magnetization of the current-carrying wire.[1]

We made figure 18.2 in order to facilitate the comprehension of their explanation. In figure 18.2 (a) there is a cylindrical wire perpendicular to the plane of the paper, with the electric current i leaving the paper. We can assume that the wire is horizontal. We will neglect in this analysis the magnetic action of the Earth on the magnetized needle. If at point A there is a North magnetic pole, then Ørsted's experience indicates that there will be a force F acting on it and pointing to the left, as represented in figure 18.2.

Biot and Savart had explained Ørsted's experiment in 1820 assuming that the current-carrying wire acted on a molecule of austral or boreal magnetism, that is, acting on the supposed North and South poles of the magnetized needle. They assumed that this action took place due to the magnetization of the wire caused by the flow of an electric current through it. In the work describing their experiment they made the following statement:[2]

[1] [Biot and Savart, 1824, pp. 769-770], [Farrar, 1826, p. 361, figure 162] and [Biot and Savart, 1885, pp. 122-123].

[2] [Biot and Savart, 1820, p. 223], [Biot and Savart, 1965b, p. 119] and [Assis and Chaib, 2006, p. 308].

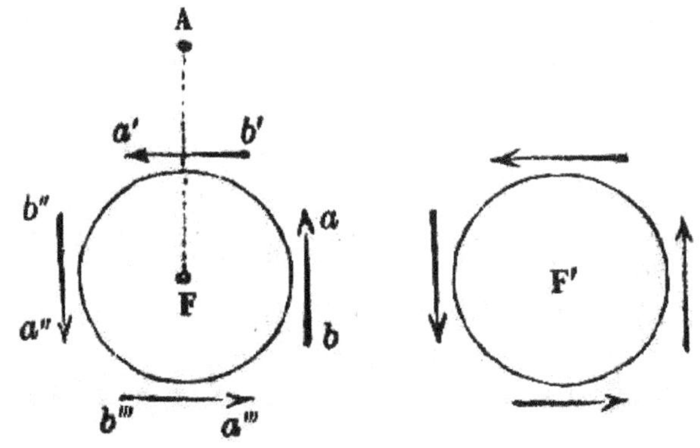

Figure 18.1: Biot and Savart's original figure utilized to explain Ørsted's experiment.

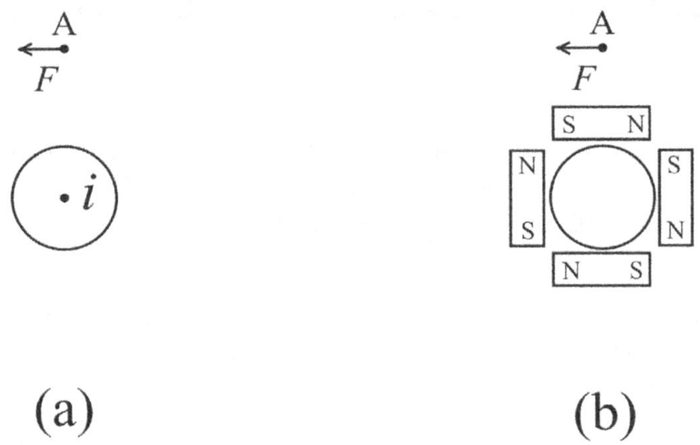

Figure 18.2: Biot and Savart's explanation of Ørsted's experiment.

> The nature of the action is the same as that of a magnetized needle which is placed on the contour of a wire in a certain constant direction in relation to the direction of the current; [...]

Biot made an analogous statement in 1824:[3]

> All experiments agree in showing that the total action of the wire upon such a particle situated in A, is exactly similar to what an infinitely small magnetic needle $a'b'$ would exert, having the direction of a tangent to the section F [of the current-carrying wire], the centre being placed upon the straight line A drawn from the magnetic particle to the centre of the wire.

Figure 18.2 (b) illustrates Biot and Savart's explanation. The force acting on a North pole situated at point A, pointing to the left, can be interpreted, according to Biot and Savart, as being due to a small magnetized needle orientated tangentially to the wire, as the small needle NS drawn above the horizontal wire of the figure. Biot drew four such magnetized needles placed tangentially around the wire, as represented in figures 18.1 and 18.2.

Ørsted's experiment, on the other hand, indicated that a supposed North pole would always be under the action of an anti-clockwise tangential force at all points around the wire, as indicated in figure 18.3 (a). The problem with Biot and Savart's explanation is that it only explains correctly the direction of the force acting on North poles placed at points A, C, E and G of figure 18.3 (b). On points B, D, F and H, on the other hand, the directions of the forces acting on a North pole would be opposite to the directions indicated by Ørsted's experiment.

By doubling the number of small magnetized needles around the current-carrying wire, we might explain correctly the direction of the force acting on a North pole placed at points $ABCDEFGH$. However, in this

[3] [Biot and Savart, 1824, pp. 769-770], [Farrar, 1826, p. 361] and [Biot and Savart, 1885, p. 123].

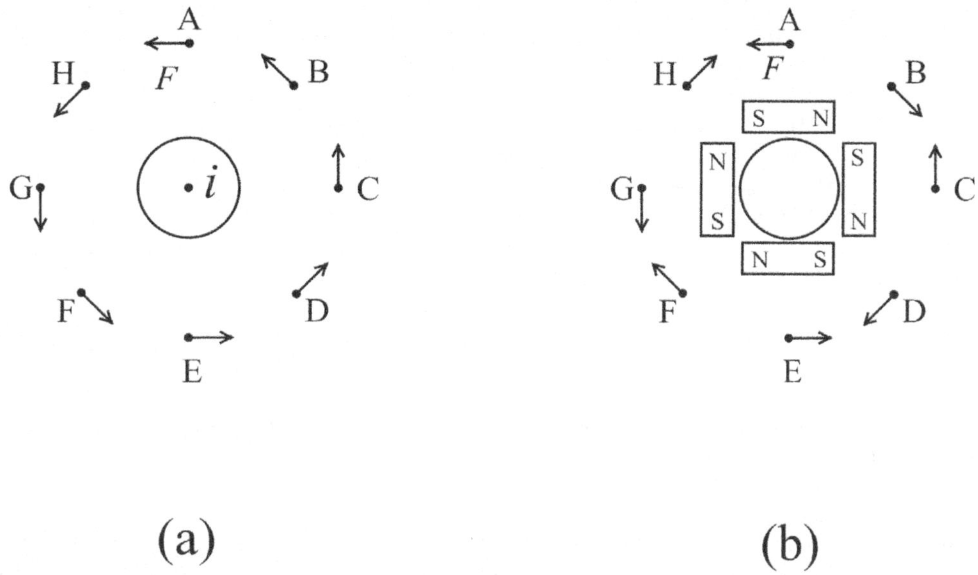

Figure 18.3: (a) Directions of the force acting on a North pole placed around a current-carrying wire according to Ørsted's experiment. (b) Directions of the force acting on a North pole placed around a current-carrying wire according to the magnetization of this wire as suggested by Biot and Savart.

case the force acting on a North pole placed at all intermediary points (like the intermediary point between A and B, the intermediary point between B and C, ...) would point in the wrong direction.

Moreover, by increasing indefinitely the number of needles around the cross section of the wire, decreasing simultaneously their individual lengths, we end up with a magnetized ring like that of Gay-Lussac and Welter, figures 9.1 and 9.3 (b). In the limit of an infinite number of small tangential magnets, this magnetized ring will not interact with an external magnetic pole nor with an external compass needle. That is, it will exert zero net force on an external magnetic pole and it will exert zero net torque on an external compass needle.

In conclusion, it is not possible to explain Ørsted's experiment in all points around the current-carrying wire by assuming a transverse magnetization of the current-carrying wire.

Figures 18.2 (b) and 18.3 (b) assumed that the magnetization of the wire was only due to the flow of the current through the wire, no matter the location of the external magnetic pole (it might be located at points A, B, C, ...). The only way to save Biot and Savart's explanation would be to assume that the location of the four magnetized needles around the wire would be due not only to the influence of the current flowing through the wire, but also on the location of the external magnetic pole which is under the action of the current-carrying wire. Figure 18.4 (a) presents their original explanation when a North pole is located at point A. Figure 18.4 (b) presents the new hypothetical orientation of the four magnetized needles assuming now that the North pole is located at point B.

Jons Jakob Berzelius (1779-1848) presented an explanation analogous to that of Biot and Savart based on the transversal magnetization of the wire due to the flow of an electric current through it. However Ørsted himself negatively criticized this explanation of his experiment.[4]

18.1.2 Quantitative Error

Ampère showed in his *Théorie* that the magnetization hypothesis of Biot and Savart also had a quantitative problem. He initially quoted Biot and Savart's statement of 1820 containing the interpretation of Ørsted's experiment by these authors based on a transverse magnetization of the wire. This quotation was repeated in the third edition of Biot's book, *Précis élémentaire de physique expérimentale*, which Ampère quoted wrongly as *Traité élémentaire de physique*.[5] Ampère then stated the following:[6]

[4] [Ørsted, 1998f, pp. 435-436].

[5] [Biot and Savart, 1820, p. 223], [Biot and Savart, 1965b, p. 119], [Assis and Chaib, 2006, p. 308], [Biot and Savart, 1824, p. 773] and [Biot and Savart, 1885, p. 126].

[6] [Ampère, 1826f, p. 188] and [Ampère, 1823c, Ampère, 1990, p. 360].

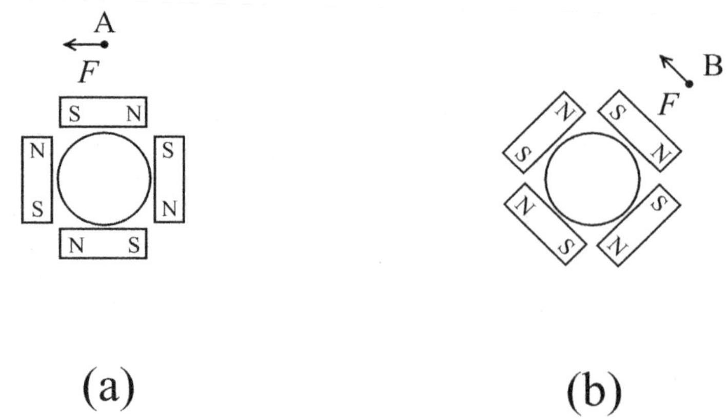

Figure 18.4: (a) Magnetization of the wire when the magnetic pole is located at point A. (b) Magnetization of the wire when the magnetic pole is located at point B.

> Finally, M. Biot repeats in the third edition of his *Traité élémentaire de physique* (volume II, page 773),[7] what he had already said in the note he published, in the *Annales de Chimie et de Physique*,[8] as regards the first experiments relative to this subject on which we are now dealing, which he had made with M. Savart, the following: that when an element of a very thin and indefinite connecting wire acts on a magnetic molecule, "the nature of the action is the same as that of a magnetized needle which is placed on the contour of a wire in a certain constant direction in relation to the direction of the voltaic current." However the action of this needle on a magnetic molecule is directed following the same straight line of the reaction of the molecule on the needle, and moreover it is easy to see that the force resulting from this action is in inverse ratio of the cube of the distance, and not [in inverse ratio] of the square of the distance, as M. Biot himself found is that [force] of the element of the wire.

Ampère here emphasized the fact that the force exerted by a magnetic dipole on another magnetic pole varies as the inverse cube of the distance between the pole and the midpoint of the dipole. Biot and Savart, on the other hand, stated that the action of a current element on a magnetic pole is inversely proportional to the square of their distance. Therefore, Biot and Savart's hypothesis of the transverse magnetization of a current-carrying wire is quantitatively incompatible with their own force law between a current element and a magnetic pole!

This aspect has also been discussed by Tricker and Hofmann.[9] Hofmann, in particular, expanded Ampère's criticism even more. He supposed the whole wire being composed of magnetic dipoles following Biot and Savart's supposition of transverse magnetization. Hofmann integrated along the whole wire the force proportional to $1/r^3$ exerted by each magnetic dipole acting on an external magnetic pole. He obtained that the integrated force of the magnetized wire acting on an external magnetic pole would vary as the inverse square of the shortest distance between the pole and the wire. This proportionality is quantitatively incompatible with Biot and Savart's own experiments, according to which the action of a long straight wire acting on an external magnetic pole should be inversely proportional to the distance between the pole and the wire!

18.2 The Principle of the Conservation of Living Forces

As discussed in Section 7.1, Faraday obtained in September 1821 continuous rotation of the extremity of a current-carrying wire around a magnet fixed relative to the ground. He also obtained continuous rotation of the extremity of a magnet around a current-carrying wire fixed relative to the ground. Ampère obtained the final value of his force law with $k = -1/2$ by developing and analyzing experiments related to this phenomenon, as seen in Chapter 7.

However, Ampère's initial interest in these experiments had another nature. He realized that Faraday's experiment could not be explained with Biot and Savart's hypothesis related to the magnetization of the wire due to the flow of an electric current through it. He showed, in particular, that this phenomenon was

[7] [Biot, 1824, p. 773].
[8] [Biot and Savart, 1820].
[9] [Tricker, 1965, p. 24] and [Hofmann, 1982, pp. 234-236].

a fatal blow to Biot and Savart's theory.[10]

In order to understand Ampère's argument, it is necessary to discuss initially the theorem of the conservation of living forces.

The theorem of the conservation of energy appeared in generic form, including chemical and thermal energies, in the decade of 1840's. By the 1820's it was not yet clear that the source of energy of the voltaic batteries was located in the chemical reactions taking place inside the batteries. In any event, for a long time scientists argued about the impossibility of perpetual motion. Faraday's discovery seemed to go against this expectation, as it indicated the possibility of a continuous macroscopic motion. Ampère was impressed by this experiment as the continuous circular motion of the extremity of a magnet or current-carrying wire took place in mercury or in acidulated water, where losses of motion by mechanical friction were inevitable.

Before presenting Ampère's quotations related to this topic it is relevant to present a brief historical digression on this subject based on the works of Lagrange,[11] Grattan-Guinness[12] and Mach.[13]

The so-called principle of the conservation of living forces was first utilized by Huygens (1629-1695), in 1673, when he solved the problem of the center of oscillation of a compound pendulum. Other scientists before Ampère developed and expanded the range of application of this principle: Jakob Bernoulli (1654-1705), also known as James or Jacques Bernoulli; l'Hôpital (1661-1704); Johann Bernoulli (1667-1748), also known as John or Jean Bernoulli; Euler (1707-1783); Daniel Bernoulli (1700-1782); d'Alembert (1717-1783); Lagrange (1736-1813); Lazare Carnot (1753-1823); etc. The name of this principle is due to Johann Bernoulli. The expression *vis viva* (*living force*) had been introduced by Leibniz (1646-1716) in 1695. It is analogous to twice the modern kinetic energy. Suppose one has a particle of mass m moving with velocity v relative to an inertial frame of reference. Its kinetic energy T is defined nowadays by $T \equiv mv^2/2$. Its *living force* as defined by Leibniz, on the other hand, is given by mv^2. Poisson, for instance, defined this concept at the time of Ampère as follows:[14]

> By the *living force*, or the *vis viva* of a material point, or more generally of a body, all the points of which are endowed with the same velocity, is meant the product of its mass by the square of this velocity.

Coriolis (1792-1843) denominated *work* the product of the force by the displacement of a particle along the direction of the force.

In modern vector notation it is possible to say that the work W exerted by a conservative force $\vec{F}$ on a point particle of mass m when it moves between points $\vec{r}_1$ and $\vec{r}_2$ of an inertial frame of reference under the action of this force is equal the variation of its kinetic energy $T = mv^2/2$:

$$W = \int_1^2 \vec{F} \cdot d\vec{r} = T_2 - T_1 \; . \tag{18.1}$$

In this equation $d\vec{r}$ represents an infinitesimal length along the curve of integration.

When the particle describes a closed trajectory, its kinetic energy is conserved. That is, its kinetic energy at the end of the cycle, T_f, will be equal to its kinetic energy at the beginning of the cycle, T_i:

$$W = \oint \vec{F} \cdot d\vec{r} = T_f - T_i = 0 \; . \tag{18.2}$$

Daniel Bernoulli, in particular, showed that when several mobile bodies are interacting with one another through conservative forces, the change in the living force of the system is only determined by the initial and final distances between the interacting particles. This fact indicated that the system could not gain kinetic energy by performing a closed loop, returning to its initial configuration. It was a known fact that central forces depending only on the distance between bodies were conservative. A particular example of a conservative force is Newton's force, varying as $1/r^2$. This behavior is valid for gravitational, electrostatic and magnetic interactions. All of these forces are central ones and vary as $1/r^2$. The frictional forces, on the other hand, are not conservative. That is, the work performed by a frictional force on a particle describing a closed loop is different from zero.

In 1811 Lagrange stated the following as regards the principle of the living forces:[15]

---

[10] [Blondel, 1982, p. 117].

[11] [Lagrange, 1788, Second Partie: "La Dynamique", Section I: "Sur les différens Principes de la Dynamique", pp. 158-188], [Lagrange, 1811, Second Partie: "La Dynamique", Section I: "Sur les différens Principes de la Dynamique", pp. 221-246] and [Lagrange, 1997, Part II: "Dynamics", Section I: "The various principles of dynamics", pp. 169-184].

[12] [Grattan-Guinness, 1990a, Chapter 16: "Work for the workers: engineering mechanics and its instruction", 1800-1830, pp. 1046-1121] and [Grattan-Guinness, 2000].

[13] [Mach, 1960, pp. 215-216, 357-365 and 433-440: "The principle of vis viva"].

[14] [Poisson, 1833, p. 29] and [Poisson, 1842, p. 23].

[15] [Lagrange, 1811, pp. 232 and 241-242] and [Lagrange, 1997, pp. 175 and 180-181].

> [...] He [Huygens] has also thought that this problem [on the center of oscillation of a rigid body under the action of terrestrial gravity] must be regarded as entirely new and since he could not resolve it by existing methods he invented a new but indirect principle which has become popular since then under the name of the **Conservation des Forces Vives**.
>
> [...]
>
> The first of these four principles, the one of the Conservation des Forces Vives was discovered by Huygens, but in a form slightly different from the one which we will presently discuss. We referred to this principle earlier when we discussed the problem of the center of oscillation. [...]
>
> Up to that time, this principle was only viewed as a simple theorem of mechanics. But when John Bernoulli adopted the distinction established by Leibnitz between the dead forces or pressures which act without actual motion and the live forces which accompany this motion as well as the measure of these latter forces by the product of the masses with the square of the velocities, he simply found in this principle a consequence of the theorem of live forces and a general law of nature according to which the sum of the live forces of several bodies is the same during the time that these bodies act upon one another with simple pressures, and is always equal to the simple live force which is the result of the action of the actual forces which put the bodies in motion. He thus gave to this principle the name of Conservation des Forces Vives and he used it successfully to solve some problems which had not yet been solved and which were difficult to resolve by means of the direct methods.

In his book of 1811 Lagrange presented a mathematical formulation of this principle and soon after stated the following:[16]

> This last equation expresses the principle known by the name of the *Conervation des Forces Vives.* [...] Thus $S((dx^2 + dy^2 + dz^2)/dt^2)m$ will be the sum of the *forces vives* of all the bodies or the *force vive* of the entire system. It is obvious from the cited equation that this *force vive* is equal to the quantity [...], which depends simply on the accelerating forces acting on the bodies and not on their mutual connections so that the *force vive* of the system has at every instant the same magnitude that the bodies would have gained if, acted upon by the same forces, they were freely moved on the line they described. This conception is what has given the name of *Conservation des Forces Vives* to this property of motion.

In the *Théorie*, Ampère presented in several forms the principle of the conservation of living forces. Here we quote one of them:[17]

> As a matter of fact, from the principle of the conservation of living forces, which is a necessary consequence of the laws of motion, it follows necessarily that when the elementary forces — which would be here attractions and repulsions in the inverse ratio of the square of the distances — are expressed by simple functions of the mutual distances of the points between which these forces are exerted, and when one part of these points are invariably connected with one another and they move only due to these forces, the other points remaining fixed, the first points cannot return to the same situation, relative to the latter points, with velocities larger than the velocities they had when they began from this same configuration.

Another formulation:[18]

> Combining this result with this rigorous consequence of the general principle of the conservation of living forces, already quoted many times in this memoir, — [namely,] that all action reducible to forces, [which are only] functions of the distances, acting between material points forming two rigid systems, one fixed and the other mobile, can never give rise to a motion which is indefinitely continuous, despite the resistances and frictions acting on the mobile system, — [...]

18.3 It is Not Possible to Explain Continuous Rotation with the Hypothesis of the Magnetization of the Wire

Faraday's experiment on continuous rotation caught Ampère's attention for the following reasons. Suppose there is a central conservative force which is a function only of the distance between the interacting bodies, like the gravitational or electrostatic forces which vary as $1/r^2$. Suppose the bodies exerting this force are at rest relative to an inertial frame of reference, while the test body can move relative to this frame. If we

[16] [Lagrange, 1811, p. 290] and [Lagrange, 1997, p. 213].
[17] [Ampère, 1826f, pp. 124-125] and [Ampère, 1823c, Ampère, 1990, pp. 296-297].
[18] [Ampère, 1826f, pp. 193-194] and [Ampère, 1823c, Ampère, 1990, pp. 365-366].

calculate the work exerted on this test body by the fixed source bodies along a closed trajectory of arbitrary shape, we obtain a zero value. Therefore the kinetic energy of the test body at the end of the motion will be equal to its initial kinetic energy. That is, there is a conservation of its *living force*. In general, when there is a set of mobile particles interacting with one another only through conservative forces, the kinetic energy of the system will be always the same at all instants in which the bodies return to the same relative positions relative to one another, no matter the trajectories they followed in order to return to this initial configuration.

Suppose now there are two forces acting on a test body, namely, a conservative force and a resistive force (like a mechanical friction). In this case, if the test body describes a closed trajectory of arbitrary shape being under the influence of these two forces, then its kinetic energy at the end of the trajectory will be smaller than its kinetic energy at the same point in the beginning of the trajectory. The mechanically resistive force will be responsible for this decrease in the kinetic energy of the test body.

Ampère seemed to be referring to something of this sort when he analyzed Faraday's experiment. As Faraday obtained continuous rotation in a mechanically resistive medium, Ampère concluded that this could only take place due to forces acting between bodies in relative motion. Conservative central forces depending only on the distance between the interacting bodies could not give rise to continuous rotation of the test body in a mechanically resistive medium.

Some scientists like Berzélius (1779-1848), Davy (1778-1829), Schweigger (1779-1857), Prechtl (1778-1854), Wollaston (1766-1828), Biot and Savart had tried to explain Ørsted's experiment supposing the magnetization of the wire due to the passage of a current through it.[19] They believed that it was possible to explain Ørsted's experiment supposing a distribution of small magnets along the length of the conductor, or distributed tangentially around its cross section. However, none of them presented a detailed study of this supposed distribution of magnetic poles on the cross section of the wire. Ampère realized that he could utilize Faraday's discovery of continuous rotation in order to show that these explanations of Ørsted's experiment based on the magnetization of the current-carrying wire must be wrong. Coulomb had already shown that the magnetic force had the same behavior as Newton's law of gravity. That is, the force between two magnetic poles is central and varies as the inverse square of their distance. Consequently, the magnetic force is conservative, just like the gravitational force. If there were no friction in Faraday's experiment, the magnet rotating around a fixed current-carrying wire would constantly increase its angular velocity. The free extremity of the magnet only maintains a continuous rotation with a constant angular velocity around the wire due to the resistive forces exerted by the fluid where it is moving.

The same happened in Faraday's experiment when the free extremity of the current-carrying wire rotated continuously around the fixed magnet. The current-carrying wire began at rest. Its free extremity was accelerated tangentially around the fixed magnet, beginning to rotate around it. It described a closed circular trajectory around the magnet. Its kinetic energy began from zero, increased, and then reached a constant value due to the friction exerted by the conducting fluid where it was moving. Were it not for the friction of the medium, then the tangential velocity of the free extremity of the wire would increase indefinitely at each turn around the fixed magnet. However, due to the theorem of the living forces, this continuous tangential acceleration could not be explained by conservative forces, like the electrostatic force or like the Coulombian force between two magnetic poles. According to Ampère, Ørsted's experiment and Faraday's continuous rotation had to be due to interactions between fluids in relative motion.

For instance, in a work presented to the Academy of Sciences of Paris on 8 April 1822, Ampère said the following:[20]

> Such are the new progresses which a branch of physics has just made, of which we had not the slightest idea only two years ago, and which caused us to realize facts more surprising than anything which the science had offered to us up to now of marvelous phenomena. A motion continuing always in the same sense, despite the frictions, despite the resistance of the medium, with this motion being produced by the mutual action between two bodies which remain constantly in the same state, is a fact unprecedented in the middle of everything we know of the properties which inorganic matter can offer. It proves that the action which emanates from voltaic conductors cannot be due to a specific distribution of certain fluids at rest within these conductors, as is the case for ordinary electric attractions and repulsions. This action can only be attributed to fluids in motion within the conductors, moving rapidly from one extremity of the pile [battery] to the other extremity.

In the *Théorie*, Ampère repeated this quotation, introducing it with the following words:[21]

---

[19] [Blondel, 1982, pp. 46-48].

[20] [Ampère, 1822r, pp. 65-66], [Ampère, 1822j, p. 205], [Ampère, 1885e, pp. 242-243] and [Hofmann, 1996, p. 297].

[21] [Ampère, 1826f, pp. 125-126] and [Ampère, 1823c, Ampère, 1990, pp. 297-298].

> At the time of the first works of the physicists on electrodynamic phenomena, several scientists believed that they could explain them through distributions of molecules, electrical or magnetic, at rest in the voltaic conductors. Since Faraday published the discovery of the first motion of continuous rotation, I saw immediately that it contradicted completely this hypothesis. I announced this observation in the following terms, while what I have said here is only the development, in the *Exposé sommaire des nouvelles experiénces eléctro-magnétiques faites par différens physiciens, depuis le mois de mars 1821*,[22] which I read at the public meeting of the *Académie royale des Sciences* on 8 April 1822: [...]

In the *Théorie* he pointed out several times that it is not possible to explain the experiments of continuous rotation supposing electric charges or magnetic poles at rest in the conductors and interacting with one another through conservative forces. For instance:[23]

> Now, in the motion of continuous rotation given to a mobile conductor by the action of a fixed conductor, all the points of the first [conductor] return to the same position with increasing velocities at each turn, up to a point in which the frictions and the resistance of the acidulated water where the crown of the conductor is submerged introduce a limit in the increase of the velocity of rotation of this conductor. This velocity becomes then constant, despite these frictions and the resistance.
>
> Therefore, it is completely demonstrated that one would not know how to explain the phenomena produced by the action of two voltaic conductors, supposing that electric molecules acting in the inverse ratio of the square of the distance were distributed on conducting wires, remaining fixed in them and, consequently, so that they might be considered as being invariably connected to one another. [...]
>
> It is only in the case in which one supposes the electric molecules at rest in the bodies in which they manifest their presence by the attractions and repulsions produced by them between these bodies, that one demonstrates that an indefinitely accelerated motion cannot result from the fact that the forces which the electric molecules exert in this state of rest depend only on their mutual distances. [...]

Ampère had already presented analogous arguments in his fundamental paper of 1822, in which he deduced the final value of his constant, namely, $k = -1/2$.[24]

[22] [Ampère, 1822j].
[23] [Ampère, 1826f, pp. 125-127] and [Ampère, 1823c, Ampère, 1990, pp. 297-299].
[24] [Ampère, 1822o, pp. 401-405], [Ampère, 1822y, pp. 296-300] and [Ampère, 1885p, pp. 272-274].

Chapter 19

The Magnetic Poles and Dipoles are Disposable Hypotheses

19.1 Ampère Against the Existence of Magnetic Poles and Dipoles

Ørsted explained his experiment of the deflection of a magnetized needle by supposing that its magnetic poles were propelled by electric charges flowing in the space around the current-carrying wire. Biot, Savart and Faraday explained the same experiment supposing a rotational, rotary, or revolutive force exerted by the current-carrying wire acting on the supposed magnetic poles of the magnetized needle.

Ampère, on the other hand, rejected the existence of these magnetic poles or magnetic fluid particles, as expressed in several of his works.[1] He was also against the existence of magnetic dipoles, that is, a particle having opposite poles separated by a small distance.

One example of Ampère's statement in 1820:[2]

> [...] However, as M. Ampère noted in his article, his explanation of the action of voltaic conductors on magnets offers a double advantage:
>
> 1st: *It is not necessary to suppose that these conductors act on magnetic particles, of which no one demonstrates their existence*, [...]

An analogous quotation appears in Ampère's work presented to the Academy of Sciences of Paris on 4 December 1820:[3]

> The theory which reduces all phenomena of the magnet to the phenomena offered by electricity, when two currents act on one another, a theory supported by many other facts, has these two advantages, as regards the phenomena which interest us here:
>
> 1st: It does not make the conducting wire AB act on magnetic particles, of which no one demonstrates their existence, *and which is a gratuitous supposition*, [...]

In a letter of 1821 he stated the following:[4]

> [...] and it seems to me that, according to the properties which I recognized in the voltaic conductors, the supposition of magnetic fluids or magnetic forces, which are different from the electric fluids or electric forces, is only a purely gratuitous conception.

In 1823, when commenting on a paper by Savary, Ampère presented similar arguments:[5]

> [...] while, in M. Ampère's theory, instead of these fluids of a particular nature,[6] of which no one proves the existence, one admits [...]

[1] [Hofmann, 1996, pp. 269-270].
[2] [Ampère, 1820d, p. 549] and [OErsted et al., 1822, pp. 82-83], our emphasis.
[3] [Ampère, 1885j, p. 133], our emphasis.
[4] [Ampère, 1821a, p. 309] and [Ampère, 1822n, p. 120].
[5] [Ampère, 1885l, p. 379], our words in the footnote.
[6] That is, molecules of austral or boreal fluids, or South and North poles, respectively.

19.2 Identification of the Magnetic Fluid with the Galvanic Fluid

There is an important letter by Ampère to Hachette, dated 14 November 1820, in which he synthesized one of the main differences between his theory of the electromagnetic phenomena and the theory of Ørsted. The latter explained the electromagnetic phenomena assuming the existence of distinct fluids, namely, magnetic and galvanic ones. He supposed the existence of electric charges flowing not only inside, but also outside the wire, with these latter charges propelling the magnetic poles of a magnetized needle. Ampère, on the other hand, explained the same phenomena utilizing only the interaction between current elements. To this end he assumed the existence of electric currents inside magnets. With these assumptions, Ampère was essentially identifying the magnetic and galvanic fluids. That is, according to Ampère's conception, there are no real magnetic poles. He assumed only the existence of current-carrying conductors interacting with one another. A relevant portion of this letter:[7]

> There is, in the first place, a clear falsehood in the statement in which M. Œrsted announced the identity of magnetic and galvanic fluids; not only he himself does not mention anything in this regard in everything that we have from him but also, in the paper in which he published his discovery, he explained the phenomena by a hypothesis according to which these fluids are totally distinct from one another. I was the first to state that this identity [between the magnetic and galvanic fluids] resulted from these phenomena, but not according to the explanation of M. Œrsted which supposed them totally distinct, but according to my own theory which is totally distinct from his and which is the only one which identifies them [that is, which identifies the magnetic and galvanic fluids].

Ampère claimed here that his theory was totally distinct from that of Ørsted. Moreover, while Ørsted assumed that the magnetic and electric fluids were totally distinct from one another, Ampère believed that only current-carrying conductors were necessary in order to explain these phenomena.

Analogous information appeared in a paper by Laumont:[8]

> M. OErsted, professor at Copenhagen University, had published, before 1807, a work in which he announced that he intended to verify *if the electricity in its most latent state does not have some action on the magnet*[9]; but it was only during the winter of 1819, that he discovered *the action of the connecting wire on the magnetized needle*, when he offered a particular explanation, which does not agree in any way with the idea that the electric and magnetic fluids are the same thing[10].

Later on, in the same article, we have the following:[11]

> All these facts, and even M. OErsted's experiments, were far from proving *the identity of electricity with the magnet*; this identity was only established by M. Ampère [...]

Sometimes, instead of talking about the identity between the magnetic and galvanic fluids, Ampère mentioned the identity between electricity and magnetism:[12]

> One then represents by a single force [between current elements], always directed along the straight line connecting the two points between which it is exerted, not only the magnetic phenomena already known for a long time, but also all circumstances of the action of a voltaic conductor on a magnet, discovered by M. Oersted, and of that [action] which I recognized between two conductors. And this [fact] seems to me to support strongly the point of view which I expressed at the time of my first works on this subject, related to the identity of electricity and magnetism. The results of the experiments which I have made since then seem to make it each time more probable.

In the *Précis* of 1824 he presented a similar point of view:[13]

> Some physicists concluded that the absolutely different phenomena which were produced, some of them by ordinary electricity and the other by dynamic electricity, should not be attributed to the same electric fluids, at rest in the first case, and in motion in the second case. It is precisely as if we had concluded, from the fact that the suspension of mercury in the barometer is a phenomenon completely different

[7] [Launay (ed.), 1943, p. 906].
[8] [Laumont, 1820, p. 535], [Laumont, 1822, p. 69] and [Hofmann, 1987, p. 326, n. 47], italics in the original.
[9] [Note by Laumont:] Chapter VIII of the translation into German by M. Marcel de Serres, published in 1807.
[10] [Note by Laumont:] *Annales de Chimie et de Physique*, volume XIV, p. 424.
[11] [Laumont, 1820, pp. 538-539] and [Laumont, 1822, p. 72], italics in the original.
[12] [Ampère, 1822q, pp. 111-112].
[13] [Ampère, 1824e, p. 60].

> from the phenomenon of sound, that we should not attribute them to the same atmospheric fluid, at rest in the first case, and in motion in the second case, in such a way that it would be necessary to admit, for two facts so different from one another, two fluids, of which one of them acted only to press the free surface of mercury, while the other transmitted the vibratory motions producing sound.
>
> It is this habit of multiplying, as it is said, the entities without necessity which during some time made one admit in physics a luminous fluid distinct from the fluid to which one attributed the phenomena of heat; it is this habit which leads one to suppose up to now two magnetic fluids different from the two electric fluids, although it has been shown that electricity, by moving around the particles of the magnetized bodies in the same way as it moves in the voltaic conductor and, consequently, exerting in this way the same action, should necessarily produce effects completely identical to the effects which are attributed to what is called *molecules of austral fluid and of boreal fluid.*

19.3 The Elementary Force Must Act between Entities of the Same Nature

This Section discusses Ampère's criticisms against the conceptions of Ørsted, Biot, Savart and Faraday. These authors believed in the existence of interactions between a current-carrying wire and a magnetic pole, that is, between magnitudes of different nature. Ampère, on the other hand, argued that the fundamental interactions took place only between entities or magnitudes of the same nature. Examples of this last kind of interaction were the forces and reaction between two massive bodies in the case of gravitation, the forces between two electrified particles in the case of electrostatics, or the forces between two current elements in the case of electrodynamics.

It is also possible to utilize Ampère's arguments against the supposition of a magnetic field acting on a magnetic pole, against the action of a magnetic field on a current element, and also against the interaction between a magnetic field and an electric charge. After all, in these three examples we have interactions between entities of different nature, namely: (i) a magnetic field and a magnetic pole, (ii) a magnetic field and a current element, and (iii) a magnetic field and an electric charge.

Ørsted explained his own experiment supposing the interaction between the imagined electric charges flowing in the space around the current-carrying conductor and the magnetic fluids he supposed to exist inside the magnet. However, he never explained how an electric charge might propel a magnetic pole, nor how their supposed interaction took place. Ampère never accepted the idea of an interaction between magnitudes of different nature.

In a very interesting letter addressed to Speyert van der Eyk, dated 15 August 1821, Ampère discussed what should be considered as the *primitive fact* from which we might deduce all the phenomena. He assumed that this primitive fact should be understood as a *fundamental interaction* from which it might be possible to explain more complex phenomena. Without mentioning Ørsted's name, Ampère rejected his points of view, according to which one should consider as fundamental the interaction between a current-carrying wire and a magnetized needle. Instead of this assumption, Ampère proposed to consider fundamental the interaction between two current-carrying conductors. He justified his choice as follows:[14]

> 1st: Every explanation in the sciences consists in discovering a primitive fact expressed by a general law which, once presented, can be utilized in order to deduce from it all other [facts or laws].
>
> 2nd: The primitive fact cannot be here the action between a voltaic conductor and a magnet, because, these two things being heterogeneous, the mutual action between them must necessarily be more complicated than that [mutual action] which takes place between two magnets, or than that [mutual action] which I discovered between two conductors; [...]

Ampère repeated these arguments in a letter of 1821, addressed to G. Moll.[15] He also insisted on this argument in a letter of 1822 addressed to Auguste de la Rive:[16]

> Is it not evident that one should look for the primitive fact in the action between two things of the same nature, like two conductors, and not in that action between two heterogeneous things, like a conductor and a magnet?

---

[14] [Launay (ed.), 1936a, pp. 570-572].
[15] [Ampère, b, *carton* 9, *chemise* 183] and [Ampère, c].
[16] [Launay (ed.), 1936a, p. 605] and [Blondel, 1982, p. 112].

In the *Théorie*, Ampère argued that the elementary force must act between magnitudes of the same nature. The elementary force might not act, for instance, between a current element and a magnetic pole. He was then going against Biot's conceptions. But his criticisms might also be directed against Ørsted's conception, according to which there was an interaction between a magnetic pole and an electric charge flowing in the space around a current-carrying wire. Here are his words:[17]

> Moreover, though M. Biot determined the value and direction of the force when an element of conducting wire acts on each particle of a magnet and defined this as the elementary force,[18] it is clear that a force cannot be regarded as truly elementary which manifests itself in the action of two elements which are not of the same nature, or which does not act along the straight line which joins the two points between which it is exerted.

Ampère utilized an analogous argument against Faraday's conceptions. As was seen in Section 13.2, in 1821 Faraday stated the following as regards the phenomenon discovered by Ampère of the attraction and repulsion between current-carrying conductors:[19]

> [...] the attractions and repulsions of M. Ampère's wires are not simple, but complicated results.

Ampère's reply to this statement came in a joint paper with Savary, followed by a French translation of this article of Faraday:[20]

> If M. Faraday, in this passage, considered only that the attractions and repulsions between electric currents are complicated facts due to the fact that they result from an infinity of actions between all infinitely small parts of these currents, then he was in agreement with M. Ampère. However, he [Faraday] considered them complicated from another point of view, as he considered as a primitive fact the rotational action and shows quite well that these attractions and repulsions can be reduced to it. But we have just seen that by considering, on the contrary, as primitive facts the attractions and repulsions between the small portions of electric currents, according to the laws given by M. Ampère, one deduces immediately the circular motions of the conducting wires and magnets around each other.

That is, instead of considering as a basic or fundamental fact the rotational action between magnitudes of different nature (like the action between a current-carrying wire and a magnetic pole), Ampère preferred to adopt as a primitive fact the central force between magnitudes of the same nature, namely, the actions and reactions between two current elements.

Ampère would certainly criticize the modern concepts of electric and magnetic fields for similar reasons. Nowadays the action and reaction between two electric charges in relative motion is not accepted. Instead of this conception, it is assumed that a charge in motion generates a magnetic field. This magnetic field propagates in space, typically at light velocity. When it reaches a second moving charge, the magnetic field exerts a force on it. If this second charge is free to move relative to an inertial frame of reference, then it will be accelerated by this magnetic field which is reaching it. Ampère would not accept the idea of a certain magnitude (like the magnetic field in this example) exerting an action on another magnitude of a different nature (like the second charge in this example).

Likewise, nowadays the textbooks do not utilize the concept of a direct action and reaction between two current elements. Instead of this action at a distance, most scientists assume that a certain current element A generates a magnetic field. They assume that this magnetic field propagates in space, typically at light velocity. When it reaches a second current element B, it generates a force on this element. Ampère would not accept the idea of a magnetic field exerting a force on a current element, as these two magnitudes have different nature.

Some questions which Ampère might ask as regards this modern conception of interaction: How a "magnetic field" might exert a force on a charge or on a current element? By which mechanism an immaterial medium, like the magnetic field, might act on a charged material particle? Faraday and Maxwell understood a magnetic field as a region of space in the neighborhood of a magnet or around a current-carrying wire, as discussed in Section 15.1. How might a source magnet produce a region of space around it? How might a region of space act on matter? How might an electric charge exert a reaction force on the magnetic field acting on it? How might a current element exert a reaction force on a magnetic field acting on it? How might a charged particle (or a current element) exert a reaction force on a region of space? Etc.

[17][Ampère, 1826f, pp. 108-109], [Ampère, 1823c, Ampère, 1990, pp. 280-281] and [Ampère, 1965b, p. 195].
[18]*Précis élémentaire de physique*, vol. II, p. 122, 2nd edit.; see [Biot, 1821a, p. 122].
[19][Faraday, 1822e, p. 79] and [Faraday, 1952d, p. 799].
[20][Savary and Ampère, 1821, pp. 373-374], [Savary and Ampère, 1822, pp. 161-162] and [Savary and Ampère, 1885, p. 187].

We list here some specific criticisms which Ampère might present against the field concept: This field concept would lead to a violation of the principle of action and reaction, as an electric charge might not exert a reaction on a magnetic field; likewise, a current element might not exert a reaction on a magnetic field; a charged particle (or a current element) might not exert a force of reaction on a region of space.

A current element and a magnetic field are distinct magnitudes. They are of different nature and have different units. They are ontologically different from one another. Therefore, Ampère would never accept the action between these two magnitudes as something fundamental or primitive.

Chapter 20

Defense of Action and Reaction along the Straight Line Connecting the Interacting Bodies

20.1 Ampère Against the Rotational Action Around a Current-Carrying Wire

Some authors, like Ørsted, defended the idea of a flux of particles in the space around a current-carrying wire. This helical or circular flux of particles would be responsible for the deflection of a magnetized needle placed close to the wire, as discussed in Section 11.1. Ørsted initially imagined a helical flux around a long and straight current-carrying wire. Later on he imagined a circular flux around this wire. Other authors, like Biot, Savart and Faraday, supposed that somehow the current-carrying wire generated a rotational, revolutive or gyratory action which would force the poles of a magnet to rotate around the wire. Modern physics has a somewhat similar concept. Nowadays it is argued that a long straight wire carrying a steady current generates a circular magnetic field around it.

Ampère, on the other hand, argued strongly against these vortices, whirls, or rotational actions around a current-carrying wire. This is the subject of this Section.

In a letter addressed to Paul Erman (1764-1851), dated 9 August 1823, he expressed his generic point of view as follows:[1]

> According to a passage in your letter, I believe there is a certain obscurity in Germany as regards this aspect of my theory. By seeing the planets revolving around the Sun, Cartesian physics imagines that they are pushed in the sense in which they move by vortices revolving in the same sense; when it [the Cartesian physics] sees a pole of a magnet carried to the right and the other [pole carried] to the left of a conducting wire, it supposes a vortex independent of the wire. Newtonian physics explains all celestial phenomena by an attraction directed along the straight line connecting the two interacting particles and the motion is a complicated result of this attraction. As regards the new [electromagnetic and electrodynamic] phenomena I make what Newton has done for celestial motions, [namely,] I explain them by attractive and repulsive forces. And the direction of the electric current, which follows a straight line when the conductor is rectilinear, is only introduced in the calculation due to the fact that the mathematical value of the exerted forces depends, as you can see in my *Recueil*,[2] on the angles formed by this direction with the direction of another current and with the direction of the straight line connecting the particles which are in mutual action.

20.1.1 Ampère Against Ørsted's Rotational Vortices

In the *Théorie*, Ampère criticized Ørsted's own interpretation of his experiment of the deflection of a compass needle close to a current-carrying wire. Ørsted's explanation was based on a helical flux of electrical particles around the wire. This explanation reminded Ampère of Descartes's explanation for the motion of the planets around the Sun, according to which there was a material vortex flowing around the Sun and propelling the

[1] [Grattan-Guinness, 1990b, pp. 1333-1334].

[2] [Ampère, 1822w].

planets along their tangential motion, like a wind propelling particles which were floating in air. Ampère, on the other hand, preferred to explain Ørsted's experiment utilizing a Newtonian interaction. That is, utilizing a force acting along the straight line connecting the interacting elements and following the principle of action and reaction:[3]

> It does not appear that this [Newtonian] approach, the only one which can lead to results which are free of all hypothesis, is preferred by physicists in the rest of Europe like it is by Frenchmen; the famous scientist [Ørsted] who first saw the poles of a magnet transported by the action of a conductor in directions perpendicular to those of the wire, concluded that electrical matter revolved about it and pushed the poles along with it, just as Descartes made "the matter of his vortices" revolve in the direction of planetary revolution. Guided by Newtonian philosophy, I have reduced the phenomenon observed by M. Oerstedt, as has been done for all similar natural phenomena, to forces acting along a straight line joining the two particles between which the actions are exerted; and if I have established that the same arrangement, or the same movement of electricity, which exists in the conductor is present also round the particles of the magnets, it is certainly not to explain their action by impulsion as with a vortex, but to calculate, according to my formula, the resultant forces acting between the particles of a magnet and those of a conductor, or of another magnet, along the lines joining the particles in pairs which are considered to be interacting, and to show that the results of the calculation are completely verified [...]

20.1.2 Ampère Against the Rotational Actions of Biot, Savart and Faraday

Faraday presented in 1821 his discovery of the rotation of one extremity of a current-carrying wire around a stationary magnet, and also the rotation of one extremity of a magnet around a stationary current-carrying wire.[4] Faraday, Biot and Savart tried to explain these new phenomena utilizing similar concepts which they had utilized in order to explain Ørsted's experiment, namely, the existence of rotational or revolutive actions exerted by the current-carrying wire acting on the poles of the magnet. As regards Faraday's experiment, this rotational action exerted by the current-carrying wire would cause an extremity of the magnet to turn around the wire. By action and reaction, the magnet would exert an opposite rotational force acting on the current-carrying wire.

Faraday's work was translated into French and published in the same year.[5] Savary and Ampère published a Note following this translation of Faraday's work.[6] In the paper of 1821 this note appears without the names of the authors. In the *Recueil* edited by Ampère, this note appears under the names of Savary and Ampère. Blondel also presented this work under both names,[7] while Hofmann presented only Savary as its author.[8] We will follow the *Recueil* and Blondel indicating Savary and Ampère as the authors of this note.

In their note, Savary and Ampère presented two main criticisms against Faraday's rotational, rotary, or revolutive action. They first pointed out that this hypothesis does not lead to a mathematical theory with which it would be possible to explain quantitatively the phenomena:[9]

> The rotational action of a conducting wire and a magnet around each other, which M. Faraday considers as a primitive fact in all portions of his article, is not enough to submit the phenomena to calculation; it would be necessary that he had determined precisely the action taking place between each element of the wire and each particle of the magnet.

The second argument contains Ampère's main criticism. He considered that the principal advantage of his theory was to be able to explain this phenomenon discovered by Faraday utilizing only central forces satisfying the principle of action and reaction, that is, Newton's third law of motion.[10] Their reasoning was expressed as follows:[11]

> [...] these facts [that is, the continuous rotations discovered by Faraday] are included in the general laws of physics, and one is not obliged to admit as a *simple and primitive fact a rotational action*, to which nature has not offered any other example, and which seems to us difficult to consider as such.

---

[3] [Ampère, 1826f, pp. 5-6], [Ampère, 1823c, Ampère, 1990, pp. 177-178] and [Ampère, 1965b, pp. 156-157].
[4] [Faraday, 1822e] and [Faraday, 1952d, pp. 805-807].
[5] [Faraday, 1821c], [Faraday, 1822c] and [Faraday, 1885].
[6] [Savary and Ampère, 1821], [Savary and Ampère, 1822] and [Savary and Ampère, 1885].
[7] [Blondel, 1982, pp. 111 and 188].
[8] [Hofmann, 1996, pp. 291 and 399].
[9] [Savary and Ampère, 1821, p. 371], [Savary and Ampère, 1822, p. 159] and [Savary and Ampère, 1885, p. 184].
[10] [Blondel, 1982, p. 112].
[11] [Savary and Ampère, 1821, p. 374], [Savary and Ampère, 1822, p. 162] and [Savary and Ampère, 1885, p. 187], our emphasis.

That is, as nature does not present any other example of a rotary action, it was difficult for Ampère to consider an action of this sort as something simple and primitive.

20.1.3 "Ampère" Against the Rotational Magnetic Field

Ampère's arguments against the rotational, rotary, or revolutive actions of Ørsted, Biot, Savart and Faraday are also relevant as regards the modern conception of a magnetic field. Ampère would certainly be against a revolutive magnetic field around a long and straight current-carrying wire. This azimuthal magnetic field is orthogonal to the wire, pointing according to the right-hand rule, as indicated in figure 20.1.

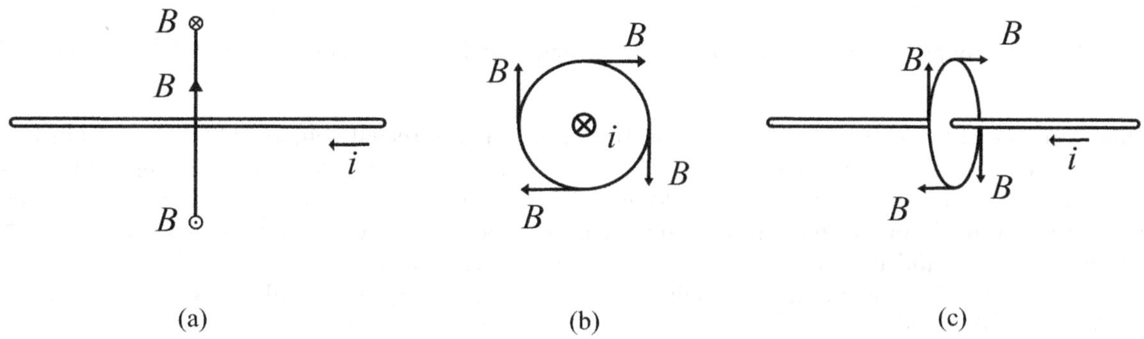

Figure 20.1: Magnetic field around a long current-carrying wire. (a) Lateral view. (b) View orthogonal to the wire. (c) View in perspective.

Figures like this appear in most textbooks, like that of Feynman, figure 20.2.[12]

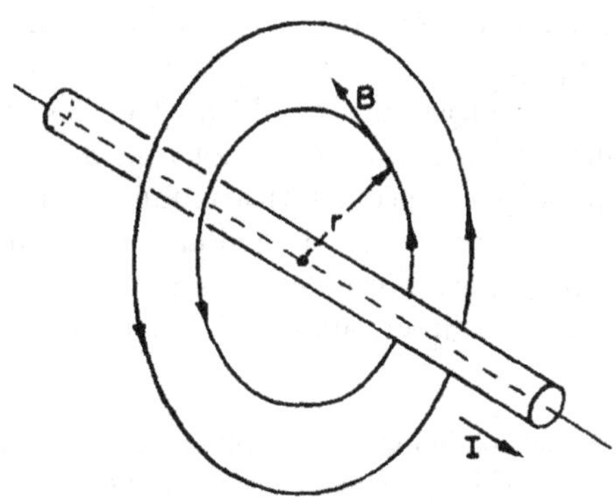

Figure 20.2: Magnetic field around a current-carrying wire, according to Feynman.

The similarity of figures 20.1 and 20.2 with Ørsted's original representation, figure 11.3, is amazing.

Ampère was against Ørsted's vortices. He was also against Biot, Savart and Faraday's rotational, rotary, or revolutive actions. He would certainly be totally against this revolutive magnetic field around a current-carrying wire.

20.2 Ampère's Criticisms Against the Primitive Couple

In this Section we present Ampère's criticisms against the primitive couples appearing in the theoretical conceptions of Biot, Savart and Faraday. He would certainly present similar arguments against the primitive couples appearing in Grassmann's force and also against the primitive couple originated in the modern concept of a magnetic field.

A couple is a pair of forces, equal in magnitude, oppositely directed, whose lines of action do not coincide, as in figure 20.3. Although there is no resultant force on the system, these opposite forces produce a torque about an axis which is normal to the plane of the forces. Suppose that a rigid test body, or a system of

[12][Feynman et al., 1964, Chapter 13, figure 13-7].

bodies rigidly connected to one another, is initially at rest relative to an inertial frame of reference. A couple acting on this test body, or on this system of bodies rigidly connected to one another, will make it begin to turn around an axis passing through its center of mass and orthogonal to the plane of the couple.

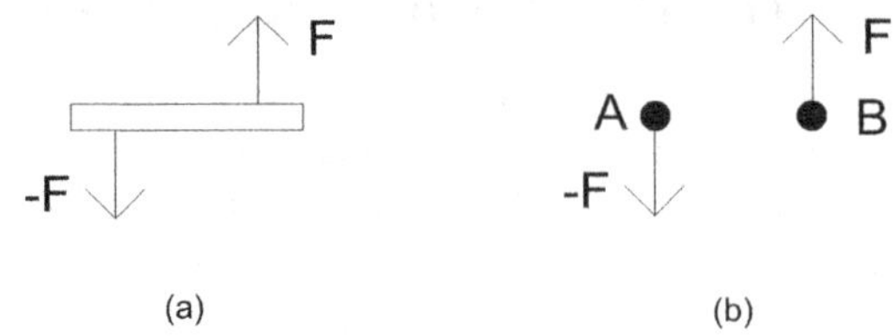

Figure 20.3: (a) A couple acting on a rigid body. (b) A couple acting on a system composed of two bodies, A and B.

What Ampère called a primitive couple was an hypothetical internal couple. That is, the pair of equal and opposite forces acting along parallel lines would be generated by internal forces acting on the system. Two particles belonging to the test body would act on one another along parallel lines, creating an internal torque on the system. Suppose that the system were initially at rest relative to an inertial frame of reference, being free to turn around its center of mass. The action of this primitive couple would make the system begin to turn around an axis passing through its center of mass and orthogonal to the plane of the couple, increasing its angular velocity.

This primitive couple might also be called a primitive torque, an internal torque or an internal couple. Ampère fought vigorously against any theory which accepted the existence of these hypothetical primitive couples.

20.2.1 Primitive Couple of Biot and Savart

In Chapter 17 we saw several criticisms Ampère presented against Biot and Savart's procedures in order to deduce their law expressing the supposed force exerted by a current element acting on a magnetic pole. Even disregarding his criticisms to their procedure to obtain the law, he was always totally against their force. One of the main aspects of his criticism was that Biot and Savart's force was not directed along the straight line connecting the two supposed interacting entities, namely, a current element and a magnetic pole. Ampère's force, on the other hand, was always directed along the straight line connecting the interacting entities, namely, the two current elements. Ampère mentioned this criticism in several works.

Biot and Savart described as follows the direction of the force exerted by a current element on a magnetic pole:[13]

> This force, like the total action, is perpendicular to the plane drawn through the longitudinal element of wire in the shortest distance between this element and the magnetic molecule which is influenced.

Biot and Savart believed that this force was due only to interactions between magnetic poles, supposing that the wire had become magnetized due to the flow of the electric current through it. As the force between magnetic poles follows the principle of action and reaction, it can be concluded that the reaction force exerted by the magnetic pole and acting on the current element would also be perpendicular to the plane drawn through the current element and through the line connecting it to the magnetic pole. Biot himself admitted explicitly the principle of action and reaction between the wire and a magnetic molecule:[14]

> Then the general law of equality which is observed between action and re-action in all the phenomena of nature, shows that the [magnetic] particle will re-act upon the wire precisely as the wire acted upon the particle; so that the wire will receive in a contrary direction all the motions which the particle exhibited.

There would then be two equal and opposite forces acting along parallel straight lines. These two opposite forces exerted between a current element and a magnetic pole would not be along the straight line connecting them.

Ampère concluded that these elementary forces of Biot and Savart, supposedly acting between a current element and a magnetic pole, would produce in this system a *primitive couple*. In several portions of the *Théorie* he criticized this property of Biot and Savart's law.[15]

[13] [Biot and Savart, 1824, p. 740], [Farrar, 1826, p. 334], [Biot and Savart, 1885, p. 113] and [Biot and Savart, 1965a, p. 134].
[14] [Biot and Savart, 1824, p. 750] and [Farrar, 1826, p. 343].
[15] [Ampère, 1826f, pp. 123, 154, 171 and 176-179] and [Ampère, 1823c, Ampère, 1990, pp. 295, 326, 343 and 348-351].

This primitive couple can be visualized in figure 20.4. In figure 20.4 (a) there is a current element ids interacting with a magnetic pole p. Let us suppose that this is a North or positive pole. Figure 20.4 (b) presents the force F exerted by the current element on the North pole. It is orthogonal to the paper, penetrating it. The force $-F$, exerted by the pole on the current element, also acts orthogonally to the paper, leaving it. Therefore, according to Biot and Savart's hypothesis, the action and reaction are directed along parallel straight lines. They do not act along the straight line connecting the pole with the midpoint of the current element.

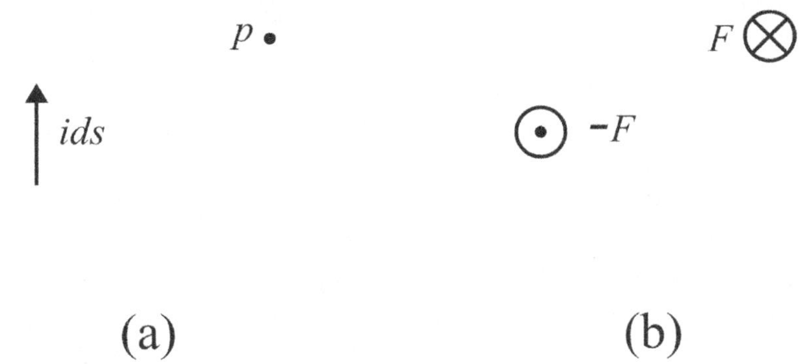

Figure 20.4: Primitive couple according to Biot and Savart's law.

These opposite forces exert a primitive couple on this system composed of the magnetic pole and the current element. Ampère was totally against this concept.

20.2.2 Primitive Couple of Faraday

Also in Faraday's conception there is a primitive couple, as seen in Section 13.1. It is illustrated in figure 20.5. There is a long straight wire normal to the paper, carrying a constant current i leaving the paper. One of the arrows indicate the force F exerted by the wire and acting on the North pole p of the magnetized needle. The other arrow indicates the force $-F$ exerted by the North pole and acting on the current-carrying wire.

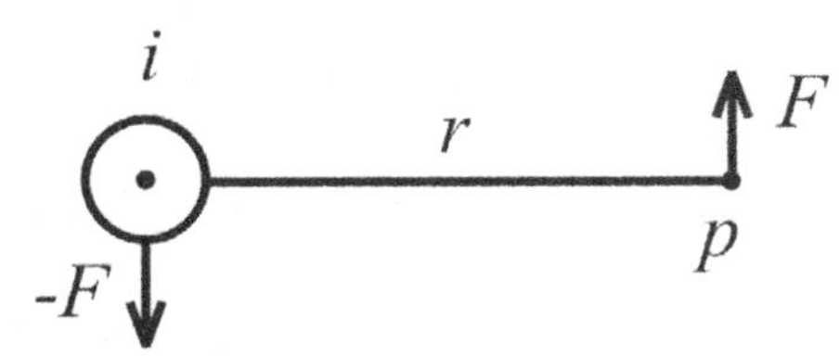

Figure 20.5: Primitive couple according to Faraday's conceptions.

These opposite forces act along parallel straight lines, generating a primitive couple on this system composed of the North pole and the current-carrying wire.

20.2.3 Primitive Couple of Grassmann

Grassmann's force is given by equations (14.7) up to (14.10). It is also possible to show with this force that there are configurations yielding a primitive force, as that shown in figure 20.6.

We have two parallel current elements making angles $\alpha = \beta = 45^o$ with the straight line connecting their midpoints. According to Ampère's force, these two elements attract one another along the straight line connecting them, figure 20.6 (a). According to Grassmann's force, there is also action and reaction in this specific example. However, Grassmann's opposite forces do not act along the straight line connecting the elements, as indicated in figure 20.6 (b).

Grassmann's action and reaction forces in this example act along parallel straight lines. Therefore they generate a primitive couple in this system.

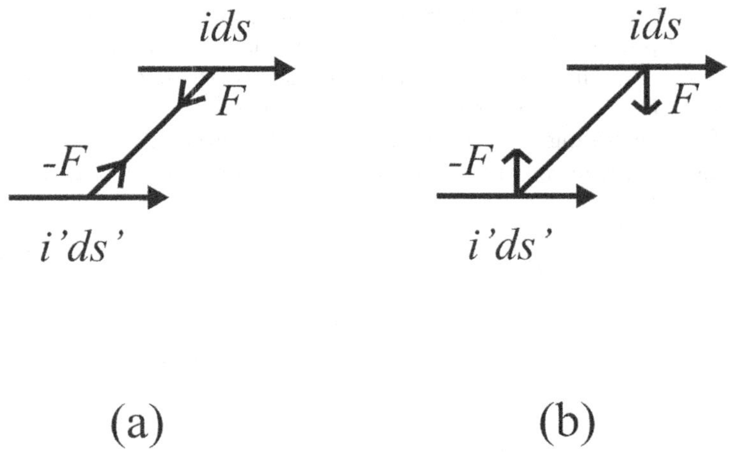

Figure 20.6: (a) Central action and reaction, according to Ampère's force. (b) Primitive couple, according to Grassmann's force.

20.2.4 Primitive Force with the Concept of a Magnetic Field

It is also possible to visualize the existence of a primitive couple utilizing the modern concept of a magnetic field, as shown in figure 20.7. There are two parallel current elements making angles $\alpha = \beta = 45^o$ with the straight line connecting their midpoints.

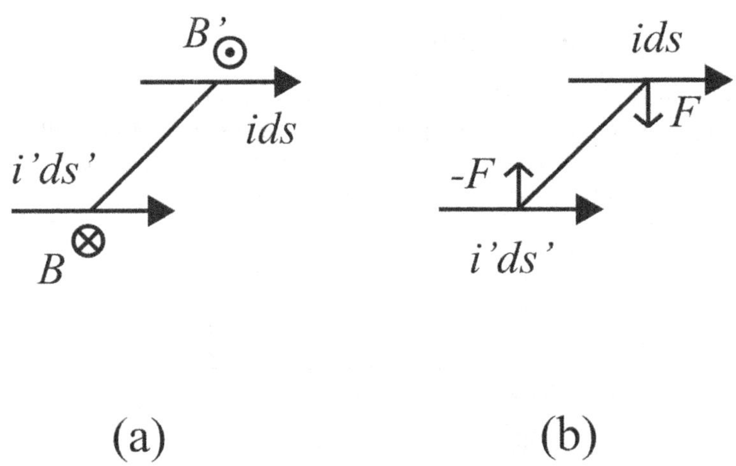

Figure 20.7: (a) Magnetic field B' generated by the current element $i'ds'$ and acting on the current element ids. There is a component of this magnetic field orthogonal to the paper, leaving it. Also shown is the magnetic field B generated by ids and acting on $i'ds'$, with a component orthogonal to the paper, penetrating it. (b) forces generated by these magnetic fields.

The magnetic fields produced by each current element are given by equation (15.2). Figure 20.7 (a) presents the magnetic field B' generated by the current element $i'ds'$ and acting on the current element ids. Also shown is the magnetic field B generated by ids and acting on $i'ds'$. Figure 20.7 (b) shows the forces generated by these magnetic fields according to equation (15.5). These opposite magnetic forces acting along parallel straight lines illustrate the primitive couple generated by the magnetic field concept.

20.2.5 Ampère Against the Primitive Couple

According to Ampère, all these conceptions in which there is a primitive couple violate Newton's third law in the strong form. In the *Théorie*, for instance, he said the following:[16]

[16] [Ampère, 1826f, p. 154] and [Ampère, 1823c, Ampère, 1990, p. 326].

> The physicists who adopted this opinion are then forced to admit a really elementary action, consisting in two equal forces acting in opposite senses along parallel straight lines, forming in this way a primitive couple, which cannot be reduced in terms of forces for which the action and reaction are opposite along the same straight line. I always considered this hypothesis of the primitive couples as absolutely contrary to the first laws of mechanics, among which one should count, with Newton, the equality of action and reaction in opposite senses along the same straight line.

He expressed the same point of view in several papers and letters. He was always against the primitive couples embedded in Biot and Savart's law, as well as in Faraday's conceptions. Likewise, he certainly would reject the primitive couple arising with Grassmann's force. He would also be totally against the primitive couple generated by the modern magnetic field concept.

It should be emphasized that Ampère's force between two current elements is the only force acting along the straight line connecting these elements which yields also Ampère's integrated force exerted by a closed circuit of arbitrary form acting on another current element.[17]

20.3 Ampère Against the Violation of the Law of Action and Reaction

In some situations Grassmann's force between current elements violates Newton's third law even in the weak form. That is, it is possible to have a force due to a current element $i'ds'$ acting on a current element ids having a magnitude which is different from the magnitude of the force exerted by ids on $i'ds'$, as illustrated in figure 20.8. In this case there are two orthogonal current elements. The current element $i'ds'$ is directed along the straight line connecting their midpoints, while the current element ids is orthogonal to this straight line.

Figure 20.8: Two orthogonal elements. Grassmann's force F exerted by ids on $i'ds'$ is different from zero, while Grassmann's force F' exerted by $i'ds'$ on ids is zero.

According to Ampère's force, equations (2.1) up to (2.20), these two current elements do not interact with one another. According to Grassmann's force given by equations (14.7) up to (14.10), the current element $i'ds'$ does not exert a force on the current element ids. The current element ids, on the other hand, exerts a force on $i'ds'$ which is orthogonal to $i'ds'$ and which is also orthogonal to the straight line connecting them, as represented in figure 20.8. Therefore, according to Grassmann's law, there is no action and reaction in this configuration, not even in the weak form. Moreover, even in this example there is also a primitive couple of this system relative to the midpoint of the current element ids.

In this configuration there is not only a primitive couple, but also a net force different from zero acting on this system of two current elements. According to Grassmann's law, this net force has an internal origin, being due to the interaction between these two current elements.

There is also a violation of Newton's third law, in the strong and weak forms, utilizing the modern concept of a magnetic field, as illustrated in figure 20.9. The magnetic field generated by a current element is given by equation (15.2). Figure 20.9 (a) shows the magnetic field B generated by ids penetrating the paper and acting on $i'ds'$. On the other hand, there is no magnetic field B' generated by $i'ds'$ acting at the location of ids. The force exerted by a magnetic field on a current element is given by equation (15.5). Figure 20.9 (b) shows the force F acting on $i'ds'$ pointing upwards. This force is due to the magnetic field generated by ids. On the other hand, there is no net force F' acting on ids due to the magnetic field generated by $i'ds'$.

The modern conception of a magnetic field leads to a violation not only of Newton's action and reaction law in the weak form, but also to a violation of its strong form, as shown in this example.

[17] [Duhem, 1886] and [Duhem, 1892, Appendice au Livre XIV: Sur la loi d'Ampère, pp. 309-332].

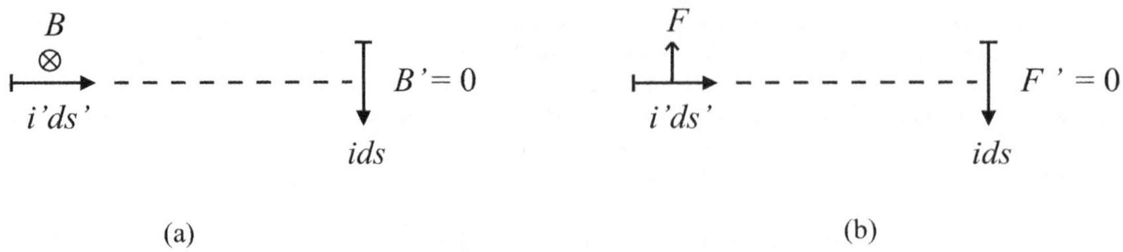

Figure 20.9: (a) Magnetic fields generated by the current elements. (b) Forces acting on the current elements.

Ampère certainly would reject completely situations like this which blatantly violate the principles of Newtonian mechanics. Here there is not only a primitive couple, but also an internal net force acting on the system! He would never accept such conceptions.

20.4 Maxwell's Appraisal of Ampère's Force between Current Elements

We close this Chapter with Maxwell's comparison of the forces between current elements of Ampère and Grassmann, together with his clear defense of action and reaction along the straight line connecting the interacting bodies.

Maxwell knew not only Ampère's force, but also Grassmann's one. In his *Treatise on Electricity and Magnetism* he made an analysis of four formulas expressing the forces between two current elements, namely, those of Ampère, Grassmann and two others of his own creation. His final judgement:[18]

> Of these four different assumptions that of Ampère is undoubtedly the best, since it is the only one which makes the forces on the two elements not only equal and opposite but in the straight line which joins them.

Maxwell's admiration for Ampère's work and for his force between current elements has been expressed in the following words:[19]

> The experimental investigation by which Ampère established the law of the mechanical action between electric currents is one of the most brilliant achievements in science. The whole, theory and experiment, seems as if it had leaped, full grown and full armed, from the brain of the 'Newton of Electricity'. It is perfect in form, and unassailable in accuracy, and it is summed up in a formula from which all the phenomena may be deduced, and which must always remain the cardinal formula of electro-dynamics.

[18] [Maxwell, 1954, vol. 2, articles 526 and 527, p. 174].
[19] [Maxwell, 1954, vol. 2, article 528, p. 175].

Chapter 21

Three Experiments Illustrating the Confrontation of These Opposing Theories

21.1 Explanations for the Interaction between Two Current-Carrying Wires

21.1.1 Ampère's Explanation

The essence of Ampère's electrodynamics is the hypothesis of attractive and repulsive forces between current elements, coupled with the supposition of electric currents flowing around the particles of magnetized bodies. It was based on these assumptions that he presented his original interpretation of Ørsted's experiment. He was then led to predict the interaction between two current-carrying conductors, a phenomenon which had never been observed. He was the first to observe it in the attractions and repulsions between flat spirals carrying steady currents, as discussed in Subsection 3.6.1. He then performed his most famous experiment in which he verified the attraction between parallel conductors carrying currents in the same direction. A repulsion was observed when the parallel currents flowed in opposite directions, Subsection 3.6.2.

The result of this experiment had been *predicted* by Ampère, based on his assumption of an interaction between current-carrying conductors. According to Ampère, this experiment always proved the existence of attractive and repulsive forces acting between current elements.

Ampère did not stop at the qualitative aspect of this experiment. He also calculated the net force exerted by a long straight wire carrying a constant current i and acting on another parallel straight wire carrying a current i'. Supposing the first wire to have an infinite length, he showed that the force acting on a length ℓ of the second wire was inversely proportional to the distance a between them. In the *Théorie*,[1] for instance, he showed that this force was proportional to:

$$\frac{ii'\ell}{a} \, . \tag{21.1}$$

That is, the force per unit length is proportional to the product of the current intensities and inversely proportional to the distance between the wires.

21.1.2 Ørsted Could Never Satisfactorily Explain the Interaction between Current-Carrying Wires

According to Ampère, the primitive fact which should be explained by the other physicists was the interaction between electric currents. He was never satisfied with Ørsted's explanations for this phenomenon.

Ørsted knew Ampère's fundamental experiment showing the attraction between parallel wires carrying currents in the same sense, together with their repulsion when the currents flowed in opposite senses. He

[1] *Théorie*, Section 29.14, "Mutual Action between Two Rectilinear Conductors", [Ampère, 1826f, pp. 65-94] and [Ampère, 1823c, Ampère, 1990, pp. 237-266].

tried to explain these interactions with his model of positive and negative electricities flowing in helices or in circles around each wire.[2] We were unable to understand his explanations for this experiment. In any event, we quote here some of his words related to this topic. As discussed in Section 11.1, he utilized for the modern concept of charged electric particles some alternative terms: electric matter, electricities, electric forces or electrical powers. Here are his words related to Ampère's experiment:[3]

> [...] Thereby he [Ampère] found that conductors through which the electric forces move in the same direction attract and, in the opposite case, repel each other. Ampère considered this to be the fundamental law for all electromagnetic effects; the author [Ørsted], on the other hand, regards it as an obvious consequence of the law he has discovered. If we imagine two parallel conductors which carry current in the same direction, the similar forces would not meet except at a couple of turning points precisely because they move in the same direction, but the opposite forces must meet and thus produce attraction. If, on the other hand, the two conductors are traversed in opposite directions, the similar forces meet and produce repulsion.
>
> The author [Ørsted] does not content himself with regarding this as a consequence of the law he has found for electromagnetism; he even tries to show that a circuit of forces is the only condition under which current-carrying conductors can exert the attractions and repulsions described here. If the forces flowed outward from the conductor in straight line from its axis, no effect would be produced when both forces balanced each other and repulsion in all cases where one of the forces was dominant. This empirical relationship is possible only when the radiating forces act along the tangents to the circumference of the conductor, or in directions which can be decomposed in such a way that only effective tangential forces remain, and when each of the forces acts from any given point along opposite tangents.

Apparently Ørsted was trying to show that with a pair of radial fluxes of positive and negative charges (or, in Ørsted's mode of expression, a pair of outward fluxes of electric forces), moving in opposite senses along straight lines orthogonal to the axes of the rectilinear wires, it would not be possible to explain the attractions and repulsions between current-carrying wires observed by Ampère.

However, this was not Ampère's conception. Ampère explained the interactions between the wires by direct actions between the current-carrying conductors. Ampère did not consider it necessary to introduce fluxes of electric charges flowing in any direction in the space outside the current-carrying wires in order to explain their attractions or repulsions.

It seems to us that Ørsted's explanations were not very clear. This lack of clarity applies not only for his interpretation of the interaction between current-carrying wires, but for electromagnetic phenomena in general. Faraday himself, when trying to describe how Ørsted explained his own experiment of the deflection of a compass needle by a current-carrying wire, stated the following:[4]

> Notwithstanding all this, I have very little to say on M. Oersted's theory, for I must confess I do not quite understand it.

Ampère's experiment showed the interaction between current-carrying conductors. Ørsted never presented a quantitative explanation for this experiment based on electric conflict existing outside the wires. The lack of a mathematical explanation left a considerable gap for a clear comprehension of how he understood the phenomena. His arguments to explain electromagnetic phenomena seem obscure and not convincing to us.

21.1.3 Problems with Biot and Savart's Explanation for the Interaction between Current-Carrying Wires

According to Ampère, the fundamental interaction took place between current elements. It was from this primitive fact that he tried to explain all other electromagnetic phenomena, like Ørsted's experiment showing the torque exerted by a current-carrying wire on a compass needle or his own experiment showing the attractions and repulsions between current-carrying wires.

Biot and Savart, on the other hand, considered the interaction between current-carrying conductors as a secondary effect. They assumed that the primary interaction took place between magnetic fluids (or between magnetic poles). They believed that they could explain Ampère's experiment on the interaction between current-carrying wires by supposing the magnetization of both wires due to the flow of electric currents through them. In 1824 Biot presented figure 21.1 in order to explain Ampère's experiment.[5]

[2] [Ørsted, 1998f, pp. 437-439] and [Ørsted, 1998e, p. 426].

[3] [Ørsted, 1998e, pp. 426-427].

[4] [Faraday, 1822b, p. 107].

[5] [Biot and Savart, 1824, pp. 769-770], [Farrar, 1826, p. 361, figure 162] and [Biot and Savart, 1885, pp. 122-123, figure 10].

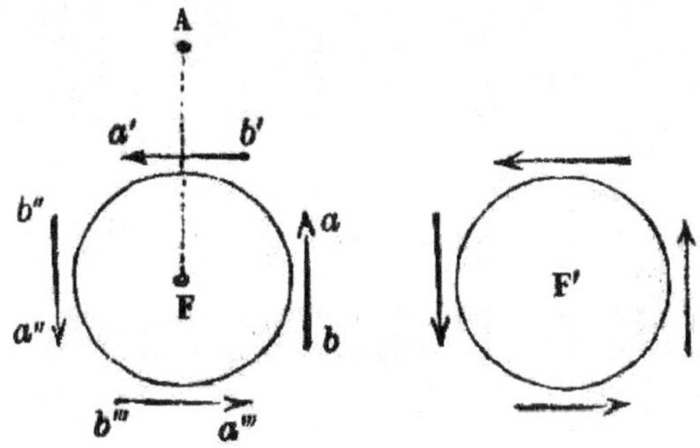

Figure 21.1: Biot and Savart's explanation of Ampère's experiment.

The two circles centered in F and F' represent the cross sections of two cylindrical wires which are orthogonal to the paper. Electric currents flow in the same sense in both wires. Let us suppose that they are leaving the paper, as in figure 21.2 (a). In this configuration, Ampère observed an attraction between the wires, supposing that it was due to a direct interaction between the current elements of both wires. Biot and Savart's explanation, on the other hand, was based on the tangential magnetization of both wires, followed by an interaction between the magnetic dipoles which were supposed to exist around the circumference of both wires:[6]

> Now, if we imagine that instead of the unknown mode of action of these wires, we substitute the tangential [magnetized] needles which are equivalent to them, it results from the figure itself that they will mutually present to each other faces the needles of which tend to attract each other; [...] Therefore, if the wires act upon magnetic bodies in virtue of a momentary magnetizing of their particles, as all reasoning from analogy leads us to suppose, we should expect these same actions to be exerted also by one wire upon the other according to the same laws.

According to Biot and Savart, the main effect produced by the flow of a current was the tangential magnetization of the wire, as represented by the arrows around the wires of figure 21.1. We can imagine each arrow as a magnetic dipole, with the North pole in the tip and the South pole in the back. The right arrow around the center F points oppositely relative to the left arrow around the center F', causing a magnetic attraction between these closest magnetic dipoles. This attraction between the dipoles would produce an attraction between the wires, as the dipoles would somehow be connected to the corresponding wires. This explanation is illustrated in figure 21.2 (b).

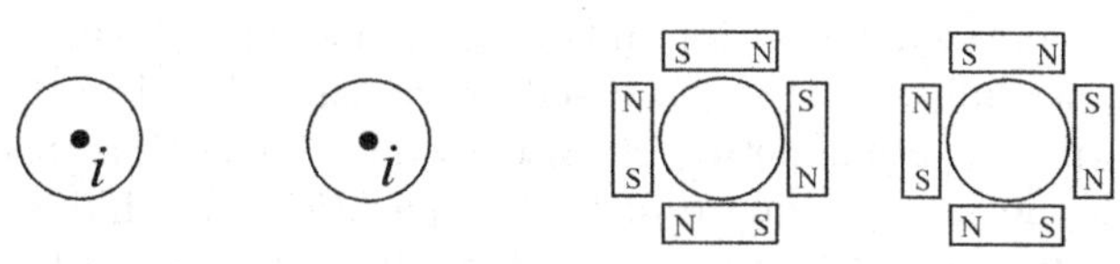

(a) (b)

Figure 21.2: (a) Two straight wires carrying currents in the same direction and attracting one another. (b) Biot and Savart's explanation of this effect supposing the magnetization of both wires and the interaction of the supposed magnets distributed around the cross sections of both wires.

Ampère also observed a repulsion between straight and parallel wires carrying currents in opposite directions, as represented in figure 21.3 (a). Biot and Savart's explanation of this repulsion is illustrated in figure 21.3 (b).

There are two problems with this explanation of Biot and Savart. Qualitatively it only works for some relative configurations of both wires, like that of figure 21.1. Consider now figure 21.4, which is the previous figure with the wire centered on F' rotated by 45° around F. We have once more two parallel straight wires with currents leaving the paper. According to Ampère's experiment, these wires attract one another. According to Biot's magnetization hypothesis, on the other hand, they should repel one another. The reason for this predicted repulsion is that the closest magnetic poles of the wires centered at F and F' are located

[6][Biot and Savart, 1824, p. 770], [Farrar, 1826, p. 361] and [Biot and Savart, 1885, pp. 123-124].

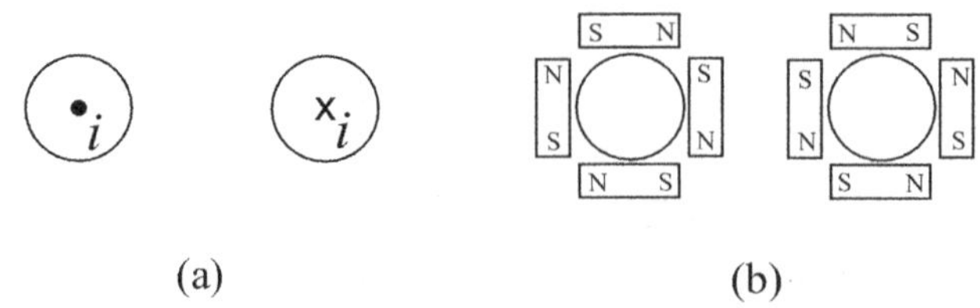

Figure 21.3: (a) Two straight wires carrying currents in opposite directions and repelling one another. (b) Biot and Savart's explanation of this effect supposing the magnetization of both wires and the interaction of the supposed magnets distributed around the cross sections of both wires.

at points a-c'' and b'-d''. The pair a-c'' is composed of magnetic poles of the same type, the same happening with the pair b'-d''. Therefore, according to Biot's assumption, these wires should repel one another, while Ampère's experiment showed that they attract one another.

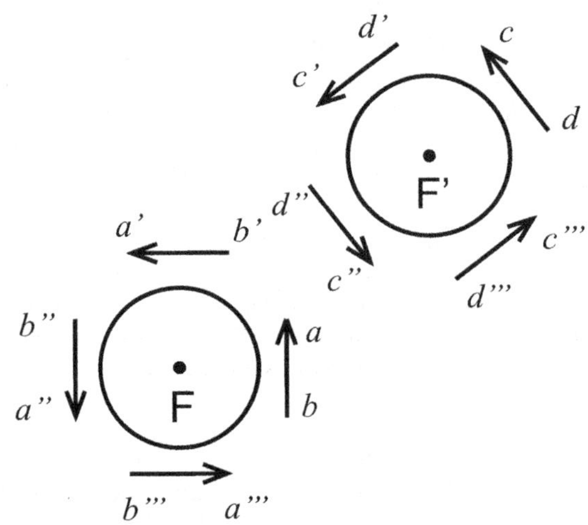

Figure 21.4: Two cylindrical wires orthogonal to the paper with the currents flowing upwards. They attract one another according to Ampère's experiment. However, they should repel one another according to Biot and Savart's magnetization hypothesis.

There is also a quantitative problem with Biot and Savart's model for the interaction between two current-carrying wires. This problem is similar to that discussed previously related to their model for the interaction between a current-carrying wire and a magnetic pole discussed in Subsection 18.1.2. Ampère's force between current elements varies with the inverse square of their separation. By integrating this force in order to obtain the net force between two parallel current-carrying wires, Ampère obtained that this last resultant force was inversely proportional to the distance between both wires. The force between two magnetic dipoles, on the other hand, varies typically as the inverse fourth power of their separation. Therefore, it cannot be equivalent to Ampère's force between two current elements. Moreover, by integrating the force exerted by an infinite straight line with magnetic dipoles spread orthogonally to it acting on another parallel and infinite straight line with magnetic dipoles spread orthogonally to it, one obtains a force inversely proportional to the cube of the separation between both wires. And this proportionality is not compatible with Ampère's experimental result.

By increasing indefinitely the number of needles around the cross section of each wire, decreasing simultaneously their individual lengths, we end up with two magnetized rings like those of Gay-Lussac and Welter, figures 9.1 and 9.3 (b). In the limit of an infinite number of small tangential magnets, these magnetized rings will not interact with each other, if we follow Biot and Savart's magnetization hypothesis. Consequently, the two parallel current-carrying wires should not interact magnetically with one another, as Ampère correctly pointed out in the *Théorie*.[7] Ampère's experiment, on the other hand, showed that they will repel or attract one another depending on the directions of the currents.

[7][Ampère, 1826f, pp. 185-186] and [Ampère, 1823c, pp. 357-358].

21.1.4 Ampère Against Faraday's Explanation for the Interaction between Current-Carrying Wires

Ampère was the first scientist to observe the attractions and repulsions between parallel and rectilinear current-carrying conductors. Faraday tried to explain this phenomenon based on what he considered as a simple fact, namely, the rotational or revolutive actions between a magnetic pole and a current-carrying wire.[8] In a paper of 1822, for instance, he said the following:[9]

> Then follow the actions between two wires, these when similarly electrified attract as M. Ampère has shewn; for then the opposite sides are towards each other, and the four powers all combine to draw the currents together forming a double attraction; but, when the wires are dissimilar they repel, because, then on both sides the wire the same powers are opposed, and cause a double repulsion.

Maybe what he had in mind was similar to Biot and Savart's explanation presented in figures 21.2 (b) and 21.3 (b). We discussed in Subsection 21.1.3 the problems pointed out by Ampère against this model.

In any event, Faraday's explanation was only qualitative. He was never able to show, for instance, that the force between two current-carrying parallel straight wires should be inversely proportional to their distance.

Ampère himself was against Faraday's explanation. Ampère had obtained the continuous rotation of an extremity of a current-carrying conductor moving in a resistive and conducting medium. Coulomb's force between magnetic poles is conservative as it is a central force varying inversely as the square of the distance between the poles. Therefore it is impossible to explain the experiment of continuous rotation in a resistive medium utilizing only this conservative force. Ampère utilized this argument against Faraday's hypothesis of the transverse magnetization of the wire. In a letter addressed to Auguste de la Rive, of October 1822, for instance, he said:[10]

> [...] in order to explain it [the mutual action between two current-carrying conductors] with the law of M. Faraday in the same way that he explained the experiment between a conductor and a magnet, it would be necessary to suppose, in one of both wires, anything similar to a magnet. It would be necessary to compare the voltaic conductors to steel wires transversely magnetized, as was done in Germany, but this identification is denied by the facts, because a mobile conductor can always turn in the same sense by the action of another conductor, and it is not possible to obtain the same result by making two steel wires acting on one another, no matter their magnetizations.
>
> This difference between the steel wires transversely magnetized and the voltaic conductors is a necessary consequence of my theory, and it is contradictory to all those [theories] which want to oppose themselves to it.

21.2 Explanations for Ampère's Bridge Experiment

Ampère obtained the final value $k = -1/2$ for the constant appearing in his force law in 1822, as discussed in Chapter 7. This negative value of k was totally unexpected for him. It meant, among other things, that two parallel and aligned current elements should repel one another if the currents flowed in the same sense, as indicated in figure 21.5.

Figure 21.5: Two parallel and aligned current elements.

Ampère designed his bridge experiment in order to test the existence of this repulsion, as discussed in Section 8.2. The experiment confirmed his prediction.

We do not know of any explanation of Ampère's bridge experiment offered by Ørsted, Biot, Savart and Faraday.

This experiment presents a challenge to Grassmann's force given by equations (14.7) up to (14.10). According to these expressions, the two current elements of figure 21.5 should exert no force on one another.

---

[8] [Williams, 1989b, p. 92].

[9] [Faraday, 1822e, p. 85] and [Faraday, 1952d, p. 801].

[10] [Launay (ed.), 1936a, p. 606].

Utilizing the magnetic field one also obtains this prediction of zero force acting on each current element of figure 21.5. Equation (15.2) shows that a current element does not produce any magnetic field along its own direction. Therefore, a current element cannot exert any force on another parallel and aligned current element.

One of the authors of this work has published many papers[11] and two books[12] devoted to inductance calculations and to a comparison between Ampère and Grassmann's formulas for the forces between current elements. One of the main subjects of these books was an analysis of Ampère's bridge experiment. It was then shown that Grassmann's force can only explain quantitatively the measured values for the net force acting on the bridge if one includes the force exerted by the bridge on itself.

This is also the only way of explaining quantitatively Ampère's bridge experiment utilizing the magnetic field concept. That is, it would be necessary to include not only the force acting on the bridge due to the magnetic field generated by the remainder of the circuit, but also the force due to the magnetic field generated by the bridge itself. When this last component of the force is not included, then the force calculated with Grassmann's expression or with the magnetic field concept turns out to be much smaller than the measured value.

Ampère would certainly be totally against these explanations of the bridge experiment utilizing Grassmann's force or the magnetic field concept. His reasoning for rejecting these explanations would be very simple, namely, they violate Newton's third law of motion.

With Ampère's force between current elements, on the other hand, it is possible to explain quantitatively the measured values of the force acting on the bridge without violating Newton's action and reaction law. After all, the force exerted by the bridge on itself is zero with Ampère's force. And his integrated force exerted only by the remainder of the circuit and acting on the bridge coincides with the measured value of this force. His explanation satisfies the principle of action and reaction. This compliance with Newton's third law has been emphasized many times by Ampère.

21.3 Explanations for the Rotation of a Magnet around Its Axis

In this Section we discuss alternative explanations for the working mechanism of Ampère's motor.[13]

As seen in Subsection 7.2.3, in 1822 Faraday stated the following after obtaining the rotation of one extremity of a magnet around a current-carrying wire:[14]

> Having succeeded thus far, I endeavoured to make a wire and a magnet revolve on their own axis by preventing the rotation in a circle round them, but have not been able to get the slightest indications that such can be the case; nor does it, on consideration, appear probable.

In 1822 Ampère published a paper discussing several new experiments on electromagnetism.[15] In the *Recueil* he introduced some Notes on this paper. In one of these Notes he stated the following:[16]

> Since I became aware, at the end of October 1821, of the memoir in which M. Faraday had published, a short time before, his important discovery of the continuous motion of rotation of a voltaic conductor around a magnet, and of a magnet around a conductor, and in which he had announced that he had not been able to make, by the action of the latter, a magnet spin around its axis, I tried to produce this kind of motion by making magnets arranged in all possible manner act on mobile conductors which I had utilized up to this time in all my experiments, with their two extremities located at the rotation axis, [...]

Ampère performed the experiment of the rotation of a magnet around its axis between November and December 1821, being the first scientist to succeed in observing this remarkable fact. He presented his result to the Academy of Sciences of Paris on 7 January 1822,[17] as discussed in Subsection 7.2.3.

[11] [Assis and Bueno, 1995], [Bueno and Assis, 1995], [Assis and Bueno, 1996], [Bueno and Assis, 1997a], [Bueno and Assis, 1997c], [Bueno and Assis, 1997b], [Bueno and Assis, 1998] and [Assis and Bueno, 2000].

[12] [Bueno and Assis, 2001] and [Bueno and Assis, 2015].

[13] [Assis and Chaib, 2012] and [Chaib and Assis, 2013].

[14] [Faraday, 1822e, p. 79] and [Faraday, 1952d, p. 798].

[15] [Ampère, 1822r], [Ampère, 1822j] and [Ampère, 1885e].

[16] [Ampère, 1822p, p. 235] and [Ampère, 1885m, pp. 267-268].

[17] [Ampère, 1822d], [Ampère, 1822e], [Ampère, 1885d] and [Blondel, 1982, pp. 114-115].

21.3.1 Faraday's Explanation for the Rotation of a Magnet around Its Axis

Ampère wrote a letter to Faraday informing him of his discovery. This letter has been lost.[18] Faraday replied on 2 February 1822, presenting an explanation for this phenomenon which was different from that of Ampère:[19]

> You mention your opinion that this experiment will be competent to decide the question whether the currents of electricity assumed by your theory exist round the axis of the magnet or round each particle from which I gather that the view you take of it differs from the one I at present have, since to me it seems a modification of the revolution of a wire round a pole. I presume much in differing from yourself on this subject and more in stating the difference to you but I do not hesitate a moment in concluding that in the true spirit of philosophy you are anxious to hear (or at last willing) even the doubt of one young in the subject if there be the smallest possibility that they will either correct or confirm previous views. [...]
>
> I regret that my deficiency in mathematical knowledge makes me dull in comprehending these subjects. I am naturally sceptical in the matter of theories and therefore you must not be angry with me for not admitting the one you have advanced immediately. The ingenuity and applications are astonishing and exact. But I cannot comprehend how the currents are produced and particularly if they be supposed to exist around each atom or particle and I wait for further proofs of their existence before I finally admit them.

That is, Faraday did not accept Ampère's explanation. Faraday presented an alternative model, in analogy with the explanation he had given for his own experiment of the rotation of the extremity of a current-carrying wire around a fixed magnet, discussed in Section 7.1. Faraday utilized once more his supposed rotational or revolutive action of a magnetic pole acting on a current-carrying wire, discussed in Section 13.1. In this letter of 2 February 1822, addressed to Ampère, Faraday explained as follows this new phenomenon of the rotation of a magnet around its axis:[20]

> The rotation of the magnet [around its axis] seems to me to take place in consequence of the different particles of which it is composed being put into the same state by the passing current of electricity as the wire of communication between the voltaic poles, and the relative position of the magnetic pole to them.
>
> Thus [see figure 21.6] the little arrows may represent the progress of the electricity; then any line of particles parallel to them except that line which passes as an axis through the pole (represented by a dot) will be in the situation of the revolving wire[21] and will endeavour to revolve round the pole and as all the lines act in the same direction or tend to go one way round the pole the whole magnet revolves [around its own axis].

Faraday's explanation for the rotation of a magnet around its axis is based only on the existence of rotational or rotatory forces acting inside the magnet. That is, the pole inside the magnet exerts a rotational force, always in the same sense (for instance, anti-clockwise when the magnet is seen from above along its axis of symmetry), acting on the electric currents flowing down inside the magnet. We can illustrate his explanation with figure 21.7.

21.3.2 Biot's Explanation for the Rotation of a Magnet around Its Axis

An analogous explanation was presented by Biot in 1824:[22]

> This experiment may be varied by immersing the wire CF in a small metallic cup filled with mercury, and fixed upon the upper pole of the bar itself, figure 25 [our figure 21.8]; the communication being completed by the lateral wire $F'Z$ as before. *In this case, the upper portion of the bar itself serves as a conductor of the voltaic current, and the lateral forces which proceed from it, act upon the magnetism of its pole A, and cause it to turn rapidly upon itself.* The effect would not be produced if the extremity F' of the second conducting wire, instead of being placed laterally a little below A, had been put directly below the opposite pole B of the bar; for then the current would traverse the bar throughout its whole

[18] [Blondel, 1982, p. 116].

[19] [Launay (ed.), 1943, p. 910].

[20] [Launay (ed.), 1943, p. 913].

[21] That is, any line will be in the situation of the revolving wire in Faraday's original experiment in which one extremity of a current-carrying wire rotated around a magnet, as represented in figures 7.1 and 7.2.

[22] [Biot and Savart, 1824, p. 754] and [Farrar, 1826, pp. 346-347], our emphasis.

Figure 21.6: Faraday's illustration of the electric currents flowing inside the magnet. The central dot represents the magnetic pole while the arrows indicate the directions of the electric currents.

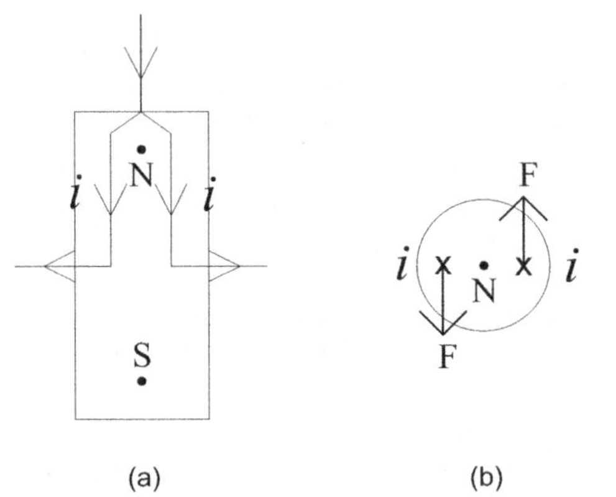

Figure 21.7: (a) Cylindrical magnet NS and two lines of current i flowing inside it. (b) Cross section of the magnet as seen from above, showing the North pole, the currents penetrating the paper, and the force F exerted by the North pole on each electric current.

> length, and would act upon its two poles with an equal energy; and as it would tend to turn them in opposite directions, the resultant of its efforts would be nothing, and the bar would no longer turn upon itself.
>
> Mr Faraday was the first who made this ingenious application of the transverse force exerted by a uniting wire when traversed by the voltaic current. [...]

In the previous quotation we emphasized the crucial section. This portion shows that Biot's explanation is essentially the same given by Faraday, although inverting the roles of the agents. According to Biot, the lateral forces exerted by the current flowing *inside* the magnet and acting on the magnetic pole of the magnet were responsible for the rotation of the magnet around its axis. According to Faraday, on the other hand, the lateral forces exerted by the magnetic pole and acting on the current flowing *inside* the magnet were responsible for the rotation of the magnet around its axis.

That is, in both explanations there are lateral forces (also called transverse, rotational, rotatory or revolutive forces) exerted by *internal* elements of the magnet, acting upon other *internal* elements of the magnet, causing the magnet to rotate around its axis relative to the ground.

21.3.3 Ampère Against the explanations of Faraday and Biot

Ampère realized immediately that these explanations of Faraday and Biot violated Newton's third law in the strong form. According to this law, a system composed of many bodies cannot exert a net torque on itself due only to mutual forces acting internally to the system. In Ampère's experiment the magnet was

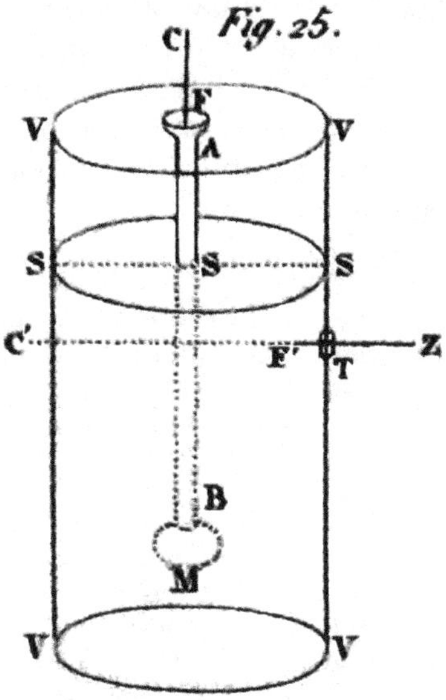

Figure 21.8: Biot's figure related to the rotation of a magnet around its axis.

spinning around its axis while floating on a resistive medium, namely, mercury. In the stationary state, it spins around its axis, relative to the ground, with a constant angular velocity. There is a net torque exerted on the magnet by the resistive mercury, which would decrease its angular velocity if it were acting alone. Therefore another opposite torque must be in action in order to keep the magnet spinning with a constant angular velocity in this resistive medium. Ampère knew that, according to Newton's third law of motion, this opposite torque could only be due to an *external* agent not belonging to the magnet. That is, this opposite torque could not be generated inside the magnet.

Ampère expressed clearly his objections relative to Faraday's explanation in a letter addressed to him dated 10 July 1822:[23]

> A fundamental and obvious principle of physics is that, the action always being equal to the reaction, it is impossible that a rigid system be put in motion in any way by a mutual action between two of its particles, as this action produces on the two particles two equal and opposite forces which tend to move the body in opposite senses. It then follows that, when the particles of a magnet traversed by an electric current which puts them in the same state of the conducting wire act on the pole or on any other part of the magnet, no motion in this body can result from this action, [...]
>
> From this observation, the rotation of a floating magnet around its axis can only be explained as I did in the memoir included in the May issue of the *Annales de Chimie et de Physique*,[24] which I sent to you recently.

In the *Théorie*, Ampère also expressed himself against Biot's explanation for the rotation of a magnet around its axis:[25]

> He [Biot] explains in particular, on page 754 of volume II of this work,[26] the motion of rotation of a magnet around its axis, when a portion of current traverses it, supposing that the magnet rotates by the action which this portion exerts on the remainder of the magnet, however, [the magnet] forms with it [that is, with the portion of current traversing it] a system of invariable form in which all parts are rigidly connected to one another. Evidently, this [explanation] supposes that the action and the reaction of this portion of the current and the remainder of the magnet produce a torque. [...]
>
> Is there any hypothesis which is more contrary to these analogies [with the other laws of attraction] than to imagine forces such that the mutual action between the several parts of a system with invariable shape can place this system in motion?

[23] [Ampère, 1885g, p. 298] and [Launay (ed.), 1936a, p. 586].
[24] [Ampère, 1822d], [Ampère, 1822e] and [Ampère, 1885d].
[25] [Ampère, 1826f, pp. 181-182 and 184] and [Ampère, 1823c, Ampère, 1990, pp. 353-354 and 356].
[26] Biot's *Précis élémentaire de Physique expérimentale*, [Biot and Savart, 1824, p. 754].

Once more Ampère was defending Newton's third law of motion and going against any hypothesis which admitted that a system of particles might exert a net torque on itself. Ampère emphasized this aspect in several points of the *Théorie* not only conceptually, but even performing specific experiments to confirm the law of action and reaction in the realm of electromagnetism. We quote here another portion of his masterpiece,[27] see Section 29.22, appearing just after the description of these experiments devised to prove the validity of action and reaction in electrodynamics:

> [...] we have then a direct proof of the principle on which rests a part of the explanations which I supplied, namely: That the portions of current traversing the magnet do not act at all on it, because the forces which might result from its action on the internal currents of the magnet, or acting on what is called magnetic molecules, as they take place between the particles belonging to a single rigid body, are necessarily destroyed by an equal and opposite reaction.
>
> I confess that this experimental proof of a principle which is only a necessary consequence of the first laws of mechanics seems completely useless to me, as it would appear to all physicists who considered this principle as one of the foundations of science. I would not have made the observation, if it had not been supposed [by others] that the mutual action between an element of a conducting wire and a magnetic molecule consisted in a primitive couple composed of two equal and parallel forces which are not directly opposite to one another, due to which a portion of current established inside a magnet might move it. This supposition is contrary to the principle which is being discussed here and is denied by the previous experiment according to which there is no action exerted on the magnet by the portions of current traversing it when it is not covered with an insulating coat, as the motion which exists in this case remains the same when the currents are impeded to traverse the magnet by covering it with this coat.

21.3.4 Ampère's Explanation for the Rotation of a Magnet around Its Axis

The arguments discussed in Subsection 21.3.3 led Ampère to conclude that the external agent responsible for the torque acting on the magnet was composed of electric currents flowing in the mercury outside the magnet. By action and reaction, Ampère concluded that when the magnet was spinning anti-clockwise, the mercury should experience an opposite torque, making it rotate clockwise relative to the ground.

In a letter addressed to van Beek in 1821, he stated the following:[28]

> But this proof [in favor of the existence of microscopic currents around the particles of the magnet] should no longer be established in the same way, after the new experiments and thoughts that they suggested to me showed that the rotation of a magnet around its axis, [a phenomenon] which I was the first to observe, and the motion of revolution of the same magnet around a vertical conducting wire, discovered by M. Faraday, are much less due to the action of this wire than to the action of the electric currents established in the mercury, the reaction of which is the cause of the rotation of the mercury in the experiment of Sir H. Davy.

Ampère's explanation for his experiment utilizes the currents represented in figure 21.9.[29] This figure corresponds to figures 7.7 and 7.8 as seen from above. The cylindrical magnet is represented by the region internal to the circle nmn'. The mercury is represented by the region between the circle nmn' and the circle cfc'. In figure 21.9 (b), we represent the current i, which originates in one of the terminals of the battery, flowing vertically downwards along the magnet and leaving horizontally along the radial direction by its lower portion floating in mercury. There are analogous currents leaving from Z and going to all points along the periphery of the circle cfc' containing a metal ring connected to one of the terminals of the battery. Current i in the Zm portion flows inside the magnet, while in the portion mM it flows in the mercury outside the magnet. The current i', along the circle nmn', represents the internal equivalent current of the magnet responsible for its magnetic properties. That is, it represents the resulting value of all molecular currents of Ampère which cancel in all internal points of the magnet, there remaining only the circular tangential current around its surface.

Ampère expressed his explanation in the following words:[30]

> Let ZM (figure 9 [our figure 21.9 (a)]) be one of these currents, the portion Zm does not act, according to what has been said before, on the electric currents of the magnet; the portion mM attracts mn' and

[27][Ampère, 1826f, p. 171] and [Ampère, 1823c, Ampère, 1990, p. 343].

[28][Ampère, 1821d, p. 455], [Ampère, 1822x, pp. 180-181], [Ampère, 1885o, p. 222] and [Blondel, 1982, p. 115].

[29][Ampère, 1822d, figure 9], [Ampère, 1822e, figure 9] and [Ampère, 1885d, figure 9].

[30][Ampère, 1822d, pp. 70-71], [Ampère, 1822e, pp. 247-248] and [Ampère, 1885d, pp. 201-202].

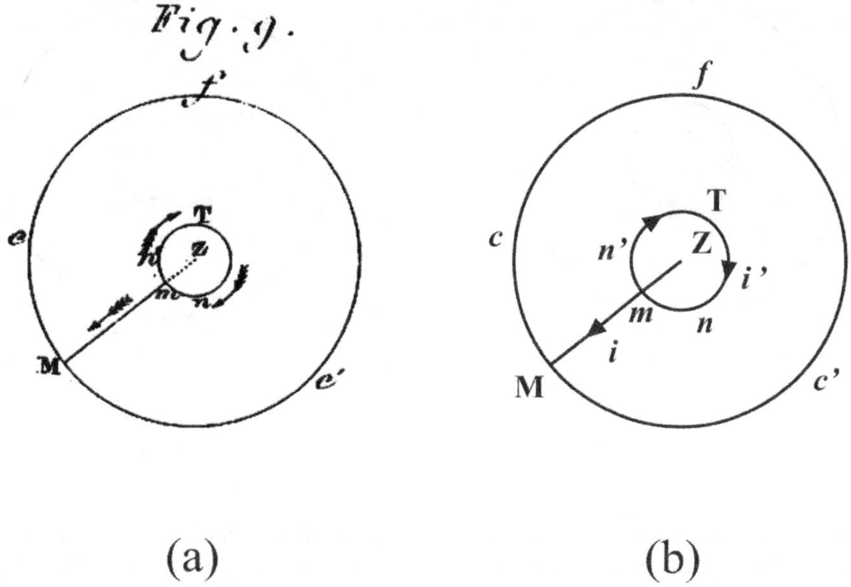

Figure 21.9: (a) Detail of the currents of figure 7.8. (b) Our representation of this figure indicating the external current i and the net value i' of the molecular currents of the magnet.

> repels mn. These two combined forces tend to make the magnet spin around itself in the sense $n'mn$. Similar forces are exerted simultaneously on all points of the magnet, [therefore] it turns around itself indefinitely.

We represent in figure 21.10 (a) the current i originating in the battery flowing externally to the magnet, from m to M. We also represent the net molecular current i' due to the magnet itself. It flows in the sense nmn'. The arrows of figure 21.10 (b) represent the forces indicated by Ampère. Current i in mM attracts the current i' in mn'. Current i in mM repels the current i' in nm. These two forces exert on the magnet a torque around its axis making it turn anti-clockwise. By action and reaction, the molecular currents i' of the magnet exert contrary forces on the current i in the mercury located outside the magnet. These two reaction forces of i' on i are also represented in figure 21.10 (b). We represent these two reaction forces acting on the midpoint of the section mM. They exert a torque on the mercury making it turn clockwise.

There will be a similar force acting in all points along the periphery $nmn'n$. These forces will generate an *external* torque acting on the magnet, making it spin anti-clockwise around its axis. By action and reaction, an opposite torque will act on the mercury, making it turn clockwise.

Ampère presented this explanation several times. In the *Recueil*, for instance, he said:[31]

> I then obtained from these considerations the following three consequences, namely:
>
> 1st: The continuous motion of a mobile conductor, with its upper extremity located in the rotation axis, can only happen when its lower extremity follows a circumference around this axis in a conducting liquid. M. Faraday obtained this rotation [see figure 7.1] utilizing mercury. I replaced it, advantageously [due to the lower friction], by acidulated water [see Subsection 7.2.1].
>
> 2nd: The liquid conductor should, by the action of the magnet, tend to rotate in the opposite sense, in agreement with what has just been observed by Sir H. Davy.[32]
>
> 3rd: It is only possible to make a magnet spin around its axis by the action of a voltaic circuit when a portion of this circuit passes through the magnet, because in this case, as there is no longer an action of this portion, the remainder of the circuit, exerting on it an equal action in opposite sense, gives it the desired motion.

In the third consequence Ampère said that "there is no longer an action of this portion". He wanted to say that this portion of the electric current flowing inside the magnet cannot make the magnet spin around

[31] [Ampère, 1822p, p. 236] and [Ampère, 1885m, p. 268].

[32] [Davy, 1823], [Davy, 1824] and [Davy, 1885]. A summary of Davy's experiment appears in [Ampère, 1822t].

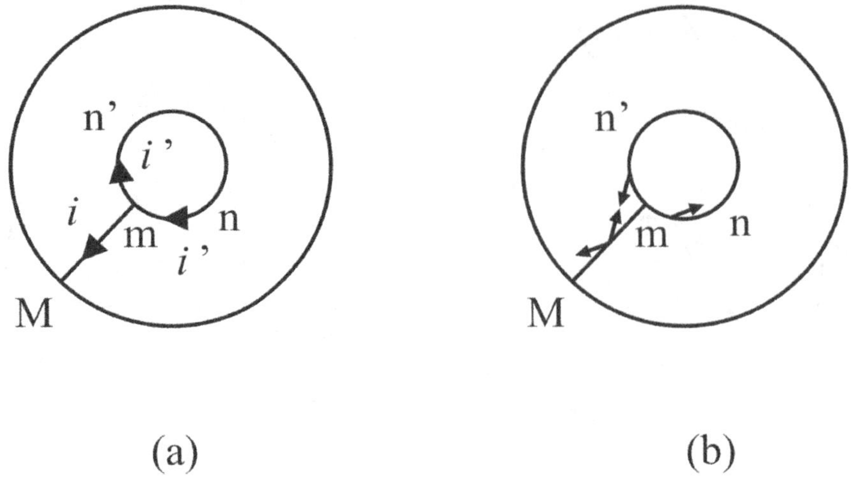

Figure 21.10: (a) Current i generated in the battery and flowing outside the magnet in the mercury located in the portion mM, together with the net molecular current i', due only to the magnet. (b) The arrow in mn' indicates the attraction of the current i located in mM acting on the current i' located in mn', the arrow in nm indicates the repulsion of the current i in mM acting on the current i' in nm, while the arrows in mM indicate the reaction forces exerted by the current i' in nmn' acting on the current i in mM.

its own axis. The reason for this conclusion is once more Newton's third law of motion. Considering the magnet and this internal portion of the current as a single system, then it cannot exert a net torque on itself due only to forces acting internally. The remainder of the circuit, in particular the external currents flowing in mercury or in acidulated water, can make the magnet spin around its axis. By action and reaction, Ampère concluded that the magnet, together with the internal current flowing through it, should exert an opposite torque acting on the remainder of the circuit outside the magnet. This reaction torque acting on the mercury makes it turn in the opposite sense relative to the ground.

These aspects were didactically explained by Ampère in another work of 1822:[33]

> These facts are easily explained in the theory of M. Ampère. He showed that they are a necessary consequence of the fact that the electric currents, excited by the pile, initially flow in the mercury and, then, in the body of the magnet, and from the fact that the portion of these currents which is inside the mercury exerts on those [currents] which he assumes to exist around the particles of the magnet, [flowing] on planes perpendicular to its axis, an action whose reaction tends to move the particles of the mercury in opposite sense, as observed in Sir H. Davy's experiment[34] (we will soon discuss it), while, *with respect to the portion of the currents located inside the body of the magnet, the action and the reaction taking place on the particles of a single body, no motion can result in this body.* Therefore, it is only necessary to consider the resultant of all the actions, attractive and repulsive, exerted by the small portions of current in the mercury acting on the small portions of currents in the magnet, according to M. Ampère's formula, and one immediately realizes that this resultant [force] is directed in the sense in which the motion really takes place [that is, it spins around its axis].

In summary, the explanations of Faraday and Biot for the rotation of a magnet around its axis violate Newton's third law, as these authors assumed that the torque on the magnet was generated by currents flowing inside the magnet. Ampère's explanation, on the other hand, utilized Newton's action and reaction law. According to Ampère, the torque on the magnet was caused by the external currents flowing in the mercury. The opposite torque generated by the magnet and its internal currents acted on the mercury, making it turn in the opposite sense. The motion of the mercury in the appropriate direction was confirmed by Davy's experiments.

[33][Ampère, 1822t, p. 22], our emphasis.

[34][Davy, 1823], [Davy, 1824] and [Davy, 1885]. A summary of Davy's experiment appears in [Ampère, 1822t].

21.3.5 Explanation for the Rotation of a Magnet around Its Axis Utilizing the Field Concept

Many modern textbooks present an explanation for the rotation of a magnet around its axis utilizing the magnetic field concept. The experimental situation is illustrated in figure 7.9. A pile or battery generates a current i flowing in a wire. It penetrates laterally on the magnet through a sliding contact between the wire and the magnet. If flows radially toward the center of the magnet, leaving along its axis, and returning to the battery through the closing wire. When there is little friction in this system, the magnet rotates around its axis in this configuration. Figure 21.11 presents some typical explanations of this phenomenon utilizing the magnetic field concept.[35]

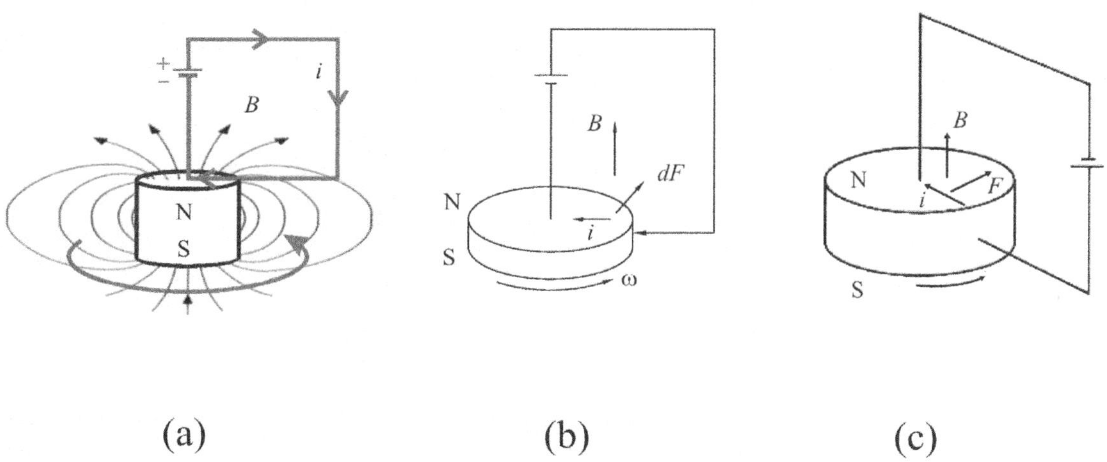

Figure 21.11: Modern explanations for the rotation of a magnet around its axis utilizing his own magnetic field B and the force exerted by this magnetic field acting on a current flowing inside the magnet.

The explanations are always the same. The authors assume that the magnetic field $\vec{B}$ generated by the magnet will act on the *internal* radial current. That is, acting on the portion of the current i penetrating the magnet radially and flowing to its axis. The authors then assume the magnetic force exerted by this magnetic field act on a current element as given by equation (15.5). The authors then assume that this force will generate a torque on the magnet, making it spin around its axis relative to an inertial frame of reference.

21.3.6 "Ampère" Against the Explanation of the Torque Utilizing the Magnetic Field

As discussed in Subsection 21.3.3, Ampère opposed the explanations of Faraday and Biot for the rotation of a magnet around its axis. His argument against Faraday was that he utilized the idea of a force exerted by a magnetic pole of the magnet acting on currents flowing *inside* the magnet would produce a torque on the magnet itself. His argument against Biot's explanation was that he utilized the idea of a force exerted by current flowing *inside* the magnet and acting on magnetic pole of the magnet would produce a net torque on the magnet. According to Ampère, these explanations blatantly violated Newton's third law of motion.

By the same reason, Ampère certainly would be against the modern explanation of this torque discussed in Subsection 21.3.5. After all, the modern explanation is similar to those given by Faraday and Biot. It assumes that the magnetic field generated by the magnet acts on currents flowing *inside* the magnet, generating a self torque which would make the magnet spin around its own axis.

Ampère assumed that in his original experiment the external currents flowing in mercury were responsible for the torque acting on the magnet, as discussed in Subsections 7.2.3 and 21.3.4. The reaction torque would be applied in the mercury, making it turn in the opposite direction.

Likewise, he certainly would utilize in the situation of figure 21.11 the *external* current flowing in the wire as being responsible for the torque acting on the magnet. His explanation might be represented by

[35] [Chiaverina, 2004], [Schlichting and Ucke, 2004b], [Schlichting and Ucke, 2004c], [Stewart, 2007], [Featonby, 2006] and [Monteiro et al., 2010].

figure 21.12. The current i in the wire flows in the sense P_0, P_1 and P_2. The magnet is represented by the region internal to the circle $P_2P_3P_4$. Current i' represents the net effect of all molecular currents of the magnet. The current i, on the other hand, has a macroscopic origin, external to the magnet, depending on the battery and on the resistance of the whole closed circuit. These two currents, i' and i, are independent from one another.

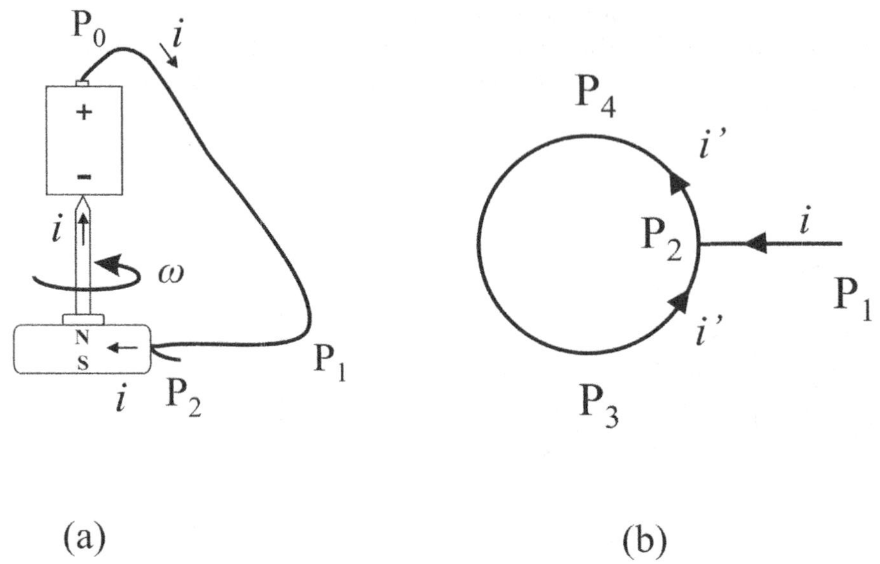

Figure 21.12: (a) Magnet and screw rotate around the common axis. (b) Configuration as seen from above indicating the current i generated by the battery and flowing in the sense P_1P_2, together with the net effect i' of all molecular currents due to the magnet itself flowing in the sense $P_3P_2P_4$.

Ampère's explanation might be similar to what has been represented in figures 21.12 and 21.13. The torque on the magnet would be essentially generated by the current i flowing externally to the magnet in the sense $P_0P_1P_2$ along the wire. The reaction torque acting on the wire would point in the opposite sense, being due to the internal molecular currents of the magnet. Suppose there are no frictions in this system. Then, if the magnet spins anti-clockwise relative to the ground as seen from above, the external wire will turn clockwise.

We present in figure 21.13 only the most relevant forces. Current i in segment P_1P_2 attracts the molecular current i' in the segment P_3P_2, while repelling the current i' in the segment P_2P_4. Current i', on the other hand, exerts opposite reaction forces on current i. These forces of action and reaction are indicated by arrows in figure 21.13.

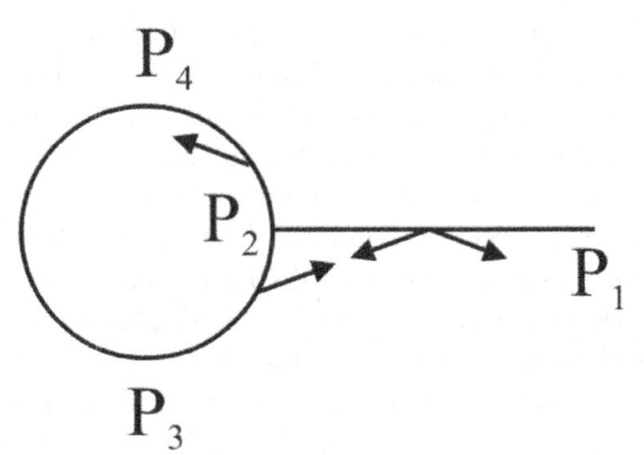

Figure 21.13: Current i in P_1P_2 attracts current i' in P_3P_2 and repels current i' in P_2P_4. The reaction forces exerted by i' on i are also shown.

The net torque on the magnet is due to the current flowing along the whole external wire in the sense P_0, P_1 and P_2 of figure 21.12. In figure 21.13 we represent only the forces on the magnet due to the horizontal component of the external wire close to the magnet, as this was the most relevant contribution for the torque in Ampère's original experiment in which the magnet was floating in mercury, figures 7.7 and 7.8. In any event, it should be kept in mind that the torque on the magnet was generated, according to Ampère, by the current which was flowing in the whole external circuit.

Several experiments prove the existence of an opposite torque acting on the *external* current i flowing in the wire outside the magnet.[36] This reaction torque acting on the external wire can be easily identified as the reaction torque which is opposed to the torque acting on the magnet. The external wire generates a torque on the magnet. By action and reaction, the magnet generates an opposite torque on the external wire.

[36] [Guala-Valverde and Mazzoni, 2001], [Guala-Valverde, 2002], [Guala-Valverde et al., 2002], [Schlichting and Ucke, 2004b], [Featonby, 2006], [Munoz, 2007], [Wong, 2009b] and [Monteiro et al., 2010].

Chapter 22

Unification of the Magnetic, Electromagnetic and Electrodynamic Phenomena

22.1 The Attempt to Explain Ørsted's Experiment Supposing Only the Interaction between Magnetic Poles Does Not Lead to the Unification of Magnetic, Electromagnetic and Electrodynamic Phenomena

As seen in Subsection 11.1, Ørsted interpreted his fundamental experiment as being caused by electric charges, which he supposed were flowing in helices or circles outside the wire, propelling the magnetic poles of the magnetized needle. Biot and Savart, on the other hand, supposed that the wire was magnetized due to the flow of an electric current through it. The deflection of the compass needle from its normal orientation would be due to the magnetic poles of this needle interacting with the magnetic poles of the magnetized wire. They expected to explain Ørsted's experiment through this supposed interaction between magnetic poles. Biot and Savart considered the interaction between magnetic poles as a primitive fact.

Ampère expressed himself against this hypothesis. One of the reasons for his rejection was that with this assumption it would not be possible to unify the interactions then known. These authors were dealing with three kinds of interaction, as discussed in Sections 1.3, 1.4 and 5.1. There were magnetic phenomena (forces and torques between magnets, or between a magnet and the Earth), electromagnetic phenomena (forces and torques between a magnet and a current-carrying conductor) and the electrodynamic phenomena (forces and torques between current-carrying conductors). Ampère realized that Biot and Savart's fundamental assumption was to try to reduce these three phenomena to interactions between magnetic poles only, that is, to reduce them to magnetostatics. The magnetostatic interaction is characterized by central forces varying as the inverse square of the distance between the interacting poles. Ampère concluded that it would not be possible to unify the three kinds of interaction utilizing Biot and Savart's hypothesis. This was one of the reasons why he rejected it.

He expressed his point of view in a Note he included in the *Recueil*, soon after the reproduction of Savary's paper of 1823 discussed in Chapter 9:[1]

> No matter the so remarkable and complete analogy between the magnets and the electrodynamic helices or cylinders imagined by M. Ampère to support his opinion on the identity between electricity and magnetism, the three laws aforementioned representing three kinds of actions, the first action being exerted between two magnets, the second between a voltaic conductor and a magnet, the third action between two conductors, were, from the mathematical point of view, independent from one another. On the other hand, it had been demonstrated that it is not possible to explain the set of phenomena of attraction and repulsion presented by the voltaic conductors attributing their properties to small magnets which would be produced by the electric action of the pile, no matter how these magnets were placed, because it is possible to produce a motion of continuous rotation, always in the same sense, by

[1] [Ampère, 1822u, pp. 355-357] and [Ampère, 1885l, pp. 377-378].

> the action either of a closed circuit, or of a magnet, and it is impossible to produce this kind of motion utilizing only magnets or rigid conductors forming closed circuits. Now, it is evident that the law given by Coulomb could only be applied to calculate the action between a conductor and a magnet, or between two conductors, by supposing, in the first case, the conductor as a set of small magnets, and adopting, in the second [case], the same supposition as regards the two conductors. In both cases we would have only actions expressed as a function of the distances between the points between which it was considered that they acted, and the acceleration of the motion of continuous rotation would be impossible, but this acceleration, verified by many experiments, follows equally from the law of M. Biot and from the law of M. Ampère. Therefore, these two laws cannot be deduced from Coulomb's law.
>
> The law of M. Biot, giving the value of the mutual action between a conductor and a magnet, could not as well lead to the law of M. Ampère relative to the action between two conductors, because it would be necessary, in order to apply it to this last case, to consider one of the two conductors as a set of small magnets on which the other conductor would act according to this law, and it would be impossible to admit this assumption for one of the conductors, without admitting it to both of them, but this would once again reduce all electrodynamic phenomena to mutual actions between sets of magnets, and would, consequently, be in direct opposition with the fact of the acceleration of the motion of continuous rotation. Therefore, the law of M. Ampère could not be deduced from the law of M. Biot, as it could not be [deduced] from the law of Coulomb. It remained to know if these two last laws could not, on the contrary, be deduced from the first [law, that is, from Ampère's force between current elements]. The solution to this question is the main subject of the memoir from M. Savary. He shows that the law of M. Ampère, applied to electric currents forming, in the magnets, closed circuits arranged as he mentioned before, reproduces the two laws, [namely, that] of Coulomb and that of M. Biot. This is a mathematical result independent of all hypothesis. The other results of the memoir of M. Savary are consequences of it offering, by their turn, another verification of these same laws, and the most complete confirmation of M. Ampère's point of view as regards the constitutions of magnets.

22.2 Ampère's Unification

As seen in Section 1.4, we are utilizing in this book the following nomenclature:

- *Electrostatic phenomena:* Forces and torques between electrified bodies which are at rest relative to one another.
- *Magnetic phenomena:* Forces and torques between magnets, together with the torques exerted by the Earth on magnets (orientation of compass and dip needles).
- *Electromagnetic phenomena:* Forces and torques between a current-carrying conductor and a magnet, together with the forces and torques exerted by the Earth on current-carrying conductors.
- *Electrodynamic phenomena:* Forces and torques between current-carrying conductors.

Magnetic phenomena were known many centuries before Ampère. These phenomena include the orientation of a compass needle by terrestrial magnetism; the orientation of dip needles by the Earth; the attractions and repulsions between magnets depending on their distances and relative orientations; the torques between magnets depending on their distances and orientations; etc.

These phenomena may be described theoretically utilizing a mathematical expression yielding the forces and torques between two magnetic dipoles. The Earth should also be considered as a magnet. An isolated magnetic pole has never been found in nature. In any event, the basic interaction between two magnetic dipoles, A and B, may be imagined to be due to four forces between their poles as given by Coulomb's law, equation (1.3). The force between each pole of A and each pole of B points along the straight line connecting them. It is proportional to the product of the pole intensities, varying as the inverse square of their distance. Poles of the same type repel one another, while poles of opposite type attract one another. The forces and torques between two magnets may be obtained utilizing Coulomb's law. The orientation of compass and dip needles by the Earth can also be obtained with Coulomb's law coupled with an appropriate distribution of magnetic dipoles on the Earth.

The discovery of electrodynamic phenomena taking place only between current-carrying wires, without the influence of any magnet, is due totally to Ampère in the period 1820-1827. He was the first person to observe the attractions and repulsions between flat spirals, the attractions and repulsions between parallel conductors, together with other related phenomena, as described in Section 3.6. He also discovered the torque between current-carrying conductors, Sections 3.8 and 3.9; the continuous rotation of a current-carrying conductor due to its interaction with another current-carrying conductor, Subsection 7.2.4; Ampère's bridge experiment, Section 8.2, etc. He also obtained many equilibrium experiments involving current-carrying conductors, as the cases of equilibrium of the sinuous wire, of anti-parallel currents, of the nonexistence of continuous rotation, of the nonexistence of tangential force, of the law of similarity, etc.

Ampère was able to explain all these electrodynamic phenomena utilizing his force between current elements, equations (2.1) up to (2.3), presented in Section 2.1. This force could also be expressed as equation (7.36). By integrating this expression he showed that a closed circuit of arbitrary form exerts a force on a current element of another circuit given by equations (10.12) up to (10.14). These formulas can be expressed in vector notation as represented in equations (10.24) or (10.28). He showed that a closed current of arbitrary form exerts a force on an external current element which is always orthogonal to this element and to a certain straight line passing through the midpoint of this element. This straight line was called directrix or normal to the directing plane. Ampère and his former student Savary introduced the concept of the electrodynamic solenoid discussed in Chapters 9 and 10. They obtained the force and torque between a current element and an electrodynamic solenoid, between a closed circuit of arbitrary form and an electrodynamic solenoid, and also between two electrodynamic solenoids.

Many scientists contributed to the discovery of electromagnetic effects. In 1820, Ørsted discovered the first electromagnetic phenomenon, namely, the deflection of a compass needle by a nearby current-carrying wire, removing the needle from its natural orientation along the magnetic meridian (see Section 1.3). Soon after he showed the inverse phenomenon, namely, the orientation of a current-carrying loop by a nearby magnet, Section 3.9. In this way he discovered the torque and counter-torque acting between a magnet and a current-carrying wire. Between 1820 and 1822 Ampère discovered many new electromagnetic phenomena: forces of attraction and repulsion between a magnet and a current-carrying wire, Sections 3.3 and 3.5; the orientation of current-carrying loops due to the influence of the Earth (situations analogous to the orientation of a compass needle and of a dip needle by terrestrial magnetism), Section 3.7; etc. Biot and Savart obtained the torque exerted by a straight wire acting on a small magnet as a function of their distance. They also obtained the torque of a bent wire acting on a small magnet as a function of the opening angle of the bent wire. In 1821 Faraday discovered a new electromagnetic phenomenon, namely, the rotation of the extremity of a magnet around a fixed current-carrying wire, together with the rotation of one extremity of a current-carrying wire around a fixed magnet, Section 7.1. Between 1821 and 1822 Ampère discovered many new phenomena related to this topic like the rotation of a current-carrying wire due to the influence of the Earth, Subsection 7.2.2; the rotation of a magnet around its axis, Subsection 7.2.3; etc. He also obtained some equilibrium experiments involving magnets and current-carrying conductors like the case of equilibrium of orthogonal currents, Section 6.3.

Ampère obtained also one of the first unifications in the history of science. He succeeded in combining magnetic, electrodynamic and electromagnetic phenomena into a single theoretical framework. When he first heard of Ørsted's experiment, he had an original and extremely fruitful insight, namely, he supposed the existence of electric currents flowing inside magnets and also inside the Earth. Moreover, he assumed that all magnetic and electromagnetic interactions were due essentially to electrodynamic forces. That is, he interpreted the electromagnetic experiments of Ørsted, himself and Faraday as being due to interactions between the electric current flowing in the wire and the supposed microscopic currents flowing around the particles of the magnets. He interpreted the terrestrial orientation of a compass needle and of a dip needle as being due to torques exerted by the supposed microscopic electric currents flowing around the particles of the Earth acting on the microscopic electric currents flowing around the particles of the magnet. Likewise, he interpreted the forces and torques acting between two magnets as being due to electrodynamic forces acting between the supposed microscopic currents of both magnets.

Utilizing his force between current elements, Ampère unified theoretically these three branches of science. He and Savary modeled a magnetic pole as the extremity of a simply indefinite electrodynamic solenoid. Through equation (10.37) they obtained the force between a current element and this solenoid. This expression is analogous to Biot and Savart's formula for the interaction between a current element and a supposed magnetic pole given by equation (10.38). Ampère also obtained the force exerted by a closed circuit of arbitrary shape carrying a steady current acting on a simply indefinite electrodynamic solenoid, as given by equation (10.42). He could then unify electromagnetism with electrodynamics through the mathematical identification of the extremity of a simply indefinite electrodynamic solenoid with a magnetic pole placed at this extremity. The mathematical proportionality between the intensity of a magnetic pole and a simply indefinite electrodynamic solenoid is expressed by equation (10.39). With these equations it is possible to explain quantitatively the electromagnetic experiments of Ørsted, Biot, Savart, Faraday and Ampère.

Ampère also obtained the analytical formula expressing the electrodynamic interaction between two simply indefinite electrodynamic solenoids, as given by equation (10.44). It represents a force pointing along the straight line connecting these two extremities and varying as the inverse square of their distance. This force is mathematically analogous to the action between two magnetic poles given by equation (1.3). Ampère could mathematically unify magnetism with electrodynamics by identifying these two pairs of magnitudes through relation (10.46). That is, a pair of interacting magnetic poles was identified with a pair of indefinite electrodynamic solenoids interacting with one another. Ampère and Savary obtained also the force and torque between two definite electrodynamic solenoids, as discussed in Subsection 10.2.4. This interaction might be reduced to four forces, each one pointing along the straight line connecting one extremity of a solenoid to one extremity of the other solenoid, varying as the inverse square of their distance. This force was proportional to the product of the current intensities of both solenoids. They were then able to identify a magnet as a definite electrodynamic solenoid.

Ampère was also able to show that any given closed circuit of arbitrary form carrying a steady current was equivalent to a set of two surfaces very close to one another, terminated by this circuit, and over which were spread the two magnet fluids of opposite type and the same intensity, the so-called magnetic shell or magnetic dipole layer. The equivalence here refers to the fact that any one of these systems (the closed circuit or the magnetic shell) exerts the same force and torque on another closed circuit or on another magnetic shell.

He also obtained the forces and torques exerted between two small planar closed loops of areas λ and λ' of arbitrary shapes, carrying constant currents of intensities i and i', respectively, supposing their typical dimensions being much smaller than the distance between their centers. Following Poisson, he also calculated the forces and torques between two small magnetic dipoles of lengths $\delta\rho$ and $\delta\rho'$, supposing their lengths are much smaller than the distance between the centers of these dipoles. Let μ and $-\mu$ be the intensities of the magnetic poles of one dipole, while μ' and $-\mu'$ are the corresponding intensities of the other dipole. Ampère showed that the forces and torques between the two current-carrying loops are equivalent to the forces and torques between the two dipoles when the loops are replaced by the dipoles, provided the axis connecting the North and South pole of one dipole was normal to the area of one loop, while the axis connecting the North and South pole of the other dipole was normal to the area of the other loop. This mathematical equivalence can be expressed by equation (10.57). Therefore, the electrodynamic equivalent of a small magnetic dipole is a small plane loop of arbitrary shape, carrying a constant current, with the plane of the loop being orthogonal to the dipole axis. In this way Ampère obtained a complete mathematical equivalence of the magnetic phenomena with the electrodynamic phenomena.

Until Ampère's time scientists explained magnetic phenomena supposing the existence of austral and boreal fluids, that is, supposing the existence of North and South poles inside magnets and also inside the Earth. One of the basic concepts was that of a magnetic dipole, that is, two opposite poles of the same intensity separated by a small distance. With Ampère's unification, these concepts of magnetic poles and dipoles became superfluous and unnecessary, as he could explain all magnetic phenomena while dealing only with the interaction of electric currents. This explanation was not only qualitative and conceptual, but also quantitative, through his expression for the force between two current elements, together with the assumption of electric currents flowing around the particles of magnetized bodies and the Earth.

What was missing was the unification of electrodynamic phenomena with electrostatic phenomena. With this last unification it would be possible to deduce, supposing only the interaction between electrified bodies, four kinds of phenomena, namely: (a) The magnetic phenomena represented by the interaction between magnets, (b) the electrodynamic phenomena represented by the interaction between current-carrying conductors, (c) the electromagnetic phenomena represented by the interaction between a current-carrying conductor and a magnet, and also (d) the electrostatic phenomena represented by the interaction between charges at rest. It would not be necessary to talk about the North and South poles of a magnet. It would also not be necessary any longer to talk about current elements. The fundamental magnitudes interacting with one another would only be electrified bodies, which might be at rest or moving relative to an inertial frame of reference, while following open or closed paths.

The fundamental unification of Ampère's electrodynamics with electrostatics was performed by Wilhelm Weber (1804-1891), in 1846.[2] He proposed a force law between electric charges depending not only on their distance, but also on their relative radial velocity and on their relative radial acceleration. He considered each current element of Ampère as composed of positive and negative charges of the same magnitude and opposite signs, in which at least one of them would be moving relative to the conductor. He could then deduce from his force law not only the electrostatic force between stationary charges, but also Ampère's force

[2][Weber, 1846], with English translation in [Weber, 2007].

between current elements. Moreover, he could also deduce from his force Faraday's law of induction of 1831. In this way it was no longer necessary to consider only the interaction of conductors carrying steady currents, as was the case of Ampère's experiments. With Weber's law it was also possible to consider experiments dealing with variable currents, that is, currents with intensities varying in time, as was the case with some of Faraday's experiments. Weber accomplished all these facts following strictly the principles of Newtonian mechanics, which were always so important for Ampère. Weber's law is a central force, acting along the straight line connecting the two interacting charges, while following Newton's action and reaction law. It complies with the principles of the conservation of linear momentum, angular momentum and energy.

We will not go into further details in this book. The interested readers can consult Weber's complete works which were published in six volumes between 1892 and 1894.[3] He wrote eight major Memoirs between 1846 and 1878 under the general title *Elektrodynamische Maassbestimmungen* (*Electrodynamic Measurements, Determination of Electrodynamic Measures* or *Electrodynamic Measure Determinations*).[4] The eighth Memoir was published only posthumously in his collected papers. Three of these eight Memoirs have already been translated into English, namely, the first, *Determinations of electrodynamic measure: Concerning a universal law of electrical action*;[5] the sixth, *Electrodynamic measurements—Sixth Memoir, relating specially to the principle of the conservation of energy*;[6] and the eighth, *Determinations of electrodynamic measure: Particularly in respect to the connection of the fundamental laws of electricity with the law of gravitation.*[7] Also published in 1848 was an abridged version of the first Memoir,[8] which has also been translated into English, *On the measurement of electro-dynamic forces.*[9] Some of his other papers have been translated to English[10] or to Portuguese.[11] Some important works are related to his introduction and measurement of the constant c connecting the electrostatic unit of charge with the electrodynamic unit of charge and also with light velocity.[12] Details about his life and work have been published by several authors, along with modern applications of Weber's law to electromagnetism and gravitation.[13]

Certainly these works of Ampère and Weber represent some of the main achievements of all time in physics.

---

[3] [Weber, 1892b, Weber, 1892a, Weber, 1893, Weber, 1894b, Weber and Weber, 1893, Weber and Weber, 1894].

[4] [Weber, 1846, Weber, 1852b, Weber, 1852a, Kohlrausch and Weber, 1857, Weber, 1864, Weber, 1871, Weber, 1878, Weber, 1894a].

[5] [Weber, 2007].

[6] [Weber, 1872].

[7] [Weber, 2008].

[8] [Weber, 1848].

[9] [Weber, 1966].

[10] [Assis, 2010d].

[11] [Weber and Kohlrausch, 2008].

[12] [Kirchner, 1957], [Rosenfeld, 1957], [Wiederkehr, 1960], [Wiederkehr, 1967], [Woodruff, 1968], [Woodruff, 1981], [Wise, 1981], [Rosenfeld, 1981], [Harman, 1982], [Jungnickel and McCormmach, 1986, volume 1, pp. 144-146 and 296-297], [Wiederkehr, 1993a], [Wiederkehr, 1994], [Hecht, 1996], [Hecht, 1997], [Assis et al., 2002], [Assis et al., 2004], [Wiederkehr, 2004] and [Mendelson, 2006].

[13] [Wiederkehr, 1960], [Wiederkehr, 1967], [Biermann, 1971], [Woodruff, 1981], [Jungnickel and McCormmach, 1986], [Atherton, 1989], [Thomson, 1885], [O'Rahilly, 1965, volume 2, Chapter 11], [Whittaker, 1973, Chapter 7], [Reich, 1977], [Wise, 1981], [Harman, 1982, pp. 32, 96 and 103–107], [Wesley, 1987], [Atherton, 1989], [Archibald, 1989], [Graneau, 1990d], [Graneau, 1990a], [Graneau, 1990c], [Graneau, 1990e], [Graneau, 1990b], [Wesley, 1990b], [Wesley, 1990c], [Wesley, 1990d], [Wesley, 1990a], [Phipps Jr., 1990a], [Phipps Jr., 1990b], [Wesley, 1991], [Assis, 1991], [Ragusa, 1992], [Phipps Jr., 1992], [Assis, 1992], [Graneau and Graneau, 1993, Chapter 3: The Riddle of Inertia], [Schreier, 1993], [Wiederkehr, 1993b], [Zylbersztajn, 1994], [Assis and Thober, 1994], [Assis, 1994], [Galeczki and Marquardt, 1995], [Kinzer and Fukai, 1996], [Phipps, Jr., 1996], [Tennenbaum, 1996], [Guala-Valverde, 1997], [Bunchaft and Carneiro, 1997], [Drury, 1997], [Tennenbaum, 1997], [Fukai and Kinzer, 1997], [Guala-Valverde, 1998], [Assis, 1998], [Guala-Valverde, 1999a], [Guala-Valverde, 1999b], [Zylbersztajn and Assis, 1999], [Assis, 1999a], [Assis, 1999b], [Gluckman, 1999], [Assis and Guala-Valverde, 2000], [Assis and Torres Silva, 2000], [Assis et al., 2000], [Guala-Valverde, 2001], [Guala-Valverde and Mazzoni, 2001], [Assis and Zylbersztajn, 2001], [Bunchaft, 2001], [Bueno and Assis, 2001], [Fukai and Assis, 2003], [Assis and Guala-Valverde, 2002], [Wesley, 2002], [Assis and Guala-Valverde, 2003], [Assis and Wiederkehr, 2003], [Fukai, 2003], [Assis, 2004], [Assis et al., 2004], [Guala-Valverde, 2004], [Guala-Valverde et al., 2005], [Assis and Hernandes, 2007], [Assis and Hernandes, 2009], [Assis et al., 2011], [Haertel, 2012a], [Haertel, 2012b], [Warkulwiz, 2013, Chapter 5: The Logic of Relational Physics], [Tajmar, 2013], [Phipps Jr., 2013], [Assis and Hernandes, 2013], [Assis, 2013], [Assis et al., 2014], [Assis, 2014], [Assis, 2015b], [Torres-Silva et al., 2015], [Bueno and Assis, 2015], [Tajmar, 2015], [Tajmar and Assis, 2015a], [Tajmar and Assis, 2015b], [Assis, 2015a], etc.

Part VI

Complete English Translation of Ampère's First Paper on Electrodynamics

Chapter 23

Ampère's Works Translated into English

We first present a list of the works and letters of Ampère which, to our knowledge, have been translated into English. We quote the original French work followed by its English translation, ordered by year.

- 1793-1804
  - [Cheuvreux, 1869] and [Cheuvreux, 1873]: The Story of His Love being the Journal and Early Correspondence of André-Marie Ampère with His Family Circle during the First Republic, 1793-1804.
- 1814
  - [Ampère, 1814] and [Ampère, 1815]: Letter from M. Ampère to Count Berthollet, on the determination of the proportions in which bodies are combined, according to the respective number and arrangement of the molecules of which their integrant molecules are composed.
- 1816
  - [Ampère, 1816a] and [Ampère, 1816b]: Essay towards a natural classification of simple bodies.
- 1820
  - [Ampère, 1820c] and [Ampère, 1965a, pp. 140-146]: Partial English translation of the first part of Ampère's first paper on electrodynamics. Chapter 24 presents a complete and commented translation of this work.
  - [Ampère, 1820f] and [Ampère, 1965a, pp. 146-154]: Partial English translation of the second part of Ampère's first paper on electrodynamics. Chapter 25 presents a complete and commented translation of this work.
- 1824
  - [Ampère, 1824d] and [Ampère, 1825f]: On the nature of the electric current.
  - [Ampère and Dulong, 1824b] and [Ampère and Dulong, 1824a]: Abstract of the report on M. Rouseau's memoir respecting a new method of measuring the power of bodies to conduct electricity.
- 1825
  - [Ampère, 1825d] and [Ampère, 1825c]: Memoir on a new electro-dynamic experiment, on its application to the formula representing the mutual action of the two elements of voltaic conductors, and on some new results deduced from that formula.
  - [Ampère, 1825b, Ampère, 1887b] and [Ampère, 1825a]: A letter from M. Ampere to M. Gerhardi on various electro-dynamic phenomena.

  - [Ampère, 1825g] and [Ampère, 1826e]: Sequel of the memoir of M. Ampere on a new electrodynamic experiment, on its application to the formula representing the mutual action of the two elements of voltaic conductors, and on new results deduced from that formula.

- 1826

  - [Ampère, 1826d] and [Ampère, 1827b]: On the action of a moving metallic disc, upon a portion of a voltaic conductor.
  - [Ampère, 1826f, Ampère, 1823c] and [Ampère, 1965b, Ampère, 2012]: Partial and complete English translations of Ampère's *Théorie*. Chapter 29 presents a complete and commented English translation of Ampère's masterpiece.

- 1832

  - [Ampère, 1832] and [Ampère, 1833]: M. Ampère's communication to the Academy of Sciences upon an experiment of M. Pixii relative to a current produced by the rotation of a magnet with an improved apparatus, Oct. 29, 1832.

- 1833

  - [Ampère, 1936] and [Ampère, 1993]: Letter 654 of Ampère to Faraday, 13 April 1833.

- 1835

  - [Ampère, 1835b] and [Ampère, 1835a]: Note by M. Ampère on heat and light considered as the results of vibratory motion.

Chapter 24

On the Effects of Electric Currents [First Part]

24.1 Translator's Introduction

This is a complete English translation of the first part of the first paper published by Ampère devoted to electrodynamics, entitled "On the effects of electric currents." This work was presented to the *Académie royale des Sciences* on 2 October 1820, containing a summary of the readings at the *Academy* on 18 and 25 September 1820.[1] This first part has been completely translated to Portuguese.[2] To our knowledge there is only a partial English translation of this first part made by O. M. Blunn.[3] The English translation presented here is based upon Blunn's version.

The English translation of the second part is presented in Chapter 25.

The footnotes by Ampère are indicated by [N. A.], while those introduced by the translators are indicated by [N. T.]. The number of the original footnotes are indicated by [N. A. m], where "m" indicates the number of the original footnote introduced by Ampère. The words between square brackets in the middle of the text have been included by the translator to facilitate the understanding of some sentences. The beginning of each original page m is also indicated between square brackets, [page m].

24.2 Translation

Dissertation presented to the Académie royale des Sciences *on 2 October 1820, containing a summary of the readings at the* Académie *on 18 and 25 September 1820, on the effects of electric currents.*

[page 59][4]

24.2.1 I. The Mutual Action of Two Electric Currents

1. The electromotive action[5] manifests itself by two types of effects which I believe should be first distinguished by a precise definition.

I will call the first effect *electric tension* and the other *electric current.*

The first effect is observed when the two bodies [page 60] between which the electromotive action[6] takes place are separated from one another[7] by non-conducting bodies, over all points of their surfaces except at

[1][Ampère, 1820c].

[2][Chaib and Assis, 2007d] and [Assis and Chaib, 2011, Chapter 23].

[3][Ampère, 1965a, pp. 140-146].

[4][N. T.] The beginning of each original page m is indicated between square brackets, [page m].

[5][N. T.] The word "electromotive" comes from Alessandro Volta. It indicates a motive force acting on the electric charges located inside his pile, which he called an *electro-motive apparatus* in his original publication, [Volta, 1800a], [Volta, 1800b], [Volta, 1964] and [Magnaghi and Assis, 2008].

[6][N. T.] In the printed version we have the expression "cette action" instead of "l'action électromotrice." But in the errata which appeared on p. 223 of the *Annales de Chimie et de Physique*, Vol. 15, 1820, it was pointed out that this last expression was the correct one.

[7][N. A. 1] When this separation is due to the simple interruption by conducting bodies, it is still a non-conducting body, air, which separates them.

those points where tension is established. The second effect, on the other hand, occurs when these bodies form part of a circuit of conducting bodies by which contact is made by points on their surface which are different from those points where this action[8] is produced.[9] In the first case, the effect of the electromotive action is to place the two bodies, or two systems of bodies, between which the electromotive action takes place, in two states of tension, the difference between which is constant while this action is constant, for example, when it is produced by the contact of two substances of different nature. This difference, on the other hand, would vary with the cause which produces it, if it were due to rubbing or pressure.[10]

This first case is the only one which can takes place when the electromotive action develops between the several parts of a single non-conducting body. An example of this case is given by the tourmaline when it changes temperature.[11]

In the second case, there is no longer electric tension, the light bodies are no longer perceptibly attracted, and the ordinary electrometer no longer indicates what [page 61] takes place in the body. However, the electromotive action continues to act; because if, for instance, water, an acid, an alkali, or a saline dissolution belong to the circuit, these bodies are decomposed, especially when the electromotive action is constant, as has been known for a long time. Moreover, as M. OErsted has just discovered, when the electromotive action is produced by the contact of metals, the magnetized needle is deflected from its direction when it is placed close to any portion of the circuit. But these effects stop, the water is no longer decomposed, and the [magnetized] needle returns to its ordinary orientation when the circuit is interrupted, when the tensions are recovered, and when the light bodies are once more attracted, which proves that these tensions are not the cause of the water decomposition, nor of the changes of direction of the magnetized needle discovered by M. OErsted. This second case is obviously the only one which can take place if the electromotive action developed between the several parts of the same conducting body. The deduced consequences, in this dissertation, of the experiences of M. OErsted will lead us to recognize the existence of this [second] circumstance as the only explanation that we need to admit at the moment in order to explain what is observed.

2. Let us consider now to what [aspect] is due the difference of these two kinds of completely distinct phenomena, of which one consists in the tension and in the attractions or repulsions which have been known for a long time, and the other phenomenon consists in the decomposition of water and in a great number of other substances, in the changes of direction of the [magnetized] needle, and in a kind of attractions and repulsions totally [page 62] different from the ordinary electric attractions and repulsions; which I believe has been first recognized by myself, and which I designated *attractions and repulsions of electric currents*,[12] in order to distinguish them from these last [ordinary electric attractions and repulsions]. When there is no continuity of conductors from one of the bodies, or system of bodies, to the other [body], between which the electromotive action is developed, and when these bodies are themselves conductors, as in Volta's battery, one can only conceive this action as carrying constantly the positive electricity in one [sense], and the negative electricity in the other [sense]. In the first moment, when nothing prevents the effect which it tends to produce, the two electricities accumulate each one in the part of the total system towards which it is carried; but this effect stops when the difference of the electric tensions[13] gives to their mutual attraction, which

---

[8][N. T.] In the printed version we have the expression "l'action électromotrice" instead of "cette action." But in the errata which appeared on p. 223 of the *Annales de Chimie et de Physique*, Vol. 15, 1820, it was pointed out that this last expression was the correct one.

[9][N. A. 2] This case includes that one in which the two bodies, or system of bodies, between which the electromotive action takes place, would be in complete communication with the common reservoir which then would form part of the circuit.

[10][N. T.] As regards pressure, Ampère was referring to piezoelectricity, which is the ability of some crystals to get electrically polarized in response to applied mechanical stress or pressure. The reverse effect is the internal generation of a mechanical strain resulting from an applied electrical tension.

[11][N. T.] Maxwell presented the following description of this phenomenon, [Maxwell, 1954, Volume I, §58, pp. 61-62]:

> 58.] Certain crystals of tourmaline, and of other minerals, possess what may be called Electric Polarity. Suppose a crystal of tourmaline to be at a uniform temperature, and apparently free from electrification on its surface. Let its temperature be now raised, the crystal remaining insulated. One end will be found positively and the other end negatively electrified. Let the surface be deprived of this apparent electrification by means of a flame or otherwise, then if the crystal be made still hotter, electrification of the same kind as before will appear, but if the crystal be cooled the end which was positive when the crystal was heated will become negative.
>
> These electrifications are observed at the extremities of the crystallographic axis. [...]

[12][N. T.] In the printed version we have the expression "voltaïques" instead of "des courants électriques." But in the errata which appeared on p. 223 of the *Annales de Chimie et de Physique*, Vol. 15, 1820, it was pointed out that this last expression was the correct one.

[13][N. A. 3] When the battery is insulated, this difference is the sum of the two tensions, one positive and the other negative. When one of its extremities communicates with the common reservoir at zero tension, the same difference has an absolute value equal to the value of the tension at the other extremity.

tends to reunite them, enough force to equilibrate the electromotive action. Then everything remains in this state, except the loss of electricity which can happen little by little through the non-conducting body, air, for instance, which interrupts the circuit; as it seems that there are no absolutely insulating bodies. When this loss takes place; the tension decreases; but when this tension decreases, the mutual attraction of the two electricities does not balance any longer the electromotive action. Therefore this last [electromotive] force, when [page 63] it is constant, carries once more the positive electricity towards one side and the negative electricity towards the other side, restoring the tensions. I designate *electric tension* to this state of a system of electromotive and conducting bodies. It is known that it continues in the two halves of this system, either when they have just been separated, or even in the case when they remain in contact after the electromotive action stopped, provided that then it [i.e., the electromotive action] happens by pressure or friction between bodies which are not both conductors. In these two cases, the tensions decrease gradually due to the loss of electricity aforementioned.

But when the two bodies, or two systems of bodies, between which the electromotive action takes place are in communication by conducting bodies between which there is not another electromotive action equal and opposite to the first one, which would maintain the state of electric equilibrium, these tensions disappear, or at least become very small, and the phenomena indicated above occur, characterizing this second case.[14] Since nothing has changed in the arrangement of bodies between which the electromotive action took place, one cannot doubt that the electromotive action continues to act, and since the mutual attraction of the two electricities, as measured by the difference of the electric tensions which has become zero, or else is considerably diminished, can no longer balance this action, it is generally accepted that this action continues to carry the two electricities in the two senses as before. [page 64] A double current thus results, the one positive electricity and the other negative electricity, beginning in opposite senses from the points where the electromotive action takes place, and meeting in the part of the circuit opposite these points.[15] These currents accelerate until the electromotive force is balanced by the inertia of the electric fluids[16] and the resistance they experience by the imperfections of even the best conductors, whereupon they continue indefinitely at a constant speed so long as this force conserves the same intensity; but they cease instantly whenever the circuit is interrupted. For the sake of simplicity I will designate *electric current* to this state of the electricity in a series of electromotive and conducting bodies. And since I shall speak continually of the two opposite senses in which the two electricities move, to avoid unnecessary repetition I shall invariably imply *positive electricity* by the words *sense of the electric current.*[17] Thus, for example, for a voltaic battery, the expression *direction of the electric current in the battery* shall designate the direction from the extremity where hydrogen is disengaged in the decomposition of water to that where oxygen is obtained; and the expression *direction of the electric current in the conductor which establishes the communication between the two extremities of the battery* shall designate, on the contrary, the direction from the extremity where oxygen is produced to that where hydrogen develops. To cover these two cases by a single definition, it may be said that what is called the direction of the electric current is the direction [page 65] followed by the hydrogen and the bases of salts when the water, or a saline substance, is part of the circuit and is decomposed by the current, whether these substances form part of the conductor in a voltaic battery, or whether they are interposed between the pairs of which the battery is composed.

The learned researches of MM. Gay-Lussac and Thenard[18] into this apparatus, a fruitful source of great discoveries in almost all branches of physical sciences, have demonstrated that the decomposition of water, salts, etc., is in no way produced by the difference in tension between the two extremities of the battery, but solely by what I have called *the electric current*, since in plunging the two conducting wires into pure water the decomposition is practically zero; whereas, without in any way changing the disposition of the rest of the

[14][N. T.] Ampère believed that the tensions in the pile disappeared or became negligible when the circuit was closed and an electrical current began to flow through the circuit. Maybe he had this point of view because his battery had a high internal resistance compared with his conducting circuit. That is, the battery terminal voltage dropped a lot under load, rising again when the load was removed. See also: [Blondel, 1982, pp. 81-82], [Benseghir, 1989] and [Blondel and Wolff, 2006].

[15][N. T.] Ampère's model of an electric current inside a current-carrying wire is somewhat similar to Ørsted's model. According to Ørsted, there would be an "electric conflict" inside the wire. Positive and negative charges would move in opposite senses relative to the wire, generating two opposite fluxes of electrified particles which would collide with one another when they met.

[16][N. T.] It is interesting to note that Ampère admitted the possibility of an inertia of the electric fluids. Many years after Ampère scientists began to suppose that each electrified elementary particle would have not only an electric charge, but also an inertial mass. This statement suggests that Ampère already envisaged this possibility.

[17][N. T.] Ampère's convention has been utilized until modern days. For instance, nowadays we assume that the ordinary current in a resistive metal wire connected to a battery is due only to the flow of negative electrons, while the positive ions remain fixed to the lattice. Despite this fact, we still assume that the sense of the electric current is opposite to the motion of the electrons relative to the wire. That is, if the electrons move to the right, we assume that the current flows to the left.

[18][N. T.] Joseph Louis Gay-Lussac (1778-1850) and Louis Jacques Thenard (1777-1857).

apparatus, if an acid or saline solution is mixed with water where the wires are plunged, this decomposition becomes very fast, because pure water is a bad conductor, whereas when it is mixed with a certain amount of these substances it conducts electricity well.

Now, it is evident that the electric tension of the extremities of the wire immersed in the liquid could not have been increased in the second case. The electric tension can only decrease according as this liquid becomes a better conductor. What increases in this case is the electric current. Therefore, it is solely due to this electric current that the decomposition of the water and of the salts occurs. It is also easy to verify that it is only the electric current that acts on the magnetized needle in [page 66] the experiments of M. OErsted. For this it is sufficient to place a magnetized needle on a horizontal battery situated roughly in the direction of the magnetic meridian; so long as its extremities are not in communication, the needle conserves its ordinary direction. But if a metal wire is attached to one extremity of the battery and the other end [of the wire] is brought into contact with the other extremity of the battery, the needle suddenly changes direction, and it remains in its new position so long as contact is made and the battery conserves its energy. It is only to the extent that the battery loses its energy, that the needle reverts to its ordinary direction; whereas if the current is made to cease by interrupting the communication, the needle returns instantly. However, it is this same communication which causes the electric tensions to cease or to decrease considerably; therefore, it cannot be these tensions, but the current alone, which influences the direction of the magnetized needle. When pure water forms part of the circuit, and the decomposition is hardly perceptible, the magnetized needle placed above or below another portion of the circuit is deflected just as slightly. When nitric acid is mixed with the water, without otherwise altering the apparatus in any way, this deviation is increased at the same time as the decomposition of the water is made more rapid.

3. The ordinary electrometer indicates the presence of tension and the intensity of this tension. There used to be no instrument for making known the presence of electric current in a battery or conductor, and which would indicate its energy and direction. Such an instrument does exist today. It is sufficient to place the battery, or any portion of the [page 67] conductor, roughly in the horizontal position in the direction of the magnetic meridian, and to place an apparatus similar to a compass, the only difference being the use to which it is put, on the battery, or well above or below this portion of conductor. As long as the circuit is interrupted, the magnetized needle remains in its ordinary position. But it deviates away from it as soon as the current is established, and more so the greater its energy. The direction of the current can be known according to this general fact, that if one imagines oneself placed along the direction of the current, so that the current flows from the feet to the head of the observer when facing the needle, it is constantly to his left that the action of the current deflects the extremity [of the magnetized needle] which was pointing to the North [ward], what I shall always call the *austral pole of the magnetized needle*, because it is the pole which is homologous to the [magnetic] austral pole of the Earth.[19] This is what I express more concisely in saying that the austral pole of the [magnetized] needle is carried to the left of the current acting on the needle.[20] To distinguish this instrument from the ordinary electrometer, I think that it ought to be given the name *galvanometer*. It is appropriate to use it in all experiments on electric currents, as one habitually adapts an electrometer with electric machines, so as to see at each instant if the current is there, and find out its energy.

The first use to which I put this device was to check that the current which exists in the voltaic battery, from the negative extremity to the positive extremity, had the same influence on the magnetized needle [page 68] as the current in a conductor which flows, on the contrary, from the positive extremity to the negative one.

It is desirable to have for this two magnetized needles, one placed on the battery and the other above or below the conductor; it is seen that the austral pole of each needle is carried to the left of the current near to which it is placed. Thus, when the second [needle] is above the conductor, it is carried to the side opposite to that towards which the needle on the battery tends, since the currents have opposite directions in these two portions of the circuit. The two needles are, on the contrary, carried to the same side, remaining roughly parallel to each other, when one is above the battery and the other below the conductor.[21] As soon as the circuit is interrupted, they immediately revert, in both cases, to their ordinary orientation.[22]

[19][N. T.] What Ampère named the austral pole of the magnetized needle is usually called nowadays the *North pole of the magnetized needle*. This is the pole which points approximately towards the geographic North of the Earth.

[20][N. T.] In this paragraph he introduced what became known as *bonhomme d'Ampère*, or Ampère's observer, [Poudensan, 1964, p. 23], [Tricker, 1965, pp. 28-29], [Kastler, 1977, p. 144] and [Blondel, 1982, pp. 73-74].

[21][N. A. 4] For this experiment to leave no doubt as to the action of the current in the battery, it is convenient to use a trough battery with zinc and copper plates soldered together over the entire interface, and not just simply over a branch of metal which can rightly be regarded as a portion of conductor.

[22][N. T.] This subject was discussed in Section 3.4.

4. Such are the differences which were known to exist, before me, between the effects produced by electricity in its two states just described, the one being, if not a state of rest, at least one of slow motion due solely to the difficulty of isolating bodies in which electric tension occurs, the other effect being the double current of positive and negative electricity along a [page 69] continuous circuit of conducting bodies. In the ordinary theory of electricity the two fluids of which it is thought to be constituted, are conceived to be perpetually separated from one another in a part of the circuit and to be carried rapidly in contrary senses into another part of the circuit where they are continually re-uniting. Though the electric current so defined might be produced by arranging a conventional [friction] machine, disposing it so as to develop the two electricities, and in joining by a conductor the two parts of the apparatus where they are produced, this current can only be obtained with a certain energy by a voltaic battery, unless very large machines are used, because the quantity of electricity produced by a friction machine is constant throughout a given period, whatever the conduction capability of the rest of the circuit, whereas that which a battery places in motion in a similar period increases indefinitely according as the two extremities are connected by a better conductor.

But the differences I just remembered are not the only ones which distinguish these two states of electricity. I discovered other more remarkable differences by disposing, in parallel directions, two straight parts of two conductors connected to the extremities of two voltaic batteries. One straight part was fixed, but the other, suspended on tips and made highly mobile by a counter-weight, free to move parallel towards it or away from it.[23] I observed that by passing an electric current through both parts at the same time, they were mutually attracted when both currents were [page 70] in the same sense, and that they repelled one another when the currents were in opposite directions.

Now, these attractions and repulsions of electric currents are essentially different from those produced by electricity at rest; in the first place, they cease, like the chemical decompositions, at the moment in which the circuit of conducting bodies is interrupted. In the second place, in the ordinary electric attractions and repulsions, the electricities which attract one another are those of opposite kinds, and those of the same type repel one another; whereas in the attractions and repulsions of electric currents, it is precisely the opposite which takes place, because it is when the two parallel conducting wires are placed in such a way that the extremities [of the two voltaic batteries] having the same names are found [connected] on the same side and very close to one another, that attraction takes place, while there is repulsion when the currents flow in opposite senses in the two parallel conductors, in such a way that the extremities of the same name are at the greatest possible distance from one another. In the third place, when attraction takes place in such a way that it is strong enough to bring the mobile conductor in contact with the fixed conductor,[24] they remain connected to one another like two magnets, and do not separate from one another immediately, as happens when two conducting bodies which are attracting one another, because one of them is electrified positively and the other negatively, come into contact.[25] Finally, and it seems that this last circumstance has the same cause as the preceding one, two electric currents attract and repel in vacuum as they do in the presence of air; this being once more contrary to what is observed in the mutual action of two conducting bodies commonly electrified.[26] [page 71] The question here is not to explain these new phenomena, the

[23][N. T.] In Ampère's figure 1 the fixed straight part is represented by AB. The mobile wire is represented by $XCDY$. The counter-weight is represented by VH. The mobile portion can turn around the horizontal axis XY. With the counter-weight it is possible to adjust the distance between the rotation axis XY and the center of gravity of the mobile system (composed of the counter-weight plus the mobile portion $XCDY$). By decreasing this distance, the sensitivity of the balance is increased. This means that the mobile system will be deflected from the vertical plane by a greater angle when it is under the action of a horizontal force exerted by the fixed portion. This force can be attractive or repulsive.

[24][N. T.] Recently J. Lühr built a replica of Ampère's balance. He had to utilize currents with intensities greater than 30A in order to bring the mobile conductor in contact with the fixed conductor, [Luhr, 2000, p. 151].

[25][N. T.] This electrostatic phenomenon of attraction, communication of electricity, and repulsion, has been called *ACR* mechanism (i.e., Attract, Communicate, Repel) by Heilbron, [Heilbron, 1999, pp. 5 and 255-258]. It was first recognized by Du Fay (1698-1739) in 1733. For a detailed discussion of this mechanism, presenting simple experiments which can be easily made with cheap materials and including many references, see [Assis, 2010b], [Assis, 2010c], [Assis, 2011b], [Boss et al., 2012] and [Assis, 2015c].

[26][N. T.] Two aspects should be observed here. The first one is that this passage suggests that Ampère performed experiments of attraction and repulsion between current-carrying conductors surrounded not only by air, but also by vacuum. We were unable to find any electrodynamic experiment in which he worked with vacuum pumps. Probably he never performed these experiments.

The second aspect to take notice of refers to the distinction between the usual attractions and repulsions between electrified bodies surrounded by air at atmospheric pressure and their interactions when they are surrounded by a vacuum. The difference pointed out by Ampère should be related to the dielectric strength. At normal pressure dry air behaves as a good insulator for electrostatic experiments. An electric field of $3 \times 10^6\ V/m$ is necessary for break down in air. When this happens, an electrified body is discharged by electric arcs or sparks. The dielectric strength of air decreases with pressure until we reach a pressure of approximately $10^{-3}\ atm$, when it then grows quickly, [Heilbron, 1982, pp. 164-165] and [Heilbron, 1999, p. 207].

attractions and repulsions which occur between two parallel currents, when they are directed in the same sense or in opposite senses, [as] these are facts given by an experiment which can be easily repeated. To prevent in this experiment the motions given to the mobile conductor by the small agitations of air, it is necessary to place the apparatus inside a glass case under which is passed, in the base which supports it, the portions of conductors which should communicate with the two extremities of the battery. The most convenient disposition of these conductors is to place one of them over two supports in a horizontal position where it remains at rest, while the other conductor is suspended by two metal wires forming part of it, to a glass axis located above the first conductor, and which rests, by very thin steel tips, over two other metal supports; these tips are welded to the two extremities of the metal wires aforementioned; in such a way that the communication is established by the supports with the help of these tips. (See this apparatus in figure 1).

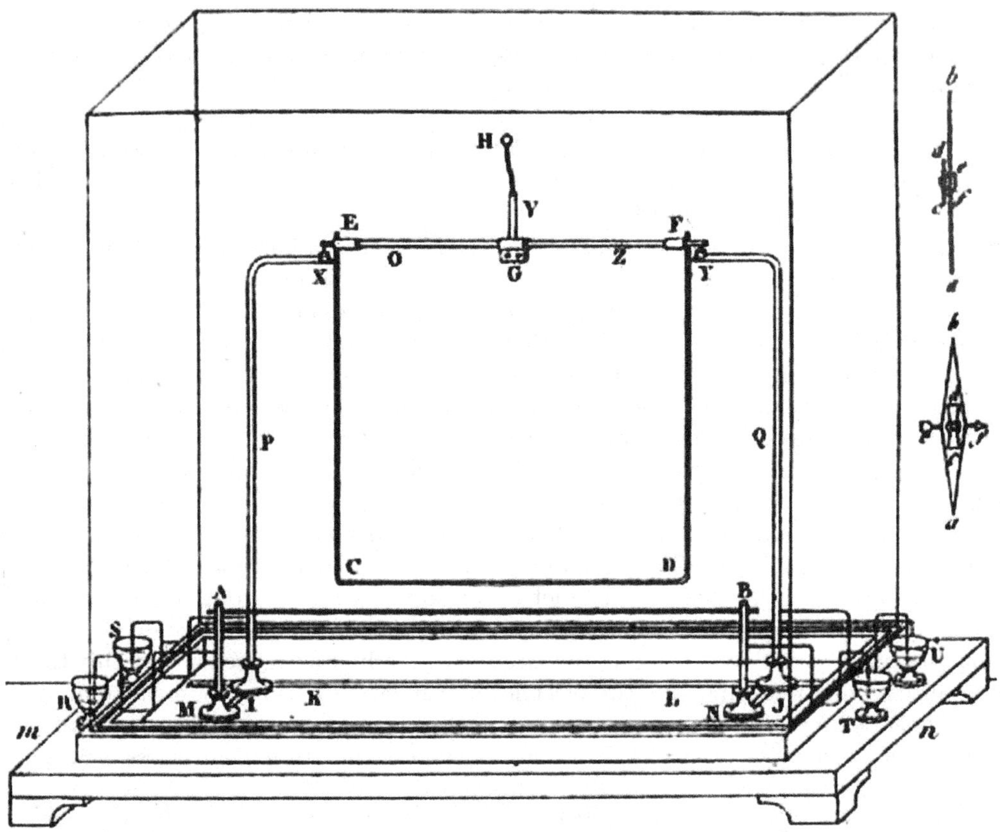

Figure 1.

In this way the two conductors are parallel to one another, and side by side in the same horizontal plane. One conductor is mobile through the oscillations it can make around the horizontal line passing through the extremities of the two steel tips. During this motion, it remains necessarily parallel to the fixed conductor.

A counter-weight is connected above and in the center of the glass axis, in order to increase the mobility of the portion [page 72] of the apparatus apt to oscillate, by raising the center of gravity.

Initially I believed it would be necessary to establish the electric current in the two conductors by means of two different batteries. But this is not necessary, it is enough that these two conductors belong to the

Paschen's law gives the breakdown voltage between two electrodes in a gas as a function of pressure and gap length. It was due to the German physicist F. Paschen (1865-1947) who discovered it at the end of the XIXth century. The vacuum pumps at the time of Ampère had not yet reached pressures so low as 10^{-3} *atm*. Therefore, the "vacuum" available in the beginning of the XIXth century behaved as a good conductor. Any electrified body placed in this "vacuum" was immediately discharged through sparks. It was then not possible to observe the ordinary attractions and repulsions between electrified bodies placed in this medium. Nowadays it is known that for pressures much smaller than 10^{-3} *atm* air behaves as an excellent insulator, so that it is possible to observe in high vacuum the ordinary attractions and repulsions between electrified bodies. They have the same behavior as those interactions taking place at atmospheric pressure.

same circuit; since the electric current exists in all its parts with the same intensity. It should be concluded from this observation that the electric tensions of the two extremities of the battery have no relation with the phenomena being described here; since certainly there is no tension in the remainder of the circuit. This conclusion is also confirmed by the possibility of making the magnetized needle move at a great distance from the battery, by means of a very long conductor, the middle part of which curves along the direction of the magnetic meridian above or below the needle. This experiment, which was completely successful, was suggested to me by the illustrious scholar[27] to whom the sciences of physics and mathematics owe their great progress in our days.

The two extremities of the fixed conductor are designated by A and B, the extremity of the mobile conductor which is on the side of A is designated by C, while that which is on the side of B is designated by D. It is clear that if one of the extremities of the battery is placed in communication with A, B with C, and D with the other extremity of the battery, the electric current will flow in the same sense along both conductors; then they will be seen attracting one another. If, on the contrary, A always communicates with one extremity of the battery, B communicates with D, and C [communicates] with the other extremity of the battery; the current will flow in opposite senses along the two conductors, [page 73] therefore, they will repel one another. Moreover, one conceives that the attractions and repulsions of the electric currents take place at all points of the circuit, so that with a single fixed conductor it is possible to attract and repel as many conductors as we wish, and change the direction of an arbitrary number of magnetized needles. I intend that two mobile conductors be constructed inside the same glass case, in such a way that by making them, together with a common fixed conductor, part of the same circuit, they both shall be alternatively attracted, both of the repelled, or one attracted and the other repelled simultaneously, according to the way in which the communications are established. After the success of the experiment suggested to me by M. Marquis de Laplace, one could, by means of as many conductors and magnetized needles as there are letters [in the alphabet], establish with the aid of a battery placed far away from these needles, and communicating alternately by its two extremities to the extremities of each conductor, form a kind of telegraph appropriate to write all details that we wished to transmit, through all kinds of obstacles, to the person in charge of observing the letters placed over the needles. By establishing over the battery a keyboard of which the keys carried the same letters and established the communication by being pressed, this kind of communication could happen very easily, and would require only the time necessary to press one side and read each letter on the other side.[28]

[page 74] If the mobile conductor, instead of being obliged to move parallel to the fixed conductor, can only turn in a plane parallel to this fixed conductor, around a common perpendicular passing through their midpoints, it is clear that, according to the law that we just recognized for the attractions and repulsions of electric currents, each half of the two conductors will attract and repel one another simultaneously, if the currents flow in the same sense or in opposite senses. Consequently, the mobile conductor will turn until it remains parallel to the fixed conductor, in such a way that the currents will be directed to the same sense. From this reasoning it follows that in the mutual action of two electric currents, the directive action and the attractive or repulsive action depend upon the same principle, and represent only different effects of the same and single action. Therefore, it is no longer necessary to establish between these two effects the distinction which is so important to make, as will be seen shortly, when we are dealing with the mutual action between an electric current and a magnet considered as usually done with respect to its axis, since, in this action, the two bodies tend to place themselves along directions perpendicular to one another.

Consider now this last action [between an electric current and a magnet] and that of two magnets acting upon one another. [page 75] It will be seen that both actions come under the same law of mutual action between two electric currents, if it is assumed that one of these currents is established at each point of a line drawn on the surface of the magnet from one pole to the other in planes perpendicular to the axis of this magnet. It hardly seems possible to me, from consideration of all the facts, to doubt that such currents do exist about the axis of magnets, or rather that magnetization is nothing other than the operation by which particles of steel are endowed with the property to produce, in the sense of the currents about which we have just been speaking, the same electromotive action as in the voltaic battery, in the oxidized zinc of mineralogists, in heated tourmaline, and even in a battery formed by wet boards and disks of the same metal at two different temperatures. But since with magnets this electromotive action develops between different

[27][N. T.] Ampère is referring to P. S. Laplace (1749-1827), as is pointed out further along in this paper.

[28][N. A. 5] After the writing of this dissertation; I was informed by M. Arago that this telegraph had already been proposed by M. Soemmering. However, instead of observing the changes in the directions of magnetized needles, which were not yet known, the author proposed to observe the decomposition of water inside the same number of vessels as there are letters [in the alphabet].

particles of one and the same body, a good conductor, it[29] can never, as pointed out above, produce any electric tension, only a continuous current like that which would occur in a voltaic battery connected to itself in a closed curve. It is quite clear from the foregoing observations that such a battery could produce in any of its points no tensions, attractions, no ordinary electric repulsions, nor any chemical phenomena, since a liquid cannot be interposed in the circuit. It is evident that any current which is established in this battery would immediately act to direct, attract, or repel another electric current or [page 76] a magnet, which, as we shall see, is only an assembly of electric currents.

It is thus that the unexpected result is reached that magnetic phenomena are due solely to electricity and there is no difference between the two poles of a magnet other then their position in regard to the currents of which the magnet is composed, the austral pole[30] being that to the right of the currents and the boreal pole to their left.

END OF THE FIRST PART OF AMPÈRE'S FIRST PAPER ON ELECTRODYNAMICS.

[29][N. T.] That is, the electromotive action.

[30][N. A. 6, with the words of the translators between square brackets:] The pole which points North [ward] in a magnetized needle; it is on the right of the currents forming the magnet because it is to the left of an external current in the same direction facing the needle.

Chapter 25

On the Effects of Electric Currents [Second Part]

25.1 Translator's Introduction

This is a complete English translation of the second part of the first paper published by Ampère devoted to electrodynamics, entitled "On the effects of electric currents." This work was presented to the *Académie royale des Sciences* on 26 December 1820,[1] containing a summary of the readings at the *Academy* on 18 and 25 September 1820.[2] This second part has been completely translated to Portuguese.[3] To our knowledge there is only a partial English translation of this second part made by O. M. Blunn.[4] The English translation presented here is based upon Blunn's version.

The translation of the first part of this paper appeared in Chapter 24.

The footnotes by Ampère are indicated by [N. A.], while those introduced by the translators are indicated by [N. T.]. The number of the original footnotes are indicated by [N. A. m], where "m" indicates the number of the original footnote introduced by Ampère. The words between square brackets in the middle of the text have been included by the translator to facilitate the understanding of some sentences. The beginning of each original page m is also indicated between square brackets, [page m].

We kept the original numbering of the figures, which follow the numbering of the first part of this paper.

Although figure 5 is not mentioned explicitly in this paper, it belongs to the original images included for this work. We therefore present it here.

Probably this figure 5 refers to the instrument Ampère mentioned on page 193 of the original article. It would represent a rectangular circuit of sides $BC = 30$ cm and $CD = 60$ cm which could turn freely around the horizontal axis GH. The battery might be connected to the goblets U and T. Supposing the current to enter at U, it would flow trough the trajectory $USABCDEFGQ$, leaving at T and connecting itself to the negative terminal of the battery. Suppose the rectangular circuit $ABCDEF$ were initially at rest in a horizontal plane, with no current along the circuit. By closing the circuit, its constant current would interact with the Earth's magnetism, generating a torque upon the circuit. The mobile circuit would acquire a new orientation of equilibrium, inclined relative to the horizontal plane. In this new equilibrium position the plane of the rectangular circuit would remain orthogonal to the axis of a dip needle. That is, the plane $ABCDEF$ would remain parallel to the equatorial plane of a magnetized dip needle. With this experiment Ampère would be reproducing electrodynamically (that is, utilizing only a current-carrying circuit, without any magnet) the magnetic behavior of a dip needle.

After this short Introduction we present the English translation of the second part of Ampère's first paper on electrodynamics.

[1][Hofmann, 1996, p. 238].
[2][Ampère, 1820f].
[3][Chaib and Assis, 2009b] and [Assis and Chaib, 2011, Chapter 24].
[4][Ampère, 1965a, pp. 146-154].

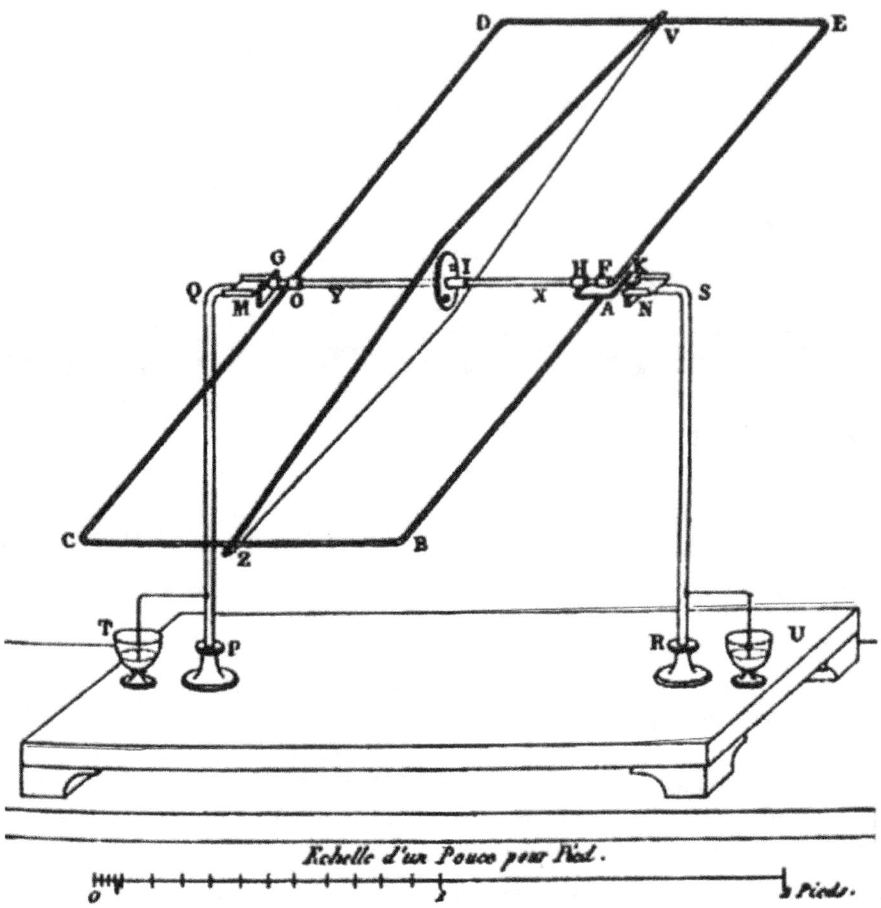

Figure 5.

25.2 Translation

Continuation of the dissertation on the mutual action between two electric currents, between an electric current and a magnet or the terrestrial globe, and between two magnets.

[page 170][5]

25.2.1 Continuation of the First §

The subject of the next paragraphs will be to present the experiments which I made on the mutual action between an electric current and the terrestrial globe or a magnet, to deduce from these experiments the consequences which, by considering the magnets as assemblies of electric currents disposed as I have said,[6] reduce this action, like the action between of the Earth upon a magnet, or the action between two magnets, to the mutual action between two electric currents. Before this presentation I should complete what I have said about this last action by exposing the new results which I obtained since the printing of what preceded, and that I communicated to the *Académie des Sciences*, in two [page 171] dissertations, one read on 9 October [1820] and the other on 6 November [1820].

The first experiment attached to those just described was made with the instrument represented in figure 2.

The electric current, arriving in this instrument by the support CA (figure 2), follows initially the conductor AB, returning by the support BDE. From this support, by the small steel goblet F, where I

[5] [N. T.] The beginning of each original page m is indicated between square brackets, [page m].

[6] [N. T.] In the first part of this work, see Chapter 24.

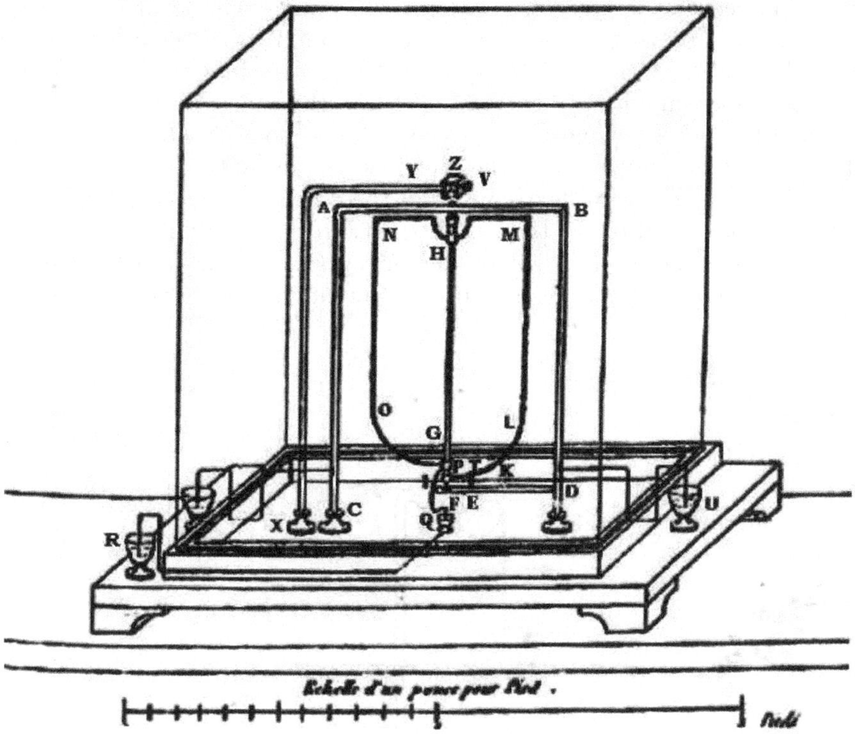

Figure 2.

placed a globule of mercury, and inside which turned the steel pivot of the glass axis GH,[7] the current communicated itself to the copper joint I and to the conductor $KLMNOPQ$, the extremity Q of which plunged into the mercury which was placed in communication with the other extremity of the battery. Everything being so arranged, it is clear that, in the situation in which this conductor is represented and in which I initially placed it, supporting it against the appendix T of the first conductor, the current of the part MN flowed in opposite sense as regards the current of AB, whereas when one made [the current] describe a semi-circumference $KLMNOPQ$, the two currents flowed in the same sense.

Then the effect I expected was produced. At the moment in which the circuit was closed, the mobile portion of the instrument turned by the mutual action between this portion and the fixed conductor AB, until the situation in which the currents, which were initially in contrary senses, remained in such a configuration in which they were parallel and [flowed] along the same sense. The acquired speed made the circuit go beyond this last position; but it returned to it, went a little beyond it, and finished at rest after a few oscillations.

[page 172] I conceive the magnet as an assembly of electric currents in planes perpendicular to the line joining its poles. This conception made me initially search to simulate its action by conductors bent into helix, in which each coil represented a current disposed like those of a magnet, and my first idea was that the obliquity of these coils could be neglected when they had a small pitch. At that moment I did not realize that as this pitch decreases, the number of these coils, for a given length, increases in the same rate and that, consequently, as I recognized later on, the effect of this obliquity remained always the same.

I announced in the dissertation read at the *Académie* on 18 September [1820], the intention to have helices built with brass wires in order to simulate the effects of the magnet, either [the effect] of a fixed magnet with a fixed helix, or [the effect] of a magnetized needle with a helix coiled around a glass tube suspended in its center above a very thin tip, like the needle of a compass.[8] I expected not only that the extremities of this helix would be attracted and repelled like the poles of a [magnetized] needle, by the extremities of a magnetized bar, but also that it would be directed by the action of the terrestrial globe. I was completely

[7][N. T.] The letters G and H do not appear in the original figure of Ampère. A figure with these letters as they are shown here appears in [Ampère, 1964, p. 36].

[8][N. A. 1] Later on I changed this kind of suspension, as explained later on.

successful as regards the action of the magnetized bar. But as regards the directive force of the Earth, the apparatus was not mobile enough, and this force acted [page 173] by a very short lever arm to produce the desired effect. I only obtained this last effect at a later time, with the aid of instruments which will be described in the following paragraphs. The brass wire which composed the helix which I made to be built, coiled around the two glass tubes ACD and BEF (figure 3), prolongs itself in both portions by returning along the interior of these tubes, and its two extremities leave [the tubes] at D and F, one extremity, DG, descends vertically, while the other extremity is curved as seen along FHK. These two extremities are terminated by steel tips which are plunged into the mercury contained in the two small goblets M and N, and placed in communication with the two extremities of the battery, the upper tip alone pressing against the bottom of goblet N. I don't need to say that, from the two extremities of this needle [made] of electric helix, the extremity which is found at the right of the currents is the one which presents, as regards the magnetized bar, the phenomena exhibited by the austral pole of a compass needle, and the other those of the boreal pole.

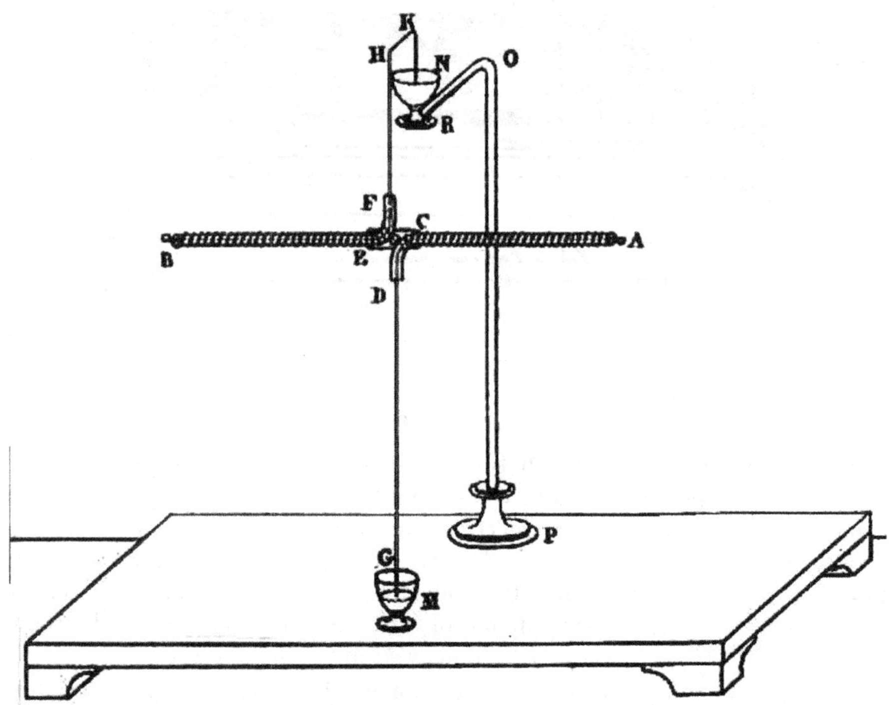

Figure 3.

Afterwards I required an instrument to be built similar to that of figure 1,[9] in which the fixed conductor and the mobile conductor were replaced by brass helices surrounding glass tubes, but whose prolongations, instead of returning inside these tubes, were placed in communication with the two extremities of the battery, as the straight conductors of figure 1.[10] It was by utilizing this instrument that I discovered a new fact which

[9][N. T.] This is the figure 1 of the first part of this paper; see our page 294.

[10][N. T.] Ampère did not present any figure to illustrate this instrument. Here we present an image which may help in the understanding of this question, [Blondel, 1982, p. 87]. The helices AB and CD are coiled in the same sense. They are parallel to one another, side by side in a horizontal plane. Helix AB is fixed in the laboratory, while the portion $ECDF$ can turn around a horizontal axis passing through EF. In this way helix CD can move towards helix AB, or away from it.

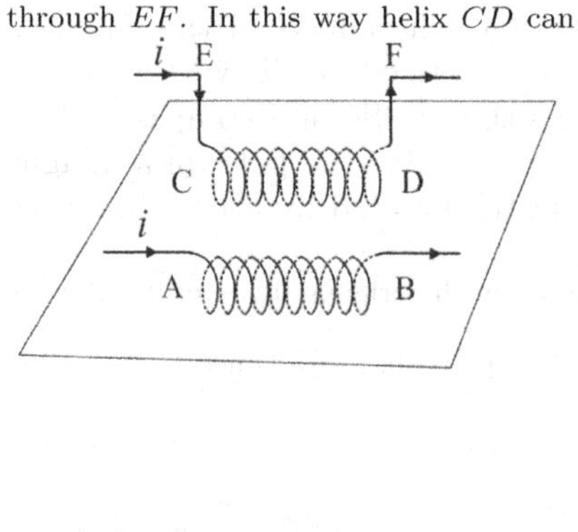

initially did not seem to agree with the other phenomena which I had until then observed in the mutual action between two electric currents, or between a current and a magnet. Later on I recognized that [page 174] this new fact did not contradict the set of these phenomena. But in order to explain it [i.e., in order to explain this new fact], it is necessary to admit as a general law of the mutual action between electric currents, a principle which I verified up to now only as regards the currents of metal wires bent into helix, but which I believe to be generally valid, as regards infinitely small portions of electric current, which one conceives as composing every current of finite length, whether [the electric current] follows a straight line or a curve.[11]

In order to have a clear idea of this law, it is necessary to conceive in space a line representing in magnitude and in direction the resultant of two forces, which are similarly represented by two other lines, and suppose, in the directions of these three lines, three infinitely small portions of electric currents, the intensities of which are proportional to their lengths.[12] The law at issue consists in the fact that the small portion of electric current, directed along the resultant, exerts, in any direction, upon another current or upon a magnet, an attractive or repulsive action equal to what would result, in the same direction, from the combination of the two portions of current directed along the components. It is easily conceived why it should be like that when one considers the current in a conducting wire bent into a helix as regards the actions it exerts parallel to the helix's axis and along the planes perpendicular to this axis, because then the ratio [*rapport*] of the resultant and of the components is the same for each infinitely small arc of this curve, together with the ratio of the actions produced by the corresponding portions of electric currents, [page 175] from which it follows that this last ratio also exists between the integrals of these actions. Moreover, if the law at issue is true for two components as regards their resultant, it must also be true for an arbitrary number of forces as regards the resultant of all these forces, as can be easily seen, by applying [this law] successively first to two of the given forces, then to their resultant and to another one of these forces, and always continuing this procedure until arriving at the resultant of all given forces. It follows from what we have just said as regards the electric currents in wires bent into helices, that the action produced by the current of each coil is composed of two other actions, of which one action would be produced by a current parallel to the helix's axis, represented in magnitude by the pitch of this coil, and the other [action would be produced] by a circular current represented by the section perpendicular to this axis along the cylindrical surface over which the helix is located. As the sum of pitches of all coils, considered parallel to the helix's axis, is necessarily equal to this axis, it follows that beyond the other action produced by the transversal circular currents, which I compared to the [action] of a magnet, the helix produces simultaneously the same action that [would be produced by] a current of equal intensity which were flowing through its axis.

When the conducting wire which forms the helix is made to return along this axis, by enclosing it inside a glass tube placed inside this helix in order to insulate it from the coils composing it, the current of this rectilinear portion of the conducting wire would be in contrary sense to the current which would correspond to the portion of the helix's action taking place [page 176] parallel to its axis, it will repel what this [equivalent rectilinear portion of the helix's current] would attract, and it will attract what it would repel. Therefore, this last action would be destroyed by the action of the rectilinear portion of the conductor, and from the joint actions of this rectilinear portion of the conductor with the helix, there would remain only the action of the transverse circular currents, [an action] perfectly similar to the action of a magnet. This assembly did take place in the instrument represented in figure 3, although I had not planned its advantages, and for this reason this instrument presented exactly the effects of a magnet, while the helices for which the rectilinear portion of conductor did not return inside the axis presented, on the contrary, the effects of a rectilinear conductor equal to the axis of these helices.[13] As the radius of the cylindrical surfaces over which the helix

---

[11][N. T.] This new principle which Ampère is introducing here, referred by him as a law, has been called *Ampère's addition law* by Hofmann, [Hofmann, 1996, pp. 246-250]. It is discussed in Subsection 4.2.1.

[12][N. T.] In this initial stage of his researches, Ampère presented some concepts which had not yet been completely clarified. In particular, he did not present a clear distinction between the length of a current element and the current intensity of this element. But in a paper published in the end of 1820, Gillet de Laumont (1747-1834) replaced Ampère's original phrase "infinitely small portions of electric currents, the intensities of which are proportional to their lengths" by a much more accurate reference to "electric currents of which the attractive or repulsive forces are proportional to their lengths," [Laumont, 1820, p. 544] and [Hofmann, 1996, p. 253].

[13][N. T.] That is, when Ampère coiled a helix around a glass tube, without returning the wire along the axis of the helix, this helix presented the same behavior as that due to a rectilinear current flowing through the axis of the helix. Ampère concluded that the action of this helix upon another current, could be represented by the action upon this other current due to two currents, namely, (a) a straight wire having the same length as the helix and (b) a current flowing azimuthally over the surface of the helix. This can be represented by:

was located was very small[14] in the helices which I utilized, it was the very effects along the longitudinal sense which were more relevant, and this phenomenon astonished me before I could discover its cause. I was still searching [for the explanation of this phenomenon], and wanted, by new experiments, to study all circumstances of this phenomenon, which I had initially observed in the action of two conductors bent into helices, and later on in the action of a conductor of this kind and a magnetized needle, when M. Arago observed [this phenomenon] in this last situation, before I had spoken to him. These helices, in which the wire returns along a straight line by the axis, will constitute an important research instrument, not only because they present the same kind of action as the magnets, when the coils have a small pitch, but also because, when they have a large pitch, we would have an approximately adynamic [*adynamique*] conductor, to carry to and fro the electric current, without [page 177] being afraid that the currents which are located in this portion of conductor would alter the effects of other portions of the circuit, of which we are observing or measuring their action.

It is then possible to simulate exactly the phenomena of the magnet utilizing a conducting wire coiled like that of figure 4, wherein there is between all portions of the conductor which lie in the direction of the axis, the same compensation which existed in the helices aforementioned, between the action of the rectilinear portion of the conductor and the action exerted by the coils in contrary sense parallel to the helix's axis.

It is seen that in this instrument the brass wire which returns inside the tube BH is the prolongation of the wire which forms the circular rings E, F, G, etc., and that each ring is connected to the next one by a small arc of a helix of which each coil would have a large pitch in comparison with the radius of the cylindrical surface over which it is located.

As the action exerted parallel to the tube's axis by these small arcs of helix which are designated in the figure by the letters M, N, O, etc., is equal and opposite to the action exerted by the portion AB of the conductor, there only remains, in this instrument, the actions in planes perpendicular to the tube's axis, and those actions produced in these planes by the small arcs M, N, O, etc. As these last actions are very weak, the effects obtained in the experiments performed with this instrument will be those due to the rings E, F, G, etc.

Ever since my first researches on this subject, I have sought to find the law governing [page 178] the attractive or repulsive action of two electric currents on variation of the distance between them and the angles which determine their position. I was soon convinced that this law could not be found by experiment, because no simple representation could be obtained except by considering portions of currents of infinitesimal length, and experiments cannot be performed on such currents. The action of currents with measurable effects is the sum of the infinitesimal actions of these elements, a sum which can only be obtained by two successive integrations, of which one must be performed over the full extent of one current for the same point of the other, whilst the other must be performed with respect to the result of the first between the limits set by the first current over the full extent of the second current. It is only the result of this last integration, taken between the limits set by the extremities of the second current, that can be compared with experimental data. Hence, as I said in my dissertation to the *Académie* on 9 October [1820], these integrations must be considered before one can determine the interaction of two currents of finite length, whether rectilinear, or curvilinear, bearing in mind that in a curvilinear current the direction of the constituent portions is determined at each point by the tangent to the curve which is its path, and that the action of an electric current on a magnet, or between two magnets, is then found by regarding, in these two later cases, the magnets as assemblies of electric currents arranged [page 179] in the way I have indicated above. From M. Biot's splendid experiment, currents which are in one and the same plane perpendicular to the axis of the magnet, must be regarded as having the same intensity, since it results from the experiment where he compared the effects produced by the action of the Earth on two similarly magnetized bars of the same size and shape, of which one was hollow and the other solid, that the motive force is proportional to the mass and that in consequence the causes to which it is due act with the same intensity on all particles of one and the same cross-section perpendicular to the axis, the intensity varying from section to section according as these sections are close to or far from the poles. When the magnet is a solid of rotation about the line joining its two poles, all the currents of one and the same section must be circles. The calculations for magnets of this shape can be simplified by first calculating the action of an infinitesimal portion of current on an assembly

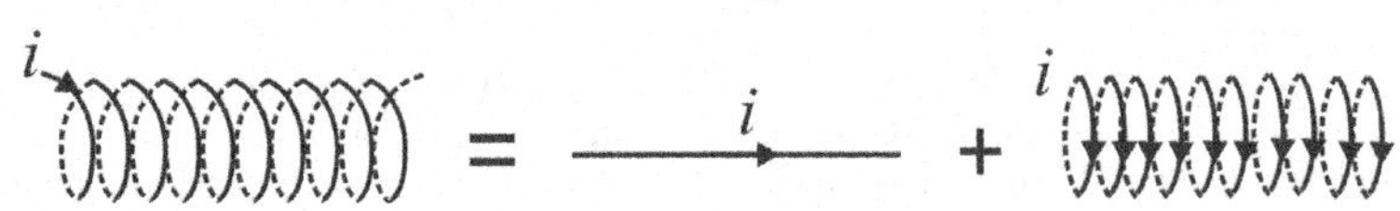

[14][N. T.] Radius very small in comparison with the length of the helix.

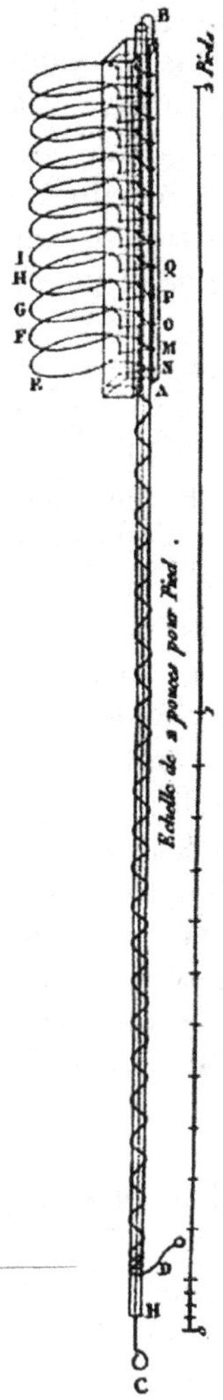

Figure 4.

of concentric circular currents occupying the entire space enclosed within the surface of a circle, such that the intensities which are attributed to them in the calculation are proportional to the infinitesimal distance of two consecutive currents measured on their radius, otherwise the result of integration would depend on the number of infinitesimal parts into which this radius was divided by the circumference representing the currents, which is absurd. Since a circular current is attracted in the portion wherever it flows in the same direction as a current acting on it, and repelled in the portion wherever it flows in the [page 180] opposite direction, the action on the surface of a circle perpendicular to the axis of a magnet consists of a resultant [force] equal to the difference between the components of the attractions and repulsions parallel to this resultant, and of a resultant couple[15] which the attractions and repulsions equally tend to produce. The value of the action is found by integrations with respect to the radii of the circular currents, which should be taken between zero and the radius of the surface for a solid magnet, and between the radii of the inside and outside surfaces for a hollow cylinder, and the result of this operation must then be multiplied:

(1) by the infinitesimal thickness of the cross-section and the overall intensity of the currents composing it, and

(2) by the intensity and the length of the infinitesimal portion of current which is assumed to be acting upon it.

The values are thus obtained of the resultant [force] and resultant couple constituting the elemental action between a circular or crown-shaped section and an infinitesimal portion of this current.

Having found this value, if it is a question of the interaction of a magnet and a current, whether curvilinear

[15] [N. T.] That is, a resultant torque.

or rectilinear of finite length, in order to obtain the mutual action, it is only necessary to perform the integrations which are required for calculation of the resultant [force] and resultant couple of all the elemental actions between each section of the magnet and each infinitesimal portion of the electric current.

But if it is a question of the mutual action of two hollow or solid cylindrical magnets, it is first necessary [page 181] to obtain the value of the interaction between a circular or crown-shaped section and an infinitesimal portion of electric current in order to deduce, by two integrations, the interaction between this section and a similar section, regarding this latter section as composed of circular currents like the first section. The resultant [force] and resultant couple of the mutual action of two infinitely minute sections are thus obtained and, by new integrations, the same can be obtained with regard to the action of two magnets under the surfaces of revolution, having on each occasion first determined by comparison of the calculated and experimental results the relationship between the distance from each section to one of the magnet poles and the intensity of the section currents. I have still not finished the calculations connected with the action of a magnet on an electric current, nor with the interaction of two magnets,[16] but only that by which I determined the mutual action [page 182] of two rectilinear currents of finite magnitude, using the hypothesis which agrees best with the observed phenomena and the general results of experiments in respect of the value of the attraction or repulsion which occurs between two infinitesimal portions of electric currents. At first I did not plan to publish this formula or its diverse applications until I had been able to compare it with the results of precise measurement. But, having considered all the circumstances associated with the phenomena, I believe I saw sufficient probability in favor of this hypothesis to give an outline of it now, and this will be the object of the following paragraphs.

I asked to be constructed, for these experiments, an instrument which I showed on 17 October [1820], to MM. Biot and Gay-Lussac, and which only differs from the apparatus represented in figure 1 in the fact that the fixed conductor of this last one was replaced by a conductor connected to a circle which turned around a horizontal axis perpendicular to the direction of the mobile conductor, by means of a return pulley [*poulie de renvoi*], and graduated in such a way that we could see over the limb the angle formed by the directions of the two currents, in the different positions which could be done successively to the conductor carried by the graduated circle.

I do not present this apparatus in the figures attached to this dissertation because, in conserving the same disposition for this last conductor, and in placing the mobile conductor in a vertical situation, I constructed the apparatus shown in figure 6, as being more appropriate than my original device for the particular measurements that I had in mind, especially as the support of the graduated circle, [page 183] besides its movement which allows the moving conductor to be brought nearer or taken further away, can now also be moved by means of an adjustable screw in two other ways, namely, vertically, and horizontally transverse to the other two movements. The first of these three movements is indispensable for measurement by the device, and originally this was the only possible movement in my first apparatus, the aim of the two additional movements being to simplify the measurements when the lines joining the midpoints of the two currents are not perpendicular to them. For this reason I think that the adjustment by hand before the experiment is preferable to the use of adjusting screws, provided that the support of the graduated circle can afterwards be fixed in a stable manner in the same position as previously.

This new instrument is represented in figure 6, for which I shall explain the construction. If I speak here of the first [instrument], it is because it was with it that I noticed, for the first time, the action of the terrestrial globe upon the electric currents, which altered the effects of the mutual action of the two conductors which I intended to measure. Therefore, I interrupted these observations, and asked to be constructed the two instruments which fully exhibit this action of the Earth, and with which I equally produced, with electric currents, the motions corresponding to the direction of the compass needle in the horizontal plane, along the line of declination, and the motion of the dip needle in the plane of the magnetic meridian. These last instruments and the experiments I made with them will be described [page 184] in the next paragraph, as they were described in the dissertation which I read to the *Académie des Sciences* on 30 October [1820]. Let us return to the apparatus represented in figure 6 designed to measure the action of two electric currents in all kinds of orientations.

The first of the three movements of the support KFG is made by the adjusting screw M, the other two

[16][N. A. 2] The calculations assume that the presence of an electric current, or of another magnet, changes nothing in the electric currents of a magnet on which they act. This is never the case with soft iron; but since tempered steel preserves the modifications which it undergoes, whether in the experiments of M. Arago on the magnetization of steel by an electric current, or whether in the procedures of ordinary magnetization, it seems to me that when magnetized steel is in precisely the same state as prior to the action of another magnet or electric current upon it, it can be inferred that the constituent currents are practically constant in direction and intensity during their action, for otherwise the modifications which they undergo would not persist after the action had ceased.

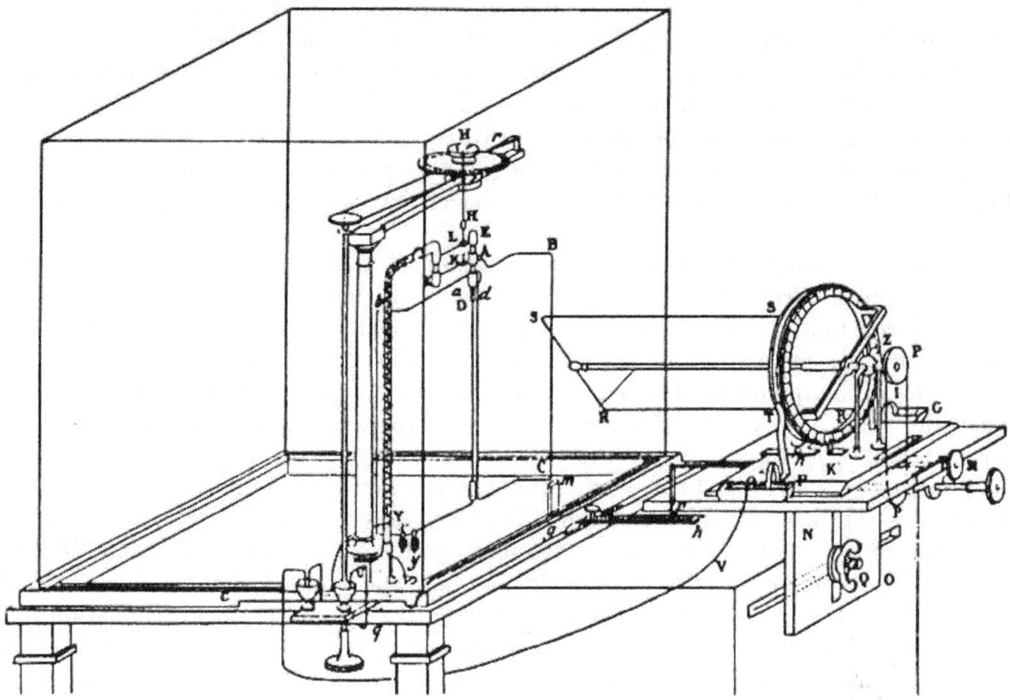

Figure 6.

movements are made by the connecting piece by which the support is fixed to the block of wood N which is free to slide horizontally and vertically on the other block of wood O at the base of the device. A horizontal slot is made in one block and a vertical slot in the other, and at the intersection of these two slots there is a screw nut Q which serves to arrest the moving piece on the fixed piece in the desired position. The graduated circle for inclining the attached portion of conducting wire at any designed angle is revolved by the two return pulleys P and P'. In order that there should be no action of the Earth on the moving conductor to combine with the action of the fixed conductor, the former is made of two equal and opposite parts $ABCd$ and $abcDE$, with the shape shown in the diagram. And in order to bring its two extremities into contact with the extremities of the battery, the moving conductor is interrupted at the angle A of the suspension piece HH' which balances with torsion the attraction or repulsion of the two currents. The branch BA continues beyond A and the branch DE continues beyond E, both terminating at tips K and L which are immersed in two small mercury-filled goblets without touching the bottom.

[page 185] The base which supports these two small goblets can be advanced or retreated by means of the screw nut a, which fixes it in the slot ef. The small goblets can be of iron or platinum. One of them is placed in communication with one of the two extremities of the battery by the conductor XU enclosed in a glass tube around which the conductor YVT is bent into helix with large coils, terminated by a kind of copper spring, which is supported in T over the circumference of the graduated circle, where it is in contact with a circle of brass wire communicating with the branch SS' of the conductor of which the part SR is destined to act upon the mobile conductor, and which branch RR' is connected to a second brass wire over which a spring ZI similar to the first one is supported in Z, and which communicates, from the side I, with the other extremity of the battery. It is clear that in turning the graduated circle around the horizontal axis which supports it, the portion SR of the conductor will turn in a vertical plane, in such a way as to make any desired angle with the direction of the portion BC of the mobile conductor, upon which it acts through the glass case inside of which is located this mobile conductor, to prevent the influences upon it due to air agitations.

In order to measure the attractions and repulsions of two conductors at different distances, when they are parallel to one another, and when the line which joins their midpoints is perpendicular to them, one turns the vertical axis to which the suspension wire is attached, in such a way that the portion BC of the mobile conductor corresponds to the zero of the scale gh. This can be obtained by placing it immediately below the bevel which terminates the copper piece m. An index [page 186] np, attached in n to the support of the graduated circle, marks upon this scale the distance of the two portions of conductor BC and SR. When the communication of the two extremities of the circuit with the extremities of the battery is established, the first portion BC moves forwards or backwards according as it is attracted or repelled by SR. But we can bring it back to the position where it was previously, by turning the axis of the suspension wire. The number of turns and portions of turn indicated by the index r upon the dial pq attached to this axis, gives the value of the attraction or repulsion between the two electric currents, as measured by the torsion of the wire.

There is no need to remind physicists who are accustomed to this type of measurement that owing to continuous variation of current intensity with the energy of the battery, it is necessary to repeat an experiment at some constant distance in between each experiment so as to know how the intensity of the currents varies and its value at each instant from the action observed each time at this constant distance and by the ordinary rules of interpolation. The same approach is to be adopted to compare the attractions and repulsions when the angle between the two currents varies if the line joining their midpoints is constantly perpendicular to them. The intermediate observations, to determine by interpolation the energy of the battery at each instant, are simplified since, with the distance between the two portions of conductor BC and SR constant, it is sufficient to turn the graduated circle in order to return SR each time in the direction parallel to BC. [page 187] Finally, if it is desired to measure the interaction of BC and SR when the line joining their midpoints is not perpendicular to their direction, the support of the graduated circle is set in the appropriate position by the screw nut Q which sets it in the desired position in relation to the rest of the apparatus and then by performing a series of experiments similar to those in the preceding case, the results obtained in each position of the conductors can be compared with those in the case when the line joining the midpoints is perpendicular, this comparison being made for one and the same shorter distance between currents and then for the various other distances. Everything necessary is thus obtained to see how and up to what point these different circumstances influence the interaction of the electric currents. There only remains to see if all the results agree with the calculation of the effects which must be produced in each arrangement from the law acknowledged to govern the attraction between two infinitesimal portions of electric current.

With the addition of another mobile conductor which has exactly the same suspension, and which is represented separately in figure 10, where the corresponding parts are designated by the same letters, I also made this instrument appropriate to measure the torque of the forces which tend to turn a conductor, by the action of another conductor which makes successively with it different angles which correspond to different torques. This mobile conductor $ABOCDEF$ has the shape seen in figure 10, and is suspended [page 188] in the middle of its horizontal superior side, where it is interrupted between the points A and F, where the two extremities of this conductor have the two steel tips M and N, which are situated along the same vertical line, and which plunge into the mercury of the two small goblets of figure 6, without touching the bottom due to the suspension of the torsion wire. In order to measure the torque produced by a rectilinear conductor, it is placed inside the glass case very close to the inferior horizontal side CD of the mobile conductor, figure 10, in such a way that it corresponds to its midpoint. Therefore this last mobile conductor turns by the action of the fixed conductor without the influence of the Earth, as there is a compensation between the actions exerted by the Earth upon the two equal and opposite halves of the mobile conductor.[17]

25.2.2 II. Orientation of Electric Currents by the Action of the Terrestrial Globe

I did not succeed[18] in moving the conducting wire of an electric current by the action of the terrestrial globe in my first experiments, perhaps not so much by the difficulty in obtaining a very mobile suspension, but by the fact that, instead of looking in the theory which reduces the phenomena of the magnet to those of electric currents, the most favorable disposition for this kind of action, I was concerned with the idea of trying to simulate [page 189] as best as I could the disposition of the electric currents of the magnet in the arrangement of the currents on which I wanted to observe the action of the Earth. Only this idea had guided

[17][N. T.] We have called this mobile rectangular conductor *Ampère's astatic coil*, Subsection 4.3.2. It is indifferent for terrestrial magnetism. The oppositely directed torques produced by the Earth's magnetic action on the two halves of the rectangle left it in equilibrium with respect to terrestrial magnetism.

[18][N. A. 3] What is contained in this paragraph was read to the *Académie royale des Sciences*, in the meeting of 30 October [1820].

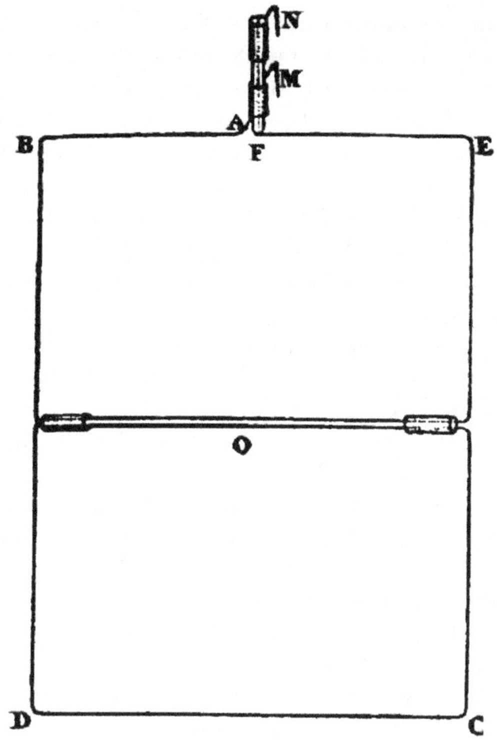

Figure 10.

me in the construction of the instrument represented in figure 3, and it prevented me from realizing that it is only through an indirect manner that this action carries the austral pole of the magnetized needle to the North and to the bottom, and carries the boreal pole to the South and to the top. [This idea also prevented me from realizing] that the immediate effect of this action is to place the planes perpendicular to the axis of the magnet, in which are located the electric currents composing it, parallel to a plane determined by the resultant action of all these [currents] of our globe, and which is, in each place, perpendicular to the dip needle. It follows from this consideration that the terrestrial action does not immediately direct a straight line, but a plane. Therefore, what needs to be simulated, is the disposition of the electricity following the equator of the magnetized needle, and this equator is a curve which returns in itself, and to see, after this, when an electric current is so disposed, whether the action of the Earth tends to place the plane where this current is located along a direction parallel to the direction where it tends to place the equator of the magnet, that is, along a direction perpendicular to the dip needle, in such a way that the current that we try to direct in this way should [flow] in the same sense as those currents of the magnetized needle which followed the action of the terrestrial globe.

The magnet receives different motions whether it can only turn in the plane of the horizon like the needle of a compass, or whether in the plane of the magnetic meridian, as the dip needle attached to [page 190] a horizontal axis and perpendicular to the magnetic meridian. To simulate these two motions by imparting analogous [motions] to an electric current, it is necessary that the plane where it [i.e., this current] is located should be, in the first case, vertical like the plane of the equator of a horizontal magnetized needle, and it should turn around the vertical which passes through its center of gravity; and, in the second case it should, like the equator of a dip needle, turn only around a line included in this plane, which must be at the same time horizontal and perpendicular to the magnetic meridian.

Initially I placed in these two positions a double copper spiral which seemed to me very appropriate to represent the electric currents of the equator of a magnet. And I saw this apparatus move when I established in it an electric current, precisely like the equator of the needle of a compass would move in the first case, and in the second case like a dip needle [would move]. But the same thing which happened to M. OErsted also happened to me. In his experiments, the directive force of the electric current which he made to act upon a magnetized needle tended to place it along a direction which made a right angle with that of the current. But he never obtained a deflection of one hundred degrees[19] leaving the conducting wire in the direction of the magnetic meridian, because the action of the terrestrial globe, combining with the action of the electric current, directed the magnetized needle along the resultant of these two actions. In the experiments performed with the double spiral, the directive force of the Earth was antagonized, in the first case, by the torsion of the wire to which this instrument was suspended; in the second case, [page 191] by its weight, because the center of gravity [of this instrument] could not be exactly situated in the horizontal line around which the double spiral turned.

I then thought that in multiplying the number of coils composing the spiral, this would not increase the effect produced by the action of the Earth, as the mass to be moved would increase proportionately to the motive force, from which I concluded that I would obtain more easily the same directive phenomena utilizing, in order to represent the equator of a magnetized needle, a single electric current returning upon itself, and forming a circuit which, if not absolutely closed as it would be impossible to establish the current in the copper wire, at least allowing only a very small interruption sufficient to make the communication between its two extremities and those of the battery.

At the same time I realized that the shape of the circuit was indifferent, provided all of its parts were in the same plane, because it was a plane that it should direct.

Therefore I ordered two instruments to be built. In one of them, the conducting wire has the shape of a circumference $ABCD$, figure 7, of which the radius is a little greater than two decimeters. The two extremities of the brass wire forming this circumference are soldered to the two copper joints E and F, attached to a glass tube Q, and which has two steel tips M and N, plunged into the mercury contained in two small platinum goblets O and P, and which only the superior point N reaches the bottom of goblet P. These two goblets are supported by the copper joints G and H, which communicate with the two extremities of [page 192] the battery, by means of two conductors made of brass wire. One [of these two brass wires] is contained inside the glass tube which supports these two last joints, and serves as a support for the instrument, and the other forms around this tube a helix of which the coils have a very great pitch in relation to the diameter of the tube, so that the actions exerted by the two portions of currents which flow through these conductors in contrary senses neutralize one another almost completely. Under the glass case which covers this instrument, to protect it from the agitations of air, I placed another circle made of brass wire, with a diameter a little larger [than the previous one], which is fixed and sustained by a support similar to the support of the mobile circle, in the situation seen in the figure. This circle also communicates with two conductors disposed in the same manner, and which serve to make the electric current pass through it when, instead of observing the action of the terrestrial globe upon the mobile circle, one wishes to see the effects due to the action of two circular currents one upon the other, whereas when one wishes to observe the action exerted by the Earth upon an electric current, this current is passed only through the mobile circle. We are dealing here only with the action of the Earth. Therefore, I will speak exclusively of the case for which only the conductors of the mobile circle are in communication with the two extremities of the battery. Therefore the fixed circle serves only to indicate precisely the vertical plane which is perpendicular to the magnetic meridian, to which the mobile circle should be carried by the action of the Earth. In the first place we then place the fixed circle in this plane by means of a compass, and the mobile circle is placed in another situation which will be, for example, that of the magnetic meridian itself. Then, [page 193] when an electric current is passed through this mobile circle, it will turn to place itself in the plane indicated by the fixed circle, but at first it will move beyond it due to the acquired speed, then it will return, stopping after a few oscillations.

The sense in which this motion takes place depends upon the sense of the electric current established in the mobile circle. In order to predict this motion in advance, one should consider a line passing through the center of this circle, and perpendicular to its plane. This line will arrive in the magnetic meridian when the mobile circle is brought in the plane which is perpendicular to it. And it will arrive in this place in such a way that the extremity which is at the right of the current considered as acting upon a point considered at will outside this circle and, consequently, at the left of an observer who, placed along the sense of the current, looked at the needle, extremity which represents the austral pole of a magnetized needle, be found in the North side. And this is enough to determine the sense of motion taken by the mobile circle.[20]

In the other apparatus, the equator of the dip needle is represented by a rectangle of brass wire having approximately 3 decimeters in width and 6 decimeters in length [(figure 5)].[21] Moreover, the suspension is

[19][N. T.] Probably Ampère is referring to 100 *grades*, that is, one fourth of a total circumference, which is equivalent to 90°, or to $\pi/2$ rad.

[20][N. T.] Ampère is referring once more to his imaginary observer, see Footnote 20 on page 292 of this book.

[21][N. T.] Probably Ampère is referring to figure 5 of this paper, reproduced on pages 298 and 310 of this book, although it

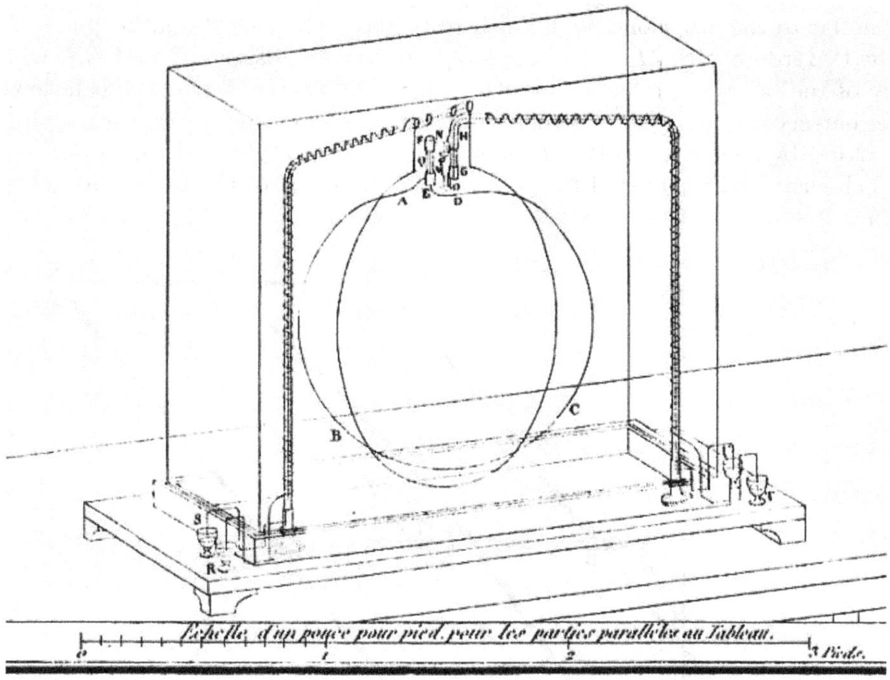

Figure 7.

equal to that of a dip needle. It was with these two instruments that I observed, in experiments repeated many times, the phenomena of orientation by the action of the Earth, much more completely than I had observed with the double spiral. With the first [instrument], the mobile circle is stopped, as I have just said, precisely in the situation in which the action of the terrestrial globe should carry it [page 194] according to the theory. With the second [instrument], the conductor constantly left a position where I had found, by making it oscillate, that the equilibrium was stable, in order to reach another situation more or less close to the situation which the equator of a magnetized needle would take in the same circumstances, and it would stop there, after a few oscillations, in equilibrium between the directive force of the Earth and the weight that acted to bend the brass wire, which lowered the center of gravity of the conductor below the horizontal axis. When the circuit was interrupted, it returned, in this last case, to its first position or, if it did not return precisely to it, even when it remained sometimes very far [from this first position], it was evident, from all circumstances of the experiment, that this was due to the aforementioned curvature [of the wire], which produced, in the position of the center of gravity, a small alteration which persisted when the current had stopped. In the two cases, I was careful to change the extremities of the conducting wires in relation to the extremities of the battery, in order to verify that the current that existed here [i.e., inside the battery] was not the cause of the produced effect, as it would then always happen in the same sense, while this effect should happen in contrary sense, according to the theory. By maintaining the same extremities in communication, I also passed, from the right to the left of the instrument, the wires which made the communication of the mobile conductor with the two extremities of the battery, in order to verify that the currents of these wires, of which the largest portion was far away from the instrument, had no perceptible influence upon its motions. I don't need to say that, in every case, the [page 195] motions take place in the sense in which the equator of a magnetized needle would move. This means that the extremity of the perpendicular to the plane of the conductor, which is at the right of the current and, consequently, to the left of a person which looks at it in the situation described in the first paragraph of this dissertation, is carried to the North in the first case, and down in the second case, as the austral pole of a magnet which is represented by this extremity would be carried. The instrument[22] with which I had made this experiment is composed of a brass wire $ABCDEFG$ soldered at A to a similar piece of wire HAK carried by the glass tube XY by means of the copper joint H, and to which is fixed a small steel axis which rests above the rim cut in bevel out of an iron blade N over which mercury is placed in contact with this axis. The portion FG of this brass wire passes through the glass tube and is soldered to the copper joint G, which supports a

was not explicitly mentioned at this point of the paper.

[22][N. T.] See figure 5 on pages 298 and 310 of this book.

small steel axis similar to the other one, and which rests above the rim of another blade M where there is also mercury. The two iron blades M and N are supported by the bases PQ and RS, which communicate with the mercury of the boxwood goblets T and U, where the two conductors which depart from the two extremities of the battery are plunged. In order to prevent the curvature of the brass wire $ABCDEF$, the glass tube XY carries, by means of another copper joint I, a very light and thin wooden lozenge ZV, the extremities of which sustain the center of the portions BC and DE of the brass wire which are parallel to the glass tube XY.

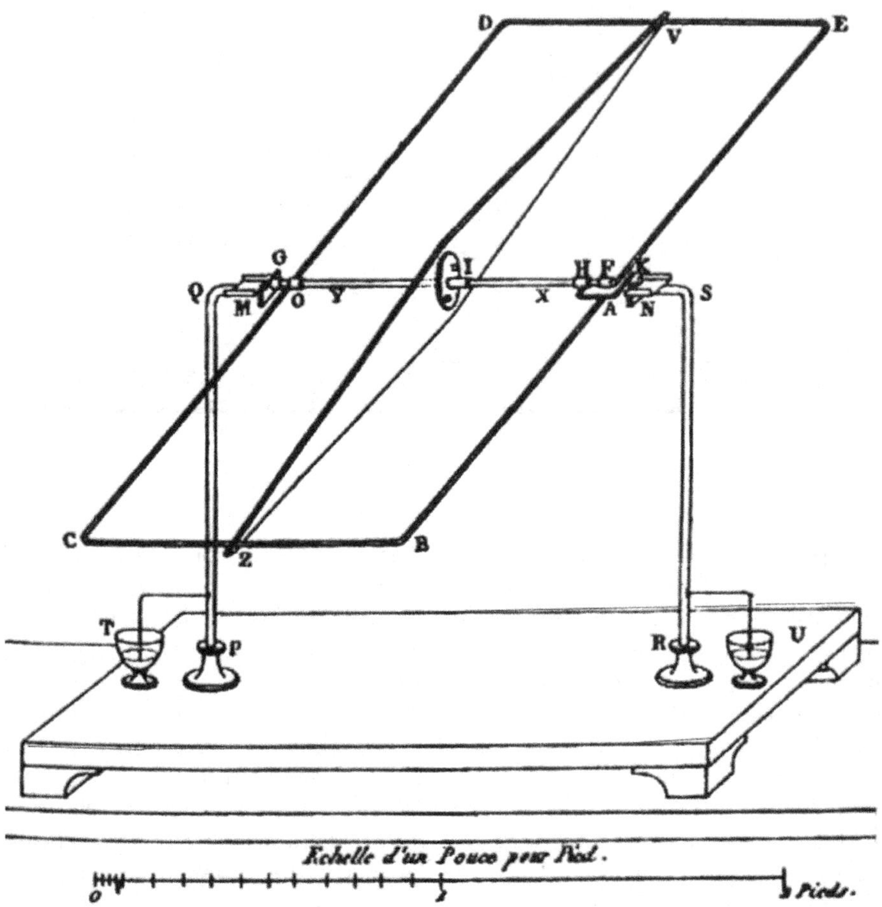

Figure 5.

The mercury interposed in this instrument, and in those which I just described, in all places where the communication should happen by parts which are not soldered, [page 196] although not being always necessary, still is the best means known to me in order to assure the success of the experiments. For instance, twice I had tried without success of an experiment which was perfectly successful when, by trying it a third time, I made a more complete contact by means of a globule of mercury.

25.2.3 III. The Interaction Between an Electrical Conductor and a Magnet

This action that M. OErsted discovered led me to look for the interaction of two electric currents, the action of the Earth on a current, and how the electricity might produce all phenomena presented by the magnets, looking for a distribution [of electric currents inside each magnet] similar to that of a conductor of electric current, with closed curves perpendicular to the axis of each magnet. These points of view, most of which have only recently been confirmed by experiment, were communicated to the *Académie* in its session of 18 September 1820. I will transcribe what I read in this session, only removing the passages which would represent a repetition of what I just said and, in particular, those passages where I described the instruments which I then proposed to be built. Since then these instruments have been built and most of them have been described in the previous paragraphs. In this way, it is possible to have a better idea of the path I followed in my research of the topics discussed here.

The experiments which I made on the mutual action of conductors which place in communication the extremities [page 197] of a voltaic battery, showed me that all facts relative to this action might be reduced to two general results, which should be initially considered as being done only by observation, expecting to reduce them to a single principle, as I shall soon try to do. I begin presenting them in the simplest and most general form.

These results consist, on the one hand, in the directive action of one of these bodies upon the other; on the other hand, in the attractive or repulsive action established between them, according to the circumstances.

Directive Action.

When a magnet and a conductor act upon one another, one of them being fixed, while the other can only turn in a plane perpendicular to the shortest distance between the conductor and the axis of the magnet, the mobile body tends to move in such a way that the directions of the conductor and the axis of the magnet form a right angle, and such that the pole of the magnet which normally points North is at the left of what is ordinarily called *the galvanic current*, a denomination which I believe should be changed to that of an electric current, and the opposite pole is at the right [of the electric current]. It should be understood that the line which measures the shortest distance between the conductor and the axis of the magnet meets the direction of this axis between the two poles. In order to keep this statement with the greatest possible generalization, it is necessary to distinguish two kinds of conductors:

1. In the first place, the battery itself, inside of which the electric current, with the meaning I give to this word, flows from the extremity where hydrogen is produced in the decomposition of water, to the extremity where oxygen [page 198] is liberated.

2. In the second place, the metal wire which connects the two extremities of the battery, and where we should then consider the same current as flowing, in the contrary, from the extremity which gives oxygen, to that where hydrogen is developed.

It is possible to combine these two cases under a single definition, by saying that one understands by electric current the direction according to which the hydrogen and the bases of the salts are transported by the action of the whole battery, by conceiving this battery composing with the conductor a single circuit, when this circuit is interrupted in order to place in it water or a saline solution which this action decomposes. Moreover, everything which I will say about this subject does not suppose in any way that there is really a current along this direction, and we can consider it as just a convenient and useful denomination which I utilize here for an electric current.

In the experiments of M. OErsted, this directive action is always combined with the action exerted by the terrestrial globe upon the magnetized needle, and sometimes it combines itself with the action which I will soon describe under the denomination of *attractive or repulsive action*. This combination leads to complicated results where it is difficult to analyze the circumstances and recognize the laws.

In order to observe the effects of the *directive action* of an electric current upon a magnet, without altering it by these diverse causes, I asked to be constructed an instrument which I denominated *astatic magnetized needle*.[23]

This instrument, represented in figure 8, consists of a magnetized needle AB fixed perpendicularly [page 199] to an axis CD, which one can place in the desired direction, by means of a motion similar to that of the base of a telescope and of two return screws E and F. The needle so disposed can only move by turning in a plane perpendicular to this axis, taking care that its center of gravity is located exactly in this plane, in such a way that before its magnetization one can be assured that the weight has no action to make it change its position. It is then magnetized, and this instrument serves to verify that, when the plane in which this needle moves is not perpendicular to the direction of the dip needle, terrestrial magnetism makes the magnetized needle to tend to take the direction of that line which, between those lines plotted over the plane, is the closest possible to the direction of the dip needle, that is to say, the projection of this direction over the same plane. Afterwards the axis [CD] is placed parallel to the direction of the dip needle. As the plane where the magnetized needle moves is perpendicular [to the direction of the dip needle], terrestrial magnetism has no longer any action to direct the magnetized needle which, therefore, remains completely

[23][N. T.] The adjective *astatic* has the meaning of "having no particular directional characteristics." In the case of figure 8, the magnetized needle AB is free to turn in a plane perpendicular to the axis CD. This axis is placed parallel to a dip needle. In this situation the needle AB will remain at rest in any position. That is, it is not oriented by terrestrial magnetism, having a neutral equilibrium as regards the magnetic action of the Earth.

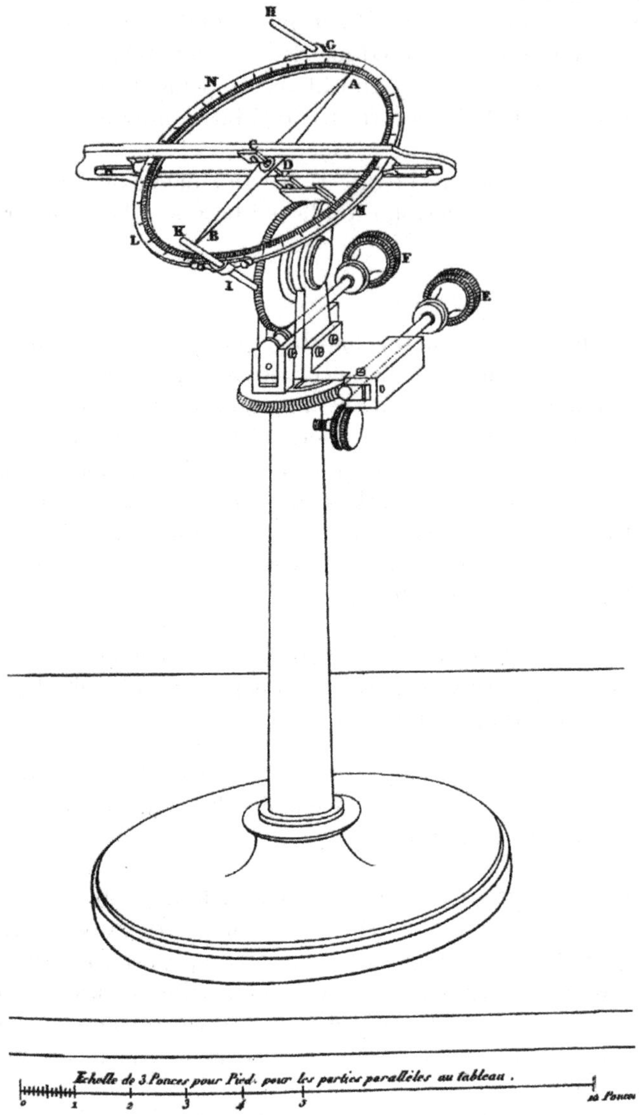

Figure 8.

astatic. The same apparatus has, in the plane of this needle, a graduated circle LMN, over which two small glass bars GH and IK are placed, in order to fix the conductive wires, the directive action of which therefore operates alone, without complications arising from the weight and from terrestrial magnetism.

The main experiment to be made with this apparatus is to show that the angle between the directions of the needle and the conductor is always a right one when the *directive action* acts alone.

Attractive or Repulsive Action.

This second general result [page 200] consists, in the first place, in the fact that a conductor connecting the two extremities of a voltaic battery, and a magnet in which the axis makes a right angle with the direction of the current which takes place in this conductor according to the previous definitions, attract one another when the austral pole is to the left of the current which acts upon it, that is to say, when the position is that configuration which the conductor and the magnet tend to acquire due to their mutual action, and repel one another when the austral pole of the magnet is to the right of the current, that is to say, when the current and the magnet are maintained in the situation opposite to that which they tend to give to one another. One can see, by the statement of these two results, that the action between the conductor and magnet is always reciprocal. In the first place I tried to verify [experimentally] this reciprocity, although it seemed to me self-evident. It would be superfluous to describe here the experiments which I made to verify this reciprocity. It is enough to say that they were completely successful.[24]

The two kinds of action between a magnet and a connective wire, which I have just presented and considered as simple results of experiments, are sufficient to explain the facts observed by M. OErsted, and to predict what should happen in analogous cases which have not yet been observed. They indicate in advance, for example, everything that should happen when an electric current acts upon the dip needle. I will not present any detail in this respect, as everything that I could say on this subject stems immediately from the previous statements. I will only say that after having deduced only the [page 201] first general result from the note of M. OErsted, I deduced from it the explanation of the magnetic phenomena, based upon the existence of electric currents in the globe of the Earth and in magnets. This explanation led me to the second general result and suggested to me, in order to verify it, an experiment which was completely successful. When I communicated it to M. Arago, he pointed out correctly that only this attraction between a magnet and an electrical conductor placed at right angles along the direction to which they tend to place themselves mutually, and this repulsion, in the opposite direction, could justify the results published by the author of the discovery [that is, Ørsted], in the case in which a galvanic conductor in a vertical position is brought near a horizontal magnetized needle, and that we could easily deduce this law, from one of M. OErsted's experiments, that one which he presented as follows: *Posito autem filo (cujus extremitas superior electricitatem à termino negativo apparatûs galvanici accipit) è regione puncto inter polum et medium acûs sito, occidentem versùs agitur.*[25]

Because this motion of the magnetized needle, indicated as taking place whether the conductor is found at the West or at the East of the needle, is in the first case an attraction, as the austral pole is at the left of the current, and in the second case it is a repulsion, as it [i.e., the austral pole] is at the right [of the current].

But by agreeing with the correctness of this observation, it seems to me that the distinction I made of the two general results of the mutual action between a magnet and a conducting wire only becomes more important [page 202] to explain what then happens by showing that, in this case, there is sometimes an attraction and sometimes a repulsion, always according to the law of the second general result which I just presented, whereas, in the experiment which M. OErsted presented immediately before in these terms: *Quando filum conjungens perpendiculare ponitur è regione polo acûs magneticæ, et extremitas superior fili electricitatem à termino negativo apparatûs galvanici accipit, polus orientem versùs movetur,*[26] this motion only takes place because the magnetized needle takes the direction, as regards the conductor, determined by the first general result, with all the circumstances which I understood in its statement and, in particular, the observation which finishes it [i.e., which finishes the statement of the first general result].

I still need to describe the instrument with which I verified the existence of this action between an electric current and a magnet, which was designated in the preceding pages by the name of *attractive or repulsive action*, and I observed the effects without the interference of the *directive action*. This instrument, represented in figure 9, is composed of a base ABC of which the arms BEG and BFH support the connective horizontal wire KL, close to which is suspended a small cylindric magnetized needle MN, at the extremity C of this base, by means of a silk thread MC, sometimes by its austral pole and sometimes by its boreal pole.

When first I wanted to find the causes of the new phenomena discovered by M. OErsted, I reflected that since the order in which two facts are discovered in no way affects any conclusions which can be drawn from the analogies they present, it might, [page 203] before we knew that a magnetized needle points constantly from South to North, have first been known that a magnetized needle has the property of being influenced by an electric current into a position perpendicular to the current, in such a way that the austral pole of the magnet is carried to the left of the current,[27] and it could then have subsequently been discovered that the extremity of the needle which is carried to the left of the current points constantly towards the North: would not the simplest idea, and the one which would immediately occur to anyone who wanted to explain the constant direction from South to North, be to postulate an electric current in the Earth in a direction such that the North would be to the left of a man who, lying on its surface facing the needle, received this

[24] [N. T.] The reciprocity of the torque had also been observed experimentally by Ørsted in 1820, [Franksen, 1981, pp. 421-424].

[25] [N. T.] *When the connecting wire is placed perpendicularly opposite to a point between the pole and the middle of the needle, and the upper extremity of the wire receives electricity from the negative end of the galvanic apparatus, the pole is moved towards the west,* [Ørsted, 1998b, pp. 415 and 419] and [Oersted, 1965, pp. 115 and 116].

[26] [N. T.] *When the connecting wire is placed perpendicularly opposite to the pole of the magnetized needle, and the upper extremity of the wire receives electricity from the negative end of the galvanic apparatus, the pole is moved towards the east,* [Ørsted, 1998b, pp. 415 and 419] and [Oersted, 1965, p. 115].

[27] [N. T.] That is, the austral or North pole of the magnet is carried to the left of Ampère's imaginary observer. This observer is placed along the current-carrying wire, facing the magnet, with the current penetrating its feet and leaving by his head. With this orientation the North pole of the magnet tends to move to the left of this observer.

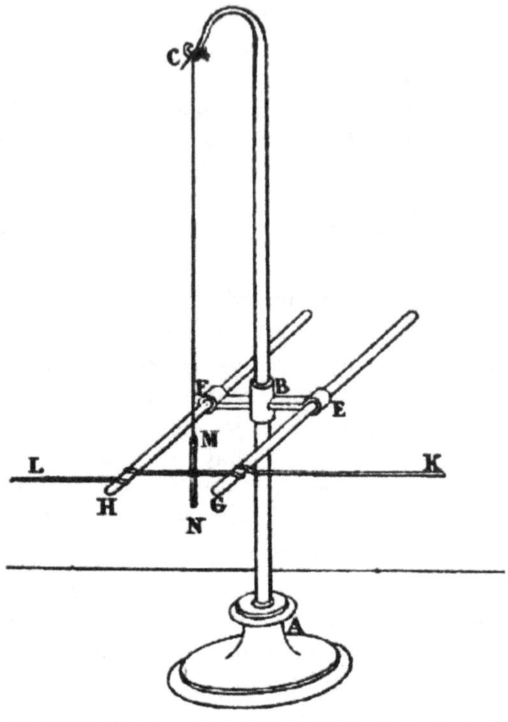

Figure 9.

current in the direction from his feet to his head, and to draw the conclusion that it takes place from East to West in a direction perpendicular to the magnetic meridian?

This hypothesis becomes even more probable when one pays attention to all known facts. This current, if it exists, should be compared to the current inside the battery which I showed to act upon the magnetized needle, as flowing from the copper extremity to the zinc extremity, when a conductor is established between them, and which would take place even if, the battery forming a closed current, they [i.e., the extremities] were reunited by a pair similar to the other ones: because there is probably nothing in our globe which looks like a continuous and homogeneous conductor. But the diverse substances which compose our globe are precisely in the situation of a voltaic battery formed by elements disposed at random and that, returning upon itself, would form like a continuous belt [page 204] around the Earth. The elements so disposed no doubt give less electric energy than they would offer if they had a regular periodic order. But it would be necessary to arrange them designedly, in a series of different substances forming a closed curve around the Earth, in such a way that no current would be flowing in one sense or in the other. It is found that, according to the arrangement of the substances of the Earth, this current takes place from East to West, and that it directs everywhere the magnetized needle perpendicularly to its own direction. This direction marks in this way over the Earth a magnetic parallel, in such a way that the pole of the needle which should be at the left of the current is found, due to this current, constantly carried towards the North, and the needle is directed according to the magnetic meridian.

I should remark in this regard that the effect produced by the English batteries, where one burns a thin metal wire even with a single pair in which the zinc and the copper plunge into an acid, proves sufficiently that it is a very narrow supposition to admit the electromotive action only between the metals, and to consider the interposed liquid merely as conductor. No doubt there is an [electromotive] action between two metals, Volta demonstrated this in a very complete way. But does there exist a reason why there is no [electromotive action] between them [i.e., the metals] and other bodies, or only between these other bodies? Probably there is [this electromotive action] in the contact between all bodies which can conduct more or less the electricity under a small tension. But this action is more perceptible in the batteries composed of metals and extended acids, not only because it seems that these are the substances in which it develops [page 205] with greater energy, but also because these are the substances which better conduct the electricity.

The several arrangements which we can give to non-metallic bodies cannot produce an electromotive action comparable to the action of a voltaic battery composed of metal disks separated alternately by liquids, due to the small length we can give to our instruments. But a battery that could make a turn around the Earth no doubt would conserve a certain intensity even if it were not composed of metals, and when the elements were arranged at random in it; because in a length so great, it would be necessary to have an arrangement done on purpose in order that the actions in one sense were exactly destroyed by the actions in the other sense.

On this subject I should observe that the electric currents in a single body cannot be independent from one another, unless these currents were separated by substances which insulated them completely along all their lengths. But even in this last case, they should influence one another, as their actions are transmitted through all bodies. When these currents coexist in a globe in which all portions are continuous, they have a stronger reason to be all directed in the same sense, following the direction which they tend to receive from the assembly of all electromotive actions of this globe. Moreover, I am far from believing that it is only in these [electromotive] actions which reside the cause of the electric currents which are indicated [in the globe of the Earth] by the direction taken by the magnetized needle in each point over the surface of the Earth. I believe, instead, that the main cause is totally different, [page 206] as I will have opportunity to say elsewhere. This cause, moreover, depending upon the rotation of the Earth, would give in each place a constant direction to the needle, which is against the observation. Therefore, I consider that the electromotive action of the substances composing the planet we inhabit, should combine with this general action, and explaining the variations [observed in the direction of the needle] as the oxidation progress in one or in the other continental region of the Earth.

As regards the diurnal variations, they can be easily explained by the alternate changes of temperature of these two regions during one rotation of the terrestrial globe. These diurnal variations can even be more easily explained since we have long known the influence of the temperature upon the electromotive action, and M. Dessaignes has made very interesting observations as regards this influence of the temperature. It should also be considered between the electromotive actions of the different parts of the Earth, the action of the magnetized minerals which the Earth contains and that should, as we have just seen, be considered as the same number of voltaic batteries. The increase of temperature inside the conductors of electric currents should also happen in the conductors of the terrestrial globe. Would this not be the cause of this internal heat recently verified by the reported experiments presented in one of the last sessions of the *Académie* by one of its members, whose works on heat have made this part of physics to enter in the domains of mathematics? And when one realizes that this increase of temperature produces, when the current is very energetic, a permanent incandescence, accompanied by the most intense light, without combustion or [page 207] loss of substance, should we not conclude that the opaques globes have this property only due to the small energy of the electric currents established in them, and to discover in the more active currents the cause of the heat and light of the globes shinning by themselves?

We know that the magnetic phenomena have been explained in the past by currents, but these currents have been supposed parallel to the axis of the magnet, and in this situation they could not exist without crossing and destroying one another.[28]

Now, if electric currents are the cause of the directive action of the Earth, then electric currents could also cause the action of one magnet on another magnet. It therefore follows that a magnet could be regarded as an assembly of electric currents in planes perpendicular to its axis, their direction being such that the austral pole of the magnet, pointing North, is to the right of these currents[29] since it is always to the left of a current placed outside the magnet, and which faces it in a parallel direction, or rather that these currents establish themselves first in the magnet along the shortest closed curves, whether from left to right, or from right to left, and the line perpendicular to the planes of these currents then becomes the axis of the magnet and its extremities make the two poles. Thus, at each pole the electric currents of which the magnet is composed are directed along closed concentric curves. I simulated this arrangement as much as possible by bending a conducting wire in a spiral. This spiral was made from brass wire terminating in two straight portions enclosed in two glass tubes[30] [page 208] so as to eliminate contact and attach them to the two extremities of the battery.

Depending on the direction of the current, such a spiral is greatly attracted or repelled by the pole of a magnet which is presented with its axis perpendicular to the plane of the spiral, according as the current of the spiral and of the pole of the magnet flow in the same or opposite directions. In replacing the magnet by

[28][N. T.] Ampère is referring to Descartes's theory of magnetism.

[29][N. T.] That is, the austral or North pole of the magnet, pointing approximately towards the North geographic pole of the Earth, is to the right of Ampère's imaginary observer. See also Tricker's book: [Tricker, 1965, p. 28].

[30][N. A. 4] I have since changed this arrangement as I shall show later.

another spiral with its current in the same direction, the same attractions and repulsions occur.[31] It is in this way that I discovered that two electric currents attract each other when they flow in the same direction and repel each other in the other case.

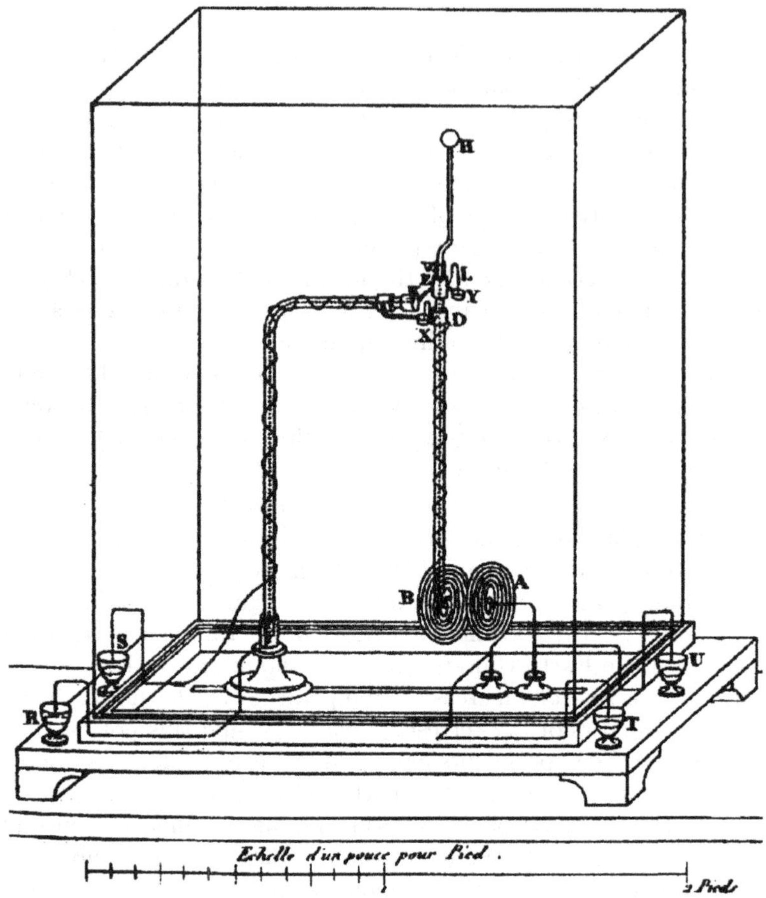

Figure 11.

Replacing the spirally wound metal wire by another magnet in the experiment on the interaction between the pole of a magnet and the current in a spiral, the effects are still the same, whether in attraction or repulsion, in conformity with the law of the known phenomena of a magnet. It is also evident that all circumstances associated with these phenomena are a necessary corollary of the arrangement of the component electric currents so that they attract or repel each other.

I built another apparatus in which the conducting wire is bent into a helix around a glass tube. According to the theory which I made of these kinds of phenomena, this conductor should present, when the electric current flowed through it, [page 209] an action similar to that of a magnetized needle or bar, in all circumstances in which these bodies act upon other bodies, or are placed in motion by terrestrial magnetism.[32] I already observed part of the expected effects coming from the utilization of a conductor bent into a helix, and I have no doubt that the more one varies the experiments based upon the analogy established by the theory between this instrument and a magnetized bar, more proofs will be obtained that the existence of electric currents in magnets is the only cause of all magnetic phenomena.

[31][N. T.] A representation of this instrument with two spirals can be seen in figure 11. This was an extremely important experiment in the history of physics. It showed for the first time attractions and repulsions between current-carrying wires. This created a new area of physics which Ampère later on called electrodynamics, [Ampère, 1822d, p. 60], [Ampère, 1822j, p. 200], [Ampère, 1885e, p. 239], [Ampère, 1822e, p. 237] and [Ampère, 1885d, p. 192]. Recently we reproduced this fundamental experiment with simple and cheap materials, [Assis et al., 2007] and [Souza Filho et al., 2007].

[32][N. A. 5] When I wrote this, I still did not know well the action exerted by the helix parallel to its axis, and I believed that we could neglect this action, but this cannot be done. But everything which I say here will be true if one understands it as being said of a helix in which this action has been destroyed by an opposite rectilinear current, established inside the glass tube which is surrounded by the coils, so that it remains only the action exerted by each coil in a plane perpendicular to the axis of the helix, as I explained in the first paragraph of this dissertation.

I completed the reading which I made at the *Académie* of what I have just transcribed in the session of 25 September [1820]. I finished this lecture by a summary in which I deduced, from the facts then presented, the following conclusions:

1. Two electric currents attract one another when they move parallel in the same sense; they repel one another when they move parallel in opposite senses. [page 210]

2. It follows from this that when the metal wires in which these currents flow can only turn in parallel planes, each one of the two currents tends to place the other in a situation in which it is parallel to the first one and directed in the same sense.

3. These attractions and repulsions are absolutely different from the ordinary electric attractions and repulsions.

4. All phenomena presented by the mutual action between an electric current and a magnet, discovered by M. OErsted, which I analyzed and reduced to two general facts in the previous dissertation, read to the *Académie* on 18 September 1820, follow from the law of attraction and repulsion of two electric currents, as this law has just been presented, by supposing that a magnet is only an assembly of electric currents which are produced by an action of the steel particles upon one another, analogous to that [action] of the elements of a voltaic battery, and which takes place in planes perpendicular to the line which connects the two poles of the magnet.

5. When the magnet is in the position which it tends to assume by the action of the terrestrial globe, these currents [of the magnet] are directed in the sense opposite to the direction of the apparent motion of the Sun. When the magnet is placed in the contrary position, in such a way that those of its poles which pointed towards the poles of the Earth are the same species as these poles of the Earth, the same currents flow in the sense of the apparent motion of the Sun.

6. The known phenomena which are observed when two magnets act upon one another are included in the same law. [page 211]

7. The same thing happens as regards the action exerted by the terrestrial globe upon a magnet, by supposing in it electric currents in planes perpendicular to the direction of the dip needle, and which move from East to West, below this direction.

8. There is nothing special in one of the poles of a magnet in comparison with the other pole. The only difference between them is that one of them is at the left and the other at the right of the electric currents which give the magnetic properties to steel.

9. When Volta proved that the two electricities, positive and negative, of the two extremities of the battery attracted and repelled one another according to the same laws followed by the two electricities produced by the procedures known before him, he still had not completely demonstrated by this means the identity of the fluids placed in action by the battery and by friction. But this identity was completely demonstrated, as far as a physical truth can be demonstrated, when he showed that two bodies, of which one had been electrified by the contact of metals and the other by friction, acted upon one another, in every circumstance, as if both of them had been electrified with the battery or with the ordinary electric machine. The same kind of proof is found here as regards the identity of the attractions and repulsions of electric currents and magnets. I have just shown to the *Académie* the mutual action between two currents. The known ancient phenomena related to the mutual action between two magnets is included in the same law. Beginning with this similarity, we would only prove that the electric and magnetic fluids are subjected to the same law, as has been known for a long time, and the only change [page 212] to be made in the ordinary theory of magnetization would be to admit that the magnetic attractions and repulsions should not be assimilated to those which result from the electric tension, but to those attractions and repulsions which I observed between two currents. The experiments of M. OErsted, in which an electric current produces further the same effects upon a magnet, prove moreover that the fluids which act in both cases are the same.

In the session of 9 October [1820], I insisted once again upon this identity of the electricity and the cause of the magnetic phenomena, by showing that the magnet only has the properties which characterize it because, in planes perpendicular to the line joining the poles, the same disposition of electricity which exists in the conductor connecting the two extremities of a voltaic battery is found. I designated this disposition by the name *electric current*, always insisting, in the dissertations which I read to the *Académie*, in this aspect, namely, that the identity between the magnetic parallels and the conductors of a battery of Volta, which I especially wished to establish, was independent of the conception, whatever it was, which we had of this electric disposition.

In order to demonstrate this identity by direct experiments, which I repeated at the *Académie des Sciences*, in the session of 9 October [1820], I asked to be constructed the apparatus of which I spoke in the first paragraph of this dissertation, and which is represented here in figure 1.[33] It is placed over a base mn to which the framework which supports the glass case intended to protect the whole apparatus from the small agitations of air is attached. Outside this case, I placed four boxwood goblets R, S, T and U, [page 213] to be filled with mercury in which the brass wires passing through the framework over which it rests plunged, and are soldered to the four supports M, N, P and Q, of which the first two carry the fixed conductor AB, which can be moved towards the other [conductor] or away from it, by slipping these supports in the slots I and J, where they are fixed as desired by means of screws placed below the base, and the other two [supports] P and Q are terminated by the steel goblets [*chapes*] X and Y, large enough to retain the globules of mercury placed there, and where two steel tips attached to the copper joints E and F are plunged, inside of which enter the two extremities of a glass tube OZ having in its middle portion another copper joint to which a copper tube V is soldered inside of which the stem of a counter-weight H is fixed by friction. This stem is bent, as seen in the figure, in order to change the position of the center of gravity of the whole mobile portion of the apparatus, by turning the bent stem upon itself inside the copper tube. One can approach or drive away these supports to or from one another by slipping them in the slot KL, where they are fixed at the desired distance, by means of screws placed below the base. At the two copper joints E and F, the two extremities of the brass wire $ECDF$ are soldered, of which the portion CD, parallel to AB, is what I named the *mobile conductor*.

When we wish to utilize this instrument, after fixing the two supports P and Q, at such a distance that the center of goblets X and Y correspond to the steel tips carried by the joints E and F, and [fixing] the supports M and N, at the most convenient distance from the two first supports; these steel tips are placed inside the [page 214] goblets, and the stem of the counter-weight H is turned, inside the cylinder V, until the mobile conductor rests by itself in the desired position, the arms EC and FD which belong to it being approximately vertical. Then, if we wish to show the attraction of two currents when they flow in the same sense, one establishes, by a brass wire passing below the instrument, the extremities of which are curved in order to plunge inside the boxwood goblets, like R and U, or S and T, the communication between the opposite extremities of the two conductors AB and CD, and one communicates the two remaining goblets with the extremities of the battery, by means of two other brass wires. When we wish to observe the repulsion, it is necessary that the first brass wire makes the communication between two goblets like R and S, or T and U, corresponding to the extremities of two conductors situated in the same side, whereas the extremities of the battery are placed in communication with the two goblets placed in the opposite side.

These goblets offer, when desired, the means of establishing the electric current in a single conductor, by plunging the two wires departing from the extremities of the battery inside the mercury of two goblets which communicate with this conductor. This disposition of four boxwood goblets arranged in this way, is found in several instruments already described or which I still need to describe, therefore I will explain it here only once in order to avoid unnecessary repetitions, and I will content myself in representing them in the figures of these instruments, without speaking about them in the text.

[page 215] To make the identity of the currents of connective wires and those which I assume in magnets evident, I obtained two small strongly magnetized needles, fixed in the center of a double brass hook, carrying an arrow which indicated the direction of the current of the magnet. I represented beside figure 1 one of these needles in front-view and the other sideways. The needle was represented by ab, the double hook by cd, and the arrow by ef. By means of the double hook, these needles adapt themselves, when we wished to place them there, in the conductors AB and CD, in a situation such that the line joining their poles is vertical, and where the currents, always parallel to the conductors, are directed as desired in the same sense or in opposite senses. These needles can be utilized as follows: after producing the attractions and repulsions between the conductors AB and CD, by passing the electric current through both of them, one passes the electric current through only one of them, and one places over the other conductor one of the magnetized needles in the situation just indicated, in such a way that the current which I suppose to exist in the needle flows at first in the same sense as that current which existed before in the conductor to which it is adapted. One then observes that the phenomenon of attraction or repulsion, which previously the two conductors

[33] [N. T.] This is the figure 1 of the first part of this paper; see our page 294.

indicated, continues to exist due to what I called the *attractive or repulsive action* in the beginning of this paragraph. Then we place the same needle in such a way that its current is directed in the opposite sense, and one obtains the inverse phenomenon, due to the same action, precisely as if we had changed the direction of the current which this needle replaces, [page 216] by communicating, in an opposite order to that order which had been established at first, the two extremities of the battery with the extremities of the conductor of this current.

Finally, when there is no current in both conductors, and by placing in each one a magnetized needle always in the same vertical situation in such a way that its axis makes a right angle with the conductor carrying it, so that its currents should still be parallel to this conductor, we have once more, according to the known action of two magnets upon one another, the same attractions and repulsions which did happen when the currents were established in the two conductors, when both of the currents in the needles flow in the same sense, or both of them in opposite sense, relative to the electric currents that they replace, and inverse phenomena [take place] when one [of the currents] is in the same sense and the other in the opposite sense. Everything agrees with the theory based upon the identity between the currents of the magnet and those which are produced with Volta's battery.

One can also verify this identity with the instrument represented in figure 2. By replacing the fixed conductor AB by a magnetized bar situated horizontally in a direction perpendicular to that of this conductor, and in such a way that the currents of this magnet flow in the same sense as the electric current established at first in the fixed conductor, one then passes the current only through the mobile conductor, and one observes that this one turns by the action of the magnet precisely as it would turn in the experiment in which the current had been established in the two conductors, and in which there was no magnetized bar. It was with the goal to attach this bar, that [page 217] I connected to this apparatus the support XY, terminated in Y by the box Z open in both ends where one fixes the magnet in the position which I have just explained by means of the pressure screw V.

As regards the apparatus represented in figure 9,[34] one can see by this figure that the means of communication with the extremities of the battery, and the mode of suspension of the mobile conductor, are more or less the same as in the instrument which is represented in figure 1, these two instruments differ only in this aspect, namely, in the apparatus of figure 9, the two conductors A and B are folded in spiral, and the mobile conductor B is suspended to a vertical glass tube CD. This tube is terminated inferiorly at the center of the spiral forming this conductor, and receives in its interior the extension of the brass wire of this spiral. This extension arrives in D, at the top of the tube, being soldered there to the copper joint E, which carries the copper tube V inside of which the counter-weight H enters by friction, and a steel tip L which one plunges into the globule of mercury in goblet [*chape*] Y, whereas the other extremity of the same brass wire, before surrounding the tube CD like a helix, is soldered to the copper joint D, to which is attached the other steel tip K meant to be also plunged into the globule of mercury placed inside the goblet X. These two goblets are made of steel, to prevent their damage by the mercury. The tips rest over their concave surface as in the instrument represented in figure 1.

Here would be the place to speak of another kind of action of electric currents upon steel, that action by which they communicate to steel the magnetic properties, and to [page 218] show that all circumstances of this action, which we owe to M. Arago, represent as many proofs of the theory exposed in this dissertation relative to the electric nature of the magnet. It seems to me that these evidences complete this theory. In order not to omit anything which is known about the mutual action between connective wires and magnets, I could also speak of the most interesting experiments communicated to the *Académie* in a dissertation that a physicist full of sagacity, M. Boisgiraud, read in the session of 9 October 1820. One of these experiments does not leave any doubt as regards an important point of the theory of the mutual action between a connective wire and a magnet, by proving that this action takes place between the connective wire and all the slices perpendicular to the line connecting the two poles of the small magnet over which it acts, and does not develop with greater energy upon the poles of this magnet, as it happens when one observes the action which the several points of the length of a magnetized bar exert upon a small needle. But the discoveries of M. Arago were exposed by himself in this journal, and I hope that M. Boisgiraud will soon publish his experiments, and that I will have the opportunity, in a second dissertation[35] in which I will occupy myself with the mathematical theory of the phenomena presented by the electric currents, to deduce from it, as

[34] [N. T.] Ampère probably was referring here to figure 11 and not to figure 9.

[35] [N. A. 6] What I have to say about the mutual action between two magnets is less composed of new facts, and more composed of calculations by means of which we reduce this action to the action between two electric currents. Therefore, it seems to me that I should include in this second dissertation the paragraph in which I proposed myself to examine the laws according to which this action is exerted, and to show that these laws are a necessary consequence of the cause which I assigned it in the conclusions which I read to the *Académie* on 25 September [1820].

evidences of the exactness of this theory, the consequences which follow naturally from the facts which he observed.

END OF THE SECOND PART OF AMPÈRE'S FIRST PAPER ON ELECTRODYNAMICS.

Part VII

Ampère's Main Book on Electrodynamics

Chapter 26

Introduction to Ampère's *Théorie*

Ampère's masterpiece was published in November 1826: *Théorie des phénomènes électro-dynamiques, uniquement déduite de l'expérience.*[1]

In the first paragraph Ampère mentioned that the *Théorie* was based on a collection of papers delivered at the Academy of Sciences of Paris at the meetings of 4 and 26 December 1820, 10 June 1822, 22 December 1823, 12 September 1825 and 21 November 1825.[2]

26.1 The Cases of Equilibrium Discussed in the *Théorie*

The *Théorie* is developed deductively. Instead of presenting his several experiments and the varied theoretical procedures he had originally utilized in order to arrive at his force between current elements, he decided to present a mature and refined work. He utilized only four cases of equilibrium in order to obtain his main results. He included long and almost literal copies of sections of his main works published between 1824 and 1825.[3]

In the *Théorie*, Ampère obtained the final expression for his force between current elements by deducing it from only four cases of equilibrium.[4] The first one was the case of equilibrium of anti-parallel currents, figure 4.19, discussed in Subsection 4.3.4. The second one was the case of equilibrium of the sinuous wire, figure 4.17, discussed in Subsection 4.3.3. This second case of equilibrium allowed him to describe the force between two current elements as a sum of the forces acting between the orthogonal components of these elements projected along three orthogonal axes. He then utilized the first case of equilibrium and the principle of symmetry, described in Subsection 4.2.2, in order to eliminate the forces between the components of these current elements which were orthogonal to one another. With these two cases of equilibrium he arrived at equation (2.1), without specifying the numerical values of the constants n and k. The third case of equilibrium discussed in the *Théorie* was that of the nonexistence of tangential force, figure 10.14, discussed in Section 10.4. This case of equilibrium allowed him to arrive at equation (7.34), relating the constant n with the constant k. The fourth and last case of equilibrium discussed in the beginning of the *Théorie* was that of the law of similarity, figure 10.19, discussed in Section 10.5. It gave him the value $n = 2$. He then obtained, utilizing equation (7.34), the value $k = -1/2$ and his final formula given by equation (7.36).

Two extremely important experiments performed previously by Ampère do not appear in any figure of the *Théorie*, namely, Ampère's balance and the case of equilibrium of the nonexistence of continuous rotation. The first experiment on the attraction and repulsion between parallel straight wires carrying steady currents was discussed in Subsection 3.6.2, appearing in our figure 3.10. This is Ampère's most famous experiment. It is the basis of the construction of the current balance (or Ampère balance) appearing in most university physics courses for the precise measurement of electric current intensities. This experiment appeared in

[1] [Ampère, 1826f].

[2] Complete references of these works: [Ampère, 1820d], [Ampère, 1820e], [Ampère, 1885j], [Ampère, 1885m], [Ampère, 1820b], [Ampère, 1821b], [Ampère, 1822o], [Ampère, 1885p], [Ampère, 1824c], [Ampère, 1824f], [Ampère, 1824e], [Ampère, 1885f], [Ampère, 1825d], [Ampère, 1825c], [Ampère, 1825e], [Ampère, 1825g], [Ampère, 1826e], [Ampère, 1826b], [Ampère, 1887c] and [Ampère, 1887f].

[3] In particular: [Ampère, 1824c], [Ampère, 1824f], [Ampère, 1824e], [Ampère, 1885f], [Ampère, 1825d], [Ampère, 1825c], [Ampère, 1825g], [Ampère, 1826e], [Ampère, 1826b], [Ampère, 1887c] and [Ampère, 1887f].

[4] [Ampère, 1826f], [Ampère, 1823c] and [Ampère, 1990].

the only figure of the first part of Ampère's first paper on electrodynamics.[5] The case of equilibrium of the nonexistence of continuous rotation, discussed in Subsection 7.3.2 and presented in figure 7.16, was the crucial experiment with which Ampère obtained for the first time the correct numerical value of his constant k, namely, $k = -1/2$.

[5] [Ampère, 1820c] and [Chaib and Assis, 2007d].

Chapter 27

Comparison of the *Théorie* Published in 1826 with the *Théorie* Published in 1827

27.1 Similarities and Differences

We now compare two versions of Ampère's masterpiece, the book *Théorie*. One of them was published as a monograph in November of 1826.[1] The other version was published in volume 6 of the *Mémoires de l'Academie Royale des Sciences* corresponding to the year 1823.[2] Despite this date, this version of Ampère's work was only published in 1827. Both works mention that a specific final section was written on 30 August 1826.[3]

This work of 1827 was reissued in 1883, 1887, 1958 and 1990.[4] The page numbering of the works published in 1827 and 1990 coincide with one another. The same does not happen with the editions of 1883, 1887 and 1958.

The title of the 1826 version reads as follows: *Théorie des phénomènes électro-dynamiques, uniquement déduite de l'expérience.* The title of the 1827 version is slightly different, including the word *mathématique*, namely: *Mémoire sur la théorie mathématique des phénomènes électro-dynamiques uniquement déduite de l'expérience.* As discussed below, we believe that the 1826 version was written later than the 1827 edition, despite the publication dates. Probably Ampère realized that in physics we normally understand the theory of a set of phenomena as a mathematical theory. As the word "mathematical" is normally implicit in the meaning of a physical theory, he decided to drop this word from the title as being unnecessary.

One difference between these two versions is that only the work of 1826 included a table of contents dividing the text into Sections, Topics, or Subjects (see Section 29.1 up to Section 29.26 of Chapter 29). This table of contents did not appear in the 1827 version. In the English translation presented here we include this table of contents. It facilitates the orientation of the reader.

The text of these two versions is exactly the same, except for the different pagination and by the termination of a single paragraph (see footnote 8 in our page 343).

A significant difference between these two versions appears in the Notes or Appendices included at the end of both works. There are five Notes in the 1826 edition and four Notes in the 1827 edition.[5] As these Notes present some differences, we decided to translate both of them, as presented in Chapters 30 and 31. The Notes published in 1826 were reprinted in 1887, with a different pagination.[6] The Notes published in 1827 were reprinted in 1858 and 1883, with a different pagination from that appearing in the 1827 edition.[7] The 1827 Notes were also reprinted in 1990, with the same pagination as the 1827 edition.[8]

[1] [Ampère, 1826f].

[2] [Ampère, 1823c].

[3] See our page 458, corresponding to pages 196 and 368 of the 1826 and 1827 versions, respectively, [Ampère, 1826f, p. 196] and [Ampère, 1823c, p. 368].

[4] [Ampère, 1883], [Ampère, 1887d], [Ampère, 1958] and [Ampère, 1990].

[5] [Ampère, 1826f, pp. 202-222] and [Ampère, 1823c, pp. 374-387].

[6] [Ampère, 1887d, pp. 174-190].

[7] [Ampère, 1958, pp. 152-164] and [Ampère, 1883, pp. 152-164].

[8] [Ampère, 1990, pp. 374-387].

We point out here the main similarities and differences between these Notes.

The first Note of 1826 is almost identical to the first Note of 1827, except for the final portions. The 1826 Note finishes with a paragraph in which Ampère described some limitations of the experiment of the case of equilibrium of the nonexistence of tangential forces discussed in our Section 10.4; see our pages 463 and 463. Nowadays these limitations are no longer relevant due to the fact that the experimental configuration of this apparatus was greatly improved by Albert von Ettingshausen by supporting the bar QO connected to the arch AA′ of figure 10.14, appearing in our page 183, by a bifilar suspension.[9] This final paragraph does not appear in the first Note of 1827. The text appearing at the end of the first Note of the 1827 edition appears in our page 479.

In any event, it is relevant to know if Ampère's final words were those appearing in the 1826 Note, recognizing the problems with this crucial experiment, or those of the 1827 Note, in which he did not mention these problems. It seems to us that his final words were those presented in 1826. In Section 27.2 we present detailed arguments supporting our point of view.

There is a difference between the last paragraphs of the second Notes of the 1826 and 1827 versions of the *Théorie*.[10] In the 1827 edition, the final portion of the second Note presents some information. This information is enlarged and developed in much more detail in a third Note appearing in the 1826 edition, as can be seen in our Section 30.3. This third Note of the 1826 version does not appear in the 1827 version. Therefore, the 1827 edition contains only four Notes, while the 1826 edition has five Notes.

The fourth Note of the *Théorie* published in 1826, Section 30.4, is identical to the third Note of the *Théorie* published in 1827, Section 31.3.

The fifth and final Note published in 1826 is almost identical to the fourth and final Note published in 1827. They differ only in the final portions. The information appearing in the 1826 version is much more detailed than the information appearing in the 1827 version. In our book the different relevant final portions appear between pages 474 and 476 for the 1826 version and between pages 485 and 487 for the 1827 version.

27.2 Ampère's Final Words

The differences between the Notes of the *Théorie* published in 1826 and 1827 raise a relevant question. Which version corresponds to Ampère's final words?

This question has not yet been solved by the specialists who studied Ampère's works. Grattan-Guinness did not reach a conclusion on which version was first written.[11] Hofmann mentioned that the initial version was written for the Memoirs of the French Academy of Science, being published in 1827, while the final version was published in November of 1826.[12] Blondel, on the other hand, believed that the initial version was the one published in 1826, while the final one was published in 1827.[13] Maybe Blondel reached this conclusion judging from the publication dates of both versions.

After considering in detail several aspects, we concluded that Ampère's final words are contained in the *Théorie* published in 1826. That is, the initial version was delayed in publication, appearing only in 1827,[14] while the final version was published earlier, in 1826.[15] We list here some arguments which led us to this conclusion.

- Ampère used to publish the same paper in a journal, as a separate brochure, and in collections of his works. He paid for the publication of these brochures out of his own pocket, printing a great number of copies that he distributed to his colleagues. Many times he included extra material in the separate brochure or in the collection of his works, although the title remained the same as in the original publication. Sometimes the texts with additional material were published faster than the original journal or periodical. This procedure created problems for the scientists studying his works. Blondel listed several examples of this procedure in her book and on the site on Ampère and the history of electricity.[16] We quoted several examples in this book, for instance: [Ampère, 1820a] and [Ampère, a]; [Ampère, 1821d] and [Ampère, 1822x]; [Ampère, 1822d] and [Ampère, 1822e]; [Ampère, 1822o] and [Ampère, 1822y];

[9][Ampère, 1887d, Note 2 by Joubert on pp. 17-18], [Blondel, 1982, Note 27, p. 148], [Ettingshausen, 1878a], [Ettingshausen, 1878b] and [Ettingshausen, 1879].

[10]See our pages 465 and 480.

[11][Grattan-Guinness, 1991, p. 126].

[12][Hofmann, 1996, p. 336].

[13][Blondel, 1982, pp. 67-68 Note 30, pp. 147-148 and p. 191 reference 57].

[14][Ampère, 1823c].

[15][Ampère, 1826f].

[16][Blondel, 1982, pp. 187-192] and [Blondel, 2005, Publications d'Ampère].

[Ampère, 1822r], [Ampère, 1822i] and [Ampère, 1822j]; [Ampère, 1822k] and [Ampère, 1822b]; [Ampère, 1822f], [Ampère, 1822g], [Ampère, 1823a] and [Ampère, 1822h]; [Ampère, 1825d] and [Ampère, 1825e].

- It was common in Ampère's time, remaining so even today, that the scientific journals and periodicals were published with delay. Even volume VI of the *Mémoires de l'Académie des Sciences de l'Institut de France* in which Ampère published his masterpiece, which corresponds to the year 1823, was only published in 1827. Moreover, although this volume is relative to the year 1823, Ampère's text refers to papers published between 1820 and 1825. At the end of the text he mentioned the date of writing as August 30, 1826; see our page 458.[17] Maybe Ampère sent this work for publication between August and November 1826, although it was only published in 1827.

- A systematic comparison between the figures of the 1826 and 1827 versions of the *Théorie* shows that most differences are improvements incorporated only in the publication which took place in 1826. These changes are detailed in the following footnotes of our translation: 105, 108, 109, 126, 167, 171 and 209 (see our pages 368, 370, 371, 377, 399, 401 and 415, respectively). We illustrate one of these improvements in figure 27.1, showing at the left side the image of the 1827 version and at the right side the image of the 1826 edition, in which the arrows were included to indicate the direction of the current.

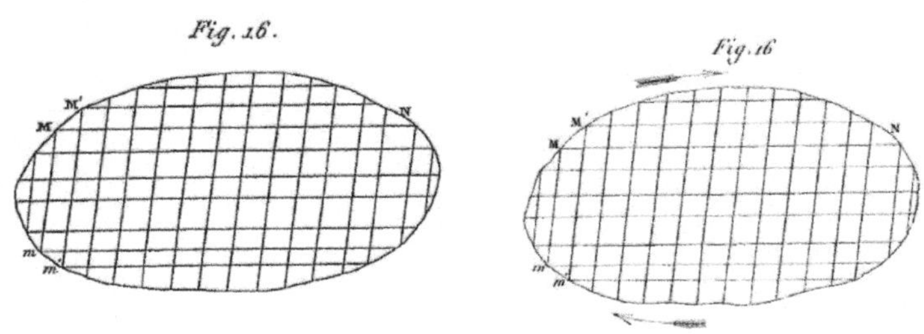

Figure 27.1: Figure 16 of the *Théorie* published in 1827 (left side) and 1826 (right side).

- The table of contents dividing the *Théorie* into sections, topics, or subjects, as shown in Sections 29.1 up to 29.26 of Chapter 29, appeared only in the 1826 version. Ampère probably included this table of contents to facilitate the orientation of the readers, after perceiving the size and complexity of his final work. It seems to us that he included this table of contents in the published version of 1826, after the work had been completed. The first version, although published only in 1827, appeared without this table of contents. This hypothesis seems more natural to us than to assume that he published the work with this table of contents in 1826, deciding to remove it in the later edition published in 1827.

- The first item in the errata of the *Théorie* published in 1827 was incorporated in the text of the version published in 1826, as can be seen on footnote 2, in our page 342.

- There are some identical typographical errors in the texts of the *Théorie* published in 1826 and 1827 which were corrected only in the errata of the 1826 edition. Probably Ampère only realized the existence of these errors after sending the first version for publication in the Memoirs of the Academy of Sciences of Paris, published in 1827. These errors are indicated in the footnotes 83, 160 and 175 of our pages 364, 393 and 404, respectively. This fact supports once more the hypothesis that the 1826 version came later than the 1827 edition.

- There is an identical typographical error in both versions of the *Théorie* relative to the closing parenthesis of an equation. This mistake was corrected in the 1827 errata. However, when correcting this mistake, the errata introduced another mistake in one of the fractions appearing in this equation. The 1826 errata corrects not only the closing parenthesis, but also then new mistake introduced in the 1827 errata, as can be seen in the footnote 144 of our page 386). Therefore the 1826 errata came later than the 1827 errata.

- There is another fact which seems to us a definitive proof in favour of the hypothesis that Ampère's final words are those of the 1826 version of the *Théorie*. Between his manuscripts,[18] we found a

[17] [Ampère, 1823c, p. 368].
[18] [Ampère, b, *carton* 9, *chemise* 172].

printers proof of the 1826 version containing the text from the last printed page of the 1827 version of the *Théorie* with handwritten corrections which were incorporated in the *Théorie* published in 1826. Figure 27.2 shows the last page of the *Théorie* published in 1827,[19] figure 27.3 shows the manuscript containing the corrections to this page,[20] while figure 27.4 shows the last page of the *Théorie* published in 1826 in which these handwritten corrections were incorporated by Ampère.[21]

We then conclude that the initial version of the *Théorie* was delayed in publication, appearing only in 1827.[22] The final version was published earlier, in 1826. Therefore, Ampère's final words on electrodynamics are contained in the *Théorie* published in 1826.[23]

[19] [Ampère, 1823c, Ampère, 1990, p. 387].
[20] [Ampère, b, *carton* 9, *chemise* 172].
[21] [Ampère, 1826f, p. 222].
[22] [Ampère, 1823c].
[23] [Ampère, 1826f].

ÉLECTRO-DYNAMIQUES. 387

perpendiculaires DI, DK, EU, EV, il est évident qu'on obtient ainsi pour la valeur de la force cherchée

$$\rho\left(\frac{1}{p_{1,2}}+\frac{1}{p_{2,1}}-\frac{1}{p_{1,1}}-\frac{1}{p_{2,2}}\right).$$

La direction perpendiculaire au plan du parallélogramme NRST suivant laquelle le pôle d'un aimant situé en B est porté par l'action du courant électrique qui parcourt le contour de ce parallélogramme, est la directrice de l'action électro-dynamique qu'il exerce au point B: d'où il suit que s'il y avait à ce point un élément de courant électrique situé dans le plan du parallélogramme, il formerait un angle droit avec la directrice, et qu'ainsi l'action de ce courant sur l'élément serait une force située dans ce plan, perpendiculaire à la direction de l'élément, et égale à celle que le même courant exercerait sur le pôle d'un aimant placé au point B multipliée par un rapport constant, qui est ici celui de ρ à $\frac{1}{2}ii'ds$, en nommant cet élément ds; en sorte que la force ainsi dirigée qui agirait sur l'élément aurait pour valeur

$$\frac{1}{2}ii'ds\left(\frac{1}{p_{1,2}}+\frac{1}{p_{2,1}}-\frac{1}{p_{1,1}}-\frac{1}{p_{2,2}}\right).$$

Lorsque l'élément situé en B n'est pas dans le plan du parallélogramme, mais forme avec ce plan un angle égal à ω, on peut le remplacer par deux éléments de même intensité, l'un dans ce plan, l'autre qui lui est perpendiculaire : l'action du courant du parallélogramme sur ce dernier étant nulle, on ne doit tenir compte que de celle qu'il exerce sur le premier; elle est évidemment dans le plan du parallélogramme, perpendiculaire à l'élément et égale à

$$\frac{1}{2}ii'ds\cos.\omega\left(\frac{1}{p_{1,2}}+\frac{1}{p_{2,1}}-\frac{1}{p_{1,1}}-\frac{1}{p_{2,2}}\right).$$

Figure 27.2: Last page of the *Théorie* published in 1827.

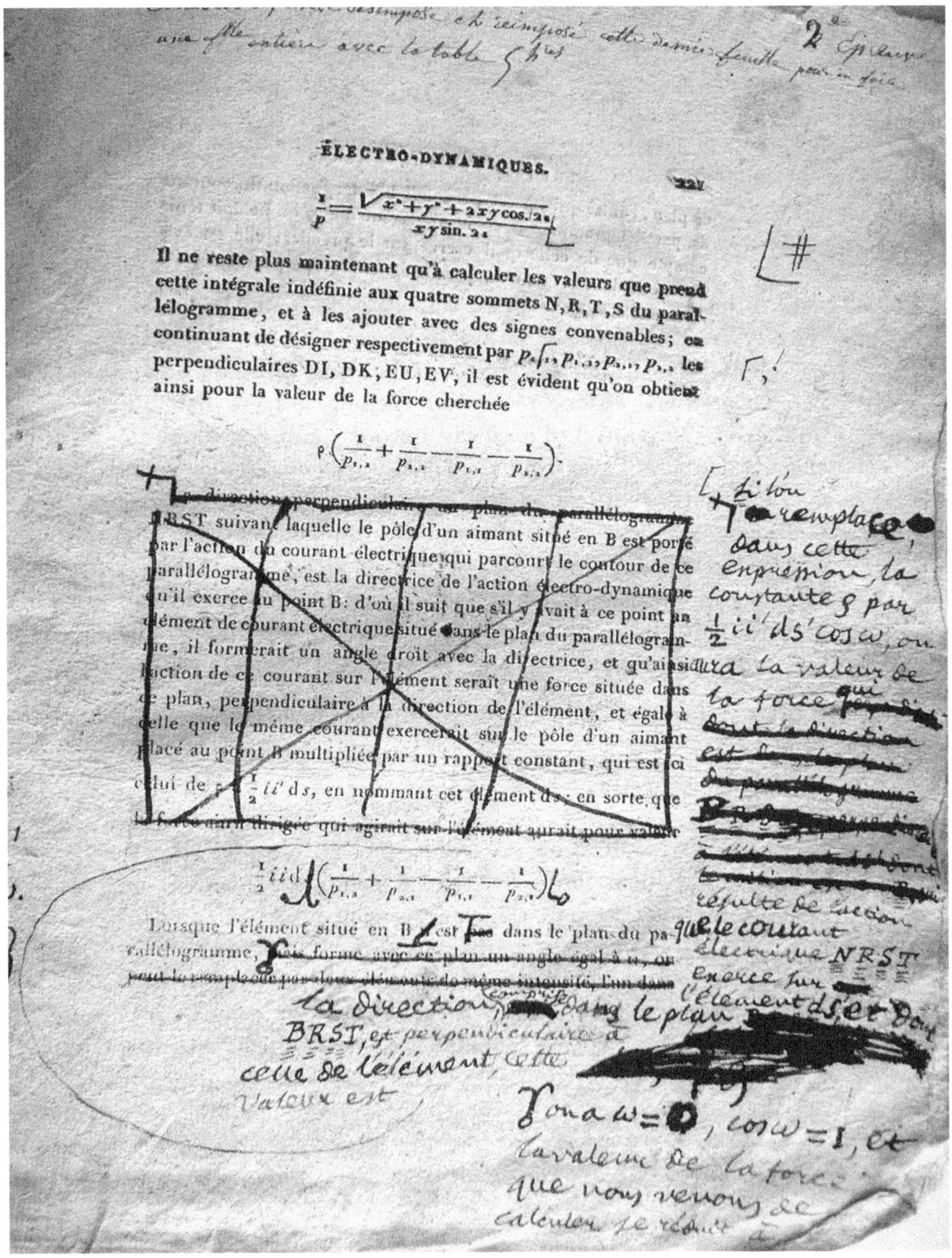

ÉLECTRO-DYNAMIQUES. 227

$$\frac{1}{p}=\frac{\sqrt{x^2+y^2+2xy\cos.2\varepsilon}}{xy\sin.2\varepsilon}$$

Il ne reste plus maintenant qu'à calculer les valeurs que prend cette intégrale indéfinie aux quatre sommets N, R, T, S du parallélogramme, et à les ajouter avec des signes convenables; en continuant de désigner respectivement par $p_{1,1}, p_{1,2}, p_{2,1}, p_{2,2}$, les perpendiculaires DI, DK; EU, EV, il est évident qu'on obtient ainsi pour la valeur de la force cherchée

$$\rho\left(\frac{1}{p_{1,1}}+\frac{1}{p_{2,2}}-\frac{1}{p_{1,2}}-\frac{1}{p_{2,1}}\right).$$

La direction perpendiculaire au plan du parallélogramme NRST suivant laquelle le pôle d'un aimant situé en B est porté par l'action du courant électrique qui parcourt le contour de ce parallélogramme, est la directrice de l'action électro-dynamique qu'il exerce au point B: d'où il suit que s'il y avait à ce point un élément de courant électrique situé dans le plan du parallélogramme, il formerait un angle droit avec la directrice, et qu'ainsi l'action de ce courant sur l'élément serait une force située dans ce plan, perpendiculaire à la direction de l'élément, et égale à celle que le même courant exercerait sur le pôle d'un aimant placé au point B multipliée par un rapport constant, qui est ici celui de $\frac{1}{2}ii'ds$, en nommant cet élément ds; en sorte que la force ainsi dirigée qui agirait sur l'élément aurait pour valeur

$$\frac{1}{2}ii'ds\left(\frac{1}{p_{1,1}}+\frac{1}{p_{2,2}}-\frac{1}{p_{1,2}}-\frac{1}{p_{2,1}}\right)$$

Lorsque l'élément situé en B est dans le plan du parallélogramme, mais forme avec ce plan un angle égal à ω, on peut le remplacer par deux éléments de même intensité, l'un dans

Si l'on remplace dans cette expression, la constante ρ par $\frac{1}{2}ii'ds'\cos\omega$, on aura la valeur de la force qui résulte de l'action que le courant électrique NRST exerce sur l'élément ds', et dont la direction comprise dans le plan BRST, et perpendiculaire à celle de l'élément, cette valeur est

Si on a ω=0, cos ω=1, et la valeur de la force que nous venons de calculer se réduit à

Figure 27.3: Handwritten corrections to the text from the last page of the *Théorie* published in 1827 shown in figure 27.2.

en nommant p la perpendiculaire PQ abaissée du point P sur BM, parce que le double de l'aire du triangle BPM est à la fois égal à $p\sqrt{x^2+y^2+2xy\cos.2\varepsilon}$ et à $xy\sin.2\varepsilon$, ce qui donne

$$\frac{1}{p}=\frac{\sqrt{x^2+y^2+2xy\cos.2\varepsilon}}{xy\sin.2\varepsilon}.$$

Il ne reste plus maintenant qu'à calculer les valeurs que prend cette intégrale indéfinie aux quatre sommets N, R, T, S du parallélogramme, et à les ajouter avec des signes convenables; en continuant de désigner respectivement par $p_{1,1}, p_{1,2}, p_{2,1}, p_{2,2}$ les perpendiculaires DI, DK, EU, EV, il est évident qu'on obtient ainsi pour la valeur de la force cherchée

$$\rho\left(\frac{1}{p_{1,2}}+\frac{1}{p_{2,1}}-\frac{1}{p_{1,1}}-\frac{1}{p_{2,2}}\right).$$

Si l'on remplace, dans cette expression, la constante ρ par $\frac{1}{2}ii'ds'$, cos. ω, on aura la valeur de la force qui résulte de l'action que le courant électrique NRST exerce sur l'élément ds', et dont la direction, comprise dans le plan BRST, est perpendiculaire à celle de l'élément; cette valeur est

$$\frac{1}{2}ii'ds'\left(\frac{1}{p_{1,2}}+\frac{1}{p_{2,1}}-\frac{1}{p_{1,1}}-\frac{1}{p_{2,2}}\right)\cos.\omega.$$

Lorsque l'élément situé en B est dans le plan du parallélogramme, on a $\omega=0$, $\cos.\omega=1$, et la valeur de la force que nous venons de calculer se réduit à

$$\frac{1}{2}ii'ds'\left(\frac{1}{p_{1,2}}+\frac{1}{p_{2,1}}-\frac{1}{p_{1,1}}-\frac{1}{p_{2,2}}\right).$$

Figure 27.4: Last page of the *Théorie* published in 1826 incorporating the corrections appearing in figure 27.3.

Chapter 28

Observations about the English Translation

To our knowledge the first complete translation of Ampère's *Théorie*[1] was made to the Portuguese language.[2] There are two English translations known to us, a partial one made by Blunn[3] and a complete one made by Godfrey.[4] The translation presented here is based on the works of Blunn and Godfrey. We are not aware of translations of the *Théorie* to any other languages.

We indicate all original page numbers between square brackets, [pages m and n], in order to facilitate the comparison with the original versions or the location of any specific passage. In this case page m corresponds to the *Théorie* published in 1826,[5] while page n corresponds to the version published in 1827, which has the same pagination as the 1990 edition of the *Théorie*.[6]

In both versions of the *Théorie* published in 1826 and 1827, the figures are not included separately throughout the text, being collected at the end of the work in two plates for each version. The two plates of the 1826 version, for instance, are shown in our figures 28.1 and 28.2, respectively.[7]

[1] [Ampère, 1826f] and [Ampère, 1823c].
[2] [Chaib, 2009] and [Assis and Chaib, 2011].
[3] [Ampère, 1965b].
[4] [Ampère, 2012].
[5] [Ampère, 1826f].
[6] [Ampère, 1823c] and [Ampère, 1990].
[7] [Ampère, 1826f].

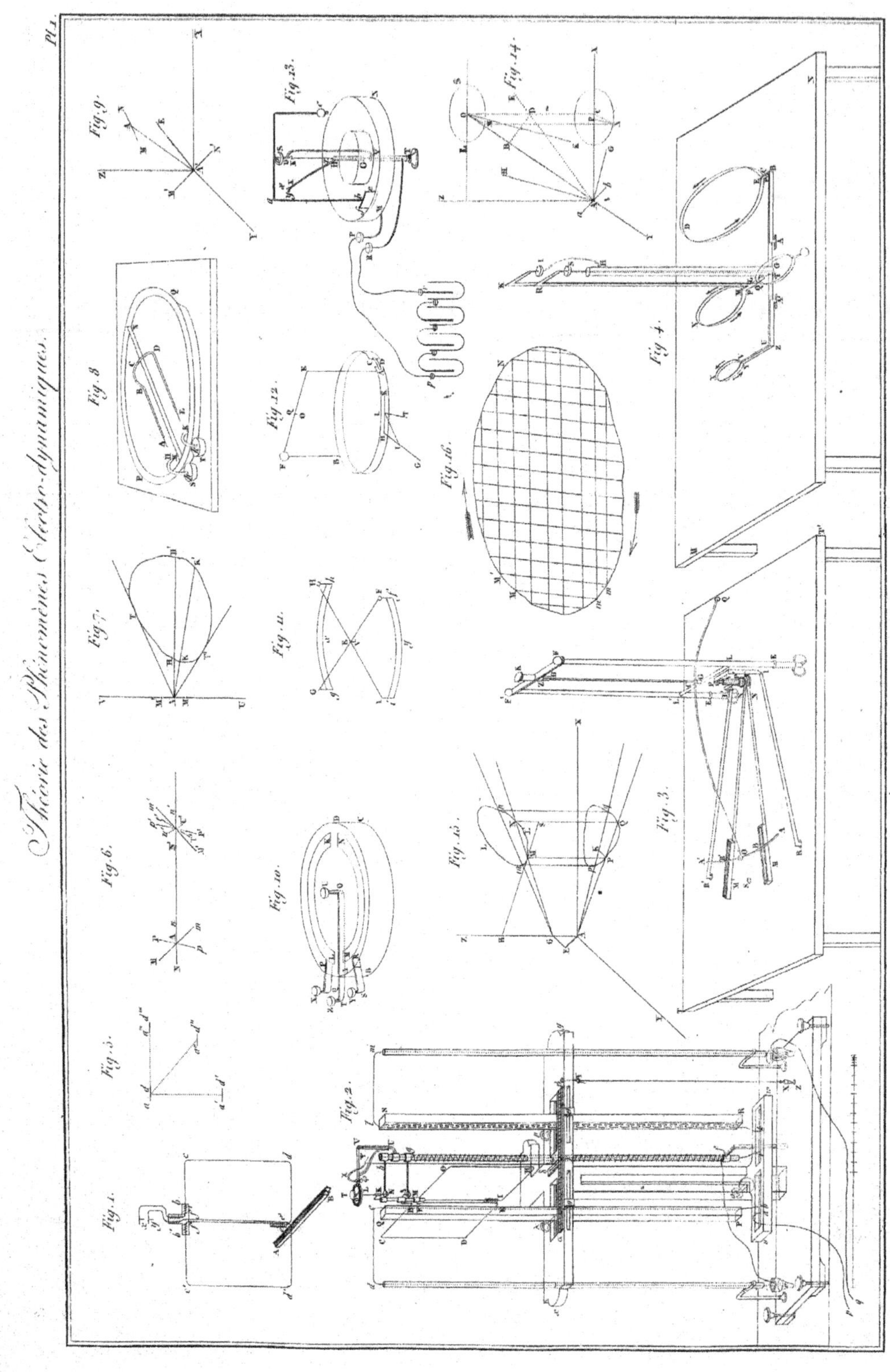

Figure 28.1: First plate of the 1826 version of the *Théorie* containing the first half of its figures.

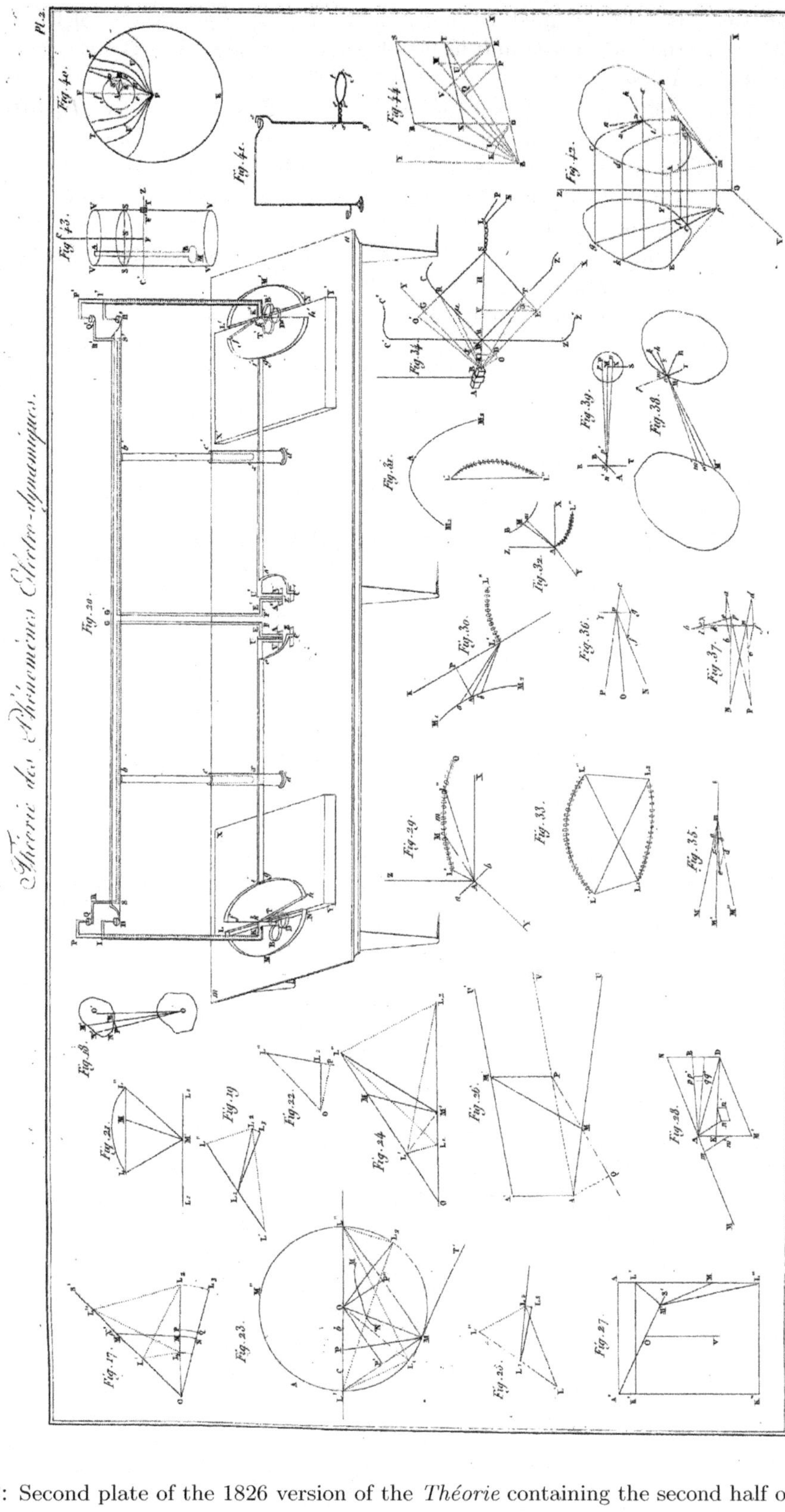

Figure 28.2: Second plate of the 1826 version of the *Théorie* containing the second half of its figures.

In our complete translation we scanned in high definition all original figures of the 1826 version of the *Théorie*. The images were then separated and inserted when first quoted in the text. In this way we hope to facilitate the reading of Ampère's text.

We made a systematic comparison between the figures of both versions of the *Théorie*, namely, those of 1826 and 1827. The differences between the analogous figures are pointed out in footnotes in the translation.

All 44 figures of the *Théorie* were also redrawn with graphic software, included in Appendix A. We could then prepare images with well-defined lines, including large and visible letters. These figures should be considered only as a complement to the original images.

All words between square brackets in the middle of the text, [ ], were included by the translators in order to facilitate the comprehension of some sentences.

The footnotes by the author, Ampère, are indicated by [N. A.], while those introduced by the translators are indicated by [N. T.].

We now present the complete and commented English translation of Ampère's masterpiece, the *Théorie*.

Part VIII

Complete and Commented English Translation of Ampère's Main Work on Electrodynamics

Chapter 29

Theory of Electrodynamic Phenomena, Uniquely Deduced from Experience

THÉORIE

DES

PHÉNOMÈNES ÉLECTRO-DYNAMIQUES,

UNIQUEMENT DÉDUITE DE L'EXPÉRIENCE,

PAR ANDRÉ-MARIE AMPÈRE,

DE L'ACADÉMIE ROYALE DES SCIENCES, DE LA SOCIÉTÉ PHILOMATIQUE, DE LA SOCIÉTÉ ROYALE D'ÉDIMBOURG, DE LA SOCIÉTÉ HELVÉTIENNE DES SCRUTATEURS DE LA NATURE, DE LA SOCIÉTÉ PHILOSOPHIQUE DE CAMBRIDGE, DE CELLE DE PHYSIQUE ET D'HISTOIRE NATURELLE DE GENÈVE, DE L'ACADÉMIE ROYALE DES SCIENCES ET BELLES-LETTRES DE BRUXELLES ET DE L'ACADÉMIE ROYALE DE LISBONNE, CHEVALIER DE LA LÉGION-D'HONNEUR, PROFESSEUR A L'ÉCOLE POLYTECHNIQUE ET AU COLLÉGE DE FRANCE.

A PARIS,

CHEZ MÉQUIGNON-MARVIS, LIBRAIRE-ÉDITEUR, RUE DU JARDINET, N° 13.

Et A BRUXELLES, AU DÉPÔT GÉNÉRAL DE LIBRAIRIE MÉDICALE FRANÇAISE.

NOVEMBRE 1826.

Figure 29.1: Cover of the *Théorie* published in 1826.

MÉMOIRES

DE

L'ACADÉMIE ROYALE DES SCIENCES

DE L'INSTITUT

DE FRANCE.

ANNÉE 1823.

TOME VI.

PARIS,

CHEZ FIRMIN DIDOT, PÈRE ET FILS, LIBRAIRES,

RUE JACOB, N° 24.

1827.

Figure 29.2: Cover of the *Mémoires de l'Académie royale des Sciences de l'Institut de France* in which the *Théorie* was published in 1827.

THÉORIE

DES PHÉNOMÈNES ÉLECTRO-DYNAMIQUES,

UNIQUEMENT DÉDUITE DE L'EXPÉRIENCE,

Ouvrage dans lequel se trouvent réunis les Mémoires que M. Ampère a communiqués à l'Académie royale des Sciences, dans les séances des 4 *et* 26 *décembre* 1820, 10 *juin* 1822, 22 *décembre* 1823, 12 *septembre et* 28 *novembre* 1825.

L'ÉPOQUE que les travaux de Newton ont marquée dans l'histoire des sciences n'est pas seulement celle de la plus importante des découvertes que l'homme ait faites sur les causes des grands phénomènes de la nature; c'est aussi l'époque où l'esprit humain s'est ouvert une nouvelle route dans les sciences qui ont pour objet l'étude de ces phénomènes.

Jusqu'alors on en avait presque exclusivement cherché les causes dans l'impulsion d'un fluide inconnu qui entraînait les particules matérielles suivant la direction de ses propres particules; et partout où l'on voyait un mouvement révolutif, on imaginait un tourbillon dans le même sens.

Newton nous a appris que cette sorte de mouvement doit, comme tous ceux que nous offre la nature, être ramenée, par le calcul, à des forces agissant toujours entre deux particules matérielles suivant la droite qui les joint, de manière que

Figure 29.3: First page of the *Théorie* published in 1826.

Theory of Electrodynamic Phenomena, Uniquely Deduced from Experience

[Pages 3 and 175][1]

Work collecting the papers delivered at the Académie royale des Sciences *by Ampère on the 4 and 26 December 1820, 10 June 1822, 22 December 1823, 12 September and 28 November*[2] *1825.*[3]

29.1 Exposition of the Path Followed in Research into the Laws of Natural Phenomena and the Forces that They Produce

The new era in the history of science marked by the works of Newton,[4] is not only the age of man's most important discoveries in the causes of natural phenomena, it is also the age in which the human spirit has opened a new highway into the sciences which have natural phenomena as their object of study.

Until Newton, the causes of natural phenomena had been sought almost exclusively in the impulsion of an unknown fluid which entrained particles of materials in the same direction as its own particles; wherever rotational motion occurred, a vortex in the same direction was imagined.[5]

Newton taught us that motion of this kind, like all motions in nature, must be reducible by calculation to forces acting between two material particles along the straight line between them such that [pages 4 and 176] the action of one upon the other is equal and opposite to that which the latter has upon the former and, consequently, assuming the two particles to be permanently associated, that no motion whatsoever can result from their interaction. It is this law, now confirmed by every observation and every calculation, which he represented in the last of the three axioms at the beginning of the *Philosophiæ naturalis principia mathematica*.[6] But it was not enough to reach this high conception, the law had to be found which governs the variation of these forces with the positions of the particles between which they act, or, what amounts to

[1][N. T.] The beginnings of the pages are indicated by "[pages m and n]." The number m refers to the *Théorie* published in 1826, [Ampère, 1826f], while page n refers to the *Théorie* published in 1827, [Ampère, 1823c]. This last pagination coincides with the 1990 edition, [Ampère, 1990].

[2][N. T.] The 1827 version of the *Théorie* presents the date "21 novembre" instead of "28 novembre." But in the errata which appears at the end of the work published in 1827 it is mentioned that this last date is the correct one. This correction was implemented in the 1826 edition.

[3][N. T.] Complete references of these works: [Ampère, 1820d], [Ampère, 1820e], [Ampère, 1885j], [Ampère, 1885m], [Ampère, 1820b], [Ampère, 1821b], [Ampère, 1822o], [Ampère, 1885p], [Ampère, 1824c], [Ampère, 1824f], [Ampère, 1824e], [Ampère, 1885f], [Ampère, 1825d], [Ampère, 1825c], [Ampère, 1825e], [Ampère, 1825g], [Ampère, 1826e], [Ampère, 1826b], [Ampère, 1887c] and [Ampère, 1887f].

[4][N. T.] Isaac Newton (1642-1727) published in 1687 his book *Mathematical Principles of Natural Philosophy*, also known by its first Latin name, *Principia*, [Newton, 1934] and [Newton, 1952a]. The *Opticks* was published in 1704, [Newton, 1952b] and [Newton, 1979]. These two books have already been translated to Portuguese, [Newton, 1996], [Newton, 1990] and [Newton, 2008].

[5][N. T.] Ampère is referring here to the conceptions of René Descartes (1596-1650). An analysis of Descartes's works on gravitation and magnetism can be found at [Whittaker, 1973, pp. 3-11 and 33-35].

[6][N. T.] Newton's third "Axiom or Law of Motion," [Newton, 1934]: "To every action there is always opposed an equal reaction: or, the mutual actions of two bodies upon each other are always equal, and directed to contrary parts."

the same thing, the value of these forces had to be expressed by a formula.

Newton was far from thinking that this law could be invented from abstract considerations, however plausible they might be. He established that this law must be deduced from observed facts, or preferably, from empirical laws, like those of Kepler,[7] which are only the generalized results of a great number of particular observations.[8]

To observe first the facts, varying the conditions as much as possible, to accompany this with precise measurement, in order to deduce general laws based solely on experience, and to deduce therefrom, independently of all hypothesis regarding the nature of the forces which produce the phenomena, the mathematical value of these forces, that is to say, to derive the formula which represents them, such was the road which Newton followed. This was the approach generally adopted by the learned men of France to whom physics owes the immense progress which has been made in recent times, and similarly it has guided me in all my research [pages 5 and 177] into electrodynamic phenomena. I have relied solely on experimentation to establish the laws of the phenomena and from them I have derived the formula which alone can represent the forces which are produced; I have not investigated the possible cause of these forces, convinced that all research of this nature must proceed from pure experimental knowledge of the laws and from the value, determined solely by deduction from these laws, of the individual forces in the direction which is, of necessity, that of a straight line drawn through the material points between which the forces act. That is why I shall refrain from discussing any ideas which I might have on the nature of the cause of the forces produced by voltaic conductors, though this is contained in the notes which accompany the *Exposé sommaire des nouvelles expériences électromagnétiques faites par plusieurs physiciens depuis le mois de mars 1821*,[9] which I read at the public session of the *Académie des Sciences*, 8 April 1822; my remarks can be seen in these notes on page 215 of my *Recueil d'Observations Électro-dynamiques*.[10] It does not appear that this approach, the only one which can lead to results which are free of all hypothesis, is preferred by physicists in the rest of Europe like it is by Frenchmen; the famous scientist[11] who first saw the poles of a magnet transported by the action of a conductor in directions perpendicular to those of the wire, concluded that electrical matter revolved about it and pushed the poles along with it,[12] just as Descartes made the matter of his vortices revolve in the direction of planetary revolution. Guided by the principles [pages 6 and 178] of Newtonian philosophy, I have reduced the phenomenon observed by M. Oerstedt, as has been done for all similar natural phenomena, to forces acting along a straight line joining the two particles between which the actions are exerted; and if I have established that the same arrangement, or the same movement of electricity which exists in the conductor, is present also around the particles of magnets, it is certainly not to explain their action by impulsion as with a vortex, but to calculate, according to my formula, the resultant forces acting between the particles of a magnet and those of a conductor, or of another magnet, along the lines joining the particles in pairs which are considered to be interacting, and to show that the results of the calculation are completely verified by (1) the experiments of M. Pouillet[13] and my own into the precise determination of the conditions which must exist for a mobile conductor to remain in equilibrium when acted upon, whether by another conductor, or by a magnet, and (2) by the agreement between these results and the laws which Coulomb[14] and M. Biot[15] have deduced by their experiments, the former relating to the interaction of two magnets, and the latter to the interaction between a magnet and a conductor.

The principal advantage of formulae which are derived in this way from general facts gained from suffi-

---

[7][N. T.] Johannes Kepler (1571-1630). He presented three laws of planetary motion which have been expressed as follows by Symon, [Symon, 1971, p. 135]: "1. The planets move in ellipses with the sun at one focus. 2. Areas swept out by the radius vector from the sun to a planet in equal times are equal. 3. The square of the period of revolution is proportional to the cube of the semimajor axis." The first two laws were discovered in the period 1601-1606, being published in 1609, while the third law was discovered in 1618, being published in 1619. An excellent discussion of Kepler's work can be found in [Koestler, 1973].

[8][N. T.] The 1826 edition of the *Théorie* ends this paragraph with the following words: "ne sont que les résultats généralisés d'un grand nombre d'observations particulières." The 1827 edition, on the other hand, had finished this paragraph as follows: "ne sont que les résultats généralisés d'un grand nombre de faits." In Tricker's book the final portion of this paragraph has been translated as follows, [Ampère, 1965b, p. 156]: "which are only the generalized results of very many facts."

[9][N. T.] This work was published originally in 1822, [Ampère, 1822r], being reprinted in [Ampère, 1822j] and [Ampère, 1885e].

[10][N. T.] *Collection of Observations in Electrodynamics*. Ampère's notes were published originally in the *Recueil* and are reprinted in Joubert's *Collection of Papers related to Physics*, [Ampère, 1822p, p. 215] and [Ampère, 1885m, p. 250], respectively. There is a partial English translation of these notes in [Caneva, 1980].

[11][N. T.] H. C. Ørsted.

[12][N. T.] See our discussion of this topic in Section 11.1.

[13][N. T.] Claude Servais Mathias Pouillet (1790-1868). Pouillet was a protegé of Jean-Baptiste Biot (1774-1862). Biot and Pouillet defended the hypothesis of the magnetization of the wire when an electric current flowed through it, [Hofmann, 1996, pp. 141, 316 and 319]. For an analysis of his works see [Taton, 1981].

[14][N. T.] Charles Augustin de Coulomb (1736-1806). His laws of force between electric charges and magnetic poles were published in 1785. They can be represented by equations (1.2) and (1.3).

[15][N. T.] The electromagnetic works of Biot were discussed in Chapter 6.

cient observations for their certitude to be incontestable, is that they remain independent, not only of the hypotheses which may have aided in the quest for these formulae, but also independent of those which may later be adopted instead. The expression for universal attraction from the laws of Kepler[16] is completely independent of the hypotheses which some writers have advanced to justify [pages 7 and 179] a mechanical cause to which they would ascribe it. The theory of heat is founded on general facts which have been obtained by direct observation; the equation deduced from these facts, being confirmed by the agreement between the results of calculation and of experiment, must be equally accepted as representative of the true laws of heat propagation by those who attribute it to the radiation of caloric molecules as by those who take the view that the phenomenon is caused by the vibration of a diffuse fluid in space; it is only necessary for the former to show how the equation results from their way of looking at heat and for the others to derive it from general formulae for vibratory motion; doing so does not add anything to the certitude of the equation, but only substantiates the respective hypotheses. The physicist who refrains from committing himself in this respect acknowledges the heat equation to be an exact representation of the facts without concerning himself with the manner in which it can result from one or other of the explanations of which we are speaking; and if new phenomena and new calculations should demonstrate that the effects of heat can in fact only be explained in a system of vibrations, the great physicist[17] who first produced the equation and who created the methods of integration to apply it in his research, is still just as much the author of the mathematical theory of heat, as Newton is still the author of the theory of planetary motion, even though the theory was not as completely demonstrated by his works as his successors have been able to do in theirs.

It is the same with the formula by which I represented [pages 8 and 180] the electrodynamic action. Whatever the physical cause to which the phenomena produced by this action might be ascribed, the formula which I have obtained will always remain the true statement of the facts. If it should later be derived from one of the considerations by which so many other phenomena have been explained, such as attractions in inverse ratio to the square of the distance, those which become imperceptible at all appreciable distance between particles between which these forces are exerted, the vibration of a fluid in space, etc., another step forward will have been made in this field of physics; but this inquiry, in which I myself have not yet been occupied, though I fully recognize its importance, will change nothing in the results of my work, since to be in agreement with the facts, the hypothesis which is eventually adopted must always be in accord with the formula which fully represents them.

From the time[18] when I noticed that two voltaic conductors interact, now attracting each other, now repelling each other, ever since I distinguished and described the actions which they exert in the various positions where they can be in relation to each other, and after I had established that the action exerted by a straight conductor is equal to that exerted by a sinuous conductor whenever the latter only deviates slightly from the direction of the former and both terminate at the same points, I have been seeking to express the value of the attractive or repulsive force between two elements, or infinitesimal parts, of conducting wires by a formula so as to be able to derive by the known methods of integration the action which takes place between two portions of conductors of the shape in question in any given conditions.

[pages 9 and 181]

The impossibility of conducting direct experiments on infinitesimal portions of a voltaic circuit makes it necessary to proceed from observations of conductors of finite dimension and to satisfy two conditions, namely that the observations be capable of great precision and that they be appropriate to the determination of the interaction between two infinitesimal portions of wires. It is possible to proceed in either of two ways: one is first to measure values of the mutual action of two portions of finite dimension with the greatest possible exactitude, by placing them successively, one in relation to the other, at different distances and in different positions, for it is evident that the interaction does not depend solely on distance, and then to advance a hypothesis as to the value of the mutual action of two infinitesimal portions, to derive the value of the action which must result for the test conductors of finite dimension, and to modify the hypothesis until the calculated results are in accord with those of observation. It is this procedure which I first proposed to follow, as explained in detail in the paper which I read at the *Académie des Sciences* on 9 October

[16][N. T.] Ampère has an interesting didactic work discussing how to deduce from Kepler's laws the fact that the gravitational force acting on a planet is inversely proportional to its distance from the Sun, [Ampère, 1829].

[17][N. T.] Jean Baptiste Joseph Fourier (1768-1830). His work *Théorie Analytique de la Chaleur* was published in 1822 and has already been translated into English, [Fourier, 1952].

[18][N. T.] The portion of the *Théorie* from this paragraph up to the middle of our page 349, corresponding to the middle of page 16 of the 1826 edition, or corresponding to the middle of page 188 of the 1827 edition, was taken from the paper of the 10th of June 1822 as presented in the *Recueil*, [Ampère, 1822y, pp. 293-303] and [Ampère, 1885p, pp. 270-277].

1820;[19, 20] though it leads to the truth only by the indirect route of hypothesis, it is no less valuable because of that, since it is often the only way open in investigations of this kind. A member[21] of this *Académie* whose works have covered the whole range of physics has aptly [pages 10 and 182] expressed this in the *Notice sur l'aimantation imprimée aux métaux par l'électricité en mouvement*,[22] which he read on 2 April 1821, designating this procedure as a kind of divination, saying that prediction of this kind was the aim of practically all physical research.[23, 24]

However, the same end can be reached more directly in the way which I have since followed: it consists in establishing by experiment that a mobile conductor remains exactly in equilibrium between equal forces, or between equal torques, these forces and these torques being produced by portions of fixed conductors of arbitrary shape and dimension without equilibrium being disturbed in the conditions of the experiment, and in determining directly therefrom by calculation what the value of the mutual action of the two infinitesimal portions must be for equilibrium to be, in fact, independent of all variations of shape and dimension compatible with the conditions.[25]

This last procedure can only be adopted when the nature of the action being studied is such that cases of equilibrium which are independent of the shape of the body are possible; it is therefore of much more restricted application than the first method which I discussed; but since voltaic conductors do permit equilibrium of this kind, it is natural to prefer the simpler and more direct method which is capable of great exactitude if ordinary precautions are taken for the experiments. There is, however, [pages 11 and 183] in connection with the action of conductors, a much more important reason for employing it in the determination of the forces which produce their action: it is the extreme difficulty associated with experiments where it is proposed, for example, to measure the forces by the number of oscillations of the body which is subjected to the actions. This difficulty is due to the fact that when a fixed conductor is made to act upon the mobile portion of a circuit, the pieces of apparatus which are necessary for connection to the battery act on the mobile portion at the same time as the fixed conductor, thus altering the results of the experiments. I believe, however, that I have succeeded in overcoming this difficulty in a suitable apparatus for measuring the mutual action of two conductors, one fixed and the other mobile, by the number of oscillations in the latter for various shapes of the fixed conductor. I shall describe this apparatus in the course of this dissertation.[26]

---

[19][N. A.] This paper has not been published separately, but the principal results are included in volume XV of the *Annales de chimie et de physique*.

[20][N. T.] See, in particular, [Ampère, 1820f, pp. 182-188 and 212-213], [Chaib and Assis, 2009b, pp. 126-128 and 139-140] and [Ampère, 1885b, pp. 17 and 29-33].

[21][N. T.] J. B. Biot.

[22][N. T.] *Notice on the magnetization of metals by electricity in motion*, [Biot, 1821b].

[23][N. A.] See *Journal des Savants*, p. 233, April 1821.

[24][N. T.] Original reference: [Biot, 1821b, p. 233]. We quote the relevant portion of this paper, our emphasis in italics:

> The apparatuses themselves by which these effects were obtained, show also the revolving character of the magnetic force acquired by the conducting wire. As this force is impressed upon the wire by the presence of the electric principles which flow through it along its length, it is therefore necessary that these principles have in their motions some transverse and revolving property in a determined sense, by means of which they excite or exert themselves the power shown by the wire. All our efforts nowadays should be devoted to discover the nature of this modification. Therefore, as I said in the beginning of this discourse, we arrived, in a subject so young, at the last stage of purely experimental research. The research which remains to be done, based upon particularities which are not immediately perceptible to our senses, can only consist in developing intellectual conceptions, in verifying conditions of cause which the imagination, guided by the usual study of nature, can only suggest. *This work, a kind of divination, is the goal of almost all physical research.* It is always necessary in chemistry, which always operates upon individual properties of the particles of bodies. Chemistry, for this reason, so positive in its details, is at the same time so exposed to the systematic approach [*sprit de système*] in its general overviews. Newton did not need to overcome this kind of difficulty when he analyzed the motions of planets and comets. These luminaries gave him in the curvatures of their orbits, in their shapes, their velocities, the perceptible and simple indications of the forces acting upon them. Therefore, he could arrive directly, without hypotheses, at the general law of these forces, from which he and his successors deduced, by calculation, all phenomena of the system of the world.

[25][N. T.] Ampère is referring here to the new method he introduced in physics for the determination of elementary forces. In this book we designated this procedure by the name *cases of equilibrium*. It is discussed mainly in Section 4.3.

[26][N. T.] Ampère described this apparatus in Section 29.13 of the *Théorie*, pages 62 and 234 of the 1826 and 1827 editions, respectively, corresponding to our page 382.

The original paper from which Ampère copied this portion has a slightly different formulation, [Ampère, 1822o, p. 401] and [Ampère, 1822y, p. 296], our emphasis in italics:

> I believe, however, that I have succeeded in overcoming this difficulty in a suitable apparatus for measuring the mutual action of two *circular concentric* conductors, one fixed and the other mobile, by the number of oscillations in the latter, *and changing the distance by means of different fixed conductors, in which the electric current will flow successively.*

It is true that the same obstacles do not arise when the action of a conducting wire on a magnet is measured in the same way; but this method cannot be employed when it is a question of determining the forces which two conductors exert upon each other, the question which must be our first consideration in the investigation of the new phenomena. It is evident[27] that if the action of a conductor on a magnet is due to some other cause than that which produces the effect between two conductors, experiments performed with respect to a conductor and a magnet can add nothing to the study of two conductors; if magnets only owe their properties to electric currents, which encircle their particles, it is necessary, in order to draw [pages 12 and 184] definite conclusions as to the action of the conducting wire on these currents, to be sure that these currents are of the same intensity near to the surface of the magnet as within it, or else to know the law governing the variation of intensity; whether the planes of the currents are everywhere perpendicular to the axis of a bar magnet, as I at first supposed,[28] or whether the mutual action of the currents of the magnet itself inclines them more to the axis when at a greater distance from this axis, which is what I have since concluded from the difference which is noticeable between the position of the poles on a magnet and the position of the points which are endowed with the same properties in a conductor of which one part is

---

We include in this footnote a figure showing Ampère's original apparatus intended to perform this experiment with *concentric circular conductors*. It was presented one year after the original text, 16 June 1823, [Ampère, 1822p, p. 224]. A discussion of this experiment can be found in [Hofmann, 1996, p. 296] and [Blondel, 1982, p. 126]. There is a mobile circular conductor surrounded by two fixed semicircular conductors. The mobile conductor can oscillate about its equilibrium position when it is acted upon by the currents in the fixed conductors.

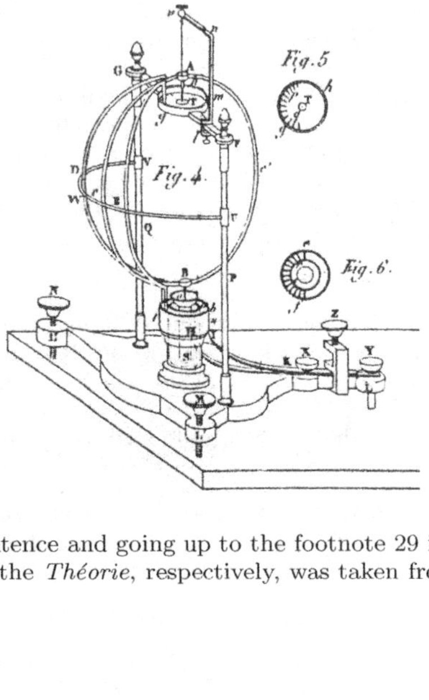

[27] [N. T.] The portion beginning in this sentence and going up to the footnote 29 in our page 347, corresponding to the pages 12 and 184 of the 1826 and 1827 editions of the *Théorie*, respectively, was taken from [Ampère, 1822o, pp. 403-404].

[28] [N. T.] See our discussion in Section 5.3.

helically wound.[29, 30]

[pages 13 and 185]

29.2 Description of the Experiments from which One Finds Four Cases of Equilibrium which Yield the Laws of Action to which the Electrodynamic Phenomena are Due

The various cases of equilibrium which I have established by precise experiment provide the laws leading directly to the mathematical expression for the [pages 14 and 186] force which two elements of conducting wires exert upon each other, in that they first make the form of this expression known and then allow the initially unknown constants to be determined, just as the laws of Kepler first show that the force which holds the planets in their orbits tends constantly towards the center of the Sun, since it varies for a particular planet in inverse ratio to the square of its distance to the solar center, so that the constant coefficient which represents its intensity has the same value for all planets. These cases of equilibrium are four in number:[31] the first demonstrates the equality in absolute value of the attraction and repulsion which is produced when a current flows alternately in opposite directions in a fixed conductor, the distance to the body on which it acts remaining constant. This equality[32] results from the simple observation that two equal portions of one and the same conductor which are covered in silk to prevent contact, whether both straight, or twisted together to form around each other two equal helices, in which the same electric current flows, but in opposite directions, exert no action on either a magnet or a mobile conductor. This can also be established by the mobile conductor which is illustrated in plate 1, figure 9 of *Annales de chimie et de physique*, volume XVIII,[33] relating to the description of the electrodynamic apparatus of mine which is reproduced here (figure 1).[34]

---

[29][N. A.] I believe I must here introduce the following note which is taken from the analysis of the works of the *Académie* in the year 1821, published 8 April 1822. (See the mathematical portion of this analysis, pp. 22 and 23.)

> "The principal difference between the manner in which a magnet acts and the action of a conductor of which one part is rolled in a helix round the other, is that the poles of the magnet are situated nearer to the midpoint of the magnet than its extremities, whereas the points which present the same properties in the helix occur exactly at the extremities of this helix: this is what must take place as the intensity of the currents of the magnet diminishes outwards from the midpoint to its extremities. But M. Ampère has since recognized that another factor can also cause this effect. Having concluded from his more recent experiments that the electrical currents of a magnet exist around each one of its particles, it was easier for him to see that it is unnecessary to assume, as he had done at first, that the planes of these currents are everywhere perpendicular to the axis of the magnet. Their interaction, however, must tend to incline the planes to the axis, particularly towards the extremities, so that the poles, instead of being exactly situated there, as they ought to be according to the calculations by the formulae of M. Ampère for the case when all the currents are assumed to be the same in intensity and in planes perpendicular to the axis, must be closer to the midpoint of the magnet according as the planes of a larger number of currents are thus inclined, and yet more so with greater inclination. That is to say, the thicker the magnet in relation to its length, the greater is this effect, which agrees with experiments. In helical conductors, where one part returns along the axis to cancel the effect of the part of the currents of each turn which acts as though they were parallel to the axis, the two circumstances which, from the foregoing remarks, do not necessarily occur in magnets, must on the contrary exist in the conductors; it may also be observed in experiments that helices have poles similar to those of magnets, but placed exactly at their extremities as calculated."

It will be seen from this note that after 1821 I concluded from the phenomena associated with magnets: (1) that in regarding each particle of a bar magnet as a magnet, the axes of those elementary magnets must be, not parallel to the axis of the total magnet as was supposed at the time, but inclined to the axis in directions determined by their interaction; (2) that this is one reason why the poles of a bar magnet are not at the extremities, but in between the extremities and its midpoint. Both these conclusions have already been demonstrated by the results of M. Poisson using the formulae by which he represented the distribution of the forces emanating from each particle of a magnet. These formulae are based on Coulomb's law and, in consequence, nothing is changed by adopting my approach to the explanation of magnetic phenomena, since this law is a corollary of my formula, as has been seen in the course of this paper.

[30][N. T.] Original reference of the work from which Ampère obtained this quotation: [Ampère, 1822s, p. 45]. The works of S. D. Poisson (1781-1840) quoted by Ampère are [Poisson, 1822a] and [Poisson, 1822b].

[31][N. T.] The four cases of equilibrium presented in the *Théorie* are not ordered according to the historical sequence in which Ampère obtained them.

[32][N. T.] The portion from this sentence until the expression "bent in any manner" in our page 349, corresponding to the middle of pages 16 and 188 of the 1826 and 1827 editions of the *Théorie*, respectively, do not appear in Ampère's original paper of 1822, [Ampère, 1822o]. This portion appeared for the first time in the *Recueil*, see [Ampère, 1822y, pp. 301-303] and [Ampère, 1885p, pp. 275-277].

[33][N. T.] See [Ampère, 1821c] and our figure 4.16 (a). This is the astatic coil discussed in Subsection 4.3.2.

[34][N. T.] This specific experiment, the first case of equilibrium of the *Théorie*, has been denominated the *case of equilibrium of anti-parallel currents* in this book. It was discussed at length in Subsection 4.3.4.

Fig. 1.

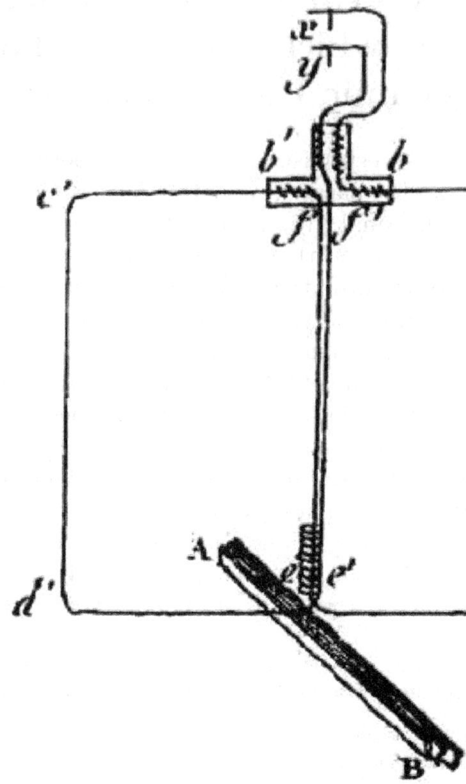

Figure 1.

A horizontal [fixed] straight conductor AB, doubled several times over,[35] is placed slightly below the lower part $dee'd'$ of this [mobile] conductor, in an arbitrary direction, [pages 15 and 187] such that its midpoint in length and thickness is in the vertical line through the tips[36] x and y about which the mobile conductor turns freely. It is seen that this [mobile] conductor stays in the position where it is placed, which proves that there is equilibrium between the actions [that is, there is equilibrium between the torques] exerted by the fixed conductor on the two equal and opposite portions of the circuit $bcde$ and $b'c'd'e'$, which differ only in that the current flows towards the fixed conductor AB in the one [portion], and away from it in the other [portion], whatever the angle between the fixed conductor and the plane of the mobile conductor. Now, considering first the two actions exerted between each portion of the circuit and the half of the conductor AB which is the nearest, and then the two actions between each of the two portions and the half of the conductor [AB] which is the furthest away, it will be seen without difficulty (1) that the equilibrium under consideration cannot occur at all angles except insofar as there is equilibrium separately between the first two actions and the last two; (2) that if one of the first two actions is attractive because current flows in the same direction along the sides of the acute angle formed by the portions of the conductors,[37] the other will be repulsive because the current flows in opposite directions along the two sides of the equal and opposite angle at the vertex, so that, for there to be equilibrium, the first two actions which tend to make the mobile conductor turn, the one in one direction, and the other in the opposite direction, must be equal [pages 16

[35][N. T.] What has been translated as *doubled several times over* appears as *plusieurs fois redoublé* in the original. It seems to us that the fixed wire has been doubled several times over in order to increase the magnitude of the current acting upon the astatic coil above it. This would increase a possible torque exerted by the fixed conductor upon the mobile astatic coil. We discussed this expression *plusieurs fois redoublé* in Subsection 4.3.4.

[36][N. T.] Original word in French: "pointes." This word was translated as "points" by Blunn, [Ampère, 1965b, p. 162].

[37][N. T.] Tricker has the following note at this point, [Ampère, 1965b, p. 163]: "Here Ampère is considering mainly the interaction between the fixed conductor AB and the two pieces of the suspended conductor, de and $d'e'$ which lie nearest to it."

and 188] to each other; and the last two actions, the one attractive and the other repulsive, because the sides of the two obtuse and opposite angles at the vertex and the supplements[38] of those [angles] about which we have just been speaking, must also be equal to each other. Needless to say, these actions [that is, these torques] are really sums of products of forces which act on each infinitesimal portion of the mobile conductor multiplied by their distance to the vertical about which this conductor is free to turn; however, the corresponding infinitesimal portions of the two arms *bcde* and $b'c'd'e'$ always being at equal distances from the vertical about which they turn, the equality of the torques makes it necessary for the forces to be equal.

I indicated the second[39] of the three[40] general cases of equilibrium towards the end of the year 1820; it consists in the equality of the actions exerted on a mobile straight conductor by two fixed conductors situated the same distance away from it, of which one is also straight, but the other bent in any manner. This was the apparatus[41] by which I verified the equality of the two actions[42] in the precise experiments, the results of which were communicated to the *Académie* in the session of 26 December 1820.[43]

The two wooden posts PQ and RS (figure 2), are slotted on the sides which mutually face each other, the straight wire *bc* being laid in the slot of PQ, and the wire *kl* in that of RS; over its entire length this wire [*kl*] is twisted in the plane perpendicular to that joining the two axes of the posts, such that [pages 17 and 189] the wire at no point departs more than a very short distance from the midpoints of the slot.

These two wires serve as conductors for the two portions of a current which is made to repel the part GH of a mobile conductor consisting of two almost closed and equal rectangular circuits BCDE and FGHI in which the current flows in opposite directions so that the effect of the Earth on these two circuits cancel out. At the two extremities of this mobile conductor there are two tips A and K which are immersed in the mercury-filled cups M and N and soldered to the extremities of the copper arms *g*M and *h*N. These arms make contact via the copper bushings *g* and *h*, the first with the copper wire *gfe*, helically wound around the glass tube *hgf*, the other with the straight wire *hi* which goes through the inside of this tube to the trough *ki* made in the piece of wood *vw* which is fixed at the desired height against the pillar *z* with the set screw *o*. In view of the experiment of which I referred above, the portion of the circuit composed of the helix *gf* and the straight wire *hi* can exert no action on the mobile conductor. For current to flow in the fixed conductors *bc* and *kl*, the connecting wires of these conductors are continued by *cde* and *lmn* in two glass tubes[44] attached to the cross-piece [pages 18 and 190] *xy*, finally terminating, the first in cup *e* and the other in cup *n*. With everything so arranged, one places mercury in all cups and inside the two troughs *ba* and *ki*, and one plunges the positive rheophore[45] *pa* in the trough *ba* which is also dug in the piece of wood *vw*, and the negative rheophore *qn* in the trough *n*. The current flows through the conductors of the

[38][N. T.] Two angles are supplementary if they add up to 180 degrees.

[39][N. T.] What Ampère denominated the second case of equilibrium in the *Théorie* has been called in this book the *case of equilibrium of the sinuous wire*, discussed in Subsection 4.3.3.

[40][N. T.] Here it should be "four" instead of "three," as four general cases of equilibrium will be discussed in the *Théorie*. This mistake is due to the fact that Ampère is reproducing here verbatim a portion of his paper of 1822 printed in the *Recueil* of 1823, [Ampère, 1822y, pp. 302-303] and [Ampère, 1885p, pp. 276-277]. In this paper of 1822, Ampère discussed only three cases of equilibrium, namely, anti-parallel currents, sinuous wire, and nonexistence of continuous rotation. This last case of equilibrium was discussed in Subsection 7.3.2. In the *Théorie* the case of equilibrium of the nonexistence of continuous rotation was replaced by what we called the *case of equilibrium of the nonexistence of tangential force*, discussed in Section 10.4. The fourth case of equilibrium of the *Théorie*, namely, what we are calling the *case of equilibrium of the law of similarity*, was only introduced in 1825; see Section 10.5.

[41][N. T.] This description beginning here and going up to our page 351 was taken verbatim from [Ampère, 1822p, pp. 215-221] or from [Ampère, 1885m, pp. 251-256].

[42][N. T.] The "actions" to which Ampère is referring here are the equal and opposite torques exerted on an astatic coil by a rectilinear and by a sinuous conductor.

[43][N. T.] See [Ampère, 1820b], [Ampère, 1821b] and [Ampère, 1885m].

[44][N. A.] These tubes are used to prevent flexure of the enclosed wires by holding them at equal distances from the two conductors *bc* and *kl*, so that their actions on GH, which reduce that of these two conductors, should reduce them equally.

[45][N. T.] *Rhéophore* in French. Rheophore is a current-carrier. It can be any wire, electrode, or connector that conducts electricity. In this case it is the wire of a voltaic apparatus which is connected to a terminal of the battery and traversed by a current. The positive (negative) rheophore is connected to the positive (negative) terminal of the battery. This word was introduced by Ampère, [de la Rive, 1822c, p. 266] and [de la Rive, 1885, p. 312]. According to Demonferrand, Ampère's student, this word was introduced in analogy with the expression *electrophore*, given to the apparatus utilized to transport static electricity, [Demonferrand, 1823, p. 10].

Ampère defined this term as follows, [Ampère, 1823b, Note 1 on p. 392], [Ampère, 1822l, Note 1 on p. 367] and [Ampère, 1885h, Note 1 on p. 386]:

> I denominate rheophores the two copper wires welded at the two extremities of the battery, utilized to *carry the electric current* in the instruments which have the purpose to observe the mutual action between the several portions of this current, and [to observe that action] between these portions and the terrestrial globe or a magnet.

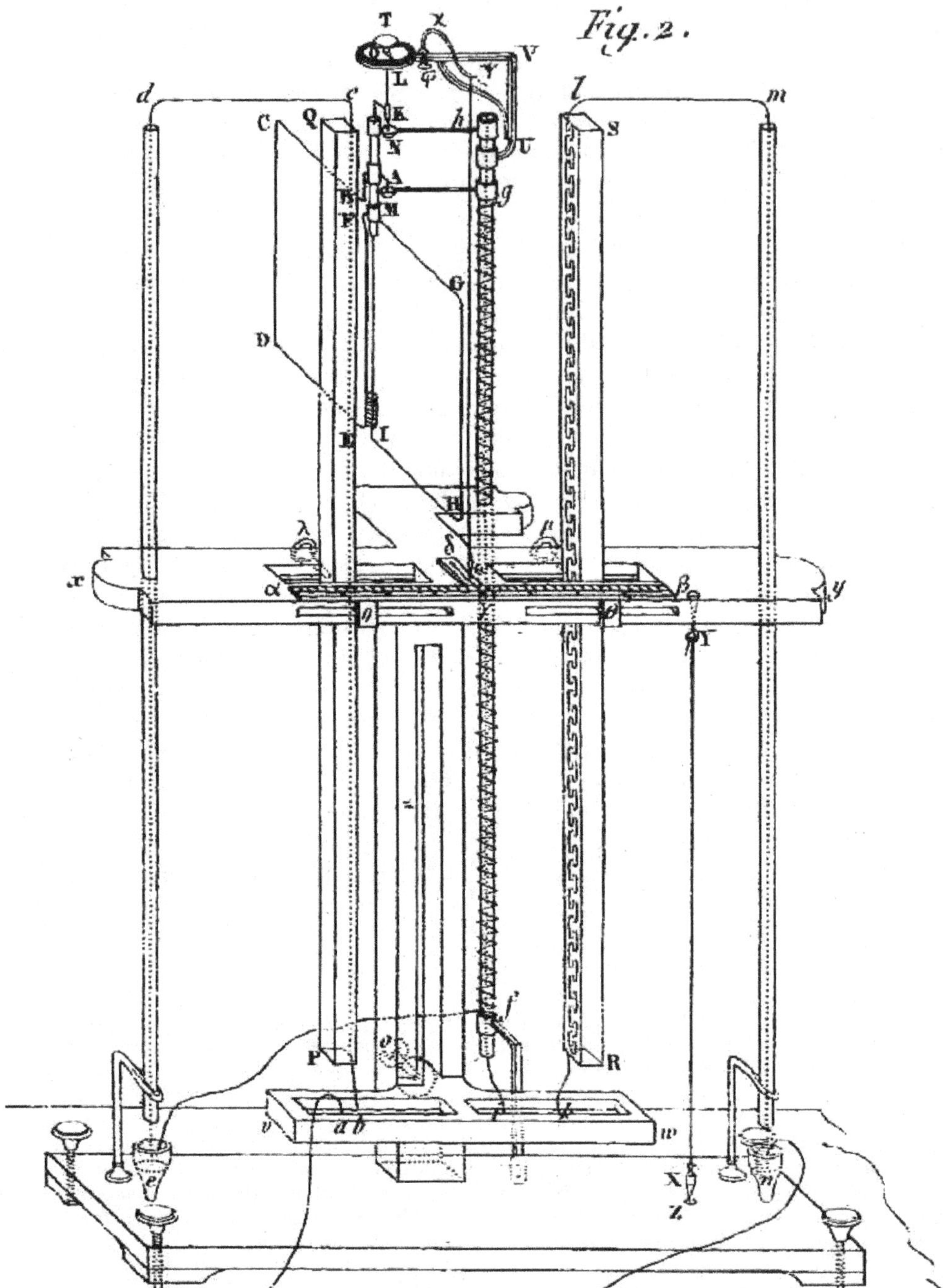

Figure 2.

apparatus in the following order: *pabcdefg*MABCDEFGHIKN*hiklmnq*. As a result, the current flows up the two fixed conductors and down that part, GH, of the mobile conductor which is acted upon in its position midway between the two fixed conductors and lies in the plane which passes through their axes. The part GH is thus repelled by *bc* and *kl*: whence it follows that if the action of these two conductors is the same at equal distances, GH must remain midway between them; this is, in fact, what happens.

It is as well to point out (1) that though the two axes of the fixed conductors are the same distance from GH, it cannot be stated rigorously that the distance is the same for all points of the conductor *kl* owing to its contours and bends. But since these bends are in a plane perpendicular to the plane through GH and through the axes of the fixed conductors, it is evident that the resulting difference in distance, is as small as possible, and as much less than half of the width of the slot RS as this half is less than the interval between the two posts (this difference, in the case when it is the largest possible, is equal to that between the radius and the secant of an arc with tangent equal to half the width of the slot, [pages 19 and 191] and belonging to a circle of which the diameter is the interval between the two slots). (2) That if each infinitesimal portion of the conductor *kl* is resolved in the same way as a force could be resolved, into two minute portions which are projections, the one along the vertical axis of the conductor and the other along horizontal lines drawn through all points of the conductor in the plane in which it is bent, the sum of the first projections (taking as negative those which, having a direction opposite to the direction of the others, should produce an action in the opposite sense), will be equal to the length of this axis; hence the total action resulting from all these projections is the same as that of a straight conductor equal to the axis, that is to say, it is equal to that of the conductor *bc* situated on the other side at the same distance from GH. The other projections will have zero effect on the mobile conductor GH since the planes erected vertically at the midpoint of each of them pass approximately through GH. The joining of these two series of projections thus produces an action on GH equal to that of *bc*. And since experience also proves that the sinuous conductor *kl* produces an action equal to that of *bc*, whatever its bends and contours, it follows that it acts in all cases like the combination of the two series of projections, which cannot occur independently of the manner in which the conductor is bent unless each part of it acts separately as the resultant of its two projections.

For this experiment to have the desired exactitude, it is necessary for the two posts to be exactly vertical and at precisely the same distance from the [pages 20 and 192] mobile conductor. To fulfill these conditions, a division[46] $\alpha\beta$ is matched to the cross-piece xy and the posts are fixed by two clamps η and θ, and two adjustable screws λ and μ, so that the posts may be moved apart or brought closer together at will, keeping the same distance from the midpoint γ of the division $\alpha\beta$. The apparatus is so constructed that the two posts are perpendicular to the cross-piece xy, and this is made horizontal by the screws at the four corners of the base of the device and the plumb line XY which corresponds exactly to a point Z as conveniently marked on the base with the cross-piece xy perfectly level.

For the conductor ABCDEFGHIK to revolve about a vertical axis at an equal distance from the two conductors *bc* and *kl*, this conductor is suspended by a very fine metal wire attached to the center of a knob T which can rotate without altering the distance between the two conductors; this knob is at the center of a small dial O,[47] on which the letter L marks the place where it is necessary to stop in order that the part GH of the mobile conductor should hang, without the suspension being twisted, at the midpoint of the interval between the two fixed conductors *bc* and *kl*, in order to be able immediately to return the needle to the position in which it should be whenever it is desired to repeat the experiment. It is checked that GH is an equal distance from *bc* and *kl* by another plumb line $\psi\omega$ which is attached to the copper arm $\varphi\chi\psi$ carried, like the dial O, on the support UVO, in which this arm $\varphi\chi\psi$ is free to revolve about the axis of the knob φ at the end of it, thus making it possible to have the tip of the plumb line ω correspond to the line $\gamma\delta$ in the middle of the division $\alpha\beta$. When the conductor is in the appropriate position, [pages 21 and 193] the three verticals $\psi\omega$, GH and CD are in the same plane, as can easily be checked by placing one's eye in this plane in front of $\psi\omega$.

The mobile conductor is thus arranged beforehand in the position where it will be in equilibrium between the repulsions of the two fixed conductors, if these repulsions are exactly equal: these actions are then produced by immersing into the trough *ba* and the cup *n* respectively the wires *ap* and *nq* which connect to the two extremities of the battery, and the conductor GH is found to remain in this position despite the great mobility associated with suspensions of this kind. If the mark L is displaced, even slightly, which brings GH into a position which is no longer equidistant between the fixed conductors *bc* and *kl*, it is seen to move as soon as communication with the battery is established, swinging away from whichever conductor is the nearest. At the time when I had this device constructed, I established in this way that the actions of the two conductors are equal from sufficient experiments with all the necessary precautions, for there to be no doubt about the result.

The same law can also be demonstrated by a simple experiment for which it is sufficient to take a silk-covered copper wire and to wind a part round the straight portion without being separated from it other than by the silk. It is then found that another portion of the wire does not affect the assembly of two portions; and since it would be the same for an assembly of two straight wires with a similar electric current in opposite direction [pages 22 and 194] (from the experiment by which the first case of equilibrium was

[46] [N. T.] That is, a partition, rule, or demarcation.

[47] [N. T.] *Cadran* in the original. It is a graduated circle with a pointer indicating the angle.

very simply established), it follows that the action of the current in the wound portion is exactly equal to that of the current in the straight part between identical extremities, because the action of both these two conductors would be counterbalanced by the action of the current in a straight portion of equal length, but in the opposite direction.[48]

The third case of equilibrium[49] is that a closed circuit of any arbitrary shape cannot produce movement in a portion of conducting wire which is in the form of an arc of a circle whose center lies on a fixed axis about which it may turn freely and which is perpendicular to the plane of the circuit of which the arc forms part.[50]

On[51] the base table TT′ (figure 3)[52] two columns EF and E′F′ are erected which are joined by the cross-pieces[53] LL′ and FF′; an upright GH is held in the vertical position between these two cross-pieces. Its two extremities G and H, ending in sharp tips, fit into two tapered holes, one in the lower cross-piece LL′, the other in the extremity of the screw KZ carried by the upper traverse FF′ which locates the upright GH without locking it. At C a support QO is fixed rigidly to this upright. At its extremity O is a hinge which engages the midpoint of the circular arc AA′ (formed by a metal wire) which remains constantly in the horizontal position. The radius of this arc is the distance from the point O to the axis GH. This arc is held in equilibrium by the counter-weight Q, thus reducing the friction of the upright GH in the tapered holes where its extremities are held.

[pages 23 and 195]

Below the arc AA′ there are two small troughs M and M′ which are filled with mercury so that the surface of the mercury, rising above the brim, just touches the arc AA′ at B and B′. These two small troughs are connected by the metallic conductors MN and M′N′ to the cups P and P′, which are full of mercury. The cup P and the conductor MN, which connects it to the trough M, are fixed to a vertical upright which is bedded in the table, but leaving it free to turn. The cup P′, to which the conductor M′N′ is connected, is

---

[48][N. T.] It is worthwhile reproducing here a note of 1887 written by Joubert when the *Théorie* was reprinted, [Ampère, 1887d, p. 17]. This note refers to the following paper by Bertrand: [Bertrand, 1874]. Joubert's note runs as follows:

> M. J. Bertrand showed (*Journal de Physique* [I], Volume III, p. 297; 1874) that the theorem of *sinuous currents*, considered by Ampère as one of the four fundamental theorems which he utilized in order to establish his formula, is a consequence of the hypothesis that the action of two current elements is directed along the straight line connecting them and of the theorem provided by the third case of equilibrium, according to which the action of a closed current upon a current element is always orthogonal to this element.
>
> "Let us suppose, writes M. Bertrand [see [Bertrand, 1874, p. 301]], that Ampère had initially verified and stated this last theorem and that, only by reasoning, he had deduced from it the theorem of sinuous currents, he could then say: If the action between two current elements is, as it seems probable to me, directed along the straight line connecting them, it would follow necessarily that a sinuous conductor would exert the same action as a rectilinear conductor following the same direction. Having experience afterwards confirmed this prediction, could it not be considered, rightly, as a very strong proof in favor of the hypothesis leading to it? Would the order in which the truths were discovered and the time in which their mutual dependence were pointed out change anything as regards their probability?"

[49][N. T.] This third case of equilibrium of the *Théorie* is being called here the *case of equilibrium of the nonexistence of tangential force*. It was discussed in Section 10.4.

[50][N. T.] See, in particular, [Ampère, 1825d] and [Ampère, 1825c]. This paper begins with the following words:

> The manner in which I have determined the relation between the two co-efficients of the formula by which I represented the mutual action of the two elements of electric currents, in the memoir which I read before the Acdemy on the 10th of June 1822 [see [Ampère, 1822o]], being liable to some difficulties, I have endeavoured to establish this relation in a more simple and direct manner. I succeeded in this very easily by means of an instrument which I shall first describe; I will then present some new results which I have deduced from this formula.

Then Ampère described the apparatus which appears in figure 3 of the *Théorie*; see our page 353. This new experiment, the case of equilibrium of the nonexistence of tangential forces, also presents some experimental problems. It has little sensitivity due to the several kinds of friction which prevent the free mobility of the instrument. These frictions appear along the axis of rotation, and between the circular arc and the mercury in the radial troughs. The motion only begins when the center of the circular arc is far away from the axis of rotation.

As discussed in Section 27.1, the experimental conditions of this instrument were greatly improved by Ettingshausen (1796-1878) when he supported the support QO of figure 3, connected to the circular arc AA′, by a bifilar suspension, [Ampère, 1887d, Note 2 by Joubert on pp. 17-18], [Blondel, 1982, Note 27, p. 148], [Ettingshausen, 1878a], [Ettingshausen, 1878b] and [Ettingshausen, 1879].

[51][N. T.] The text describing this experiment beginning in this sentence and going up to the expression "prominent part in the effect" in our page 354, corresponding to pages 25 and 197 of the 1826 and 1827 editions of the *Théorie*, respectively, was taken from [Ampère, 1825d, pp. 382-385], with English translation in [Ampère, 1825c].

[52][N. T.] In the original figure the letter *E′* of the left column appears as *E*.

[53][N. T.] In the original version we have the following text: *traverses LL′, EF′*. The correct expression, according to Ampère's original figure, should be *traverses LL′, FF′*.

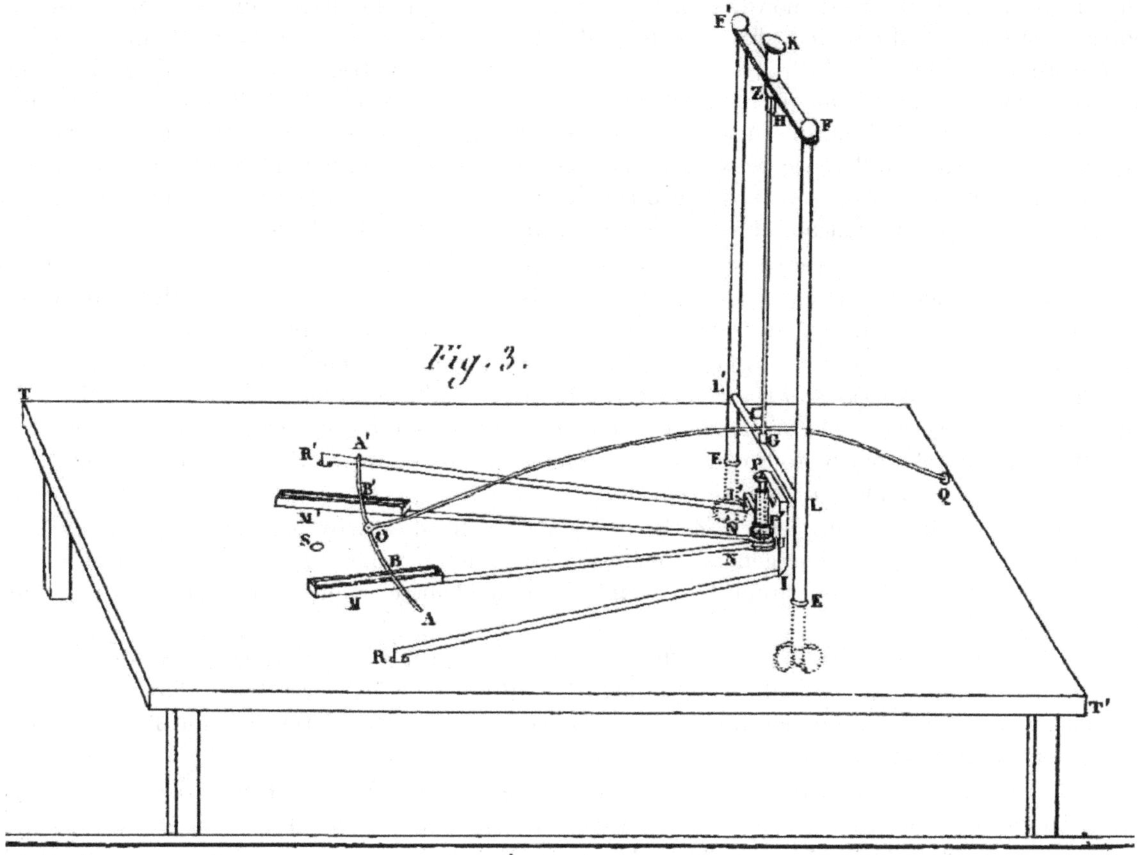

Figure 3.

traversed by the same upright, about which it, too, can revolve independently. The cup is insulated from the upright by the glass tube V which envelopes it, and by the glass ring U which separates it from the conductor of the trough M so as to be able to arrange the conductors MN and M′N′ at any desired angle.

Two other conductors IR and I′R′, attached to the table, are immersed respectively in the cups P and P′ and connect them to the cavities R and R′ which are made in the table and filled with mercury. Finally, a third cavity S, likewise full of mercury, is situated in between the other two.

This apparatus is used in the following way: one of the rheophores,[54] say the positive rheophore, is immersed in the cavity R, whilst the negative rheophore is immersed in cavity S, which is made to communicate with the cavity R′ by a curvilinear conductor of arbitrary shape. The current follows the conductor RI, passes into the cup P, and thence to the conductor NM, the trough M, the conductor M′N′, the cup P′, the conductor I′R′ and finally from the cavity R′ into the curvilinear conductor which connects to the mercury of the cavity S, where the negative rheophore is immersed.

[pages 24 and 196]

With this arrangement the total circuit is formed by:

1. the arc BB′ and the conductors MN and M′N′;

2. the circuit consisting of the parts RIP and P′I′R′ of the apparatus, the curvilinear conductor from R′ to S and the battery itself.

This latter circuit must act as a closed circuit since it is only interrupted by the glass which insulates the two cups P and P′. It is therefore sufficient to observe its action on the arc BB′ to determine the action of a closed circuit on an arc in different positions in relation to each other.

When by means of the hinge O the arc AA′ is positioned such that its center lies outside the axis GH, the arc moves and slides on the mercury of the troughs M and M′ owing to the action of the closed curvilinear

[54][N. T.] See footnote 45 on page 349.

current flowing from R′ to S. If, however, its center is on the upright, it remains stationary; hence, the two portions of the closed circuit which tend to make it turn in opposite directions about the axis, exert [opposite] torques on this arc which are equal in absolute value, no matter what the magnitude of the part BB′, as determined by the opening of the angle of the conductors MN and M′N′. If, therefore, two arcs BB′ are taken in succession which hardly differ from each other, since the torque is zero for both of them, it will also be zero for the slight difference between them, and, in consequence, it is likewise zero for any element of circumference with center on the axis; hence the direction of the action exerted by the closed circuit on the element is along the upright and it is necessarily perpendicular to the element.

When the arc AA′ is positioned so that its center is [pages 25 and 197] on the upright, the portions MN and M′N′ of the conductor exert equal and opposite repulsive actions on the arc BB′ with the result that no effect is produced; since no movement occurs, it is certain that no torque is produced by the closed circuit.

When the arc AA′ moves in the other situation which we initially envisaged [that is, with its center outside the axis GH], the actions of the conductors MN and M′N′ are no longer equal; it could be thought that the movement was due solely to this difference if the movement did not increase, or decrease, according as the curvilinear circuit from R′ to S comes nearer or moves further away, which leaves no doubt that the closed circuit plays a prominent part in the effect.

This result, occurring for any length of the arc AA′,[55] will necessarily occur for each of the elements of which the arc is composed. The general conclusion may therefore be drawn that the action of a closed circuit, or of an assembly of closed circuits, on an infinitesimal element of an electric current is perpendicular to this element.[56]

It is by the fourth case of equilibrium,[57] about which I have still to speak, that the constant coefficients occurring in my formula may be finally determined without recourse, as I first had to have, to experiments where a magnet and a conductor interact.[58] The device by which this determination may rest solely on observation of the interaction of two conductors is shown in figure 4.[59]

A cavity A is made in the table MN (figure 4).[60] The cavity is filled with mercury and from it runs the fixed conductor ABCDEFG made from a sheet of copper. The part CDE is circular, [pages 26 and 198] and the parts CBA and EFG are insulated from each other by a silk covering. At G this conductor is soldered to the copper tube GH, which carries the cup I which is in contact with the tube by means of the copper support HI. The mobile conductor IKLMNPQRS, of which the part MNP is circular, starts from the cup I; the parts MLK and PQR are insulated by a silk covering. The conductor is held horizontal by the counterweight a fixed on the circumference of a circle formed around the tube GH by the continuation

[55][N. T.] The original French text has the expression "quelle que soit la longueur de l'axe AA′." We believe the correct expression in French should be "de l'arc AA′" and not "de l'axe AA′."

[56][N. T.] A detailed discussion of this experiment and the relevance of this conclusion has been presented in Section 10.4.

[57][N. T.] This fourth case of equilibrium of the *Théorie* is being called in this book the *case of equilibrium of the law of similarity*. It was discussed in Section 10.5.

[58][N. T.] Ørsted designated *electromagnetic interactions* the mutual torques between a magnet and a current-carrying wire, [Ørsted, 1998c, p. 421], [Ørsted, 1998f, p. 426], [Grattan-Guinness, 1990a, p. 920] and [Grattan-Guinness, 1991, p. 116]. Ampère emphasized here that he will obtain the values of all the constant coefficients which appear in his force law utilizing only mutual actions between current-carrying conductors, which he called *electrodynamic interactions*. That is, he will not need to utilize magnets in his experiments.

In the original paper in which he introduced the case of equilibrium of the law of similarity, Ampère expressed himself as follows, [Ampère, 1826b, pp. 37-38] and [Ampère, 1887f, p. 205]:

> There are two determinations for the value of [the constant] n. One of them is deduced from the experiments of M. Biot, utilizing the number of oscillations made by a small magnet due to the action of an unlimited rectilinear conductor when the small magnet is at different distances from the conductor. The other determination of n is based upon the experiments due to MM. Gay-Lussac and Velter, which shows the null action [upon a small magnet] due to a steel ring in which all points are magnetized with the same intensity [along the direction of the ring]. As these two determinations [of the value of n] are due to experiments which utilize magnets, they cannot be extended, speaking rigorously, to the mutual action between two conductors. It would be important to discover the value of n utilizing observations made directly with conducting wires. This can be obtained very simply utilizing this last result just obtained. To this end it is sufficient to build an instrument composed of three similar circuits, circular for instance, in which the homologous dimensions form a continuous proportion, with all of them in the same plane, in such a way that they can be located between the sides of an angle which are simultaneously tangent to the three circumferences.

[59][N. T.] This instrument was described in [Ampère, 1826b] and [Ampère, 1887f]. The portion which begins in the next paragraph, going up to the expression "form a system similar to that of the circles O′ and O″" in our page 355, corresponding to the final of pages 26 and 198 of the 1826 and 1827 editions of the *Théorie*, respectively, was taken with a few modifications from [Ampère, 1826b, pp. 38-39] and [Ampère, 1887f, pp. 206-207].

[60][N. T.] Ampère used the letters MN to indicate the table and also the mobile central conductor.

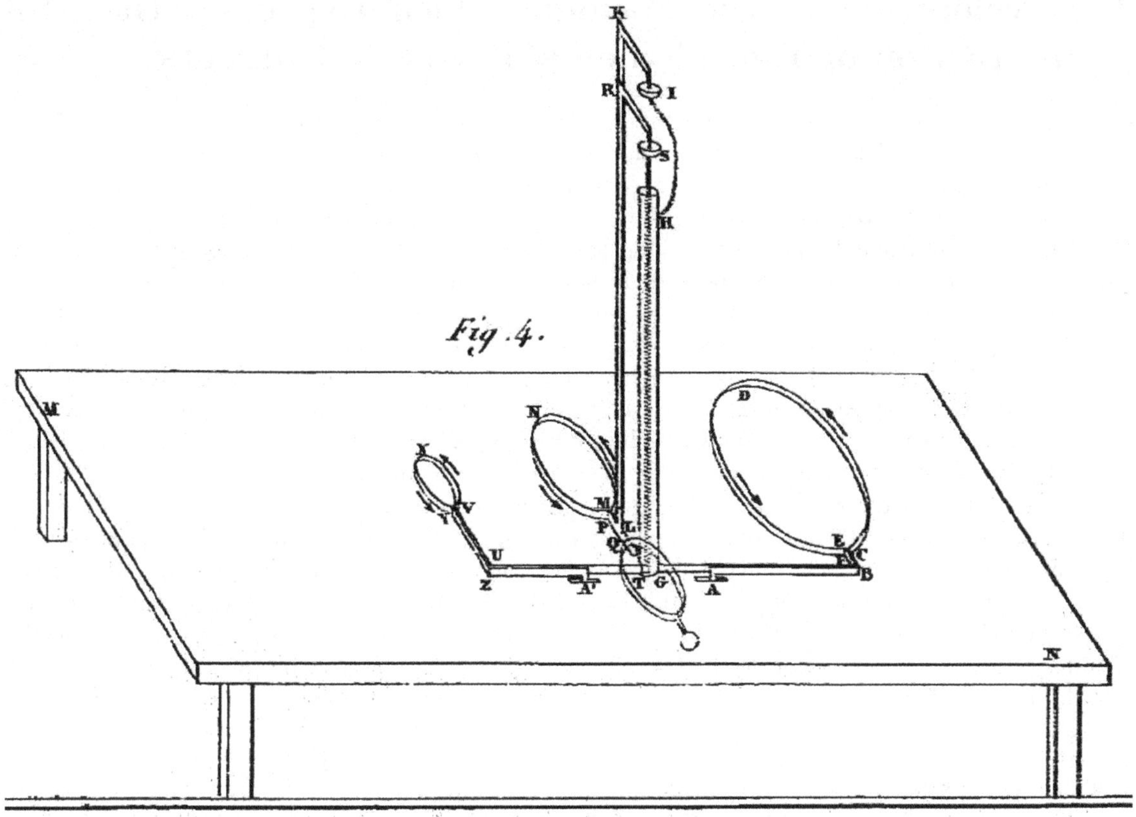

Figure 4.

bcg of the sheet constituting the mobile conductor.[61] The cup S is supported by the rod ST which has the same axis as GH, but from which it is insulated by a resinous substance which is poured into the tube. The base of the rod ST is soldered to the fixed conductor TUVXYZA′, which passes out of the tube GH through an opening large enough for the resin to insulate it as completely at this place as in the rest of the tube GH with regard to ST. At the outlet from the tube this conductor is covered with silk to prevent contact between the portions TUV and YZA′. The portion VXY is circular and the extremity A′ is immersed in the mercury-filled cavity A′ in the table.

The centers[62] O, O′ and O″ of the three circular [horizontal] portions are in a straight line; the radii of the circles which they form are in continuous geometric proportion and the mobile conductor is first placed in such a way that the distances OO′ and O′O″ bear the same relation to each other as consecutive terms in this proportion; hence the circles O and O′ form a system similar to that of the circles O′ and O″. The positive rheophore is then immersed in A with the negative rheophore in A′, and the current flows in succession through the circles with centers at O, O′ and O″, which repel each other in pairs, because [pages 27 and 199] the current flows in the opposite directions in neighboring parts.

The purpose of the experiment is to prove that the mobile conductor remains in equilibrium in the position where the ratio of OO′ to O′O″ is the same as that of the radii of two consecutive circles, and that if it is moved away from this position, it returns to it after oscillating about it.

[61][N. T.] The letters *a* and *bcg* did not appear in figure 4 of the 1826 and 1827 versions of the *Théorie*. We included these letters in our reproduction of this figure drawn with graphic software; see figure A.4 in our page 497.

[62][N. T.] In figure 4 of the 1826 *Théorie* reproduced here these centers do not appear. They appear in figure 4 of the 1827 *Théorie*; see figure 10.19 on page 187. In this case center O″ belongs to the left circle, O′ belongs to the middle circle, while O belongs to the right circle. We included these centers in our diagram made with graphic software; see figure A.4 on page 497.

29.3 Development of the Formula which Expresses the Mutual Interaction of Two Elements of Voltaic Currents

I will now explain how to deduce rigorously from these cases of equilibrium the formula by which I represent the mutual action of two elements of voltaic current, showing that it is the only force which, acting along the straight line joining their midpoints, can agree with the facts of the experiment. First of all, it is evident that the mutual action of two elements of electric current is proportional to their length; for, assuming them to be divided into infinitesimal equal parts along their lengths, all the attractions and repulsions of these parts can be regarded as directed along one and the same straight line, so that they necessarily add up. This action must also be proportional to the intensities of the two currents. To express the intensity of a current as a number, suppose that another arbitrary current is chosen for comparison, that two equal elements are taken from each current, and that the ratio is required of the actions which they exert at the same distance on a similar element of any other current if it is parallel to them, or if its direction is perpendicular to the straight lines which join its midpoint with the midpoints of two other elements. This ratio will be the measure of the intensity of one current, assuming that the other is unity.

Let us put i and i' for the ratios of the intensities of two given currents to the intensity of the reference current taken [pages 28 and 200] as unity, and put ds and ds' for the lengths of the elements which are considered in each of them; their mutual action, when they are perpendicular to the line joining their midpoints, parallel to each other and situated a unit distance apart, is expressed by $ii'dsds'$; we shall take the sign + when the two currents, flowing in the same direction, attract, and the sign − in the other case.

If it is desired to relate the action of the two elements to gravity, the weight of a unit volume of suitable matter could be taken for the unit of force. But then the current taken as unity would no longer be arbitrary; it would have to be such that the attraction between two of its elements ds and ds', situated as we have just said, could support a weight which would bear the same relation to the unit of weight as $dsds'$ bears to 1. Once this current was determined, the product $ii'dsds'$ would denote the ratio of the attraction of two elements of arbitrary intensity, still in the same situation, to the weight which would have been selected as the unit of force.

Suppose[63] we now consider two elements placed arbitrarily; their mutual action will depend on their lengths, on the intensities of the currents of which they are part, and on their relative position. This position can be determined by the length r of a straight line joining their midpoints, the angles θ and θ' between a continuation of this line and the directions of the two elements in the same direction as their respective currents, and finally the angle ω between the planes drawn through each of these directions and the straight line joining the midpoints of the elements.[64]

Consideration of the diverse attractions and repulsions observed in nature led me to believe that the force which I was seeking to represent, acted in some inverse ratio [pages 29 and 201] to distance; for greater generality, I assumed that it was in inverse ratio to the nth power of this distance, n being a constant to be determined.[65] Then, putting ρ for the unknown function of the angles θ, θ' and ω, I had $\rho ii'dsds'/r^n$ as the general expression for the action of two elements ds and ds' of the two currents with intensity i and i', respectively. It remained to determine the function ρ. For that I shall first consider two elements ad and $a'd'$ (figure 5), parallel to each other, perpendicular to the straight line joining their midpoints, and a distance r apart;

---

[63] [N. T.] What Ampère is presenting here was discussed in several of his paper, namely: [Ampère, 1820d], [Ampère, 1820e] and [Ampère, 1885j] (4 December 1820), and also [Ampère, 1822r], [Ampère, 1822j] and [Ampère, 1885e] (8 April 1822).

[64] [N. T.] These angles are represented in figure 2.15 (b).

[65] [N. T.] Note written by Joubert when the 1887 edition of the *Théorie* was printed, [Ampère, 1887d, p. 24], referring to Verdet's work of 1872, [Verdet, 1872, p. 144]:

> With all rigor, nothing proves:
>
> 1. That the function of the distance has the form $1/r^n$.
> 2. That this form should be the same when the element occupies the position $a''d''$ and the position ad (figure 5), and that, consequently, there should exist a constant ratio k, independent of the distance, between the actions exerted in both cases.

In Note 1 at the end of this Dissertation [see Sections 30.1 and 31.1], Ampère presents a more general method which does not suppose anymore, *a priori*, that the action should be in inverse ratio of a power of the distance. In the *Works* of Verdet (*Conférences à l'École Normale*, Volume I, p. 144) an even more general demonstration can be found, due to Blanchet, and in which two different functions are considered to represent the action between two elements of current located in positions ad and $a''d''$.

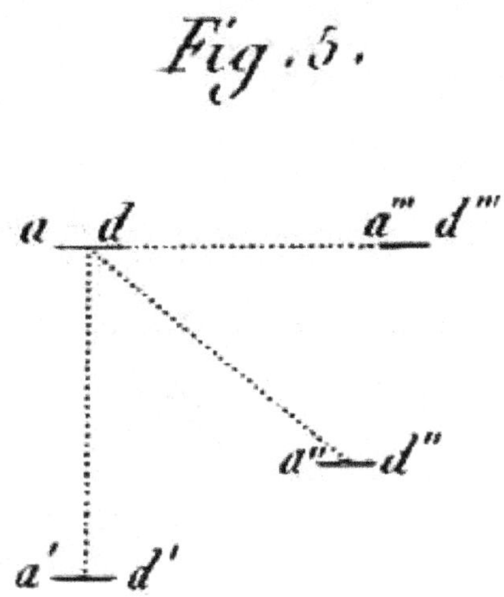

Figure 5.

their action being represented in accordance with the foregoing remarks by $ii'dsds'/r^n$, I assumed that ad remained fixed and that $a'd'$ was transported parallel to itself in such a way that its midpoint was always the same distance from the midpoint of ad; ω being always zero, the value of their mutual action could depend only on the angles represented above by θ and θ' and which, in this case, are equal, or supplementary of one another, according as the currents flow in the same or opposite direction; in this way I obtained the value $ii'dsds'\varphi(\theta,\ \theta')/r^n$. By putting k for the positive or negative constant to which $\varphi(\theta,\ \theta')$ is reduced when the element $a'd'$ is at $a'''d'''$ on the continuation of ad and in the same direction, I obtained $kii'dsds'/r^n$ to represent the action of ad on $a'''d'''$. In this expression the constant k represents the ratio of the action of ad on $a'''d'''$[66] to that of ad on $a'd'$, a ratio which is independent of the distance r, the intensities i and i', and of the lengths ds and ds' of the two elements under consideration.

These values of the electrodynamic action [pages 30 and 202] are sufficient, in the two simplest cases, for finding the general form of the function ρ, by reason of the experiment which shows that the attraction of an infinitely small rectilinear element is the same as that of any other sinuous element, terminating at the ends of the first, and the theorem which I will establish, namely:[67] that an infinitely small portion of current exerts no action on another infinitesimal portion of a current which is situated in a plane which passes through its midpoint and which is perpendicular to its direction.[68] In fact, the two halves of the first element produce equal actions on the second, the one attractive and the other repulsive, because the current tends to approach the common perpendicular in one of these halves and to move away from it in the other. The two equal forces form an angle which tends to two right angles according as the element tends to zero. Their resultant is therefore infinitesimal in relation to these forces and in consequence it can be neglected in the calculations. Let[69] Mm = ds and M$'m'$ = ds' (figure 6)[70] represent two elements of electric currents with midpoints A and A$'$;

[66][N. T.] In the original text we read ad'''. This was a typographical mistake. What is being discussed here is the action between the current elements ad and $a'''d'''$.

[67][N. T.] Ampère will present now the theorem of the nonexistence of interaction between orthogonal current elements discussed in Subsection 4.2.2. The justification of this theorem in the *Théorie* is different from the justification presented in Subsection 4.2.4.

[68][N. T.] These two cases are represented in figures 4.15 (a) and (b). The plane of the paper passes through the center of ds'. The current element ds' does not exert any force on ds, according to this theorem of Ampère.

[69][N. T.] Ampère will present here the vector decomposition of the current elements. We discussed this topic in Subsection 4.2.1.

[70][N. T.] By a typographical mistake the original text had Mm = ds and M$'m$ = ds' instead of Mm = ds and M$'m'$ = ds'. In the following paragraph we have the correct expression M$'m'$ = ds'. In the *Théorie* published in 1827 the letter A$'$ was included in the middle of M$'m'$. We included this letter A$'$ in our diagram made with graphic software; see figure A.6 on page 498.

Fig. 6.

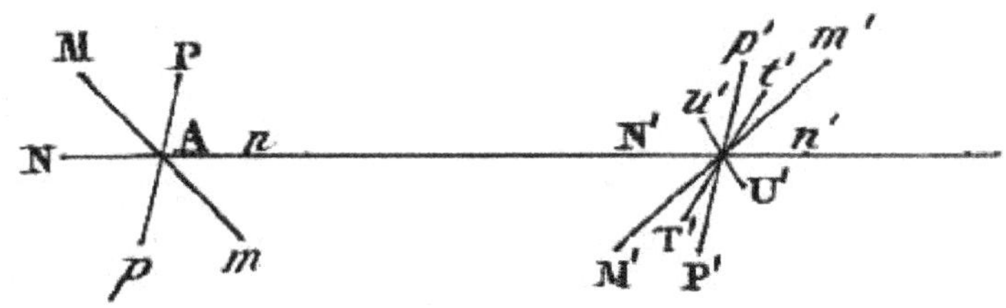

Figure 6.

suppose that the plane MA$'m$ passes along the straight line AA$'$ which joins them, and through the element Mm. We replace the portion of current ds which flows through this element by its projection N$n = ds\cos\theta$ on the straight line AA$'$ and its projection P$p = ds\sin\theta$ on the perpendicular erected at A to this straight line in the plane MA$'m$; we then replace the portion of current ds' which flows through M$'m'$ by its projection[71] N$'n' = ds'\cos\theta'$ on the straight line AA$'$ and its projection P$'p' = ds'\sin\theta'$ on the perpendicular to AA$'$ drawn through the point A$'$ on AA$'$ in the plane M$'$Am'; finally, we replace the latter by its projection T$'t' = ds'\sin\theta'\cos\omega$ in the plane [pages 31 and 203] MA$'m$ and its projection U$'u' = ds'\sin\theta'\sin\omega$ on the perpendicular to this plane through the point A$'$; according to the foregoing law, the two elements ds and ds' exert the same action as the two portions of current $ds\cos\theta$ and $ds\sin\theta$ exert together on the three portions $ds'\cos\theta'$, $ds'\sin\theta'\cos\omega$ and $ds'\sin\theta'\sin\omega$; since the latter has its midpoint in the plane MAm to which it is perpendicular, no action occurs between it and the two portions $ds\cos\theta$ and $ds\sin\theta$ which are in this plane. For the same reason, there can be no action between the portions $ds\cos\theta$ and $ds'\sin\theta'\cos\omega$, nor between the portions $ds\sin\theta$ and $ds'\cos\theta'$, since, imagining a plane through the straight line AA$'$ perpendicular to the plane MA$'m$, [then] $ds\cos\theta$ and $ds'\cos\theta'$ are in this plane and the portions $ds'\sin\theta'\cos\omega$ and $ds\sin\theta$ are perpendicular to it with their midpoints in this same plane. The action of the two elements ds and ds' therefore reduces to the two joint remaining actions, namely the interaction between $ds\sin\theta$ and $ds'\sin\theta'\cos\omega$, and [the action between] $ds\cos\theta$ and $ds'\cos\theta'$, these two actions both being along the straight line AA$'$ joining the midpoints of the currents between which they are exerted, and it suffices to add them to obtain the mutual action of the two elements ds and ds'. Now the portions $ds\sin\theta$ and $ds'\sin\theta'\cos\omega$ are in one and the same plane and both are perpendicular to the straight line AA$'$. Accordingly, their mutual action along this straight line is

$$\frac{ii'dsds'\sin\theta\sin\theta'\cos\omega}{r^n} ,$$

and that of the two portions $ds\cos\theta$ and $ds'\cos\theta'$ along [pages 32 and 204] the same line AA$'$ is

$$\frac{ii'kdsds'\cos\theta\cos\theta'}{r^n} .$$

Thus the interaction of the two elements ds and ds' is necessarily represented by

$$\frac{ii'dsds'}{r^n}(\sin\theta\sin\theta'\cos\omega + k\cos\theta\cos\theta') .$$

[71] [N. T.] In the original text we have N$'n' = ds'\cos\theta$. This was a typographical mistake. What is being discussed here is the angle θ' between ds' and the straight line AA$'$. This angle θ' is not necessarily equal to the angle θ.

This formula is simplified by introducing ε for the angle between the two elements in place of ω; for, by considering the spherical triangle[72] with sides θ, θ' and ε, we have

$$\cos\varepsilon = \cos\theta\cos\theta' + \sin\theta\sin\theta'\cos\omega \ .$$

Hence

$$\sin\theta\sin\theta'\cos\omega = \cos\varepsilon - \cos\theta\cos\theta' \ .$$

Substituting this in the foregoing formula and putting $k - 1 = h$, we get

$$\frac{ii'dsds'}{r^n}(\cos\varepsilon + h\cos\theta\cos\theta') \ .$$

It is as well to point out that a change of sign occurs when one of the currents, say that of the element ds, takes the diametrically opposite direction, for at that time $\cos\theta$ and $\cos\varepsilon$ change sign, whilst $\cos\theta'$ remains the same. This value of the mutual action of the two elements has only been obtained by the substitution of projections for the element itself; but it may be inferred without difficulty that an element can be replaced by an arbitrary polygonal contour, and then by an arbitrary curve [pages 33 and 205] which terminates at the same extremities, provided that all the dimensions of this polygon or curve are infinitesimal.

Suppose, in fact, that ds_1, ds_2, ..., ds_m are different sides of the infinitesimal polygon which is substituted for ds. The direction AA$'$ may always be regarded as in the same direction as the lines joining the respective midpoints of the sides with A$'$.

Let θ_1, θ_2, ..., θ_m be the angles which they form respectively with AA$'$, and ε_1, ε_2, ..., ε_m be those which they form with M$'m'$. Using Σ to denote a sum of terms of like form, the sum of the actions of the sides ds_1, ds_2, ..., ds_m on ds' is

$$\frac{ii'ds'}{r^n}\left(\Sigma ds_I\cos\varepsilon_I + h\cos\theta'\Sigma ds_I\cos\theta_I\right) \ .$$

Now $\Sigma ds_I \cos\varepsilon_I$ is the projection of the polygonal contour on the direction of ds' and, in consequence, it is equal to the projection of ds on the same direction, that is to say, it is equal to $ds\cos\varepsilon$. Likewise $\Sigma ds_I\cos\theta_I$ is equal to the projection of ds on AA$'$ which is $ds\cos\theta$. The action exerted on ds' by the polygonal contour terminated at the extremities of ds may therefore be represented as

$$\frac{ii'ds'}{r^n}\left(ds\cos\varepsilon + hds\cos\theta\cos\theta'\right) \ ,$$

and it is the same as that of ds on ds'.

Since this conclusion is independent of the number of sides ds_1, ds_2, ..., ds_m, it also applies to an infinitesimal arc of a curve.

It could likewise be proved that the action of ds' on ds can be replaced by that which an infinitesimal curve, having the same extremities as [pages 34 and 206] those of ds', would exert on each element of the small curve which we have just substituted for ds, and which would therefore be exerted on this small curve itself. Thus, the formula which we have obtained expresses the fact that a curvilinear element produces the same effect as an infinitesimal portion of rectilinear current with the same extremities, whatever the values of the constants n and h. The experiment by which this result has been reached cannot therefore help in the determination of these constants.

We shall therefore have to utilize two of the other cases of equilibrium which we have discussed. But first we shall transform[73] the foregoing expression for the action of two elements of voltaic currents by introducing the partial differentials of the distance of these two elements.

Let x, y and z be the coordinates of the first point [namely, the center of the element ds], and x', y' and z' those of the second. We get:[74]

---

[72][N. T.] The spherical triangle was discussed in Subsection 7.4.1.

[73][N. T.] This transformation was first presented by Ampère in the dissertation read to the *Académie* on 10 June 1822, [Ampère, 1822o] and [Ampère, 1885p]. We discussed it in detail in Section 7.4. The relevance of this transformation is that with it Ampère could calculate the action of an arbitrary closed circuit on a current element. He could also establish a mathematical relation between the constants n and k.

[74][N. T.] By two typographical mistakes the next equations appeared in the original version of the *Théorie* as:

$$\cos\theta = \frac{x - x'}{r}\frac{dx}{ds} + \frac{y - y}{r}\frac{dy}{ds} + \frac{z - z'}{r}\frac{dz}{ds} \ ,$$

$$\cos\theta = \frac{x-x'}{r}\frac{dx}{ds} + \frac{y-y'}{r}\frac{dy}{ds} + \frac{z-z'}{r}\frac{dz}{ds} ,$$

and

$$\cos\theta' = \frac{x-x'}{r}\frac{dx'}{ds'} + \frac{y-y'}{r}\frac{dy'}{ds'} + \frac{z-z'}{r}\frac{dz'}{ds'} .$$

But we also have:

$$r^2 = (x-x')^2 + (y-y')^2 + (z-z')^2 .$$

Therefore, by successively taking the partial differential coefficients with respect to s and s',

$$r\frac{dr}{ds} = (x-x')\frac{dx}{ds} + (y-y')\frac{dy}{ds} + (z-z')\frac{dz}{ds} ,$$

$$r\frac{dr}{ds'} = -(x-x')\frac{dx}{ds'} - (y-y')\frac{dy}{ds'} - (z-z')\frac{dz}{ds'} ,$$

[pages 35 and 207]
therefore

$$\cos\theta = \frac{dr}{ds} , \quad \text{and} \quad \cos\theta' = -\frac{dr}{ds'} .$$

To obtain the value of $\cos\varepsilon$, note that

$$\frac{dx}{ds} , \frac{dy}{ds} , \frac{dz}{ds} , \quad \text{and} \quad \frac{dx'}{ds'} , \frac{dy'}{ds'} , \frac{dz'}{ds'} ,$$

are the cosines of the angles formed by ds and ds' with the three axes, and it follows that

$$\cos\varepsilon = \frac{dx}{ds}\frac{dx'}{ds'} + \frac{dy}{ds}\frac{dy'}{ds'} + \frac{dz}{ds}\frac{dz'}{ds'} .$$

Now, differentiating with respect to s' the foregoing equation which gives $r\frac{dr}{ds}$, it is found that

$$r\frac{d^2r}{dsds'} + \frac{dr}{ds}\frac{dr}{ds'} = -\frac{dx}{ds}\frac{dx'}{ds'} - \frac{dy}{ds}\frac{dy'}{ds'} - \frac{dz}{ds}\frac{dz'}{ds'} = -\cos\varepsilon .$$

If in the formula for the mutual action of two elements ds and ds' we substitute for $\cos\theta$, $\cos\theta'$ and $\cos\varepsilon$ the values which have just been obtained, and putting $k = 1 + h$, the formula for the mutual action of the two elements ds and ds' becomes

$$-\frac{ii'dsds'}{r^n}\left(r\frac{d^2r}{dsds'} + k\frac{dr}{ds}\frac{dr}{ds'}\right) ,$$

which can be written as

---

and

$$\cos\theta' = \frac{x-x'}{r}\frac{dx'}{ds} + \frac{y-y'}{r}\frac{dy'}{ds'} + \frac{z-z'}{r}\frac{dz'}{ds'} .$$

We now present Ampère's results in modern vector notation. Let $(x,\ y,\ z)$ be the coordinates of the center of the element $d\vec{s} = dx\hat{\imath} + dy\hat{\jmath} + dz\hat{k}$ and $(x',\ y',\ z')$ those of the center of the element $d\vec{s}\,' = dx'\hat{\imath} + dy'\hat{\jmath} + dz'\hat{k}$. The distance between these centers is given by $r = \sqrt{(x-x')^2 + (y-y')^2 + (z-z')^2}$. The unit vector pointing from the center of $d\vec{s}\,'$ to the center of $d\vec{s}$ is given by $\hat{r} = [(x-x')\hat{\imath} + (y-y')\hat{\jmath} + (z-z')\hat{k}]/r$. As θ is the angle between $\hat{r}$ and $d\vec{s}$, while θ' is the angle between $\hat{r}$ and $d\vec{s}\,'$, we can utilize the scalar product to obtain the following relations:

$$\hat{r}\cdot d\vec{s} = \cos\theta ds = \frac{x-x'}{r}dx + \frac{y-y'}{r}dy + \frac{z-z'}{r}dz , \tag{29.1}$$

and

$$\hat{r}\cdot d\vec{s}\,' = \cos\theta' ds' = \frac{x-x'}{r}dx' + \frac{y-y'}{r}dy' + \frac{z-z'}{r}dz' . \tag{29.2}$$

We have thus obtained with modern notation the equations presented by Ampère.

$$-\frac{ii'dsds'}{r^n}\frac{1}{r^{k-1}}\frac{d(r^k\frac{dr}{ds})}{ds'} ,$$

or[75]

$$-ii'r^{1-n-k}\frac{d(r^k\frac{dr}{ds})}{ds'}dsds' .$$

[pages 36 and 208]

It could also be given the following form:

$$-\frac{ii'}{1+k}r^{1-n-k}\frac{d^2(r^{1+k})}{dsds'}dsds' .$$

29.4 Relation Given by the Third Case of Equilibrium between the Two Unknown Constants which Enter in This Formula

Let us now examine the result of the third case of equilibrium which shows that the component of the action of a closed circuit on an element in the same direction as this element is always zero, whatever the form of the circuit.[76] Putting ds' for the element in question, the action of an element ds of the closed circuit on ds' is, according to the foregoing,

$$-ii'ds'r^{1-n-k}\frac{d(r^k\frac{dr}{ds'})}{ds}ds .$$

Or, substituting $-\cos\theta'$ for dr/ds',

$$ii'ds'r^{1-n-k}\frac{d(r^k\cos\theta')}{ds}ds .$$

The component of this action along ds' is obtained by multiplying this expression by $\cos\theta'$:

$$ii'ds'r^{1-n-k}\cos\theta'\frac{d(r^k\cos\theta')}{ds}ds .$$

This differential, integrated over the whole circuit s, yields the total tangential component and it must be zero whatever the form of the circuit. Integrating it by parts, having written it thus

$$ii'ds'r^{1-n-2k}r^k\cos\theta'\frac{d(r^k\cos\theta')}{ds}ds ,$$

we shall have [pages 37 and 209]

$$\frac{1}{2}ii'ds'\left[r^{1-n}\cos^2\theta'-(1-n-2k)\int r^{-n}\cos^2\theta'dr\right] .$$

The first term $r^{1-n}\cos^2\theta'$ vanishes at the limits. As for the integral $\int r^{-n}\cos^2\theta'dr$, it is very easy to imagine a closed circuit for which it does not reduce to zero. In fact, if this circuit is cut by very close spherical surfaces with the center at the midpoint of the element ds', the two points at which each of these spheres cuts the circuit give the same value for r and equal values and opposite signs for dr; but the values of $\cos^2\theta'$ may be different and the squares of all the cosines corresponding to the points situated on one side of the extreme points of the circuit may be made less than those relative to the corresponding points on the other side in an infinite number of ways. Now, in this case, the integral does not vanish; and as the above expression must be zero, whatever the form of the circuit, the coefficient $1-n-2k$ of this integral must therefore be zero, which gives our first relation between n and k:

[75] [N. T.] Due to a misprint, the $-$ sign did not appear in the original text in front of the next equation.

[76] [N. T.] The reasoning discussed in this Section was first presented in [Ampère, 1825d, pp. 385-388], see also [Ampère, 1825c].

$$1 - n - 2k = 0 \,.^{77}$$

Before looking for a second equation for determining these two constants, we start by proving that k is negative, and, as a consequence, that $n = 1 - 2k$ is greater than 1; we will use a fact that is easily experimentally determined,[78] namely, that a rectilinear indefinite conductor attracts a closed circuit, when the electric current in this circuit, in the portion which is closest to the conductor, flows in the same sense as the current in the conductor, and repels in the opposite case.

Let UV (figure 7) be the indefinite rectilinear conductor; assume [pages 38 and 210] for simplicity that the closed circuit THKT′K′H′ is in the same plane as the conducting wire UV, and look for the action exerted by an arbitrary element MM′ of this latter [rectilinear conductor]. To this end draw from the midpoint A of this element the radius vectors connected to all the points of the circuit, and look for the action perpendicular to UV exerted by this element on the circuit.

The component perpendicular to UV of the action exerted by MM′$= ds'$ on an element KH $= ds$ will be obtained by multiplying the expression of this action by $\sin\theta'$; it will therefore be, observing that $1 - n - 2k = 0$,

---

[77][N. T.] In Section 7.5 we showed that in the paper of 10 June 1822, [Ampère, 1822o] and [Ampère, 1885p], Ampère had deduced this same relation beginning with the fact that a circular closed circuit does not exert any torque upon a circuit of arbitrary form which can turn freely around the axis of the circle when the two extremities of the mobile circuit are located on this axis. That is, Ampère originally deduced this relation utilizing the case of equilibrium of the nonexistence of continuous rotation discussed in Subsection 7.3.2. Here in the *Théorie* this relation has been obtained from the case of equilibrium of the nonexistence of tangential force discussed in Section 10.4. Moreover, the method presented in the *Théorie* leads directly to the resut $n = 2$ when it is combined with the case of equilibrium of the law of similarity discussed in Section 10.5.

Tricker presented the following note at this point of the partial English translation of the *Théorie*, [Tricker, 1965, p. 181]:

> "The derivation of Ampère's formula is virtually complete at this point. The value of n may be determined very simply by the method of dimensions, utilizing the results of the fourth experiment which shows that the forces exerted by one circuit on another is independent of the linear dimensions of the circuits. In the expression
>
> $$\frac{ii'dsds'}{r^n}(\sin\theta\sin\theta'\cos\omega + k\cos\theta\cos\theta')\ ,$$
>
> the numerator is of two dimensions in length. It follows that the denominator must also possess two dimensions in length and thus the value of the exponent n must be 2. This then gives $k = -1/2$.
>
> Apparently Ampère did not discover this line of reasoning until after the writing of his Mémoire and it is given only in a note at the end [see Sections 30.1 and 31.1]. The method occurred to him after considering similar conclusions drawn by Laplace from some of Biot's experiments."

Maybe for this reason the partial English translation of the *Théorie* which appears in Tricker's book, see [Ampère, 1965b], continues after this note in the point corresponding to the second half of page 94 of the *Théorie* published in 1826, [Ampère, 1826f], equivalent to page 266 of the *Théorie* published in 1827, [Ampère, 1823c]. That is, pages 38 to 94 of the 1826 edition (or pages 210 to 266 of the 1827 edition) were not translated in Tricker's book.

In any event, this point of view presented by Tricker that Ampère only discovered this line of reasoning after writing the *Théorie* in 1826 is not correct. As a matter of fact, already in 28 November 1825, Ampère read to the Academy of Sciences of Paris a dissertation in which he presented the case of equilibrium of the law of similarity, discussed in Section 10.5, see [Ampère, 1826b, pp. 37-38] and [Ampère, 1887f, pp. 205-207]. Moreover, in the paper of 1825 he utilized directly this argument mentioned by Tricker, which Ampère also included in the first Note at the end of the *Théorie*; see Sections 30.1 and 31.1.

We present here some words of Ampère written at the end of 1825, [Ampère, 1826b, p. 39] and [Ampère, 1887f, pp. 207-208]. He was then considering a force inversely proportional to the nth power of the distance r between two current elements. His words:

> Moreover, it should be noted that the procedure just indicated for the determination of the value of n might had been concluded from this [argument, namely,] that the mutual action between two elements of electric currents, being necessarily proportional to the product of the lengths of these elements, and represented by this product multiplied by a function of the angles which determine their orientations and divided by the nth power of their distance, the number of dimensions of the values of the double integrals expressing the resultant forces of the mutual action between two circuits is, necessarily, $2 - n$. Therefore, when it is supposed that all dimensions of the two circuits increase or decrease in the same ratio [*rapport*], without changing the angles, it follows necessarily that the action must be, as we have just seen otherwise, proportional to the $2 - n$ power of this ratio. Hence, if the action remains the same, it necessarily follows that $n = 2$.

That is, Ampère showed in 1825 that if the resultant force between two circuits does not change when all linear dimensions are altered in the same proportion, it follows necessarily that the force between the current elements must be inversely proportional to the square of their distance.

It seems to us that Ampère opted consciously to include this line of reasoning only in the first Note at the end of the *Théorie*, see Sections 30.1 and 31.1. That is, instead of being a late discovery, as suggested by Tricker, probably this procedure of including it at the end of his work was a methodological choice that Ampère made.

[78][N. T.] Ampère is referring to the experiment of May 1822 showing that $n > 1$ or that $k < 0$. This experiment was discussed in Section 8.3.

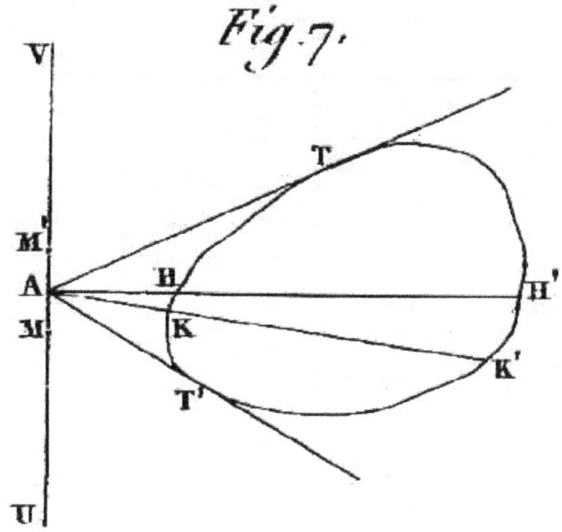

Figure 7.

$$ii'ds'\sin\theta' r^k \frac{d(r^k\cos\theta')}{ds}ds\ ,$$

or

$$\frac{1}{2}ii'ds'\tan\theta' \frac{d(r^{2k}\cos^2\theta')}{ds}ds\ ,$$

an expression which should be integrated over the entire extent of the circuit. Integration by parts yields

$$\frac{1}{2}ii'ds'\left(r^{2k}\sin\theta'\cos\theta' - \int r^{2k}d\theta'\right)\ .$$

The first term vanishes at the limits; there remains only

$$-\frac{1}{2}ii'ds'\int r^{2k}d\theta'\ .$$

Now considering the two elements [of the circuit] KH and K′H′ contained between the same two consecutive radii, $d\theta'$ is the same of both, but must be taken with the opposite sign, in such a way that, making AH $= r$ and AH′$= r'$, one has for the joint action of the two elements

$$-\frac{1}{2}ii'ds'\left[\int (r'^{2k} - r^{2k})d\theta'\right]\ ,$$

[pages 39 and 211]
where we assumed that r' is greater than r. The term in this integral which results from the action of the part THT′, convex toward UV, will dominate over that [term] which is produced by the action of the concave part TH′T′ if k is negative; the reverse will hold if k is positive, and there will be no action if k is zero. Occurring the same results for all the elements of UV, it follows that the convex part toward UV will have more influence on the movement of the circuit than the concave part, if $k < 0$, equal influence if $k = 0$, and less [influence] for $k > 0$. Now, experience proves that it has a larger influence. One then takes $k < 0$, and it follows that $n > 1$, because $n = 1 - 2k$.

From this reasoning this remarkable consequence follows, namely, that the parts of the same rectilinear current repel each other; as a matter of fact, by making $\theta = 0$ and $\theta' = 0$, the formula for the attraction between two elements becomes $kii'dsds'/r^n$; and as it is negative, because k is negative, there is repulsion. This is what I verified by the experiment which I now describe.[79]

[79][N. T.] This experiment was made originally in Geneva between the end of August and the beginning of September 1822; see Section 8.2.

Fig. 8

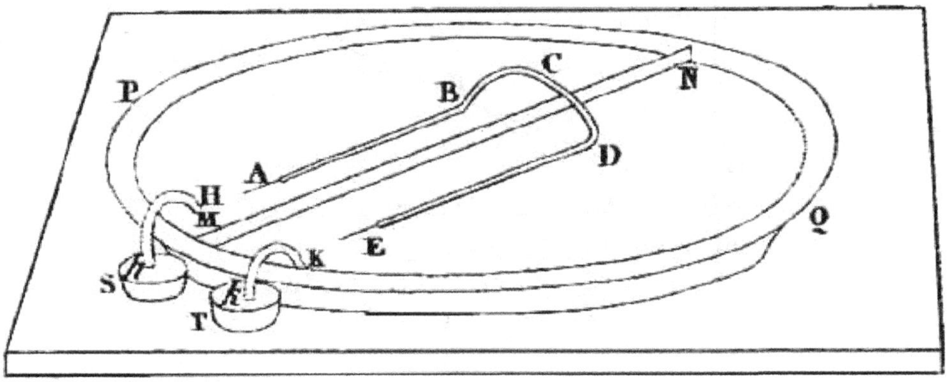

Figure 8.

One takes a glass container PQ (figure 8) separated by an [insulating] partition MN into two equal compartments filled with mercury, one places in it a silk-covered copper wire ABCDE, with the branches AB and ED, situated parallel to the partition MN, floating on the mercury which communicates with the bare extremities A and E of these portions. By placing the rheophores in the capsules S and T, where the mercury communicates with that [mercury] of the vase PQ by the portions *h*H and *k*K of the conductor, one establishes two currents, in which each one has as conductor one part of the mercury and a part solid: whatever the direction of the current, one sees always the two wires AB and ED moving away from the points H and K parallel to the partition MN, which indicates a repulsion [pages 40 and 212] for each wire between the current established in the mercury and its extension in the wire itself. Depending upon the sense of the current, the motion of the wire is more or less quick, because, in one case, the action exerted by the [terrestrial] globe on the portion BCD of this wire, adds to the obtained effect, while in the other case, on the contrary, it decreases the obtained effect and should be subtracted from it.[80]

29.5 General Formulas which Represent the Action of a Closed Voltaic Circuit, or of a System of Closed Circuits, on an Electric Current Element

We examine now the action exerted by an electric current which forms a closed circuit,[81] or a system of currents which also form closed circuits, on an element of electric current.[82]

Take the origin of the coordinate [system] at the center[83] A′ (figure 9)[84] of the considered element M′N′,

[80][N. T.] The force exerted by the current in the mercury acting on the bridge ABCDE is always repulsive, no matter the sense of the current in the bridge. This repulsive force moves the bridge away from the points H and K, in such a way that it approaches the point N. The force exerted by the Earth on the bridge, on the other hand, depends upon the sense of the current. Let us suppose that when the current flows in the sense ABCDE the force exerted by the Earth on the bridge acts in the sense from M to N. In this case it will add to the force exerted on the bridge by the current in the mercury. The net force on the bridge has then a larger intensity than the repulsive force exerted only by the mercury on the bridge. In the opposite case, when the current flows in the sense EDCBA, the force exerted by the Earth on the bridge will be in the opposite direction, from N to M. Therefore, in this case the net force on the bridge will be smaller than the force exerted only by the mercury on the bridge.

[81][N. T.] By a typographical mistake, in the original text we read "formant un circuit formé" instead of "formant un circuit fermé."

[82][N. T.] The following text which goes until the final equation for U at the end of our page 374, corresponding to pages 53 and 225 of the 1826 and 1827 versions of the *Théorie*, respectively, were taken from Ampère's paper of 22 December 1823, [Ampère, 1824c], [Ampère, 1824f], [Ampère, 1824e] and [Ampère, 1885f].

[83][N. T.] Original French sentence in the 1826 and 1827 versions of the *Théorie*: "Prenons l'origine des coordonnées au milieu A de l'élément proposé M′N′, [...]." However, as indicated in figure 9 appearing on page 40 of the 1826 edition of the *Théorie*, and on page 212 of the 1827 edition, the center of the element M′N′ is the point A′. There was a typographical mistake in the French sentence. Only in the *Théorie* of 1826 we have the errata informing that the center of the element M′N′ is the point A′.

[84][N. T.] We call attention here that Ampère always utilized a left-handed coordinate system.

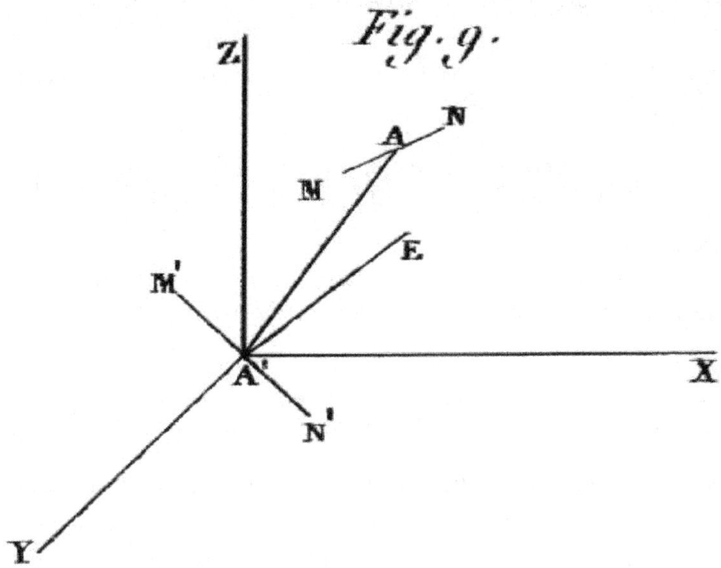

Figure 9.

and let λ, μ and ν, be the angles which it makes with the three axes. Let MN be any element of a current forming a closed circuit, or of one of the currents forming equally closed circuits composing the considered system of currents, and naming ds' and ds the elements M′N′ and MN, r the distance AA′ of their centers and θ' the angle of the current M′N′ with AA′, the formula that we previously found expressing the mutual action of two elements becomes, by replacing[85] dr/ds' by $-\cos\theta'$:[86]

$$ii'ds'r^k\frac{d(r^k\cos\theta')}{ds}ds\ .$$

The angles which AA′ makes with the three axes have for cosines x/r, y/r and z/r; [therefore] one has

$$\cos\theta' = \frac{x}{r}\cos\lambda + \frac{y}{r}\cos\mu + \frac{z}{r}\cos\nu\ ;$$

by substituting this value for $\cos\theta'$,[87] and multiplying by x/r, we find as the expression of the component [of the force] following the x axis,
[pages 41 and 213]

$$ii'ds'r^{k-1}xd(r^{k-1}x\cos\lambda + r^{k-1}y\cos\mu + r^{k-1}z\cos\nu)\ ,$$

the sign d referring only, except in the factor ds', to the differentials taken when varying only s, this expression can be written as[88]

$$= ii'ds'\left[\cos\lambda r^{k-1}xd(r^{k-1}x) + \frac{x\cos\mu}{y}r^{k-1}yd(r^{k-1}y) + \frac{x\cos\nu}{z}r^{k-1}zd(r^{k-1}z)\right]$$

[85][N. T.] By a typographical mistake, in the original text we read dr/ds. The correct expression should be dr/ds', as Ampère was referring here to the current M′N′, corresponding to the element ds' and the angle θ' it makes with AA′.

[86][N. T.] In the original text the portion in the parenthesis appeared as $(r^n\cos\theta')$. This was a typographical mistake. The index n should not appear in this place, as this equation came from

$$-ii'ds'r^{1-n-k}\frac{d(r^k\frac{dr}{ds'})}{ds}ds\ .$$

Substituting n by $1-2k$, we obtain the equation included in this English translation.

[87][N. T.] By a typographical mistake, in the original text we read $\cos\theta$. This section of the text is discussing the element ds', which makes an angle θ' with the straight segment of length r connecting the two elements. For this reason the correct expression should be $\cos\theta'$.

[88][N. T.] In the original text the last fraction in the third line was written as $d\frac{x}{y}$. This was a typographical mistake. The correct expression should be $d\frac{x}{z}$, as we wrote in this English translation.

$$= \frac{1}{2} ii'ds' \left[\cos\lambda d(r^{2k-2}x^2) + \frac{x}{y}\cos\mu d(r^{2k-2}y^2) + \frac{x}{z}\cos\nu d(r^{2k-2}z^2) \right]$$

$$= \frac{1}{2} ii'ds' \left(d\frac{x^2\cos\lambda + xy\cos\mu + xz\cos\nu}{r^{n+1}} - \frac{y^2\cos\mu}{r^{n+1}} d\frac{x}{y} - \frac{z^2\cos\nu}{r^{n+1}} d\frac{x}{z} \right)$$

$$= \frac{1}{2} ii'ds' \left(d\frac{x\cos\theta'}{r^n} + \frac{xdy - ydx}{r^{n+1}}\cos\mu - \frac{zdx - xdz}{r^{n+1}}\cos\nu \right) ,$$

by replacing $2k - 2$ by its value $-n - 1$.

If one represents by r_1, x_1, θ'_1, and r_2, x_2, θ'_2, the values of r, x and θ', at the two extremities of the arc s, and by X the resultant along the x axis of all the forces exerted by the elements of this arc on ds', one obtains

$$X = \frac{1}{2} ii'ds' \left(\frac{x_2\cos\theta'_2}{r_2^n} - \frac{x_1\cos\theta'_1}{r_1^n} + \cos\mu \int \frac{xdy - ydx}{r^{n+1}} - \cos\nu \int \frac{zdx - xdz}{r^{n+1}} \right) .$$

If this arc forms a closed circuit, r_2, x_2 and θ'_2, will be equal to [pages 42 and 214] r_1, x_1 and θ'_1, and the value of X reduces to

$$X = \frac{1}{2} ii'ds' \left(\cos\mu \int \frac{xdy - ydx}{r^{n+1}} - \cos\nu \int \frac{zdx - xdz}{r^{n+1}} \right) .$$

By designating Y and Z the forces along the y and z axes resulting from the action of the same elements on ds', one finds by a similar calculation

$$Y = \frac{1}{2} ii'ds' \left(\cos\nu \int \frac{ydz - zdy}{r^{n+1}} - \cos\lambda \int \frac{xdy - ydx}{r^{n+1}} \right) ,$$

$$Z = \frac{1}{2} ii'ds' \left(\cos\lambda \int \frac{zdx - xdz}{r^{n+1}} - \cos\mu \int \frac{ydz - zdy}{r^{n+1}} \right) ,$$

and by taking

$$\int \frac{ydz - zdy}{r^{n+1}} = A , \quad \int \frac{zdx - xdz}{r^{n+1}} = B , \quad \text{and} \quad \int \frac{xdy - ydx}{r^{n+1}} = C ,$$

it becomes

$$X = \frac{1}{2} ii'ds'(C\cos\mu - B\cos\nu) ,$$

$$Y = \frac{1}{2} ii'ds'(A\cos\nu - C\cos\lambda) ,$$

$$Z = \frac{1}{2} ii'ds'(B\cos\lambda - A\cos\mu) .$$

By multiplying the first of these equations by A, the second by B and the third by C, one finds $AX + BY + CZ = 0$; and if one conceives at the origin [A′] a straight line A′E [see figure 9] which makes with the [Cartesian orthogonal] axes angles whose cosines are given by, respectively,[89]

$$\frac{A}{D} = \cos\xi_1 , \quad \frac{B}{D} = \cos\eta_1 , \quad \frac{C}{D} = \cos\zeta_1 ,$$

by supposing, for brevity,[90]
[pages 43 and 215]

$$\sqrt{A^2 + B^2 + C^2} = D ,$$

[89][N. T.] The first equation in the original text was written as "$\frac{A}{B} = \cos\xi_1$." This was a typographical mistake, as the hypotenuse here is the length $D = \sqrt{A^2 + B^2 + C^2}$.

[90][N. T.] By a typographical mistake, in the original text we read $\sqrt{A^2 + B^2 + C^2 = D}$.

it[91] will be perpendicular to the resultant R of the three forces X, Y and Z, which make with the [orthogonal Cartesian] axes angles whose cosines are[92]

$$\frac{X}{R}\,,\ \frac{Y}{R}\,,\ \frac{Z}{R}\,,$$

since one has, by virtue of the preceding equation,[93]

$$\frac{A}{D}\frac{X}{R}+\frac{B}{D}\frac{Y}{R}+\frac{C}{D}\frac{Z}{R}=0\ .$$

It should be remarked that the straight line [A′E] which we have determined is completely independent of the direction of the element M′N′; because it is immediately deduced from the integrals A, B and C, which depend only on the closed circuit and on the position of the coordinate planes, and which are the sums of the projections on the coordinate planes of the areas of the triangles which have their vertex at the center of the element ds', and as bases the different elements of the closed circuit s,[94] all these areas being divided by the power $n+1$ of the radius vector r.[95] The resultant is perpendicular to this straight line A′E, which I will call directrix;[96] it is located, regardless of the direction of the element [ds'], in the plane raised at the point A′ perpendicular to the [straight line] A′E; I will call this plane the directing plane.[97] If one forms the sum of the squares of X, Y and Z, one finds as the value of the resultant of the action [on ds' exerted by] the single circuit, or by the system of circuits which one has considered, [the following value:][98]

$$R=\frac{1}{2}Dii'ds'\left[(\cos\zeta_1\cos\mu-\cos\eta_1\cos\nu)^2+(\cos\xi_1\cos\nu-\cos\zeta_1\cos\lambda)^2\right.$$
$$\left.+\,(\cos\eta_1\cos\lambda-\cos\xi_1\cos\mu)^2\right]^{1/2}\,,$$

or, naming ε the angle of the element ds' with the directrix, [one obtains:][99]

$$R=\frac{1}{2}Dii'ds'\sin\varepsilon\ .$$

[pages 44 and 216]

It is easy to determine the component of this action in a given plane passing through the element ds' and making an angle φ with the plane formed by ds' and the directrix. In effect, the resultant R being perpendicular to this last plane, its component on the given plane will be

$$R\sin\varphi\ ,\quad\text{or}\quad\frac{1}{2}Dii'ds'\sin\varepsilon\sin\varphi\ .$$

Now, $\sin\varepsilon\sin\varphi$ is equal to the sine of the angle ψ which the directrix makes with the given plane. This is what one deduces immediately from the trihedral angle formed by ds', by the directrix and by its projection on the given plane. The component in this plane will therefore have as its expression

$$\frac{1}{2}Dii'ds'\sin\psi\ .$$

[91][N. T.] That is, the straight line A′E.

[92][N. T.] That is, the resultant force R has three components X, Y and Z along the orthogonal Cartesian axes, such that $R=\sqrt{X^2+Y^2+Z^2}$. The cosines of the angles which the component forces X, Y and Z make with the three orthogonal Cartesian axes are given by, respectively: X/R, Y/R and Z/R.

[93][N. T.] By a typographical mistake, the second fraction of the second term was written as $\frac{Y}{X}$.

[94][N. T.] In the original text we read "des circuits fermés s."

[95][N. T.] This is an extremely important result. Ampère obtained this result at the end of 1823, just after the publications of Savary; see Chapter 9 and references [Ampère, 1824c], [Ampère, 1824f], [Ampère, 1824e] and [Ampère, 1885f].

[96][N. T.] The word "directrice" appeared in italics in [Ampère, 1887d, p. 38]. Its meaning was discussed in Section 10.1.

[97][N. T.] The expression "plan directeur" appeared in italics in [Ampère, 1824c, p. 142], [Ampère, 1885f, p. 398] and also in the 1887 edition of the *Théorie*, [Ampère, 1887d, p. 38]. The first name which Ampère gave to the *directrix* was *normal to the directing plane*, "normale au plan directeur," [Ampère, 1826b, p. 41], [Ampère, 1885f, p. 398], [Ampère, 1887c, p. 195] and [Ampère, 1887f, p. 210]. This expression appeared also in the *Théorie*, see pages 97, 98, 104 and 105 of the 1826 edition, or pages 269, 270, 276 and 277 of the 1827 edition.

[98][N. T.] By a typographical mistake, the last term inside the square root of the original text did not appear elevated to 2.

[99][N. T.] This extremely important result was discussed in Section 10.1.

The angle ε defined here between ds' and the directrix due to a system of closed circuits should not be confused with the angle ε between two elements of current introduced on pages 32 and 204 of the 1826 and 1827 editions of the *Théorie*, respectively, which correspond to page 359 of this book.

This expression can be put in another form by observing that ψ is the complement[100] of the angle which the directrix makes with the normal to the plane in which one considers the action. One has therefore, naming ξ, η and ζ the angles that this last straight line forms with the three axes,

$$\sin\psi = \frac{A}{D}\cos\xi + \frac{B}{D}\cos\eta + \frac{C}{D}\cos\zeta \ ,$$

and the expression of the action becomes

$$\frac{1}{2}ii'ds'(A\cos\xi + B\cos\eta + C\cos\zeta) \ ,$$

or

$$\frac{1}{2}Uii'ds' \ ,$$

by using [the following definition:]

[pages 45 and 217]

$$U = A\cos\xi + B\cos\eta + C\cos\zeta \ .$$

One sees that this action is independent of the direction of the element in the plane that one has considered; we designate it under the name of the action exerted in this plane,[101] and we conclude that it [i.e., this action] remains the same when one gives successively to the element various directions in the same plane;[102] [we also conclude] that if that [action] which the Earth exerts on a mobile conductor in a fixed plane is produced by electric currents forming closed circuits, the distances of which to the [mobile] conductor being sufficiently large to be considered as constants during the time that it [i.e., this conductor] moves in this plane, it [i.e., this action] will always have the same value in the various positions which the conductor successively takes, because the actions exerted on each of the elements of which it is composed always resting the same and always being perpendicular to these elements, their resultant cannot vary in its size nor in its direction relative to the conductor. This direction will change moreover in the fixed plane by following there the motion of this conductor: it is in effect what one observes with respect to a conductor which is mobile in a horizontal plane, and which one directs successively in various azimuths.

29.6 Experiment by which One Verifies a Consequence of These Formulas

One can verify this result by the following experiment:[103] within a wooden disc ABCD (figure 10), one carves a circular channel KLMN in which one places two copper vessels KL and MN of the same form, and which each occupy nearly the half-circumference of the channel in a manner such that there remains between them two intervals KN and LM which one fills with an insulating putty; on each of these vessels one solders the two copper plates PQ and RS, embedded in the [pages 46 and 218] disc and which have goblets X and Y, designed to allow, through the mercury which they contain, the communication of the vessels KL and MN with the rheophores of a very strong battery; in the disc there is embedded another plate TO, carrying the goblet Z, where one also places a small amount of mercury; this plate TO is soldered at the center O of the disc to a vertical rod to which is soldered a fourth goblet U, which has its bottom covered by a piece of glass or agate to make more mobile the crossed framework[104] which we will discuss, but whose edges are sufficiently high so as to be in communication with the mercury which one places in this goblet; it [i.e., this goblet U] receives the tip V (figure 11)[105] which forms the pivot of the crossed framework FGHI, whose branches EG

[100][N. T.] The complement of an angle φ_1 is another angle φ_2 such that $\varphi_1 + \varphi_2 = \pi/2$.

[101][N. T.] The expression "action exercée dans ce plan" appeared in italics in [Ampère, 1887d, p. 39].

[102][N. T.] The portion beginning in this point and going up to the expression "which should be verified" in our page 370, which corresponds to pages 47 and 219 of the 1826 and 1827 editions of the *Théorie*, respectively, was taken from [Ampère, 1824c, pp. 138-140].

[103][N. T.] See [Ampère, 1824c, p. 139 and figure 2].

[104][N. T.] *Sautoir* in the original.

[105][N. T.] The tip V below the central point E did not appear in the original figure 11 of the 1827 edition of the *Théorie*. This was corrected in the 1826 edition of the *Théorie*, as indicated here. This tip V will remain in electric contact with the mercury of goblet U of figure 10. This tip V had already appeared in the first presentation of this experiment, see [Ampère, 1824c, p. 139 and figure 3].

Fig. 10.

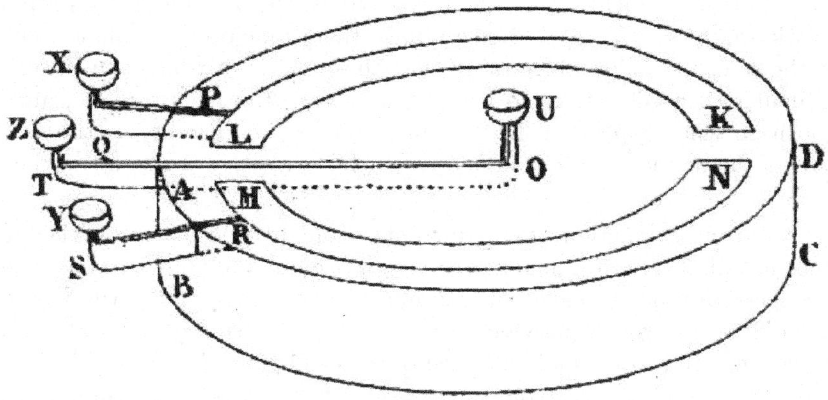

Figure 10.

and EI are mutually equal and soldered at G and I to the laminas gxh and iyf which are submerged in the acidulated water of the vases KL and MN [of figure 10], when the tip V rests over the bottom of goblet U, and which are attached by their other extremities h and f by arms EH and EF, without communicating with them. These two laminas are equal, similar, and folded in an approximately 90° circular arc.

Fig. 11.

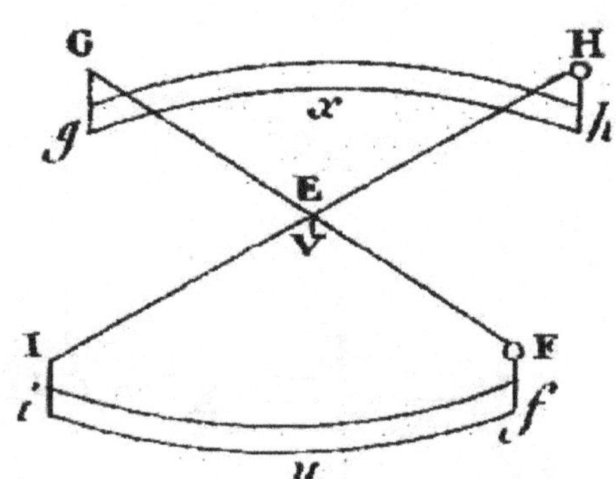

Figure 11.

When one inserts the rheophores, one in the goblet Z, the other in one of the two goblets X or Y, the current only passes through one of the arms of the crossed framework, and one sees it [i.e., this crossed framework] turn on the tip V due to the Earth's action, from East to West by the South when the current goes from the circumference to the center, and in the opposite direction when it [i.e., when the current] goes from the center to the circumference, conforming to the explanation of this phenomenon that I have given, and which one can see in my *Recueil d'Observations électro-dynamiques*, on page 284.[106] But when one inserts them [i.e.,

[106][N. T.] Ampère is referring here initially to a work of Auguste de la Rive which was published in 1822 and reprinted in the *Recueil* with some modifications due to Ampère, [de la Rive, 1822b, p. 46], [de la Rive, 1822a, p. 47], [de la Rive, 1822c, p. 284] and [de la Rive, 1885, p. 326]. These works were followed by Notes written by Ampère himself; see [Ampère, 1822z] and [Ampère, 1822a]. In the *Recueil* Auguste de la Rive's paper came after a Preface written by Ampère, [Ampère, 1822c] and

the rheophores] in the goblets X and Y, the current flows in the opposite directions through the two arms EG and EI, [and] the crossed framework remains stationary in any situation it was placed, [pages 47 and 219] as when, for example, one of the arms is parallel and the other perpendicular to the magnetic meridian, and this [lack of motion of the crossed framework] takes place even when one knocks lightly the disc ABCD, increasing the mobility of the instrument by the small vibrations which result. By slightly bending the arms of the crossed framework around the point E, one can make them take different angles, and the result of the experiment is always the same. It follows inevitably that the force with which the Earth acts on a portion of conductor, perpendicular to its direction, to make it move in a horizontal plane, and, by consequence, in a plane given in position with respect to the system of terrestrial currents, is the same, no matter the direction, in this plane, of the portion of the conductor, which is precisely the result of the calculation it was meant to verify.

It should be remarked that the action of the currents of acidulated water on their extensions on the laminas gh and if does not disturb in any manner the equilibrium of the device; since it is easy to see that the action which is in question tends to cause the lamina gh to turn about the tip V in the direction hxg, and the lamina if in the direction fyi, from which results, due to the equality of these laminas, two torques that cancel one another, since they are equal and with opposite signs.

One knows that it is M. Savary who is responsible for the experiment by which one found this action;[107] this experiment can be made easier by replacing the copper wire spiral in the device, which was first used, by a circular lamina of the same metal. This lamina ABC (figure 12)[108] forms a circular arc nearly equal to a complete circumference; but its extremities A and C are separated from each other by a piece D of insulating material.

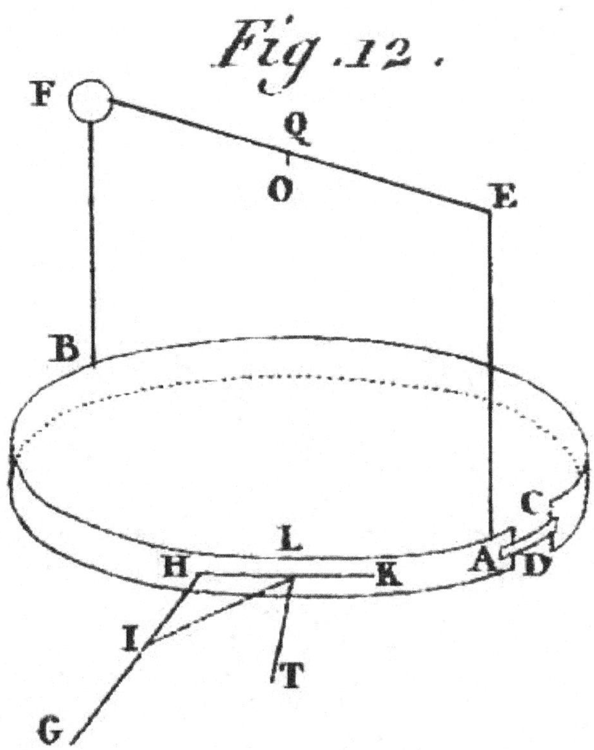

Figure 12.

One makes [pages 48 and 220] one of its extremities A, for instance, in communication with one of the

[Ampère, 1885c].

[107][N. T.] Ampère mentioned this experiment performed by Savary in [Ampère, 1822d, p. 66], [Ampère, 1822e, p. 243] and [Ampère, 1885d, p. 198].

[108][N. T.] The tip O below letter Q at the center of the wire EF did not appear in figure 12 of the *Théorie* published in 1827. This was corrected in figure 12 of the *Théorie* published in 1826. This tip allows an electric contact with one of the poles of the battery connected to the mercury located inside cup S of figure 13; see our page 371.

rheophores at the tip O which one places in the cup S (figure 13)[109, 110] filled with mercury. This [cup] is joined by the metal wire STR to the cup R in which one of the rheophores is immersed. This tip O (figure 12) connects with the extremity A by the copper wire AEQ whose extension QF supports at F the lamina ABC by a ring of insulating material, which covers the copper wire at this point. When the tip O rests on the bottom of the cup S (figure 13), the lamina ABC (figure 12) is immersed in the acidulated water contained in the copper vessel MN (figure 13) which communicates with the cup P which contains the other rheophore;

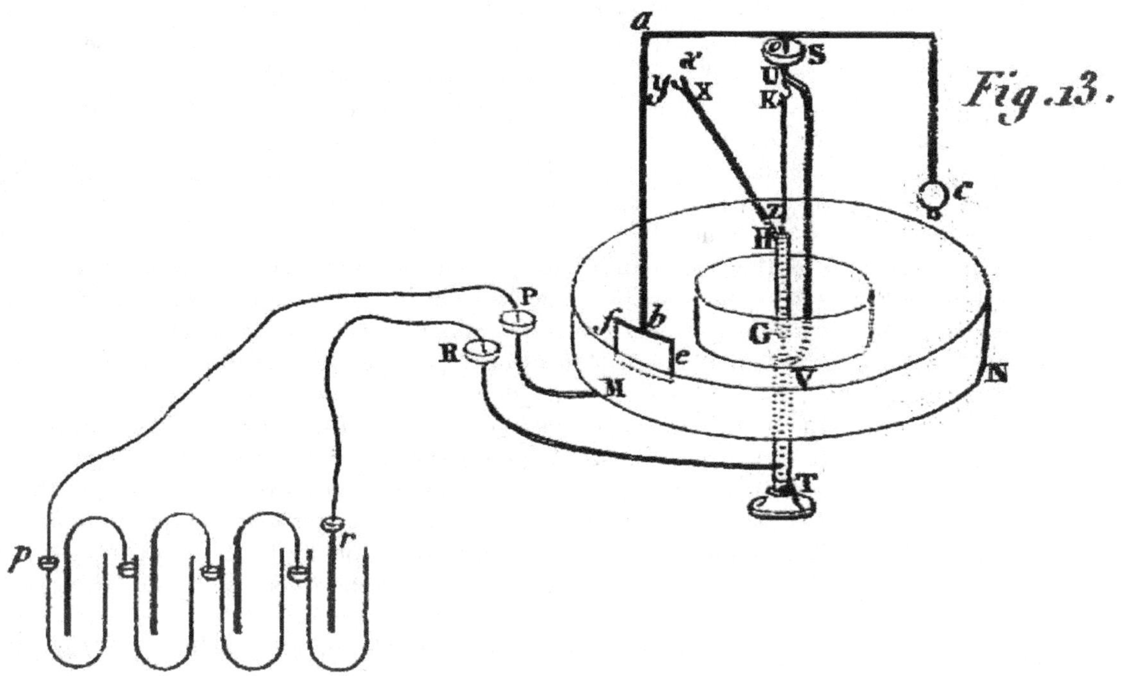

Figure 13.

one sees therefore turning of this lamina in the direction CBA, and provided that the battery is strong enough, the movement continues in this direction until one reverses the communications with the battery, by reciprocally changing the two rheophores of the cup P with the cup R, thus proving that this movement is not at all due to the action of the Earth[111] and only derive from the action that the currents in the acidulated water exert on the current flowing in the circular lamina ABC (figure 12), an action which is always repulsive, because if GH represents one of the acidulated water currents which extends to HK in the

[109][N. T.] In figure 13 of the *Théorie* published in 1827 the letter O in cup S did not appear. This was corrected in the 1826 edition presented here. Tip O below letter Q appeared in figure 12; see our page 370.

[110][N. T.] The figure included in this footnote is a simpler device than that shown in figure 13 of the *Théorie*. It was published in 1824, [Ampère, 1824c, figure 6]. It has been inserted here to facilitate the comprehension of Ampère's experiment. Figure 13 of the *Théorie* is more complex than this device due to the discussion which Ampère will make in our page 449, corresponding to page 174 of the *Théorie* published in 1826 and to page 346 of the *Théorie* published in 1827. As the figures were collected in plates at the end of the book, a specific figure might have rich details which were utilized in several discussions. Sometimes this made these figure reasonably complex, as is the case with figure 13 of the *Théorie*.

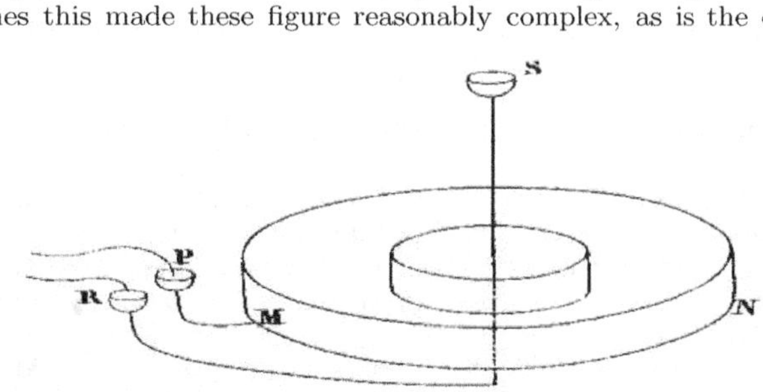

[111][N. T.] Ampère himself had shown that the force or torque exerted by the Earth on a current-carrying wire reverses its direction by inverting the sense of the current in the conductor. In this experiment, on the other hand, the action exerted on the mobile conductor does not change its direction by inverting the sense of the current flowing though it.

lamina ABC, regardless of the direction of this current, it will obviously travel one of the sides of the angle GHK while approaching, and the other [side] while flowing away from the vertex H. But it is necessary, so that the movement which one observes in this case can take place, that the repulsion between two elements, one in I and the other in L, take place along the line IL, oblique to the arc ABC, and not following the perpendicular LT at the element situated in L, since the direction of this perpendicular encounters the vertical drawn through the point O around which the mobile part of the device is allowed to turn, a force [pages 49 and 221] directed along this perpendicular could not impart any rotational movement.

I have just said that, when one wants to be assured that the movement of this device is not produced by the action of the Earth, by establishing that it continues to happen in the same direction when one reverses the connections to the battery by changing the rheophores at the cups, it is necessary to use a battery of sufficient strength; it is effectively impossible, in this arrangement of a mobile conductor, to avoid the Earth's action on the vertical wire AE moving it to the West, when the current there is ascending, to the East when the current is descending, and on the horizontal wire EQ, in order to make it turn about the vertical passing through the point O, in the sense directly East, South, West, when the current goes from E to Q, while approaching the rotational center, and in a retrograde Western, Southern, Eastern direction, when it goes from Q to E, while following the same center.[112, 113] The first of these actions is hardly observable, at least when one gives to the vertical wire AE a length only sufficient for the stability of the mobile conductor at its tip O; but the second [action] is determined by the dimensions of the device; and since it changes direction when one reverses connections with the battery, it is added in one order of the connections with the action exerted by the acidulated water currents, and it reduces in the other; this is why the observed movement is always more rapid in one case than in the other; this difference is more pronounced [pages 50 and 222] when the current produced by the battery is weaker, because as its intensity decreases, the electrodynamic action being, all other things remaining the same, as the product of the intensities of the two portions of the currents which act on each other, this action between the acidulated water currents and those of the lamina ABC, decreases as the square of their intensity, while the intensity of the terrestrial currents remaining the same, their action, on those of the lamina, will only decrease proportionally to the same intensity: as the intensity of the battery diminishes, the action of the Earth becomes more and more able to destroy that [action] of the acidulated water currents in the arrangement of the connections with the battery where these actions are opposed, and one sees, when this energy becomes very weak, the device will stop in this case, and the movement then appears in the contrary direction; thus the experiment leads to a conclusion opposite that which was expected to be established, since the action of the Earth became dominant one could ignore the existence of those from the acidulated water currents. For the rest, the first of these two actions is always null on the circular lamina ABC, because the Earth acts like a system of closed currents, the force that they exert on each element being perpendicular to the direction of this element, passes through the vertical set by the point O, and cannot, as a consequence, tend to cause rotation of the mobile conductor around the vertical.

29.7 Application of the Preceding Formulas to a Circular Circuit

We will, to serve as an example, apply the preceding formulas to the case where the system reduces to a single closed circular current.[114]

When the system is only composed of a single current, [pages 51 and 223] traversing a circular circumference of any radius m, one simplifies the calculation, by taking, for the xy plane, the plane through the coordinate origin, that is to say, through the center A of the element ab (figure 14)[115] parallel to that of the circle; and for the xz plane, the one that goes perpendicularly through the plane of the circle by the same origin [A] and by the center O [of the circular current].

---

[112] [N. A.] Note for these two kinds of actions exerted by the Earth, what is said in my *Recueil d'Observations électro-dynamiques*, pages 280, 284.

[113] [N. T.] Ampère is referring here initially to a work of Auguste de la Rive published in 1822 and reprinted in the *Recueil* including some modifications made by Ampère, [de la Rive, 1822b, pp. 42 to 46], [de la Rive, 1822a, pp. 44-47], [de la Rive, 1822c, pp. 280 to 284] and [de la Rive, 1885, pp. 323 to 327]. These works were followed by Notes written by Ampère himself, [Ampère, 1822z] and [Ampère, 1822a]. In the *Recueil* the work of Auguste de la Rive was preceded by a Preface written by Ampère, [Ampère, 1822c] and [Ampère, 1885c].

[114] [N. T] The following text has been taken essentially from the following works: [Ampère, 1824c, pp. 148-151] and [Ampère, 1824e, pp. 16-20].

[115] [N. T.] The figure presented in this footnote is much simpler than figure 14 of the *Théorie*. This illustration was published in 1824, [Ampère, 1824c, table 2, figure 7]. Figure 14 of the *Théorie* is more complex than the figure of this footnote due to the discussion which Ampère will present in our pages

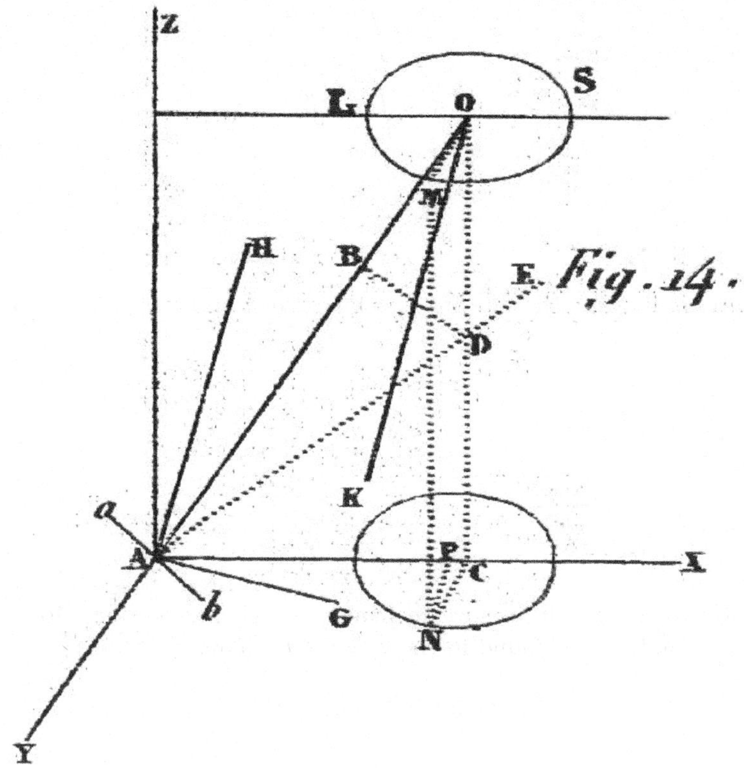

Figure 14.

Let p and q be the coordinates of this center O;[116] suppose that the point C is the projection of O on the xy plane, N that of any point M of the circle, and name ω the angle ACN; if one projects NP perpendicularly on AX, the three coordinates x, y and z of the point M will be MN, NP and AP, and one easily finds for their values:

$$z = q \ , \ y = m \sin \omega \ , \quad \text{and} \quad x = p - m \cos \omega \ .$$

The magnitudes that we have designated[117] by A, B and C being respectively equal to

$$\int \frac{ydz - zdy}{r^{n+1}} \ , \ \int \frac{zdx - xdz}{r^{n+1}} \ , \quad \text{and} \quad \int \frac{xdy - ydx}{r^{n+1}} \ ,$$

we have:

469 and 481, corresponding to pages 214 and 380 of the *Théorie* published in 1826 and 1827, respectively.

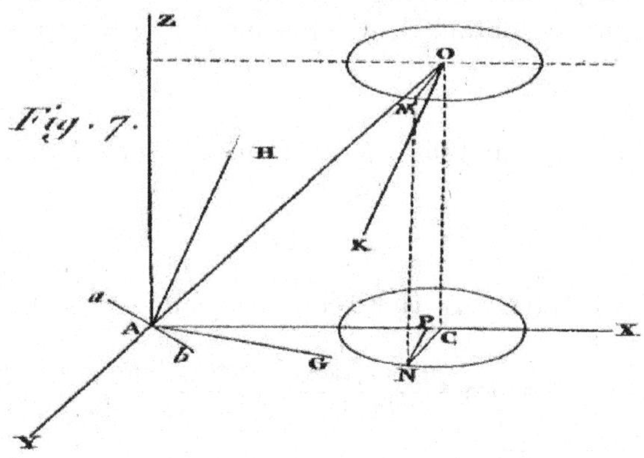

[116][N. T.] That is, according to figure 14, the coordinates (x_O, y_O, z_O) of the center O are given by $(x_O, y_O, z_O) = (p, 0, q)$, respectively.

[117][N. T.] On pages 42 and 214 of the 1826 and 1827 versions of the *Théorie*, corresponding to our page 366.

$$A = -mq \int \frac{\cos \omega d\omega}{r^{n+1}} ,$$

$$B = mq \int \frac{\sin \omega d\omega}{r^{n+1}} ,$$

and

$$C = mp \int \frac{\cos \omega d\omega}{r^{n+1}} - m^2 \int \frac{d\omega}{r^{n+1}} .$$

If one integrates by parts those of these terms which contain $\sin \omega$ and $\cos \omega$, while noting that

[pages 52 and 224]

$$r^2 = x^2 + y^2 + z^2 = q^2 + p^2 + m^2 - 2mp \cos \omega ,$$

gives

$$dr = \frac{mp \sin \omega d\omega}{r} ,$$

if one removes the terms which are null because their integrals should be taken from $\omega = 0$ to $\omega = 2\pi$, and one sets the values of A, B and C thus found in the value of U, [namely,][118]

$$U = A \cos \xi + B \cos \eta + C \cos \zeta ,$$

one obtains[119]

$$U = m^2 \left[(n+1)(p^2 \cos \zeta - pq \cos \xi) \int \frac{\sin^2 \omega d\omega}{r^{n+3}} - \cos \zeta \int \frac{d\omega}{r^{n+1}} \right] .$$

Now, the angle ξ can be expressed by means of ζ; since, by designating h the perpendicular OK projected from the center O on the plane bAG for which one calculates the value of U, one obtains $h = q \cos \zeta + p \cos \xi$, and this value becomes

$$U = m^2 \left\{ (n+1) \left[(p^2 + q^2) \cos \zeta - hq \right] \int \frac{\sin^2 \omega d\omega}{r^{n+3}} - \cos \zeta \int \frac{d\omega}{r^{n+1}} \right\} .$$

29.8 Simplification of the Formulas when the Diameter of the Circular Circuit is Very Small

The evaluation is quite simple in the case where the radius m is very small when compared to the distance l of the origin A to the center O; since, if one develops in series [the value of U] following the powers of m, one has that when one neglects the powers of m higher than 3, the terms in m^3 disappear between the limits 0 and 2π, and those that are in m^2 are obtained by replacing r by $l = \sqrt{p^2 + q^2}$; it only remains therefore to calculate the values of

$$\int \sin^2 \omega d\omega \quad \text{and of} \quad \int d\omega \ \text{ from } \omega = 0 \text{ to } \omega = 2\pi ;$$

[pages 53 and 225]

which gives π for the first [integral] and 2π for the second; the value of U therefore reduces to

$$U = \pi m^2 \left[\frac{(n-1) \cos \zeta}{l^{n+1}} - \frac{(n+1)hq}{l^{n+3}} \right] .$$

[118][N. T.] The magnitude U had been introduced in pages 45 and 217 of the 1826 and 1827 versions of the *Théorie*; corresponding to our page 368.

[119][N. T.] In the original texts of the versions published in 1826 and 1827 we have incorrectly the letter m instead of m^2 multiplying the square brackets of this equation. Joubert corrected this misprint; see [Ampère, 1887d, p. 46].

29.9 Application to a Planar Circuit which Forms a Closed Curve of Arbitrary Shape, at First in the Case where the Dimensions are All Very Small, and then when They are of Any Size Whatsoever

For increased generality, we will now assume that the closed current, instead of being circular, has any form, but still remains plane and very small.[120]

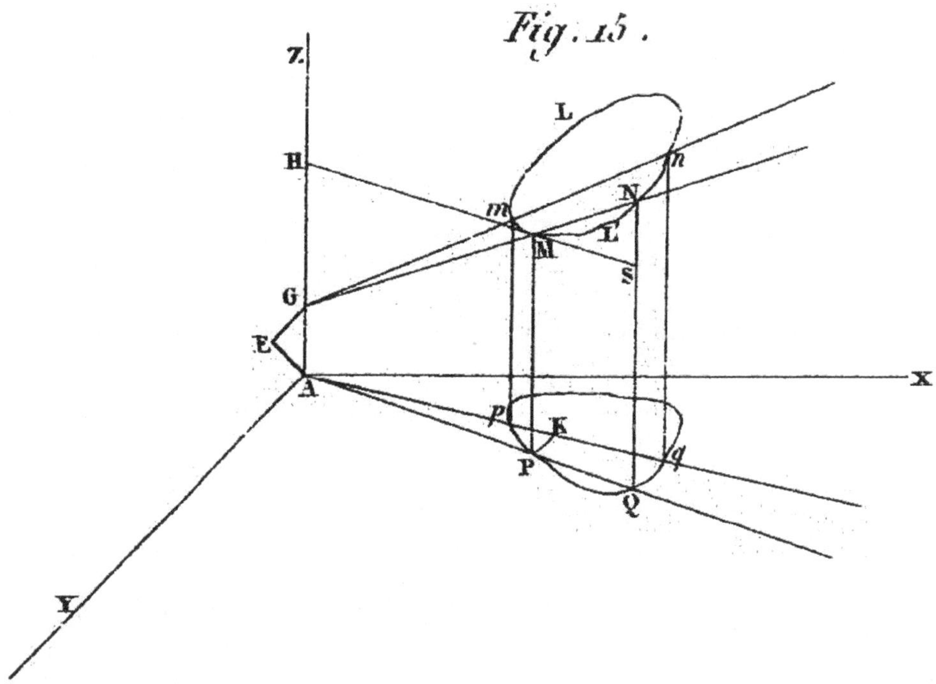

Figure 15.

Let MNL (figure 15) be a very small closed and planar circuit of which the area is λ and which acts on an element placed at the origin A. Partition its surface into infinitely small elements, by planes passing by the z axis, and let APQ be the trace of one of these planes [projected on the xy plane], with M and N being its meeting points with the circuit λ, projected on the xy plane in P and Q. Extend the chord MN to the z axis in G; draw from A a perpendicular AE $= q$ to the plane of the circuit, and join EG. Let Apq be the trace of a plane infinitely close to the first [plane], making with it an angle $d\varphi$; let AP $= u$ and PQ $= \delta u$. The action of the circuit on the element in A depends, as we have seen, on three integrals designated by A, B and C, which we will calculate. Consider first C, whose value is

$$C = \int \frac{xdy - ydx}{r^{n+1}} = \int \frac{u^2 d\varphi}{r^{n+1}} \,.$$

This integral is relative to all the points of the circuit, and if one considers simultaneously the two elements included between the two adjoining planes AGNQ and AGnq, related to the equal values and opposite signs of $d\varphi$, one will see that the actions of these two elements should be subtracted from each other, and that the [action of] one of the elements which is the closest [pages 54 and 226] to A produces the stronger action. Observing that to have the action from the farthest [element], it is necessary to replace u and r by $u + \delta u$ and $r + \delta r$, one finds

$$C = \int \frac{u^2 d\varphi}{r^{n+1}} - \int \frac{(u+\delta u)^2 d\varphi}{(r+\delta r)^{n+1}} \,,$$

120[N. T.] Ampère will show that the action of a planar and very small closed circuit acting on a current element of another circuit is independent of the shape of the closed circuit, being proportional to its area. This demonstration can be found at the end of a memoir of 12 September 1825, [Ampère, 1825e, pp. 38-41]. This demonstration did not appear in the text published originally in the *Annales de Chimie et de Physique*, [Ampère, 1825d] and [Ampère, 1825c].

these two integrals being taken between the two values of φ relative to the extreme points L and L′ between which the circuit is included.

The difference between these two integrals can be considered as the variation of the first taken with the sign reversed; if one neglects all the powers of the circuit dimensions whose exponents are greater than unity, it becomes

$$C = -\delta \int \frac{u^2 d\varphi}{r^{n+1}} = \int \left[\frac{(n+1)u^2 \delta r}{r^{n+2}} - \frac{2u\delta u}{r^{n+1}} \right] d\varphi \, .$$

Now

$$r^2 = u^2 + z^2 \, ,$$

from where

$$\delta r = \frac{u\delta u + z\delta z}{r} \, ;$$

also the angle ZAE being equal to ζ, one has

$$\text{AG} = \frac{q}{\cos\zeta} \quad \text{and} \quad \text{GH} = z - \frac{q}{\cos\zeta} \, ,$$

and, due to the similar triangles MHG and MSN,

$$\text{MH} : \text{MS} :: \text{GH} : \text{NS} \, ,$$

that is to say

$$u : \delta u :: z - \frac{q}{\cos\zeta} : \delta z \, ;$$

by extracting from this proportion the value of δz and carrying it into that of δr, one obtains

$$\delta z = \frac{z\cos\zeta - q}{u\cos\zeta}\delta u \quad \text{and} \quad \delta r = \frac{(u^2+z^2)\cos\zeta - qz}{ur\cos\zeta}\delta u = \frac{r^2\cos\zeta - qz}{ur\cos\zeta}\delta u \, ,$$

[pages 55 and 227]

and by substituting this value into C, it becomes:[121]

$$C = \int \left[\frac{(n+1)(r^2\cos\zeta - qz)}{r^{n+3}\cos\zeta} - \frac{2}{r^{n+1}} \right] u\delta u d\varphi$$
$$= \int \left[\frac{n-1}{r^{n+1}} - \frac{(n+1)qz}{r^{n+3}\cos\zeta} \right] u\delta u d\varphi \, .$$

Since the circuit is very small, one can consider the values of r and z as constants and equal, for example, to those that occur at the center of gravity of the area of the circuit, to ensure that the third-order terms vanish, and representing these values [of r and z] by l and z_1, the preceding integral takes this form

$$C = \left[\frac{n-1}{l^{n+1}} - \frac{(n+1)qz_1}{l^{n+3}\cos\zeta} \right] \int u d\varphi \delta u \, .$$

But $ud\varphi$ is the arc PK given by A as center with the radius u and PQ $= \delta u$; therefore $ud\varphi\delta u$ is the infinitely small area PQqp, and the integral $\int ud\varphi du$ gives the total area of the projection of the circuit, that is, $\lambda\cos\zeta$, since ζ is the angle of the plane of the circuit with the xy plane; one obtains therefore finally

$$C = \left[\frac{(n-1)\cos\zeta}{l^{n+1}} - \frac{(n+1)qz_1}{l^{n+3}} \right] \lambda \, .$$

One obtains analogous values for B and A, namely:

[121][N. T.] In the first integral we read in Ampère's original *Théorie* the expression $(r^2\cos\xi - qz)$ instead of $(r^2\cos\zeta - qz)$. This was a misprint.

$$B = \left[\frac{(n-1)\cos\eta}{l^{n+1}} - \frac{(n+1)qy_1}{l^{n+3}} \right] \lambda \ ,$$

$$A = \left[\frac{(n-1)\cos\xi}{l^{n+1}} - \frac{(n+1)qx_1}{l^{n+3}} \right] \lambda \ .$$

One knows thus the angles that the directrix makes with the axes, since one has for their cosines A/D, B/D, A/D, by making

[pages 56 and 228]

$$D = \sqrt{A^2 + B^2 + C^2} \ .$$

As for the force produced by the action of the circuit on the element situated at the origin, it will have, as we saw above, the expression $\frac{1}{2}ii'ds'D\sin\varepsilon$, ε being the angle which this element makes with the directrix, to which this force is perpendicular as it is perpendicular to the direction of the element.[122]

In the case where the small circuit that is considered is in the same plane[123] as the element ds' on which it acts, one has, by taking this plane as the one for the xy,

$$q = 0, \ \cos\zeta = 1, \ \cos\eta = 0, \ \cos\xi = 0 \ ,$$

and consequently

$$A = 0, \ B = 0, \ C = \frac{n-1}{l^{n+1}}\lambda \ ;$$

therefore, D reduces to C; ε is equal to $\pi/2$, and the action of the circuit on the element[124] ds' becomes[125]

$$\frac{n-1}{2}\frac{ii'ds'\lambda}{l^{n+1}} \ .$$

I will now present a new manner of considering the action of planar circuits of any form and size.

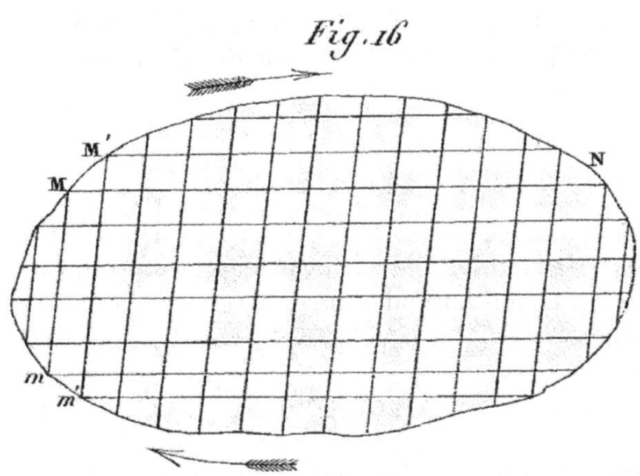

Figure 16.

Consider any planar circuit MNm (figure 16);[126] partition its surface into infinitely small elements by parallel lines cut by a second system of parallels making right angles with the first ones, and imagine, around each of these infinitely small areas, currents directed in the same sense as the current MNm. All the parts

[122] [N. T.] That is, this force is perpendicular to the directrix and also to the direction of the element.

[123] [N. T.] See the work of 21 November 1825, [Ampère, 1826b] and [Ampère, 1887f].

[124] [N. T.] By a misprint in the original text we had here ds instead of the correct ds'.

[125] [N. T.] In the original text we read $iids'\lambda$. This was another misprint. After all, Ampère was discussing the interaction between two elements with current intensities given by i and i'. The correct expression should be $ii'ds'\lambda$.

[126] [N. T.] In the *Théorie* published in 1827 the two arrows indicating the sense of the current do not appear. They were included in figure 16 of the 1826 version reproduced here.

of these currents which are found following these straight lines will be destroyed, because there will be two [currents of] contrary signs which follow the same line; [pages 57 and 229] and there will only remain the curved parts of these currents, such as MM′ and mm', which will form the complete circuit MNm.

It follows from this [argument] that the three integrals A, B and C will be obtained for the planar circuit of finite size, by substituting in the values which we obtained for these three quantities, in place of λ any element of the area of the circuit that we can represent by $d^2\lambda$ and integrating in all the extent of this area.

When, for example, the element is situated in the same plane as the circuit, and one considers this plane as that of the xy, one has

$$A = 0,\ B = 0,\ C = (n-1)\int\int\frac{d^2\lambda}{l^{n+1}}\,,$$

and the value of the force becomes

$$\frac{n-1}{2}ii'ds'\int\int\frac{d^2\lambda}{l^{n+1}}\,;$$

from which it follows that, if at each of the points of the area of the circuit one raises a perpendicular equal to $1/l^{n+1}$, the volume of the prism which has as its base the circuit and which is terminated on the surface formed by the extremities of these perpendiculars, will represent the value of [127] $\int\int\frac{d^2\lambda}{l^{n+1}}$; and this volume multiplied by $\frac{n-1}{2}ii'ds'$ will express the sought action.

It is important to observe that as the question was directed to the cubature of a solid,[128] one could adopt the system of coordinates and the division of the area of the circuit into elements which will lead to the simpler calculations.

29.10 Mutual Interaction of Two Closed Circuits Located in the Same Plane, First Assuming that All the Dimensions are Very Small, and then in the Case where the Two Circuits are of Any Form and Size Whatsoever

Let us proceed to the mutual action of two very small circuits [pages 58 and 230] O and O′ (figure 18)[129] situated in the same plane. Let MN be an arbitrary element ds' of the second [circuit]. The action of the circuit O on ds' is, after the preceding,[130]

$$\frac{n-1}{2}\frac{ii'ds'\lambda}{r^{n+1}}\,.$$

Calling $d\varphi$ the angle MNO and drawing the arc MP between the sides of this angle, one can replace the small current MN by two currents MP and NP of which the lengths are respectively $rd\varphi$ and dr; the action of the circuit O on the element MP, which is normal to its direction, is obtained by replacing in the preceding expression ds' by MP, and becomes

$$\frac{n-1}{2}\frac{ii'\lambda d\varphi}{r^{n}}\,;$$

the action on NP, perpendicular to its direction, becomes similarly

$$\frac{n-1}{2}\frac{ii'\lambda dr}{r^{n+1}}\,.$$

This last [equation] integrated over the entire closed circuit O′ is null, [therefore,] it suffices to consider the first [force component] which is directed toward the point O, from which it already results that the action of these two small circuits is directed along the line which joins them.

---

[127] [N. T.] In the 1826 and 1827 versions of the *Théorie* the expression which appears in the integrals was given by l^{n+2}. The errata appearing at the end of these works informed that the correct expression should be l^{n+1}.

[128] [N. T.] That is, calculation of the volume of a solid.

[129] [N. T.] Some figures appear out of order in the *Théorie*. One example is this figure 18 appearing in the text before figure 17.

[130] [N. T.] Due to a misprint in the 1826 and 1827 versions of the *Théorie*, the numerator of the second fraction appeared as $ii'ds'\lambda d\varphi$ instead of the correct expression $ii'ds'\lambda$. This mistake was corrected in [Ampère, 1887d, p. 51].

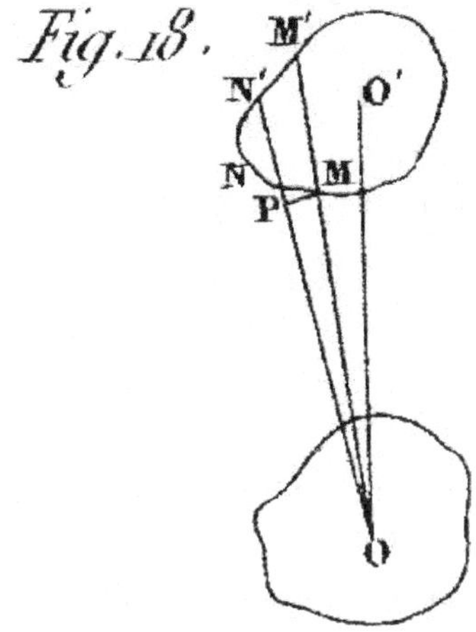

Figure 18.

Extend the rays OM and ON until they encounter the curve in M′ and N′; the action of M′N′ should be subtracted from that of MN, and the resulting action is obtained by taking as before the variation of that of MN with the sign reversed; this gives

$$\frac{n(n-1)}{2}\frac{ii'\lambda d\varphi\delta r}{r^{n+1}} \quad \text{or} \quad \frac{n(n-1)}{2}\frac{ii'\lambda r d\varphi\delta r}{r^{n+2}} .$$

Now, $rd\varphi\delta r$ is the measure of the infinitely small segment MNN′M′. [pages 59 and 231] Taking the sum of all the analogous expressions relative to different elements of the circuit O′ and considering r as constant and equal to the distance between the centers of gravity of the areas λ and λ' of the two circuits, one obtains for the action which one exerts on the other

$$\frac{n(n-1)}{2}\frac{ii'\lambda\lambda'}{r^{n+2}} ,$$

and this action will be directed along the line OO′. Therefore the mutual action of two finite circuits situated in the same plane is obtained by considering their areas as partitioned into elements, infinitely small in all respects, and supposing that these elements act on one another along the line that joins them, in direct proportion of their surfaces and in inverse proportion of the power $n+2$ of their distance.

As the mutual action of closed currents is a function only of the distance, one draws this important consequence, that there can never result from this action a continuous rotational motion.

29.11 Determination of the Two Unknown Constants which Enter into the Fundamental Formula

The formula which we just found for obtaining the mutual action of two closed and planar circuits based on the actions of the elements of the areas of these circuits, leads to the determination of the value of n. In effect, if one considers two similar systems composed of two closed and planar circuits, the similar elements of their areas will be proportional to the square of the homologous lines, and the distances of these elements will be proportional to the first powers of the same lines. Calling m the ratio of homologous lines of the two systems, the actions of two elements of the first system and their correspondents in the second will be, [pages 60 and 232] respectively,

$$\frac{n(n-1)}{2}\frac{ii'\lambda\lambda'}{r^{n+2}} \quad \text{and} \quad \frac{n(n-1)}{2}\frac{ii'\lambda\lambda' m^4}{r^{n+2}m^{n+2}} .$$

their ratio, and hence the ratio between the total actions, will thus be m^{2-n}. However, we have described previously an experiment by which one can prove directly that these two actions are equal;[131] therefore, it is necessary that[132]

$$n = 2 ,$$

and, due to the equation $1 - n - 2k = 0$, [it is also necessary] that

$$k = -\frac{1}{2} .$$

These values of n and k reduce to a very simple form the expression[133]

$$-\frac{ii'}{1+k} r^{1-n-k} \frac{d^2(r^{1+k})}{dsds'} dsds' ,$$

of the mutual action between ds and ds'; this expression becomes

$$-\frac{2ii'}{\sqrt{r}} \frac{d^2(\sqrt{r})}{dsds'} dsds' .$$

It also follows from the value $n = 2$ that, in the case where the directions of the two elements stay the same, this action becomes proportional to the inverse square of their distance. One knows that M. de La Place established the same law, based on an experiment of M. Biot, in the case of the mutual action of an element of a voltaic conductor and of a magnetic molecule: but this result could only be extended to the action of two elements of conductors, by assuming that the actions of the magnets is due to electric currents; while the experimental demonstration that I just gave is independent of all the hypotheses that one could make about the constitution of the magnets.

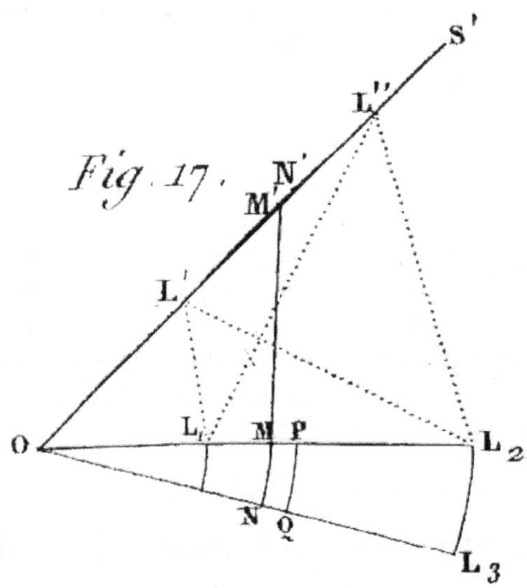

Figure 17.

[131][N. T.] Ampère was referring here to the case of equilibrium of the law of similarity, mentioned on pages 25 and 197 of the *Théorie* published in 1826 and 1827, respectively, corresponding to our page 354.
[132][N. T.] Utilizing that $m^{2-n} = 1 = m^0$ in order to have the equality of the two forces.
[133][N. T.] Due to a misprint in the original versions of the *Théorie* this formula appeared as

$$-\frac{1+k}{ii'} r^{1-n-k} \frac{d^2(r^{1+k})}{dsds'} .$$

As Ampère was discussing the previous expression for the force between two current elements, the correct expression should be the one presented here.

29.12 Action of a Conducting Wire which Forms a Segment of a Circle on a Rectilinear Conductor Passing Through the Center of the Segment

Let MON (figure 17) be a circuit forming a sector whose sides comprise an infinitely small angle, and look for the action that it exerts on a rectilinear conductor OS′ passing through [pages 61 and 233] the center O of the sector,[134] and calculate first that [action] of an element MNQP of the area of the sector on an element M′N′ of the conductor OS′. Let OM $= u$, MP $= du$, OM′ $= s'$, MM′$= r$, S′ON$= \varepsilon$ and NOM $= d\varepsilon$. The torque of MNQP in order to cause M′ to turn about O will, by observing that the area MNQP has as expression $udud\varepsilon$, be given by

$$\frac{1}{2}ii's'ds'\frac{udud\varepsilon}{r^3} ,$$

and the torque of the sector on the conductor s' will be obtained by integrating this expression with respect to u and s'. One has

$$r^2 = s'^2 + u^2 - 2us'\cos\varepsilon ,$$

from which

$$r\frac{dr}{du} = u - s'\cos\varepsilon, \quad r\frac{dr}{ds'} = s' - u\cos\varepsilon ,$$

and, by differentiating a second time,

$$r\frac{d^2r}{duds'} + \frac{dr}{ds'}\frac{dr}{du} = -\cos\varepsilon ,$$

or, by substituting for dr/ds' and dr/du their values,

$$r\frac{d^2r}{duds'} + \frac{(u - s'\cos\varepsilon)(s' - u\cos\varepsilon)}{r^2} = -\cos\varepsilon ,$$

which becomes, by carrying out the calculation and simplifying,

$$r\frac{d^2r}{duds'} + \frac{us'\sin^2\varepsilon}{r^2} = 0 ,$$

from which one extracts

$$\frac{us'}{r^3} = -\frac{1}{\sin^2\varepsilon}\frac{d^2r}{duds'} ;$$

substituting this value in the elementary torque, one has [pages 62 and 234] for the expression of the total torque[135]

$$\frac{1}{2}ii'd\varepsilon\int\int\frac{us'duds'}{r^3} = -\frac{1}{2}ii'\frac{d\varepsilon}{\sin^2\varepsilon}\int\int\frac{d^2r}{duds'}duds' .$$

By considering the portion L′L″ of the current s′, and the portion L_1L_2 of the sector, and by making $L'L_1 = r_1'$, $L''L_1 = r_1''$, $L'L_2 = r_2'$ and $L''L_2 = r_2''$, the value of this integral is evidently

$$\frac{1}{2}ii'\frac{d\varepsilon}{\sin^2\varepsilon}(r_2' + r_1'' - r_2'' - r_1') .$$

As it is from the center O that start [both] the sector and the conductor s', the distance $r_1' = 0$; and if one makes $OL_2 = a$, $OL'' = b$ and $L''L_2 = r$, one finds that their mutual action is expressed by

$$\frac{1}{2}ii'\frac{d\varepsilon}{\sin^2\varepsilon}(a + b - r) .$$

When the conductor L′L″ (figure 19) has for midpoint the center L_1 of the sector, and when its length is double the radius a of this sector, one has $a = b$, and by making $L'L_1L_2 = 2\theta = \pi - \varepsilon$,

[134] [N. T.] With the rectilinear conductor situated in the plane of the sector.

[135] [N. T.] Due to a misprint, the right hand side of the following equation had ii instead of the correct value ii'.

$$r_1' = r_1'' = a,\ r_2' = 2a\sin\theta,\ r_2'' = 2a\cos\theta,\ d\varepsilon = -2d\theta\ ,$$

in such a way that the value of the torque becomes

$$aii'\frac{d\varepsilon}{\sin^2\varepsilon}(\sin\theta - \cos\theta) = \frac{1}{2}\frac{aii'd\theta\,(\cos\theta - \sin\theta)}{\sin^2\theta\cos^2\theta}\ .$$

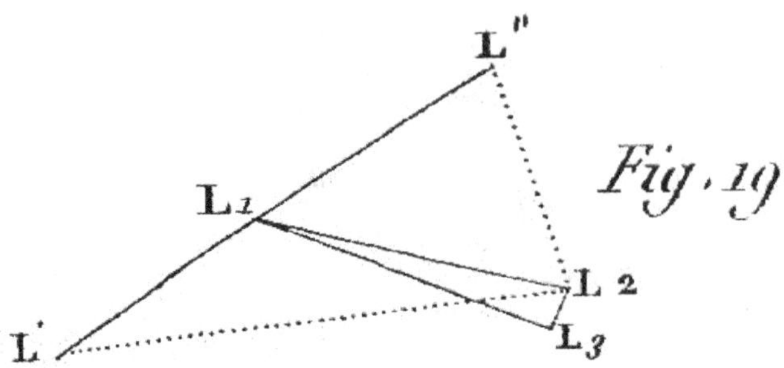

Figure 19.

One can deduce from this result a means of verifying my formula utilizing an instrument which I will now describe.[136]

29.13 Description of an Instrument Designed to Verify the Results of the Theory for Conductors of This Form

At the two points a and a' (figure 20)[137] of the table mn stand two supports ab and $a'b'$ of which the upper parts cb and $c'b'$ are insulated; they support a copper strip HdeH$'d'e'$ [pages 63 and 235] folded in half along the line HH$'$, and which is terminated by two cups H and H$'$ where one places mercury. At points A, C, A$'$ and C$'$ on the table are four cavities also filled with mercury. From A starts a copper conductor AEFGSRQ, supported by HH$'$ and terminated by a cup Q; from A$'$ there starts a second [conductor] A$'$E$'$F$'$G$'$S$'$R$'$Q$'$ symmetric to the first; they are both enclosed in silk, in order to be insulated from each other and from the conductor HH$'$. In the cup Q the tip of a mobile conductor QPONMLKIH returning to itself from K to I is inserted, and having in this part its two branches PO and KI enclosed in silk; it is terminated by a second tip inserted in the cup H; [the branch] NML forms a semicircle of which LN is the diameter, and K the center; the stem PKp is vertical, and terminated at p by a tip held by three horizontal circles B, D and T, which can turn about their centers and are designed to reduce friction.

[In this figure 20,] XY is a fixed shelf which receives in a groove a conductor VU$ifkhgo$ZC returning on itself from g to o and covered by silk in this part; $ifkhg$ is a sector of a circle which has as its center the point k; the parts Ui and go are rectilinear; they traverse at x the support ab, in which there is an opening for this purpose, and which separates at o in order to plunge respectively in the cavities A and C. To the right of FG there is an assemblage of conductors both fixed and mobile exactly similar to those that we just described, and when one inserts the positive rheophore of the battery into C, and the negative into C$'$, the electric current passes through the conductors CZ$oghkfi$UV and AEFGSRQ; from there it passes into the

[136][N. T.] This experiment was mentioned in the work of 12 September 1825, [Ampère, 1825d, p. 398], [Ampère, 1825c], [Ampère, 1825e], [Ampère, 1825g], [Ampère, 1826e] and [Ampère, 1887c]. The instrument designed to perform this experiment was completely described in a memoir of 21 November 1825, [Ampère, 1887f, p. 208, figure 2] and [Ampère, 1826b, pp. 39-41]. In the last paragraph of the *Théorie* Ampère informed that up to that moment he had not yet constructed this instrument. It was apparently never built and the experiment never performed, as discussed in our page 460. The goal of the instrument was to determine, by the number of oscillations of a mobile conductor under the action of another fixed conductor, the value of the torque exerted on the mobile conductor by the fixed conductor. An earlier version of this instrument appears in footnote 26 in our page 345.

[137][N. T.] In figure 20b we present only half of figure 20, so that the letters appearing in it can be read; see our page 384. In the 1826 version of the *Théorie* the letters d, e, p, O, d', e', p' and O', were removed. These letters appeared in the 1827 version of the *Théorie*. The letters i' and g' of the 1826 version presented here appeared in the 1827 version as i and g, respectively. In both versions the letter f' appeared as f. In the 1827, version the letters d' and e' do not correspond to the analogous letters d and e. In the illustration we prepared utilizing graphic software, figure A.20 on page 504, we maintained the letters d, e, p, O, d', e', p' and O' as they appeared in the 1827 version of the *Théorie*, although reversing the positions of the letters d' and e'.

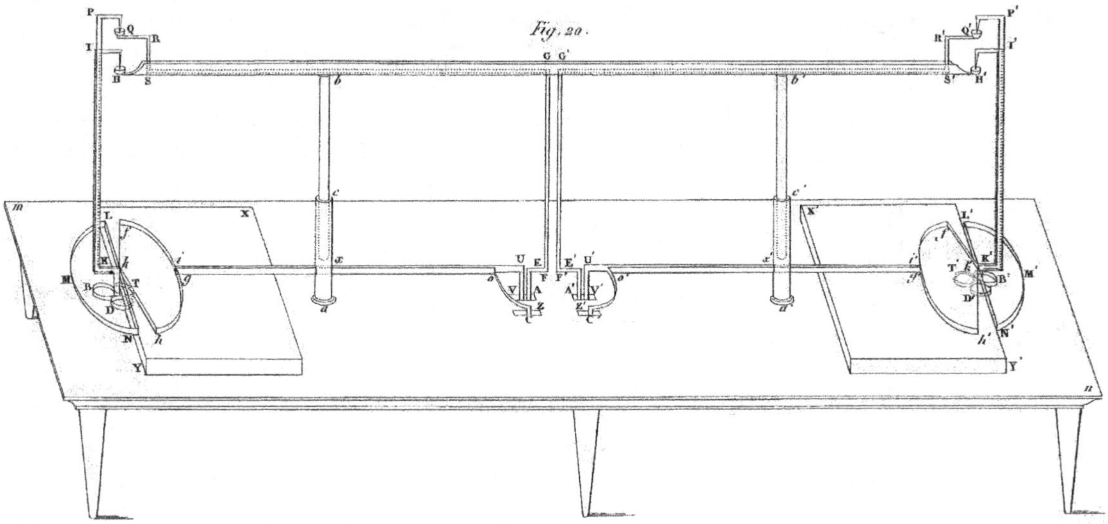

Figure 20.

mobile conductor QPONMLKIH, and connects to H′ by HH′; it then goes through the symmetric mobile conductor [pages 64 and 236] H′I′K′L′M′N′O′P′Q′, arriving at Q′, follows the conductor Q′R′S′G′F′E′A′ which conducts it into the cavity A′, where it connects to C′ by the conductor V′U′$i'f'k'h'g'o'$Z′C′, and from there to the negative rheophore.[138]

The current flowing in the direction LN in the diameter LN, and from h to k, then from k to f, in the rays hk and kf, there is a repulsion between the rays and the diameter; also, the closed circuit $ghkfi$ does not produce any action on the semi-circle LMN whose center is found in the fixed axis pH, the mobile conductor can only be set in motion by the action of the sector $ghkfi$ on the diameter LN, seeing that in all the other parts of the apparatus two opposed currents flow whose actions cancel each other. Equilibrium will be obtained when the diameter LN makes equal angles with the rays kf and kh; and if one departs from this position, there will be oscillation solely due to the action of the sector $ghkfi$ on the diameter.

Letting 2η be the angle at the center of the sector, one obtains at the equilibrium position[139]

$$2\theta = \frac{\pi}{2} + \eta \quad \text{or} \quad \theta = \frac{\pi}{4} + \frac{\eta}{2} \ ,$$

from which one concludes

$$\cos\theta - \sin\theta = \cos\theta - \cos\left(\frac{\pi}{2} - \theta\right) = 2\sin\frac{\pi}{4}\sin\left(\frac{\pi}{4} - \theta\right)$$

$$= -\sqrt{2}\sin\frac{1}{2}\eta \ ,$$

and

$$\sin\theta\cos\theta = \frac{1}{2}\sin 2\theta = \frac{1}{2}\cos\eta \ .$$

But it is easy to see that when one displaces, from its equilibrium position, the conductor[140] L′L″ by an amount equal to $2d\theta$, the torque of the forces which tend to restore it are composed of those [pages 65 and 237] which produce two small sectors whose angles are equal to this displacement, and whose actions are equal, a torque whose value, after that which we have seen just now, is

[138] [N. T.] In a Note published in [Ampère, 1825d, p. 401], see also [Ampère, 1825c, p. 385], Ampère justified the construction of the symmetrical instrument as follows:

> By comparing among themselves measures determined by successive operations, we also completely avoid the inaccuracy produced by the variations of the energy of the pile, which necessarily alter the results deduced from experiment.

[139] [N. T.] Due to a misprint, the right hand side of the last equation appeared in the published versions as $\theta = \pi/4 + \eta$.

[140] [N. T.] The conductor L′L″ at figure 19 of the *Théorie* corresponds to the conductors LN and L′N′ at figure 20.

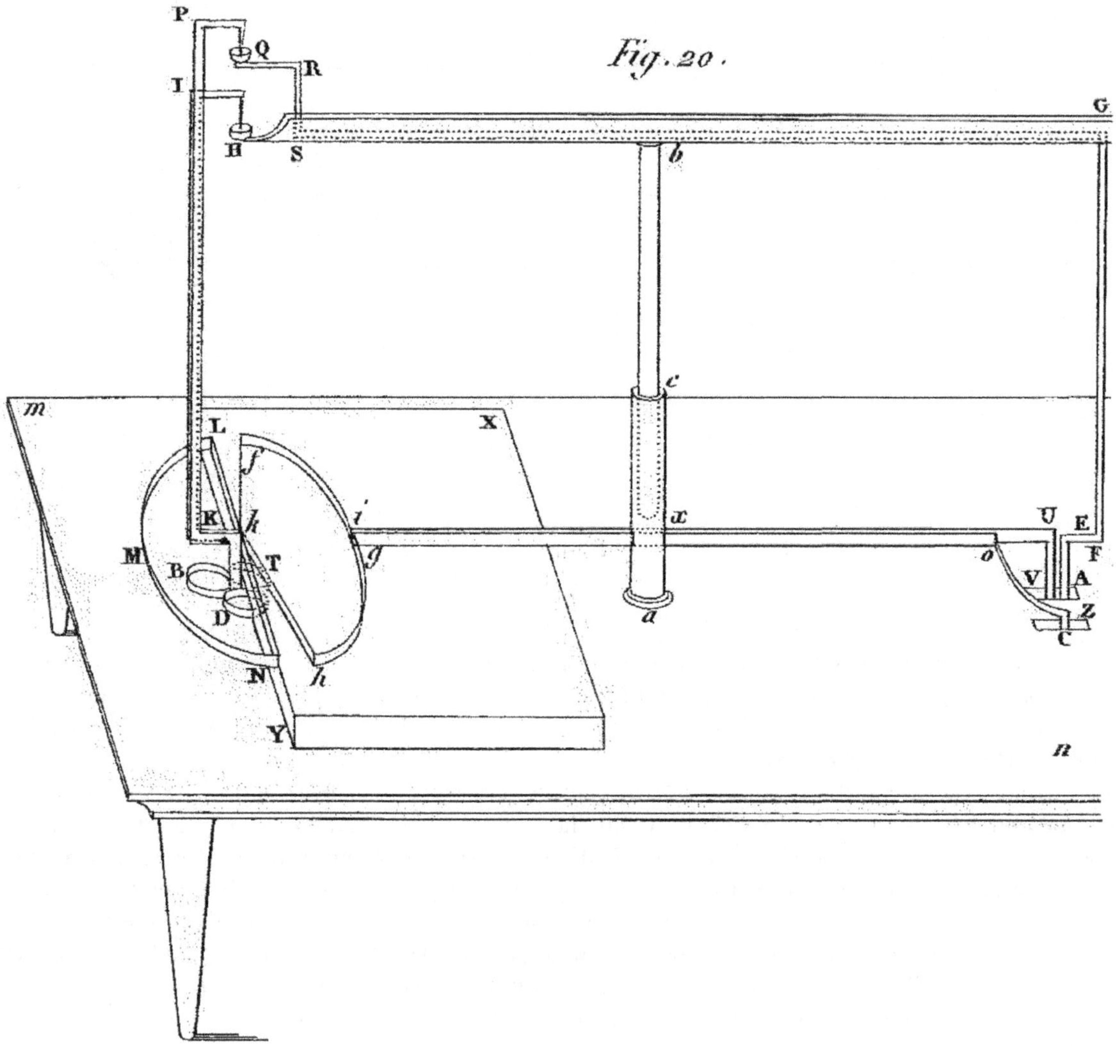

Figure 20b. It shows half of Figure 20 of the *Théorie*, so that the letters can be read.

$$\frac{1}{2}\frac{aii'(\cos\theta-\sin\theta)}{\sin^2\theta\cos^2\theta}d\theta=-\frac{2aii'\sqrt{2}\sin\frac{1}{2}\eta}{\cos^2\eta}d\theta\;.$$

From which it follows that the duration of these oscillations will be, for the same diameter, proportional to

$$\frac{\sqrt{\sin\frac{1}{2}\eta}}{\cos\eta}\;.$$

Therefore by causing simultaneous oscillations of the mobile conductors in the two symmetric parts of the apparatus, supposing the angles of the sectors are different, one will have currents of the same intensity, and one will observe if the numbers of oscillations in the same time, are proportional to the two expressions

$$\frac{\sqrt{\sin\frac{1}{2}\eta}}{\cos\eta}\quad\text{and}\quad\frac{\sqrt{\sin\frac{1}{2}\eta'}}{\cos\eta'}\;;$$

calling the two angles at the center of the two sectors 2η and $2\eta'$.

29.14 Interaction of Two Rectilinear Conductors

We now examine the mutual action of two rectilinear conductors; and we recall first that by calling β the angle comprised between the direction of the element ds' and that of the line r, the value of the action[141] that the two elements of electric current ds and ds' exert on each other has already been put in the form

$$ii'ds'r^k d\left(r^k\cos\beta\right) \ ,$$

by multiplying and dividing by $\cos\beta$, and noting that $k=-1/2$ gives $r^{2k}=1/r$, we see that one can [pages 66 and 238] write it as:

$$\frac{ii'ds'}{\cos\beta}r^k\cos\beta d\left(r^k\cos\beta\right)=\frac{1}{2}\frac{ii'ds'}{\cos\beta}d\left(\frac{\cos^2\beta}{r}\right) \ ,$$

from which it is easy for us to conclude that the component of this action following the tangent of the element ds', is equal to

$$\frac{1}{2}ii'ds'd\left(\frac{\cos^2\beta}{r}\right) \ ,$$

and that the component normal to the same element, is as

$$\frac{1}{2}ii'ds'\tan\beta d\left(\frac{\cos^2\beta}{r}\right) \ ,$$

an expression which can be put in the form

$$\frac{1}{2}ii'ds'\left[d\left(\frac{\sin\beta\cos\beta}{r}\right)-\frac{d\beta}{r}\right] \ .$$

These values of the two components can be found on page 331 of my *Recueil d'Observations électro-dynamiques*, published in 1822.[142]

Apply this last [normal component] to the case of two rectilinear parallel currents, situated at a distance a from each other.[143]

One then has

$$r=\frac{a}{\sin\beta} \ ,$$

and the normal component becomes

$$\frac{1}{2}ii'ds'\left[\frac{d(\sin^2\beta\cos\beta)}{a}-\frac{\sin\beta d\beta}{a}\right] \ .$$

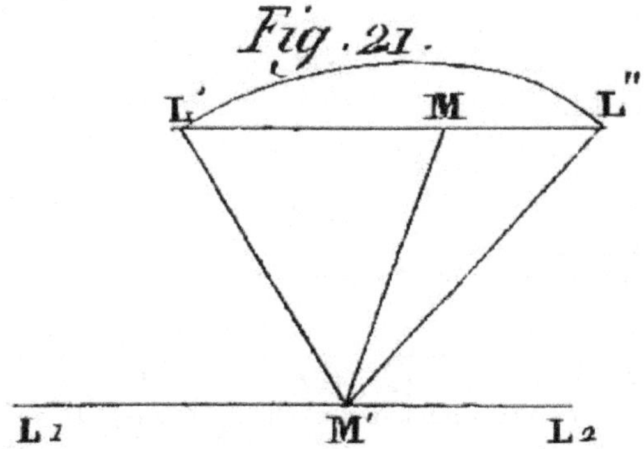

Figure 21.

[141] [N. T.] That is, the value of the force.

[142] [N. T.] See [Ampère, 1822h, p. 331] and [Ampère, 1823a, p. 9].

[143] [N. T.] The next calculations relative to the force between two rectilinear conductors are reproduced from the memoir of 12 September 1825, [Ampère, 1825d, pp. 402-404], [Ampère, 1825c] and [Ampère, 1825e].

Let M′ (figure 21) be any point of a current flowing in the line L_1L_2; and let β' and β'' be the angles L′M′L_2 and L″M′L_2 formed [pages 67 and 239] with L_1L_2 by the extreme radius vectors M′L′ and M′L″; one obtains the action of ds' on L′L″ by integrating the preceding expression between the limits β' and β'', which gives

$$\frac{1}{2a}ii'ds'\left(\sin^2\beta''\cos\beta''+\cos\beta''-\sin^2\beta'\cos\beta'-\cos\beta'\right)\ ;$$

but one has at each limit, by representing the values of s by b' and b'',

$$s'=b''-a\cot\beta''=b'-a\cot\beta',\quad ds'=\frac{ad\beta''}{\sin^2\beta''}=\frac{ad\beta'}{\sin^2\beta'}\ ;$$

substituting these values and integrating anew between the limits β_1', β_2' and β_1'', β_2'', one obtains for the value of the sought force,

$$\frac{1}{2}ii'\Bigg(\sin\beta_2''-\sin\beta_1''-\sin\beta_2'+\sin\beta_1'$$

$$-\frac{1}{\sin\beta_2''}+\frac{1}{\sin\beta_1''}+\frac{1}{\sin\beta_2'}-\frac{1}{\sin\beta_1'}\Bigg)\ ,$$

or[144]

$$\frac{1}{2}ii'\left(\frac{a}{r_2''}-\frac{a}{r_1''}-\frac{a}{r_2'}+\frac{a}{r_1'}+\frac{r_1''+r_2'-r_2''-r_1'}{a}\right)\ .$$

If the two conductors are of the same length and perpendicular to the straight lines which join the two extremities of the same side, one has

$$r_1'=r_2''=a\quad\text{and}\quad r_2'=r_1''=c\ ,$$

naming c the diagonal of the rectangle formed by these two straight lines and the two directions of the currents, the preceding expression then becomes

$$ii'\left(\frac{c}{a}-\frac{a}{c}\right)=\frac{ii'l^2}{ac}\ ;$$

[pages 68 and 240]

naming l the length of the conductors, and when this rectangle becomes a square, one has $ii'/\sqrt{2}$ for the value of the force; finally, if one supposes that one of the conductors is indefinite in the two directions,[145] and that l is the length of the other, the terms in which r_1', r_2', r_1'' and r_2'' are found in the denominator will disappear; we will have

$$r_2'+r_1''-r_2''-r_1'=2l\ ,$$

and the expression for the force will become

$$\frac{ii'l}{a}\ ,$$

[144] [N. T.] In the 1826 and 1827 versions of the *Théorie* we had

$$\frac{1}{2}ii'\left(\frac{a}{r_2''}-\frac{a}{r_1''}-\frac{a}{r_2'}+\frac{a}{r_1'}\right)+\frac{r_1''+r_2'-r_2''-r_1'}{a}\ .$$

The errata appearing at the end of the 1827 edition requested to change this equation to:

$$\frac{1}{2}ii'\left(\frac{a}{r_2''}-\frac{a}{r_1''}-\frac{a}{r^{2'}}+\frac{a}{r_1'}+\frac{r_1''+r_2'-r_2''-r_1'}{a}\right)\ .$$

The errata appearing at the end of the 1826 edition, on the other hand, corrected not only the printed versions of the 1826 and 1827 editions, but also the 1827 errata. This 1827 errata had corrected the closing parenthesis but, at the same time, introduced a mistake in the third fraction. The 1826 errata indicated that the correct equation is that one reproduced here in this English translation of the *Théorie*.

[145] [N. T.] That is, by supposing an infinite straight conductor.

which reduces to ii' when the length l is equal to the distance a.

With regard to the action of two currents parallel to the s' direction, it can be obtained no matter the form of the current s. In effect the component which follows ds' is

$$\frac{1}{2}ii'ds'd\left(\frac{\cos^2\beta}{r}\right) ,$$

the total action exerted by ds' in this direction on the current L′L″ (figure 21) has as its value

$$\frac{1}{2}ii'ds'\left(\frac{\cos^2\beta''}{r''}-\frac{\cos^2\beta'}{r'}\right) ,$$

and it is remarkable that it only depends on the position of the extremities L′ and L″ of the conductor s; it is therefore the same, whatever the form of the conductor, which can be folded following any line.

If one names a' and a'' the perpendiculars from the [pages 69 and 241] two extremities of the portion of the conductor L′L″, which one considers as being mobile, on the rectilinear conductor on which one calculates the action [exerted by the mobile conductor] parallel to its direction, one obtains

$$r''=\frac{a''}{\sin\beta''} , \quad r'=\frac{a'}{\sin\beta'} ,$$

$$ds'=-\frac{dr''}{\cos\beta''}=\frac{a''d\beta''}{\sin^2\beta''}=-\frac{dr'}{\cos\beta'}=\frac{a'd\beta'}{\sin^2\beta'} ,$$

and as a consequence

$$\frac{ds'}{r''}=\frac{d\beta''}{\sin\beta''} , \quad \frac{ds'}{r'}=\frac{d\beta'}{\sin\beta'} ,$$

from which it is easy to conclude that the sought integral is[146]

$$-\frac{1}{2}ii'\int\left(\frac{\cos^2\beta''d\beta''}{\sin\beta''}-\frac{\cos^2\beta'd\beta'}{\sin\beta'}\right)$$

$$=-\frac{1}{2}ii'\left(\ln\frac{\tan\frac{1}{2}\beta''}{\tan\frac{1}{2}\beta'}+\cos\beta''-\cos\beta'+C\right) .$$

It is necessary to take this integral between the limits determined by the two extremities of the rectilinear conductor; naming β'_1, β'_2 and β''_1, β''_2 the values of β' and β'' relative to these limits, one has immediately the value of the force exerted by the rectilinear conductor, and this last value only depends on the four angles β'_1, β''_1, β'_2 and β''_2.

When one wants the value of this force for the case where the rectilinear conductor extends indefinitely in both directions, it is necessary to set $\beta'_1=\beta''_1=0$ and $\beta'_2=\beta''_2=\pi$: it appears, at first glance, that they become null, which would be contrary to experience; but one sees easily that the part of the integral where the cosines of these four angles enter is the [pages 70 and 242] only place where they vanish in this case, and that the rest of the integral

$$\frac{1}{2}ii'\left(\ln\frac{\tan\frac{1}{2}\beta''_1}{\tan\frac{1}{2}\beta'_1}-\ln\frac{\tan\frac{1}{2}\beta''_2}{\tan\frac{1}{2}\beta'_2}\right)$$

$$=\frac{1}{2}ii'\ln\frac{\tan\frac{1}{2}\beta''_1\cot\frac{1}{2}\beta''_2}{\tan\frac{1}{2}\beta'_1\cot\frac{1}{2}\beta'_2} ,$$

becomes, because one has $\beta''_2=\pi-\beta''_1$ and $\beta'_2=\pi-\beta'_1$,

146[N. T.] We are utilizing here the modern notation for the natural logarithm of a number x, namely, $\ln x$. This notation appeared only at the end of the XIXth century. Ampère represented the natural logarithm by the letter "l". The following equation, for instance, appeared in the *Théorie* as follows:

$$-\frac{1}{2}ii'\int\left(\frac{\cos.^2\beta''d\beta''}{\sin.\beta''}-\frac{\cos.^2\beta'd\beta'}{\sin.\beta'}\right)$$

$$=-\frac{1}{2}ii'\left(\mathrm{l}\frac{\mathrm{tang}.\frac{1}{2}\beta''}{\mathrm{tang}.\frac{1}{2}\beta'}+\cos.\beta''-\cos.\beta'+C\right) .$$

$$\frac{1}{2}ii'\ln\frac{\tan^2\frac{1}{2}\beta_1''}{\tan^2\frac{1}{2}\beta_1'} = ii'\ln\frac{\tan\frac{1}{2}\beta_1''}{\tan\frac{1}{2}\beta_1'} = ii'\ln\frac{a''}{a'} \ .$$

This value shows that the sought force can only depend on the ratio of the two perpendiculars a'' and a',[147] drawn onto the indefinite rectilinear conductor from the two extremities of the portion of the conductor on which it acts; which is also independent of the form of this portion, and only becomes null, as it should be, when the two perpendiculars are equal to each other.

To obtain the distance of this force[148] to the rectilinear conductor whose direction is parallel to its own, it is necessary to multiply each one of the elementary forces composing it[149] by its distance to the conductor, and integrate the result with respect to the same limits; one will thus obtain the torque which must be divided by the force in order to obtain the sought distance.

One easily finds, using the values above, that the elementary torque has as its value

$$\frac{1}{2}ii'ds'r\sin\beta d\frac{\cos^2\beta}{r} \ .$$

This value can only be integrated when one has substituted in this expression one of the variables r or β for its value as a function of the other [variable], taken from the equations which determine the form of the mobile portion [pages 71 and 243] of the conductor; this is very simple when this portion is on a line drawn by an arbitrary point of the rectilinear conductor which one considers as fixed, perpendicular to its direction, because by taking this point for the origin of the s', one has

$$r = \frac{s'}{\cos\beta} \ ,$$

and [one has] that s' is a constant relative to the differential

$$d\frac{\cos^2\beta}{r} \ .$$

The value of the elementary torque then becomes

$$\frac{1}{2}ii'ds'\frac{\sin\beta}{\cos\beta}d(\cos^3\beta) = -\frac{3}{2}ii'ds'\sin^2\beta\cos\beta d\beta \ ,$$

whose integral between the limits β'' and β' is

$$-\frac{1}{2}ii'ds'\left(\sin^3\beta'' - \sin^3\beta'\right) \ .$$

By replacing ds' by the values of this differential found above, and by integrating again, one has, between the limits determined by the rectilinear conductor,

$$\frac{1}{2}ii'\left[a''\left(\cos\beta_2'' - \cos\beta_1''\right) - a'\left(\cos\beta_2' - \cos\beta_1'\right)\right] \ .$$

If one supposes that the conductor extends indefinitely in both directions, it would be necessary to give to β_1', β_1'', β_2' and β_2'', the values which we have already assigned to them in this case, and one obtains

$$-ii'\left(a'' - a'\right)$$

for the sought value of the torque, which will be, as a consequence, [pages 72 and 244] proportional to the length $a'' - a'$ of the mobile conductor, and will not change at all while this length remains the same, whatever the values of the other distances of the extremities of his last conductor to the conductor which is considered as fixed.

Let us calculate now[150] the action exerted by a curve of arbitrary form NM in order to turn an arc of the circle L_1L_2 about its center.

---

[147][N. T.] Due to a misprint in the 1826 and 1827 versions of the *Théorie*, this sentence appeared with a' and a', instead of a'' and a'.

[148][N. T.] Ampère wishes now to calculate the torque exerted by this force. To this end he needs to calculate the length of the arm of the lever.

[149][N. T.] That is, composing the net or resultant force.

[150][N. T.] According to Joubert, the following calculations belonged to Ampère's manuscript memoir of 21 November 1825. They were published for the first time only here in the *Théorie*, [Ampère, 1887d, note 1 on p. 65].

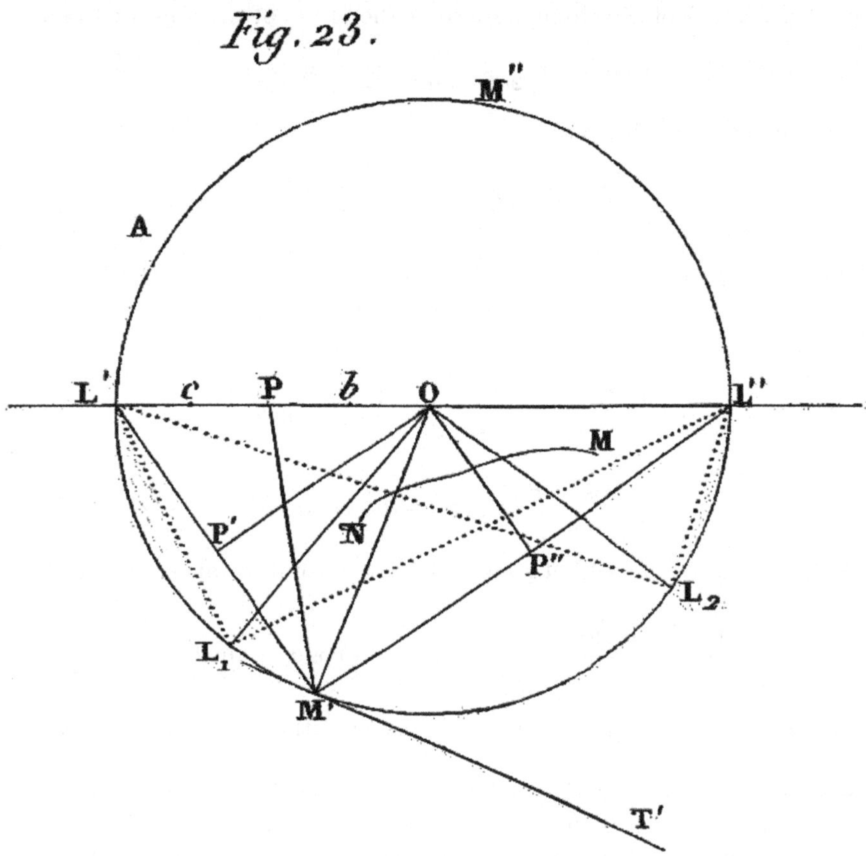

Figure 23.

Let M′ (figure 23)[151] be the midpoint of an arbitrary element ds' of the arc L_1L_2, and a the radius of the circle. The torque of an element ds of MN in order to cause ds' to turn about the center O can be obtained by multiplying the tangential component [of the force acting] on M′ by its distance a to the fixed point, which gives

$$\frac{1}{2}aii'ds'd\frac{\cos^2\beta}{r} .$$

Taking β', β'' and r', r'' to be the values of β and r relative to the limits M and N, one has for the torque of ds'

$$\frac{1}{2}aii'ds'\left(\frac{\cos^2\beta''}{r''}-\frac{\cos^2\beta'}{r'}\right) ,$$

a result which only depends on the position of the extremities M and N.

We carry out the calculation by assuming that the line MN is a diameter L′L″ of the same circle.

Take 2θ to be the angle M′OL′; M′T′ being the tangent at M′, the angles L′M′T′ and L″M′T′ will be, respectively, β' and β'', and one obviously obtains

$$\cos\beta' = -\cos\theta,\ \cos\beta'' = \sin\theta,\ r' = 2a\sin\theta,\ r'' = 2a\cos\theta .$$

The action of the diameter L′L″ to turn [around center O] the element situated at M′ [pages 73 and 245] will then be

$$\frac{1}{4}ii'ds'\left(\frac{\sin^2\theta}{\cos\theta}-\frac{\cos^2\theta}{\sin\theta}\right) .$$

[151] [N. T.] In figure 23 of the *Théorie* published in 1827 there is a letter p between points c and P. This letter p is not mentioned in the text of the *Théorie* related to this figure. For this reason it was removed in figure 23 of the 1826 version of the *Théorie*.

If one takes an arbitrary point A of the circumference as the origin of the arcs, and make AL$'$ $= C$, one has

$$s' = C + 2a\theta \quad \text{and} \quad ds' = 2ad\theta \ ,$$

which changes the preceding expression into

$$\frac{1}{2}aii'\left(\frac{\sin^2\theta d\theta}{\cos\theta} - \frac{\cos^2\theta d\theta}{\sin\theta}\right) ,$$

which must be integrated over the entire extent of the arc L_1L_2, in order to have the torque of this arc about its center.

However one has

$$\int\frac{\sin^2\theta d\theta}{\cos\theta} = \ln\tan\left(\frac{\pi}{4}+\frac{1}{2}\theta\right) - \sin\theta + C_1 \ ,$$

$$\int\frac{\cos^2\theta d\theta}{\sin\theta} = \ln\tan\frac{1}{2}\theta + \cos\theta + C' \ .$$

If then one calls $2\theta_1$ and $2\theta_2$ the angles L$'$OL_1 and L$'$OL_2, the total torque of the arc L_1L_2 becomes

$$\frac{a}{2}ii'\left\{\ln\frac{\tan\left(\frac{\pi}{4}+\frac{1}{2}\theta_2\right)\tan\frac{1}{2}\theta_1}{\tan\frac{1}{2}\theta_2\tan\left(\frac{\pi}{4}+\frac{1}{2}\theta_1\right)} - \sin\theta_2 - \cos\theta_2 + \sin\theta_1 + \cos\theta_1\right\} \ .$$

This expression, changing sign, gives the value of the torque of the diameter L$'$L$''$ due to the action of the arc L_1L_2.

In an apparatus which I described previously,[152] a conductor which has the form of a circular sector, acts on another conductor composed of a diameter and of a half-circumference which is mobile about an axis passing through [pages 74 and 246] the center of this semi-circumference and perpendicular to its plane. The action which it[153] experiences from the part of the sector is destroyed by the resistance of the axis, since the contour which forms the sector is closed; there only remains the action on the diameter. We have already calculated that [action] of the arc [on the diameter]; it remains for us in addition to obtain those [actions] of the radii of this sector on the same diameter.

For determining these, we will look for the torque which results from the mutual action of two rectilinear currents situated in the same plane, and which tend to cause them to turn in contrary senses about the point of intersection of their directions.

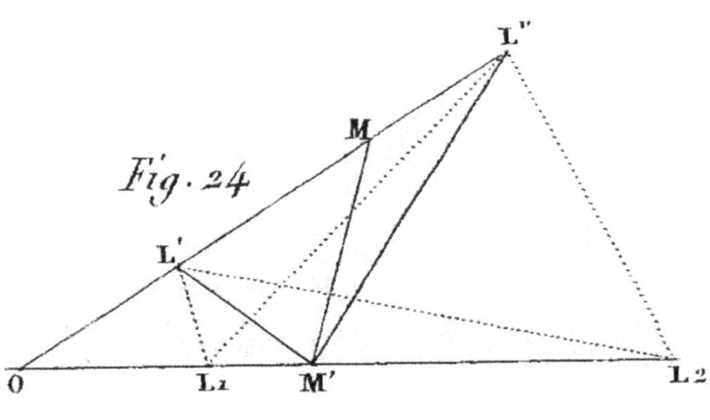

Figure 24.

The component [of the force] normal to the element ds' located in M$'$ (figure 24), is, as we have seen previously,

[152][N. T.] See Section 29.13 of the *Théorie*.

[153][N. T.] In the original French text we have here the word *elle*, referring to the semi-circumference, *demi-circonférence*, which is a feminine word in French.

$$\frac{1}{2} ii' ds' \left(d\frac{\sin\beta\cos\beta}{r} - \frac{d\beta}{r} \right) .$$

The torque of ds which causes rotation of ds' about O, is obtained by multiplying this force by s'; one then obtains, naming M the total torque,[154]

$$\frac{d^2M}{dsds'} dsds' = \frac{1}{2} ii' s' ds' \left(d\frac{\sin\beta\cos\beta}{r} - \frac{d\beta}{r} \right) ,$$

from which, by integrating with respect to s,

$$\frac{dM}{ds'} ds' = \frac{1}{2} ii' s' ds' \left(\frac{\sin\beta\cos\beta}{r} - \int \frac{d\beta}{r} \right) .$$

But, following the manner in which the angles were determined in the calculation of the formula which represents the mutual action of two elements of voltaic conductors, the angle $\mathrm{MM'L_2} = \beta$ is exterior of the triangle OMM′; and, naming ε the angle MOM′ which is between the directions of the two currents, one finds that the third angle OMM′ is equal to $\beta - \varepsilon$, which gives

[pages 75 and 247]

$$r = \frac{s' \sin\varepsilon}{\sin(\beta - \varepsilon)} ;$$

one has, therefore,

$$\frac{dM}{ds'} ds' = \frac{1}{2} ii' \frac{ds'}{\sin\varepsilon} \left[\cos\beta \sin\beta \sin(\beta - \varepsilon) + \cos(\beta - \varepsilon) + C \right] .$$

By replacing in this expression $\cos(\beta - \varepsilon)$ by

$$\cos^2\beta \cos(\beta - \varepsilon) + \sin^2\beta \cos(\beta - \varepsilon) ,$$

one easily sees that it reduces to

$$\frac{dM}{ds'} ds' = \frac{1}{2} ii' \frac{ds'}{\sin\varepsilon} \left[\cos\varepsilon \cos\beta + \sin^2\beta \cos(\beta - \varepsilon) + C \right] ,$$

which must be taken between the limits β' and β''; one has thus the difference between two functions of the same form, one with β'', the other with β', which must be newly integrated to obtain the torque which is sought: it suffices to make this second integration on just one of these two quantities: let then a'' be the distance OL″ which corresponds to β''; one has, in the triangle OM′L″,[155]

$$s' = \frac{a'' \sin(\beta'' - \varepsilon)}{\sin\beta''} = a'' \cos\varepsilon - a'' \sin\varepsilon \cot\beta'' \quad \text{and} \quad ds' = \frac{a'' \sin\varepsilon d\beta''}{\sin^2\beta''} ;$$

and the quantity[156] that we first proposed to integrate becomes

$$\frac{1}{2} a'' ii' \left[\frac{\cos\varepsilon \cos\beta'' d\beta''}{\sin^2\beta''} + \cos(\beta'' - \varepsilon) d\beta'' \right] ,$$

whose integral taken between the limits β_1'' and β_2'' is

$$\frac{1}{2} a'' ii' \left[\sin(\beta_2'' - \varepsilon) - \sin(\beta_1'' - \varepsilon) - \frac{\cos\varepsilon}{\sin\beta_2''} + \frac{\cos\varepsilon}{\sin\beta_1''} \right] .$$

[154] [N. T.] Ampère probably chose the letter M to represent the total torque because this is the first letter of the French expression "moment total."

[155] [N. T.] In the 1826 and 1827 editions of the *Théorie* the last fraction was written as $ds' = \frac{a'' \sin r d\beta''}{\sin^2\beta''}$. This was a misprint, as Ampère was differentiating the expression of s' in relation to β''. Therefore, the variable r should not appear.

[156] [N. T.] That is, the torque around O.

Designating by p''_2 and p''_1[157] the perpendiculars [pages 76 and 248] from the point O on the distances $\mathrm{L}''\mathrm{L}_2 = r''_2$ and $\mathrm{L}''\mathrm{L}_1 = r''_1$, one has obviously

$$a'' \sin(\beta''_2 - \varepsilon) = p''_2 \ , \ a'' \sin(\beta''_1 - \varepsilon) = p''_1 \ ,$$

$$\frac{a''}{\sin\beta''_2} = \frac{r''_2}{\sin\varepsilon} \ , \ \frac{a''}{\sin\beta''_1} = \frac{r''_1}{\sin\varepsilon} \ ,$$

and the preceding integral becomes

$$\frac{1}{2} ii' \left[p''_2 - p''_1 - (r''_2 - r''_1)\cot\varepsilon\right] \ .$$

If one pays attention that when designating the distance OL′ by a', one has also, in the triangle OM′L′,

$$s' = \frac{a' \sin(\beta' - \varepsilon)}{\sin\beta'} = a'\cos\varepsilon - a'\sin\varepsilon\cot\beta' \ , \ ds' = \frac{a'\sin\varepsilon d\beta'}{\sin^2\beta'} \ ,$$

[then] it is easy to see that the integral of the other quantity is formed from the one that we have obtained, by there changing p''_2, p''_1, r''_2 and r''_1 into p'_2, p'_1, r'_2 and r'_1; which gives for the value of the torque which is the difference of the two integrals,

$$\frac{1}{2} ii' \left[p''_2 - p''_1 - p'_2 + p'_1 - (r''_2 - r''_1 - r'_2 + r'_1)\cot\varepsilon\right] \ .$$

This value reduces to the one that we found above in the case where the angle ε is a right angle, because then $\cot\varepsilon = 0$.

When[158] one assumes that the two currents leave the point O, and that their lengths OL″ and OL_2 (figure 22)[159] are represented respectively by a and b, the perpendicular OP by p, and the distance $\mathrm{L}''\mathrm{L}_2$ by r, one has $p''_2 = p$, $p''_1 = p'_2 = p'_1 = 0$, $r''_2 = r$, $r''_1 = a$, $r'_2 = b$, $r'_1 = 0$, and

$$\frac{1}{2} ii' \left[p + (a + b - r)\cot\varepsilon\right] \ ,$$

for the value which is taken by the torque.

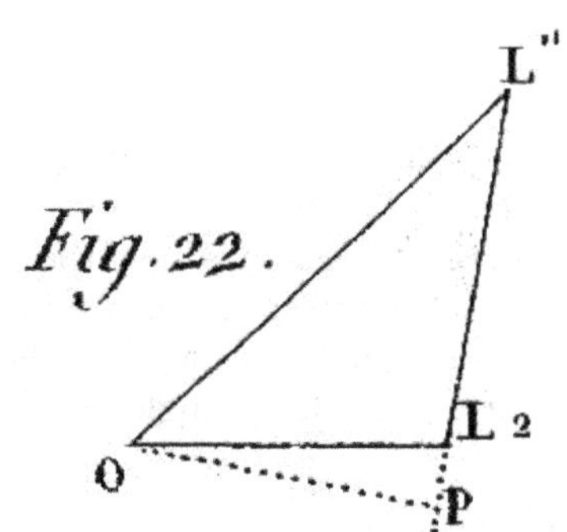

Figure 22.

[pages 77 and 249]

The quantity $a + b - r$, excess of the sum of two sides of a triangle on the third, is always positive: from which it follows that the torque is larger than the value $ii'p/2$ which it takes when the angle ε between the

[157] [N. T.] Due to a typographical mistake, in the 1826 and 1827 versions of the *Théorie* we had here p''_2 and p'_2. In the sequence Ampère associates p''_2 with r''_2 and p''_1 with r''_1. This misprint appeared also in the 1887 and 1958 editions of the *Théorie*, [Ampère, 1887d, p. 68] and [Ampère, 1958, p. 57].

[158] [N. T.] In the 1826 and 1827 editions of the *Théorie* we had here the French word "quant" instead of "quand." But in the errata appearing at the end of the work it was pointed out that the correct word should be "quand".

[159] [N. T.] In the 1826 and 1827 editions of the *Théorie*, we had here "figure 24." But the errata corrected this expression to figure 22.

two conductors is a right one, while $\cot\varepsilon$ is positive, that is, while this angle is acute; but it becomes smaller [than the value $ii'p/2$] when the same angle is obtuse, because then $\cot\varepsilon$ is negative. It is evident moreover that its value increases as the value of ε decreases, and that it goes to infinity as $\cot\varepsilon$ when ε approaches zero; but it is good to demonstrate that it stays always positive, no matter how close the angle may be to two right angles.

It is sufficient for this [proof] to pay attention that by naming α the angle of the triangle $\mathrm{OL''L_2}$ between the sides a and r, and β [the angle] between the sides b and r, one has

$$\cot\varepsilon = -\cot(\alpha+\beta),\ p = a\sin\alpha = b\sin\beta,\ r = a\cos\alpha + b\cos\beta\ ,$$

and as a consequence

$$\begin{aligned} a+b-r &= a\,(1-\cos\alpha) + b\,(1-\cos\beta) \\ &= p\tan\frac{1}{2}\alpha + p\tan\frac{1}{2}\beta\ , \end{aligned}$$

and

$$\frac{1}{2}ii'\left[p+(a+b-r)\cot\varepsilon\right] = \frac{1}{2}ii'p\left(1-\frac{\tan\frac{1}{2}\alpha+\tan\frac{1}{2}\beta}{\tan(\alpha+\beta)}\right)\ ,$$

a value which stays always positive, however small the angles α and β become, since $\tan(\alpha+\beta)$, for angles less than $\pi/4$, is always larger than $\tan\alpha+\tan\beta$, and with stronger reason larger than $\tan\frac{1}{2}\alpha+\tan\frac{1}{2}\beta$. This value [pages 78 and 250] surely tends toward the limit $ii'p/4$ as the angles α and β approach zero; they vanish with p when these angles become null.

Recall now the general value of the torque in which enter only the distances[160] $\mathrm{OL''} = a''$ (figure 24) and $\mathrm{OL'} = a'$, and the various angles, the value of which is

$$\frac{1}{2}ii'\Big[a''\sin(\beta_2''-\varepsilon) - a''\sin(\beta_1''-\varepsilon) - a'\sin(\beta_2'-\varepsilon)$$

$$+\,a'\sin(\beta_1'-\varepsilon) - \frac{a''\cos\varepsilon}{\sin\beta_2''} + \frac{a''\cos\varepsilon}{\sin\beta_1''} + \frac{a'\cos\varepsilon}{\sin\beta_2'} - \frac{a'\cos\varepsilon}{\sin\beta_1'}\Big]\ ,$$

and apply this [expression] to the case where one of the conductors $\mathrm{L'L''}$ (figure 25) is rectilinear and mobile about its center $\mathrm{L_1}$, and when the other [conductor] starts from this center. Taking $\mathrm{L'L''} = 2a$, one has

$$a'' = a,\ a' = -a,\ \beta_1' = \pi+\varepsilon,\ \beta_1'' = \varepsilon,\ \sin\beta_1' = -\sin\beta_1''\ ,$$

and by designating as before the perpendiculars drawn from $\mathrm{L_1}$ onto[161] $\mathrm{L'L_2}$ and $\mathrm{L''L_2}$, the expression for the torque becomes

$$\frac{1}{2}ii'\left(p_2''+p_2' - \frac{a\cos\varepsilon}{\sin\beta_2''} - \frac{a\cos\varepsilon}{\sin\beta_2'}\right)\ .$$

Now

$$\sin\beta_2'' : a :: \sin\varepsilon : r_2'' \quad \text{and} \quad -\sin\beta_2' : a :: \sin\varepsilon : r_2'\ ,$$

and the values of r_2'' and of r_2' taken from these proportions and substituted into the preceding expression change it to

$$\frac{1}{2}ii'\left[p_2''+p_2'+\cot\varepsilon(r_2'-r_2'')\right]\ .$$

When one assumes $\mathrm{L_1L_2}$ to be infinite, one has $p_2'' = p_2' = a\sin\varepsilon$, $r_2'-r_2'' = 2a\cos\varepsilon$, and this value of the torque reduces to

[160][N. T.] In the 1826 and 1827 versions of the *Théorie* we read here "$\mathrm{OL''} = a''$ and $\mathrm{OL'} = a'$." However, only the 1826 errata stated that we should have here "$\mathrm{OL''} = a''$ (figure 24) and $\mathrm{OL'} = a'$." This figure 24 appears in our page 390.

[161][N. T.] In the 1826 and 1827 versions of the *Théorie* we have $\mathrm{L'L_2L''L_2}$. The errata of this work corrects this expression to $\mathrm{L'L_2}$ and $\mathrm{L''L_2}$.

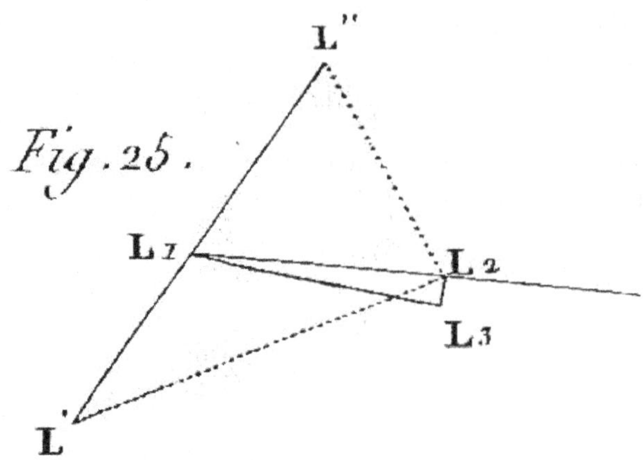

Figure 25.

$$\frac{1}{2}aii'\left(2\sin\varepsilon+\frac{2\cos^2\varepsilon}{\sin\varepsilon}\right)=\frac{aii'}{\sin\varepsilon}\ ;$$

[pages 79 and 251]

it is therefore inversely proportional to the sine of the angle between the two currents, and proportional to the length of the finite current.

When $\mathrm{L}_1\mathrm{L}_2 = \mathrm{L}'\mathrm{L}''/2 = a$ and if one represents the angle $\mathrm{L}'\mathrm{L}_1\mathrm{L}_2$ by 2θ, one has $p_2'' = a\sin\theta$, $p_2' = a\cos\theta$, $r_2' = 2a\sin\theta$, $r_2'' = 2a\cos\theta$ and $\cot\varepsilon = -\cot 2\theta$, and the torque becomes

$$\frac{1}{2}aii'\left[\cos\theta+\sin\theta+2\cot 2\theta\left(\cos\theta-\sin\theta\right)\right]\ ,$$

by replacing $2\cot 2\theta$ by its value

$$\frac{1-\tan^2\theta}{\tan\theta}=\frac{\cos^2\theta-\sin^2\theta}{\sin\theta\cos\theta}=\frac{(\cos\theta+\sin\theta)\,(\cos\theta-\sin\theta)}{\sin\theta\cos\theta}\ ,$$

one finds that the value of this torque is equal to

$$\frac{1}{2}aii'\left(\cos\theta+\sin\theta\right)\left[1+\frac{(\cos\theta-\sin\theta)^2}{\sin\theta\cos\theta}\right]$$

$$=\frac{1}{2}aii'\left(\cos\theta+\sin\theta\right)\left(\frac{1}{\sin\theta\cos\theta}-1\right)\ .$$

To obtain the sum of the actions of two radii between which is an infinitely small sector of which the arc is $d\varepsilon$, it is necessary to pay attention that these two radii being traversed [by currents flowing] in opposite senses, this sum becomes equal to the differential of the preceding expression; one thus finds that it is represented by

$$\frac{1}{2}aii'\left[(\cos\theta-\sin\theta)\left(\frac{1}{\sin\theta\cos\theta}-1\right)\right.$$
$$\left.-\frac{(\cos\theta+\sin\theta)\,(\cos^2\theta-\sin^2\theta)}{\sin^2\theta\cos^2\theta}\right]d\theta$$

$$=\frac{1}{2}aii'\left(\cos\theta-\sin\theta\right)\left(\frac{1}{\sin\theta\cos\theta}-1-\frac{(\cos\theta+\sin\theta)^2}{\sin^2\theta\cos^2\theta}\right)d\theta$$

$$=-\frac{1}{2}aii'\left(\cos\theta-\sin\theta\right)\left(\frac{1}{\sin^2\theta\cos^2\theta}+\frac{1}{\sin\theta\cos\theta}+1\right)d\theta\ .$$

But the action of the arc L_2L_3 on the diameter L′L″ is equal and [pages 80 and 252] opposed to that which the diameter exerts on the arc to cause it to turn on its center; the torque of this action, following that which we just saw, is therefore equal to

$$\frac{1}{2}aii'\left(\frac{\cos^2\theta}{\sin\theta}-\frac{\sin^2\theta}{\cos\theta}\right)d\theta=\frac{1}{2}aii'\left(\cos\theta-\sin\theta\right)\left(\frac{1}{\sin\theta\cos\theta}+1\right)d\theta\ ;$$

by adding it to the preceding [result], one has for the torque resulting from the action of the infinitely small sector on the diameter L′L″ [the following value:]

$$-\frac{1}{2}aii'\left(\cos\theta-\sin\theta\right)\frac{d\theta}{\sin\theta\cos\theta}\ .$$

This value only differs in the sign of the one we have already found for the same torque, the difference comes obviously from [the fact] that we obtained this last [expression] from the formula relative to the action of a very small closed circuit on an element, where we have changed the sign of C to make it positive.

We will examine now the action that two rectilinear currents,[162] which are not in the same plane, exert on one another, whether to move in parallel with their common perpendicular, or to turn about this line.

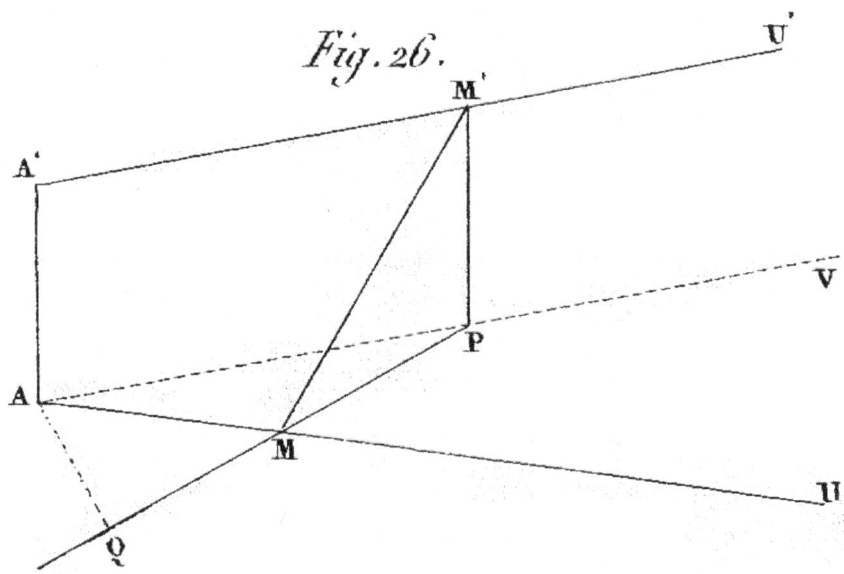

Figure 26.

Let AU and A′U′ be the two currents (figure 26); $AA'=a$ their common perpendicular; AV a parallel to A′U′: the action[163] of two elements located in M and M', if one sets $n=2$ and $h=k-1=-3/2$ in the general formula

$$\frac{ii'dsds'}{r^n}\left(\cos\varepsilon+h\cos\theta\cos\theta'\right),$$

becomes

$$\frac{1}{2}\frac{ii'dsds'\left(2\cos\varepsilon+3\frac{dr}{ds}\frac{dr}{ds'}\right)}{r^2}\ ,$$

because[164]

[162][N. T.] The most important steps in the following calculations leading to the result obtained in our page 404, corresponding to pages 94 and 265 of the 1826 and 1827 versions of the *Théorie*, respectively, were published in [Ampère, 1825g], see also [Ampère, 1826e]. Here in the *Théorie* Ampère developed the original calculations, as he had promised previously, [Ampère, 1825g, p. 41] and [Ampère, 1826e].

[163][N. T.] That is, the force.

[164][N. T.] See our page 360, corresponding to pages 35 and 207 of the 1826 and 1827 versions of the *Théorie*, respectively.

$$\cos\theta = \frac{dr}{ds} \ , \quad \cos\theta' = -\frac{dr}{ds'} \ ;$$

[pages 81 and 253]

but on making AM $= s$, A′M′ $= s'$ and VAU $= \varepsilon$, one has

$$r^2 = a^2 + s^2 + s'^2 - 2ss'\cos\varepsilon \ ,$$

from where

$$r\frac{dr}{ds} = s - s'\cos\varepsilon, \ r\frac{dr}{ds'} = s' - s\cos\varepsilon, \ r\frac{d^2r}{dsds'} + \frac{dr}{ds}\frac{dr}{ds'} = -\cos\varepsilon \ ;$$

and as

$$\frac{d\frac{1}{r}}{ds} = -\frac{\frac{dr}{ds}}{r^2} \ ,$$

$$\frac{d^2\frac{1}{r}}{dsds'} = -\frac{r\frac{d^2r}{dsds'} - 2\frac{dr}{ds}\frac{dr}{ds'}}{r^3} = \frac{\cos\varepsilon + 3\frac{dr}{ds}\frac{dr}{ds'}}{r^3} \ ,$$

the value of the action of the two elements becomes

$$\frac{1}{2}ii'dsds'\left(\frac{\cos\varepsilon}{r^2} + r\frac{d^2\frac{1}{r}}{dsds'}\right) .$$

In order to have the component parallel to AA′, it is necessary to multiply this expression by the cosine of the angle MM′P which makes MM′ with M′P parallel to AA′, that is to say, by M′P/M′M, or a/r, which gives

$$\frac{1}{2}aii'dsds'\left(\frac{\cos\varepsilon}{r^3} + \frac{d^2\frac{1}{r}}{dsds'}\right) \ ;$$

and by integrating over the total extent of the two currents, one finds for the total action

$$\frac{1}{2}aii'\left(\frac{1}{r} + \cos\varepsilon\int\int\frac{dsds'}{r^3}\right) .$$

If the two currents form between them a right angle, one has $\cos\varepsilon = 0$, and the action parallel to AA′ is reduced, by taking [pages 82 and 254] the integral between the suitable limits, and by employing the same notation as above, to

$$\frac{1}{2}ii'\left(\frac{a}{r''_2} - \frac{a}{r''_1} - \frac{a}{r'_2} + \frac{a}{r'_1}\right) .$$

This expression is proportional to the shortest distance of the two currents and becomes as a consequence null when they are in the same plane, as should be obvious.

If the currents are parallel, one has $\varepsilon = 0$ and

$$r^2 = a^2 + (s - s')^2 \ ,$$

from which

$$\int\int\frac{dsds'}{r^3} = \int ds'\int\frac{ds}{[a^2 + (s-s')^2]^{\frac{3}{2}}}$$

$$= \int ds'\frac{s - s'}{a^2\sqrt{a^2 + (s-s')^2}} = -\frac{\sqrt{a^2 - (s-s')^2}}{a^2} = -\frac{r}{a^2} \ ,$$

which means, between the limits of integration,

$$\frac{r_2' + r_1'' - r_1' - r_2''}{a^2} \ ;$$

and since $\cos\varepsilon = 1$, the total action becomes

$$\frac{1}{2} ii' \left(\frac{a}{r_2''} - \frac{a}{r_2'} - \frac{a}{r_1''} + \frac{a}{r_1'} + \frac{r_1'' + r_2' - r_2'' - r_1'}{a} \right) \ .$$

We will see later how to carry out the integration in the case where the angle ε is arbitrary.

We search now the torque about the common perpendicular: for this [calculation] it is necessary to know first the component along MP, and multiply it by the perpendicular AQ from A onto MP, which amounts [pages 83 and 255] to multiplying the force along MM′ by $\frac{\text{MP}}{\text{MM}'}\text{AQ}$, or by $ss'\sin\varepsilon/r$; one thus obtains

$$\frac{1}{2} ii' \sin\varepsilon \left(ss' \frac{d^2\frac{1}{r}}{dsds'} dsds' + ss' \frac{\cos\varepsilon dsds'}{r^3} \right) \ ;$$

setting $ss'/r = q$, one obtains

$$\frac{dq}{ds} = \frac{s'}{r} + \frac{ss'd\frac{1}{r}}{ds} \ ,$$

and

$$\frac{d^2q}{dsds'} = \frac{1}{r} - \frac{s'}{r^2}\frac{dr}{ds'} - \frac{s}{r^2}\frac{dr}{ds} + ss'\frac{d^2\frac{1}{r}}{dsds'}$$

$$= \frac{1}{r} - \frac{s'(s' - s\cos\varepsilon) + s(s - s'\cos\varepsilon)}{r^3} + ss'\frac{d^2\frac{1}{r}}{dsds'} \ ;$$

and by simplifying

$$\frac{d^2q}{dsds'} = \frac{a^2}{r^3} + \frac{ss'd^2\frac{1}{r}}{dsds'} \ ,$$

from which one extracts

$$ss'\frac{d^2\frac{1}{r}}{dsds'} = \frac{d^2q}{dsds'} - \frac{a^2}{r^3} \ .$$

Now, we have previously found

$$r\frac{d^2r}{dsds'} + \frac{dr}{ds}\frac{dr}{ds'} = -\cos\varepsilon \ ;$$

or

$$r\frac{d^2r}{dsds'} + \frac{(s - s'\cos\varepsilon)(s' - s\cos\varepsilon)}{r^2} = -\cos\varepsilon \ ;$$

carrying out the multiplication and replacing $s^2 + s'^2$ by its value obtained from

$$r^2 = a^2 + s^2 + s'^2 - 2ss'\cos\varepsilon \ ,$$

[pages 84 and 256]

one obtains by simplification

$$\frac{d^2r}{dsds'} + \frac{ss'\sin^2\varepsilon + a^2\cos\varepsilon}{r^3} = 0 \ ,$$

from which

$$\frac{ss'}{r^3} = -\frac{1}{\sin^2\varepsilon}\left(\frac{d^2r}{dsds'} + \frac{a^2\cos\varepsilon}{r^3}\right) \ .$$

Substituting this value and also that of $ss'\frac{d^2\frac{1}{r}}{dsds'}$ into the expression for the torque of the element, it becomes[165]

$$\begin{aligned}&\frac{1}{2}ii'\sin\varepsilon dsds'\left[\frac{d^2q}{dsds'}-\frac{a^2}{r^3}-\frac{\cos\varepsilon}{\sin^2\varepsilon}\left(\frac{d^2r}{dsds'}+\frac{a^2\cos\varepsilon}{r^3}\right)\right]\\ &=\frac{1}{2}ii'dsds'\left(\sin\varepsilon\frac{d^2q}{dsds'}-\frac{a^2\sin\varepsilon}{r^3}-\cot\varepsilon\frac{d^2r}{dsds'}-\frac{\cos^2\varepsilon\, a^2}{\sin\varepsilon\ r^3}\right)\\ &=\frac{1}{2}ii'dsds'\left(\sin\varepsilon\frac{d^2q}{dsds'}-\cot\varepsilon\frac{d^2r}{dsds'}-\frac{1}{\sin\varepsilon}\frac{a^2}{r^3}\right)\ ;\end{aligned}$$

and integrating with respect to s and s', one has for the total torque

$$\frac{1}{2}ii'\left(q\sin\varepsilon-r\cot\varepsilon-\frac{a^2}{\sin\varepsilon}\int\int\frac{dsds'}{r^3}\right)\ ;$$

the calculation is thus reduced, as before, to finding the value of the double integral $\int\int\frac{dsds'}{r^3}$.

If the currents are in the same plane, one has $a=0$, and the torque reduces to

$$\frac{1}{2}ii'\left(q\sin\varepsilon-r\cot\varepsilon\right)\ ,$$

a result which coincides with that which we have obtained in directly treating two currents situated in the same [pages 85 and 257] plane. Since q is nothing but ss'/r, and r being MP, one has

$$q\sin\varepsilon=\frac{ss'\sin\varepsilon}{r}=\frac{\mathrm{MP}\cdot\mathrm{AQ}}{\mathrm{MP}}=\mathrm{AQ}\ ;$$

and we have found by the other procedure,

$$\frac{1}{2}ii''(p-r\cot\varepsilon)\ ;$$

where p designates the perpendicular AQ; the two results are therefore identical. The integration carried out between the limits gives

$$\frac{1}{2}ii'\left[p_2''-p_1''-p_2'+p_1'+\cot\varepsilon(r_1''+r_2'-r_2''-r_1')\right]\ ;$$

if the angle ε is a right angle, this torque reduces to

$$\frac{1}{2}ii'(p_2''-p_1''-p_2'+p_1')\ .$$

When $\varepsilon=\pi/2$, but a is not null, the previous torque becomes

$$\frac{1}{2}ii'\left(q-a^2\int\int\frac{dsds'}{r^3}\right)\ .$$

The integral that needs to be calculated in this case is

$$\int ds'\int\frac{ds}{r^3}=\int ds'\int\frac{ds}{(a^2+s^2+s'^2)^{\frac{3}{2}}}=\int\frac{s}{(a^2+s'^2)\sqrt{a^2+s^2+s'^2}}ds'\ ,$$

which must be integrated over again with respect to s'; it becomes

$$\int\frac{sds'}{(a^2+s'^2)\sqrt{a^2+s^2+s'^2}}=\int\frac{(a^2+s^2)sds'}{(a^4+a^2s'^2+a^2s^2+s^2s'^2)\sqrt{a^2+s^2+s'^2}}=$$

$$\int\frac{s(a^2+s^2)\frac{ds'}{\sqrt{a^2+s^2+s'^2}}}{a^2(a^2+s^2+s'^2)+s^2s'^2}=\int\frac{\frac{s(a^2+s^2)ds'}{(a^2+s^2+s'^2)^{\frac{3}{2}}}}{a^2+\frac{s^2s'^2}{a^2+s^2+s'^2}}=\int\frac{\frac{dq}{ds'}ds'}{a^2+q^2}=\frac{1}{a}\arctan\frac{q}{a}+C\ .$$

[165] [N. T.] Due to a misprint the first line of this equation appeared originally as

$$\frac{1}{2}ii'\sin\varepsilon dsds'\left[\frac{d^2q}{dsds'}-\frac{a^2}{r^3}-\frac{\cos\varepsilon}{\sin^2}\left(\frac{d^2r}{dsds'}+\frac{a^2\cos\varepsilon}{r^3}\right)\right]\ .$$

[pages 86 and 258]

Let M be the value of the torque when the two electric currents, of which the lengths are s and s', start from the points where their directions meet the line which measures the shortest distance between them, one obtains

$$M = \frac{1}{2}ii'\left(q - a\arctan\frac{q}{a}\right) ,$$

an expression which reduces, when $a = 0$, to $M = ii'q/2$, which accords with the value $M = ii'p/2$ which we have already found for this case, because then q becomes the perpendicular which we have designated by p.[166] If one assumes a infinite, M becomes null, as it should be, as follows from

$$a\arctan\frac{q}{a} = q .$$

If one names z the angle of which the tangent is

$$\frac{ss'}{a\sqrt{a^2+s^2+s'^2}} ,$$

it follows

$$M = \frac{1}{2}ii'q\left(1 - \frac{z}{\tan z}\right) ;$$

this is the value of the torque which is produced by a force equal to

$$\frac{1}{2}ii'\left(1 - \frac{z}{\tan z}\right) ,$$

acting along the line which joins the two extremities of the conductors opposed to those [extremities] where they are met by the line which measures the shortest distance.

It is sufficient to quadruple these expressions to have the torque [pages 87 and 259] produced by the mutual action of two conductors such that one is mobile about the line which measures their shortest distance, in the case where this line touches the two conductors at their centers, and where their lengths are respectively represented by $2s$ and $2s'$.

It is, for the rest, easy to see that if, instead of assuming that the two currents start from the point where they encounter the line, one had made the calculation for arbitrary limits, one should obtain a value of [the torque] M composed of four terms of the form of those that we obtained in this particular case, two of these terms are positive and the two others are negative.

Consider now two rectilinear currents A′S′ and L′L″ (figure 27),[167] not situated in the same plane and whose directions form a right angle.

Let A′A be their joint perpendicular, and find the action of L′L″ which turns A′S′ about a parallel OV to L′L″ drawn at the distance A′O $= b$ from A.

Let M and M′ be two arbitrary elements of these currents; the general expression for the component of their action [i.e., force] parallel with the common perpendicular AA′, becomes,[168] by making $\varepsilon = \pi/2$,

$$\frac{1}{2}aii'\frac{d^2\frac{1}{r}}{dsds'}dsds' ;$$

its torque in relation to the point O is therefore, by taking A′ as the origin of the s', equal to

$$\frac{1}{2}aii'(s'-b)\frac{d^2\frac{1}{r}}{dsds'}dsds' ;$$

[166] [N. T.] In our page 398, corresponding to pages 85 and 257 of the 1826 and 1827 versions of the *Théorie*, respectively, Ampère had designated p the perpendicular AQ of figure 26.

[167] [N. T.] In figure 27 of the 1827 version of the *Théorie*, the line segment connecting M′ with S′ did not appear. This was corrected in figure 27 of the 1826 version reproduced here.

[168] [N. T.] Due to a misprint, the next equation appeared in the original text as follows:

$$\frac{1}{2}aii'\frac{d^2\frac{1}{r}}{dsds'}ds'ds' .$$

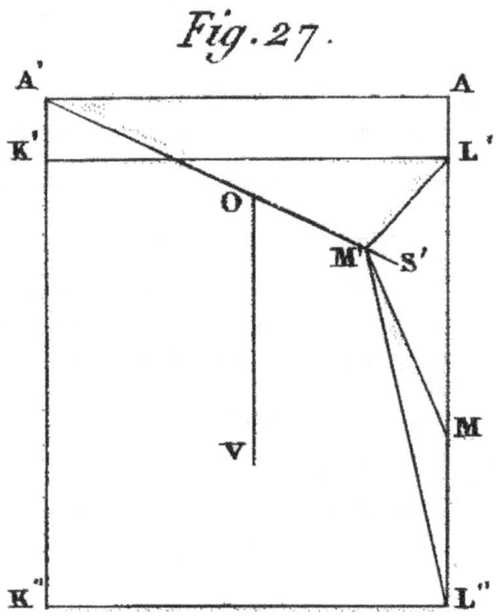

Figure 27.

[pages 88 and 260]

by integrating with respect to s, it becomes

$$\frac{1}{2}aii'(s'-b)\frac{d\frac{1}{r}}{ds'}ds' \ ;$$

and naming r' and r'' the distances M$'$L$'$ and M$'$L$''$ of M$'$ to the points L$'$ and L$''$, and integrating between these limits the action of L$'$L$''$ to turn the element M$'$, is

$$\frac{1}{2}aii'(s'-b)ds'\left(\frac{d\frac{1}{r''}}{ds'}-\frac{\frac{1}{r'}}{ds'}\right) \ ,$$

expression which must be integrated with respect to s'. Now

$$\frac{1}{2}aii'\int(s'-b)d\frac{1}{r''}=\frac{1}{2}aii'\left(\frac{s'-b}{r''}-\int\frac{ds'}{dr''}\right) \ ,$$

and it is also easy to see that in naming c the value AL$''$ of s which corresponds to r'', and which is a constant in the actual integration, one has A$'$L$''$ $=\sqrt{a^2+c^2}$, from which it follows that

$$r''=\frac{\sqrt{a^2+c^2}}{\sin\beta''}, \quad s'=-\sqrt{a^2+c^2}\cot\beta'', \quad \text{and} \quad ds'=\frac{\sqrt{a^2+c^2}}{\sin^2\beta''}d\beta'' \ ;$$

therefore

$$\int\frac{ds'}{r''}=\int\frac{d\beta''}{\sin\beta''}=\ln\frac{\tan\frac{1}{2}\beta''_2}{\tan\frac{1}{2}\beta''_1} \ ;$$

the second term integrates in the same manner, and one obtains finally for the torque to be found

$$\frac{1}{2}aii'\left(\frac{s'_2-b}{r''_2}-\frac{s'_1-b}{r''_1}-\frac{s'_2-b}{r'_2}+\frac{s'_1-b}{r'_1}-\ln\frac{\tan\frac{1}{2}\beta''_2\tan\frac{1}{2}\beta'_1}{\tan\frac{1}{2}\beta''_1\tan\frac{1}{2}\beta'_2}\right) \ .$$

In the case where the axis of rotation parallel to the line L$'$L$''$ or s passes by the point of intersection A$'$ of the lines a and s', one has $b=0$; and if one assumes, furthermore, that the current which [pages 89 and 261] flows in s' starts at this point of intersection, one obtains in addition

$$s_1' = 0,\ \beta_1' = \frac{\pi}{2},\ \beta_1'' = \frac{\pi}{2}\ ,$$

so that the value of the torque reduces to

$$\frac{1}{2}aii'\left(\frac{s_2'}{r_2''} - \frac{s_2'}{r_2'} - \ln\frac{\tan\frac{1}{2}\beta_2''}{\tan\frac{1}{2}\beta_2'}\right)\ .$$

I will now search for the action of a conducting wire folded following the perimeter of a rectangle K′K″L′L″ for turning a rectilinear conductor A′S′ $= s_2'$, perpendicular to the plane of this rectangle, and mobile about one of its sides K′K″ that it meets at the point A′: the torque produced by this action of the side K′K″ being then obviously null,[169] it will be necessary [in order to obtain the total torque,] to add, to that [torque] due to the action of L′L″ and which we have just calculated its value, the torque produced by K′L′ in the same direction as that [torque] of L′L″, and to subtract that [torque] which is [produced] by K′L″ whose action tends to cause the turning of A′S′ in a contrary direction; now, following the preceding calculations, by naming g and h the shortest distances A′K′ and A′K″, from AS′ to the lines K′L′ and K″L″ which are both equal to a, one has for the absolute values of these torques

$$\frac{1}{2}ii'\left(q' - g\arctan\frac{q'}{g}\right) \quad\text{and}\quad \frac{1}{2}ii'\left(q'' - h\arctan\frac{q''}{h}\right)\ ,$$

by setting

$$q' = \frac{as_2'}{\sqrt{g^2 + a^2 + s'^2}} = \frac{as_2'}{r_2'}\ ,\ q'' = \frac{as_2'}{\sqrt{h^2 + a^2 + s'^2}} = \frac{as_2'}{r_2''}\ ,$$

that [value] of the total torque is then

$$\frac{1}{2}ii'\left(h\arctan\frac{q''}{h} - g\arctan\frac{q'}{g} - a\ln\frac{\tan\frac{1}{2}\beta_2''}{\tan\frac{1}{2}\beta_2'}\right)\ .$$

[pages 90 and 262]

Such is the value of the torque resulting from the action of a conductor having as its form the perimeter of a rectangle, and acting on a mobile conductor around one of the sides of the rectangle, when the direction of this conductor is perpendicular to the plane of the rectangle, whatever besides its distance to the other sides of the rectangle and its dimensions. By determining by experiment the instant at which the mobile conductor is in equilibrium with respect to the opposing actions of two rectangles situated in the same plane, but of various sizes and at various distances from the mobile conductor, one has a quite simple means to obtain some verifications of my formula which are susceptible of great precision; it is this [experiment] that one can carry out easily with the aid of an instrument of which it is too easy to conceive the construction for it to be necessary to explain it here.

Integrate now the expression[170] $\int\int \frac{dsds'}{r^3}$ in the extent of two rectilinear currents not situated in the same plane, and making between them an arbitrary angle ε, in the case where these currents start at the common perpendicular; the other cases can be deduced immediately.

Let A (figure 28)[171] be the point where the common perpendicular meets the direction AM of the current s, AM′ a parallel drawn by this point to the current s', and mm' the projection on the plane MAM′ of the line which joins the two elements ds and ds'.

Draw by A a line An parallel and equal to mm', and form in n a small parallelogram nn' having its sides parallel to the lines MAN and AM′, and equal to ds and ds'.

If one repeats the same construction for all the elements, the parallelograms so formed will compose the entire parallelogram NAM′D, and, their surfaces having for their extent [pages 91 and 263] $dsds'\sin\varepsilon$, one obtains the proposed integral multiplied by $\sin\varepsilon$, in searching for the volume having as its base NAM′D, and terminated on the surface whose ordinates drawn to various points of this base have for value $1/r^3$; r being

[169][N. T.] Due to a zero-length arm.

[170][N. T.] In the original versions of the *Théorie* we had here due to a misprint the following expression: $\int\int \frac{dsds}{r^3}$.

[171][N. T.] In figure 28 of the 1827 version of the *Théorie* we had a letter m' between M and A. This letter was corrected to m in figure 28 of the 1826 version of the *Théorie*.

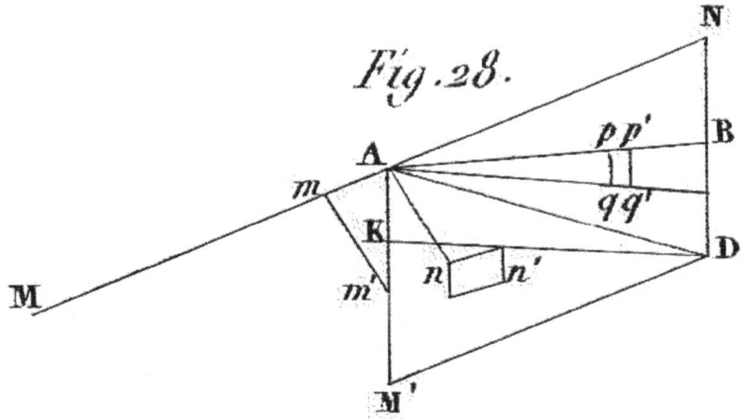

Figure 28.

the distance of the two elements of currents, which correspond, after our construction, to all these points of the surface NAM′D.[172]

Now, to calculate this volume, we can partition the base into triangles having for common vertex the point A.

Let Ap be a line connected to any of the points of the surface of the triangle AND, and $pqq'p'$ the surface included between the two infinitely close lines Ap and Aq' and the two arcs of a circle described by A with the radii A$p = u$ and A$p' = u + du$: because the angle NAM′ $= \pi - \varepsilon$ and naming φ the angle NAp, we have

$$\sin\varepsilon \int\int \frac{dsds'}{r^3} = \int\int \frac{ududφ}{r^3} \ .$$

Now, if a designates the common perpendicular to the direction of the two conductors, and s and s' [designate] the distances counted from A on the two currents, one has

$$r = \sqrt{a^2 + u^2}, \ u = \sqrt{s^2 + s'^2 - 2ss'\cos\varepsilon} \ ;$$

therefore, by integrating first from $u = 0$ up to $u = \text{AR} = u_1$,

$$\sin\varepsilon \int\int \frac{dsds'}{r^3} = \int\int \frac{udud\varphi}{(a^2 + u^2)^{\frac{3}{2}}} = \int d\varphi \left(\frac{1}{a} - \frac{1}{\sqrt{a^2 + u_1^2}} \right) .$$

It remains to integrate this last expression with respect to φ: for this we will calculate u_1 as a function of φ by the proportion AN : AR :: $\sin(\varphi + \varepsilon) : \sin\varepsilon$, or $s : u_1 :: \sin(\varphi + \varepsilon) : \sin\varepsilon$; [pages 92 and 264] and by substituting $a^2 + u_1^2$ by the value taken from this proportion, we will have to calculate

$$\int d\varphi \left[\frac{1}{a} - \frac{1}{\sqrt{a^2 + \frac{s^2\sin^2\varepsilon}{\sin^2(\varphi+\varepsilon)}}} \right] = \frac{\varphi}{a} - \int \frac{d\varphi \sin(\varphi + \varepsilon)}{\sqrt{s^2\sin^2\varepsilon + a^2\sin^2(\varphi + \varepsilon)}}$$

$$= \frac{\varphi}{a} + \frac{1}{a} \int \frac{d\cos(\varphi + \varepsilon)}{\sqrt{\frac{a^2+s^2\sin^2\varepsilon}{a^2} - \cos^2(\varphi + \varepsilon)}}$$

$$= \frac{1}{a} \left[\varphi + \arcsin \frac{a\cos(\varphi + \varepsilon)}{\sqrt{a^2 + s^2\sin^2\varepsilon}} + C \right] .$$

Name μ and μ' the angles NAD and M′AD, and take the preceding integral between $\varphi = 0$ and $\varphi = \mu$, it then becomes

$$\frac{1}{a} \left[\mu + \arcsin \frac{a\cos(\mu + \varepsilon)}{\sqrt{a^2 + s^2\sin^2\varepsilon}} - \arcsin \frac{a\cos\varepsilon}{\sqrt{a^2 + s^2\sin^2\varepsilon}} \right] ,$$

[172][N. T.] Due to a misprint, in the original French text we have NAMD.

and, since $\mu + \varepsilon = \pi - \mu'$, it changes to

$$\frac{1}{a}\left[\mu - \arcsin\frac{a\cos\mu'}{\sqrt{a^2 + s^2\sin^2\varepsilon}} - \arcsin\frac{a\cos\varepsilon}{\sqrt{a^2 + s^2\sin^2\varepsilon}}\right] ;$$

now

$$\cos\mu' = \frac{\mathrm{AK}}{\mathrm{AD}} = \frac{s' - s\cos\varepsilon}{\sqrt{(s' - s\cos\varepsilon)^2 + s^2\sin^2\varepsilon}} = \frac{s' - s\cos\varepsilon}{\sqrt{s^2 + s'^2 - 2ss'\cos\varepsilon}} ,$$

from which one extracts for the integral the following expression:

$$\frac{1}{a}\left[\mu - \arcsin\frac{a(s' - s\cos\varepsilon)}{\sqrt{a^2 + s^2\sin^2\varepsilon}\sqrt{s^2 + s'^2 - 2ss'\cos\varepsilon}}\right.$$

$$\left. - \arcsin\frac{a\cos\varepsilon}{\sqrt{a^2 + s^2\sin^2\varepsilon}}\right] ,$$

or, in changing from the sine to the tangent for the two arcs,

$$\frac{1}{a}\left[\mu - \arctan\frac{a(s' - s\cos\varepsilon)}{s\sin\varepsilon\sqrt{a^2 + s^2 + s'^2 - 2ss'\cos\varepsilon}} - \arctan\frac{a\cot\varepsilon}{\sqrt{a^2 + s^2}}\right] ,$$

and since one finds the integral relative to the triangle M′AD [pages 93 and 265] in changing in this expression μ to μ' and s to s', one has for the total integral, because $\mu + \mu' = \pi - \varepsilon$,

$$\frac{1}{a}\left(\pi - \varepsilon - \arctan\frac{a(s' - s\cos\varepsilon)}{s\sin\varepsilon\sqrt{a^2 + s^2 + s'^2 - 2ss'\cos\varepsilon}}\right.$$

$$- \arctan\frac{a\cot\varepsilon}{\sqrt{a^2 + s^2}} - \arctan\frac{a(s - s'\cos\varepsilon)}{s'\sin\varepsilon\sqrt{a^2 + s^2 + s'^2 - 2ss'\cos\varepsilon}}$$

$$\left. - \arctan\frac{a\cot\varepsilon}{\sqrt{a^2 + s'^2}}\right) .$$

In calculating the tangent of the sum of the two arcs whose values contain s and s', one changes this expression to

$$\frac{1}{a}\left(\pi - \varepsilon - \arctan\frac{a\sin\varepsilon\sqrt{a^2 + s^2 + s'^2 - 2ss'\cos\varepsilon}}{ss'\sin^2\varepsilon + a^2\cos\varepsilon}\right.$$

$$\left. - \arctan\frac{a\cot\varepsilon}{\sqrt{a^2 + s^2}} - \arctan\frac{a\cot\varepsilon}{\sqrt{a^2 + s'^2}}\right) ;$$

and since[173]

$$\frac{\pi}{2} - \arctan\frac{a\sin\varepsilon\sqrt{a^2 + s^2 + s'^2 - 2ss'\cos\varepsilon}}{ss'\sin^2\varepsilon + a^2\cos\varepsilon}$$

$$= \arctan\frac{ss'\sin^2\varepsilon + a^2\cos\varepsilon}{a\sin\varepsilon\sqrt{a^2 + s^2 + s'^2 - 2ss'\cos\varepsilon}} ,$$

one has,[174] by dividing by $\sin\varepsilon$,

$$\int\int\frac{dsds'}{r^3} = \frac{1}{a\sin\varepsilon}\left(\arctan\frac{ss'\sin^2\varepsilon + a^2\cos\varepsilon}{a\sin\varepsilon\sqrt{a^2 + s^2 + s'^2 - 2ss'\cos\varepsilon}}\right.$$

$$\left. - \arctan\frac{a\cot\varepsilon}{\sqrt{a^2 + s^2}} - \arctan\frac{a\cot\varepsilon}{\sqrt{a^2 + s'^2}} + \frac{\pi}{2} - \varepsilon\right) ;$$

[173][N. T.] In the 1827 version of the *Théorie* we had $2ss\cot\varepsilon$ in the denominator of the right hand side of the equality, instead of $2ss'\cos\varepsilon$. This misprint was corrected in the 1826 edition of the *Théorie*.

[174][N. T.] In his work of 12 September 1825, Ampère presented the value of the next integral, without indicating the calculations leading to it; see: [Ampère, 1825d], [Ampère, 1825g], [Ampère, 1825e], [Ampère, 1825c] and [Ampère, 1826e].

an expression which, when one assumes that $\varepsilon = \frac{\pi}{2}$, reduces to

$$\frac{1}{a}\left(\arctan\frac{ss'}{a\sqrt{a^2+s^2+s'^2}}\right) ,$$

as we have found previously.

[pages 94 and 266]

One can remark that the first term of the value that we have just found in the general case is the indefinite integral of[175]

$$\frac{dsds'}{(a^2+s^2+s'^2-2ss'\cos\varepsilon)^{\frac{3}{2}}} ,$$

as one can verify by differentiation, and that the three other [terms] are obtained by successively applying into this indefinite integral:

$$\text{1st} \quad s'=0; \quad \text{2nd} \quad s=0; \quad \text{3rd} \quad s'=0 \quad \text{and} \quad s=0 .$$

If the currents do not start from the common perpendicular, one would obtain an integral still composed of four terms which would all be of the same form as the indefinite integral.

29.15 Action Exerted on an Element of Conducting Wire by an Assembly of Closed Circuits of Very Small Dimensions, which Received the Name of Electrodynamic Solenoid

Until now we have considered the mutual action of electric currents in the same plane and rectilinear currents situated arbitrarily in space; it still remains to consider the mutual action of curvilinear currents which are not in the same plane. First we shall assume that these currents describe planar and closed curves with all their dimensions infinitesimal.[176] As we have seen, the action of a current of this kind depends on the three integrals A, B and C, whose values are[177]

---

[175] [N. T.] In the 1826 and 1827 versions of the *Théorie* this equation appeared as

$$\frac{dsds'}{(a^2+s^2+s'^2-2ss'\cos\varepsilon)^{\frac{5}{2}}} .$$

However, only in the 1826 edition we have the errata informing that the correct expression is the one presented in this English translation.

[176] [N. T.] The following paragraphs related to the theory of the solenoid are a reproduction of [Ampère, 1824c, pp. 151-162], [Ampère, 1824f, pp. 246-258] and [Ampère, 1824e, sections III, IV and V]. In the papers of 1824 Ampère worked with solenoids having circular areas of cross section represented by πm^2. In the *Théorie*, on the other hand, these calculations have been generalized to take into account solenoids having cross sections of arbitrary shape having areas represented by λ.

[177] [N. T.] See our page 366 up to 377, corresponding to pages 42 up to 55 of the 1826 version of the *Théorie*, and to pages 214 up to 227 of the 1827 version of the *Théorie*. In the previous pages Ampère was working with an arbitrary value of his constant n. In the next calculations he utilized the specific value $n = 2$.

In the paper published in 1824, Ampère mentioned the following, soon after the formula for the magnitude U appearing in our page 374, corresponding to pages 53 and 225 of the 1826 and 1827 versions of the *Théorie*, respectively, and immediately before considering the action of a solenoid on a current element, [Ampère, 1824c, p. 150-151]; see also [Ampère, 1824e, p. 19] and [Ampère, 1887d, note 1 on p. 85]:

> The results presented in these two paragraphs [that is, the general properties of the three integrals A, B and C] are independent of the value which is given to the exponent [n] of the power of the distance which one supposes that the electrodynamic action is reciprocally proportional when this distance varies, while the interacting elements of electric current do not change their directions. The same [conclusion] does not take place in the results with which I am dealing in the remainder of my paper, which are related to the case in which the system of currents forming closed circuits, of which we have just examined their properties, becomes an electrodynamic solenoid as I defined above. These results take place in only two cases, [namely,] in the case of nature, that is, when the electrodynamic action is inversely proportional to the square of the distance when only this distance varies, and in the case in which it would be supposed that [the action or force] is directly proportional to the distance. These [results] are due to M. Savary, who obtained them initially for an electrodynamic cylinder [see [Savary, 1823a], [Savary, 1822], [Savary, 1823b], [Savary, 1823c] and [Savary, 1885b]] and, then, for an arbitrary solenoid [that is, when the centers of the small circles composing this solenoid follow an arbitrary curve in space, so that they do not need to follow a straight line as was the case with the electrodynamic cylinder, see [Anonymous, 1823] and

$$A = \lambda \left(\frac{\cos\xi}{l^3} - \frac{3qx}{l^5} \right) ,$$

$$B = \lambda \left(\frac{\cos\eta}{l^3} - \frac{3qy}{l^5} \right) ,$$

$$C = \lambda \left(\frac{\cos\zeta}{l^3} - \frac{3qz}{l^5} \right) .$$

[pages 95 and 267]

Now imagine in space any line MmO (figure 29),[178] which the electric currents encircle forming very small closed circuits around this line, in infinitely close planes which are perpendicular to this line, such that the areas occupied by these circuits are all equal to each other and represented by λ, that their centers of gravity are on MmO, and that these planes have the same distance, measured along this line, between two consecutive planes. Putting g for this infinitesimal distance between neighboring planes, the number of currents found to correspond to an element ds of the line MmO, will be ds/g; and it is necessary to multiply by this number the values of A, B and C which have just been found for a single circuit so as to obtain the values which refer to the circuits of the element ds; by integrating over the arc s from one extremity L$'$ to the other extremity L$''$ of this arc, one obtains the values of A, B and C relative to the assembly of all circuits which encircle it, an assembly which I have called an *electrodynamic solenoid*, from the Greek word $\sigma\omega\lambda\eta\nu o\varepsilon\iota\delta\eta\zeta$,[179] which means that which forms a canal, that is to say, the surface of this form on which all the circuits are located.

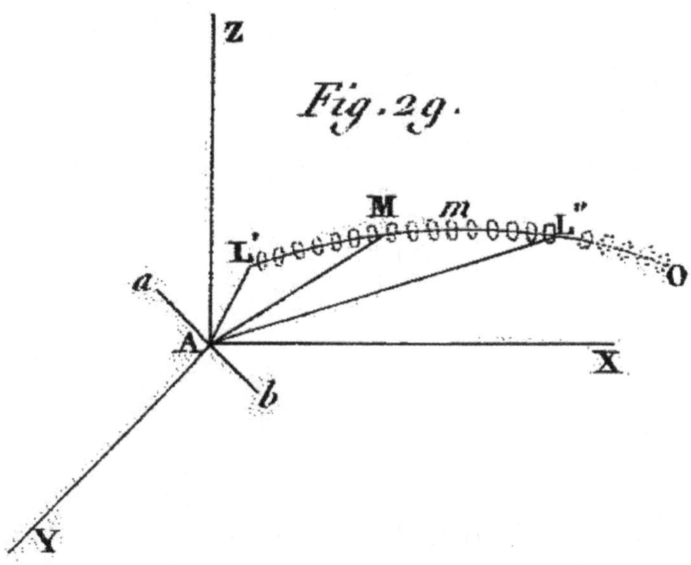

Figure 29.

Thus, for the entire solenoid,

$$A = \frac{\lambda}{g} \int \left(\frac{\cos\xi ds}{l^3} - \frac{3qxds}{l^5} \right) ,$$

$$B = \frac{\lambda}{g} \int \left(\frac{\cos\eta ds}{l^3} - \frac{3qyds}{l^5} \right) ,$$

[Savary, 1885a]]. The new demonstration which I present [for these results] can be applied directly to the solenoid and includes therefore, the case in which it is a cylinder, which is only a particular kind of solenoid.

[178][N. T.] In the 1826 edition of the *Théorie* the loops between L$''$ and O are dotted. In the 1827 edition they appeared as all the other loops, namely, drawn with continuous lines.

[179][N. T.] *Solenoid.* Ampère introduced this denomination in 1824, see Section 10.2 and [Ampère, 1824c, p. 135].

$$C = \frac{\lambda}{g} \int \left(\frac{\cos\zeta ds}{l^3} - \frac{3qzds}{l^5} \right) .$$

[pages 96 and 268]

Now, since the direction of the line g which is perpendicular to the plane of λ, is parallel to the tangent to the curve s, it follows that

$$\cos\xi = \frac{dx}{ds} , \quad \cos\eta = \frac{dy}{ds} , \quad \text{and} \quad \cos\zeta = \frac{dz}{ds} .$$

Moreover, q is evidently equal to the sum of the projections of the three coordinates x, y and z on its direction;[180] thus

$$q = \frac{xdx + ydy + zdz}{ds} = \frac{ldl}{ds} ,$$

since[181] $l^2 = x^2 + y^2 + z^2$. Substituting these values into the expression which has just been found for C, it becomes[182]

$$C = \frac{\lambda}{g} \int \left(\frac{dz}{l^3} - \frac{3zdl}{l^4} \right) = \frac{\lambda}{g} \left(\frac{z}{l^3} + C \right) .$$

Putting x', y', z', l' and x'', y'', z'', l'' for the respective values of x, y, z, l at the two extremities L$'$ and L$''$ of the solenoid, we have

$$C = \frac{\lambda}{g} \left(\frac{z''}{l''^3} - \frac{z'}{l'^3} \right) .$$

Likewise, finding similar expressions for the two other integrals A and B, the values for the three quantities which it is proposed to calculate for the entire solenoid are:

$$A = \frac{\lambda}{g} \left(\frac{x''}{l''^3} - \frac{x'}{l'^3} \right) ,$$

$$B = \frac{\lambda}{g} \left(\frac{y''}{l''^3} - \frac{y'}{l'^3} \right) ,$$

[pages 97 and 269]

$$C = \frac{\lambda}{g} \left(\frac{z''}{l''^3} - \frac{z'}{l'^3} \right) .$$

For a solenoid with a closed curve as its directrix[183] one would have $x'' = x'$, $y'' = y'$, $z'' = z'$, $l'' = l'$, and therefore $A = 0$, $B = 0$ and $C = 0$; if it did extend to infinity in both directions,[184] all the terms of the values of A, B and C would be zero separately, and it is evident that in these two cases the action exerted by the solenoid [on a current element] will be reduced to zero. Assuming that it only extends to infinity on one side, which I shall indicate by referring to it as an indefinite solenoid in a single direction,[185] it is only

[180][N. T.] The magnitude q was defined on pages 51 and 53 of the 1826 version of the *Théorie*, corresponding to pages 223 and 225 of the 1827 version of the *Théorie*; see our pages 373 and 375, respectively.

[181][N. T.] Due to a misprint, the original French text had the following expression: $l^2 = x^2 + y^2 + z_2$.

[182][Note by R. A. R. Tricker, [Ampère, 1965b, p. 183]:] Ampère uses the same symbol C, for the arbitrary constant of integration and the expression on the left hand side of the equation. The two are, of course, not the same.

[183][N. T.] The meaning of the word *directrix* in this context was discussed in Section 10.2. The expression *directrix of a solenoid* refers to a straight or curved line passing through the centers of all loops composing the solenoid. This directrix of a solenoid should not be confused with the *directrix* or *normal to the directing plane* discussed in Section 10.1. Maybe Ampère utilized only the expression *normal to the directing plane* in the next few pages of the *Théorie* when referring to this last meaning of the word *directrix* in order to avoid confusion with the meaning of the word *directrix* which he was utilizing in this portion of his text.

[184][N. T.] Original French text: "s'il s'étendait à l'infini dans les deux sens." Blunn translated this sentence as, [Ampère, 1965b, p. 183]: "if they extend to infinity in both directions."

[185][N. T.] Original French text: "solénoïde indéfini dans un seul sens." It might also be translated as a "semi-infinite solenoid."

necessary to consider the extremity with coordinates x', y' and z' of finite value, because the other extremity is assumed to be infinitely remote and the first terms of the values which have just been found for A, B and C are necessarily zero; thus

$$A = -\frac{\lambda x'}{gl'^3}\,, \quad B = -\frac{\lambda y'}{gl'^3}\,, \quad \text{and} \quad C = -\frac{\lambda z'}{gl'^3}\,,$$

and therefore $A : B : C :: x' : y' : z'$; hence the normal to the directing plane[186] which passes through the origin and forms angles to the axes with cosines

$$\frac{A}{D}\,, \frac{B}{D}\,, \frac{C}{D}\,,$$

where $D = \sqrt{A^2 + B^2 + C^2}$, also passes through the extremity of the solenoid[187] with the coordinates x', y' and z'.

As we have seen, in the general case, the total resultant is perpendicular to this normal;[188] thus the action of an indefinite solenoid on an element is perpendicular to the straight line joining the midpoint of this element to the extremity of the solenoid; [pages 98 and 270] and since it[189] is likewise perpendicular to the element, it follows that it is also perpendicular to the plane drawn through this element and through the extremity of the solenoid.

Its direction being determined, it only remains to find its value; now, according to the analysis for the general case, this value is

$$-\frac{Dii'ds' \sin \varepsilon'}{2}\,,$$

where ε' is the angle between the element ds' and the normal to the directing plane; and since $D = \sqrt{A^2 + B^2 + C^2}$, it is easily found that

$$D = -\frac{\lambda}{gl'^2}\,,$$

which gives for the value of the resultant[190]

$$\frac{\lambda ii'ds' \sin \varepsilon'}{2gl'^2}\,.$$

It is therefore seen that the action exerted by an indefinite solenoid with its extremity at L′ (figure 29) on the element ab is normal at A to the plane bAL′, proportional to the sine of the angle bAL′, and is inversely proportional to the square of the distance AL′, and it always remains the same, whatever the shape and direction of the indefinite curve L′L″ on which all the centers of gravity of the currents composing the indefinite solenoid are assumed to lie.[191]

If it should be desired to consider a definite solenoid with its two extremities situated at two given points L′ and L″, it is sufficient to assume a second indefinite solenoid commencing at the point L″ of the first and coinciding with it from this point to infinity, with currents [pages 99 and 271] opposite in direction,[192] but equal in intensity, the action of the latter being opposite in sign to that of the first indefinite solenoid from L′, and destroying its action over the part extending from L″ to infinity in the direction L″O where they are superposed. The action of the solenoid L′L″ will therefore be the same as that which would be

[186] [N. T.] That is, the directrix discussed in our Section 10.1, see the footnote 96 in our page 367.

[187] [N. T.] That is, the directrix at a given point due to a simply indefinite solenoid is the straight line connecting this point to the extremity of the solenoid, see our Subsection 10.2.1.

[188] [N. T.] Ampère is pointing out here that in the general case the total force exerted by a closed circuit on a current element is perpendicular to the directrix due to the closed circuit and passing through the midpoint of the current element. As pointed out by Tricker, Ampère's directrix possesses the same direction as what is now called the magnetic induction, see [Ampère, 1965b, note on p. 184].

[189] [N. T.] That is, the net force acting on the current element.

[190] [N. T.] Due to a misprint in the 1826 and 1827 versions of the *Théorie*, we had here $\frac{\lambda ii'ds' \sin \varepsilon}{2gl'^2}$. This mistake appeared also in Tricker's book, [Tricker, 1965, p. 185], and in the French editions of 1887 and 1958, [Ampère, 1887d, p. 88] and [Ampère, 1958, p. 76]. The correct expression should have the angle ε', namely, $\frac{\lambda ii'ds' \sin \varepsilon'}{2gl'^2}$.

[191] [N. T.] This force is also orthogonal to the current element ab and has its application point in the center A of this element; see figure 29 of the *Théorie*.

[192] [N. T.] If the currents in a solenoid are flowing clockwise in the small closed loops, the currents in the other solenoid would be flowing anti-clockwise.

exerted by joining the two indefinite solenoids and, in consequence, it consists of the force which has just been calculated and another force which acts in the opposite direction, passing likewise through point A, perpendicular to the plane bAL$''$, and having for value

$$\frac{\lambda ii'ds'\sin\varepsilon''}{2gl''^2},$$

where ε'' is the angle bAL$''$, and l'' is the distance AL$''$. The total action of the solenoid L$'$L$''$ is the resultant of these two forces and, like them, it passes through point A.

29.16 Action on a Solenoid Exerted by an Element or by a Finite Portion of a Conducting Wire, by a Closed Circuit, or by a System of Closed Circuits

Since the action of a definite solenoid can be deduced directly from that of an indefinite solenoid, we shall in all that remains to be said on the subject proceed from the indefinite solenoid. This simplifies the calculations and conclusions can readily be drawn for a definite solenoid.

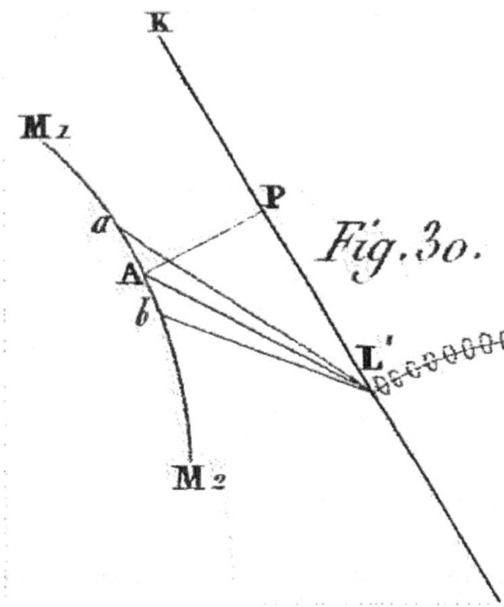

Figure 30.

Let L$'$ (figure 30), be the extremity of an indefinite solenoid, A the midpoint of an arbitrary element ba of an electric current M_1AM_2 and L$'$K an arbitrary and fixed straight line through the point L$'$; we put θ for the variable angle KL$'$A, μ for the inclination of the planes bAL$'$ and AL$'$K to each other, and l' for the distance L$'$A. Since the action of the element ba on the solenoid is equal and opposite to that which the solenoid exerts on the element, for its determination it is necessary to consider a point situated in A which is permanently associated with the solenoid, and which is influenced by a force which, ignoring the sign, may be represented as[193]

[pages 100 and 272]

$$\frac{\lambda ii'ds'\sin b\text{AL}'}{2gl'^2} \quad \text{or} \quad \frac{\lambda ii'd\nu}{gl'^3},$$

where $d\nu$ is the area aL$'b$ equal to

[193][N. T.] The force exerted by the current element ab on the indefinite solenoid has its application point at the center A of the element; see figure 30 of the *Théorie*. This point of the application of the force should be considered as invariably and rigidly connected to the solenoid.

$$\frac{l'ds'\sin b\text{AL}'}{2} \; .$$

Since this force is normal at A to the plane AL$'b$, to obtain its torque about the axis L$'$K, it is necessary to find the component which is perpendicular to AL$'$K and to multiply it by the perpendicular[194] AP dropped from point A on to the straight line L$'$K. The angle between the planes AL$'b$ and AL$'$K being μ, this component is obtained by multiplying the foregoing expression by $\cos\mu$; but $d\nu\cos\mu$ is the projection of the area $d\nu$ on the plane AL$'$K, whence it follows that in representing this projection by du, the value of the required component is

$$\frac{\lambda ii'du}{gl'^3} \; .$$

Now, the projection of the angle aL$'b$ on AL$'$K can be regarded as the infinitesimal difference between the angles KL$'a$ and KL$'b$; it will therefore be $d\theta$, and we will have

$$du = \frac{l'^2 d\theta}{2} \; ;$$

which reduces the previous expression to

$$\frac{\lambda ii'd\theta}{2gl'} \; ;$$

but since AP $= l'\sin\theta$, for the required torque we have

$$\frac{\lambda ii'}{2g}\sin\theta d\theta \; .$$

[pages 101 and 273]

This expression, integrated over the curve M_1AM_2, yields the torque of this current to make the solenoid revolve about L$'$K; now, if the current is closed, the integral, which is in general

$$C - \frac{\lambda ii'\cos\theta}{2g} \; ,$$

vanishes between the limits, and the torque is zero in respect of any straight line L$'$K through the point L$'$.

Hence, in the action of a closed circuit, or of any system of closed circuits, on an indefinite solenoid, all the forces applied to the various elements of the system, about an arbitrary axis, produce the same torques as if they were [acting] at the extremity of the solenoid; their resultant passes through this extremity and in no case can the forces tend to impart a rotational motion to the solenoid about a straight line through its extremity, which is in agreement with the results of the experiments.[195] If the current represented by the curve M_1AM_2 were not closed, its torque for rotation of the [simply indefinite] solenoid about L$'$K, putting θ'_1 and θ'_2 for the extreme values of θ in respect of point L$'$ and of the extremities M_1 and M_2 of the curve M_1AM_2, would be

$$\frac{\lambda ii'}{2g}(\cos\theta'_1 - \cos\theta'_2) \; .$$

[194][N. T.] In the 1826 and 1827 versions of the *Théorie* this sentence appeared as follows: "et la multiplier par la perpendiculaire à AP abaissé du point A sur la droite L$'$K." The word "à" which changes the meaning of the sentence seems to have been introduced due to a misprint. In Ampère's original article of 1824 this sentence appeared without the word "à"; see [Ampère, 1824c, p. 157]. In our translation we utilized the sentence as written in 1824: "to multiply it by the perpendicular AP," instead of "to multiply it by the perpendicular to AP." This perpendicular AP will be the arm of the lever relative to the axis L$'$K as regards the torque exerted by the force acting on A.

[195][N. T.] On 8 April 1822, Ampère mentioned the following; see [Ampère, 1822r, p. 64], [Ampère, 1822j, p. 203] and [Ampère, 1885e, p. 241]:

> Ever since I became aware of M. Faraday's work announcing the motion of revolution that a magnet produces always in the same sense to a voltaic conductor, it was easy for me to realize that if this effect had not been observed earlier, it was due to the fact that [up to that moment] one had utilized conductors making circuits which were almost closed, in which the electromagnetic action cannot produce this kind of motion [...].

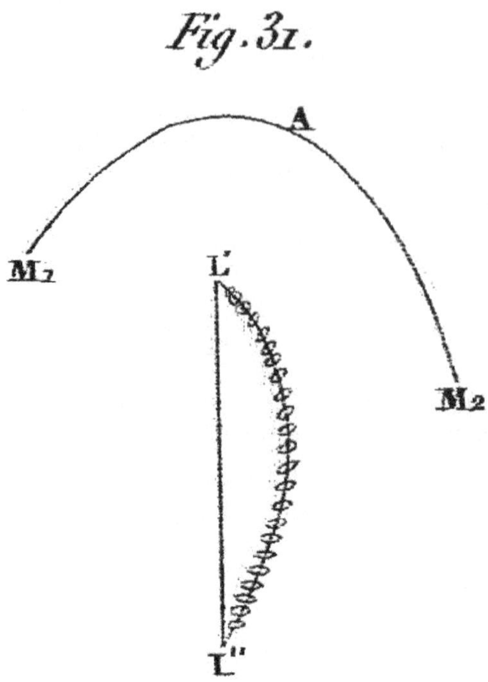

Figure 31.

Consider now the definite solenoid L′L″ (figure 31) which may only revolve about an axis through its two extremities. We shall again be able to replace it by two indefinite solenoids; the sum of the actions of the current $\mathrm{M_1AM_2}$ on each of them is equivalent to its action on L′L″. The torque of the first has just been found; [pages 102 and 274] putting θ_1'' and θ_2'' for the angles corresponding to θ_1' and θ_2', but in respect of the extremity L″, one obtains for that of the second

$$-\frac{\lambda ii'}{2g}(\cos\theta_1'' - \cos\theta_2'') ;$$

the total torque produced by the action of $\mathrm{M_1AM_2}$ for rotation of the solenoid about its axis L′L″ therefore is

$$\frac{\lambda ii'}{2g}(\cos\theta_1' - \cos\theta_1'' - \cos\theta_2' + \cos\theta_2'') .$$

This torque is independent of the shape of the conductor $\mathrm{M_1AM_2}$, of its magnitude and of its distance from the solenoid L′L″, and it remains the same so long as any such variation entails no change in the four angles θ_1', θ_2', θ_1'' and θ_2''; it is zero not only when the current $\mathrm{M_1M_2}$ forms a closed circuit, but also when the current is assumed to extend to infinity in both directions, because in that event, the two extremities of the current being infinitely remote from the extremities of the solenoid, the angle θ_1' becomes equal to θ_1'', and the angle θ_2' to θ_2''.

All the torques about straight lines drawn through the extremity of an indefinite solenoid being zero, this extremity is the point at which the resultant for the forces exerted on the solenoid is applied by a closed circuit,[196] or by a system of currents forming more than one closed circuit; it may therefore be assumed that all these forces are transported there and this point may be taken as the origin of coordinates A (figure 32): suppose that BM is a portion of one of the currents acting on the solenoid. From the foregoing the force due to some element Mm of BM is normal to the plane AMm and represented as

$$\frac{\lambda ii' d\nu}{gr^3} ,$$

where $d\nu$ is the area AMm and r is the variable distance AM.

[196] [N. T.] The extremity of an indefinite solenoid is the electrodynamic analog of a magnetic pole, as discussed in our Section 9.2.

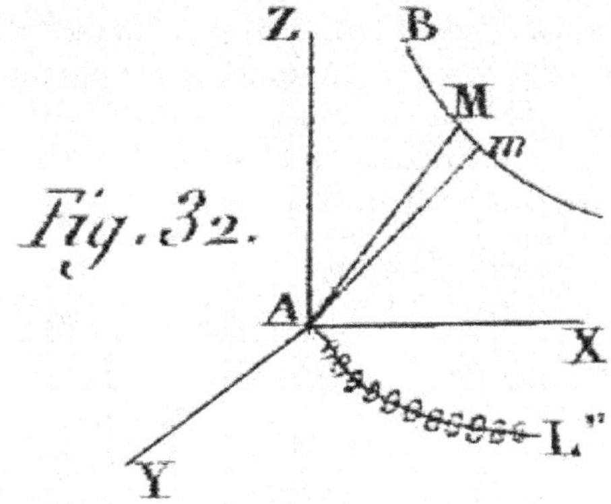

Figure 32.

[pages 103 and 275]

To obtain the component of this action along AX, it has to be multiplied by the cosine of the angle which it forms with AX, which is the same as the angle between the planes AMm and ZAY; but $d\nu$, multiplied by this cosine, is the projection of AMm on ZAY, which is equal to

$$\frac{ydz - zdy}{2} .$$

If therefore it is desired to find the action exerted along AX by currents forming closed circuits, it is necessary to take the following integral over the entire range of the currents[197]

$$\frac{\lambda ii'}{2g} \int \frac{ydz - zdy}{r^3} , \quad \text{which is equal to} \quad \frac{\lambda ii' A}{2g} ,$$

the quantity A being the same as before, where[198] $n+1$ was replaced by its value 3; likewise the action along AY is

$$\frac{\lambda ii' B}{2g} ,$$

and that along AZ

$$\frac{\lambda ii' C}{2g} .$$

The resultant of these three forces, which is the total action exerted by a number of closed circuits on an indefinite solenoid, is therefore equal to

$$\frac{\lambda ii' D}{2g} ,$$

where $D = \sqrt{A^2 + B^2 + C^2}$; and the cosines of the angles which it forms with the x, y and z axes are:

[pages 104 and 276]

$$\frac{A}{D} , \frac{B}{D} , \frac{C}{D} ,$$

[197][N. T.] Due to a misprint in the original French text, we have $ydz - rdy$ in the numerator of the integral, instead of the correct expression $ydz - zdy$.

[198][N. T.] In the original French text we have "on a remplacé n par sa valeur 3." However, the correct expression should be "on a remplacé $n+1$ par sa valeur 3." We utilized this correction in our English translation. The expression $n+1$ is the power of the distance r appearing in the calculation of quantity A; see our page 366, corresponding to pages 42 and 214 of the 1826 and 1827 versions of the *Théorie*, respectively.

which are exactly the values of the cosines of the angles between the same axes and the normal to the directing plane as if the action of the circuits on an element at A were considered.[199] Now this element would be transported by the action of the system in a direction contained within the directing plane; hence the remarkable conclusion is reached that when a system of closed circuits acts alternately on an indefinite solenoid and on an element situated at the extremity of this solenoid, the respective directions in which the element and the extremity of the solenoid are carried, are mutually perpendicular. If the element is situated in the directing plane, the action exerted upon it by the system is at its maximum and equal to

$$\frac{ii'Dds'}{2} .$$

The action which this system exerts on the solenoid was found just now to be

$$\frac{\lambda ii'D}{2g} .$$

These two forces are therefore always in a constant ratio for a particular element and a particular solenoid which is equal to

$$ds' : \frac{\lambda}{g} ;$$

that is to say, the forces are in the same relation as the length of the element bears to the area of the closed curve described by one of the currents of the solenoid divided by the distance between two consecutive currents; this ratio is independent of the form and magnitude of the currents of the system acting on the element and solenoid.

[pages 105 and 277]

29.17 Interaction of Two Solenoids

If the system of closed circuits is itself an indefinite solenoid, the normal to the directing plane through point A is, as we have just seen, a straight line joining point A to the extremity of the solenoid; hence the mutual action of two indefinite solenoids takes place along the straight line joining the extremity of one solenoid to the extremity of the other; in order to determine its value, we put λ' for the area of the circuits formed by the currents of this new solenoid, g' for the distance between the planes of two of these consecutive circuits, l for the distance between the extremities of the two indefinite solenoids, and we get $D = -\lambda'/(g'l^2)$, which yields for their interaction

$$\frac{\lambda ii'D}{2g} = -\frac{\lambda\lambda' ii'}{2gg'l^2} ,$$

which is inversely proportional to the square of the distance l. When one of the solenoids is definite, it can be replaced by two indefinite solenoids and the action is then made up of two forces, one attractive and the other repulsive, along the lines which join the two extremities of the first solenoid to the extremity of the other. Finally, if two definite solenoids $\mathrm{L'L''}$ and $\mathrm{L_1L_2}$ (figure 33) interact with each other, there are four forces along the respective straight lines $\mathrm{L'L_1}$, $\mathrm{L'L_2}$, $\mathrm{L''L_1}$ and $\mathrm{L''L_2}$ which join the extremities in pair; and if, for example, there is repulsion along $\mathrm{L'L_1}$, there will be attraction along $\mathrm{L'L_2}$ and $\mathrm{L''L_1}$, and repulsion along $\mathrm{L''L_2}$.

In order to justify the manner in which I have conceived magnetic phenomena, regarding magnets as assemblies of electric currents forming minute circuits round their particles, it should be shown from consideration of the formula by which I have represented the interaction [pages 106 and 278] of two elements of current, that certain assemblies of little circuits result in forces which depend solely on the situation of two determinate points of this system, and [it should be also shown that] these are endowed with all the

[199] [N. T.] The total force exerted by these closed circuits on a simply indefinite solenoid having its extremity at point A acts along this directrix D. The directrix due to these closed circuits passes through point A. That is to say, the force exerted by these closed circuits acts on the extremity of the simply indefinite solenoid along this directrix. These closed circuits generate the same directrix D when they act on a current element placed at A. However, as Ampère will mention, the force exerted by these closed circuits on this current element does not act along the directrix D. In this case the force acts along a direction which is orthogonal to D and which is also orthogonal to the direction of this current element. That is, this force will be along the directing plane passing through A. This directing plane is orthogonal to the directrix D.

Fig. 33.

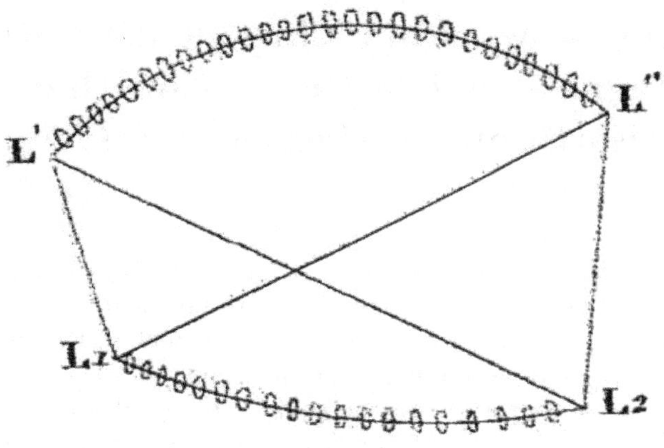

Figure 33.

properties of the forces which may be attributed to what are called molecules of austral fluid and of boreal fluid, whenever these two fluids are used to explain magnetic phenomena, whether in the mutual action of magnets, or in the action of a magnet on a conducting wire.[200] Now, the physicists who prefer explanations based on the existence of such molecules[201] to the explanation which I have deduced from the properties of electric currents, are known to admit that each molecule of austral fluid has a corresponding molecule of boreal fluid of the same intensity in each particle of the magnetized body.[202] In saying that the assembly of these two molecules, which may be regarded as the two poles of the element, is a magnetic element,[203] an explanation of the phenomena associated with the two kinds of action in question requires: (1) that the mutual action of magnetic elements should be made up of four forces, two attractive and two repulsive, acting along straight lines joining the two molecules of one of these elements to the two molecules of the other, with intensity in inverse ratio to the square of these lines; (2) that when one of these elements acts on an infinitesimal portion of conducting wire, two forces result, perpendicular to the planes passing through the two molecules of the element and through the direction of the small portion of wire, and proportional to the sines of the angles between the wire and the straight lines joining the wire to the two molecules, and which are inversely proportional to the squares of these distances. So long as my concept of the behavior of a magnet is disputed, [pages 107 and 279] and so long as the two types of force are attributed to molecules of austral and boreal fluid, it will be impossible to reduce them to a single principle; yet no sooner than my way of looking at the constitution of magnets is adopted, it is seen from the foregoing calculations that these two kinds of actions and the values of the resulting forces are deducible directly from my formula. To determine their values it is sufficient to replace the assembly of two molecules, the one of austral and the other of boreal fluid, by a solenoid with extremities that are the two determinate points on which the forces in question depend, and which are situated at precisely the same points where it is assumed that the molecules of the two fluids are placed.

---

[200][N. T.] Ampère satisfied these two requirements by considering the interaction between two solenoids.

[201][N. T.] Ampère was referring to Ørsted, Biot, Savart, Faraday, Poisson and most other scientists of his time. Ampère himself, on the other hand, fought against the existence of these magnetic molecules, as discussed in Section 19.1.

[202][N. T.] That is, each particle of a magnetized body would be composed of a small magnetic dipole with the opposite fluids separated by a small distance.

[203][N. T.] As discussed in Section 10.3, Poisson denominated "magnetic element" to what is nowadays called a magnetic dipole, that is, a North pole and a South pole of the same intensity separated by a small distance ℓ. There would be a magnetic dipole in each particle of the magnetized body. Ampère also adopted this nomenclature of Poisson from 1824 onwards.

29.18 Identity of Solenoids and Magnets when the Action Exerted on Them is from Conducting Wires, or by Other Solenoids or Other Magnets. Discussion of the Consequences that can be Drawn from This Identity, Relative to the Nature of Magnets and of the Action that one Observes between the Earth and a Magnet or a Conducting Wire

From the above, two systems of very small solenoids act on each other, according to my formula, like two magnets composed of as many magnetic elements as there are assumed to be solenoids in the two systems. One of these systems will also act on an element of electric current in the same way as a magnet.[204] In consequence, inasmuch as all calculations and explanations are based either on the attractive and repulsive forces of the molecules in inverse proportion to the squares of the distances, or on the rotational forces between a molecule and an element of electric current, the law governing which I have just indicated as accepted by physicists who do not accept my theory, they are necessarily the same whether the magnetic phenomena in these two cases is explained in my way by electric currents, or whether the hypothesis of two fluids is preferred. [pages 108 and 280] Objections to my theory, or proofs in its favor, therefore, are not to be found in such calculations or explanations. The demonstration on which I rely results above all from the fact that my theory explains with a single principle three sorts of actions[205] that all the associated phenomena prove are due to one common cause, and this cannot be done otherwise. In Sweden, Germany and England it has been thought possible to explain the phenomena by the interaction of two magnets as determined by Coulomb; the experiments which produce continuous rotational motion are manifestly at variance with this idea. In France, those who have not adopted my theory, are obliged to regard the three kinds of action which I have brought under one law, as three kinds of phenomena absolutely independent from one another. It should be remarked, in this context, that one can deduce from the law proposed by M. Biot for the interaction of an element of a conducting wire and that of what he termed a magnetic molecule, the law that Coulomb established for the action of two magnets if one accepts that one of these magnets is composed of small electric currents, like those which I have suggested; but then how can it be objected that the other is not likewise composed, thereby accepting all of my point of view?

Moreover, though M. Biot determined the value and direction of the force when an element of conducting wire acts on each particle of a magnet and defined this as the elementary force,[206, 207] it is clear that a force cannot be regarded [pages 109 and 281] as truly elementary which manifests itself in the action of two elements which are not of the same nature, or which does not act along the straight line which joins the two points between which it is exerted. However, in the mémoire which this gifted physicist communicated

[204][N. T.] Tricker introduced at this point, [Ampère, 1965b, note on p. 193], the footnote 29 which Ampère had introduced on pages 12 and 184 of the 1826 and 1827 versions of the *Théorie*, respectively; see our page 347.

[205][N. T.] Namely, the electrodynamic actions between two current-carrying conductors, the electromagnetic actions between a current-carrying conductor and a magnet, and the magnetic actions between two magnets or between a magnet and the Earth.

[206][N. A.] *Précis élémentaire de physique*, volume II, p. 122 of the second edition.

[207][N. T.] See [Biot, 1821a, p. 122]. In the third edition of his book, published in 1824, Biot continued to call the interaction between a current element and a magnetic pole as an *elementary force*, [Biot, 1824, p. 740], [Biot and Savart, 1885, p. 113], [Farrar, 1826, p. 334] and [Biot and Savart, 1965a, pp. 133-134]:

> The action of an indefinite and rectilinear connecting wire on a magnetic element, such as that obtained in the foregoing experiments, is still nothing but a composite result; for by imagining the length of wire to be divided into infinitely many fine sections of very low height, it is seen that each section must act on the [magnetized] needle with a different energy according to its distance and direction [to the center of the needle] and the direction in which its action is exerted. Now these elementary forces are just the simple result which it is specially important to know, for the total force exerted by the wire is nothing other than the arithmetic sum of these effects.

to the *Académie* on 30 October and 18 December 1820,[208, 209] he regards [pages 110 and 282] the force which an element of conducting wire exerts on a molecule of austral or boreal fluid as elementary, that is to say, the action exerted on the pole of a magnetic element, and he considers there the mutual action of two elements of voltaic conductors as a compound phenomenon.[210] Now, one can easily conceive that if in effect there exist magnetic particles, [pages 111 and 283] their mutual action can be considered as the elementary force: this was the point of view of the Swedish and German physicists, that could not support the test of

[208][N. A.] Since the latter mémoire has not been published separately, the formula for the force is only known to me from the following passage in the second edition of the *Précis élémentaire de physique*, volume II, pp. 122 and 123.

> "By imagining the length of the connecting wire Z′C′ (figure 34 [see this figure in our page 416]) to be divided into infinitely many very fine sections, it is seen that each section must act on the needle with a different energy according to its distance and direction. Now, these elementary forces are just the simple result which it is especially important to know; for the total force exerted by the complete wire is nothing other than the sum of their individual actions. However, calculation is sufficient to analyze from the resultant the simple action. This is what Laplace did. He deduced from our observations that the individual law of the elementary forces exerted by each section of the connecting wire was inversely proportional to the square of the distance, that is to say, it is precisely the same as what is known to exist in ordinary magnetic actions. This analysis showed that to complete our knowledge of the force, it remained to determine whether the action of each section of the wire was the same in all directions at the same distance, or whether the energy was greater in some directions than in others. To decide this question, in the vertical plane I bent a long copper wire ZMC at M (figure 34), in such a way that the two [rectilinear] arms ZM and MC were at the same angle of the horizontal MH. In front of this wire I stretched another piece Z′M′C′ of the same material, the same in diameter and of the same grade; this piece I set up vertically, being separated from the first piece at MM′ only by a strip of very fine paper. I then suspended our magnetized needle AB in front of this system at the height of the points M and M′, and observed its oscillations at various distances whilst passing current successively through the bent and straight wires. In this way I found that the action was inversely proportional to the distance to the points M and M′; but the absolute intensity was weaker for the oblique wire than for the straight wire in the same proportion that the angle ZMH is to unity. An analysis of this result appears to indicate to me that the action of each element μ of the oblique wire on each molecule m of austral or boreal magnetism is inversely proportional to the square of its distance μm to this molecule and [directly] proportional to the sine of the angle $m\mu$M between the distance μm and the length of the wire."

It is remarkable that this law, which is a corollary of the formula by which I have represented the interaction of two elements of conducting wires when, according to my theory, each magnetic element [that is, each magnetic dipole] is replaced by a very small electrodynamic solenoid, was first found through a mathematical error; indeed, for the law to be valid, the *absolute intensity* of the force ought to have been proportional, not to the angle ZMH, but to the tangent of half this angle, as demonstrated later by M. Savary in his dissertation at the *Académie*, 3 February 1823, and which has meanwhile been published in the *Journal de physique*, volume XCVI, pp. 1-25 and subsequent. It appears, besides, that M. Biot recognized this error, for in the third edition which has just appeared, he describes, without reference to the Mémoire where it had first been corrected, new experiments where the intensity of the total force is, in accordance with the calculation of M. Savary, proportional to the tangent of half the angle ZHM, and he concludes therefrom, with more reason than he had with his first experiments, that the force which he calls elementary, is proportional for equal distance to the sine of the angle between the direction of the element of conducting wire and the direction of the straight line joining its midpoint to the magnetic molecule. (*Précis élémentaire de physique expérimentale*, third edition, volume II, pp. 740-745).

[209][N. T.] See also [Biot, 1821a, pp. 122-123], [Biot, 1885a], [Biot, 1824, pp. 740-745], [Farrar, 1826, pp. 334-339], [Biot and Savart, 1965a, pp. 133-139] and [Ampère, 1965b, note on pp. 195-199]. In figure 34 of the 1827 version of the *Théorie*, the letters appearing below the left straight wire of the image were Z and Z′. These letters were corrected in the 1826 edition of the *Théorie* presented here to Z′ and Z′.

[210][N. T.] The following discussion, up to the material included in our page 419, corresponding to pages 116 and 288 of the 1826 and 1827 versions of the *Théorie*, respectively, came from Note B of the *Précis*, [Ampère, 1824e, Note B, pp. 49-63]. We reproduce here the beginning of this Note, which was not included in the *Théorie*:

> In the present state of physics, the cause of the observed phenomena which might be considered with certainty as being truly primitive is not known. In chemistry we consider all substance which cannot be decomposed in other [substances] as a simple body or element. Likewise, in physics one should consider as an elementary force all force which cannot be reduced to other [forces]. It is evident that the force observed in the mutual action between two conducting wires or in the action between one of these wires and a magnet cannot be reduced to the attractions and repulsions which are simple functions of the distances of the particles between which they are exerted, because one obtains, by one as by the other [that is, one obtains in the action between two conductors and also in the action between a conductor and a magnet], motions of accelerated rotation always in the same sense. [Ampère is referring here to the experiments of continuous rotation performed by Faraday and by himself; see Chapter 7. These experiments cannot be explained utilizing central forces depending only on the distance between the interacting bodies.] Therefore, it is necessary to look for the elementary force, either in the action between two elements of conducting wire, as I have made since my first experiments on this subject, or in the action that an element [of current] exerts on the two poles of a particle of the magnet, poles which are called *magnetic molecules*, when one admits the hypothesis of two [magnetic] fluids, as M. Biot admitted in the memoirs communicated to the *Académie* on 30 October and 18 December 1820, in which he considers as elementary the force that an element of a conductor exerts on a magnetic molecule, that is to say, on the pole of a particle of the magnet, and he considers as a compound phenomenon the mutual action between two elements of voltaic

experiments,[211] because this force being proportional to a function of the distance, it can never give rise to a motion which is always accelerated in the same sense,[212] at least when, as they supposed, the magnetic molecules are considered fixed at determinate points of the conducting wires which they view as assemblies of small magnets, and therefore the two other types of actions would be compound phenomena, since the voltaic element was also [considered compound]. We also conceive that the mutual action of two conducting wire elements is the action which offers the elementary force: in this case the mutual action of two magnetic elements,[213] and the action which one of these elements exerts exerts on an infinitely small portion of the voltaic conductor, are compound actions, since the magnetic element should, in this case, be considered as compound. But how can it be conceived that the elementary force is that which is manifested between a magnetic element and an infinitely small voltaic conductor, in other words between two bodies actually of a very small volume, but such that one is necessarily compound, regardless of the two ways of interpreting the phenomena that we have just discussed?

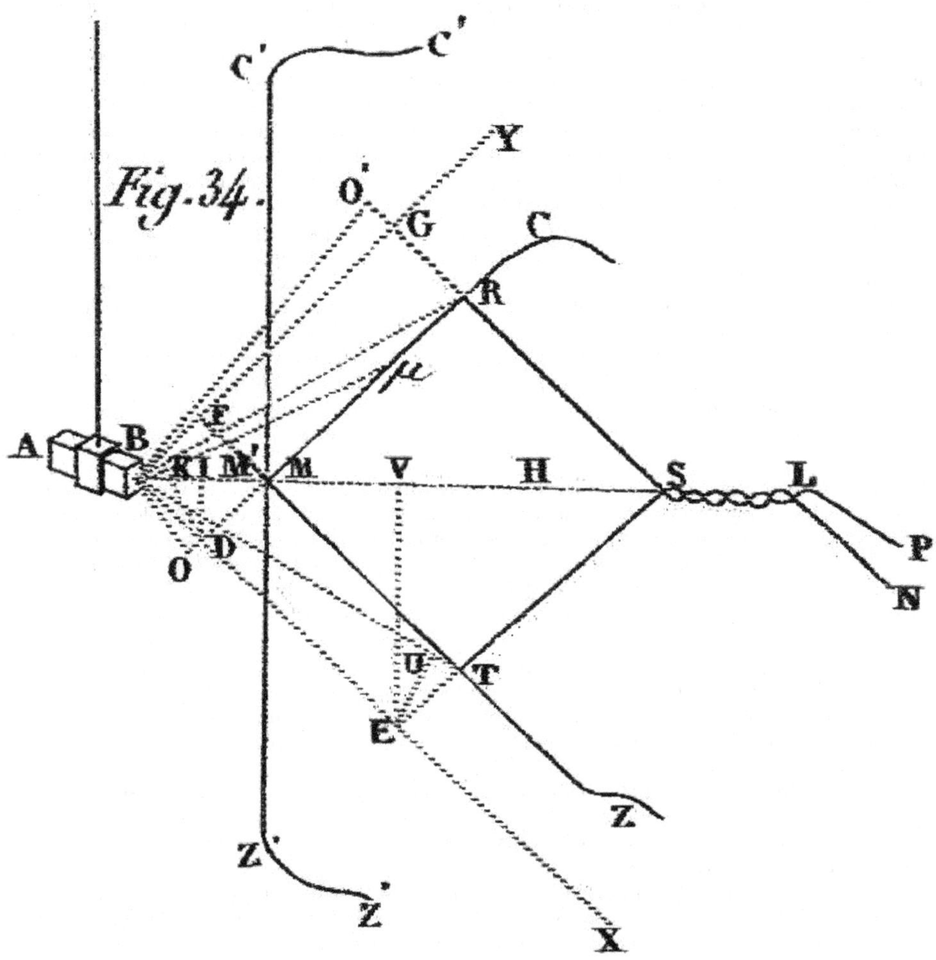

Figure 34.

The circumstance which presents the force exerted by an element of a conducting wire on a pole of a magnetic element, acting in a perpendicular direction to the line which joins the two points between which the force is determined, while the mutual action of two conductor elements follows the line which joins them, is not a proof less [pages 112 and 284] convincing than the one that the first of these two forces is a compound phenomenon. Whenever two points act on one another, whether due to an inherent force, or due to a force

---

conductors.

[211][N. T.] That is, this point of view was refuted by the experimental tests.

[212][N. T.] Ampère is referring here to a motion always accelerated clockwise or anti-clockwise around a fixed point, as in Faraday's experiments of continuous rotation discussed in Section 7.1.

[213][N. T.] That is, the interaction between two magnetic dipoles.

which comes from some other cause, such as a chemical phenomenon, a decomposition, or recomposition of the neutral fluid resultant from the reunion of the two electricities, one cannot conceive this force other than as a tendency of these two points to move closer or farther from each other along the line which joins them, with speeds reciprocally proportional to their masses, and this [behavior should take place] even when this force is only transmitted from one of the material particles to the other by means of an interposed fluid, like the mass of the bullet of a cannon[214] which is only carried in advance at a certain speed, due to the air released by the powder, by the same amount as the cannon mass is carried backwards along the same line, passing though the centers of inertia[215] of the bullet and cannon, with a speed which is to that of the bullet, as its mass is to the mass of the cannon.

This [conclusion] is a necessary result of the inertia of matter which Newton showed to be one of the principal foundations of the physical theory of the universe, in the last of the three axioms that he gave at the beginning of the *Philosophiæ naturalis principia mathematica*,[216] by stating that the action is always equal and opposite to the reaction; since two forces which impart to two masses speeds inverse to these masses, are the forces that would produce equal pressures against obstacles which would prevent invincibly that they[217] should put themselves in motion, in other words, equal forces. In order for this principle to be applicable in the case of the mutual action [pages 113 and 285] of two material particles which are traversed by an electrical current, while one assumes that this action is transmitted by an elastic fluid which fills the space, and whose vibrations constitute light,[218, 219] it must be admitted that this fluid has no appreciable inertia, as air with respect to the bullet and the cannon; but this is what one cannot doubt, since it does not oppose any resistance to the motion of the planets. The phenomenon of the rotation of an electric mill[220] has led many physicists to admit an appreciable inertia in the two electric fluids, and as a consequence in what results from their combination; but this supposition is in conflict with all that we otherwise know of these fluids, and with the fact that the planetary motions show no resistance due to the ether; there is not otherwise any motive to admit [this supposed great inertia in the electric fluids], since I have shown that the rotation of the electric mill is due to an electrodynamic repulsion produced between the tip of the mill and the ambient air particles, by the electrical current which escapes from this tip.[221, 222]

---

[214][N. T.] Ampère had discussed this example of the bullet of a cannon in a letter to Roux-Bordier dated 21 February 1821, [Launay (ed.), 1936a, pp. 565-567] and [Blondel, 1982, p. 100]. Ampère supports his points of view with Newton's third law of motion in the strong form, that is, when the action and reaction between two particles have not only the same magnitude and opposite direction, but act along the straight line connecting them. In this case we have conservation of linear and angular momentum.

[215][N. T.] That is, passing though the centers of mass.

[216][N. T.] Ampère was referring here to Newton's *Mathematical Principles of Natural Philosophy*: [Newton, 1934], [Newton, 1990], [Newton, 1999], [Newton, 2008] and [Newton, 2010].

[217][N. T.] That is, the masses.

[218][N. A.] This fluid can only be that which results from the combination of the two electricities. In order to avoid repeating the phrase for this, I think that one should employ, like Euler, the name ether, to mean the fluid as defined here.

[219][N. T.] Ampère is referring here to Leonhard Euler (1707-1783).

[220][N. T.] *Moulinet électrique* in French. This instrument was invented by Hamilton, professor of philosophy at Dublin University, around 1750, [Wilson, 1760, pp. 905-906]. This paper by Wilson was the first publication mentioning this apparatus. We present here two figures illustrating this instrument, [Jefimenko, 1973, Chapter 3: Electric wind motors]:

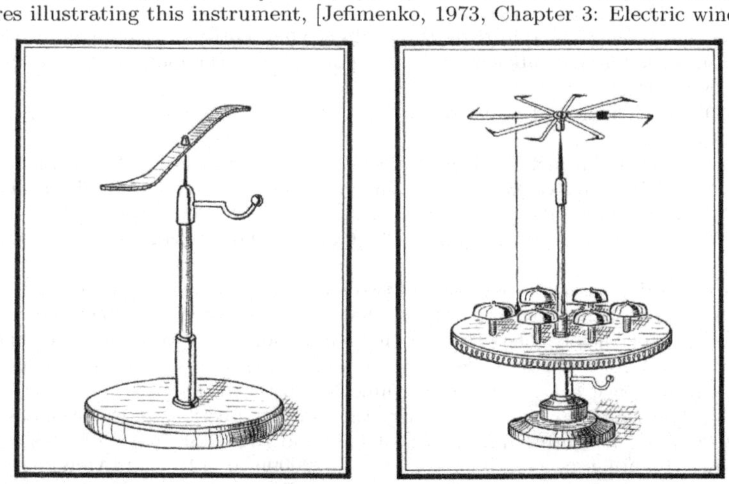

The apparatus normally turns around a vertical axis when connected to a source of high tension. It works based on the so-called electric wind and does not rotate in vacuum. There is a long controversy related to its working mechanism, [Robinson, 1962].

[221][N. A.] See the *Note* that I read before the *Académie*, on 24 June 1822, and which is included in the *Annales de chimie*, volume XX, pp. 419-421, and in my *Recueil d'observations électro-dynamiques*, pp. 316-318.

[222][N. T.] See [Ampère, 1822o, pp. 419-421], [Ampère, 1822y, pp. 316-318], [Ampère, 1885p, pp. 287-289] and [Ampère, 1885k,

When M. OErsted discovered the action which a conductor exerts on a magnet, it really ought to have been suspected that there could be interaction between two conductors; but this was in no way a [pages 114 and 286] necessary corollary of the discovery of this famous scientist, since a bar of soft iron acts on a magnetized needle, but there is no interaction between two bars of soft iron.[223] Inasmuch as it was only known that a conductor deflects a magnetized needle, could it have been concluded that electric current only imparted to this wire the property to be influenced by a needle in the same way as soft iron is so influenced by the same needle, which would be sufficient for the wire to act on the needle, without requiring interaction between two conductors when they are beyond the influence of a magnetized body? Only experiments could decide the question; I performed these in the month of September 1820, and the mutual action of voltaic conductors was demonstrated.

As regards the action of our Earth on a conducting wire, the analogy between the Earth and a magnet is doubtless sufficient to most probably produce this action, and I do not see why several of the most skilled European physicists think that the effect did not exist; not only, like M. Erman,[224] before I had made the experiment which showed it,[225] but [also] after this experiment had been presented at the *Académie des Sciences*, in the session of 30 October 1820, and repeated several times, during November of the same year, in the presence of several [pages 115 and 287] of their members and a great number of other physicists, who permitted me, at this time, to cite them as testifying to the motions produced by the action of the Earth on the movable parts of the equipment described and shown in the *Annales de chimie et de physique*,[226] volume XV, pages 191-196, plate 2, figure 5, and plate 3, figure 71, and also in my *Recueil d'observations électro-dynamiques*,[227] pages 43-48, since nearly one year later, the English physicists still raised doubts about the results of experiments so complete and which were demonstrated before a large number of witnesses.[228, 229] One cannot deny the importance of these experiments, and one cannot refuse to admit that the discovery of the action of the Earth on the conducting wires belongs to me as much as the [discovery of the] mutual action between two conductors. It was of little value that I should merely have discovered these two kinds of actions and verified them by experiment; it was more important:

(1) To find the formula for the interaction of two elements of current.

(2) To show by virtue of the law thus formulated (which governs the attraction of currents flowing in the same direction and repulsion of currents flowing in contrary senses, whether the currents are parallel or at an arbitrary angle),[230, 231] that the action of the Earth on conducting wires [pages 116 and 288] is identical in all respects to the action which would be exerted on the same wires by a system of electric currents flowing in the East-West direction, when situated in the South of Europe,[232] where the experiments which confirm this action were performed.

(3) To calculate first, from consideration of my formula and the manner in which I have explained magnetic phenomena by electric currents forming very small closed circuits round the particles of a magnetized body, the interaction between two particles of magnets regarded as two little solenoids[233] each equivalent to two

---

p. 291]. The subject of this Note by Ampère was discussed in our Chapter 8.

[223][N. T.] This argument was due to Arago, as discussed in Section 3.6.

[224][N. T.] Paul Erman (1764-1851), German physicist.

[225][N. A.] In a very remarkable memoir, printed in 1820, this well known scientist said that conducting wire had this advantage over the magnetic needle which is used for delicate experiments, [namely,] that the motion that it will take in these experiments will not be influenced at all by the action of the Earth.

[226][N. T.] See [Ampère, 1820f] and [Chaib and Assis, 2009b].

[227][N. T.] See [Ampère, 1822v, pp. 43-48].

[228][N. A.] See M. Faraday's memoir, published on 11 September 1821. The translation of this memoir appears in the *Annales de chimie et de physique*, volume XVIII, pages 337-370, and in my *Recueil d'observations électro-dynamiques*, pages 125-158. Due to a mistake in printing it shows the date 4 September 1821, instead of 11 September 1821.

[229][N. T.] See [Faraday, 1821c], [Faraday, 1822e, pp. 95-96], [Faraday, 1952d, pp. 805-807], [Faraday, 1822c] and [Faraday, 1885].

[230][N. A.] The experiments which demonstrate the action of two rectilinear currents in both cases, were communicated to the *Académie* in the session of 9 October 1820. The apparatuses that I employed are described and drawn in volume XV of the *Annales de chimie et de physique*, specifically: (1st) that for the mutual action of two parallel currents, page 72, plate 1, figure 1, and in more detail in my *Recueil d'observations électro-dynamiques*, pages 16-18; (2nd) that for the mutual action of two currents forming an arbitrary angle, page 171 of the same volume XV of the *Annales de chimie et de physique*, plate 2, figure 2, and in my *Recueil*, page 23. The figures in my *Recueil* carry the same numbers as in the *Annales*.

[231][N. T.] See [Ampère, 1820c, plate I, figure 1, pp. 71-72], [Chaib and Assis, 2007d, pp. 98-99], [Ampère, 1822v, pp. 16-18 and 23], [Ampère, 1820f, plate I, Figure 2, p. 171], [Chaib and Assis, 2009b, p. 121] and [Ampère, 1885b].

[232][N. T.] Original text in French: *situés an midi de l'Europe*. Normally the word *midi* is translated as *South*. In the partial English translation made by Blunn, this sentence was translated as *situated in the middle of Europe*, [Ampère, 1965b, p. 196].

[233][N. T.] Each one of these particles of a magnet should be considered as a small magnetic dipole, that is, as a system composed of a North and a South pole of equal magnitudes separated by a small distance. The electrodynamic analog of each one of these magnetic dipoles is a small definite solenoid, with its axis of symmetry along the straight line connecting the opposite poles of the supposed magnetic dipole.

magnetic molecules, the one of austral and the other of boreal fluid; and the action which one of these particles exerts on an element of conducting wire; then to check that these calculations give exactly for the two kinds of actions, in the first case the law established by Coulomb for the action of two magnets, and in the second case, the law which M. Biot has proposed for the forces which develop between a magnet and a conducting wire.[234] It is thus that I reduced both kinds of action to a single principle and also that [action] which I discovered exists between two conducting wires. Doubtless it was simple, [pages 117 and 289] having assembled all the facts, to conjecture that these three kinds of action depended on a single cause. But it was only by calculation that this conjecture could be substantiated, and this is what I have done. I draw no premature conclusions as to the nature of the force which two elements of conducting wires exert on each other, for I have sought only to obtain the analytical expression of this force from experimental data. By taking this as my starting point I have demonstrated that the values of the other two forces given by the experiment (the one between an element of conducting wire and what is called a magnetic molecule, the other between two of these molecules) can be deduced purely mathematically by replacing, in one or the other case, as is necessary, according to my conception of the constitution of magnets, each magnetic molecule by one of the two extremities of an electrodynamic solenoid. Thereafter, all that can be deduced from these values of the forces is necessarily contained in my manner of considering the effects which are produced and it becomes a corollary of my formula, and that alone should be sufficient to demonstrate that the interaction of two conductors is, in fact, the simplest case and that from which it is necessary to proceed in order to explain all other cases. The following considerations seem to me appropriate to furnish a complete confirmation of these general results of my work; they are founded on the simplest notions about the composition of forces, and are related to the mutual action of two systems of infinitely close points in the various cases which can arise—whether these systems only contain [pages 118 and 290] points of the same type, that is to say, points which attract or repel similar points of the other system, or whether one of the systems, or both, contains points of the two opposite types, of which those of one type attract what those of the other repel, and repel what they attract.

Initially suppose[235] that both of the two systems are made up of molecules of the same type, that is to say those that act on the other entirely by attraction or entirely by repulsion, with forces proportional to their masses; let M, M′, M″, etc. (figure 35), be the molecules which compose the first [system], and m any [molecule] of those which compose the second [system]: in successively composing all the actions ma, mb, md, etc., exerted by M, M′, M″, etc., one obtains the resultants mc, me, etc., such that the last is the action of the system MM′M″ on the point m, and passes near the center of inertia of the system.[236] By the same reasoning relative to the other molecules of the second system, one finds that the corresponding resultants all pass very close to the center of inertia of the second system, and will have a general resultant which passes also close to the center of inertia of the second: we call the *centers of action* the two points very close to the respective centers of inertia of the two systems by which the general resultant passes; it is evident that it will tend, due to the small distances that they are from the centers of inertia,[237] to impart to each system only a translational motion.

Suppose, in a second case, that the molecules of the second system are all of the same type, some [particles] of the first [system] are attractive and the others repulsive with respect to the molecules of the second system, the first [particles of the first system] will give a resultant of (figure 36), passing by their

[234][N. T.] It finishes here Ampère's copy of a portion of Note B of his *Précis* of 1824, [Ampère, 1824e]. The remainder of this Note B of the *Précis* was reproduced by Joubert, [Ampère, 1887d, note 1 on p. 104].

[235][N. T.] The portion beginning in this paragraph and going up to our page 421, corresponding to the beginning of pages 122 and 294 of the 1826 and 1827 versions of the *Théorie*, respectively, was taken from Note C of the *Précis* from 1824, [Ampère, 1824e, note C, pp. 63-67].

We present here the portion just before this part of Note C of the *Précis*, [Ampère, 1824e, p. 63]:

> We will examine here the action and reaction between two particles considered, each one, as [being composed of] a system of points or of molecules which attract or repel the points of the other system, supposing that the distance between the two systems is very large in comparison with the spaces which they occupy, and admitting, following the third axiom of the *Mathematical Principles of Natural Philosophy*, an axiom on which the whole of mechanics is based, that the action and reaction of each one of the molecules of one [system] on a molecule of the other [system] are directed along the straight line connecting them, and that they are equal and opposite.

[236][N. T.] Apparently Ampère's reasoning went as follows: The force exerted by M on m is represented by ma, acting along the straight line connecting Mm, the force of M' on m is represented by mb, acting along the line $M'm$ and the force of M'' on m is represented by md, acting along the line $M''m$. The joint force of M and M' acting on m is represented by mc, being given by the rule of the parallelogram. That is, mc is the diagonal of the parallelogram of sides ma and mb. The total force exerted by M, M' and M'' acting on m is represented by me. This line is also given by the rule of the parallelogram. That is, me is the diagonal of the parallelogram having mc and md as its sides.

[237][N. T.] That is, due to the small distance between the center of action and the corresponding center of inertia.

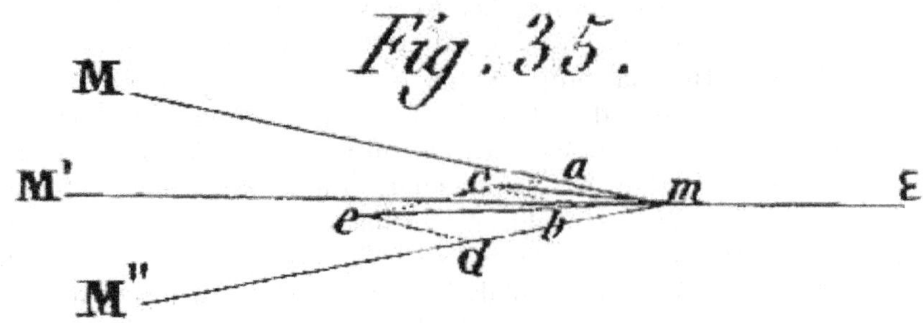

Figure 35.

center of action [pages 119 and 291] N, and by the center of action o of the other system; similarly, the repulsive particles [of the first system] yield a resultant oe, passing by their center of action P and by the same point o: the general resultant is then the diagonal og; and since it passes close to the center of inertia of the second system, it will tend to impart to it only a translational motion. Moreover, this resultant is in the same plane with the three centers of action o, N and P; and when the attracting molecules are of the same number as the repulsing ones, and acting with the same intensity, its direction is, moreover, perpendicular to the line oO which divides the angle PoN in two equal parts.

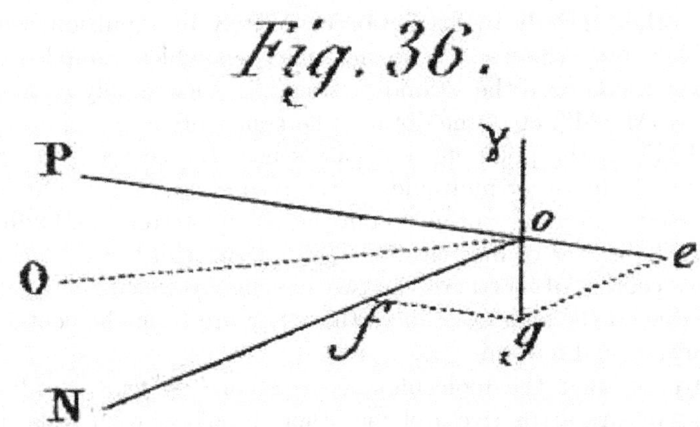

Figure 36.

Consider finally the case where the two systems are both composed of molecules of differing types. Let N and P (figure 37) be the centers of action respectively of the attractive and negative molecules of the first [system], let n and p be the corresponding centers of the second, of a type such that they have attraction between N and p, as well as between n and P, and they have repulsion between N and n, as well as between P and p. The combined actions of N and P on p give a resultant following the diagonal pe: similarly, the actions of N and P on n give a resultant nf. To obtain the general resultant, one extends these two lines until they meet in o, and taking[238] $oh = pe$, and $ok = nf$, [then] the diagonal ol will be the sought resultant which gives the action produced by the system PN on the system pn. But since the point o does not take part in the system pn, it will be necessary to conceive that it is connected to this system in an invariable manner without being [connected] to the first system PN; and the force ol will tend in general, due to this connection, to generate on pn a [pages 120 and 292] translational motion and a rotational motion about its center of inertia.

[238][N. T.] In the 1826 and 1827 versions of the *Théorie*, we had here *on* instead of *oh*. The errata at the end of the work corrected this mistake.

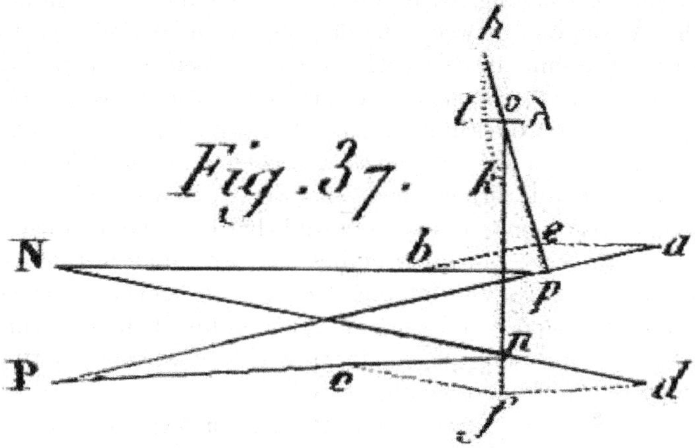

Figure 37.

Now examine the reaction exerted by the second system on the first: following the fundamental axiom of mechanics, that the action and reaction of two particles on each other are equal and directly opposed, it is necessary in order to obtain this [that is, to obtain the reaction] to successively compose the equal and directly opposed forces to those which the particles of the first system exert on the particles of the second, and it is evident that the total reaction that is thus found will always be equal and directly opposed to the total action.

Therefore, in the first case, the reaction is represented by the line $m\varepsilon$ (figure 35), equal and opposite to the resultant me, and that which one can assume is applied at the center of action of the first system which is located in that direction; from which it follows, always neglecting the small difference of the position of the center of action and the center of inertia, that one will have only translational motion.

In the second case, the reaction will likewise be represented by the line $o\gamma$ (figure 36), equal and opposite to og. But, since the point o is not part of the first system, and since in general it will not be traversed by the direction $o\gamma$, it will be necessary to conceive that this point o must be invariably connected to the first system, without being connected to the second; and, by this connection, the force $o\gamma$ will tend in general to induce on the system PN a double motion, translational and rotational. In addition, this force $o\gamma$ is in the plane PoN; and when the attractive molecules are of the same number as the repulsive ones and act with the same intensity, its direction is, like that of og, perpendicular to oO.

Finally, in the third case, the reaction is represented [pages 121 and 293] by the line $o\lambda$ (figure 37), equal and opposite to the resultant ol, and applied like it at the point o. To find the action of ol on pn, we conceived just now that this point o was located in this second system pn without being [connected] to the first PN. In order to now have the reaction exerted on this one, we consider the force $o\lambda$ applied at a point located in o, and connected in the first system PN without being [connected] to the second. This force tends in general to generate on PN a double motion of translation and rotation.

If one compares these results with the experimental measurements, relative to the directions of the forces which occur in the three types of actions that we distinguished above, it is easy to see that the three cases that we examined correspond with them exactly. When two voltaic conductor elements act on each other, the action and reaction are, as in the first case, determined by the line which connects these two elements; when it concerns the force which acts between an element of a conducting wire and a particle of a magnet[239] containing two poles of opposite types, which act in opposite senses with equal intensities, the action and reaction are, as in the second case, directed perpendicularly to the line which connects the particle and the element; and two particles of a magnetized bar, which are only two very small magnets, exert on each other a more complex action, resembling that of the third case, and thus one cannot draw a conclusion without considering it as the result of four forces, two attractive and two repulsive: it is thus easy to conclude that there is only the element of a conducting wire such that one can suppose that all the points exert the same type of action, [pages 122 and 294] and to judge that it is, of the three kinds of force which are here considered, the one that one can view as the simplest.[240]

[239][N. T.] Each particle of a magnet will be considered here as a magnetic dipole.

[240][N. T.] Here finishes the Note C of the *Précis* of 1824, [Ampère, 1824e]. That is, the simplest case is the situation represented

But from the fact that the force taking place between two elements of conducting wires is the simplest one, and that those which develop, one between one of these elements and a particle of a magnet[241] where there always exist two poles of the same intensity, the other between two such particles, are results more or less complicated, thus it is necessary to conclude that the first of these forces should be considered as truly elementary? I have always been far from thinking like that, and for this reason, after long consideration in the *Notes sur l'exposé sommaire des nouvelles expériences électro-magnetiques*, published in 1822,[242, 243] I sought to find an explanation for it through the reaction of the fluid distributed in space, and whose vibrations produce the phenomena of light: I have only said that one should consider it as *elementary*, in the sense that chemists arrange in the class of simple bodies all those [bodies] they cannot further decompose, whatever the conjectures based on analogy which might lead us to believe, on the other hand, that they are really composed, and because one has determined from it the value of the experiments and the calculations shown in this Treatise, it was starting from this single value that it was necessary to calculate those [values] of all the forces that are manifest in the more complicated cases.

But even when it were due, either to the reaction of a fluid of rarity such that one cannot suppose that it reacts due to its mass, whether to a combination of inherent forces of the two electric fluids, it would still follow [pages 123 and 295] that the action would still always be opposed to the reaction following the same line; because, as one has seen in the considerations just discussed, this circumstance necessarily enters into all the complex actions, since it occurs for all the really elementary forces which make up the complex action. By applying the same principle on the force which acts between that which one calls a magnetic molecule and an element of a conducting wire, one sees that if this force, considered as acting on the element, passes through its midpoint, the reaction of the element on the molecule should also be directed in a manner to pass through this midpoint and not through the molecule.[244] This consequence of a principle which has until now been admitted by all physicists, does not appear easy to demonstrate by experiment, in the case of the force which we are discussing, because in all the experiments where one causes a portion of a conducting wire which forms a closed circuit to act on a magnet, the result that one obtains for the total action is the same, whether one supposes that this force passes through the element of the conducting wire or through the magnetic molecule, as we have seen in this Treatise; it is this that has led many physicists to suppose that only the action exerted on the element of the conducting wire would pass through this element,[245] and that the reaction which is opposed and parallel would not be directed along the same line, [with these physicists supposing] that it would pass through the molecule and would form with the first force that which they called a primitive couple.[246]

The following calculations will provide me the occasion to examine in detail this singular hypothesis. One will see, by this examination, that it not only opposed to one of the fundamental principles of mechanics,[247]

by the force between two current elements acting along the straight line connecting them. After all, from this basic interaction it is possible to deduce the three kinds of interaction, namely: the interaction between two current-carrying conductors, the interaction between a current-carrying conductor and a magnet, and the interaction between two magnets. It is not possible to deduce the first two kinds of interactions beginning only with the force between magnetic poles. It is also not possible to deduce the first and the third interactions beginning only with the force between a magnetic pole and a current element.

In the *Précis* the last sentence was written as follows, [Ampère, 1824e, p. 67]:

> It is possible to easily judge, after this simple consideration, which one of these three kinds of action should be considered as the elementary force, following the meaning which should be given to this word in physics, according to what we said in the beginning of the previous Note.

The meaning of the expression *elementary force* to which Ampère was referring here was presented at the footnote 210 in our page 415. That is, an elementary force is that force which cannot be reduced to other known forces.

[241] [N. T.] Each particle of a magnet will be considered as a small magnetic dipole.

[242] [N. A.] *Recueil d'observations électro-dynamiques*, p. 215.

[243] [N. T.] See [Ampère, 1822p, p. 215] and [Ampère, 1885m, p. 250].

[244] [N. T.] That is, if the net force exerted by a magnetic pole on a current element passes through the midpoint of the current element, then the reaction force exerted by the current element on the magnetic pole must also pass through the midpoint of the current element, having the same intensity and opposite direction. In this case one must consider that the reaction force is acting on a point at the center of the current element which is rigidly connected to the magnetic pole, without being connected to the current element. In this way this reaction might be transmitted to the magnetic pole.

[245] [N. T.] Original French text (with our emphasis in italics): "c'est ce qui a porté plusieurs physiciens à supposer que l'action exercée *par l'élement* de fil conducteur passait seule par cet élément, [...]." It seems to us that there was a misprint here. Instead of *par l'élement*, "by the element," the correct expression should had been *sur l'élement*, "on the element." The correct sentence would then be as we wrote here, namely, "it is this that has led many physicists to suppose that only the action exerted on the element of the conducting wire would pass through this element, [...]."

[246] [N. T.] We discussed the primitive couple or the primitive torque in our Section 20.2. A primitive couple is a pair of forces, equal in magnitude, oppositely directed, whose lines of action do not coincide. Although there is no resultant force on the system, these opposite forces produce a torque about an axis which is normal to the plane of the forces.

[247] [N. T.] Namely, Newton's third law of motion in its strong form. In this case the forces of action and reaction have not

but that it is otherwise completely useless for explaining the observed facts, [pages 124 and 296] and a false interpretation of these facts could only be adopted by the physicists who do not admit that magnets obtain their properties from the action of electric currents around their particles.

The phenomena produced by the two electric fluids moving in voltaic conductors seemed so different from those which manifest their presence when they are stationary on the electrified bodies in an ordinary manner, that it was also claimed that the first [phenomena] should not be attributable to the same fluids as [those fluids which are responsible for] the second. It is exactly as if one concluded from this [fact, namely,] that the suspension of mercury in a barometer is a phenomenon entirely different from that of sound, that one should not attribute them to the same atmospheric fluid, at rest in the first case and in motion in the second; but that it was necessary to admit, for two facts so different, two fluids, one of which would act solely to press on the open surface of the mercury, and the other would transmit the vibratory motions which produce the sound.

Moreover, nothing proves that the force expressed by my formula cannot result from the attractions and repulsions of molecules of the two electric fluids,[248] in the inverse ratio of the square of the distance between the molecules. The fact of a rotational motion continually accelerating until the friction and resistance of the liquid in which the magnet or voltaic conductor is immersed that presents this kind of motion renders the velocity constant, appears at first absolutely opposed to this kind of explanation of electrodynamic phenomena. In effect, from the principle of conservation of living forces,[249] which is a necessary consequence of the laws of motion, it necessarily follows that when the [pages 125 and 297] elementary forces, which are here the attractions and repulsions in inverse ratio of the square of the distances, are expressed by simple functions of the mutual distances between the points between which they act, and that one part of these points are invariably fixed to each other and only move by virtue of these forces, the others remaining fixed, the first cannot return to the same situation, as regards the second, with velocities larger than those they had when they began from this same situation.[250] Now, in the motion of continuous rotation given to a mobile conductor by the action of a fixed conductor, all the points of the first [conductor] return to the same position with greater and greater velocities for each revolution, until the friction and the resistance of the acid in which the crown of the conductor is immersed introduce a limit in the increase of the rotational velocity of the conductor: it then becomes constant, despite the frictions and the resistance.

It is thus completely demonstrated that one would not know how to explain the phenomena produced by the action of two voltaic conductors, by supposing that electric molecules acting in inverse ratio of the square of the distance were distributed over the conducting wires, so as to remain fixed there and to be able to, in consequence, be regarded as invariably fixed between them. One must conclude that these phenomena are due to the two electric fluids flowing[251,252] continuously through the conducting wires, with [pages 126 and 298] extremely rapid motion, alternately joining and separating in the intervals between the particles of these wires.[253] It is because the phenomena here in question cannot be produced but by electricity in motion,

---

only the same intensity and opposite senses, but are also directed along the straight line connecting the interacting particles.

[248][N. T.] That is, nothing proves that Ampère's formula between current elements cannot result from the attractions and repulsions exerted between positively and negatively charged particles, as we would say nowadays.

[249][N. T.] We discussed the principle of the conservation of living forces in our Section 18.2.

[250][N. T.] For instance, when a rock is thrown upwards on the Earth, it will return to the departure point with the same speed it had at the beginning of its journey, supposing that only the gravitational force of the Earth is acting on the rock. We are here neglecting frictional forces, the action of the wind, etc.

[251][N. A.] During the first works by physicists on electrodynamic phenomena, many scientists believed they could explain them by the distribution of molecules, whether electric, or magnetic, residing at rest in the voltaic conductors. Since the discovery of the first motion of continuous rotation by M. Faraday was published, I saw immediately that it completely overthrew this hypothesis, and here are the terms in which I announced this observation, while that which I say here is just the development, in the *Exposé sommaire des nouvelles experiénces eléctro-magnétiques* made by various physicists since the month of March 1821, which I read in the public session of the *Académie royale des Sciences* on 8 April 1822.

> "Such are the new progresses which are becoming a branch of physics, of which we did not even suspect their existence only two years ago, and has already revealed to us facts more astounding than everything that science has up to the present offered of marvelous phenomena. A motion which continues forever in the same direction, despite frictions, despite resistance of the medium, and this motion produced by the mutual action of two bodies which remain constantly in the same state, is a fact without example in all that we know of the properties offered by inorganic matter; it demonstrates that the action which emanates from the voltaic conductors cannot be due to a particular distribution of certain fluids at rest in these conductors, as are the ordinary electric attractions and repulsions. One can only attribute this action to fluids in motion in the conductor where they flow going rapidly from one extremity of the battery to the other."

See the *Journal de physique* where this exposition was inserted during this time, volume XCIV, page 65, and my *Recueil d'observations électro-dynamiques*, page 205.

[252][N. T.] See [Ampère, 1822r, p. 65], [Ampère, 1822j, p. 205] and [Ampère, 1885e, pp. 242-243].

[253][N. T.] Ørsted had a somewhat similar conception of the electric current inside a wire. As he said in 1814 (see [Oersted, 1814,

that I believe they should be designated by the name *electrodynamic phenomena*;[254] those of *electromagnetic phenomena*,[255] as they were named [pages 127 and 299] until now, were well-suited, provided one were dealing only with the action discovered by M. OErsted between a *magnet* and an *electric current*, but it could only be seen as a false idea after I had found that one produces phenomena of the same kind without a *magnet*, and by only the mutual action of two *electric currents*.

It is only in the case where one assumes the electric molecules at rest within the bodies where they manifest their presence by the attractions or repulsions produced by them between these bodies, that one demonstrates that an indefinitely accelerated motion cannot result from the fact that the forces which excite the electric molecules in this state of rest depend only on their mutual distances. When on the contrary one considers that, set in motion in the conducting wires by the action of the battery, they change continually their location, recombining there at each instant into a neutral fluid, separating again, and then promptly reuniting with other fluid molecules of opposite nature, it is no longer contradictory to admit that from these actions in inverse ratio of the square of the distances exerted by each molecule, a force which depends not only on their distance can result between two elements of conducting wires, but also on the directions of the two elements [of conducting wires] following which the electric molecules move, reuniting with molecules of opposite type, and separating in the next instant to unite with others. Now, it is exactly and uniquely from this distance and from these directions that the force which is then developed depends, and of which the experiments and the calculations shown in this Treatise have given me the value. In order to make a clear idea of what happens in the [pages 128 and 300] conducting wire, it is necessary to pay attention that between the metal molecules of which it is composed, a fluid composed of positive fluid and negative fluid is spread, not just in the proportions which constitute a neutral fluid, but with an excess of the fluid which is naturally opposed to the proper electricity of the metal molecules, and which conceals this electricity, as I have explained in a letter that I wrote to M. Van-Beek at the beginning of 1822:[256, 257] it is in this intermolecular electric fluid that all the motion takes place, all the decompositions and recombinations which constitute the electric current.[258]

As the liquid interposed between the plates of the battery is, without comparison, a worse conductor than the metal wire which joins their extremities, it takes a time, very short in fact, but even so appreciable, during which the inter-molecular electricity, supposed at first to be in equilibrium, is decomposed in each one of the intervals between two molecules of this wire. This decomposition gradually augments until the positive electricity in one interval reunites with the negative electricity of the following interval in the sense of the current, and its negative electricity [reunites] with the positive electricity from the previous interval. This reunion can only be as instantaneous as the discharge of a Leyden jar;[259] and the action between the

p. 341], [Marazzini and Tucci, 1983] and [Martins, 1986, p. 96]):

> Electricity thus does not flow, through the conductors, as a liquid through a channel; but it spreads by a kind of continuous decomposition and recomposition, or better by an action which disturbs the equilibrium in each moment, and restores it in the next moment. We could express this succession of opposing forces, which exists in the transmission of electricity, by saying that *electricity always spreads in an undulatory manner.*

[254][N. T.] Ampère first defined the concept of *electrodynamic phenomena* in 1822, see our Section 1.4.

[255][N. T.] This nomenclature was due to Ørsted, as discussed in Section 1.4.

[256][N. A.] *Jounal de physique*, volume XCIII, pages 450-453, and *Recueil d'observations électro-dynamiques*, pages 174-177.

[257][N. T.] See [Ampère, 1821d, pp. 450-453], [Ampère, 1822x, pp. 174-177] and [Ampère, 1885o, pp. 217-219]. Abertus van Beek (1787-1856) was a Dutch physicist. Information about his life, work and relationship with Ampère can be found in [Snelders, 1975].

[258][N. T.] It is interesting to quote here Williams, [Williams, 1981, p. 145], when referring to Ampère's paper of 1821, [Ampère, 1821d]:

> There was no doubt that Ampère took his electrodynamic molecule [that is, the hypothesis of currents of electricity around each molecule] seriously and expected others to do so too. In an answer to a letter from the Dutch physicist Van Beck, published in the *Journal de physique* in 1821, Ampère argued eloquently for his model, insisting that it could be used to explain not only magnetism but also chemical combination and elective affinity. In short, it was to be considered the foundation of a new theory of matter. This was one of the reasons why Ampère's theory of electrodynamics was not immediately and universally accepted. To accept it meant to accept as well a theory of the ultimate structure of matter itself.

[259][N. T.] The Leyden jar was an old type of capacitor or condenser; see [Magnaghi and Assis, 2008, note 1] and [Heilbron, 1999, chapter 13: The Invention of the Condenser]. Ewald Jürgen von Kleist (c. 1700-1748) was the first scientist to build a capacitor in 1745. The Leyden jar was created by Andreas Cunaeus (1712-1788) in 1745, when developing some experiments performed by Pieter van Musschembroek (1692-1761). These instruments were initially composed of a glass bottle partially filled with water. A metal plate was placed over the exterior surface of the glass, being normally grounded, that is, connected to the Earth. The glass worked as an insulator, while the water and the external metal worked as conductors. This capacitor was charged by a wire connected to the water and to an electrostatic machine generating charges by friction. The condenser could then store opposite electric charges in the water and in the external conductor. When the water and the external plate were connected

conducting wires, that develops, while it takes place, [that is, while the reunion takes place,] in the contrary direction from that [action] which they then exerted during the decomposition, cannot by consequence diminish the effect of this [action], since the effect produced by a force is composed of its intensity [pages 129 and 301] and the time during which it acts; now here the intensity should be the same, whether the two electric fluids separate or reunite: but the time during which they are separating is without comparison much larger than that during their reunion.

As the action varies with the distance between the molecules of the two electric fluids during this separation, it would be necessary to integrate, with respect to time and during the entire period of separation, the value of the force occurring at each instant, and divide afterwards, by this period, the integral so obtained. Without making this calculation, for which the data are required, which are still missing, over the manner in which the distances of the electric molecules vary, with time, during each inter-molecular interval of the conducting wire, it is easy to see that the forces produced in this manner, between two elements of this wire, should depend on the direction of the electric current in each of these elements.

If it were possible, starting from this consideration, to find that the mutual action of two elements [of electric currents] is in effect proportional to the formula by which I have represented it, this explanation of the fundamental fact of all the theory of electrodynamic phenomena would clearly be preferred to all others; but it would require research that I have not had time to carry out,[260] nor [have I had the time for that] research even more difficult which would have to be carried out in order to find if the contrary explanation, where one attributes the electrodynamic phenomena to motions given to the ether by the electric currents, could lead to the same formula. Whatever it may be of these hypotheses and of other [pages 130 and 302] suppositions which one could make to explain these phenomena, they will always be represented by the formula that I have deduced from the results of experiments, interpreted by calculation; and it will remain demonstrated mathematically, when considering magnets to be assemblies of electric currents arranged about their particles as I have said, the values of the forces which are, in each case, given by experiment, and all the circumstances of the three types of actions which occur, one between two magnets, another between a conducting wire and a magnet, and the third between two conducting wires, are derived from a single force, acting between two electric current elements along the line which joins their midpoints.

As for the expression of this force, it is one of the most simple among those that do not depend only on the distance, but also on the direction of the two elements; because these directions only enter in [the fact] that it contains the second derivative of the square root of the distance between the two elements, taken by varying alternately the two arcs of electric currents of which this distance is a function, a differential which itself depends on the directions of the two elements, and which enters moreover in the value given by my formula in a very simple manner, as one has the second derivative for this value, multiplied by a constant coefficient and divided by the square root of the distance; by observing that the force is repulsive when the second derivative is positive, and attractive when it is negative. This [fact] is expressed by the – sign which is found in front of the general expression

$$-\frac{2ii'}{\sqrt{r}}\frac{d^2\sqrt{r}}{dsds'}dsds'$$

[pages 131 and 303]

of this force, following the usage where one takes the attractions as positive forces, and the repulsions as negative forces.

through a conductor, like a wire or the human body, a discharge took place, being very fast, almost instantaneous. When the discharge happened through the human body, a large shock might be felt.

[260][N. T.] The unification of Ampère's electrodynamics with electrostatics was performed by Wilhelm Weber (1804-1891), as discussed with references in our Section 22.2.

29.19 Identity of the Exerted Actions, either on the Pole of a Magnet, or on the Extremity of a Solenoid, by a Closed Voltaic Circuit and by an Assembly of Two Very Closely Spaced Surfaces Terminated by This Circuit, and on which are Spread and Fixed Two Fluids, such as the Two Supposed Magnetic Fluids, Austral and Boreal, in a Manner such that the Magnetic Intensity Is Everywhere the Same

Throughout history,[261] whenever hitherto unrelated phenomena have been reduced to a single principle, a period has followed in which many new facts have been discovered, because a new approach in the conception of causes suggests a multitude of new experiments and explanations. It is thus that Volta's[262] demonstration of the identity of galvanism and electricity was accompanied by the construction of the electric battery with all the discoveries which have sprung from this admirable device.[263] Judging from the important results of the work of M. Becquerel[264] on the influence of electricity in chemical compounds, and that of MM. Prévost and Dumas on the causes of muscular contraction,[265] it may be hoped that their discovery of new knowledge over the past four years and its reduction to a single principle of the laws of attractive and repulsive forces between electric conductors, will also lead to a host of other results which will establish the links between physics, on the one hand, and chemistry and even physiology, on the other, for which there has been a long-felt need, though we cannot flatter ourselves for having taken so long to realize it.

It still remains to consider the actions exerted by a closed circuit of arbitrary shape, magnitude and position on a solenoid, or on some other circuit of arbitrary shape, magnitude and position;[266] the principal result from such inquiries is the similarity which exists between the forces produced by this circuit, whether acting on another closed circuit or on [pages 132 and 304] a solenoid, and the forces which would have been exerted by points whose action were precisely that which is attributed to molecules of what is called austral and boreal fluid, when one assumes that these points are distributed in the manner which I will explain over surfaces terminated by the circuits, and the extremities of the solenoid are replaced by two magnetic molecules of opposite types. This analogy seems at first to be so complete that all electrodynamic phenomena appear to be reduced to the theory associated with these fluids. It is soon seen, however, that this only applies to conductors which form rigid and closed circuits, that it is only phenomena which are produced by conductors forming such circuits that may be explained in this way, and that in the end it is only the forces which my formula represents that fit all the facts. Moreover, it will be from this same analogy that I will deduce a demonstration of an important theorem which can be expressed as: the mutual action between two rigid and closed circuits, or that of a rigid and closed circuit[267] and a magnet, can never produce a continuous movement with a velocity which accelerates indefinitely up to a point in which the resistances and the frictions in the apparatus will render this velocity constant.

Finally in order to leave nothing out on this subject, I will start by giving the formulas relative to the mutual action of two conducting wires a more general and symmetric form. For this take s and s' any two curves that are assumed traversed by electric currents of which we continue to designate the intensities by i and i'.

Let $ds = \mathrm{M}m$ (figure 38) be an element of the first curve, $ds' = \mathrm{M}'m'$ an element of the second; x, y,

---

[261] [N. T.] This portion of Ampère's *Théorie* is contained in the final paragraph of [Ampère, 1824f, pp. 257-258] and [Ampère, 1824e, pp. 43-44].

[262] [N. T.] Alessandro Volta (1745-1827).

[263] [N. T.] Volta's original paper of 1800, in French, English and Portuguese, can be found at: [Volta, 1800a], [Volta, 1800b], [Volta, 1964] and [Magnaghi and Assis, 2008].

[264] [N. T.] Antoine César Becquerel (1788-1878), father of the physicist Alexandre Edmond Becquerel (1820-1891) and grandfather of Antoine Henri Becquerel (1852-1908), Nobel prize in physics in 1903 for his discovery of radioactivity.

[265] [N. T.] Jean-Louis Prevost (1790-1850) and Jean-Baptiste-André Dumas (1800-1884). For information of their lives and works, see [Pilet, 1981] and [Kapoor, 1981].

[266] [N. T.] The following results were presented to the Academy of Sciences of Paris in the meeting of 28 November 1825, [Ampère, 1826b]. The text read to the Academy was published in [Ampère, 1887c]. The summary published in 1825 can be found in [Ampère, 1887f].

[267] [N. T.] We translated Ampère's expression *circuit solid* by *rigid circuit.* He defined this concept as follows, [Ampère, 1822u, p. 356] and [Ampère, 1885l, p. 377]:

> One understands, by this expression, that all the parts of the portion of the conductor which forms a closed or almost closed circuit are invariably connected to one another, and cannot change their relative position.

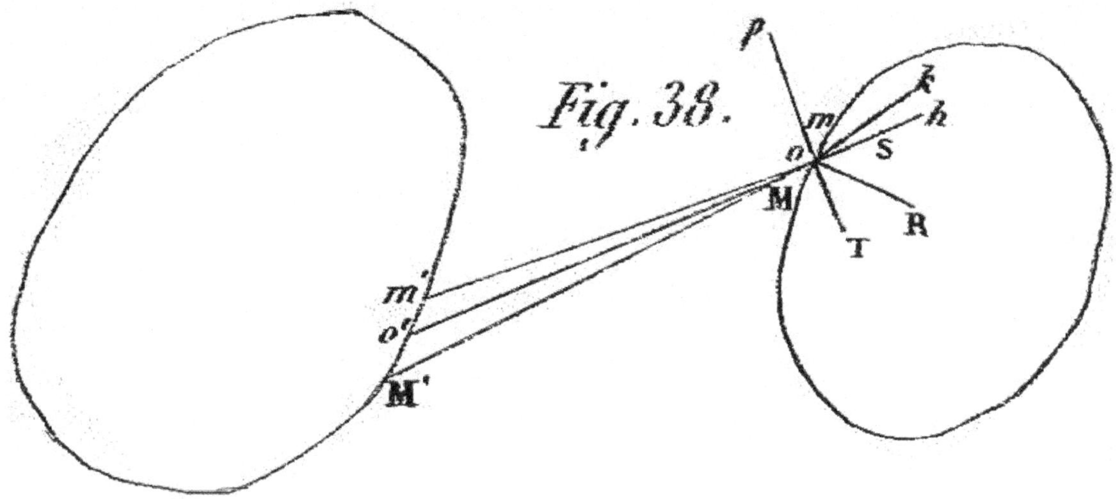

Figure 38.

z and x', y', z' the [pages 133 and 305] coordinates of their midpoints o and o', and r the line oo' which joins them, which should be considered as a function of two independent variables s and s' which represent the arcs of the two curves evaluated from two fixed points taken on them. The mutual action of the two elements ds and ds' is, as we have seen above,[268] a force directed following the straight line r, and having the value

$$-ii'dsds'r^k\frac{d(r^k\frac{dr}{ds})}{ds'} .$$

One can write this [equation] more simply as:[269]

$$-ii'r^kd'(r^kdr) ,$$

distinguishing by the characteristics d and d' the differentials relative only to the variation of the coordinates x, y and z of the element ds, from those obtained by varying only the coordinates x', y' and z' of the element ds'; a distinction which we will use in all cases where we consider differentials obtained the ones from one of these two forms, and the others from the other.

This force being attractive, it is necessary, for having that one of its components which is parallel to the x axis, to multiply the value by $(x-x')/r$ or by $-(x-x')/r$, when it is considered as acting on the element ds' or on the element ds;[270] in this last case, the component is then equal to

$$ii'r^{k-1}(x-x')d'(r^kdr) .$$

One can put this expression in another form by making use of the value that one obtains for udv, with u and v [pages 134 and 306] representing any magnitude, when one adds, side by side, the [following] two identities

$$udv + vdu = d(uv) ,$$

$$udv - vdu = u^2d\left(\frac{v}{u}\right) ,$$

[268][N. T.] See the first equation of Section 29.4 of the *Théorie* combined with the equation $1-n-2k=0$ which was also obtained in that Section.

[269][N. T.] Comparing these two equations we conclude that Ampère defined the magnitude d by $ds\frac{d}{ds}$, while the magnitude d' was defined by $ds'\frac{d}{ds'}$. In modern notation and utilizing the symbol $\equiv$ as representing a definition, Ampère's magnitudes d and d' would be written nowadays as $\partial \equiv ds\frac{\partial}{\partial s}$ and $\partial' \equiv ds'\frac{\partial}{\partial s'}$, respectively. It seems that Ampère first utilized this notation in his article of 10 June 1822, when he performed the calculations related to the case of equilibrium of the nonexistence of continuous rotation, [Ampère, 1822o].

[270][N. T.] That is, in order to obtain the x component of the force, the resultant force should be multiplied by $(x-x')/r$ or by $-(x-x')/r$.

this value is

$$udv = \frac{1}{2}d(uv) + \frac{1}{2}u^2 d\frac{v}{u} \ ,$$

and by making

$$u = r^{k-1}(x - x') \quad \text{and} \quad v = r^k dr \ ,$$

one concludes

$$r^{k-1}(x - x')d'(r^k dr) = \frac{1}{2}d'[r^{2k-1}(x - x')dr] +$$

$$\frac{1}{2}r^{2k-2}(x - x')^2 d'\frac{rdr}{x - x'} = \frac{1}{2}d'\frac{(x - x')dr}{r^n} + \frac{1}{2}\frac{(x - x')^2}{r^{n+1}}d'\frac{rdr}{x - x'} \ ,$$

since $2k + n = 1$, which gives

$$2k - 1 = -n, \quad 2k - 2 = -n - 1 \ .$$

But

$$r^2 = (x - x')^2 + (y - y')^2 + (z - z')^2 \ ,$$

and as a consequence

$$\frac{rdr}{x - x'} = dx + \frac{y - y'}{x - x'}dy + \frac{z - z'}{x - x'}dz \ ,$$

where[271]

$$d'\frac{rdr}{x - x'} = \frac{(z - z')dx' - (x - x')dz'}{(x - x')^2}dz - \frac{(x - x')dy' - (y - y')dx'}{(x - x')^2}dy \ .$$

[pages 135 and 307]

The component parallel to the x axis has therefore the value

$$\frac{1}{2}ii'd'\frac{(x - x')dr}{r^n} + \frac{1}{2}ii'\left[\frac{(z - z')dx' - (x - x')dz'}{r^{n+1}}dz \right.$$
$$\left. - \frac{(x - x')dy' - (y - y')dx'}{r^{n+1}}dy\right] \ .$$

The two terms of this expression can be considered separately as two forces whose sum is equivalent to the sought force. Now, it is easy to see that when the curve s' forms a closed circuit, all the forces such that they have for expression the part $\frac{1}{2}ii'd'\frac{(x-x')dr}{r^n}$, originating from the action of all the elements ds' of the circuit s' on the same element ds, mutually cancel each other. In effect, all the forces are applied at the same point o, midpoint of the element ds, along the same straight line parallel to the x axis; it is necessary, then, to have the force produced following this line due to the action of any portion of the conductor s', to integrate $\frac{1}{2}ii'd'\frac{(x-x')dr}{r^n}$ from one of the extremities of this portion to the other [extremity], and one finds[272]

$$\frac{1}{2}ii'\left[\frac{(x - x_2')dr_2}{r_2^n} - \frac{(x - x_1')dr_1}{r_1^n}\right] \ ;$$

[271][N. T.] In the *Théorie* published in 1990, [Ampère, 1990, p. 306], this equation appeared with a misprint in the first term, namely:

$$\frac{rdr}{x - x'} = \frac{(z - z')dx' - (x - x')dz'}{(x - x')^2}dz - \frac{(x - x')dy' - (y - y')dx'}{(x - x')^2}dy \ .$$

This mistake did not appear in the 1826 and 1827 original versions of the *Théorie*, [Ampère, 1826f, p. 134] and [Ampère, 1823c, p. 306].

[272][N. T.] Due to a misprint, the last fraction of the original work appeared as $\frac{x - x_1' dr_1}{r_1^n}$.

naming x'_1, r_1 and dr_1, the quantities corresponding to one extremity, and x'_2, r_2 and dr_2 those that are relative to the other, this value becomes evidently null when, the circuit being closed, its two extremities are at the same point.

Thus, when the conductor s' forms a closed circuit, it is therefore necessary, in order to obtain more simply the action that it produces on the element ds parallel to the x axis, to remove, [pages 136 and 308] in the expression of the component parallel to this axis, the part $\frac{1}{2}ii'\frac{d'(x-x')dx}{r^n}$, and only consider the other part

$$\frac{1}{2}ii'\left[\frac{(z-z')dx'-(x-x')dz'}{r^{n+1}}dz-\frac{(x-x')dy'-(y-y')dx'}{r^{n+1}}dy\right] ,$$

which we will represent by X.

By applying the same considerations to the other two components of the same force which are parallel to the y and z axes, we will substitute them by the forces Y and Z, having values given by

$$Y=\frac{1}{2}ii'\left[\frac{(x-x')dy'-(y-y')dx'}{r^{n+1}}dx-\frac{(y-y')dz'-(z-z')dy'}{r^{n+1}}dz\right] ,$$

$$Z=\frac{1}{2}ii'\left[\frac{(y-y')dz'-(z-z')dy'}{r^{n+1}}dy-\frac{(z-z')dx'-(x-x')dz'}{r^{n+1}}dx\right] .$$

Thus, when a closed circuit is considered, the resultant R of the three forces X, Y and Z, to which the components of the force $-ii'r^kd'(r^kdr)$ are reduced, replace this force; and the ensemble of all the forces R is equivalent to that of all the forces exerted by each of the elements ds' of the closed circuit s', and represents the total action of this circuit on the element ds. Let us see now what is the value and the direction of this force R.

Let u, v and w be the projections of the line r on the yz, xz and xy planes, making respectively the angles φ, χ and ψ, with the y, z and x axes. Consider the sector M$'om'$ (figure 38), which has as base the element ds', and for vertex the point o, midpoint of ds, the coordinates of which are x, y and z. Call λ, μ and ν the angles which the normal to the plane of this sector makes with the axes, and θ' the angle made between [pages 137 and 309] the directions of ds' and of r. Twice the area of this sector is $rds'\sin\theta'$, and its projections on the planes of the coordinates are

$$u^2d'\varphi=rds'\sin\theta'\cos\lambda=(y'-y)dz'-(z'-z)dy' ,$$

$$v^2d'\chi=rds'\sin\theta'\cos\mu=(z'-z)dx'-(x'-x)dz' ,$$

$$w^2d'\psi=rds'\sin\theta'\cos\nu=(x'-x)dy'-(y'-y)dx' .$$

We can then give this new form to the values of the forces X, Y and Z:[273]

$$X=\frac{1}{2}ii'\left(\frac{v^2d'\chi}{r^{n+1}}dz-\frac{w^2d'\psi}{r^{n+1}}dy\right)=\frac{1}{2}\frac{ii'dsds'\sin\theta'}{r^n}\left(\frac{dz}{ds}\cos\mu-\frac{dy}{ds}\cos\nu\right) ,$$

$$Y=\frac{1}{2}ii'\left(\frac{w^2d'\psi}{r^{n+1}}dx-\frac{u^2d'\varphi}{r^{n+1}}dz\right)=\frac{1}{2}\frac{ii'dsds'\sin\theta'}{r^n}\left(\frac{dx}{ds}\cos\nu-\frac{dz}{ds}\cos\lambda\right) ,$$

$$Z=\frac{1}{2}ii'\left(\frac{u^2d'\varphi}{r^{n+1}}dy-\frac{v^2d'\chi}{r^{n+1}}dx\right)=\frac{1}{2}\frac{ii'dsds'\sin\theta'}{r^n}\left(\frac{dy}{ds}\cos\lambda-\frac{dx}{ds}\cos\mu\right) .$$

Now these values give

$$X\frac{dx}{ds}+Y\frac{dy}{ds}+Z\frac{dz}{ds}=0 ,$$

$$X\cos\lambda+Y\cos\mu+Z\cos\nu=0 ;$$

[273] [N. T.] Due to a misprint, the original version of the *Théorie* contained the following expression: $X=\frac{1}{2}ii(...)$ instead of $X=\frac{1}{2}ii'(...)$.

that is to say that the direction of the force R makes with that [direction] of the element $m\mathrm{M} = ds$, and with the normal op to the plane of the sector $\mathrm{M}'om'$, the angles for which the cosines are zero, so that this force is both in the plane of the sector and perpendicular to the element ds.[274] As to its intensity, one has by known formulas[275]

$$R = \sqrt{X^2 + Y^2 + Z^2} = \frac{1}{2}\frac{ii'dsds' \sin\theta' \sin pom}{r^n}$$

$$= \frac{1}{2}\frac{ii'dsds' \sin\theta' \cos mok}{r^n} ;$$

[pages 138 and 310]

ok being the projection of om on the plane of the sector $\mathrm{M}'om'$. One can decompose this force in the plane of the same sector into two others, one [force] S directed along the line $oo' = r$, the other T perpendicular to this line. This last one is

$$T = R \cos ToR = R \cos hok = \frac{1}{2}\frac{ii'dsds' \sin\theta' \cos mok \cos hok}{r^n} ;$$

and since the trihedral angle formed by the directions of om, ok and oh gives

$$\cos mok \cos hok = \cos moh = \cos\theta ,$$

one has

$$T = \frac{1}{2}\frac{ii'dsds' \sin\theta' \cos\theta}{r^n} .$$

The force S along oh is

$$S = R \sin hok = T \tan hok .$$

But by designating ω the inclination of the plane moh on the plane hok, which is the plane of the sector $\mathrm{M}'om'$, one has

$$\tan hok = \tan\theta \cos\omega ;$$

thus

$$S = \frac{1}{2}\frac{ii'dsds' \sin\theta \sin\theta' \cos\omega}{r^n} .$$

If one integrates the expressions for X, Y and Z for the full range of the closed circuit s', one obtains the three components of the action exerted by all this circuit on the element ds; by replacing n by its value 2, the three components become

[pages 139 and 311]

[274][N. T.] Moreover, the point of application of the net force R exerted by the closed circuit s' on the element $ds = Mom$ is the midpoint o of the element Mom; see figure 38 of the *Théorie*.

[275][N. T.] It is worthwhile reproducing here the following Note by Joubert, [Ampère, 1887d, p. 123], referring to a work by Reynard, [Reynard, 1870], related to the next equation obtained by Ampère:

> This formula was reproduced by M. Reynard (*Annales de Chimie et de Physique* [4], volume XIX, p. 272; 1870), and, frequently, is designated as *Reynard's formula.* From the analytical point of view, it is the simplest [expression] with which one can represent the reciprocal action between two current elements. By adding to each one of its components an exact differential of the coordinates x', y' and z', one will obtain a formula which will also satisfy the experiments, as the integrals performed along the closed circuit will yield zero values for these terms. When one imposes to the new force so obtained the condition that it should be directed along the straight line connecting the two elements, one obtains that this term is that [term] which had been neglected previously, and we return to Ampère's formula. Therefore, it is the only one satisfying the imposed condition. [That is, Ampère's force is the only formula compatible with the experiments and satisfying the requirement of action and reaction along the straight line connecting the two current elements.]

$$\frac{1}{2}ii'\left(dz\int\frac{v^2d'\chi}{r^3}-dy\int\frac{w^2d'\psi}{r^3}\right)\,,$$

$$\frac{1}{2}ii'\left(dx\int\frac{w^2d'\psi}{r^3}-dz\int\frac{u^2d'\varphi}{r^3}\right)\,,$$

$$\frac{1}{2}ii'\left(dy\int\frac{u^2d'\varphi}{r^3}-dx\int\frac{v^2d'\chi}{r^3}\right)\,.$$

The similar forces applied to all the elements ds of the curve s will provide the total action exerted by the circuit s' on the circuit s. They will be obtained by integrating again the preceding expressions over all the extent of this last circuit.

Imagine now two surfaces chosen at random σ and σ', terminated by two contours s and s', such that all the points are invariably connected with one another and with all those of the corresponding surface, and [imagine] on these surfaces infinitely thin layers of the same magnetic fluid which is held there by a sufficiently coercive force so that it cannot be displaced. Considering on these two surfaces two portions infinitely small of the second order that we will represent by $d^2\sigma$ and $d^2\sigma'$,[276] whose positions are determined by the coordinates x, y, z for the first, x', y', z' for the second, and whose distance [from one another] is r, their mutual action will be a repulsive force directed along the line r and represented by[277] $-\mu\varepsilon\varepsilon' d^2\sigma d^2\sigma'/r^2$, that we consider as acting on the element $d^2\sigma$; ε and ε' designating here that which one calls the thickness of the magnetic layer on each surface;[278] μ is a constant, such that $\mu\varepsilon\varepsilon'$ represents the repulsive action that would occur, if one connected in two points located at a distance equal to unity, on one side all the fluid spread on an area equal to unity of the surface, where the thickness would [pages 140 and 312] be constant and equal to ε, on the other [side] all the fluid spread on another area equal to unity of the surface, where the thickness would also be constant and equal to ε'.[279]

In decomposing this force parallel to the three axes, one obtains the three components

$$\frac{\mu\varepsilon\varepsilon' d^2\sigma d^2\sigma'(x-x')}{r^3}\,,\quad\frac{\mu\varepsilon\varepsilon' d^2\sigma d^2\sigma'(y-y')}{r^3}\,,\quad\frac{\mu\varepsilon\varepsilon' d^2\sigma d^2\sigma'(z-z')}{r^3}\,.$$

Imagine now a new surface terminated by the same contour s which limits the surface σ, and such that all portions of the normals to the surface σ comprised between it and the new surface are very small. Suppose that on this last surface a magnetic fluid is distributed of contrary type to that from the surface σ,[280] in such a way that there is on the portion of the new surface circumscribed by the normals drawn by all points of the boundary of the element of surface $d^2\sigma$ a quantity equal to that of the fluid distributed on $d^2\sigma$. Taking h as the length of the small portion of the normal to the surface σ, drawn by the point which has coordinates x, y, z, and included between the two surfaces, which measures over the full extent of the area infinitely small $d^2\sigma$ the distance of its points to the corresponding points of the other surface, and designating by ξ, η, ζ the angles which this normal makes with the three axes, the three components of the mutual action between the

---

[276] [N. T.] That is, considering two infinitesimal areas $d^2\sigma$ and $d^2\sigma'$.

[277] [N. T.] In the 1826 and 1827 versions of the *Théorie*, we had here only the formula $\frac{\mu\varepsilon\varepsilon' d^2\sigma d^2\sigma'}{r^2}$. The errata at the end of this work mentioned that the $-$ sign should be included before it, together with an explanation which should be included after this formula:

> $-\frac{\mu\varepsilon\varepsilon' d^2\sigma d^2\sigma'}{r^2}$, que nous considérons comme agissant sur l'élément ds.

Certainly there was another misprint in the errata, which should have $d^2\sigma$ instead of ds. In the text at the end of our translation we replaced "that we consider as acting on the element ds" by "that we consider as acting on the element $d^2\sigma$". This mistake in the errata was pointed out by Joubert, [Ampère, 1887d, p. 125, note 1].

[278] [N. T.] Original French text: "ε, ε' désignent ici ce qu'on appelle l'épaisseur de la couche magnétique sur chaque surface."

[279] According to Coulomb and equation (1.3), the force between two magnetic poles is proportional to $-pp'/r^2$, where p and p' are the intensities of the two magnetic poles (magnetic pole-strengths) separated by a distance r. Consider now that the magnetic fluid p, instead of being concentrated in a point, is spread over an infinitesimal volume of thickness ε and area $d^2\sigma$, with a volume density of magnetic fluid given by ρ, which is the amount of magnetic fluid per unit volume. In this case $p=\rho\varepsilon d^2\sigma$ would represent the amount of magnetic fluid in the infinitesimal volume $dV=\varepsilon d^2\sigma$. Likewise, $p'=\rho'\varepsilon' d^2\sigma'$ would represent the amount of magnetic fluid in the infinitesimal volume $dV'=\varepsilon' d^2\sigma'$. When these two volume elements were separated by a distance r, the force between them would be proportional to $-pp'/r^2=-\rho\rho'\varepsilon\varepsilon' d^2\sigma d^2\sigma'/r^2$. In this case Ampère's constant μ would be then equal to $\rho\rho'$. The constant μ might also be proportional to $\rho\rho'$, depending on the chosen system of units.

[280] [N. T.] This distribution of two magnetic fluids, North and South, of the same intensity, on the two sides of a surface, separated by a small distance, is sometimes called a magnetic shell, a magnetic surface, or a magnetic dipole layer, as discussed in Sections 10.3 and 10.7. If the surface were closed, we would have, for instance, the North fluid distributed along the internal surface, while the South fluid would be distributed along the external surface.

element $d^2\sigma'$ and the small portion of the new surface circumscribed as we just said, which is always equal to $d^2\sigma$ when h is very small and one neglects in the calculation, as we do here, the powers of h greater than the first, one will obtain [an expression similar to the previous one,] by replacing in the expression which we just found, [pages 141 and 313] x, y and z by $x + h\cos\xi$, $y + h\cos\eta$ and $z + h\cos\zeta$. And since the two fluids spread on the two areas equal to $d^2\sigma$ are of opposite type, it will be required to subtract the new values of these components from the values found previously; it will reduce, as we are neglecting the powers of h greater than the first, to differentiate these values, replacing in the result the differentials of x, y and z by $h\cos\xi$, $h\cos\eta$ and $h\cos\zeta$, and by changing the sign. These differentials having been taken by passing from the first surface σ to the other one, we will designate them by δ, following the notation of the calculus of variations; we have thus for the component [of the force] parallel to x what becomes $-\mu\varepsilon\varepsilon' d^2\sigma d^2\sigma'\delta\frac{x-x'}{r^3}$, when one there replaces δx by $h\cos\xi$, that is to say

$$-\mu\varepsilon\varepsilon' d^2\sigma d^2\sigma' h\cos\xi\left(\frac{3(x-x')\frac{\delta r}{\delta x}}{r^4} - \frac{1}{r^3}\right) .$$

We will now determine the form and the position of the element $d^2\sigma$.

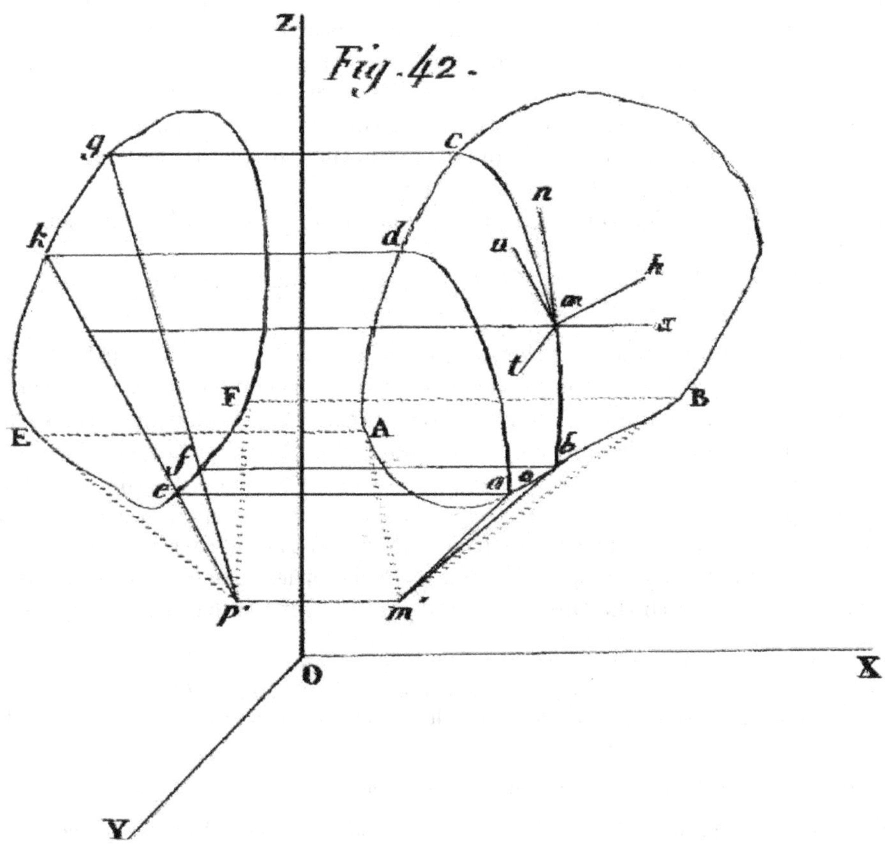

Figure 42.

Designate as before by u, v and w the projections of the line r on the yz, zx and xy planes, and by φ, χ and ψ, the angles which these projections make with the y, z and x axes, respectively. Decompose the first surface σ into an infinity of zones infinitely narrow, such as $abcd$ (figure 42), by a sequence of planes perpendicular to the yz plane connected by the coordinate $m'p' = x$ of the point m'.[281] Each zone ending at the two edges of the contour s of the surface σ, will have as projection on the yz plane an area decomposable itself into infinitely small quadrangular elements, which will correspond the same amount [pages 142 and 314] of surface elements σ on the zone being considered here. These are the elements which one should consider

[281][N. T.] The element $d^2\sigma'$ is located at the point m'.

as the values of $d^2\sigma$. That one whose position, with respect to the element $d^2\sigma'$, is determined by the polar coordinates r, u and φ, is equal to its projection $ududφ$ on the yz plane divided by the cosine of the angle ξ made between this plane and the plane tangent to the surface σ with which the element $d^2\sigma$ coincides. It is thus necessary to replace $d^2\sigma$ by $\frac{ududφ}{\cos\xi}$ in the preceding formula, and it will be

$$\mu h\varepsilon\varepsilon' d^2\sigma' ududφ \left(\frac{3(x-x')\frac{\delta r}{\delta x}}{r^4} - \frac{1}{r^3}\right) .$$

In order to calculate the value of $(x-x')\frac{\delta r}{\delta x}$, take mx as the prolongation of the coordinate $mp = x$ of the point m where is situated the element $d^2\sigma$,[282] [take] mu a parallel to the yz plane drawn on the plane $pmm'p'$, and mt perpendicular to the last plane at the point m. It is easy to see that the line mn, following which $pmm'p'$ cuts the plane tangent in m, on the surface σ, makes with the three lines mx, mu and mt, which are mutually perpendicular, the angles the cosine of which are respectively

$$\frac{dx}{\sqrt{dx^2+du^2}} , \frac{du}{\sqrt{dx^2+du^2}} \quad \text{and} \quad 0 ,$$

and that the normal mh makes with the same directions angles whose cosines are[283]

$$\frac{\delta x}{\sqrt{\delta x^2+\delta u^2+\delta t^2}} , \frac{\delta u}{\sqrt{\delta x^2+\delta u^2+\delta t^2}} , \frac{\delta t}{\sqrt{\delta x^2+\delta u^2+\delta t^2}} ,$$

δt serving as the projection of mh on mt. One obtains therefore[284]

[pages 143 and 315]

$$\frac{dx\delta x + du\delta u}{\sqrt{dx^2+du^2}\sqrt{\delta x^2+\delta u^2+\delta t^2}} ,$$

for the cosine of the angle between the line mn and the normal mh, and as this is a right angle, $dx\delta x + du\delta u = 0$, from where $dx/du = -\delta u/\delta x$. But the equation

$$r^2 = (x-x')^2 + u^2 ,$$

gives

$$r\delta r = (x-x')\delta x + u\delta u ,$$

and

$$rdr = udu + (x-x')dx ,$$

from which one deduces

$$\frac{\delta r}{\delta x} = \frac{x-x'}{r} + \frac{u}{r}\frac{\delta u}{\delta x} ,$$

and

$$\frac{dr}{du} = \frac{u}{r} + \frac{x-x'}{r}\frac{dx}{du} = \frac{u}{r} - \frac{x-x'}{r}\frac{\delta u}{\delta x} ;$$

by eliminating $\delta u/\delta x$ between these two equations, it becomes

$$(x-x')\frac{\delta r}{\delta x} + u\frac{dr}{du} = \frac{(x-x')^2}{r} + \frac{u^2}{r} = r .$$

[282][N. T.] It should be observed that the point p is the projection of the point m on the yz plane. Although this projection was drawn in figure 42, the letter p was not represented in this figure. We included this letter p in our figure A.42 on page 513 drawn with graphic software.

[283][N. T.] Due to a misprint, the term δu^2 in the first formula appeared as δ^2.

[284][N. T.] Due to a misprint, in the numerator of the original equation we had $dx\delta x + \delta u\delta u$.

If we now extract from this equation the value of $(x-x')\frac{\delta r}{\delta x}$ to substitute it in that of the force parallel to the x axis, we obtain

$$\mu h\varepsilon\varepsilon' ududφ\left(\frac{3r-3u\frac{dr}{du}}{r^4}-\frac{1}{r^3}\right)=$$

$$\mu h\varepsilon\varepsilon' d\varphi\left(\frac{2udu}{r^3}-\frac{3u^2dr}{r^4}\right)=\mu h\varepsilon\varepsilon' d\varphi d\frac{u^2}{r^3}\ .$$

[pages 144 and 316]

The height h and the thickness ε of the infinitely thin fluid layer over the surface σ can vary from one point of the surface to another; and to obtain the goal that we propose to represent with the aid of magnetic fluids, the actions of voltaic conductors, it is necessary to assume that these two quantities ε and h vary in inverse relation one to the other, in a manner such that their product[285] $h\varepsilon$ maintains the same value over the entire extent of the surface σ. Calling g the constant value of this product, the previous expression becomes

$$\mu g\varepsilon' d^2\sigma' d\varphi d\frac{u^2}{r^3}\ ,$$

and it can be immediately integrated. Its integral $\mu g\varepsilon' d^2\sigma' d\varphi\left(\frac{u^2}{r^3}-C\right)$ expresses the sum of the forces parallel to the x axis which act on the elements $d^2\sigma$ of the zone of the surface σ enclosed by the two planes drawn through $m'p'$ which include the angle $d\varphi$. The surface σ being terminated by the closed contour s, it is necessary to take this integral between the limits determined by the two elements ab and cd of this contour which are included in the angle $d\varphi$ of the two planes of which we have just spoken, such that by taking u_1, r_1 and u_2, r_2 to be the values of u and of r relative to these two elements, one obtains

$$\mu g\varepsilon' d^2\sigma' d\varphi\left(\frac{u_2^2}{r_2^3}-\frac{u_1^2}{r_1^3}\right)\ ,$$

as the sum of all the forces exerted by the element $d^2\sigma'$, on the zone parallel to the x axis.

If the surface σ, instead of being terminated by a contour, encloses from all sides a space of any shape,[286] the zone of this surface contained in the dihedral angle φ would be [pages 145 and 317] closed, and one would have $u_2=u_1$ and $r_2=r_1$; ensuring that the action exerted on this zone parallel to the x axis would be null, and as a consequence that [action] which the element $d^2\sigma'$ would exert on the entire surface σ made up then of similar zones also [would be null]. And since the same thing would take place relative to the forces parallel to the y and z axes, one sees that the assemblage of two surfaces very close to each other, enclosing from all sides a space of any shape, and covered, in the manner that we just described, one of austral fluid, the other of boreal fluid, is without action on a magnetic molecule, in whatever position it may be placed, and as a consequence [it does not act] on a body magnetized in whatever manner. Recall the preceding expression

$$\mu g\varepsilon' d^2\sigma'\left(\frac{u_2^2d\varphi}{r_2^3}-\frac{u_1^2d\varphi}{r_1^3}\right)\ ,$$

and it will be easy for us to see that, in order to have the total sum of the forces parallel to the x axis that the element $d^2\sigma'$ exerts on the entire surface σ, one must integrate, with respect to φ, the two parts that make up this expression, respectively in the two portions[287] AabB and BcdA of the contour s determined by the two tangent planes $p'm'$A and $p'm'$B, drawn by the line $m'p'$. But it amounts to the same to integrate $\mu g\varepsilon' d^2\sigma'\frac{u^2d\varphi}{r^3}$ over the entire range of the circuit s; since if one puts for u and φ their values as functions of r deduced from the equations of the curve s, one sees that in passing from the part AabB to the part BcdA, $d\varphi$ changes sign, and that as a consequence the elements of one of these parts are of the contrary sign from the others.

[285][N. T.] Sometimes this product is called *magnetic power of the magnetic layer*, [Ampère, 1887d, note on p. 129].

[286][N. T.] That is, if this surface were closed, covering a volume of arbitrary shape.

[287][N. T.] In the 1826 and 1827 versions of the *Théorie*, we had here AabB and BabA. The final errata corrected these expressions to AabB and BcdA. These letters are related to figure 42 in our page 432.

Following this, if we designate by X the sum of the forces [pages 146 and 318] parallel to x which the element $d^2\sigma'$ exerts on the assemblage of the two surfaces terminated by the same contour s, we have

$$X = \mu g\varepsilon' d^2\sigma' \int \frac{u^2 d\varphi}{r^3} \,,$$

or, which is the same thing,

$$X = \mu g\varepsilon' d^2\sigma' \int \frac{(y-y')dz - (z-z')dy}{r^3} \,,$$

x, y and z being only relative to the contour s.

One will have, just the same, designating by Y and Z the sums of the forces parallel to the y and to z which act on the same assembly of surfaces,[288]

$$Y = \mu g\varepsilon' d^2\sigma' \int \frac{v^2 d\chi}{r^3} = g\mu\varepsilon' d^2\sigma' \int \frac{(z-z')dx - (x-x')dz}{r^3} \,,$$

$$Z = \mu g\varepsilon' d^2\sigma' \int \frac{w^2 d\psi}{r^3} = g\mu\varepsilon' d^2\sigma' \int \frac{(x-x')dy - (y-y')dx}{r^3} \,.$$

Since all the elementary forces which the element $d^2\sigma'$ exerts on these surfaces pass through the point m' where it is situated, one sees that all these forces have a unique resultant the direction of which passes through the same point m', and whose components parallel to the axes are X, Y and Z. Therefore, the torques of this resultant with respect to the same axes are given by

$$Yz' - Zy' \,, \quad Zx' - Xz' \quad \text{and} \quad Xy' - Yx' \,.$$

Suppose now that instead of these forces one applies [pages 147 and 319] at the middle of each of these elements ds of the contour s a force equal to $\mu g\varepsilon' d^2\sigma' \frac{ds \sin\theta}{r^2}$, and perpendicular to the plane of the sector which has ds as its base, the point m' as its vertex, and such that the area is $\frac{1}{2} r ds \sin\theta$. The three components of this force being respectively equal to

$$\mu g\varepsilon' d^2\sigma' \frac{u^2 d\varphi}{r^3} \,, \quad \mu g\varepsilon' d^2\sigma' \frac{v^2 d\chi}{r^3} \quad \text{and} \quad \mu g\varepsilon' d^2\sigma' \frac{w^2 d\psi}{r^3} \,,$$

parallel to those [components] which pass through the element[289] $d^2\sigma'$ and directed in the same sense, we will have the same values for the three forces X, Y and Z which tend to move the circuit s; but the sums of the torques resulting from this [force], instead of being represented by[290]

$$\mu g\varepsilon' d^2\sigma' \left(z' \int \frac{v^2 d\chi}{r^3} - y' \int \frac{w^2 d\psi}{r^3} \right) \,,$$

$$\mu g\varepsilon' d^2\sigma' \left(x' \int \frac{w^2 d\psi}{r^3} - z' \int \frac{u^2 d\varphi}{r^3} \right)$$

and

$$\mu g\varepsilon' d^2\sigma' \left(y' \int \frac{u^2 d\varphi}{r^3} - x' \int \frac{v^2 d\chi}{r^3} \right) \,,$$

will be [represented] by[291]

$$\mu g\varepsilon' d^2\sigma' \left(\int \frac{z v^2 d\chi}{r^3} - \int \frac{y w^2 d\psi}{r^3} \right) \,,$$

---

[288] [N. A.] It is unnecessary to remark that these X, Y and Z express forces entirely different from those that we have already designated by the same symbols, when we were referring to the mutual action between two elements of voltaic circuits.

[289] [N. T.] Due to a misprint, in the original text of the *Théorie* we had: $d^2\sigma$.

[290] [N. T.] Due to a misprint in the 1990 edition of the *Théorie*, the closing parenthesis of the second equation was missing, [Ampère, 1990, p. 319]. This mistake did not take place in the 1826 and 1827 versions of the *Théorie*.

[291] [N. T.] Due to a misprint in the 1990 edition of the *Théorie*, the closing parenthesis of the second equation was missing, [Ampère, 1990, p. 319]. This mistake did not take place in the 1826 and 1827 versions of the *Théorie*.

$$\mu g\varepsilon' d^2\sigma' \left(\int \frac{xw^2 d\psi}{r^3} - \int \frac{zu^2 d\varphi}{r^3} \right)$$

and

$$\mu g\varepsilon' d^2\sigma' \left(\int \frac{yu^2 d\varphi}{r^3} - \int \frac{xv^2 d\chi}{r^3} \right) .$$

It appears at first that this change should give a [torque] to the action exerted on the contour s, but it is not so provided that this contour forms a closed circuit, since if one subtracts the first sum of the torques, relative to the x axis for example, from the fourth which refers to the same axis, while noting that x', y' and z' should be considered as constants in these integrations, one will obtain

$$\mu g\varepsilon' d^2\sigma' \int \frac{(z-z')v^2 d\chi - (y-y')w^2 d\psi}{r^3} =$$

[pages 148 and 320]

$$\mu g\varepsilon' d^2\sigma' \int \frac{(z-z')^2 dx - (z-z')(x-x')dz - (y-y')(x-x')dy + (y-y')^2 dx}{r^3} =$$

$$\mu g\varepsilon' d^2\sigma' \int \frac{\left[(z-z')^2 + (y-y')^2\right] dx - (x-x')\left[(z-z')dz + (y-y')dy\right]}{r^3} =$$

$$\mu g\varepsilon' d^2\sigma' \int \frac{\left[r^2 - (x-x')^2\right] dx - (x-x')\left[rdr - (x-x')dx\right]}{r^3} =$$

$$\mu g\varepsilon' d^2\sigma' \int \left[\frac{rdx - (x-x')dr}{r^2}\right] = \mu g\varepsilon' d^2\sigma' \left(\frac{x_2 - x'}{r_2} - \frac{x_1 - x'}{r_1} \right) ,$$

naming x_1, x_2, and r_1, r_2 the values of x and of r at the two extremities of the arc s for which one calculates the value of the difference of the two torques. If this arc forms a closed circuit, it is evident that $x_2 = x_1$ and $r_2 = r_1$, which results in the integral so obtained to be null; therefore one obtains[292]

$$\mu g\varepsilon' d^2\sigma' \int \frac{zv^2 d\chi - yw^2 d\psi}{r^3} = \mu g\varepsilon' d^2\sigma' \left(z' \int \frac{v^2 d\chi}{r^3} - y' \int \frac{w^2 d\psi}{r^3} \right) .$$

We find by a similar calculation that the torques relative to the other two axes are the same, for a closed circuit, if one supposes that the directions of the forces

$$\mu g\varepsilon' d^2\sigma' \frac{u^2 d\varphi}{r^3} , \quad \mu g\varepsilon' d^2\sigma' \frac{v^2 d\chi}{r^3} \quad \text{and} \quad \mu g\varepsilon' d^2\sigma' \frac{w^2 d\psi}{r^3} ,$$

pass through the element $d^2\sigma'$ or through the middle of ds; hence it follows that in these two cases the action on the contour s is exactly the same, this contour being invariably connected to the two very close surfaces which it terminates: the action exerted on these two surfaces by the element $d^2\sigma'$ thus reduces, provided that the contour s is a closed curve, to forces applied as we just said to each of the elements of this contour, that which acts on the element ds having the value

$$\mu g\varepsilon' d^2\sigma' \frac{ds \sin\theta}{r^2} .$$

[pages 149 and 321]

The force applied on the center o of the element $ab = ds$ [of figure 42], which is proportional to $ds \sin\theta$ divided by the square of the distance r of this element to the point m',[293] whose direction is perpendicular to

[292][N. T.] Due to a misprint in the 1826 and 1827 versions of the *Théorie*, we have in the original text $d^2 s'$ instead of $d^2\sigma'$ at the right hand side of this equation. This mistake also happens in [Ampère, 1887d, p. 132] and [Ampère, 1958, p. 113].

[293][N. T.] The element $d\sigma'$ is located at the point m'.

the plane which passes through the element ab and through the point m', is precisely that [force] which, as we have seen, the extremity of an indefinite electrodynamic solenoid exerts on the element ds, if one places this extremity at the point m'; this is also that [force] produced, based on the last experiments by M. Biot, by the mutual action between the element ab and a magnetic molecule located in m'.

But in giving to this force the same value and the same direction perpendicular to the plane $m'ab$, which one should give it when it is determined, as I have done, by replacing the magnetic molecule by the extremity of an indefinite solenoid, M. Biot assumes that it is in m' that its point of application is located, or rather that [point of application] of the force equal and opposite that the element ds exerts on the point m', since it is at this last [configuration] with which the experiments which he made are related;[294] whereas the direction of the force exerted by this element on the extremity situated in m' of an indefinite solenoid should pass through the point o,[295] like that [force] which the solenoid exerts on the element, when one determines this force from my formula. Thus, keeping the notation which we have used, and by representing, to abbreviate, the constant coefficient $g\mu\varepsilon' d^2\sigma'$ by ρ, the sums of the torques, following the manner in which M. Biot places the points of application of the forces, would be for the three axes, and by changing the signs, since it refers to the forces acting on the point m',

$$-\rho\int\frac{z'v^2d\chi - y'w^2d\psi}{r^3}\,,$$

[pages 150 and 322]

$$-\rho\int\frac{x'w^2d\psi - z'u^2d\varphi}{r^3}\,,$$

$$-\rho\int\frac{y'u^2d\varphi - x'v^2d\chi}{r^3}\,;$$

whereas in taking the points of application as I have found them, one obtains for the sums of the torques

$$-\rho\int\frac{zv^2d\chi - yw^2d\psi}{r^3}\,,$$

$$-\rho\int\frac{xw^2d\psi - zu^2d\varphi}{r^3}\,,$$

$$-\rho\int\frac{yu^2d\varphi - xv^2d\chi}{r^3}\,.$$

But we have just seen that these last values are respectively equal to the three preceding ones when the portion of the conductor forms a closed circuit; hence it follows that in this case, the experiment cannot decide if the point of application of the forces is really at the point m' or in the center o of the element ds.[296] And as, in those [experiments] made by the skillful physicist[297] to whom we owe the experiments which are here in question, there was in effect a completely closed circuit, composed of two rectilinear portions forming an angle to which he gave successively different values, [composed also of] the remainder of the conducting wire and of the battery, that he caused to act on a small magnet, in order to deduce the ratio of the corresponding forces for various values of this angle [from] the number of oscillations of the small magnet, during a given time, which corresponded to these various values; therefore, the results of these experiments made in this manner should be identically the same, whether one assumes the point of application of the

[294] [N. T.] That is, according to Biot's conception, the action and reaction between a current element and a magnetic pole act along parallel straight lines, one of them passing through the center of the current element, while the other line passes through the magnetic pole. These opposite forces acting along parallel straight lines would generate a primitive couple. Ampère was always against these primitive couples, as discussed in Section 20.2.

[295] [N. T.] In the 1826 and 1827 versions of the *Théorie* we have here the expression "doit passer par le point m." It seems to us that this was a misprint. We believe the correct expression should be "doit passer par le point o" as we wrote in our translation. The point o is the midpoint of the current element $ds = aob$ of figure 42 of the *Théorie*.

[296] [N. T.] In the 1826 and 1827 versions of the *Théorie* we have here the expression "ou au milieu m de l'élément ds." It seems to us that this was a misprint. We believe the correct expression should be "ou au milieu o de l'élément ds" as we wrote in our translation. The point o is the midpoint of the current element $ds = aob$ of figure 42 of the *Théorie*.

[297] [N. T.] Namely, J. B. Biot.

[pages 151 and 323] forces in o or in m', [they] cannot serve to decide which of these two hypotheses should be preferred, this question about the location of the point of application can only be resolved by other considerations; this is why I think that it is necessary, before going further, to examine it in some detail.

I communicated to the *Académie*, in the memoir that I read at the meeting of 4 December 1820, the fundamental formula of all the theory explained in this memoir, a formula which gives the value of the mutual action of two conducting wires expressed as:[298, 299]

$$\frac{ii'dsds'\left(\sin\theta\sin\theta'\cos\omega+k\cos\theta\cos\theta'\right)}{r^2},$$

k being a constant, for which I have since determined the value,[300] by proving, by other experiments, that it is equal to $-1/2$.

29.20 Examination of the Three Hypotheses that are Proposed Concerning the Nature of the Interaction of an Element of a Conducting Wire and what is Called *a Magnetic Molecule*

Sometime later, in the meeting of the 18th of the same month,[301] M. Biot read a paper in which he described the experiments which he conducted on the oscillations of a small magnet under the action of an angular conductor, and where he concluded from these experiments, due to the error in calculation shown above,[302] that the action of each element of the conductor on that which is called a magnetic molecule,[303] is represented by a force perpendicular to the plane drawn by the molecule and by the element, inversely proportional to the square of their distance, and proportional to the sine of the angle determined by the line which measures [pages 152 and 324] this distance and the direction of the element. One sees from the preceding calculations that this force is exactly that which my formula gives for the mutual action of an element of conducting wire and the extremity of a [simply indefinite] electrodynamic solenoid, and which is also that which results from Coulomb's law, in the hypothesis of two magnetic fluids, when one looks for the action which takes place between a magnetic molecule and the contour elements which terminate two infinitely close surfaces, one covered with austral fluid, the other with boreal fluid, by supposing that the molecules of these fluids are distributed on the two surfaces as I have just explained.

In these two [last] ways of conceiving these things,[304] one finds the same values for the three components, parallel to the three axes chosen at will, of the resultant of all the forces exerted by the contour elements, and, for each of these forces, the action is opposed to the reaction following the straight lines which join, pair by pair, the points between which they are exerted; it is the same for the resultant itself and for its reaction. But in the first case,[305] the point O (figure 36) represents the extremity of the solenoid to which the points P and N belong, and o being that [point] where the element is situated, the two equal and opposite forces og and $o\gamma$ pass through this element; in the second case,[306] on the contrary, it is in O that one must conceive to place the contour element of the surfaces covered by magnetic molecules P and N, and in o the molecule on which these surfaces act [should be conceived], so that the two equal and opposite forces pass by the molecule. As long as one admits that there could be no action of one material point on another, without it reacting on the first with a force [pages 153 and 325] equal and directed in the contrary sense following the same line, which leads to the same condition relative to the action and to the reaction of two systems of points invariably connected, one has to choose between these two hypotheses. And since the experiment by M. Faraday on the rotation of a piece of conducting wire about a magnet is, as I will shortly explain, in

298 [N. A.] *Journal de Physique*, volume XCI, pp. 226-230.

299 [N. T.] This memoir, read on 4 December 1820, was published in [Ampère, 1820e]. The manuscript containing the material of this work was published by Joubert, [Ampère, 1885j]. A related paper of the same date was published in [Ampère, 1820d].

300 [N. T.] The value $k=-1/2$ was determined by Ampère in 1822, being published in [Ampère, 1822o]. We discussed this determination of the value of the constant k in Chapter 7.

301 [N. T.] That is, in the meeting of 18 December 1820.

302 [N. T.] See [Ampère, 1826f, p. 110], [Ampère, 1823c, Ampère, 1990, p. 282] and [Ampère, 1887d, p. 99].

303 [N. T.] That is, a particle containing a magnetic pole. Nowadays it would be called a magnetic monopole.

304 [N. T.] That is, by Ampère's hypothesis utilizing the force between a closed current and one extremity of an electrodynamic solenoid, and by the hypothesis of a magnetic dipole layer interacting with a magnetic pole according to Coulomb's law.

305 [N. T.] That is, with Ampère's hypothesis.

306 [N. T.] That is, with the hypothesis of a magnetic dipole layer interacting with a magnetic pole according to Coulomb's law.

manifest contradiction with the first [hypothesis],[307] there should be no further difficulty in viewing, with me, as the only one admissible the [hypothesis] that passes through the center of the element the line along which the two forces are directed. But many physicists[308] imagined that, in the mutual action of one element AB (figure 39)[309] of a conducting wire and of a magnetic molecule M, the action and reaction, though equal and directed in opposite directions, were not directed along the same line, but followed two parallel lines, such that, the molecule M, acting on the element AB, would tend to move it along the line OR drawn through the midpoint[310] O of the element AB perpendicularly to the plane MAB, and that the action which this element would exert on the molecule M would tend to carry it, with an equal force, in the direction MS, parallel to OR.

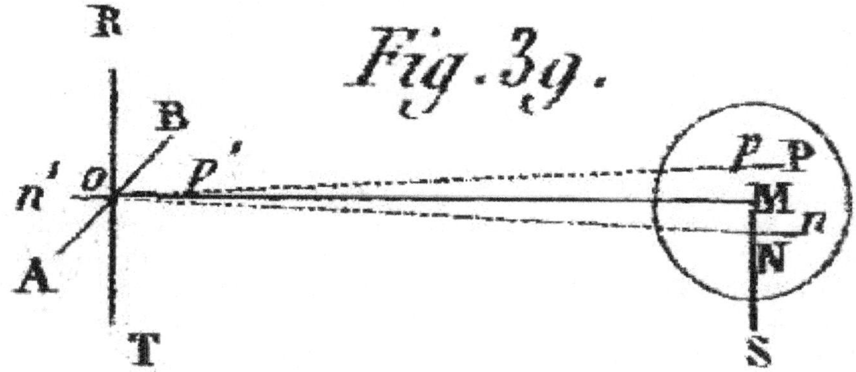

Figure 39.

It would result from this singular hypothesis, if it were true, that it would be mathematically impossible to reduce the phenomena produced by the mutual action of a conducting wire and a magnet, to forces acting, like all those that one has known up to the present to exist in nature, in such a manner that the action and the reaction are equal and opposite in the direction of the lines which join in pairs the points between which they act; because, [pages 154 and 326] in all cases where this condition is satisfied for any elementary forces, it is also evidently satisfied, following the principle of the composition of forces, for their resultants. Therefore, the physicists who have adopted this opinion are forced to admit a really elementary action, made up of two equal forces directed in opposite directions following two parallel lines, and thus forming a primitive couple,[311] which cannot be reduced to forces for which the action and the reaction would be opposed following the same line. I have always considered this hypothesis of primitive couples as absolutely contrary to the first laws of mechanics, among which one should count, with Newton, the equality of action and reaction in opposite directions following the same straight line; and I have reduced the phenomena which one observes when a conducting wire and a magnet react on each other, as all the other electrodynamic phenomena, to an action between two current elements, from which two equal and opposite forces result, both directed along the line which joins the two elements. This first characteristic of other forces observed in nature is thereby justified; and as regards that [characteristic] which consists in that the forces that one considers as really elementary are further simply functions of the distances of the points between which they are exerted, nothing opposes, as I have already remarked, that the force, of which I have determined the value by precise experiments, cannot be reduced one day to elementary forces which also satisfy this second condition, provided one includes into the calculation the continuous motion,

[307][N. T.] According to Ampère, Faraday's experiments of continuous rotation could not be explained utilizing central forces depending only on the distance between the magnetic poles, as was the case with Coulomb's force.

[308][N. T.] Ampère will discuss now the third hypothesis, due to Biot and Savart. In this case there is an interaction between a current element and a magnetic pole. According to Biot and Savart, there is also action and reaction in this case, although these two opposite forces would not act along the straight line connecting the center of the element to the magnetic pole. Therefore, the action and reaction would be directed along parallel straight lines, one of them passing through the element, while the other would pass through the magnetic pole.

[309][N. T.] In figure 39 of the 1826 and 1827 versions of the *Théorie*, the midpoint of the element AB was represented by the letter *o*. However, it should have been represented by the letter O due to the fact that in the present paragraph and in the next ones Ampère always referred to this point by the letter O. In our diagram made with graphic software, figure A.39, the midpoint of the element AB is represented by the letter O.

[310][N. T.] See the footnote 309 of our page 439.

[311][N. T.] See footnote 246 on our page 422.

in the conducting wires, of the electric molecules for which these last forces would be inherent. [pages 155 and 327] The consideration of these motions would necessarily introduce in the value of the force which would result between two [current] elements, in addition to their distance, the angles which determine the directions following which the electric molecules move, and which depend on the directions themselves of these elements; these are exactly the only angles, or, what leads to the same, the differentials of the distance of the two elements considered as a function of the arcs formed by the conducting wires, which enter with the distance in my formula. It should not be forgotten that, in the only manner of conceiving the things which seems to me admissible, two equal and opposite forces[312] OR and OT are the resultants of an infinity of forces equal and opposite pair by pair; [that is,] OR is that [resultant] of the forces On', Op', etc., which all pass through the point O, such that their resultant OR also passes there, while OT is the resultant of the forces Nn, Pp, etc., exerted by the element AB on points such as N, P, etc., invariably connected to the extremity M of the electrodynamic solenoid by which I suppose that was replaced what one calls a magnetic molecule. These points are very close to M when this solenoid is very small, but they are always distinct, and for this reason their resultant OT does not pass through the point M, but through the point O toward which all the forces Nn, Pp, etc., are directed.[313]

One sees, by all that we have said, that maintaining in the two equal forces that result from the mutual action of a conducting wire and of a magnet, and which act, one on the wire on which the element AB forms a part, and the other on the magnet of which the point M is a part, the same value, and the same direction perpendicular to the plane MAB, one can form [pages 156 and 328] three hypotheses on the point of application of these forces: in the first,[314] one supposes that the two forces pass through the point M; in the second, which is the one which results from my formula, the two forces pass through the center O of the element; in the third,[315] where the forces are OR and MS, that [force] which acts on the element is applied at the point O, and the other [force which acts on the magnet is applied] at the point M. These three hypotheses are entirely in agreement, (1) with regard to the value of these forces which are equally, in all three [hypotheses], in inverse ratio of the square of the distance MO, and in direct ratio of the sine of the angle MOB which the line OM which measures this distance makes with the element AB; (2) with regard to the direction of these same forces, always perpendicular to the plane MAB which passes through the molecule and through the direction of the element: but with regard to their points of application, they are placed differently for the two forces, in the first two hypotheses; and there is identity between the first and the third only for the forces which act on the magnet, and between the second and the third only for the forces which act on the conductor.

By virtue of the identity of the values and the directions of the forces which exist in the three hypotheses, the components of their resultants, taken parallel to three arbitrary axes, will be the same; but the torques, which depend in addition on the points of application of these forces, will only be the same, in general, with regard to the forces which tend to move the magnet, for the first and the third [hypotheses], and, as regards the forces acting on the conducting wire, [the torques will be equal] only for the second and the third [hypotheses].

We have seen that in the case where it is a question of the action of a portion of a conducting wire, forming a [pages 157 and 329] closed circuit, the values of the torques are the same, whether one takes, for each element, the point of application of the forces in O or in M; in this case, therefore, there will be, furthermore, identity between the values of the torques in the three hypotheses.

---

[312][N. T.] See the representation of these forces in figure 39 and also the footnote 309 in our page 439.

[313][N. T.] That is, the resultant of the forces exerted by the current element AOB acting on all points of the electrodynamic solenoid should be considered as passing through the center O of the element AOB. In this case one should consider this point O of the application of the net force acting on the solenoid as being rigidly connected to the solenoid, so that this force may be transmitted to the solenoid. According to Ampère's calculations, one could not consider this resultant force as passing through the extremity M of the electrodynamics solenoid.

[314][N. T.] In the first hypothesis the closed conductor carrying a steady current is replaced by a magnetic dipole layer enclosing the circuit and interacting with a magnetic pole located at point M.

[315][N. T.] This third hypothesis is due to Biot.

29.21 Impossibility of Producing an Indefinitely Accelerating Movement Due to the Interaction of a Closed Rigid Circuit and a Magnet, or of such a Circuit and an Electrodynamic Solenoid

The motion of a body, such that all the parts are invariably linked together, can depend only on the three components [of the force] parallel to three axes taken arbitrarily, and the three torques about the same axes; from where it follows that there is complete identity in the three hypotheses for the movement produced, whether in the magnet, whether in the conductor, as long as this [conductor] forms a rigid and closed circuit. That is why the impossibility of an indefinitely accelerated motion, being in general a necessary consequence of the first hypothesis, since the elementary forces are here simply functions of the distances of the points within which they interact, it follows evidently that this movement is equally impossible, in the two other hypotheses, only when the conductor forms a rigid and closed circuit.

It is easy to see, for the rest, that the demonstration thus obtained of the impossibility to produce an indefinitely accelerating movement by the mutual action of a rigid and closed electric circuit, and of a magnet, is not only a necessary result of my theory, but it results also, in the hypothesis of primitive couples, from the single value given by M. Biot for the force perpendicular to the plane MAB, as I have directly demonstrated, with all the details that could be desired, in a letter that I wrote on this subject to M. doctor Gherardi.[316] If therefore one could produce an accelerating movement by making a conductor which forms a rigid and closed circuit act on a magnet, [pages 158 and 330] it would be not only my formula which would be defective, but also that [formula] given by M. Biot, which was completely demonstrated by all observations made since then,[317] whose exactness was never contested by the physicists who admit the hypotheses of the primitive couples.

29.22 Examination of the Different Cases where an Indefinitely Accelerating Movement Can Result from the Action that a Voltaic Circuit, of Which a Part is Movable with Respect to the Rest of the Circuit, Exerts on a Magnet or on an Electrodynamic Solenoid

When[318] one makes a portion of the voltaic circuit mobile, one should distinguish three cases: that where it forms a nearly closed circuit;[319] that where, when it can only turn about one axis, it has its two extremities in this axis; that where the mobile portion does not form a closed circuit, and where one of its extremities travels at least within a sufficient space while it moves: this last case includes that where this portion is formed by a conducting liquid.

We have just seen that, in the first of the three cases, the movement that the mobile portion acquires by the action of a magnet, is identically the same in the three hypotheses, and can never indefinitely accelerate, but tends only to bring the mobile portion to a determinate position where it stops in equilibrium after having for some time oscillated about this position by virtue of the acquired speed.

It is the same for the second [case], which only differs from the first in appearance: since if one added in the axis, a current, [pages 159 and 331] which connected the two extremities of the mobile portion, one would have a closed circuit without having changed anything in the torque about this axis, because the torques of the forces acting on the added current would be obviously null; from which it follows that the movement of the mobile portion would be identically the same as that for the closed circuit so obtained.[320]

---

[316][N. T.] Silvestro Gherardi (1802-1879). He graduated in mathematics and natural sciences in Bologna in 1822. He held the physics chair and was director of the scientific cabinet of the Turin University. Ampère sent the mentioned letter in the beginning of 1826. It was published in [Ampère, 1827c] and [Ampère, 1887a].

[317][N. T.] That is, Ampère's formula between current elements was completely demonstrated by all observations.

[318][N. T.] In this Section Ampère will discuss two kinds of actions, namely, that exerted by a magnet or by an electrodynamic solenoid acting on the mobile portion of a voltaic circuit, and that exerted by a voltaic circuit on a mobile magnet.

[319][N. A.] The circuit formed by a mobile portion of conducting wire is never rigorously closed, because it is necessary that its two extremities communicate separately with those of the battery; but it is easy to make the interval which separates them sufficiently small so that one can consider it as if it were exactly closed.

[320][N. T.] As an example we can quote the case of equilibrium of the nonexistence of continuous rotation discussed in Section 7.3.

But when the mobile portion does not form a closed circuit, and if its two extremities are not in the axis about which it is subject to turning, the torques produced by the action, whether of a magnetic molecule, whether of the extremity of an indefinite solenoid, are the same only in the second and third hypotheses, and have a different value in the first. Taking the x axis as the line about which one assumes that the mobile portion is connected in a manner such that it can only rotate about this line, and in preserving the designations that we have used in the preceding calculations, we would conclude that the value of the torque produced by the forces which act on the mobile portion would be

$$\rho \int \frac{z' v^2 d\chi - y' w^2 d\psi}{r^3} ,$$

in the first hypothesis, and

$$\rho \int \frac{z' v^2 d\chi - y' w^2 d\psi}{r^3} + \rho \left(\frac{x_2 - x'}{r_2} - \frac{x_1 - x'}{r_1} \right) ,$$

in the two others.

It is due to this difference in the values of the torques that we have the possibility to prove by experiment that the first hypothesis is in contradiction with the [pages 160 and 332] facts. Because if one considers a magnet to be reduced to two magnetic molecules of an infinite force placed at its two poles, and that after having placed a vertical line joining them, one subjected a portion of a conducting wire to be turned around this line taken as the x axis, the two torques relative to the two poles will be given by the preceding formula, replacing in it x', y' and z' by x_1', y_1' and z_1', for one of the poles, and x_2', y_2' and z_2', for the other, taking care to change the sign in one of the torques, the first, for example, since the two poles are necessarily of opposite nature, one austral and the other boreal.

When the two poles are, as we assume here, situated on the x axis, one has $y_1' = 0$, $y_2' = 0$, $z_1' = 0$, $z_2' = 0$, and the two torques about the x axis become null in the first hypothesis: which was easy to foresee, since under this hypothesis the directions of all the forces applied to the mobile conductor pass through one of the poles, meeting there the fixed axis, which renders necessarily null the torques of these forces.

In the two other hypotheses, on the contrary, where the directions of the forces pass through the midpoints of the elements, the parts of the torques equal to those of the first hypothesis are the only ones which disappear; and when after they are deleted, one combines that which remains of each torque, one obtains[321]

$$\rho \left(\frac{x_2 - x_2'}{r_{2,2}} - \frac{x_1 - x_2'}{r_{1,2}} - \frac{x_2 - x_1'}{r_{2,1}} + \frac{x_1 - x_1'}{r_{1,1}} \right) ,$$

designating by $r_{2,2}$; $r_{1,2}$; $r_{2,1}$ and $r_{1,1}$ the distances of the points such that their abscissas are respectively x_2, x_2'; x_1, x_2'; x_2, x_1' and x_1, x_1'.

[pages 161 and 333]

It is easy to see that the four terms of the quantity which is contained between the parentheses in this expression are precisely the cosines of the angles which form with the x axis the lines which measure the distances $r_{2,2}$; $r_{1,2}$; $r_{2,1}$ and $r_{1,1}$: making the value we just found for the torque produced by the action of the two poles on the mobile conductor identical to that [torque] which we have already obtained resulting from the action on the same conductor due to a solenoid whose extremities would be located at these poles,[322] and whose electric currents would have an intensity i and respective distances such that one had

$$\frac{\lambda i i'}{2g} = \rho ,$$

i' being the intensity of the current in the [mobile] conductor.

The torque being always null under the first hypothesis, the mobile portion of the voltaic circuit will never turn by the action of a magnet, situated as we have said,[323] around the axis of this magnet; under the other two hypotheses, it should on the contrary turn due to the torque whose value we have just calculated, always the same, under these two hypotheses. M. Faraday, who first produced this movement, a necessary

[321] [N. T.] Due to a misprint, in the original text the sign between the first two fractions did not appear.

[322] [N. T.] See Ampère's result obtained in our page 410, corresponding to pages 102 and 274 of the *Théorie* published in 1826 and 1827, respectively.

[323] [N. T.] That is, with its two poles located on the x axis, which is the axis of rotation.

consequence of the laws which I have established on the mutual action of voltaic conductors, and of the manner in which I considered magnets as assemblages of electric currents, demonstrated [by this motion] that the direction of the action exerted by the pole of a magnet on the element of a conducting wire passes in effect through the middle of the element, conforming to the explanation that I have given of this action, and not through the pole of the magnet. Therefore the ensemble of [pages 162 and 334] electrodynamic phenomena can no longer be explained by the substitution of the action of austral and boreal magnetic molecules, distributed in the manner that I just explained on two very close surfaces terminated by the conducting wires of the voltaic circuit, in the place of the action, expressed by my formula, which the currents of these wires exert. This substitution can only take place in the case of the action of rigid and closed circuits, and its principal utility is to demonstrate the impossibility of an indefinitely accelerating motion, whether by the mutual action of two rigid and closed conductors, or whether by that of such a conductor and a magnet.

When the magnet is mobile, it is also necessary to distinguish three cases: one where all the parts of the voltaic circuit which can act on this magnet are fixed; one where some parts of the circuit are mobile, but without connection with the magnet, these portions could also be formed by a metallic wire, or by a liquid conductor; finally one where one part of the circuit passes through the magnet, or through a portion of a conductor connected to the magnet.

In the first case, the total circuit composed of the conductors and the battery is necessarily closed; and as all its parts are immovable, the three sums of the torques of the forces acting on the points of the magnet which is considered, whether as molecules of austral or boreal fluid, whether as extremities of electrodynamic solenoids, are identical in the three hypotheses, and so are the resultants of the forces; so that the movements caused to the magnet, and all the circumstances of these movements, are precisely the same, whichever [pages 163 and 335] of these hypotheses one adopts. It is this [configuration] which applies, for example, to the duration of oscillations of a magnet, under the influence of this closed and immobile circuit; and it is for this reason that the last experiments of M. Biot, from which it results that the force which produces these oscillations is proportional to the tangent of the quarter of the angle formed by the two branches of the conductor that he uses, are as consistent with this consequence of my theory that the directions of the forces which act on the magnet pass through the middle of the elements of the conducting wire, as with the hypothesis which he adopted and in which he admits that the directions [of the forces] pass through the points of the magnet where he placed the magnetic molecules.

The identity which exists in this case between the three hypotheses shows at the same time the impossibility that the movement of the magnet could accelerate indefinitely, and proves that the action of the voltaic circuit can only tend to direct it to a determined position of equilibrium.

It seems, at first glance, that the same impossibility should be present in the second case, which is contrary to experience, at least when one part of the circuit is formed by a liquid. It is evident, in effect, that the mobility of a portion of a conductor does not prevent that this portion acts at each instant as if it were fixed at the position it occupies at that instant; and we do not see at first how this mobility can change the conditions of movement of the magnet, so that it becomes susceptible to indefinite acceleration whose impossibility is demonstrated when all the parts of the voltaic circuit are fixed.

But, after one examines with some attention what [pages 164 and 336] should happen, following the laws of mutual action of a conducting body and a magnet, when the conductor is liquid, when a vertical magnetic cylinder floats in this liquid, and when the surface of the cylinder is covered with an insulating varnish so that the current cannot pass through it, which would give rise to the third case, one recognizes quickly how there results from the mobility of the liquid portion of the voltaic circuit that the floating magnet acquires a movement which indefinitely accelerates: it is only necessary for this [conclusion] to apply to this case the explanation that I have given, in the *Annales de Chimie et de Physique* (volume XX, pp. 68-70),[324] of the same movement, when one assumes that the magnet not being varnished, the currents in the liquid where it floats traverse it freely.

In effect, this explanation being based on that the portions of the currents which are within the magnet cannot have any action on it, and that those [portions of the currents] that are within the liquid outside the magnet act entirely to accelerate its movement always in the same direction, it follows necessarily that all that happens in this case should also occur when the insulating substance, which covers the magnet, removes only exactly the portions of the current which produce no action, and when it allows to subsist and to act, always in the same manner, those [portions of the currents] that, being outside the magnet, all tend to accelerate its movement constantly in the same direction. In order that one can better judge that, in effect, there is nothing that needs changing in the explanation that I just discussed, I should repeat it

[324][N. T.] See [Ampère, 1822d, pp. 68-70], [Ampère, 1822e, pp. 245-247] and [Ampère, 1885d, pp. 199-201].

here, as applied to the case where the magnet is covered with an insulating substance. I will assume, for simplicity in the explanation, that one substitutes [pages 165 and 337] for the magnet an electrodynamic solenoid, whose extremities are at the poles of the magnet, although, according to my theory, it should be considered as a bundle of solenoids. This assumption does not change the effects that are produced, because the currents of mercury act in the same manner and in the same direction on all the solenoids in the bundle, they impose a movement similar to that which they would give to a single one of the solenoids, and one can always assume that the electric currents of this [solenoid] have sufficient intensity so that its motion will be substantially the same as that of the bundle.

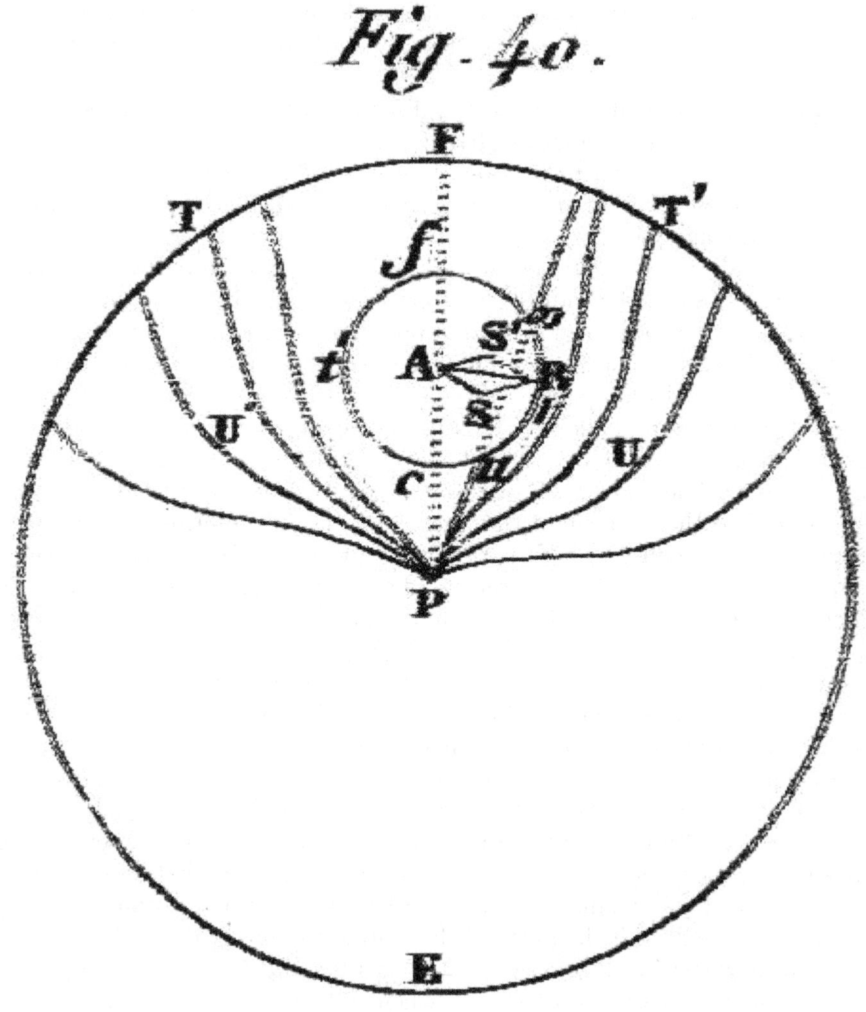

Figure 40.

Let then ETFT′ (figure 40)[325] be the horizontal section of a glass jar filled with mercury in contact with a ring of copper which provides the interior border [of the jar] and which communicates with one of the rheophores, the negative rheophore for example, while one inserts the positive rheophore at P; then there form currents in the mercury which flow from the center P of the ring ETFT′ to its circumference.

Represent the horizontal section of the solenoid by the small circle $etft'$, whose center is at A and whose circumference $etft'$ is one of the electric currents of which it is composed; by assuming that this current moves in the direction $etft'$, it will be attracted by the [electric] currents of mercury such as PUT, which is located, in the figure, to the right of $etft'$, because the semi-circumference etf, in which the current goes in the same direction, is closer of it than $ft'e$, in which it goes in the contrary direction. Let AS be this

[325] [N. T.] In the 1827 version of the *Théorie*, the letters T and T′ appeared with their positions inverted when compared with this figure 40 of the 1826 version of the *Théorie*. The correct representation should be the 1827 figure, due to the fact that Ampère will soon discuss the electric currents flowing in mercury, such as PUT, as being at the right of the circumference $etft'$. We corrected the position of these letters in our diagram made with graphic software; see figure A.40 on page 512.

attraction equal to the difference of the forces exerted by the currents PUT on the two semi-circumferences, and which necessarily passes by their center A, as it results from the forces which these currents exert on all [pages 166 and 338] the elements of the circumference $etft'$ perpendicular to them, and are, by consequence, directed following the rays of this circumference. The same current $etft'$ of the solenoid is, on the contrary, repelled by the currents which, like PU′T′, are, in the figure, to the left of this current $etft'$, because they are in opposite direction in the semi-circumference $ft'e$ the closest to PU′T′. Let AS′ be the repulsion which results from the difference of the actions exerted by the currents PU′T′ on the two semi-circumferences $ft'e$ and etf; it will be equal to AS, and will form, with the ray PAF, the angle FAS′ = PAS, since everything is equal on both sides of this radius; the resultant AR of these two forces will be then perpendicular to it, and since it will pass through the center A, together with its two components AS and AS′, the solenoid will have no tendency to turn on its axis, as one observes in effect with respect to the floating magnet which represents this solenoid; but it will tend, at each instant, to move following the perpendicular AR of the ray PAF, and since, when one conducts this experiment with a floating magnet, the resistance of the mercury cancels at each instant the acquired speed, one sees this magnet describe the curve perpendicular to all the lines which pass as PAF by the point P, that is to say, [this magnet will describe] the circumference ABC of which this point is the center.

This outstanding experiment, by M. Faraday, has been explained by the physicists who do not admit my theory,[326] by attributing the movement of the magnet to the rheophore immersed in the mercury at P, to which one ordinarily gives a direction perpendicular to the surface of the mercury. It is true that, in this case, the current in this rheophore tends to carry the magnet in the direction that it actually moves; but it is easy to establish, [pages 167 and 339] by comparative experiments, that it acts with a force very much too weak to overcome the resistance of the mercury, and to produce, despite this resistance, the movement that one observes. I was at first surprised to see that these physicists did not take account of the action that the currents in the mercury should exert in their own theory, my surprise was augmented when I found the cause in a manifest error which is found stated in these terms in the publication already cited:[327, 328]

> "The transverse action of this imaginary wire (the electric current which is in the mercury) upon the austral magnetism of A (figure 43), will therefore tend constantly to impel A from the right to the left of an observer having his head at C′, and his feet at Z. But a contrary action will be exerted upon the pole B, of equal intensity, if the horizontal line C′FF′Z is situated at the precise height of the center of the [magnetized] bar; so that the sum of these actions will produce no motion of translation. It will, therefore, be the single force exerted by CF, which will produce the rotation of the bar AB."

How did not the author see that the actions that the *imaginary wire*, placed as he said, exerts on the two poles of the bar AB, tend to carry it in the same direction, and that they add instead of subtract, because being of contrary types, these poles are found at the two opposite sides of the wire?[329]

It is important to remark on this subject, that if the portions of the currents, forming a part of those [currents] of the mercury, could be found in the interior of the small circle $etft'$ and act [pages 168 and

[326] [N. T.] Biot, in particular, as Ampère will mention in the following discussion.

[327] [N. A.] *Précis élémentaire de physique expérimentale*, third edition, volume II, page 753.

[328] [N. T.] Ampère was quoting here Biot's book, see [Biot, 1824, p. 753] and [Farrar, 1826, p. 346].

[329] [N. T.] Ampère pointed out correctly Biot's mistake. In the image of this footnote we have the current in the mercury flowing from Z to F. In the upper portion of the magnetized bar, represented by the letter A by Biot, we have a North pole N. In the lower portion of the magnetized bar, represented by the letter B by Biot, we have a South pole S. According to Biot and Savart's law, equations (6.2) up to (6.4), the force F exerted by this *imaginary wire*, both in the North pole and in the South pole of the magnet, will be penetrating the paper. That is, these two forces will point from the right to the left of an observer whose head was at C', his feet at Z, in such a way that he was looking towards the North pole N located at point A.

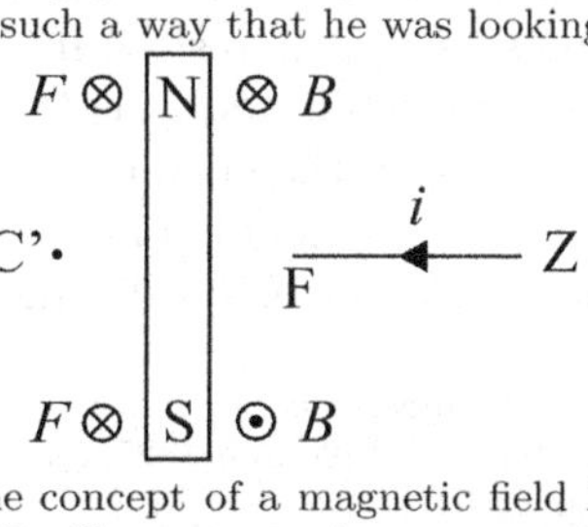

We can reach the same conclusion utilizing the concept of a magnetic field B. According to equation (15.2), the magnetic field generated by the current flowing through ZF will penetrate the paper at the location of the North pole of the magnet. At the location of the South pole of the magnet, on the other hand, the magnetic field will be leaving the paper, as represented in the figure of this footnote. According to equation (15.5), the forces F exerted by this magnetic field on the two poles of this magnetized bar will be penetrating the paper, as represented in this figure. Therefore these two forces should combine with one another, as pointed out correctly by Ampère. Biot, on the other hand, concluded wrongly that these two forces should cancel one another.

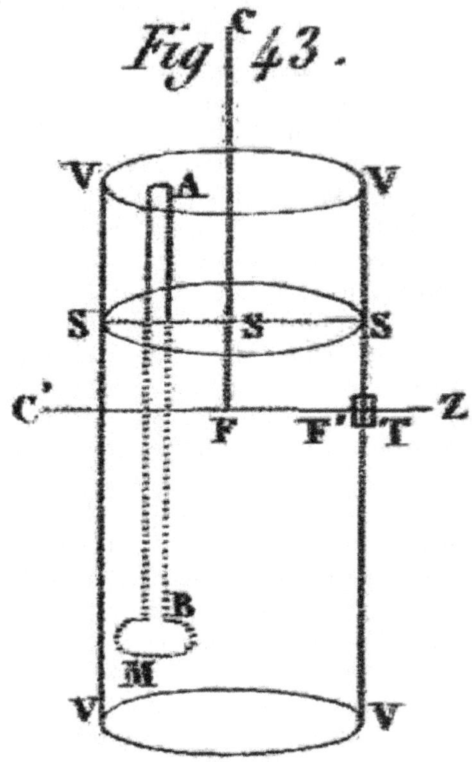

Figure 43.

340] on it,[330] they would tend to cause its rotation about the point P in the contrary direction, and with a force which, instead of being the difference of the actions exerted on the two semi-circumferences etf and $ft'e$, would be their sum, because if uv represents one of these portions, it is evident that it will attract the arc utv and will repel the arc $vt'u$, from which result two forces which together will move $etft'$ in the direction AZ opposite to AR.[331] This circumstance obviously cannot take place with the floating magnet which occupies all the interior of the small circle $etft'$, because it excludes the currents when it is covered by insulating material, and because, in the contrary case, the portions of the currents contained in this circle, occurring in the particles of the magnet invariably connected to those on which they act, the action that they produced is canceled by an equal and opposite reaction; so that there only remains, in the two cases, the forces exerted by the currents of the mercury, which all tend to move the magnet following AR.

It is uniquely for this reason that it turns about the point P in this direction, as one is assured by replacing the magnet by a mobile conductor $xzetft'sy$ (figure 41),[332] formed of a quite thin copper wire, covered by silk, whose intermediate part $etft'$ is wound in a circle, and whose two extreme portions, tied together from e to z, will, the one ezx go to x in a cup of mercury communicating with one of the rheophores, and the other $t'sy$ immerse in P (figure 40)[333] in the mercury which communicates, as we have said, with the other rheophore: we suspend this mobile conductor in a manner such that the circle $etft'$ (figure 41) is very close to the mercury surface, and one sees that it rests immobile, by virtue of the equilibrium which is established between the forces exerted by [pages 169 and 341] the portions of the currents contained in the circle $etft'$, and those [forces] that are [exerted] by the currents and current portions outside this circle. But as soon as you remove the portions of the currents included in the space $etft'$ (figure 40), by inserting in the mercury below the circle $etft'$ (figure 41) a cylinder of insulating material whose base is such as to imitate

[330][N. T.] That is, if these currents could act on the mercury.

[331][N. T.] The letter Z did not appear in figure 40 of the 1826 and 1827 versions of the *Théorie*. We included it in the diagram made with graphic software; see figure A.40 on page 512.

[332][N. T.] In figure 41 of the 1826 and 1827 versions of the *Théorie*, the letters appearing on the central column, from top to bottom, were x, χ (this letter is not easily visible in the 1826 version of the *Théorie*, as one can observe in this image), s and y. A little to the right of this central column we have, from top to bottom, the letters e and e'. However, this figure can only be compatible with the following text utilizing the letter z instead of χ, and t' instead of e'. We utilized this correction in our diagram made with graphic software; see figure A.41 in our page 513.

[333][N. T.] See our page 444.

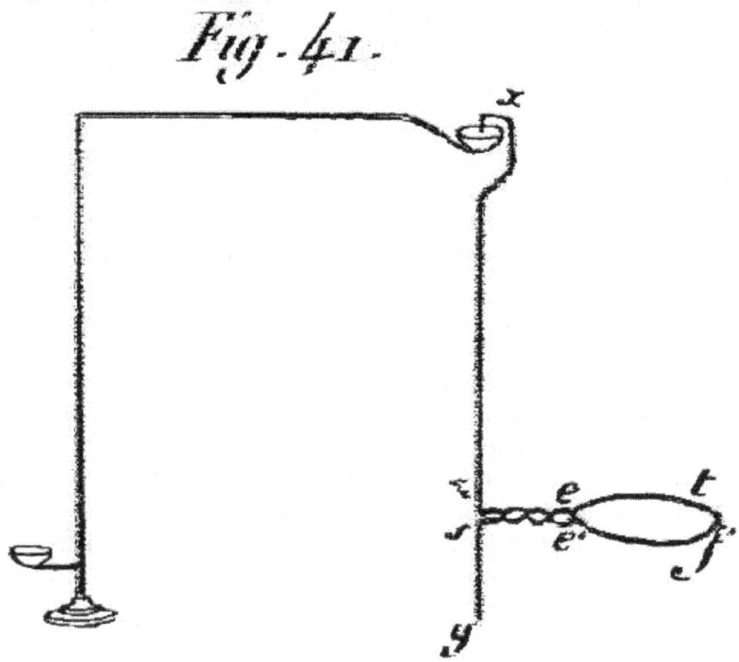

Figure 41.

that [base] which happens to the floating magnet, one sees it moving,[334] like this magnet, in the direction AR. When one leaves the cylinder of insulating material where first was the circle $etft'$, it does not turn indefinitely like the magnet, but stops after a few oscillations, in a position of equilibrium; the difference comes from the fact that the floating magnet allows, behind it, to be filled with mercury the space which it occupied at first, and drives the mercury successively from the various places to which it is transported. It is this change in the situation of a part of the mercury which causes a [change] in the electric currents, and causes, even when the total voltaic current is closed, the continuous motion of the magnet, [a motion] which is impossible by the action of a rigid and closed circuit, [although this continuous motion] still takes place in this case where the closed circuit changes shape by the movement of the magnet itself. To produce this movement by using, instead of the magnet, the mobile conductor as described above, it is necessary — since one has established that it will only move if one removes, by the cylinder of insulating material, the portions of the current interior to the small circle $etft'$, and that, [as we have also established,] when one leaves this cylinder in the same place, it stops at a specific position of equilibrium after having oscillated about it — to imitate that [continuous motion] which takes place when one is dealing with a floating magnet, by sliding the cylinder of insulating material at the bottom of the [pages 170 and 342] vase, in a manner such that it is always under the circle $etft'$ (figure 41), and such that its center always corresponds vertically to that [center] of the circle, the mobile conductor therefore starts to turn indefinitely about the point P (figure 40) like the magnet.

It is, in general, when substituting for magnets mobile conductors wound in a circle, that one can form a correct idea of the causes of the various movements of the magnets, when one wishes to analyze these movements by experiment without recourse to calculation, because this substitution provides the means to vary the condition in various manners, which are very often impossible to obtain with magnets, and only [this substitution] can clarify the difficulties which are presented by often so complicated phenomena. It is thus, for example, that in what we have just said, it is impossible, with a magnet, to verify this result of the theory, [namely,] that if the portions of currents of mercury could traverse the magnet and, despite this, could act on it conserving the intensity and direction they have in the mercury when one removes the magnet, it would not turn about the point P, and that the verification becomes easy when one substitutes for it, as we have said, the mobile conductor shown here (figure 41).

The identity of the action that one constantly observes between the movements of a mobile conductor and that of a magnet, in all cases that they are found in the same circumstances, does not permit any

[334][N. T.] That is, one sees the conductor moving.

doubt, when one has done the preceding experiment, that the magnet will also remain immobile, when it is traversed by the portions of currents interior to the circle $etft'$, if these portions could act on it; and as one sees, on the contrary, that when it is not covered [pages 171 and 343] by an insulating material, and when the currents freely traverse it, it moves exactly as when it is [covered by an insulating material] and that no portions of currents can penetrate into the interior of this magnet, one has a direct proof of the principle which rests a part of the explanations that I have given, namely: that the portions of currents which traverse the magnet do not act in any manner on it, because the forces which would result from their action on the currents proper to the magnet, or on those that one calls the magnetic molecules, by occurring between the particles of the same rigid body, are necessarily destroyed by an equal and opposite reaction.

I confess that this experimental proof of a principle which is nothing else but a necessary consequence of the first laws of mechanics, appears to me completely useless, as it should have been clear to all the physicists who considered this principle one of the foundations of science. I would not have made this observation, if it had not been assumed [by others] that the mutual action of one element of a conducting wire and of a magnetic molecule, consisted in a primitive couple composed of two forces equal and parallel without being directly opposed, by virtue of which a portion of current which is located inside a magnet might move it;[335] [this] supposition is contrary to the principle which is being discussed here, and is denied by the previous experiment from which there is no action exerted on the magnet by the portions of currents which traverse it when it is not covered by an insulating envelope, since the movement which takes place in this case stays the same if one prevents the currents from traversing the magnet, by enclosing it in this [insulating] envelope.

[pages 172 and 344]

It is from this principle that one must start in order to see what phenomena should yield a mobile magnet under the influence of a voltaic current, in the third case which still remains to be considered, where a portion of the current passes through the magnet, or through a portion of a conducting wire which is rigidly bound to it. We have just seen that when there is a motion of revolution of a magnet about a conducting wire, the movement should be the same, as is in effect, whether the current traverses or does not traverse the magnet. But this is not the case when it is a question of continuous rotational movement of a magnet about the line that joins its two poles.

I have demonstrated by theory and by various experiments of diverse kinds whose results always confirmed those of the theory, that the possibility or impossibility of this movement is solely due to [the fact] that a portion of the total voltaic circuit is in all its points separated from the magnet, or that it passes, whether inside the magnet, or in a portion of conductor bound invariably with it. In effect, in the first case, the assembly of the battery and the conducting wires forms a circuit, always closed and in which all the parts act equally on the magnet, whether they are fixed or mobile; in this last case, they exert, at each instant, precisely the same forces as if they were fixed in the position where they are at that instant. Now, we have demonstrated, first synthetically with the aid of considerations which figures 30 and 31 provided us,[336] then calculating directly the torques, [showing] that a closed circuit cannot impress on a magnet a continuous motion about the line which joins its two [pages 173 and 345] poles, whether one considers them, conforming to my theory, as the two extremities of a solenoid equivalent to the magnet, or as two magnetic molecules whose intensity is sufficiently large so that the actions exerted stay the same when one substitutes them for all those [molecules] of which one regards the magnet as composed under the hypothesis of the two fluids. The impossibility of rotational movement of the magnet about its axis, while the totally closed circuit is everywhere separated from it, is thus found completely demonstrated, not only with the application of my formula to the currents of a solenoid substituted for the magnet, but also by taking into consideration a force which would exist between an element of conducting wire and a magnetic molecule perpendicular to the plane which passes through this molecule and by the direction of the element, in inverse ratio of the square of the distance, and which would be proportional to the sine of the angle composed between the line that measures this distance and the direction of the element. But if one assumes, in this last case, that the force passes through the center of the element, whether it acts on it or reacts on the magnetic molecule, as it would happen, following my theory, with respect to the solenoid, the same movement becomes possible provided a portion of the current passes through the magnet, or by a portion of conductor invariably bound with it; because all the actions exerted by this portion on the particles [of the magnet] being destroyed by the

[335] [N. T.] Ampère was referring here, in particular, to the suppositions of Biot, Savart and Faraday. We discussed this topic in Section 21.3. We clarified the meaning of the primitive couple in footnote 246 in our page 422.

[336] [N. T.] These figures appear in our pages 408 and 410, respectively.

equal and opposite reactions that these same particles exert on it,[337] there only remains the actions exerted by the rest of the total circuit which is no longer closed, and can as a consequence rotate the magnet.

In order to fully understand all that relates to this sort [pages 174 and 346] of movement, conceive that the rod TVUS (figure 13),[338] which supports the small cup S in which is inserted the tip *o* of the mobile conductor *oab*, is folded in V and U as one sees in the figure, in such a manner as to leave free the portion VU of the straight line TS taken as the axis of rotation, so that one may suspend the cylindrical magnet GH, by a very thin wire ZK, from the hook K attached to U on this rod, and so that the mobile conductor *oab*, maintained in the position where one sees it in the figure by the counter weight *c*, is terminated at *b* by a copper plate *bef*, which is inserted into the acidulated water with which one has filled the vase MN, so that the conductor communicates with the rheophore *p*P inserted into mercury in the cup P, while the other rheophore *r*R is in communication with the rod TVUS through the mercury which one puts in the cup R, and that the battery *pr* closes the total circuit.

At the instant when one establishes the current in the apparatus, one sees the mobile conductor turn about the line TS; but the magnet is only led to a determined position about which it oscillates for some time, and where it then comes to rest. By the principle of the equality of action and reaction, which applies with regard to the torques about a common axis as with regard to the forces, if one represents by M the torque produced, by the action of the magnet, on the mobile conductor *oab*, the reaction of this [torque] will tend necessarily to cause the magnet to rotate about its own axis with the torque −M, equal to M, but acting in the contrary direction.

The immobility of the magnet obviously comes from the fact that if the mobile conductor *oab* acts on it, the rest *b*MP*pr*RTS of the total circuit cannot fail to act equally; the [pages 175 and 347] torque of the action that it exerts on the magnet,[339] combined with that of *oab*, yields the torque of the closed circuit *oab*MP*pr*RTS, which is null; from which it follows that the torque of *b*MP*pr*RTS is M, equal and opposite to −M.

But if one connects the magnet GH to the mobile conductor *oab*, there results a system of invariable form, in which the action and reaction that they exert on each other mutually cancel; and the system would obviously remain immobile, if the part *b*MP*pr*RTS did not act as before on the magnet in order to cause it to turn by imparting to it the torque M. It is due to this torque that the magnet and the mobile conductor, combined in a system of invariable form, turn about the line TS; and since this torque is, as we just saw, of the same value and the same sign as that which the magnet exerted on the conductor *oab* when this conductor was separated from it [i.e., from the magnet] and turned alone, one sees that these two movements necessarily take place in the same direction, but with speeds which are reciprocally proportional to the moment of inertia of the conductor and to the sum of this moment of inertia and that of the magnet.

I disregarded, in the preceding considerations, the action exerted by the portion *b*MP*pr*RTS of the total circuit on the mobile conductor *oab*, whether in the case where the conductor is separated from the magnet, or in the case where it is connected to it, not only because it is very small relative to that [action] which the magnet exerts, but because it tends uniquely to carry the mobile conductor into the position determined by the mutual repulsion of the elements of these two portions of the total circuit, and only contributes, as a consequence, [pages 176 and 348] in the two cases, as regards the motions of rotation of *oab*, to change a little its speed, which without this [action] would be constant.

In order to easily unify and separate alternately the magnet and the mobile conductor, without interrupting the experiments, it is convenient to fix to the hook Z by which the magnet is suspended by the wire ZK, a piece of copper wire ZX terminated at X by a fork whose two branches Xx and Xy tighten the mobile conductor *oab*, which is trapped between them, when one properly folds the rod ZX; by bending it in the contrary direction, one places it where it is shown in the figure,[340] and the conductor again becomes free.

I have explained in detail this experiment because it seems, more than any other, to support the hypothesis of the primitive couple, when one does not analyze it as I have just done. In effect, one accepts as I do, in this hypothesis, that the forces exerted by the magnet GH,[341] on the elements of the mobile conductor *oab*, pass through these elements, and assuming they are all in the vertical plane TS*ab*, drawn by the straight line TS, the forces are normal to this plane, therefore they will tend to rotate *oab* always in the same direction about TS: these forces are, after the law M. Biot proposed, precisely the same — in size, in direction and

[337][N. T.] Original French text: "par les réactions égales et opposées qu'exercent sur elles ces mêmes particules." It seems to us that the correct expression should be "par les réactions égales et opposées qu'exercent sur elle ces mêmes particules." The word *elle* here should refer to the portion of electric current which was interacting with the particles of the magnet.

[338][N. T.] See our page 371.

[339][N. T.] That is, the torque exerted by the remainder of the closed circuit acting on the magnet.

[340][N. T.] Ampère was still referring to figure 13 of the *Théorie*; see our page 371.

[341][N. T.] Magnet GH of figure 13; see our page 371.

relatively at their application points — as the forces given by my formula; they produce therefore the same torque M by virtue of which the conductor *oab* moves if it is free. But, following the physicists[342] who accept the hypothesis here in question, the forces due to the reaction of the elements of the conductor [pages 177 and 349] on the magnet will only be the same in size and in the fact that they are perpendicular to the plane TS*ab*; they[343] think that these forces are applied to the magnetic molecules, or, that which comes to the same, to the two poles of the magnet GH which are on the line TS; thus their torques are null relative to this line. They attribute the immobility of the magnet to this cause when it is not attached to any portion of the voltaic circuit; but to explain the rotational motion of the magnet in the case in which it is connected to the mobile conductor *oab*, with the aid of the rod ZX, they assume that the connection of these two bodies into a system of invariable form, does not prevent the magnet to always act to impose on the mobile conductor the same torque M, without that the conductor reacts on the magnet in a manner to prevent movement of the system, which should turn as a consequence in the same direction as the mobile conductor turned before having been rigidly attached to the magnet, but with a smaller speed in the reciprocal ratio of the moments of inertia of the conductor alone and of the conductor attached to the magnet.

It is thus that one finds in this hypothesis the same results as when one assumes that the action opposed to the reaction follows the same line, and one takes account of the action exerted on the magnet by the rest *b*MP*pr*RTS of the voltaic circuit. It results from all that has been demonstrated in this Treatise, that this identity of the produced effects and of the values of the forces that we have found, in the case that we have examined, between the manner that I have explained the phenomena and the hypothesis of the primitive couple, is a necessary consequence of the fact that the voltaic circuit that one has made to act on [pages 178 and 350] the magnet is always closed, and that, provided we are dealing with a closed circuit, not only the three forces parallel to three axes which result from the action that such a circuit exerts on a magnet, but also the three torques about these three axes, are the same in the two ways of conceiving these things, just as the motion of the magnet, which can only depend on these six quantities.

The same identity is found, as a consequence, in all the experiments of the same type, and it is not, neither from these experiments, neither from the measurement of the forces that develop between the conducting wires and the magnets, that such a question can be decided; it should be [decided]:

1. By the necessity of the principle, that the mutual action of the diverse parts of a system of invariable form cannot, in any case, impose on this system an arbitrary movement; a principle which is only a consequence of the very concept that we have of forces and of the inertia of matter.

2. From this circumstance, [namely,] that the hypothesis of the primitive couple was only imagined, by those who proposed it, because they believed that the phenomena from which they departed could not be explained otherwise, failing to take into account the action exerted on the magnet by the totality of the voltaic circuit; because they have not paid attention to the fact that the circuit is always closed, and that they did not deduce, as I did, from the law proposed by M. Biot, this rigorous consequence that, for a closed circuit, the forces and the torques are identically the same, whether one assumes that the directions of the forces acting on the magnet pass through the magnetic molecules or through the centers of the elements of the conducting wires.

3. On this [argument], when one accepts that the phenomena with which we are [pages 179 and 351] concerned can be produced, in the final analysis, by forces expressed as functions of the distances exerted by molecules of the two electric fluids,[344] and that one attributes also [these forces] to the two magnetic fluids when one views them as the cause of the phenomena, purely electric as I think, posed by the magnets, [therefore] one can well conceive that if these molecules are in movement in the conducting wires, there then results between their elements forces that do not only depend on the distances of these elements, but also on the directions according to which the movement of the electric molecules traversing them takes place, precisely such as the forces given by my formula, provided that these forces satisfy the condition that the action and the reaction are directed following the same line, whereas it is contradictory to assume that the forces, whatever were otherwise their values as functions of distances, directed along the straight lines connecting the molecules between which they are exerted, may produce, by any combination whatsoever, even when these molecules are in movement, forces for which the action and reaction are not directed following the same line, but following two parallel lines, as in the hypothesis of the primitive couple.

[342][N. T.] Physicists like Biot, Savart and Faraday, who accepted the concept of a primitive or internal couple.

[343][N. T.] That is, physicists like Biot, Savart and Faraday.

[344][N. T.] That is, by forces depending on the distance and being exerted between positively and negatively charged particles, as we would say nowadays.

One knows, in effect, that even when electric or magnetic molecules are in movement, they act at each instant as if they were at rest at the point where they are at that instant. If therefore one considers two systems of molecules, such that each molecule of one [system] exerts on each molecule of the other a force equal and opposite, along the line that joins them, to the force exerted by the second molecule on the first, and stopping [pages 180 and 352] all these molecules in their current location at that instant, one assumes that they are all rigidly connected at this location, there will necessarily be equilibrium in the rigid system, composed of two others, which results from this assumption, since there will be equilibrium between the elementary forces taken pairwise. The resultant of all the forces exerted by the first system on the second will therefore be equal and opposite, following the same line, to that [resultant] of all the forces exerted by the second on the first; and these two resultants can never produce a couple[345] capable of turning the total system, when all of its parts are rigidly bound together, as [however] is assumed by those who, while adopting the hypothesis of a couple in the mutual action of one magnetic molecule and of one element of conducting wire, pretend that this action results from [the assumption] that the element only acts on the molecule because it is itself an assemblage of magnetic molecules,[346] whose actions on that [magnetic molecule] which one considers are such that as Coulomb established them, that is to say, directed along the lines which join them to this last [molecule], and in inverse ratio of the square of the distances.

It suffices to read with some care that which M. Biot has written on the phenomena that occupy us, in the ninth book of the third edition of his *Traité élémentaire de physique expérimentale*,[347] to see that after having considered carefully the forces that the elements of conducting wires exert on magnets, as applied to magnetic molecules perpendicularly to the planes passing through each element and each molecule, he then assumes, when he speaks of the movement of conducting wires about [pages 181 and 353] magnets, that the forces exerted by the magnetic molecules on the elements of the wires, pass through these elements in directions parallel to those of the forces exerted on the magnet, and forming, as a consequence, couples with the first, instead of being opposed along the same straight line; he explains in particular on page 754, volume II of this work,[348] the rotational movement of a magnet about its axis, when a portion of the current traverses it, by assuming that the magnet turns due to the action that this portion itself exerts on the rest of the magnet, which forms with it, however, a system of fixed form[349] such that all the parts are invariably connected to one another:[350] [pages 182 and 354] which obviously implies that the action and the reaction of this portion of current and the rest of the magnet form a couple. How from that position can it be imagined that the physicist who admits such a supposition, can express in these terms on page 769 of the same book:[351]

> "If we calculate the action which a magnetic needle of an infinitely small length, and nearly molecular, would exert at a distance, it will be readily seen that we may form assemblages of such needles which shall

[345][N. T.] See footnote 246 in our page 422.

[346][N. T.] That is, these authors like Biot, etc. assumed that the current element was composed of an assemblage of magnetic poles. Two magnetic poles exert an equal and opposite force on one another along the straight line connecting them. Despite this fact, authors like Biot assumed that an assemblage of magnetic poles might act on an external magnetic pole with equal and opposite forces along parallel lines. These opposite forces acting along parallel straight lines would produce a primitive or internal couple. Ampère is here fighting against this conclusion. According to Ampère, even accepting this assumption about the constitution of a current element, the net action and reaction between an assemblage of magnetic poles and an external magnetic pole would point along the same straight line.

[347][N. T.] The correct title of Biot's book is: *Précis élémentaire de physique expérimentale*, [Biot, 1824].

[348][N. T.] See [Biot, 1824, p. 754].

[349][N. T.] That is, the magnet and the portion of the current traversing it form a system of rigid form.

[350][N. A.] I do not know if it is necessary to recall on this subject that which I have already remarked elsewhere, namely, that the electric fluids, after all the facts, above all after the nullity of the action on the lightest bodies of electricity which move in a vacuum, should be considered as incapable of action due to their mass which one can consider as infinitely small with respect to those of ponderable bodies, and so that therefore all attraction or repulsion exerted between these bodies and the electric fluids can well put these in motion, but not the ponderable bodies. For these last to move, it is necessary, when it comes to ordinary electric attractions and repulsions, that the electricity is retained on their surface, so that the force that overcomes the inertia of one, applies, if one can so express it, on the inertia of the other. It is necessary all the same, for the mutual action of two conducting wires to put these wires in motion, that the decompositions and recombinations of the neutral fluid which is present at each instant in all the elements along the length of the two wires, determine between their ponderable particles the forces capable of overcoming the inertia of their particles in imparting to the two wires the speeds proportionally reciprocal to their masses. When one speaks of the mutual action of two electric currents, one never understood, and it is evident that one cannot understand, other than those of the conductors that they traverse: the physicists who accept magnetic molecules acting on the elements of a conducting wire, conforming to the law proposed by M. Biot, accept without doubt also that this action only moves the wire because the magnetic molecule is retained by the ponderable particles of the magnet which constitutes the magnetic element of which it is a part; and it is therefore evident that in assuming that the magnet is moved by the action of the portion of the electric current which traverses it, one necessarily assumes that its movement results from the mutual action which exists between each of those of its particles which traverse the current and all the other particles of the same body.

[351][N. T.] See [Biot, 1824, p. 769], [Farrar, 1826, p. 360] and [Biot and Savart, 1885, p. 122].

> exert transverse forces. The only difficulty, but no doubt a very great one, is to combine such systems so as to produce for the laminae of a uniting wire of sensible dimensions, the precise laws of transverse action with which experiment has made us acquainted, and which have been explained above."

Without doubt from the action of two systems of small magnets, whose austral and boreal molecules attract or repel in inverse ratio of the square of their distances, along the lines which join them in pairs, *transversal actions* could result, but not *actions which are not equal and opposite to the reactions directed following the same lines*, as those [actions] supposed by M. Biot.

In one word, the value of the action of two elements of [pages 183 and 355] conducting wires, which I have deduced uniquely from experiments, depends on the angles which determine the respective directions of the two elements; following the law proposed by M. Biot, the force which develops between an element of conducting wire and a magnetic molecule, depends also on the angle which determines the direction of the element. If I called *elementary* the force which I determined the value, it was because it is exerted between two elements of conducting wires and because it has not yet been reduced to simpler forces; he has also called *elementary* the force that he assumes between a magnetic molecule and an element of conducting wire. Up to here all is similar with respect to these two sorts of forces; but for those [forces] that I have admitted, the action and the reaction are opposed following the same line and nothing prevents conceiving that they result from attractions and repulsions inherent in the molecules of the two electric fluids, provided that one assumes these molecules in motion in the conducting wires, to explain [in this way] the influence of the direction of the elements of these wires on the value of the force; whereas M. Biot, by admitting a force for which the action and the reaction are not directed in contrary directions on the same line, but on lines parallel and forming a couple, places himself in the absolute impossibility to reduce this force to attractions and repulsions directed along the lines which join pairwise the magnetic molecules, such as those admitted by all the physicists who use this [assumption] to explain the mutual action of two magnets. It is not evident that it is from this hypothesis of M. Biot, on the rotational forces for which the action and the reaction are not opposed following a single line, that one should say what he said (page 771)[352] on the subject of the [pages 184 and 356] mutual action of two elements of conducting wires, such as I have determined by my experiments and the calculations that I derived, namely: that a similar supposition *is in the first place, completely opposed in itself to all the analogy observed in the other laws of attraction?*[353] Does there exist a hypothesis more contrary to these similarities, than to imagine forces such that the mutual action of the diverse parts of a system of invariable form can set this system in motion?

It was not walking away in this way from one of the laws that Newton considered as a foundation of the physical theory of the universe that, after having discovered a great number of facts that no one had observed before me, I determined, solely by experiment and following the path traced by this great person, first the laws of electrodynamic action, then the analytic expression of the force that develops between two elements of conducting wires, and finally I deduced from this expression all the consequences expressed in this Treatise. M. Biot, by citing the names of a group of physicists who had observed new facts or invented instruments which were useful in science, mentioned neither the means by which I came to render mobile the portions of conducting wires, by suspending them on steel tips in cups filled with mercury, a method without which one can learn nothing of the actions exerted on these wires, whether by other conductors, or whether by the Earth or by magnets; nor the apparatus which I constructed to make evident all the circumstances which display these actions, and to precisely determine the cases of equilibrium from which I deduced the laws to which they are subject; nor these laws themselves as determined by my experiments; [pages 185 and 357] nor the formula that I obtained from them; nor the applications that I made of this formula. And as regards the facts that I was the first to observe, he cites just one, that of the mutual attraction of two conducting wires; and as he cites it, it is to give an explanation which had been first proposed by several foreign physicists,[354] at a time when one had not done the experiments which have demonstrated since long ago that it[355] was completely inadmissible. This explanation consists, as one knows, of assuming that two conducting wires act on each other, as they would [act] by virtue of the mutual action of infinitely small magnetic needles, tangent to circular sections that one can make in all the length of the wires assumed to be cylindrical; the ensemble of small needles of the same section forming thus a magnetic ring, similar to that

[352][N. T.] See [Biot, 1824, p. 771], [Farrar, 1826, p. 362] and [Biot and Savart, 1885, pp. 124-125].

[353][N. T.] The portion in italics is a sentence by Biot when criticizing Ampère's theory, see [Biot, 1824, p. 771], [Farrar, 1826, p. 362], [Biot and Savart, 1885, pp. 124-125] and our Section 12.2.

[354][N. T.] Biot's explanation was discussed in our Subsection 21.1.3.

[355][N. T.] That is, this explanation proposed by foreign scientists.

which MM. Gay-Lussac[356] and Velter[357] made use of to carry out, in 1820, a decisive experiment on the subject of the explanation which is here in question.[358] This experiment proved, as one knows, that such a ring exerts absolutely no action,[359] while it forms a complete circumference, even though it is so strongly magnetized that, being formed from an appropriate steel in order to preserve, when one breaks it, all of its magnetism, one finds, in breaking it, that all of the pieces are strongly magnetized.

Sir H. Davy[360] and M. Erman[361] obtained the same result as regards a steel ring of any shape. It is, for the remainder, a necessary consequence of the theory of two magnetic fluids as of mine, as it is easy to be convinced by a calculation entirely similar to that by which I demonstrated, in this Treatise, the nullity of action of a solenoid forming a closed curve, conforming to what [pages 186 and 358] M. Savary first found, by a calculation which does not differ essentially from mine, and that one can see, either in the addition which is found at the end of the memoir on the application of calculations to electrodynamic phenomena, that he published in 1823, or in the *Journal de Physique*, volume XCVI, pages 295 and following.[362] In giving once again this explanation, M. Biot shows that he does not know either the MM. Gay-Lussac and Velter [Welter] experiment, nor the calculation of M. Savary.

In addition, the small needles tangent to the circumferences of the sections of conducting wires, are considered by M. Biot as the particles themselves of the surface of the conducting wire, magnetized by the electric current which separates in these particles the austral and boreal fluid, by carrying them in contrary directions, without that the molecules of these fluids can leave the wire particles where they were originally found combined in the neutral fluid. Therefore, when the current is established after some time in the fluid and continues indefinitely, the distribution of the magnetic molecules in the conducting wires can no longer change; it is thus as if there existed in these wires a multitude of determined points that would not change position as long as the current continued with the same intensity, and which would emanate attractive and repulsive forces due to the magnetic molecules, and as a consequence reciprocally proportional to the squares of the distances.

Thus two conducting wires would not act on each other except by virtue of forces expressed as a function of the distances between the points fixed in one of the wires and the other points equally fixed in the other wire; but then one of the wires, assumed stationary, could only bring the other [pages 187 and 359] into a situation of equilibrium where the integral of the living forces, which is always obtained as a function of the coordinates of the points of the mobile wire when the forces are functions of the distances, would attain its maximum value. Never could such forces produce a rotational movement whose speed would always go on increasing in the same direction, up to a point in which this speed would become constant, because of frictions, or of resistance of the liquid in which it is necessary to insert the mobile conductors to maintain the [electric] communications. Now, I have obtained this [continuous] rotational movement by making a spiral conductor, forming nearly a circle, act on a rectilinear conducting wire, turning about one of its extremities situated at the center of the circle, while its other extremity was located quite close to the spiral conductor.[363]

This experiment, where the movement is very rapid and can last several hours, when one uses a battery of sufficient strength, is in manifest contradiction to the point of view of M. Biot; and if it is not [in contradiction] with the point of view that the action of two conducting wires results from attractive and repulsive forces inherent in the molecules of the two electric fluids, it is because these molecules do not remain circumscribed, like those that one assumes compose the two magnetic fluids, in the very small space where their distribution is determined by a permanent cause, but on the contrary they travel all the length of each wire by a sequence of compositions and decompositions, which succeed each other in very short intervals: from which, as I have observed, movements always continuing in the same direction can result, [movements which are] incompatible with the supposition that the points [pages 188 and 360] from which the attractive and repulsive forces emanate do not change position in the wires.

Finally, M. Biot repeats in the third edition of his *Traité élémentaire de physique* (volume II, page 773),[364] what he already said in the note which he published, in the *Annales de Chimie et de Physique*,[365]

---

[356][N. T.] Joseph-Louis Gay-Lussac (1778-1850).

[357][N. T.] Ampère was referring to Jean Joseph Welter (1763-1852), a chemist who collaborated with Gay-Lussac.

[358][N. T.] We discussed this experiment in Section 9.1.

[359][N. T.] This ring exerts no action on another magnet which does not belong to the ring.

[360][N. T.] Humphrey Davy (1778-1829), English chemist.

[361][N. T.] Paul Erman (1764-1851), German physicist.

[362][N. T.] See [Savary, 1823b], [Savary, 1823c], [Savary, 1885b] and [Savary, 1885a].

[363][N. T.] Section 7.2.4 discussed one of these experiments of continuous rotation which did not employ magnets or terrestrial magnetism, but only current-carrying conductors.

[364][N. T.] Correct title of Biot's work: *Précis élémentaire de physique expérimentale*, see [Biot and Savart, 1824, p. 773] and [Biot and Savart, 1885, note on pp. 125-126].

[365][N. T.] See [Biot and Savart, 1820, p. 223], [Biot and Savart, 1885, p. 126], [Biot and Savart, 1965b, p. 119] and

on the first experiments relative to the subject which we are addressing, which he made with M. Savart, namely: that when an element of a very thin and indefinite connective wire acts on a magnetic molecule, "the nature of its action is the same as that of a magnetized needle which is placed on the contour of a wire in a certain constant direction in relation to the direction of the current." However the action of this needle on a magnetic molecule is directed following the same line as the reaction of the molecule on the needle, and it is anyhow easy to see that the force which results is in inverse ratio of the cube, and not [in inverse ratio] of the square of the distance, as M. Biot himself found is that [force] of the element of the wire.

29.23 Identity of the Mutual Interaction between Two Closed Voltaic Circuits with the Mutual Interaction between Two Assemblies, Each Composed of Two Very Closely Spaced Surfaces Terminated by the Circuit Corresponding to Each Assemblage, and on Which the Two Magnetic Fluids, Austral and Boreal, are Distributed and Fixed in such a manner that the Magnetic Intensity is Everywhere the Same

It remains for me to extend to the mutual action of two closed circuits, of arbitrary size and shape, the considerations relative to surfaces terminated by these circuits and whose points act as the so-called molecules of austral and of boreal fluid, which I have previously applied to the mutual action of an arbitrary closed circuit and of an element of conducting wire. I have found[366] that the action of the element $d^2\sigma'$ on the two surfaces terminated by the contour s, was expressed by the three forces

$$\mu g\varepsilon' d^2\sigma' \frac{u^2 d\varphi}{r^3} \ , \quad \mu g\varepsilon' d^2\sigma' \frac{v^2 d\chi}{r^3} \quad \text{and} \quad \mu g\varepsilon' d^2\sigma' \frac{w^2 d\psi}{r^3} \ ,$$

applied to each element ds of this contour, I will [pages 189 and 361] now apply to the circuit s' what I have done before with regard to the circuit s. For this consider a new surface terminated on all sides, like the surface σ', by the closed curve s', and which is such that the portions of the normals of the surface σ' comprised between it and this new surface, are everywhere very small. Assume, on the new surface, fluid of type contrary to that of the surface σ', in such a manner that the quantities of the two fluids in the corresponding parts of the two surfaces are the same. Designating by ξ', η' and ζ' the angles that the normal at the point m', whose coordinates are x', y' and z' forms with the three axes, and by h' the small portion of this normal which is located between the two surfaces, we can, as we have done for the element $d^2\sigma'$, reduce the action of the element of the new surface which is represented by $d^2\sigma'$, on the ensemble of the two surfaces which are terminated by the contour s, to the forces applied, as one has seen, on page 147,[367] on the various elements of the contour; that [force] which is relative to the element ds and parallel to the x [axis] is obtained by substituting in the expression which we found for this force

$$\mu g\varepsilon' d^2\sigma' \frac{u^2 d\varphi}{r^3} \ ,$$

or

$$-\mu g\varepsilon' d^2\sigma' \frac{(y'-y)dz - (z'-z)dy}{r^3} \ ,$$

the new coordinates $x' + h'\cos\xi'$, $y' + h'\cos\eta'$ and $z' + h'\cos\zeta'$ in place of x', y' and z'. Since the forces thus obtained act in the sense contrary to the first, it is necessary to subtract them, which results in, if one neglects in the calculation the powers of h greater than the first, differentiating

$$-\mu g\varepsilon' d^2\sigma' \frac{(y'-y)dz - (z'-z)dy}{r^3} \ ,$$

[Assis and Chaib, 2006, p. 308].

366 [N. T.] See pages 144 and 316 of the 1826 and 1827 versions of the *Théorie*, respectively, corresponding to our page 434.

367 [N. T.] Page 147 of the *Théorie* published in 1826, corresponding to page 319 of the 1827 version of the *Théorie* and to our page 435.

[pages 190 and 362]

by varying x', y' and z', replacing $\delta x'$, $\delta y'$ and $\delta z'$ by $h'\cos\xi'$, $h'\cos\eta'$ and $h'\cos\zeta'$, [respectively,] and changing the sign of the result, while x, y, z, and dx, dy, dz, should be considered as constants since they belong to the element ds.

The formula into which one should substitute $h'\cos\xi'$, $h'\cos\eta'$ and $h'\cos\zeta'$ in the place of $\delta x'$, $\delta y'$ and $\delta z'$ is therefore

$$\mu g\varepsilon'\left(dzd^2\sigma'\delta'\frac{y'-y}{r^3} - dyd^2\sigma'\delta'\frac{z'-z}{r^3}\right) ,$$

which must be integrated after this substitution over all the extent of the surface σ' to have the total action of this surface and of that with which it is united over the assemblage of the two surfaces terminated by the contour s. One can carry out this double integration separately on each of the two terms in which this expression is composed. We carry out first that [integration] which is relative to the first term

$$\mu g\varepsilon' dzd^2\sigma'\delta'\frac{y'-y}{r^3} .$$

For this, decompose the surface σ' into an infinity of infinitely narrow zones by a series of planes perpendicular to the xz plane drawn by the coordinate y of the center o of the element ds. We take, on one of these zones, for $d^2\sigma'$ the element of the surface σ' which is expressed as

$$\frac{vd'vd'\chi}{\cos\eta'} ,$$

and we have thus to integrate the quantity

$$\mu g\varepsilon' dz\frac{vd'vd'\chi}{\cos\eta'}\delta'\frac{y'-y}{r^3} ,$$

which changes, by a transformation exactly similar to that [pages 191 and 363] which we used above relative to

$$d^2\sigma = \frac{udud\varphi}{\cos\xi} ,$$

into this

$$-\mu gdzh'\varepsilon' d'\chi d'\frac{v^2}{r^3} .$$

Assuming, as we did for the surface σ, that the quantities h' and ε' vary together in a manner such that their product maintains a constant value g', one integrates this last expression, assuming the angle χ to be constant, within all the length of the zone enclosed on the surface σ' between the two planes which include the angle $d'\chi$ from one of the borders of the contour s' up to the other. This first integration can be done immediately and gives

$$-\mu gg'dzd'\chi\left(\frac{v_2^2}{r_2^3} - \frac{v_1^2}{r_1^3}\right) ,$$

where r_1, v_1 and r_2, v_2 represent the values of r and of v for the two borders of the contour s'. The two parts of this expression must now be integrated with respect to χ respectively in the two portions of the contour s' determined by the two planes tangent to this contour drawn by the ordinate y of the element ds; and after our remark on page 145,[368] with respect to the value of the force parallel to x in the calculation relative to the two surfaces terminated by the contour s, it is easy to see that we will have here

$$-\mu gg'dz\int\frac{v^2d'\chi}{r^3} ,$$

in taking this integral over the total extent of the closed contour s'; the variables r, v and χ are only relative to this contour.

[368][N. T.] Pages 145 and 317 of the 1826 and 1827 versions of the *Théorie*, respectively, corresponding to our page 434.

[pages 192 and 364]

One will execute in the same manner the double integration of the other term which is equal to

$$-\mu g \varepsilon' dy d^2\sigma' \delta' \frac{z' - z}{r^3} ,$$

over the entire extent of the surface σ'. It is necessary, for this, to divide this surface into an infinity of zones, by planes drawn by the coordinate z of the center of the element ds, and take, on one of these zones, for $d^2\sigma'$ an infinitely small area which has as its expression $wd'wd'\psi/\cos\zeta'$. The formula, after having been transformed as previously, will be integrated first over all the length of the zone; the integral will only contain quantities relative to the contour s'. Then the second integration with respect to ψ over the extent of the closed contour s', will give

$$\mu g g' dy \int \frac{w^2 d'\psi}{r^3} .$$

Finally, bringing together the two results obtained by these double integrations, one will obtain

$$\mu g g' \left(dy \int \frac{w^2 d'\psi}{r^3} - dz \int \frac{v^2 d'\chi}{r^3} \right) ,$$

for the value of the force parallel to [the axis of the] x, whose direction passes through the middle of the element ds, and which comes from the action of the two surfaces terminated by the contour s' on the two surfaces terminated by the contour s.

One obtains similarly, parallel to the two other axes, the forces

$$\mu g g' \left(dz \int \frac{u^2 d'\varphi}{r^3} - dx \int \frac{w^2 d'\psi}{r^3} \right) ,$$

[pages 193 and 365]

[and]

$$\mu g g' \left(dx \int \frac{v^2 d'\chi}{r^3} - dy \int \frac{u^2 d'\varphi}{r^3} \right) .$$

Thus, by assuming the application to each element ds of the contour s of the forces which we just determined, one obtains the action which results from the attractions and repulsions of the two magnetic fluids, distributed and fixed on the two assemblies of surfaces terminated by the two contours s and s'.

29.24 Impossibility of Producing an Indefinitely Accelerating Movement by the Interaction of Two Rigid and Closed Voltaic Circuits and, Consequently, by the Interaction of Any Two Assemblages of Circuits of This Kind

But these forces applied to elements ds only differ in sign from those that we have obtained on page 139,[369] for the action of two circuits s and s', by assuming they are traversed by electric currents, provided we have $\mu g g' = ii'/2$. This [sign] difference arises because, in the calculation that we gave, the differentials $d'\varphi$, $d'\chi$ and $d'\psi$ were assumed to have the same sign as the differentials $d\varphi$, $d\chi$ and $d\psi$, whereas they should be taken with opposite signs when the two currents move in the same direction; in that case the forces produced by the mutual action of these currents are exactly the same as those which result from the action of the two surfaces σ' on the two surfaces σ, and it is thus completely demonstrated that the mutual action of two rigid and closed circuits, carrying electric currents, can be replaced by those of two assemblages each composed of surfaces having as contours these two circuits, and on which are fixed molecules of austral and boreal fluids attracting and repelling following the straight lines which joins them, in inverse ratio of the squares of the

[369][N. T.] Pages 139 and 311 of the 1826 and 1827 versions of the *Théorie*, respectively, corresponding to our page 431.

distances. Combining this result with this rigorous consequence of the general principle of the conservation of living forces, already referred to several times in this Treatise, [namely,] that [pages 194 and 366] all action which is reducible to forces, functions of distances, acting between the material points which form two rigid systems, one fixed, the other mobile, can never give rise to a movement which continues indefinitely, despite the resistances and the frictions which affect the mobile system, we thus conclude, as we did when one was dealing with a magnet and a closed and rigid voltaic circuit, that this sort of movement can never result from the *mutual action of two rigid and closed circuits.*

Instead of substituting for each circuit two surfaces very close to each other, one covered by austral fluid and the other by boreal fluid, these fluids being distributed as stated above, one could replace each circuit by a single surface on which are uniformly distributed magnetic elements[370] such as were defined by M. Poisson, in the memoir read to the *Académie des Sciences* on 2 February 1824.[371]

The author of this memoir, in calculating the formulas by which he included into the field of analysis all issues related to the magnetization of bodies, whatever the cause that one assigns to them, has given[372, 373] the values of the three forces exerted by a magnetic element on a molecule of austral or boreal fluid;[374] these values are identical to those that I deduced from my formula, for the three quantities A, B and C, in the case of a very small closed and planar circuit, when one assumes that the constant coefficients are the same,[375] and it is easy to conclude from this [result] a theorem according to which one sees immediately:

1. That the action of an electrodynamic solenoid, [pages 195 and 367] calculated from my formula, is, in all cases, the same as that [action] of a series of magnetic elements of the same strength, distributed uniformly along a straight or curved line which encloses all the small circuits of the solenoid, by giving, to each of its points, to the axes of the elements, the same direction of this line.[376]

2. That the action of a rigid and closed voltaic circuit, also calculated following my formula, is precisely that [action] which would be exerted by magnetic elements of the same strength, distributed uniformly on an arbitrary surface terminated by the circuit, when the axes of the magnetic elements are everywhere normal to this surface.

The same theorem leads also to this consequence, that if one imagines a surface enclosing on all sides a very small space; assuming, on one part, molecules of austral fluid and of boreal fluid in equal quantities distributed on this small surface, as they should be [distributed] so that they constitute the magnetic element such as that considered by M. Poisson, and, on the other part, the same surface covered by electric currents, forming on this surface small closed circuits in planes parallel and equidistant, and when one calculates the action of these currents from my formula, the forces exerted, in the two cases, whether on an element of conducting wire, whether on a magnetic molecule, are precisely the same, independent of the form of the small surface, and proportional to the volume that they enclose, the axes of the magnetic elements being represented by the line perpendicular to the planes of the circuits.

The identity of these forces once demonstrated, one could consider as being only simple corollaries, all the [pages 196 and 368] results that I have given in this Treatise, on the possibility of substituting for magnets, without changing the produced effects, assemblies of closed electric currents around its particles. I think that it would be easy for the reader to deduce this consequence, and the theorem on which it is based, from the preceding calculations; I have developed it elsewhere in another essay[377] where I discussed at the same time, under this new point of view, all that is relative to the mutual action of a magnet and a voltaic conductor.

---

[370] [N. T.] Each "magnetic element" is composed of an austral pole and a boreal pole, of the same intensity, separated by a small fixed distance. That is, each magnetic element is a magnetic dipole.

[371] [N. T.] See [Poisson, 1822a]. We discussed Poisson's work in Section 10.3.

[372] [N. A.] *Mémoire sur la théorie du magnétisme*, by M. Poisson, page 22.

[373] [N. T.] See [Poisson, 1822a, p. 268].

[374] [N. T.] The three components of the force exerted by a magnetic dipole acting on another magnetic pole calculated by Poisson are given by equations (10.48) up to (10.50).

[375] [N. T.] See the equations for the magnitudes A, B and C in our pages 377 and 405, which are equivalent to pages 55 and 94 of the 1826 version of the *Théorie*, and to pages 227 and 266 of the 1827 version of the *Théorie*, respectively.

[376] [N. T.] We can imagine this line as passing through the centers of all the small circuits composing the solenoid.

[377] [N. T.] See [Ampère, 1827a], [Ampère, 1828] and [Ampère, 1887e]. This work is followed by Ampère's letter to Gherardi, [Ampère, 1827c], [Ampère, 1887a] and [Ampère, 1887b].

29.25 Experiment which Has Just Confirmed the Theory which Attributes the Properties of Magnets to Electric Currents, Proving that a Spiral or Helical Conducting Wire Carrying a Current, Suffers, from a Moving Metallic Disc, an Action Totally Similar to that Discovered by M. Arago between This Disc and a Magnet

While I was writing this [Treatise], M. Arago discovered a new type of action on magnets.[378] This discovery, equally as important as unexpected, consists of the mutual action which develops between a magnet and a disk or ring of any substance, when the relative positions continually change. M. Arago had the idea that one should be able, in this experiment, to substitute a conductor wound in a helix for the bar magnet and he engaged me to verify this conjecture by an experiment the success of which could not be doubted. Defects in the equipment that I used, with M. Arago, to verify the existence of this action prevented us from obtaining a decisive result; but, M. Colladon[379] having agreed to improve the equipment that we used, I verified with him in a complete manner, today 30 August 1826, M. Arago's idea, by use of a very short double helix which has turns of about two inches in diameter.[380]

This experiment completes the identity of the effects produced, whether by magnets, or by assemblies of [pages 197 and 369] rigid and closed voltaic circuits;[381, 382] it demonstrates that the series of decompositions and recompositions of the [pages 198 and 370] neutral fluid, which constitutes the electric current, suffices to produce, in this case as in all the others, the effects that one ordinarily explains by the action of two fluids different from electricity, and that one designates by the names *austral fluid* and *boreal fluid.*

---

[378][N. T.] François Arago (1786-1853) presented his discovery at the Academy of Sciences of Paris in the meetings of 22 November 1824, 7 March 1825 and 3 July 1826, [Arago, 1824], [Arago, 1825] and [Arago, 1826]. These papers were also included in his complete works, [Arago, 1854b].

[379][N. T.] Jean-Daniel Colladon (1802-1893), Swiss physicist.

[380][N. T.] See [Ampère, 1826d] and [Ampère, 1827b].

[381][N. A.] It seems at first that this identity ought only to take place with respect to closed circuits of very small diameter; but it may readily be seen that it is also true of circuits of arbitrary magnitude since, as we have seen, they may be replaced by magnetic elements [that is, they may be replaced by magnetic dipoles] distributed uniformly over surfaces terminated by these circuits, whilst the number of surfaces that a particular circuit circumscribes can be multiplied as you please. The set of surfaces may be regarded as a system of magnets which are equivalent to the circuit. The same consideration proves that without in any way affecting the resulting forces, the infinitesimal currents which encircle the particles of a bar magnet can always be replaced by currents of finite dimension, these currents forming closed circuits about the axis of the bar when those of the particles are distributed symmetrically about this axis. For this it is sufficient to imagine surfaces within the bar terminating at the surface of the magnet and cutting the lines of magnetization everywhere at right angles and passing through the magnetic elements which can always be assumed to be placed at the points where these lines are met by the surfaces. Then, if all the elements of a particular surface are of equal intensity on equal areas, they can be replaced by a single current flowing through the curve formed by the intersection of this surface and that of the magnet. If they should vary, increasing in intensity from the surface to the axis of the magnet, they would first be replaced by a current at this intersection such as ought to be according to the *minimum* intensity of the particulate currents of the surface normal to the lines of magnetization under consideration, and then, for each line circumscribing the portions of this surface where the little currents become more intense, a new current should be imagined which is concentric to the previous one as required by the difference in intensity of the adjacent currents, some outward and the others inward of this line. If the intensity of the particulate currents decreases from the surface to the axis of the bar, a corresponding concentric current should be imagined on the separation line in the opposite sense. Finally, an increase of intensity which might follow the decrease would require a new concentric current directed as in the first case.

These comments are only given here so as not to omit a remarkable conclusion which may be drawn from the results of this Treatise; they are in no way intended to corroborate the supposition that the electric currents of magnets form closed circuits about their axes. Having at first hesitated between this supposition and the other way of regarding currents as encircling the particles of magnets, I have recognized for a long time that this latter concept best fitted all the facts and in this respect my opinion has not changed at all.

Moreover, this conclusion is useful in that it identifies the actions produced by an electrodynamic helix, on the one hand, or by a magnet, on the other, just as completely from the point of view of theory as when verified by experiments, and in the aspect that it justifies the explanations in which one substitutes, as I did in that explanation given above on the motion of revolution of a floating magnet, the magnet being considered by a single closed circuit.

[382][N. T.] In this note Ampère utilized the expression *courants particulaires* to designate the molecular currents around the particles of the magnet. This expression has been translated here as particulate current due to the reasons discussed in our Sections 5.2 up to 5.4. Blunn, on the other hand, utilized the following translation: particular currents, [Ampère, 1965b, note on pp. 199-200].

29.26 General Consequences of these Experiments and Calculations Relative to Electrodynamic Phenomena

After long reflection about these phenomena and after the ingenious explanation that M. Poisson has recently given[383] for the new kind of action discovered by M. Arago, I think that what we can accept as most likely in the present state of science, consists of the following propositions.

1. Without our being allowed to reject explanations based on the reaction of the ether set in motion by electric currents, there is no need, up to now, to resort to them.

2. The molecules of the two electric fluids, distributed [pages 199 and 371] on the surfaces of conductors, on the surface or in the interior of non conducting bodies, and at rest at points of these bodies where they are located, whether in equilibrium in the first case, or whether due to the fact that they are held fixed in the second case by the coercive force of the non conducting bodies, produce, by their attractions and repulsions reciprocally proportional to the square of the distances, all the phenomena of ordinary electricity.[384]

3. When the same molecules move in conducting wires, meeting there in neutral fluid and separating there at every moment, there results from their mutual action forces that depend first on the duration of extremely short periods between two consecutive meetings or separations, next on the directions following which these compositions and decompositions of the neutral fluid take place. The forces so produced are constant as soon as this dynamic state of the electric fluids in the conducting wires becomes permanent; it is these [forces] that produce all the phenomena of attraction and repulsion that I have discovered between two such wires.

4. The action, whose existence I found, between the Earth and voltaic conductors, makes it difficult to doubt that there are currents, similar to those of conducting wires, in the interior of our Earth. One can presume that these currents are the cause of its internal heat; that they occur mainly there where the oxidized layer surrounding it on all sides rests on a metallic core, in accordance with the explanation that Sir H. Davy gave of volcanoes, and it is they that magnetize magnetic minerals and bodies exposed under the right circumstances for the [pages 200 and 372] electrodynamic action of the Earth. However, any unanswerable proof that terrestrial currents are not solely established around the particles of the globe does not exist, and cannot exist, according to the identity of effects explained in the previous note.

5. The same permanent electrodynamic state consisting of a series of decompositions and recompositions of the neutral fluid which take place in conducting wires, exists around the particles of magnetized bodies, and produces there actions similar to those exerted by these wires.

6. In calculating these actions according to the formula that represents the [force between] two elements of voltaic currents, we find specifically, for the forces that result either when a magnet acts on a wire, or when two magnets interact with each other, the values that were produced from the latest of M. Biot's experiments in the first case,[385] and those of Coulomb in the second.

7. This identity, purely mathematical, confirms the most comprehensive view, based also on the body of all the facts, that the properties of magnets are actually due to the continual movement of the two electric fluids around their particles.

8. When the action of a magnet, or of a conducting wire, creates this movement around the particles of a body, the molecules of positive electricity and of negative electricity, which must be in the permanent electrodynamic state from which result the actions which it then exerts, whether on a wire or on a magnetized body, can only reach this state after a very short time, but which is never zero, and whose duration depends in general on the [pages 201 and 373] resistance that the body opposes to the movement of the electric fluids which it contains. During this movement, before reaching a state of constant motion, or when this state stops, they must exert forces that most probably produce the

[383][N. T.] See [Poisson, 1826] and [Poisson, 1823].

[384][N. T.] That is, they produce all common electrostatic phenomena.

[385][N. T.] In Biot's experiments a current-carrying wire acted on a magnet. He did not study the action of a magnet on a current-carrying wire.

> singular effects that M. Arago has discovered.[386] This explanation is, moreover, that of M. Poisson, applied to my theory, because an electric current forming a very small closed circuit acting precisely as two molecules, one of austral fluid, another of boreal fluid, located on its axis, on either side of the plane of the small current, at distances of these planes which are equal to each other, and as much larger when the electric current has more intensity, we must necessarily find the same values for the forces that develop, whether when it is assumed that the current gradually arises or ceases to exist, whether when one imagines that the magnetic molecules, initially grouped in neutral fluid, separate from one another, successively moving away to greater distances, then approaching to meet again.

In finishing this Treatise I think that I should observe that I have not had time to build the instruments shown in figure 4 of the first plate and figure 20 of the second plate.[387] Therefore, the experiments for which they are intended have not yet been done,[388] but since these experiments are only designed to verify results obtained by other means and that, on the other hand, it would be useful to perform them as a counterproof of those that provided these results, I have not thought it necessary to remove the description.

[386][N. T.] This was a very important insight by Ampère, namely, that the phenomena associated with Arago's disc should be associated with the acceleration of charges.

[387][N. T.] These figures appear in our pages 355 and 383, respectively.

[388][N. T.] See our discussion of this topic in Section 10.5.

Chapter 30

Notes [of the *Théorie* Published in 1826] on Different Subjects Considered in This Treatise

[Page 202][1]

30.1 On the Method of Demonstrating Using the Four Cases of Equilibrium Explained at the Beginning of This Treatise, that the Value of the Mutual Action of Two Elements of Conducting Wires is $-\frac{2ii'}{\sqrt{r}}\frac{d^2\sqrt{r}}{dsds'}dsds'$

Following[2] in order the transformations that I successively applied to this expression, one finds first, due to the first two cases of equilibrium,[3] that the expression is

$$\frac{ii'(\sin\theta\sin\theta'\cos\omega + k\cos\theta\cos\theta')dsds'}{r^n};$$

one deduces from the third [case of equilibrium],[4] between n and k, the relation $n + 2k = 1$, and from the fourth[5] [one deduces that] $n = 2$, from which $k = -1/2$; this fourth case of equilibrium is therefore the one employed in the last place for the determination of the value of the force which develops between two elements of conducting wires: but one can follow a different path using a consideration provided by M. de Laplace, as he concluded from M. Biot's first experiments, on the mutual action between a magnet and an indefinite rectilinear conductor, which showed that the force exerted by an element of this wire on one of the poles of the magnet varies inversely with the square of the distance, if the distance only changes in value and the angle between the measured straight line and the direction of the element stays the same. [page 203] In applying this consideration to the mutual action of two elements of conducting wires, it is easy to see, independently of any preliminary research on the value of the resulting force, that this force is also inversely proportional to the square of the distance when only it is varied, and the angles that determine the relationship between the elements are unchanged. In effect, based on the consideration developed at the beginning of this Treatise, the force in question here is necessarily directed along the line r, and has the value

[1][N. T.] As discussed in Section 27.2, we believe that these Notes represent Ampère's final points of view, although published in 1826. The earlier version of these Notes appeared in the *Théorie* published in 1827. They can be found in Chapter 31.

[2][N. T.] Due to a misprint, the formula in the title of this Section appeared as $-\frac{2ii'}{\sqrt{r}}\frac{dsds'}{d^2r}dsds'$. We included the correct expression of this formula in the title of this Section, namely, $-\frac{2ii'}{\sqrt{r}}\frac{d^2\sqrt{r}}{dsds'}dsds'$.

[3][N. T.] Namely, the case of equilibrium of the anti-parallel currents and the case of equilibrium of the sinuous wire.

[4][N. T.] Namely, the case of equilibrium of the nonexistence of tangential force.

[5][N. T.] Namely, the case of equilibrium of the law of similarity.

$$ii'f(r,\ \theta,\ \theta',\ \omega)dsds'\ ;$$

from which it follows, defining α, β and γ to be the angles that this straight line forms with the three [orthogonal Cartesian] axes, its three components are expressed by

$$ii'f(r,\ \theta,\ \theta',\ \omega)\cos\alpha dsds'\ ,$$

$$ii'f(r,\ \theta,\ \theta',\ \omega)\cos\beta dsds'\ ,$$

$$ii'f(r,\ \theta,\ \theta',\ \omega)\cos\gamma dsds'\ ,$$

and the three forces parallel to the three axes which result [from these components] between two circuits [will be obtained] by the double integrals of these expressions, i and i' being constants.

Now it follows from the fourth case of equilibrium, by replacing the three circles by any similar curves such that their homologous dimensions are in continuous geometric proportion, that these three forces have equal values in two similar systems; it is thus necessary that the integrals which express them have null dimension relative to all the lines which enter there, following the remark by M. de Laplace which I just remembered and that, by consequence, the same should also happen for the differentials of which they are composed, considering ds and ds' among the lines which are included [in these expressions], because the number of these differentials, though [page 204] infinite of second order,[6] should be considered as the same in the two systems.

Now the product $dsds'$ is two dimensional: it is then necessary that[7] $f(r,\theta,\theta',\omega)\cos\alpha$, $f(r,\theta,\theta',\omega)\cos\beta$ and $f(r,\theta,\theta',\omega)\cos\gamma$ are of dimension -2; and since the angles θ, θ', ω, α, β and γ are expressed by numbers which contribute nothing in the dimensions of the values of the differentials, and since $f(r,\ \theta,\ \theta',\ \omega)$ only contains the single line r, it is necessary that this function is proportional to $1/r^2$, so that the force applied from one to the other of the two elements of the conducting wires is given by

$$\frac{ii'\varphi(\theta,\ \theta',\ \omega)}{r^2}dsds'\ .$$

The first two cases of equilibrium then determine the function φ, where only k remains unknown, and one has

$$\frac{ii'\left(\sin\theta\sin\theta'\cos\omega + k\cos\theta\cos\theta'\right)}{r^2}dsds'\ ,$$

for the value of the sought force: it is, as is known, in this form that I presented it in the memoir that I read before the *Académie* on 4 December 1820.[8] By replacing then $\sin\theta\sin\theta'\cos\omega$ and $\cos\theta\cos\theta'$ by their values[9]

$$-\frac{rd^2r}{dsds'}\quad\text{and}\quad -\frac{dr}{ds}\frac{dr}{ds'}\ ,$$

one obtains[10]

[6][N. T.] The line integral along the closed circuit s can be considered as a sum of an infinite number of terms, each one proportional to the infinitesimal length ds. The same reasoning can be applied to the line integral along the circuit s' as regards the infinitesimal length ds'. The force of a closed circuit s acting on another closed circuit s' will be then given by a product between an infinite sum of terms proportional to ds and another infinite sum of terms proportional to ds'. This will yield the "infinite of second order" mentioned by Ampère.

[7][N. T.] Due to a misprint in the 1826 version of the *Théorie*, the last cosine appeared as $\cos\alpha$.

[8][N. T.] This work, read on 4 December 1820, was published in [Ampère, 1820e]. The manuscript containing the material that Ampère presented in this date was published by Joubert, [Ampère, 1885j]. Ampère presented another version of this work in [Ampère, 1820d].

[9][N. T.] Due to a misprint, the original version of the *Théorie* stated that the expression $\sin\theta\sin\theta'\cos\omega$ was equivalent to $-\frac{rd^2r}{dsds'}dsds'$.

[10][N. T.] Due to a misprint, the first line of the following equation appeared in the original text as follows:

$$-\frac{ii'}{r^2}\left(\frac{d^2r}{dsds'} + k\frac{dr}{ds}\frac{dr}{ds'}\right)dsds' =$$

$$-\frac{ii'}{r^2}\left(\frac{rd^2r}{dsds'}+k\frac{dr}{ds}\frac{dr}{ds'}\right)dsds'$$

$$=-\frac{ii'(rdd'r+kdrd'r)}{r^2}=-\frac{ii'r^kdd'r+kr^{k-1}drd'r}{r^{k+1}}$$

$$=-\frac{ii'd(r^kd'r)}{r^{k+1}}=-\frac{ii'dd'(r^{k+1})}{(k+1)r^{k+1}}\ ,$$

[page 205]

and simplifying by substituting $k+1=m$, one has this very simple expression for the sought force:

$$-\frac{ii'dd'(r^m)}{mr^m}\ ,$$

and there only remains to prove that $r^m=\sqrt{r}$, that is, that the constant number m is equal to 1/2.

The experiment[11] described on pages[12] 22–25, which I utilized in this Treatise to determine the value of k and, consequently, the value of $m=k+1$, is not much susceptible of precision due to the friction of the arc AA′ (figure 3)[13] with the mercury contained in the two troughs M and M′, and [due] to the difficulty that one faces in preventing that the repulsion taking place between the arc and the mercury, when the electric current traverses them, do not separate them well enough from each other to interrupt the [electric] communication.[14] Initially I deduced the value of k from another experiment[15] which did not present the same inconveniences, due to the fact that the mobile portion of the voltaic circuit had its two extremities along the vertical axis around which it was subject to rotate, the friction of the mercury took place only against the surface of two tips rotating around themselves, which rendered it essentially null and, moreover, the tips could not separate themselves from the mercury in which they were immersed; this experiment has also the advantage that it does not require a particular device, but only that [instrument] used to make all the other electrodynamic experiments, and which is described and represented in the work which I published in 1825, by the publisher Bachelier, [at the address] *quais des Augustins* number 55, with the title *Description d'un appareil électro-dynamique*, 2nd edition.[16] One can see in this description, pages 19 and 20, how one performs the experiment which is in question here; its goal is to verify that a mobile portion of conducting wire whose two extremities are located along the vertical axis around which it turns freely, [page 206] cannot move continuously around this axis due to the action of a horizontal circular conductor whose center is located in the same axis. Later on I abandoned this way of determining the value of k, because the calculation which I utilized to deduce its value supposed valid for each one of the elements of the circular conductor what the experiment only demonstrated as regards the totality of this conductor. Later on I recognized that beginning with the nullity of action of the circular conductor [acting] on a rectangular conductor whose two sides are vertical, which is the most convenient form for the experiment, one can, by means of a transformation, which will be the subject of the next Note, determine directly the value of m and, consequently, that of $k=m-1$; what dispenses the use of the instrument presented in figure 3 of plate I, and [dispenses also the use of] the experiment not much susceptible of precision to which it was intended.

---

[11] [N. T.] This final paragraph did not appear in Note 1 of the 1827 version of the *Théorie*. What appeared in the earlier version of 1827 goes from our page 479 up to our page 479.

[12] [N. T.] Pages 22–25 of the *Théorie* published in 1826, corresponding to pages 194–197 of the 1827 version of the *Théorie*, and to our pages 352–354.

[13] [N. T.] This figure appeared in our page 353, being reproduced here in our page 464.

[14] [N. T.] The experimental conditions of this device were greatly improved by Ettingshausen (1796-1878) when he replaced the bar or support QO of figure 3, connected to the arc AA′, by a bifilar suspension, [Ampère, 1887d, Note 2 by Joubert on pp. 17-18], [Blondel, 1982, Note 27, p. 148], [Ettingshausen, 1878a], [Ettingshausen, 1878b] and [Ettingshausen, 1879].

[15] [N. T.] Namely, the case of equilibrium of the nonexistence of continuous rotation, see [Ampère, 1822o], [Ampère, 1822y], [Ampère, 1885p] and our Section 7.5.

[16] [N. T.] See [Ampère, 1824a], [Ampère, 1824b] and [Ampère, 1826a].

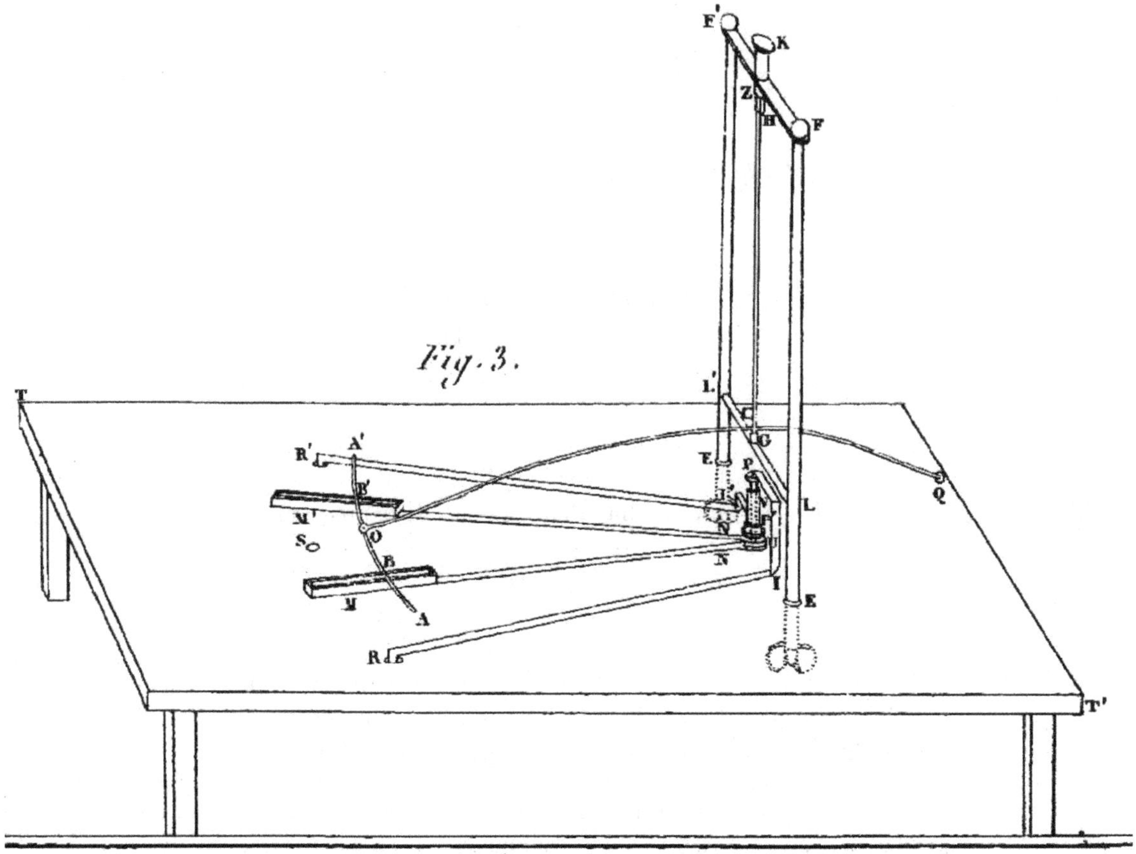

Figure 3.

30.2 On a Proper Transformation which Simplifies the Calculation of the Mutual Action between Two Rectilinear Conductors

When the two conductors are rectilinear, the angle formed by the directions of the two elements is constant and equal to that [angle] between the same directions of the two conductors; it is therefore supposed to be known, and one has, designating it by ε, [from] page 35,[17]

$$r\frac{d^2r}{dsds'}+\frac{dr}{ds}\frac{dr}{ds'}=-\frac{dx}{ds}\frac{dx'}{ds'}-\frac{dy}{ds}\frac{dy'}{ds'}-\frac{dz}{ds}\frac{dz'}{ds'}=-\cos\varepsilon\ ,$$

from which it follows that[18]

$$\frac{dd'(r^m)}{mr^m}=\frac{(m-1)drd'r+rdd'r}{r^2}=\frac{(m-2)drd'r-\cos\varepsilon dsds'}{r^2}\ .$$

By designating p some other arbitrary exponent, one has equivalently

[page 207]

$$\frac{dd'(r^p)}{pr^p}=\frac{(p-2)drd'r-\cos\varepsilon dsds'}{r^2}\ ,$$

[17] [N. T.] Page 35 of the 1826 version of the *Théorie*, corresponding to page 207 of the 1827 version, and to our page 360.

[18] [N. T.] Due to a misprint in the original version, the second and third terms of this equation appeared as $\frac{(m-1)drdr'+rdd'r}{r^2}=\frac{(m-2)drdr'-\cos\varepsilon dsds'}{r^2}$.

and, by eliminating[19] $\frac{drd'r}{r^2}$ between these two equations, one obtains

$$\frac{(p-2)dd'(r^m)}{mr^m} - \frac{(m-2)dd'(r^p)}{pr^p} = \frac{(m-p)\cos\varepsilon ds ds'}{r^2} ,$$

from which

$$\frac{dd'(r^m)}{mr^m} = \frac{m-2}{p-2}\frac{dd'(r^p)}{pr^p} + \frac{m-p}{p-2}\frac{\cos\varepsilon ds ds'}{r^2} .$$

By multiplying[20] the two members of this equation by $-ii'$, one obtains an expression for the mutual action between two elements of voltaic conductors in which one can assign the value that one wants to the indeterminate constant p; this expression is

$$-ii'\frac{dd'(r^m)}{mr^m} = -ii'\left[\frac{m-2}{p-2}\frac{dd'(r^p)}{pr^p} + \frac{m-p}{p-2}\frac{\cos\varepsilon ds ds'}{r^2}\right] .$$

30.3 Application of This Transformation to the Determination of the Constant m which Appears in the Formula by which One Expresses the Force which Two Elements of Conducting Wires Exert on Each Other, and to the Determination of the Value of This Force which should be Utilized when One Wishes to Calculate the Effects Produced by the Mutual Action between Two Rectilinear Conductors

Initially[21] one should apply the formula just obtained to the determination of the value of [the constant] m, starting from the experiment which proves that a rectangular mobile conductor whose two sides are vertical does not acquire any motion when it is under the action of a horizontal circular conductor and when it can only turn around the axis of the circle [page 208] whose circumference is formed by this last conductor. To this end, by performing one of the two differentiations indicated in the value just found for the force exerted by the element ds of the mobile conductor on the element ds' of the circular conductor, one puts it[22] in this form[23]

$$-ii'\left[\frac{m-2}{p-2}\frac{d(r^{p-1}d'r)}{r^p} + \frac{m-p}{p-2}\frac{\cos\varepsilon ds ds'}{r^2}\right] ;$$

then one takes its component along the tangent of the circular conductor by multiplying it by $\cos\theta'$, and one replaces[24] $d'r$ by its value $-ds'\cos\theta'$, yielding, for the expression of this component, [the following value:]

$$ii'ds'\left[\frac{m-2}{p-2}r^{-p}\cos\theta' d(r^{p-1}\cos\theta') - \frac{m-p}{p-2}\frac{\cos\theta'\cos\varepsilon ds}{r^2}\right] ,$$

where p can have the value one wishes.

Multiplying the expression of the component by the radius of the circumference, which I will call a, according to which the fixed conductor is bent, one will have that [expression] of the torque of the action which the element ds would exert on ds' to make it turn around the axis, if this last element were mobile, from which it follows that by changing the sign of the product one will obtain the value of the torque resulting from the action of ds' to make ds turn around the same axis. As [the value of] p can be taken at will, one will make it simpler by assuming $-p = p-1$, or $p = 1/2$; one then has

---

[19][N. T.] Due to a misprint, this term appeared in the original version of the *Théorie* as $\frac{drdr'}{r^2}$.

[20][N. T.] The final portion of this second Note of the *Théorie* published in 1826 is different from the final portion of the second Note of the *Théorie* published in 1827; see our page 480.

[21][N. T.] This third Note of the *Théorie* published in 1826 was not included in the Notes of the *Théorie* published in 1827.

[22][N. T.] That is, one puts the value of the force just found in this form.

[23][N. T.] Due to a misprint in the 1826 version of the *Théorie*, the first fraction of this term appeared in the form $\frac{m-2}{p-2}\frac{d(r^{p-1}dr')}{r^p}$.

[24][N. T.] Due to a misprint in the original text, we had here dr' instead of $d'r$.

$$r^{-p}\cos\theta' d(r^{p-1}\cos\theta') = \frac{\cos\theta'}{\sqrt{r}} d\frac{\cos\theta'}{\sqrt{r}} = \frac{1}{2} d\frac{\cos^2\theta'}{r} ,$$

and the expression of the torque takes the form

$$aii'ds'\left(\frac{m-2}{3}d\frac{\cos^2\theta'}{r} - \frac{2m-1}{3}\frac{\cos\theta'\cos\varepsilon ds}{r^2}\right) .$$

By integrating it in relation to the differentials assigned by the [page 209] symbol d, which are relative to the mobile portion of the conducting wire, and denominating by r_1, r_2, $\cos\theta_1'$ and $\cos\theta_2'$, the values of r and of $\cos\theta'$ at the two extremities of this portion, one will obtain, for that [value] of the torque by means of which it [that is, the mobile portion] tends to turn around the axis due to the action of the element ds', [the following expression:][25]

$$aii'ds'\left[\frac{m-2}{3}\left(\frac{\cos^2\theta_2'}{r_2} - \frac{\cos^2\theta_1'}{r_1}\right) - \frac{2m-1}{3}\int\frac{\cos\theta'\cos\varepsilon ds}{r^2}\right] .$$

As the straight lines drawn from all the points of the axis to the center of the element ds' of the circular conductor are orthogonal to the direction of this element, it is evident that one will have, when the two extremities of the mobile conductor are on the axis, $\cos\theta_1' = 0$ and $\cos\theta_2' = 0$, and thus the preceding value reduces to[26]

$$-\frac{(2m-1)aii'ds'}{3}\int\frac{\cos\theta'\cos\varepsilon ds}{r^2} = \frac{(2m-1)aii'ds'}{3}\int\frac{dr}{ds'}\frac{\cos\varepsilon ds}{r^2} .$$

The integral appearing in this expression should be taken for the whole contour of the rectangle formed by the mobile conductor, that is to say, for the four portions of this conductor which are the four sides of the rectangle; but initially, for the two vertical portions, the angle ε comprised between the directions of the horizontal element ds' and of the elements of which they [that is, the vertical portions] are composed is evidently a right angle, therefore the factor $\cos\varepsilon$ goes to zero, this also nullifies the integral itself in these two portions, and there only remains, consequently, to calculate the parts of the integral relative to the two horizontal portions.

Suppose that the circumference L′M′L″M″ (figure 23)[27] represents

[page 210]

the horizontal conductor, that the axis is drawn from the center O of this circle perpendicularly to the plane of the figure, that the two horizontal portions of the mobile conductor are projected from b to c on the radius OL,[28] and that P is the projection of the centers, located in the same vertical, of two equal elements of each of these portions, both represented by ds and situated at a distance OP $= s$ from the center O, considering in the two portions the origin of the s [as being] the points where their directions are met by that [origin] of the axis. Instead of calculating the integral

$$\frac{2m-1}{3}aii'ds'\int\frac{dr}{ds'}\frac{\cos\varepsilon ds}{r^2}$$

separately for each one of these two portions and combining the two results, it is better to make only once, from $s = 0$ and $b = s_1$ to $s = 0$ and $c = s_2$, the integral of the sum of the two torques of the forces exerted by the element ds' on the two elements represented by ds. Naming γ the angle L′OM′, one has $s = a\gamma + C$ and $ds' = ad\gamma$; and since the radius OM is perpendicular to the element ds', and the two horizontal portions of the mobile conductor are traversed in opposite senses by the electric current, it is evident that for that

[25] [N. T.] Due to a misprint, we had r in the denominator of the second fraction instead of r_2.

[26] [N. A.] The same reduction takes place when the mobile conductor forms a closed circuit, because then $r_2 = r_1$ and $\theta_2' = \theta_1'$, yielding

$$\frac{\cos^2\theta_2}{r_2} - \frac{\cos^2\theta_1}{r_1} = 0 .$$

[27] [N. T.] This figure appears in our page 389, being reproduced here in our page 467.

[28] [N. T.] We should have here "on the radius OL′.

Fig. 23.

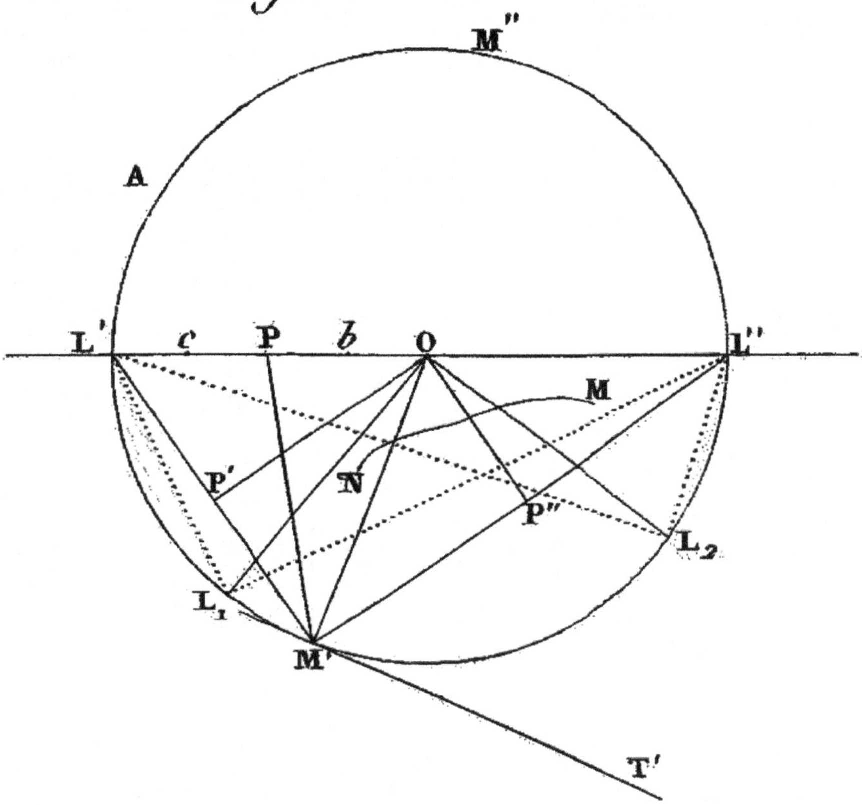

Figure 23.

[portion] where it [the current] is directed towards the axis one should make $\varepsilon = \frac{\pi}{2} - \gamma$, and for the other [portion where the current is directed away from the axis one should make] $\varepsilon = \frac{\pi}{2} + \gamma$.

Denoting by r and r' the distances of the element ds' to the two elements of these portions represented by ds, one will obtain [the following value] for the torque resulting from the action of ds' on that [element] of the first portion when $\cos\varepsilon = \sin\gamma$,

$$\frac{(2m-1)a^2ii'd\gamma}{3}\frac{dr}{ds'}\frac{\sin\gamma ds}{r^2} ,$$

and for the torque impressed, by the same action, to the element of the second [portion] when $\cos\varepsilon = -\sin\gamma$ [one will obtain the following value:]

$$-\frac{(2m-1)a^2ii'd\gamma}{3}\frac{dr'}{ds'}\frac{\sin\gamma ds}{r'^2} .$$

[page 211]

Let h and h' be the distances to the plane of the circular conductor of the two horizontal portions of the mobile conductor, one will have

$$r^2 = h^2 + a^2 + s^2 - 2as\cos\gamma \quad \text{and} \quad r'^2 = h'^2 + a^2 + s^2 - 2as\cos\gamma ;$$

thus

$$r\frac{dr}{ds'}ds' = r'\frac{dr'}{ds'}ds' = as\sin\gamma d\gamma ;$$

and since $ds' = ad\gamma$, one has

$$\frac{dr}{ds'} = \frac{s \sin\gamma}{r} \quad \text{and} \quad \frac{dr'}{ds'} = \frac{s \sin\gamma}{r'} \ .$$

Substituting these values into those that we just obtained for the two torques, one finds that their sum is equal to

$$\frac{(2m-1)a^2ii'}{3}\left(\frac{\sin^2\gamma}{r^3} - \frac{\sin^2\gamma}{r'^3}\right) sdsd\gamma \ .$$

The total torque resulting from the action of the mobile conductor on the circular conductor is equal to the double integral of this expression taken from $\gamma = 0$ to $\gamma = 2\pi$, and then from $s = s_1$ to $s = s_2$; being arbitrary the order in which these two integrations are performed, this [total] torque is then expressed by

$$\frac{(2m-1)a^2ii'}{3}\int_{s_1}^{s_2} sds \int_0^{2\pi}\left(\frac{\sin^2\gamma}{r^3} - \frac{\sin^2\gamma}{r'^3}\right) d\gamma \ ;$$

and since experience proves that it is null, it is necessary that the double integral

$$\int_{s_1}^{s_2} sds \int_0^{2\pi}\left(\frac{\sin^2\gamma}{r^3} - \frac{\sin^2\gamma}{r'^3}\right) d\gamma = 0 \ ,$$

or that $2m - 1 = 0$, this [last expression] yielding for m the value $1/2$, which we proposed to demonstrate to be in fact that [value] of the constant m.

[page 212]

Therefore, there only remains to show that this double integral can never be zero, which is easy to do, since initially the two terms $\frac{\sin^2\gamma}{r^3}$ and $\frac{\sin^2\gamma}{r'^3}$ cannot change their signs no matter the value of γ, because the two distances r and r' should always be considered as positive; then, as these two distances are those of the same element ds' of the circular conductor to two elements equal to ds located in the same vertical at each one of the horizontal portions of the mobile conductor, it is evident that if one supposes, to fix the ideas, that r refers to element ds of that [portion] of these two portions which is [located] at a smaller distance of the plane of the circular conductor, and r' [refers] to the other [element], one will always have $r < r'$ and, consequently,

$$\left(\frac{\sin^2\gamma}{r^3} - \frac{\sin^2\gamma}{r'^3}\right) d\gamma$$

[will be] positive.

All the elements of the first integral being positives, this integral taken from $\gamma = 0$ to $\gamma = 2\pi$ will also be [positive], its product with sds will have the same sign as ds as long as s is positive, that is to say, as long as the rectangle formed by the mobile conductor is entirely on the same side of the axis, as we suppose it here. The sign of ds is determined by the sense of the current in the two horizontal portions of this conductor, and as we assigned different signs to $\cos\varepsilon$ in each portion, ds has necessarily the same sign in one [portion] and in the other; therefore, all the elements on which the second integral is composed from $s = s_1$ to $s = s_2$ have also the same sign, and consequently this integral can never be zero; it is necessary therefore, from what we have just seen, that one has $m = 1/2$ [and] that the mutual action between two elements of electric currents has the value

$$-\frac{2ii'}{\sqrt{r}}\frac{d^2\sqrt{r}}{dsds'}dsds' \ ,$$

[page 213]

and that the torque due to the action of a circular conductor on a conductor mobile around the axis of the circle formed by the first [circular conductor], be always zero when the mobile conductor has its two extremities along this axis or when it forms a closed circuit, which is, as one knows, confirmed by experience, regardless of the form of the contour following which it is bent.[29]

Now that the value of m has been determined, one can substitute 1/2 in place of m in the transformation found, on page 207,[30] and suppose once again an arbitrary value for p; one thus obtains, for the mutual action of two elements ds and ds', the expression[31]

$$-\frac{2ii'dd'\sqrt{r}}{\sqrt{r}}=\frac{\frac{3}{2}ii'}{p-2}\frac{dd'(r^p)}{pr^p}-\frac{\left(\frac{1}{2}-p\right)ii'}{p-2}\frac{\cos\varepsilon dsds'}{r^2} ,$$

and one can, in this expression, assign to p the value we want. That [value] which yields a more convenient result for the calculation is $p=-1$; adopting it, one has

$$-\frac{2ii'dd'\sqrt{r}}{\sqrt{r}}=\frac{1}{2}ii'rdd'\frac{1}{r}+\frac{1}{2}\frac{ii'\cos\varepsilon dsds'}{r^2}=\frac{1}{2}ii'dsds'\left(\frac{\cos\varepsilon}{r^2}+r\frac{d^2\frac{1}{r}}{dsds'}\right) .$$

I have already obtained in another way, on page 81,[32] this expression for the force exerted between two elements of conducting wires; one can only utilize it, to simplify the calculations, when the conductors are rectilinear, since it is only in this configuration that the angle ε is constant and known; but in this case, it is by it [that is, by this expression] that one determines in the easiest way the forces and the torques resulting from the mutual action of two conductors of this type. If I utilized in this Treatise other means to calculate their values, it was because at the time when I wrote it I still did not know the transformation of my formula which I just explained.

[page 214]

30.4 On the Direction of the Straight Line which I Designated under the Name of *Directrix of the Electrodynamic Action at a Given Point*, when This Action is That of a Closed and Planar Circuit in which All of the Dimensions are Very Small

The straight line[33] which I have named *directrix of the electrodynamic action at a given point*[34] is that [line] which forms with the three axes the angles whose cosines are proportional respectively to the three quantities A, B and C; the values of these three quantities, found on page 55,[35] become

$$A=\lambda\left(\frac{\cos\xi}{r^3}-\frac{3qx}{r^5}\right) ,$$

$$B=\lambda\left(\frac{\cos\eta}{r^3}-\frac{3qy}{r^5}\right) ,$$

$$C=\lambda\left(\frac{\cos\zeta}{r^3}-\frac{3qz}{r^5}\right) ,$$

when one substitutes the number 2 for n; therefore, when one assumes the small circuit of arbitrary form located as in figure 14,[36] that is to say, after having placed the origin A of the coordinates at the given point, one takes as the z axis the perpendicular AZ drawn from the point A on the plane of the small circuit, and for the xz plane that [plane] which passes by this perpendicular and by the center of inertia O of the area

[29][N. T.] That is, no matter the shape of the path followed by the mobile conductor, provided it has its two extremities along the axis of the circular conductor or when it forms a closed circuit.

[30][N. T.] Page 207 of the *Théorie* published in 1826, corresponding to our page 465.

[31][N. T.] Due to a misprint, the term at the left hand side of this equation appeared in the form $-\frac{2iidd'\sqrt{r}}{\sqrt{r}}$.

[32][N. T.] Page 81 of the 1826 version of the *Théorie*, corresponding to page 253 of the 1827 version, and to our page 396.

[33][N. T.] The goal of this Note was to obtain the mapping of terrestrial magnetism, as we discussed in Section 10.6.

[34][N. T] See Section 10.1 and the footnote 96 in our page 367.

[35][N. T.] Page 55 of the 1826 version of the *Théorie*, corresponding to page 227 of the 1827 version, and to our page 377.

[36][N. T.] See our pages 373 and 470.

LMS to which is related the x, y and z which enter into the values of A, B and C,[37] it is evident that one has $y = 0$, $q = z$, $\xi = \eta = \pi/2$, $\zeta = 0$, and that these values reduce as a consequence to

$$A = -\frac{3\lambda xz}{r^5} \ , \ B = 0 \ \text{ and } \ C = \lambda\left(\frac{1}{r^3} - \frac{3z^2}{r^5}\right) = \frac{\lambda(x^2 - 2z^2)}{r^5} \ ,$$

because $r^2 = x^2 + z^2$. Since B is null, the directrix AE is necessarily in the xz plane determined as we have just said; it forms with the x axis an angle EAX whose tangent

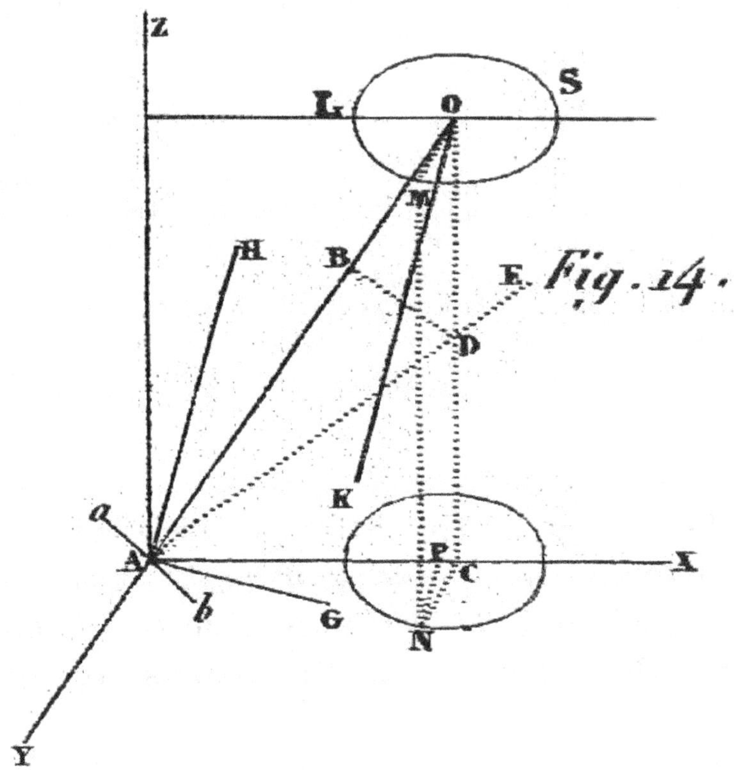

Figure 14.

[page 215]

is equal to C/A, that is, is equal to $(2z^2 - x^2)/3xz$; and as that [tangent] of the angle OAX is [equal] to z/x, one finds, for the value of the tangent of OAE

$$\tan \mathrm{OAE} = \frac{\frac{z}{x} - \frac{2z^2 - x^2}{3xz}}{1 + \frac{2z^2 - x^2}{3x^2}} = \frac{(z^2 + x^2)x}{(2x^2 + 2z^2)z} = \frac{1}{2}\frac{x}{z} = \frac{1}{2}\tan \mathrm{COA} \ .$$

From which it follows that, if one takes OB = OA/3, and one draws on OA at the point B a plane perpendicular to AO which intercepts in D the normal OC on the plane of the small circuit, the straight line ADE determined by the points A and D will be the directrix of the action exerted at the point A by the electric current flowing in it,[38] since one will have

$$\mathrm{AB} = 2\mathrm{OB} \ , \quad \tan \mathrm{BDA} = 2 \tan \mathrm{BDO} \ ,$$

and

$$\tan \mathrm{OAE} = \cot \mathrm{BDA} = \frac{1}{2}\cot \mathrm{BDO} = \frac{1}{2}\tan \mathrm{COA} \ .$$

[37] [N. T.] That is, the coordinates of the center of inertia O of the area LMS are given by $(x,\ y,\ z)$.

[38] [N. T.] That is, flowing through the small closed circuit.

This construction gives in the simplest manner the direction of the straight line AE along which we saw, on page 104,[39] that the pole of a magnet placed in A would be moved by the action of this current. It is to be remarked that it[40] is orientated as regards the plane LMS of the small circuit that it[41] describes, just as the direction of the dip needle is generally [orientated] with respect to the magnetic equator; because the point O being considered as the center of the Earth, the planes LMS and OAC like those of the magnetic equator and meridian, [respectively,] and the straight line AE like the direction of the dip needle, it is evident that the angle OAE comprised between the terrestrial ray OA and the direction AE of the magnetized needle is the complement of the inclination,[42] and that the angle COA is the complement of the magnetic latitude LOA; therefore the preceding equation becomes:[43]

$$\cot \text{incl.} = \frac{1}{2} \cot \text{lat.} \ ,$$

or[44]

$$\tan \text{incl.} = 2 \tan \text{lat.}$$

[page 216]

30.5 On the Value of the Force that an Indefinite Angular Conductor Exerts on the Pole of a Small Magnet, and on the Value of the Force that a Parallelogrammic Conductor Situated in the Same Plane Exerts on This Pole

Whether one considers the pole B (figure 34)[45] of the small magnet AB as the extremity of an electrodynamic solenoid or as a magnetic molecule, there is agreement, in both views, with respect to the expression of the force exerted on this pole by each element of the angular conductor CMZ; one finds in general that in drawing the perpendicular BO $= b$ of the point B on one of its branches CμM extended toward O, setting O$\mu = s$, BM $= a$, B$\mu = r$, the angle BμM $= \theta$, the angle CMH $=$ BMO $= \varepsilon$, and designating by ρ a constant coefficient, the force which is exerted on the pole B by the element ds situated at μ is equal to

$$\frac{\rho \sin \theta ds}{r^2} \ ,$$

which must be integrated from $s =$ OM $= a \cos \varepsilon$ to $s = \infty$, or, what amounts to the same thing, from $\theta = \varepsilon$ to $\theta = 0$; but, in the triangle BOμ, whose side OB $= b = a \sin \varepsilon$, one has

$$r = \frac{a \sin \varepsilon}{\sin \theta} \ , \ s = a \sin \varepsilon \cot \theta \ , \ ds = -\frac{a \sin \varepsilon d\theta}{\sin^2 \theta} \quad \text{and} \quad \frac{ds}{r^2} = -\frac{d\theta}{a \sin \varepsilon} \ ,$$

thus

$$\frac{\rho \sin \theta ds}{r^2} = -\frac{\rho \sin \theta d\theta}{a \sin \varepsilon} \ ,$$

whose integral is

$$\frac{\rho}{a \sin \varepsilon} (\cos \theta + \mathrm{C}) \ ,$$

[39][N. T.] Page 104 of the 1826 version of the *Théorie*, corresponding to page 276 of the 1827 version, and to our page 412.

[40][N. T.] That is, the straight line AE.

[41][N. T.] That is, the current.

[42][N. T.] The complement of an angle φ_1 is the angle φ_2 such that $\varphi_1 + \varphi_2 = \pi/2$.

[43][N. T.] Savary was the first scientist to obtain these relations utilizing Ampère's electrodynamic theory; see [Savary, 1823b, p. 25], [Savary, 1823c, p. 26] and [Savary, 1885b, p. 369].

[44][N. T.] This mathematical relation obtained by Ampère is analogous to our equation (10.56) connecting the inclination angle ζ of a dip needle relative to the horizon with the latitude angle L where the needle is located. According to this expression, which agrees with observational data, the dip needles at the surface of the Earth become orientated as indicated in figure 10.23.

[45][N. T.] See our page 416.

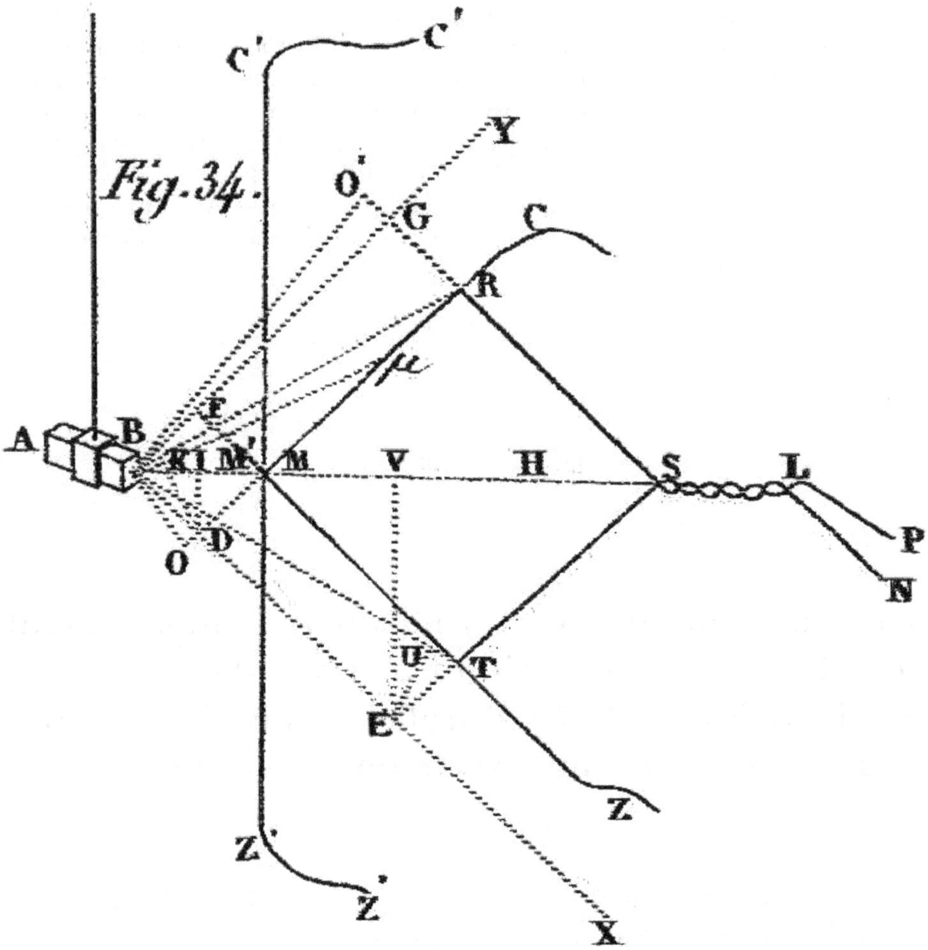

Figure 34.

or, considering it between the limits determined above,

$$\frac{\rho\,(1-\cos\varepsilon)}{a\sin\varepsilon}=\frac{\rho}{a}\tan\frac{1}{2}\varepsilon\ ,$$

[page 217]

a value which is necessary to double in order to obtain the force exerted on the pole B by the indefinite angular conductor CMZ; this force, in inverse ratio of BM $= a$, is therefore, for the same value of a, proportional to the tangent of half the angle CMH, and not to this angle itself, although it may have been claimed that the value

$$\frac{\rho\sin\theta ds}{r^2}$$

of the force exerted by the element ds on the pole B, was found *analyzing by calculation*[46] the supposition that the force produced by the conducting wire CMZ was proportional to the angle CMH. It is not possible to doubt that there was some error in this calculation; but it would be equally curious to know it,[47] a calculation intended to determine the value of a differential beginning with the value of the definite integral obtained between given limits, but it does not seem to me that, up to now, any mathematician had considered [this calculation] something possible to be done.

[46][N. T.] See Section 17.2 in which we discussed Biot's work and Ampère's criticism of it.

[47][N. T.] That is, to know this wrong calculation.

Since one cannot, in practice, make the branches MC and MZ of the angular conductor really infinite, nor extend the prolongations of the wire composing it and which connect these branches with the two extremities of the battery, at a sufficiently great distance from the small magnet AB so that they[48] will effect absolutely no action on it,[49] one should, rigorously, consider the value that we just obtained only as an approximation. In order to have an exact value which can be verified by experiment, it is necessary to calculate the force exerted on the pole B of the small magnet by a conducting wire PSRMTSN, whose portions SP and SN, which communicate with the two extremities of the battery, are covered in silk and twisted together, as one sees in SL, up to close to the battery, so that the actions that they produce cancel each other, and whose remainder [of the circuit] form a lozenge SRMT situated in such a way that the direction of the diagonal SM of this lozenge passes by the point B. To this end, by preserving the preceding names [page 218] and making moreover the angle BRM $= \theta_1$, the angle BRO$' = \theta_1'$, the distance BS $= a'$ and the perpendicular BO$' = b' = -a' \sin\varepsilon$ because the angle BSO$' = -\varepsilon$, one will easily see that the action of the portion RS of the conducting wire on the pole B is equal to

$$-\frac{\rho\,(\cos\varepsilon - \cos\theta_1')}{b'}\ ,$$

as well as, on account of $b = a\sin\varepsilon$, one would have found

$$\frac{\rho\,(\cos\theta_1 - \cos\varepsilon)}{b}\ ,$$

for that [action] that the portion MR exerts on the same pole B, by taking the preceding integral from $\theta = \varepsilon$ to $\theta = \theta_1$; and then, it is sufficient to combine these two expressions, and double the sum, to get the total action of the lozenge contour MRST, yielding

$$2\rho\left(\frac{\cos\theta_1}{b} - \frac{\cos\varepsilon}{b} + \frac{\cos\theta_1'}{b'} - \frac{\cos\varepsilon}{b'}\right)\ .$$

This value is susceptible to another form which one obtains by relating the positions of the four angles of the lozenge to two axes BX and BY drawn by the point B parallel to these sides and which join them at the points D, E, F and G; if one sets BD $=$ BF $= g$ and BE $=$ BG $= h$, one has

$$b = \mathrm{BO} = g\sin 2\varepsilon\ ,\quad b' = \mathrm{BO}' = h\sin 2\varepsilon\ ,$$

$$\cos\theta_1 = \frac{\mathrm{OR}}{\mathrm{BR}} = \frac{h + g\cos 2\varepsilon}{\sqrt{g^2 + h^2 + 2gh\cos 2\varepsilon}}\ ,$$

$$\cos\theta_1' = \frac{\mathrm{O'R}}{\mathrm{BR}} = \frac{g + h\cos 2\varepsilon}{\sqrt{g^2 + h^2 + 2gh\cos 2\varepsilon}}\ ,$$

and using these expressions one finds, for that [expression] of the force exerted on the pole B,

$$2\rho\left(\frac{h + g\cos 2\varepsilon}{g\sin 2\varepsilon\sqrt{g^2 + h^2 + 2gh\cos 2\varepsilon}} + \frac{g + h\cos 2\varepsilon}{h\sin 2\varepsilon\sqrt{g^2 + h^2 + 2gh\cos 2\varepsilon}}\right.$$
$$\left. - \frac{\cos\varepsilon}{g\sin 2\varepsilon} - \frac{\cos\varepsilon}{h\sin 2\varepsilon}\right)$$
$$= \rho\left(\frac{2\sqrt{g^2 + h^2 + 2gh\cos 2\varepsilon}}{gh\sin 2\varepsilon} - \frac{1}{g\sin\varepsilon} - \frac{1}{h\sin\varepsilon}\right)\ ,$$

[page 219]

[48] [N. T.] Ampère utilized here the word "ils." He was referring to the prolongations of the wire (*prolongement* is a masculine word in French). On page 383 of the Notes of the 1827 version of the *Théorie*, equivalent to our page 484, Ampère utilized in the same place the word "elles" instead of "ils." The reason for this replacement was that in 1827 he was referring to the branches of the wire (*branche* is a feminine word in French).

[49] [N. T.] That is, so that the prolongations effect no action on the magnet.

by reducing the first two terms to the same denominator, and by replacing in the two other [terms] $\sin 2\varepsilon$ by its value $2\sin\varepsilon\cos\varepsilon$.

Now draw from the point D the perpendiculars DI and DK on the straight lines BM and BR; the first [perpendicular] will be obviously equal to $g\sin\varepsilon$, and the second is obtained by noticing that in multiplying it by $BR = \sqrt{g^2+h^2+2gh\cos 2\varepsilon}$ one has a product equal to twice the surface of the triangle BDR, that is, $gh\sin 2\varepsilon$, so that by naming these perpendiculars $p_{1,1}$ and $p_{1,2}$, it becomes

$$\frac{1}{p_{1,1}} = \frac{1}{g\sin\varepsilon} \quad \text{and} \quad \frac{1}{p_{1,2}} = \frac{\sqrt{g^2+h^2+2gh\cos 2\varepsilon}}{gh\sin 2\varepsilon} \; ;$$

by drawing from the point E the two perpendiculars EU and EV on the straight lines BT and BS, and representing them by $p_{2,1}$ and $p_{2,2}$, the first [perpendicular] will be equal to DK due to the equality of the triangles BDR and BET, and the second will have the value $h\sin\varepsilon$, in such a way that the expression for the force exerted by the contour of the lozenge MRST on the pole B can be written as:

$$\rho\left(\frac{1}{p_{1,2}} + \frac{1}{p_{2,1}} - \frac{1}{p_{1,1}} - \frac{1}{p_{2,2}}\right) .$$

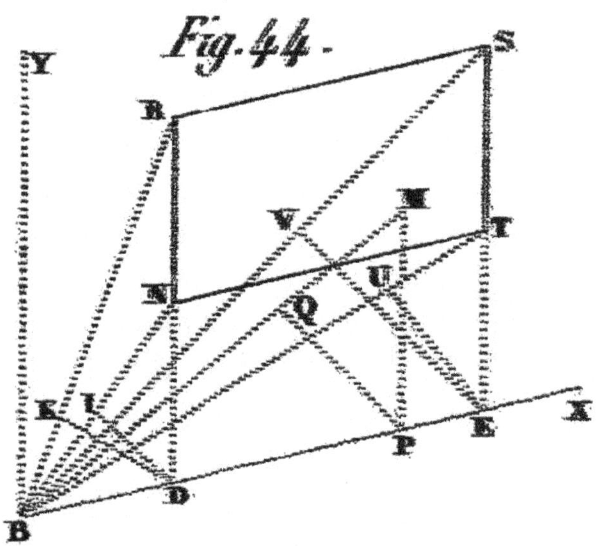

Figure 44.

In this form it applies not only to a lozenge in which one diagonal is directed so as to pass by the point B, but [also] to an arbitrary parallelogram NRST (figure 44) whose perimeter carries an electric current which acts on the pole of a magnet located in the plane of this parallelogram. It results,[50] in effect, as was already said, on page 57,[51] that by calculating the magnitudes designated by A, B, C and $D = \sqrt{A^2+B^2+C^2}$, in relation to a planar and closed voltaic circuit, such as that forming the perimeter of the parallelogram NRST, and [in relation] to a point B situated in the same plane, one has

$$A = 0\,, \quad B = 0 \quad \text{and} \quad C = D = \int\!\!\int \frac{d^2\lambda}{r^3}\,,$$

[page 220]

when one represents by $d^2\lambda$ an element of the area of this circuit and when one replaces the exponent n by its value 2. [The magnitudes] A and B being null, the directrix of the electrodynamic action exerted on the

[50] [N. T.] The text beginning in this paragraph and going up to our page 476 in this Note 5 of the 1826 version of the *Théorie*, is different from the text beginning in our page 485 and going up to our page 487 relative to the Note 4 of the 1827 version of the *Théorie*.

[51] [N. T.] Page 57 of the 1826 version of the *Théorie*, corresponding to page 229 of the 1827 version, and to our page 378.

point B by the current which we are considering is the perpendicular elevated at this point on the plane of the parallelogram, from which it follows:

(1) That the force it[52] exerts on an element ds' of electric current, whose center was located at B, is, in this plane, perpendicular to the direction of the element, and has the value, page 43,[53]

$$\frac{1}{2}Dii'ds'\sin\varepsilon = \frac{1}{2}ii'ds'\cos\omega\int\int\frac{d^2\lambda}{r^3} ,$$

naming ω the inclination of the element ds' on the plane BRST, an inclination which is the complement of the angle ε formed by the direction of this element and that of the directrix;

(2) That, after what was said, on page 103,[54] if one had placed at point B the extremity of an indefinite solenoid, the force exerted on this extremity, by the same electric current, would be perpendicular to the plane BRST and would have the value

$$\frac{\lambda' ii'D}{2g} = \frac{\lambda' ii'}{2g}\int\int\frac{d^2\lambda}{r^3} ,$$

naming λ' the area of the small circuits composing the solenoid and g the distance between the planes of two consecutive circuits [of the solenoid];

(3) That the pole of a magnet located at B would experience from the circuit NRST an action directed along the same perpendicular and expressed by

$$\rho\int\int\frac{d^2\lambda}{r^3} ,$$

ρ being a constant coefficient.

In order to obtain the value of $\int\int\frac{d^2\lambda}{r^3}$, relative to the voltaic circuit represented by the perimeter of the parallelogram NRST, [page 221] one relates all the points, such as M, of its area to two axes BX and BY drawn by the point B parallel to its sides, and by naming x and y the coordinates BP and PM, one will have

$$d^2\lambda = dxdy\sin 2\varepsilon \quad \text{and} \quad r = \sqrt{x^2+y^2+2xy\cos 2\varepsilon} ;$$

the total force, impinging on pole B of the small magnet AB, will then be

$$\rho\sin 2\varepsilon\int\int\frac{dxdy}{(x^2+y^2+2xy\cos 2\varepsilon)^{\frac{3}{2}}} .$$

Now we have seen, on page 94,[55] that the indefinite integral of

$$\frac{dsds'}{(a^2+s^2+s'^2-2ss'\cos\varepsilon)^{\frac{3}{2}}}$$

is

$$\frac{1}{a\sin\varepsilon}\arctan\frac{ss'\sin^2\varepsilon+a^2\cos\varepsilon}{a\sin\varepsilon\sqrt{a^2+s^2+s'^2-2ss'\cos\varepsilon}} ,$$

or[56]

$$-\frac{1}{a\sin\varepsilon}\arctan\frac{a\sin\varepsilon\sqrt{a^2+s^2+s'^2-2ss'\cos\varepsilon}}{ss'\sin^2\varepsilon+a^2\cos\varepsilon} ,$$

by removing the constant $\pi/2$. When $a=0$, this quantity takes the form $0/0$; but since the arc should be replaced by its tangent, the null factor $a\sin\varepsilon$ vanishes, and one has[57]

---

[52][N. T.] That is, the current-carrying circuit.

[53][N. T.] Page 43 of the 1826 version of the *Théorie*, corresponding to page 215 of the 1827 version, and to our page 367.

[54][N. T.] Page 103 of the 1826 version of the *Théorie*, corresponding to page 275 of the 1827 version, and to our page 411.

[55][N. T.] Page 94 of the 1826 version of the *Théorie*, corresponding to page 266 of the 1827 version, and to our page 404, respectively.

[56][N. T.] Due to a misprint, the next equation appeared as follows:

$$-\frac{1}{a\sin\varepsilon}\arctan\frac{a\sin\varepsilon\sqrt{a^2+s^2+s'^2-2ss\cos\varepsilon}}{ss'\sin^2\varepsilon+a^2\cos\varepsilon} ,$$

[57][N. T.] Due to a misprint, the denominator of the integral appeared as: $(s+s'^2-2ss'\cos\varepsilon)\frac{3}{2}$.

$$\int\int \frac{dsds'}{(s^2+s'^2-2ss'\cos\varepsilon)^{\frac{3}{2}}} = -\frac{\sqrt{s^2+s'^2-2ss'\cos\varepsilon}}{ss'\sin^2\varepsilon},$$

which is easy to verify by differentiation. One concludes immediately that the expression of the force that we have calculated, considered as an indefinite integral, is

$$-\frac{\rho\sqrt{x^2+y^2+2xy\cos 2\varepsilon}}{xy\sin 2\varepsilon} = -\frac{\rho}{p},$$

[page 222]

defining p to be the perpendicular PQ drawn from point P on BM, because the double of the area of the triangle BPM is both equal to $p\sqrt{x^2+y^2+2xy\cos 2\varepsilon}$ and to $xy\sin 2\varepsilon$, which gives

$$\frac{1}{p} = \frac{\sqrt{x^2+y^2+2xy\cos 2\varepsilon}}{xy\sin 2\varepsilon}.$$

There only remains now to calculate the values taken by this indefinite integral at the four vertices N, R, T and S of the parallelogram, and to sum them with convenient signs; continuing to designate respectively by $p_{1,1}$, $p_{1,2}$, $p_{2,1}$ and $p_{2,2}$ the perpendiculars DI, DK, EU and EV, it is evident that one thus obtains for the value of the force sought:

$$\rho\left(\frac{1}{p_{1,2}}+\frac{1}{p_{2,1}}-\frac{1}{p_{1,1}}-\frac{1}{p_{2,2}}\right).$$

If one replaces, in this expression, the constant ρ by $\frac{1}{2}ii'ds'\cos\omega$, one will have the value of the force resulting from the action that the electric current NRST exerts on the element ds' and whose direction, comprised in the plane BRST, is perpendicular to that [direction] of the element; this value is

$$\frac{1}{2}ii'ds'\left(\frac{1}{p_{1,2}}+\frac{1}{p_{2,1}}-\frac{1}{p_{1,1}}-\frac{1}{p_{2,2}}\right)\cos\omega.$$

When the element situated at B is in the plane of the parallelogram, one has $\omega = 0$ and $\cos\omega = 1$, and the value of the force that we just calculated reduces to

$$\frac{1}{2}ii'ds'\left(\frac{1}{p_{1,2}}+\frac{1}{p_{2,1}}-\frac{1}{p_{1,1}}-\frac{1}{p_{2,2}}\right).$$

[End of the Treatise
Theory of Electrodynamic Phenomena, Uniquely Deduced from Experience.]

Chapter 31

Notes [of the *Théorie* Published in 1827] Containing Some New Developments on the Subjects Considered in the Preceding Treatise

[Page 374][1]

31.1 On the Method of Demonstrating Using the Four Cases of Equilibrium Explained at the Beginning of This Treatise, that the Value of the Mutual Action of Two Elements of Conducting Wires is $-\frac{2ii'}{\sqrt{r}}\frac{d^2\sqrt{r}}{dsds'}dsds'$

Following[2] in order the transformations that I successively applied to this expression, one finds first, due to the first two cases of equilibrium,[3] that the expression is

$$\frac{ii'\left(\sin\theta\sin\theta'\cos\omega + k\cos\theta\cos\theta'\right)dsds'}{r^n};$$

one deduces from the third [case of equilibrium],[4] between n and k, the relation $n + 2k = 1$, and from the fourth[5] [one deduces that] $n = 2$, from which $k = -1/2$; this fourth case of equilibrium is therefore the one employed in the last place for the determination of the value of the force which develops between two elements of conducting wires: but one can follow a different path using a consideration provided by M. de Laplace, as he concluded from M. Biot's first experiments, on the mutual action between a magnet and an indefinite rectilinear conductor, which showed that the force exerted by an element of this wire on one of the poles of the magnet varies inversely with the square of the distance, if the distance only changes in value and the angle between the measured straight line and the direction of the element stays the same. [page 375] In applying this consideration to the mutual action of two elements of conducting wires, it is easy to see, independently of any preliminary research on the value of the resulting force, that this force is also inversely proportional to the square of the distance when only it is varied, and the angles that determine the relationship between the elements are unchanged. In effect, based on the consideration developed at the

---

[1][N. T.] As discussed in Section 27.2, we believe that these Notes represent only Ampère's initial points of view, although published in 1827. The improved version of these Notes was presented in the 1826 version of the *Théorie*. They can be found in Chapter 30.

[2][N. T.] Due to a misprint, the formula in the title of this Section appeared as $-\frac{2ii'}{\sqrt{r}}\frac{dsds'}{d^2r}dsds'$. We included the correct expression of this formula in the title of the Section, namely, $-\frac{2ii'}{\sqrt{r}}\frac{d^2\sqrt{r}}{dsds'}dsds'$.

[3][N. T.] Namely, the case of equilibrium of the anti-parallel currents and the case of equilibrium of the sinuous wire.

[4][N. T.] Namely, the case of equilibrium of the nonexistence of tangential force.

[5][N. T.] Namely, the case of equilibrium of the law of similarity.

beginning of this Treatise, the force in question here is necessarily directed along the line r, and has the value

$$ii'f(r,\ \theta,\ \theta',\ \omega)dsds'\ ;$$

from which it follows, defining α, β and γ to be the angles that this straight line forms with the three [orthogonal Cartesian] axes, its three components are expressed by

$$ii'f(r,\ \theta,\ \theta',\ \omega)\cos\alpha dsds'\ ,$$

$$ii'f(r,\ \theta,\ \theta',\ \omega)\cos\beta dsds'\ ,$$

$$ii'f(r,\ \theta,\ \theta',\ \omega)\cos\gamma dsds'\ ,$$

and the three forces parallel to the three axes which result [from these components] between two circuits [will be obtained] by the double integrals of these expressions, i and i' being constants.

Now it follows from the fourth case of equilibrium, by replacing the three circles by any similar curves such that their homologous dimensions are in continuous geometric proportion, that these three forces have equal values in two similar systems; it is thus necessary that the integrals which express them have null dimension relative to all the lines which enter there, following the remark by M. de Laplace which I just remembered and that, by consequence, the same should also happen for the differentials of which they are composed, considering ds and ds' among the lines which are included [in these expressions], because the number of these differentials, though [page 376] infinite of second order,[6] should be considered as the same in the two systems.

Now the product $dsds'$ is two dimensional: it is then necessary that[7] $f(r,\theta,\theta',\omega)\cos\alpha$, $f(r,\theta,\theta',\omega)\cos\beta$ and $f(r,\theta,\theta',\omega)\cos\gamma$ are of dimension -2; and since the angles θ, θ', ω, α, β and γ are expressed by numbers which contribute nothing in the dimensions of the values of the differentials, and since $f(r,\ \theta,\ \theta',\ \omega)$ only contains the single line r, it is necessary that this function is proportional to $1/r^2$, so that the force applied from one to the other of the two elements of the conducting wires is given by

$$\frac{ii'\varphi(\theta,\ \theta',\ \omega)}{r^2}dsds'\ .$$

The first two cases of equilibrium then determine the function φ, where only k remains unknown, and one has

$$\frac{ii'\left(\sin\theta\sin\theta'\cos\omega+k\cos\theta\cos\theta'\right)}{r^2}dsds'\ ,$$

for the value of the sought force: it is, as is known, in this form that I presented it in the memoir that I read before the *Académie* on 4 December 1820.[8] By replacing then $\sin\theta\sin\theta'\cos\omega$ and $\cos\theta\cos\theta'$ by their values[9]

$$-\frac{rd^2r}{dsds'}\quad\text{and}\quad-\frac{dr}{ds}\frac{dr}{ds'}\ ,$$

one obtains[10]

---

[6] [N. T.] The line integral along the closed circuit s can be considered as a sum of an infinite number of terms, each one proportional to the infinitesimal length ds. The same reasoning can be applied to the line integral along the circuit s' as regards the infinitesimal length ds'. The force of a closed circuit s acting on another closed circuit s' will be then given by a product between an infinite sum of terms proportional to ds and another infinite sum of terms proportional to ds'. This will yield the "infinite of second order" mentioned by Ampère.

[7] [N. T.] Due to a misprint in the 1827 version of the *Théorie*, the last cosine appeared as $\cos\alpha$.

[8] [N. T.] This work, read on 4 December 1820, was published in [Ampère, 1820e]. The manuscript containing the material that Ampère presented in this date was published by Joubert, [Ampère, 1885j]. Ampère presented another version of this work in [Ampère, 1820d].

[9] [N. T.] Due to a misprint, the original version of the *Théorie* stated that the expression $\sin\theta\sin\theta'\cos\omega$ was equivalent to $-\frac{rd^2r}{dsds'}dsds'$.

[10] [N. T.] Due to a misprint, the first line of the following equation appeared in the original text as follows:

$$-\frac{ii'}{r^2}\left(\frac{d^2r}{dsds'}+k\frac{dr}{ds}\frac{dr}{ds'}\right)dsds'=$$

$$-\frac{ii'}{r^2}\left(\frac{rd^2r}{dsds'}+k\frac{dr}{ds}\frac{dr}{ds'}\right)dsds' =$$

$$-\frac{ii'(rdd'r+kdrd'r)}{r^2} = -\frac{ii'r^k dd'r + kr^{k-1}drd'r}{r^{k+1}} =$$

$$-\frac{ii'd(r^k d'r)}{r^{k+1}} = -\frac{ii'dd'(r^{k+1})}{(k+1)r^{k+1}} ,$$

[page 377]

and simplifying by substituting $k+1=m$, one has this very simple expression for the sought force:

$$-\frac{ii'dd'(r^m)}{mr^m} .$$

Therefore,[11] there only remains to determine m from the case of equilibrium which shows that the sum of the components of the forces that a conducting wire exerts on an element of a conducting wire, taken in the direction of the element, is always null when the conducting wire forms a closed circuit. This case of equilibrium, which I considered in this Treatise as the third, should be like the fourth, since it is the last which one uses in the full determination of the sought force. In replacing $d'r$ by $-\cos\theta' ds'$ in the value

$$-\frac{ii'd(r^{m-1}d'r)}{r^m}$$

of the force that the two elements exert on each other, one has, for its component in the direction of the element ds',

$$\frac{ii'ds'\cos\theta' d(r^{m-1}\cos\theta')}{r^m} = \frac{1}{2}\frac{ii'ds'd(r^{2m-2}\cos^2\theta')}{r^{2m-1}} ,$$

in which it is necessary that the integral relative to the differentials which depend on ds be null at all times that the curve s is closed; but it is easy to see, by integration by parts, that it is equal to

$$\frac{1}{2}ii'ds'\left[\frac{\cos^2\theta_2'}{r_2}-\frac{\cos^2\theta_1'}{r_1}+(2m-1)\int\frac{\cos^2\theta' dr}{r^2}\right] .$$

The first part of this expression vanishes when the curve s is closed because $r_2=r_1$ and $\cos\theta_2'=\cos\theta_1'$; with regard to the second [part], one shows easily, as we have done on page[12] 209, that $\int\frac{\cos^2\theta' dr}{r^2}$ cannot vanish, whatever the form of the [page 378] closed curve s; it is therefore necessary that one has $2m-1=0$, [that is,] $m=1/2$, and that the value of the force due to the mutual action of the two elements ds and ds' is[13]

$$-\frac{ii'dd'(r^m)}{mr^m} = -\frac{2ii'dd'\sqrt{r}}{\sqrt{r}} .$$

31.2 On a Proper Transformation which Simplifies the Calculation of the Mutual Action between Two Rectilinear Conductors

When the two conductors are rectilinear, the angle formed by the directions of the two elements is constant and equal to that [angle] between the same directions of the two conductors; it is therefore supposed to be known, and, designating it ε, one has [from] page[14] 207,

[11] [N. T.] This final portion of the Note 1 published in 1827 is different from the final paragraph of the Note 1 published in the 1826 version of the *Théorie*; see our page 463 up to page 463.

[12] [N. T.] Page 209 of the 1827 version of the *Théorie*, corresponding to our page 361.

[13] [N. T.] Due to a misprint in the original version of the *Théorie*, we had in the left side of this equation the following expression: $-\frac{ii'dd'(r^m)}{m\varepsilon^m}$.

[14] [N. T.] From page 207 of the 1827 version of the *Théorie*, corresponding to page 35 of the 1826 version of the *Théorie* and to our page 360.

$$r\frac{d^2r}{dsds'}+\frac{dr}{ds}\frac{dr}{ds'}=-\frac{dx}{ds}\frac{dx'}{ds'}-\frac{dy}{ds}\frac{dy'}{ds'}-\frac{dz}{ds}\frac{dz'}{ds'}=-\cos\varepsilon\ ,$$

from which it follows that[15]

$$\frac{dd'(r^m)}{mr^m}=\frac{(m-1)drd'r+rdd'r}{r^2}=\frac{(m-2)drd'r-\cos\varepsilon dsds'}{r^2}\ .$$

By designating p some other exponent, one has equivalently

$$\frac{dd'(r^p)}{pr^p}=\frac{(p-2)drd'r-\cos\varepsilon dsds'}{r^2}\ ,$$

and, by eliminating[16] $\frac{drd'r}{r^2}$ between these two equations, one obtains

$$\frac{(p-2)dd'(r^m)}{mr^m}-\frac{(m-2)dd'(r^p)}{pr^p}=\frac{(m-p)\cos\varepsilon dsds'}{r^2}\ ,$$

from which

$$\frac{dd'(r^m)}{mr^m}=\frac{m-2}{p-2}\frac{dd'(r^p)}{pr^p}+\frac{m-p}{p-2}\frac{\cos\varepsilon dsds'}{r^2}\ .$$

[page 379]

By substituting[17] 1/2 for m in this equation, and multiplying the two members of what results from this substitution by $-ii'$, one has the value of the action of two elements of conducting wires transformed as[18]

$$-\frac{2ii'dd'\sqrt{r}}{\sqrt{r}}=\frac{\frac{3}{2}ii'}{p-2}\frac{dd'(r^p)}{pr^p}-\frac{\left(\frac{1}{2}-p\right)ii'}{p-2}\frac{\cos\varepsilon dsds'}{r^2}\ ,$$

and one can in this expression assign any value to p. The one that provides the most convenience for calculation is $p=-1$; by adopting it one has

$$-\frac{2ii'dd'\sqrt{r}}{\sqrt{r}}=\frac{1}{2}ii'rdd'\frac{1}{r}+\frac{1}{2}\frac{ii'\cos\varepsilon dsds'}{r^2}=\frac{1}{2}ii'dsds'\left(\frac{\cos\varepsilon}{r^2}+r\frac{d^2\frac{1}{r}}{dsds'}\right)\ .$$

I have already found by another means, page 253,[19] this expression of the force that two elements of conducting wires exert on one another; one can only use it, for simplification of calculations, when the conductors are rectilinear, because it is only then that the angle ε is constant and known; but in this case, it is this [expression] that gives in the simplest manner the values of the forces and the torques which result from the mutual action of two conductors of this type. If I have in this Treatise used other means to calculate these values, it is because at the time that I wrote it I did not yet know this transformation of my formula.

---

[15] [N. T.] Due to a misprint in the original version, the second and third terms of this equation appeared as $\frac{(m-1)drdr'+rdd'r}{r^2}=\frac{(m-2)drdr'-\cos\varepsilon dsds'}{r^2}$.

[16] [N. T.] Due to a misprint, this term appeared in the original version of the *Théorie* as $\frac{drdr'}{r^2}$.

[17] [N. T.] The final portion of this second Note of the *Théorie* published in 1827 is different from the final portion of the second Note of the *Théorie* published in 1826; see our page 465.

[18] [N. T.] Due to a misprint in the original work, the term at the left hand side of this equation appeared in the following form: $-\frac{2iidd'\sqrt{r}}{\sqrt{r}}$.

[19] [N. T.] Page 253 of the 1827 version of the *Théorie*, corresponding to page 81 of the 1826 version, and to our page 396.

31.3 On the Direction of the Straight Line Designated in this Treatise under the Name of *Directrix of the Electrodynamic Action at a Given Point*, when This Action is That of a Closed and Planar Circuit in which All of the Dimensions are Very Small

The straight line[20] which I have named *directrix of the electrodynamic action at a given point*[21] [page 380] is that [line] which forms with the three axes the angles whose cosines are proportional respectively to the three quantities A, B and C; the values of these three quantities, found on page 227,[22] become

$$A = \lambda \left(\frac{\cos \xi}{r^3} - \frac{3qx}{r^5} \right) ,$$

$$B = \lambda \left(\frac{\cos \eta}{r^3} - \frac{3qy}{r^5} \right) ,$$

$$C = \lambda \left(\frac{\cos \zeta}{r^3} - \frac{3qz}{r^5} \right) ,$$

when one substitutes the number 2 for n; therefore, when one assumes the small circuit of arbitrary form located as in figure 14,[23] that is to say, after having placed the origin A of the coordinates at the given point, one takes as the z axis the perpendicular AZ drawn from the point A on the plane of the small circuit, and for the xz plane that [plane] which passes by this perpendicular and by the center of inertia O of the area LMS to which is related the x, y and z which enter into the values of A, B and C,[24] it is evident that one has $y = 0$, $q = z$, $\xi = \eta = \pi/2$, $\zeta = 0$, and that these values reduce as a consequence to

$$A = -\frac{3\lambda xz}{r^5} , \ B = 0 \ \text{and} \ C = \lambda \left(\frac{1}{r^3} - \frac{3z^2}{r^5} \right) = \frac{\lambda(x^2 - 2z^2)}{r^5} ,$$

because $r^2 = x^2 + z^2$. Since B is null, the directrix AE is necessarily in the xz plane determined as we have just said; it forms with the x axis an angle EAX whose tangent is evidently equal to C/A, that is, is equal to $(2z^2 - x^2)/3xz$; and as that [tangent] of the angle OAX is [equal] to z/x, one finds, for the value of the tangent of OAE

$$\tan \text{OAE} = \frac{\frac{z}{x} - \frac{2z^2 - x^2}{3xz}}{1 + \frac{2z^2 - x^2}{3x^2}} = \frac{(z^2 + x^2)x}{(2x^2 + 2z^2)z} = \frac{1}{2}\frac{x}{z} = \frac{1}{2} \tan \text{COA} .$$

[page 381]

From which it follows that, if one takes OB = OA/3, and one draws on OA at the point B a plane perpendicular to AO which intercepts in D the normal OC on the plane of the small circuit, the straight line ADE determined by the points A and D will be the directrix of the action exerted at the point A by the electric current flowing in it,[25] since one will have

$$\text{AB} = 2\text{OB} , \quad \tan \text{BDA} = 2 \tan \text{BDO} ,$$

and

$$\tan \text{OAE} = \cot \text{BDA} = \frac{1}{2} \cot \text{BDO} = \frac{1}{2} \tan \text{COA} .$$

[20] [N. T.] The goal of this Note was to obtain the mapping of terrestrial magnetism, as we discussed in Section 10.6.

[21] [N. T.] See Section 10.1 and the footnote 96 in our page 367.

[22] [N. T.] Page 227 of the 1827 version of the *Théorie*, corresponding to page 55 of the 1826 version, and to our page 377.

[23] [N. T.] See our pages 373 and 482.

[24] [N. T.] That is, the coordinates of the center of inertia O of the area LMS are given by (x, y, z).

[25] [N. T.] That is, flowing through the small closed circuit.

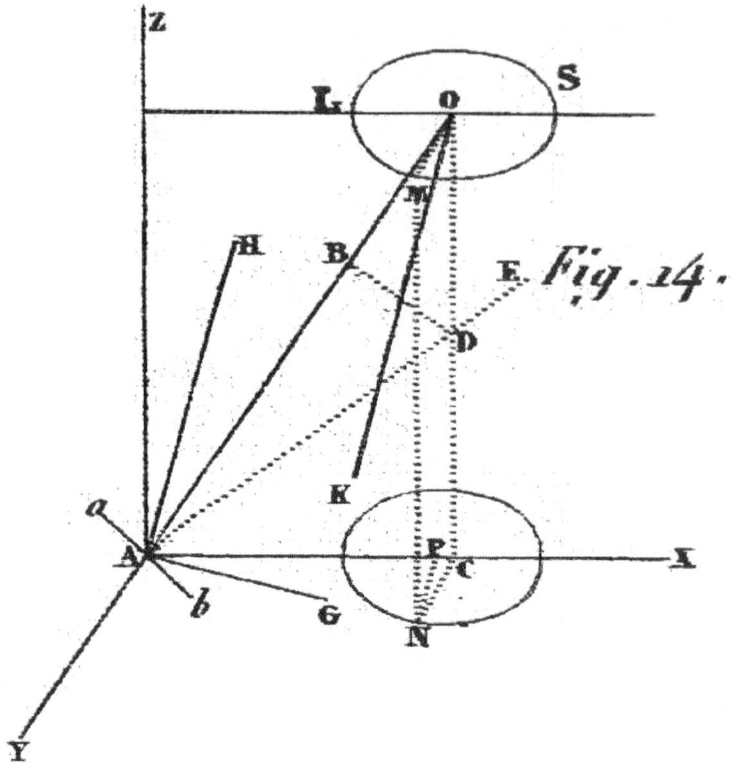

Figure 14.

This construction gives in the simplest manner the direction of the straight line AE along which we saw[26] that the pole of a magnet placed in A would be moved by the action of this current. It is to be remarked that it[27] is orientated as regards the plane LMS of the small circuit that it[28] describes, just as the direction of the dip needle is generally [orientated] with respect to the magnetic equator; because the point O being considered as the center of the Earth, the plane OAC like that [plane] of the magnetic meridian, and the straight line AE like the direction of the dip needle, it is evident that the angle OAE comprised between the terrestrial ray OA and the direction AE of the magnetized needle is the complement of the inclination,[29] and that the angle COA is the complement of the magnetic latitude LOA; therefore the preceding equation becomes:[30]

$$\cot \text{incl.} = \frac{1}{2} \cot \text{lat.} \ ,$$

or[31]

$$\tan \text{incl.} = 2 \tan \text{lat.}$$

[26][N. T.] See pages 104 and 276 of the 1826 and 1827 versions of the *Théorie*, respectively, corresponding to our page 412.

[27][N. T.] That is, the straight line AE.

[28][N. T.] That is, the current.

[29][N. T.] The complement of an angle φ_1 is the angle φ_2 such that $\varphi_1 + \varphi_2 = \pi/2$.

[30][N. T.] Savary was the first scientist to obtain these relations utilizing Ampère's electrodynamic theory; see [Savary, 1823b, p. 25], [Savary, 1823c, p. 26] and [Savary, 1885b, p. 369].

[31][N. T.] This mathematical relation obtained by Ampère is analogous to our equation (10.56) connecting the inclination angle ζ of a dip needle relative to the horizon with the latitude angle L where the needle is located. According to this expression, which agrees with observational data, the dip needles at the surface of the Earth become orientated as indicated in figure 10.23.

31.4 On the Value of the Force that an Indefinite Angular Conductor Exerts on the Pole of a Small Magnet

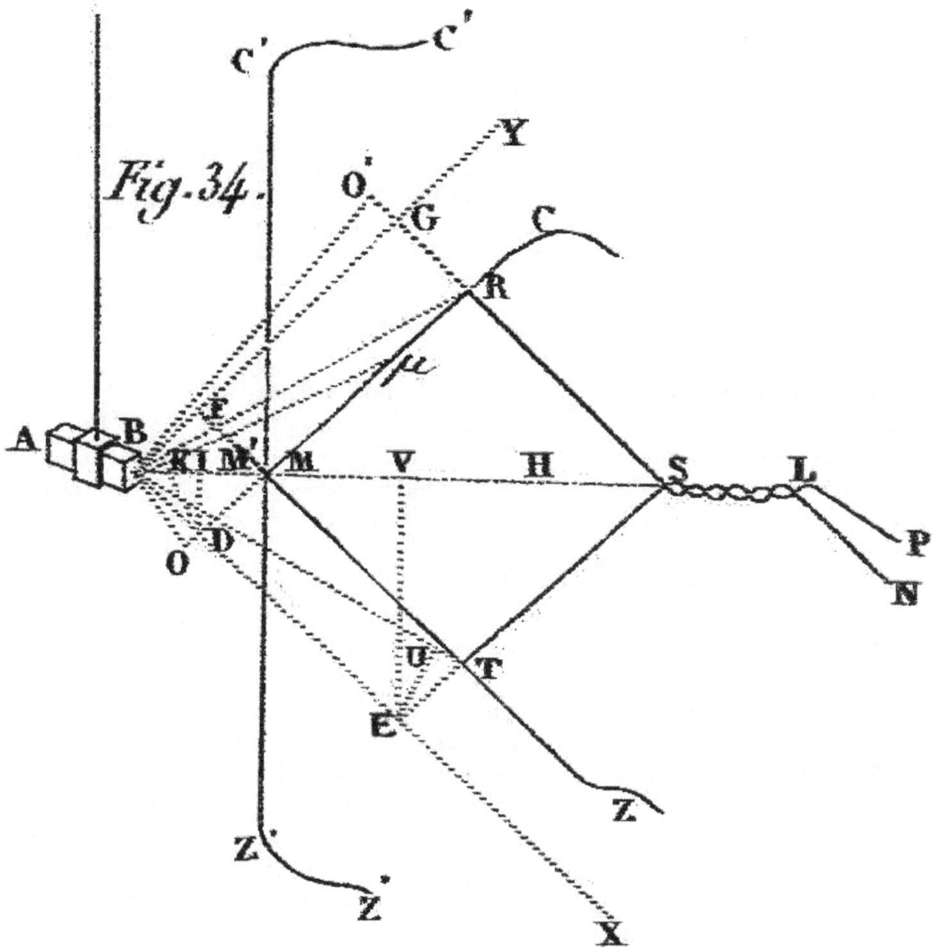

Figure 34.

Whether one considers the pole B (figure 34)[32] of the small magnet AB as the extremity of an electrodynamic solenoid or as a magnetic molecule, there is agreement, in both views, with respect to the expression of the force exerted [page 382] on this pole by each element of the angular conductor CMZ; one finds in general that in drawing the perpendicular BO $= b$ of the point B on one of its branches CμM extended toward O, setting O$\mu = s$, BM $= a$, B$\mu = r$, the angle BμM $= \theta$, the angle CMH $=$ BMO $= \varepsilon$, and designating by ρ a constant coefficient, the force which is exerted on the pole B by the element ds situated at μ is equal to

$$\frac{\rho \sin\theta ds}{r^2} ,$$

which must be integrated from $s =$ OM $= a\cos\varepsilon$ to $s = \infty$, or, what amounts to the same thing, from $\theta = \varepsilon$ to $\theta = 0$; but, in the triangle BOμ, whose side OB $= b = a\sin\varepsilon$, one has

$$r = \frac{a\sin\varepsilon}{\sin\theta} , \quad s = a\sin\varepsilon\cot\theta , \quad ds = -\frac{a\sin\varepsilon d\theta}{\sin^2\theta} \quad \text{and} \quad \frac{ds}{r^2} = -\frac{d\theta}{a\sin\varepsilon} ,$$

thus

[32][N. T.] See our page 416.

$$\frac{\rho \sin\theta ds}{r^2} = -\frac{\rho \sin\theta d\theta}{a \sin\varepsilon} ,$$

whose integral is

$$\frac{\rho}{a \sin\varepsilon} (\cos\theta + \mathrm{C}) ,$$

or, considering it between the limits determined above,

$$\frac{\rho (1 - \cos\varepsilon)}{a \sin\varepsilon} = \frac{\rho}{a} \tan\frac{1}{2}\varepsilon ,$$

a value which is necessary to double in order to obtain the force exerted on the pole B by the indefinite angular conductor CMZ; this force, in inverse ratio of BM $= a$, is therefore, for the same value of a, proportional to the tangent of half the angle CMH, and not to this angle itself, although it may have been claimed that the value

$$\frac{\rho \sin\theta ds}{r^2}$$

of the force exerted by the element ds on the pole B, was found *analyzing by calculation*[33] the supposition that the force produced [page 383] by the conducting wire CMZ was proportional to the angle CMH. It is not possible to doubt that there was some error in this calculation; but it would be equally curious to know it,[34] a calculation intended to determine the value of a differential beginning with the value of the definite integral obtained between given limits, but it does not seem to me that, up to now, any mathematician had considered [this calculation] something possible to be done.

Since one cannot, in practice, make the branches MC and MZ of the angular conductor really infinite, nor extend the prolongations of the wire composing it and which connect these branches with the two extremities of the battery, at a sufficiently great distance from the small magnet AB so that they[35] will effect absolutely no action on it,[36] one should, rigorously, consider the value that we just obtained only as an approximation. In order to have an exact value which can be verified by experiment, it is necessary to calculate the force exerted on the pole B of the small magnet by a conducting wire PSRMTSN, whose portions SP and SN, which communicate with the two extremities of the battery, are covered in silk and twisted together, as one sees in SL, up to close to the battery, so that the actions that they produce cancel each other, and whose remainder [of the circuit] form a lozenge SRMT situated in such a way that the direction of the diagonal SM of this lozenge passes by the point B. To this end, by preserving the preceding names and making moreover the angle BRM $= \theta_1$, the angle BRO$'$ $= \theta_1'$, the distance BS $= a'$ and the perpendicular BO$'$ $= b' = -a' \sin\varepsilon$ because the angle BSO$'$ $= -\varepsilon$, one will easily see that the action of the portion RS of the conducting wire on the pole B is equal to

$$-\frac{\rho (\cos\varepsilon - \cos\theta_1')}{b'} ,$$

as well as, on account of $b = a \sin\varepsilon$, one would have found

$$\frac{\rho (\cos\theta_1 - \cos\varepsilon)}{b} ,$$

[page 384]

for that [action] that the portion MR exerts on the same pole B, by taking the preceding integral from $\theta = \varepsilon$ to $\theta = \theta_1$.

By combining these two expressions, and doubling the sum, one gets the total action of the lozenge contour MRST,

[33] [N. T.] See Section 17.2 in which we discussed Biot's work and Ampère's criticism of it.

[34] [N. T.] That is, to know this wrong calculation.

[35] [N. T.] Ampère utilized here the word "elles." He was referring to the branches of the wire (*branche* is a feminine word in French). On page 217 of the Notes of the 1826 version of the *Théorie*, equivalent to our page 473, Ampère utilized in the same place the word "ils" instead of "elles." The reason for this replacement was that in 1826 he was referring to the prolongations of the wire (*prolongement* is a masculine word in French).

[36] [N. T.] That is, so that the branches effect no action on the magnet.

$$2\rho\left(\frac{\cos\theta_1}{b}-\frac{\cos\varepsilon}{b}+\frac{\cos\theta_1'}{b'}-\frac{\cos\varepsilon}{b'}\right) .$$

This value is susceptible to another form which one obtains by relating the positions of the four angles of the lozenge to two axes BX and BY drawn by the point B parallel to these sides and which join them at the points D, E, F and G; if one sets BD = BF = g and BE = BG = h, one has

$$b = \mathrm{BO} = g\sin 2\varepsilon\ , \quad b' = \mathrm{BO}' = h\sin 2\varepsilon\ ,$$

$$\cos\theta_1 = \frac{\mathrm{OR}}{\mathrm{BR}} = \frac{h+g\cos 2\varepsilon}{\sqrt{g^2+h^2+2gh\cos 2\varepsilon}}\ ,$$

$$\cos\theta_1' = \frac{\mathrm{O'R}}{\mathrm{BR}} = \frac{g+h\cos 2\varepsilon}{\sqrt{g^2+h^2+2gh\cos 2\varepsilon}}\ ,$$

and using these values, that [value] of the force exerted on the pole B becomes

$$2\rho\left(\frac{h+g\cos 2\varepsilon}{g\sin 2\varepsilon\sqrt{g^2+h^2+2gh\cos 2\varepsilon}}+\frac{g+h\cos 2\varepsilon}{h\sin 2\varepsilon\sqrt{g^2+h^2+2gh\cos 2\varepsilon}}\right.$$

$$\left.-\frac{\cos\varepsilon}{g\sin 2\varepsilon}-\frac{\cos\varepsilon}{h\sin 2\varepsilon}\right)$$

$$=\rho\left(\frac{2\sqrt{g^2+h^2+2gh\cos 2\varepsilon}}{gh\sin 2\varepsilon}-\frac{1}{g\sin\varepsilon}-\frac{1}{h\sin\varepsilon}\right)\ ,$$

by replacing in the two last terms $\sin 2\varepsilon$ by its value $2\sin\varepsilon\cos\varepsilon$.

Now draw from the point D the perpendiculars DI and DK on the straight lines BM and BR; the first [perpendicular] will be obviously equal to $g\sin\varepsilon$, and the second is obtained by noticing that in multiplying it by BR $=\sqrt{g^2+h^2+2gh\cos 2\varepsilon}$ one has a product equal to twice the surface of the triangle BDR, that is, $gh\sin 2\varepsilon$, [page 385] so that by naming these perpendiculars $p_{1,1}$ and $p_{1,2}$, it becomes

$$\frac{1}{p_{1,1}}=\frac{1}{g\sin\varepsilon} \quad\text{and}\quad \frac{1}{p_{1,2}}=\frac{\sqrt{g^2+h^2+2gh\cos 2\varepsilon}}{gh\sin 2\varepsilon}\ ;$$

by drawing from the point E the two perpendiculars EU and EV on the straight lines BT and BS, and representing them by $p_{2,1}$ and $p_{2,2}$, the first [perpendicular] will be equal to DK due to the equality of the triangles BDR and BET, and the second will have the value $h\sin\varepsilon$, in such a way that the expression of the force exerted by the contour of the lozenge MRST on the pole B can be written as:

$$\rho\left(\frac{1}{p_{1,2}}+\frac{1}{p_{2,1}}-\frac{1}{p_{1,1}}-\frac{1}{p_{2,2}}\right) .$$

In this form it applies not only to a lozenge in which one diagonal is directed so as to pass by the point B, but [also] to an arbitrary parallelogram NRST (figure 44) whose perimeter carries an electric current which acts on the pole of a magnet located in the plane of this parallelogram. It results,[37] in effect, as was already said, pages 229 and 276,[38] that the effect of NRST on the pole B is the same as if all the elements $d^2\lambda$ which compose its surface were acting on this pole with a force equal to $\rho d^2\lambda/r^3$; from which it follows, labeling by x and y the coordinates referring to the axes BX and BY, and to the origin B of an arbitrary point M of the area of the parallelogram which gives

$$d^2\lambda = dxdy\sin 2\varepsilon \quad\text{and}\quad r=\sqrt{x^2+y^2+2xy\cos 2\varepsilon}\ ,$$

the total force, impinging on pole B, will then be

[37][N. T.] In Note 5 of the 1826 version of the *Théorie*, there is a more detailed description of the contents included in the previous and in the next paragraphs; see our pages 474 up to 476.

[38][N. T.] Pages 229 and 276 of the 1827 version of the *Théorie*, corresponding to pages 57 and 104 of the 1826 version, and to our pages 378 and 412, respectively.

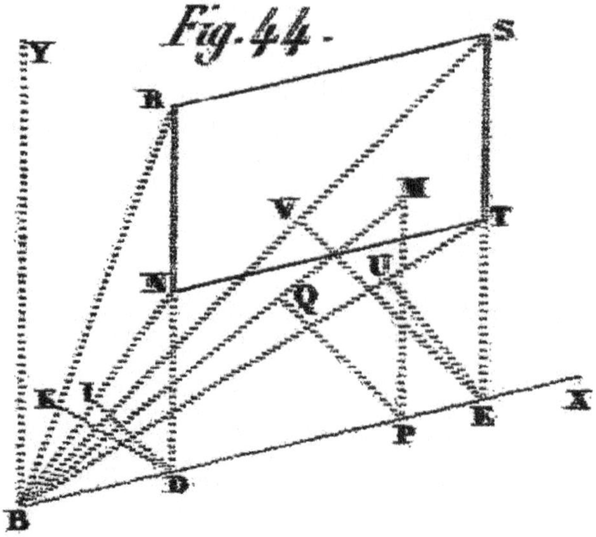

Figure 44.

$$\rho \sin 2\varepsilon \int\int \frac{dxdy}{(x^2+y^2+2xy\cos 2\varepsilon)^{\frac{3}{2}}} \, .$$

Now we have seen, page 266,[39] that the indefinite integral of

[page 386]

$$\frac{dsds'}{(a^2+s^2+s'^2-2ss'\cos\varepsilon)^{\frac{3}{2}}}$$

is

$$\frac{1}{a\sin\varepsilon}\arctan\frac{ss'\sin^2\varepsilon+a^2\cos\varepsilon}{a\sin\varepsilon\sqrt{a^2+s^2+s'^2-2ss'\cos\varepsilon}} \, ,$$

or[40]

$$-\frac{1}{a\sin\varepsilon}\arctan\frac{a\sin\varepsilon\sqrt{a^2+s^2+s'^2-2ss'\cos\varepsilon}}{ss'\sin^2\varepsilon+a^2\cos\varepsilon} \, ,$$

by removing the constant $\pi/2$. When $a=0$, this quantity takes the form $0/0$; but since the arc should be replaced by its tangent, the null factor $a\sin\varepsilon$ vanishes, and one has[41]

$$\int\int \frac{dsds'}{(s^2+s'^2-2ss'\cos\varepsilon)^{\frac{3}{2}}} = -\frac{\sqrt{s^2+s'^2-2ss'\cos\varepsilon}}{ss'\sin^2\varepsilon} \, ,$$

which is easy to verify by differentiation. One concludes immediately that the expression of the force that we have calculated, considered as an indefinite integral, is

$$-\frac{\rho\sqrt{x^2+y^2+2xy\cos 2\varepsilon}}{xy\sin 2\varepsilon} = -\frac{\rho}{p} \, ,$$

[39][N. T.] Page 266 of the 1827 version of the *Théorie*, corresponding to page 94 of the 1826 version, and to our page 404, respectively.

[40][N. T.] Due to a misprint, we had inside the square root sign the following expression: $2ss\cos\varepsilon$.

[41][N. T.] Due to a misprint, the denominator of the integral appeared as: $(s+s'^2-2ss'\cos\varepsilon)\,\frac{3}{2}$.

defining p to be the perpendicular PQ drawn from point P on BM, because the double of the area of the triangle BPM is both equal to $p\sqrt{x^2+y^2+2xy\cos 2\varepsilon}$ and to $xy\sin 2\varepsilon$, which gives

$$\frac{1}{p}=\frac{\sqrt{x^2+y^2+2xy\cos 2\varepsilon}}{xy\sin 2\varepsilon}\,.$$

There only remains now to calculate the values taken by this indefinite integral at the four vertices N, R, T and S of the parallelogram, and to sum them with convenient signs; continuing to designate respectively by $p_{1,1}$, $p_{1,2}$, $p_{2,1}$ and $p_{2,2}$ the [page 387] perpendiculars DI, DK, EU and EV, it is evident that one thus obtains for the value of the force sought:

$$\rho\left(\frac{1}{p_{1,2}}+\frac{1}{p_{2,1}}-\frac{1}{p_{1,1}}-\frac{1}{p_{2,2}}\right)\,.$$

The direction perpendicular to the plane of the parallelogram NRST, along which the pole of magnet located in B is carried by the action of the electric current which follows the contour of this parallelogram, is the directrix of the electrodynamic action which it exerts at the point B, from which it follows that if there were at this point an element of electric current situated in the plane of the parallelogram, it would form a right angle with the directrix, and thus the action of this current on the element would be a force located in this plane, perpendicular to the direction of the element, and equal to that [action] which the same current would exert on the pole of a magnet placed at the point B multiplied by a constant ratio, which is here that of ρ to $ii'ds/2$, naming this element ds; so that the force thus directed which would act on the element would have the value[42]

$$\frac{1}{2}ii'ds\left(\frac{1}{p_{1,2}}+\frac{1}{p_{2,1}}-\frac{1}{p_{1,1}}-\frac{1}{p_{2,2}}\right)\,.$$

When the element located at B is not in the plane of the parallelogram, but forms with this plane an angle equal to ω, one can replace it by two elements of the same intensity, one in this plane, the other which is perpendicular to it: the action of the current of the parallelogram on this last [element] being null, one should only take into account that [action] which it exerts on the first [element]; it is evidently in the plane of the parallelogram, perpendicular to the element, and equal to

$$\frac{1}{2}ii'ds\cos\omega\left(\frac{1}{p_{1,2}}+\frac{1}{p_{2,1}}-\frac{1}{p_{1,1}}-\frac{1}{p_{2,2}}\right)\,.$$

[End of the Treatise
Theory of Electrodynamic Phenomena, Uniquely Deduced from Experience.]

[42][N. T.] Due to a misprint, we had in the beginning of this equation the following expression: $\frac{1}{2}iids$.

Part IX

Conclusion

The *Théorie* was Ampère's masterpiece. We conclude this book presenting once more some evaluations of his experimental and theoretical work. We subscribe to them.

Maxwell comparing four formulas for the force between current elements, namely, those of Ampère, Grassmann and two others of his own creation:[43]

> Of these four different assumptions that of Ampère is undoubtedly the best, since it is the only one which makes the forces on the two elements not only equal and opposite but in the straight line which joins them.

Maxwell's appraisal of Ampère's force between current elements:[44]

> The experimental investigation by which Ampère established the laws of the mechanical action between electric currents is one of the most brilliant achievements in science. The whole, theory and experiment, seems as if it had leaped, full grown and full armed, from the brain of the 'Newton of electricity.' It is perfect in form, and unassailable in accuracy, and it is summed up in a formula from which all the phenomena may be deduced, and which must always remain the cardinal formula of electro-dynamics.

Whittaker:[45]

> [Ampère] published his collected results in one of the most celebrated memoirs in the history of natural philosophy.

Williams[46] comparing Ampère's main work[47] with Newton's masterpiece of 1687, *Mathematical Principles of Natural Philosophy*:[48]

> Having established a noumenal foundation for electrodynamic phenomena, Ampère's next steps were to discover the relationship between the phenomena and to devise a theory from which these relationships could be mathematically deduced. This double task was undertaken in the years 1821-1825, and his success was reported in his greatest work, the *Mémoire sur la théorie mathématique des phénomènes électrodynamique, uniquement déduite de l'expérience* (1827). In this work, the *Principia* of electrodynamics, Ampère first described the laws of action of electric currents, which he had discovered from four extremely ingenious experiments.

Tricker:[49]

> At the beginning of the year 1820 nothing was known of the magnetic action of an electric current. By 1826 the theory for steady currents had been completely worked out. Since then, though newer methods may have made the handling of the mathematical apparatus simpler and more concise, nothing fundamental has been changed.
>
> [...]
>
> In the theory of gravitation, Newton was already provided with a knowledge of a range of the phenomena, mainly through the medium of Kepler's laws. Ampère had to discover the laws as well as provide the theory, and thus do the work of Tycho Brahe, Kepler and Newton rolled into one.

[43][Maxwell, 1954, vol. 2, article 527, p. 174].
[44][Maxwell, 1954, vol. 2, article 528, p. 175].
[45][Whittaker, 1973, p. 83].
[46][Williams, 1981, p. 145]
[47][Ampère, 1826f] and [Ampère, 1823c].
[48][Newton, 1934], [Newton, 1990], [Newton, 1999], [Newton, 2008] and [Newton, 2010].
[49][Tricker, 1965, pp. vii and 36].

Part X

Appendix

Appendix A

Figures of the *Théorie* Drawn with Graphic Software

We present in this Appendix all figures of the *Théorie* drawn with graphic software. Our goal is to show images with more defined lines and with more readable letters. These figures may help in the comprehension of Ampère's masterpiece.

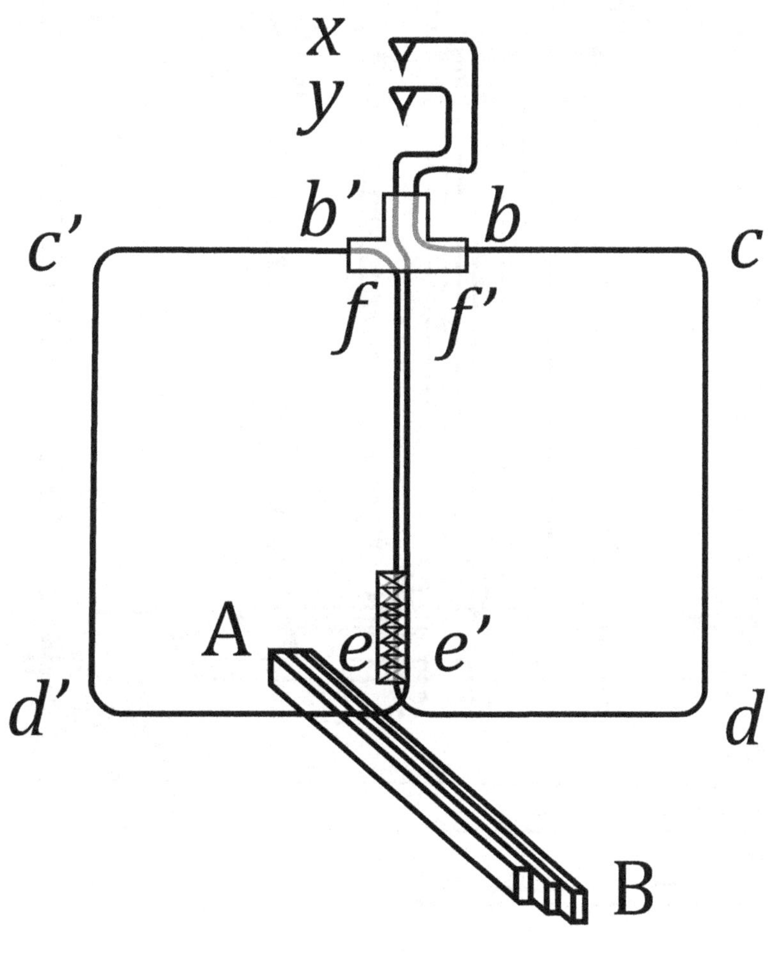

Figure A.1: Figure 1.

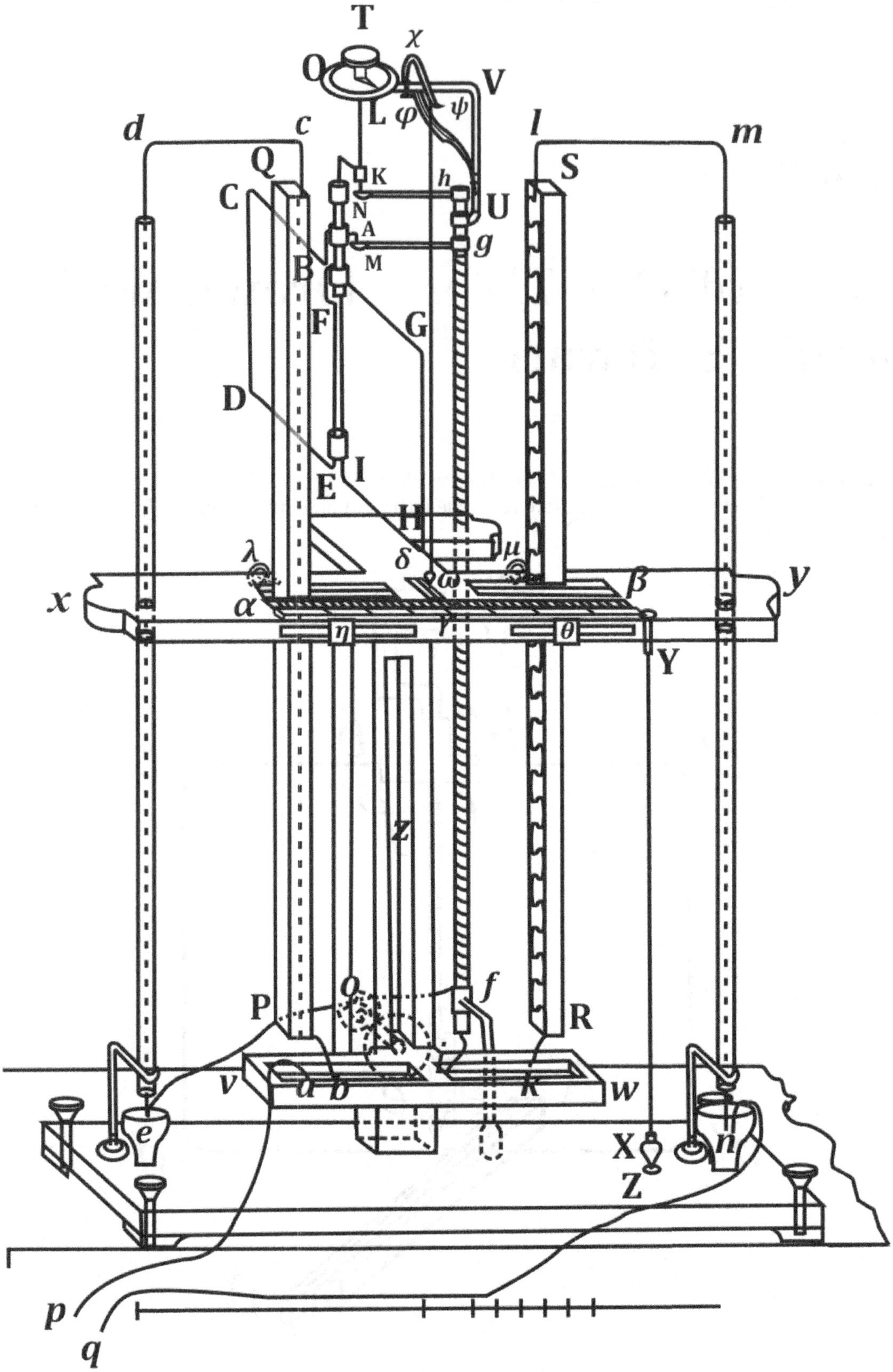

Figure A.2: Figure 2.

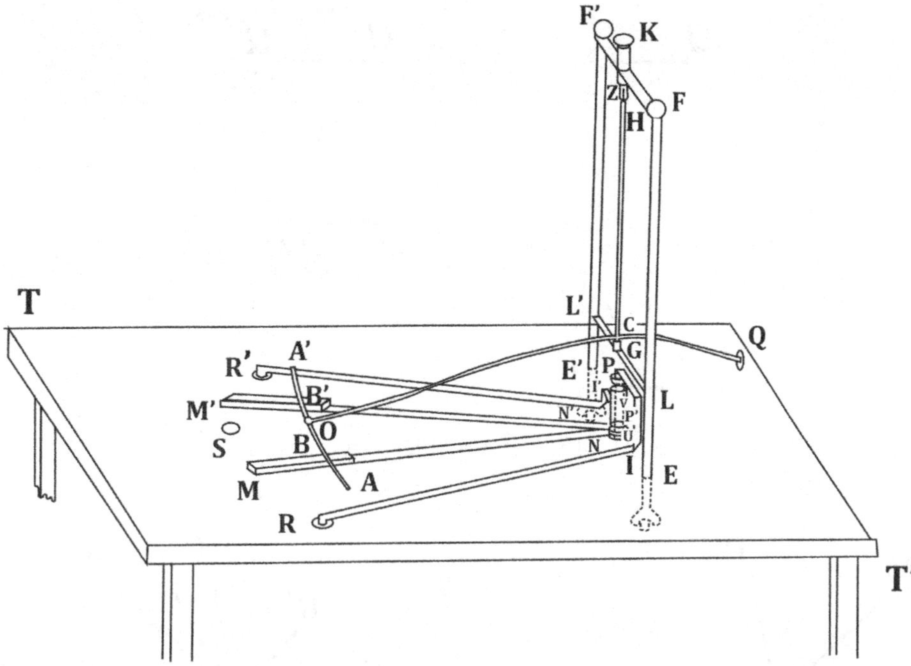

Figure A.3: Figure 3. Letter E' at the bottom of the left column appeared as letter E in figure 3 of the 1826 and 1827 versions of the *Théorie*.

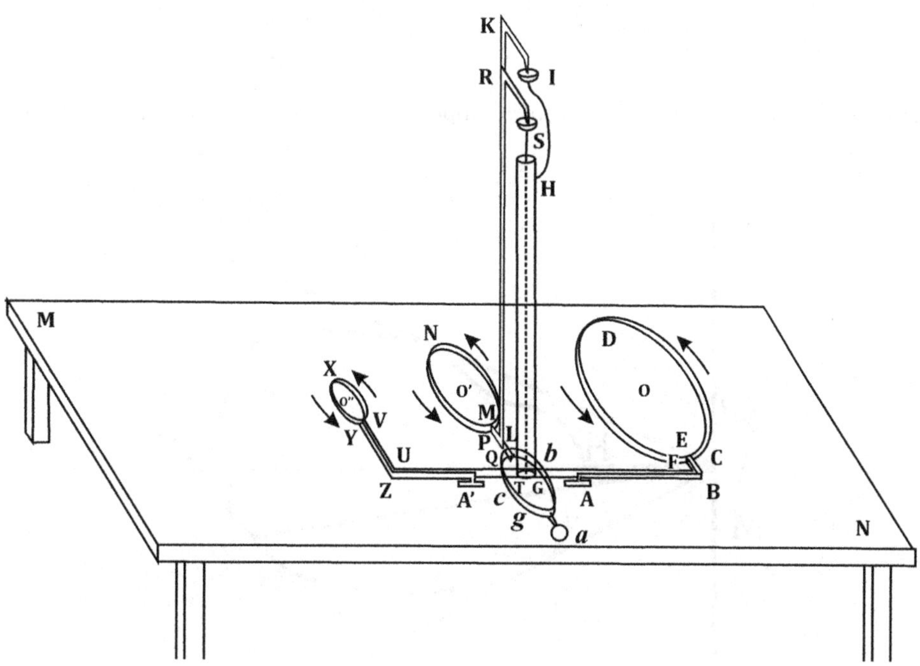

Figure A.4: Figure 4. Letter a representing the counterweight did not appear in figure 4 of the 1826 and 1827 versions of the *Théorie*. Letters O″, O′ and O representing the centers of the circles did not appear in the 1826 version of the *Théorie*, although they appeared in the 1827 version. As they are mentioned in Ampère's text, we included them here.

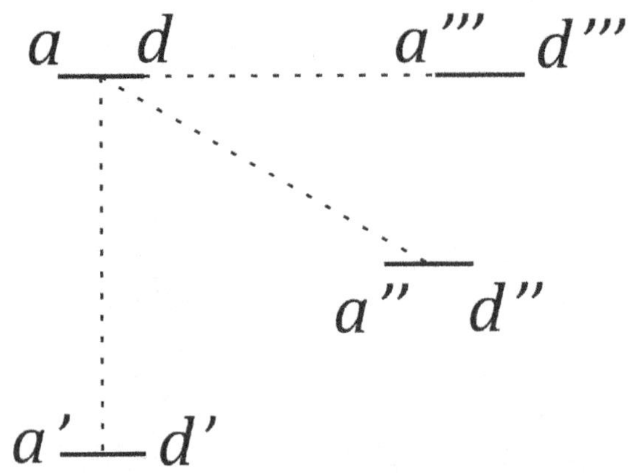

Figure A.5: Figure 5.

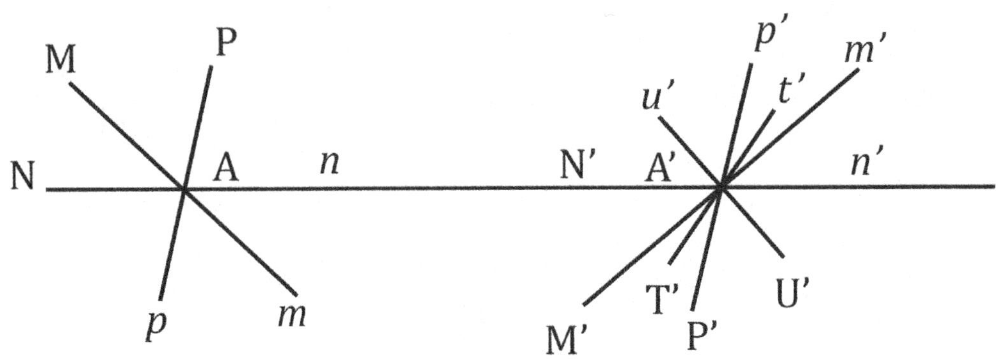

Figure A.6: Figure 6. Letter A$'$ at the center of M$'m'$ appeared in the 1827 version of the *Théorie*, but not in the 1826 version.

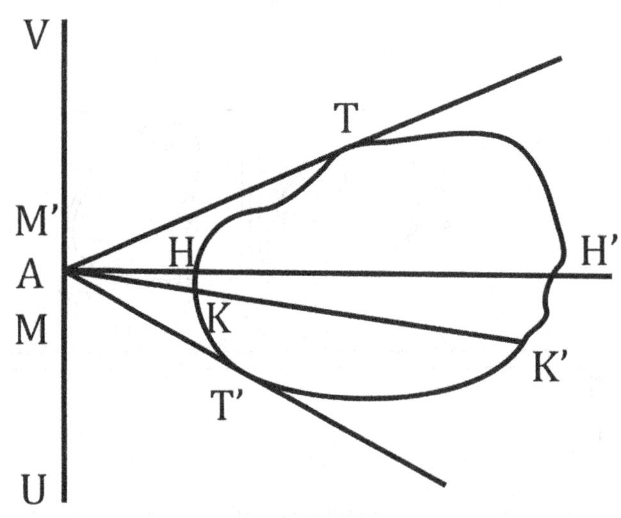

Figure A.7: Figure 7.

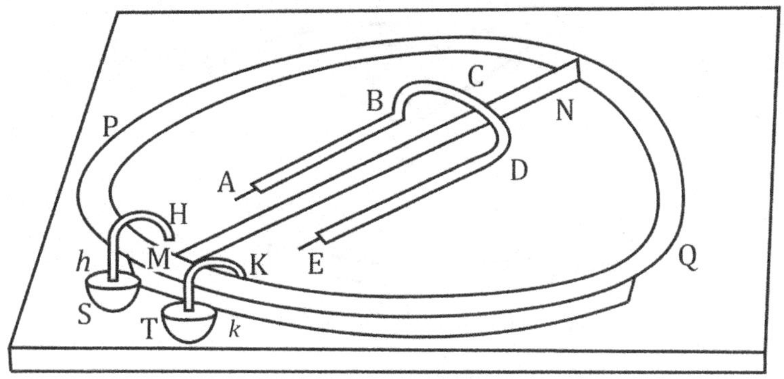

Figure A.8: Figure 8.

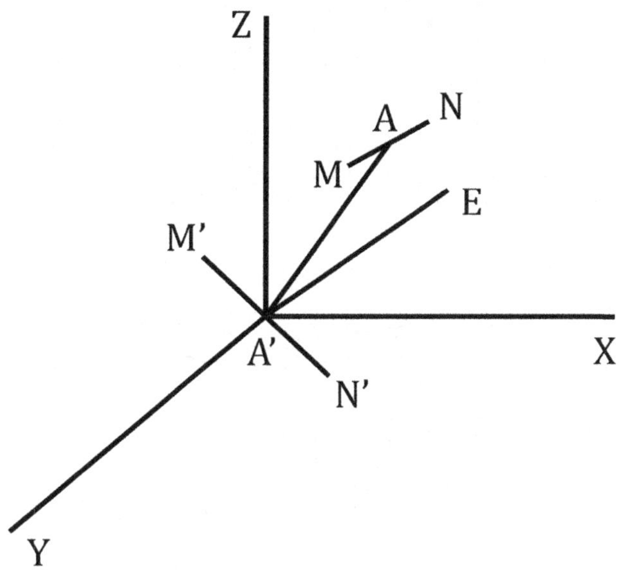

Figure A.9: Figure 9.

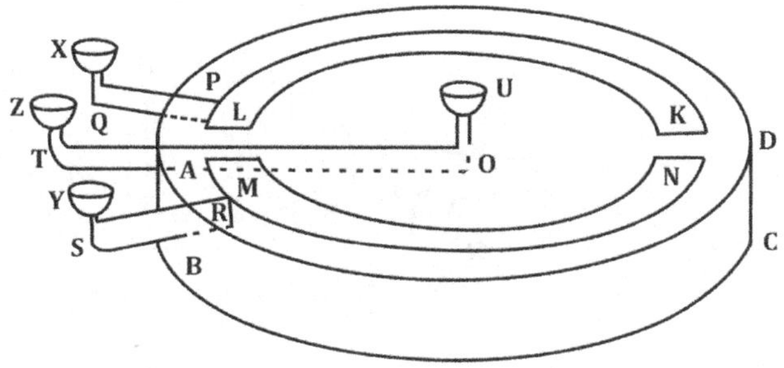

Figure A.10: Figure 10.

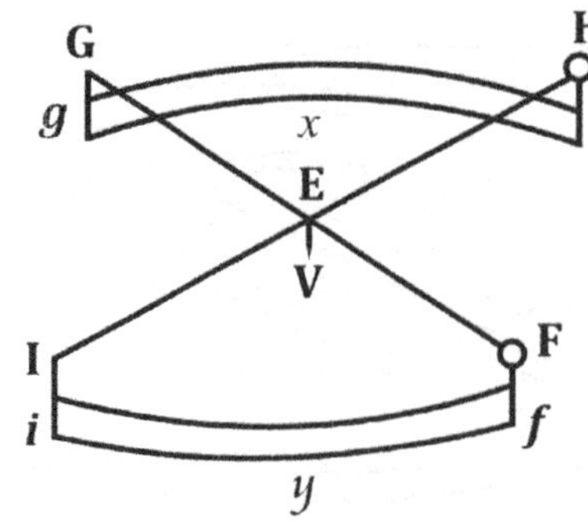

Figure A.11: Figure 11.

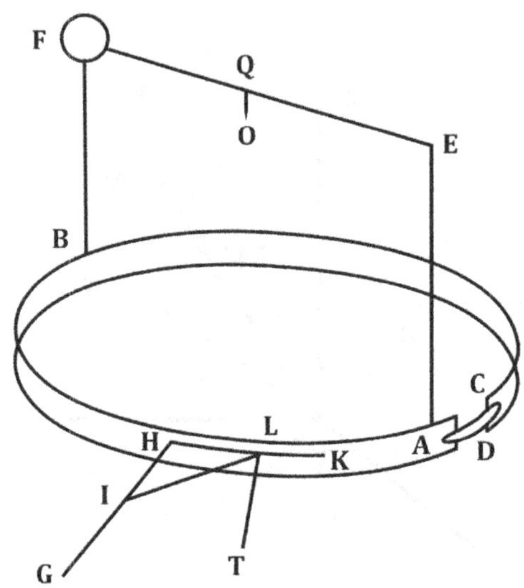

Figure A.12: Figure 12.

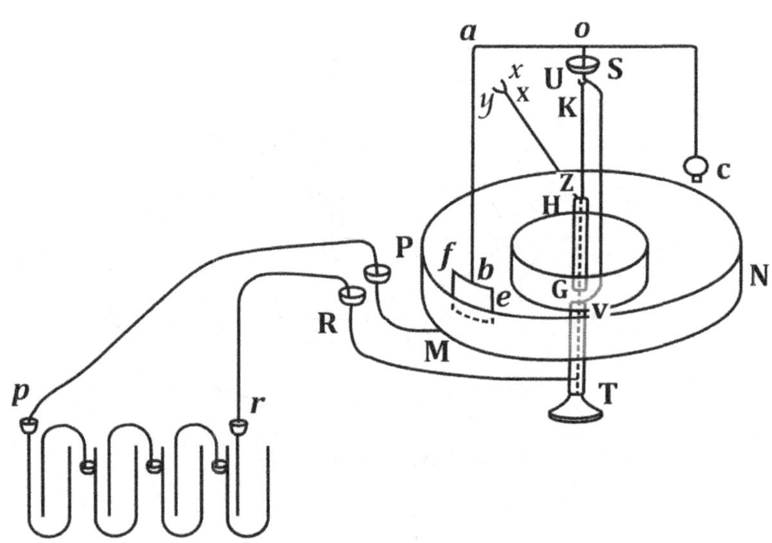

Figure A.13: Figure 13.

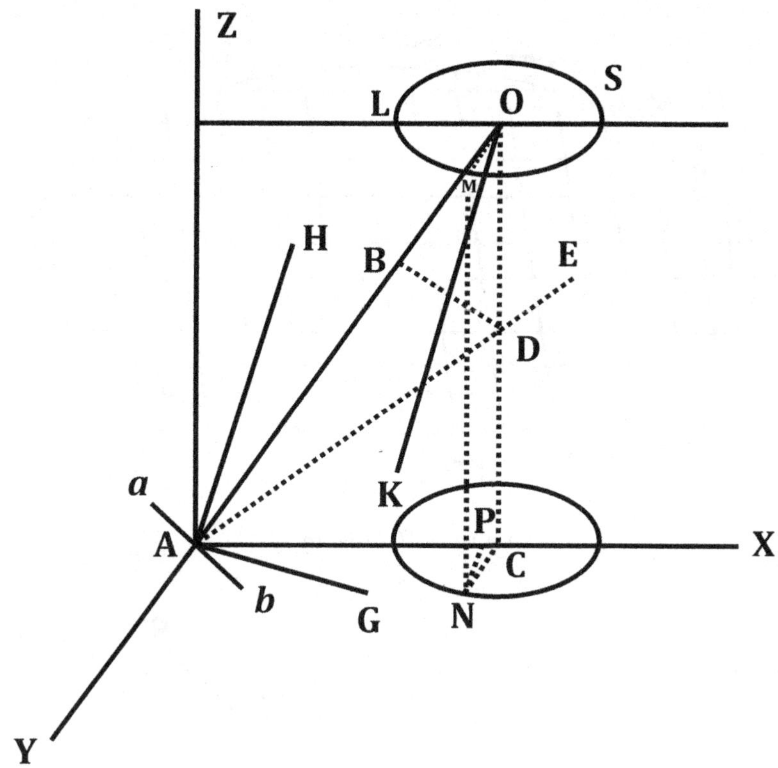

Figure A.14: Figure 14.

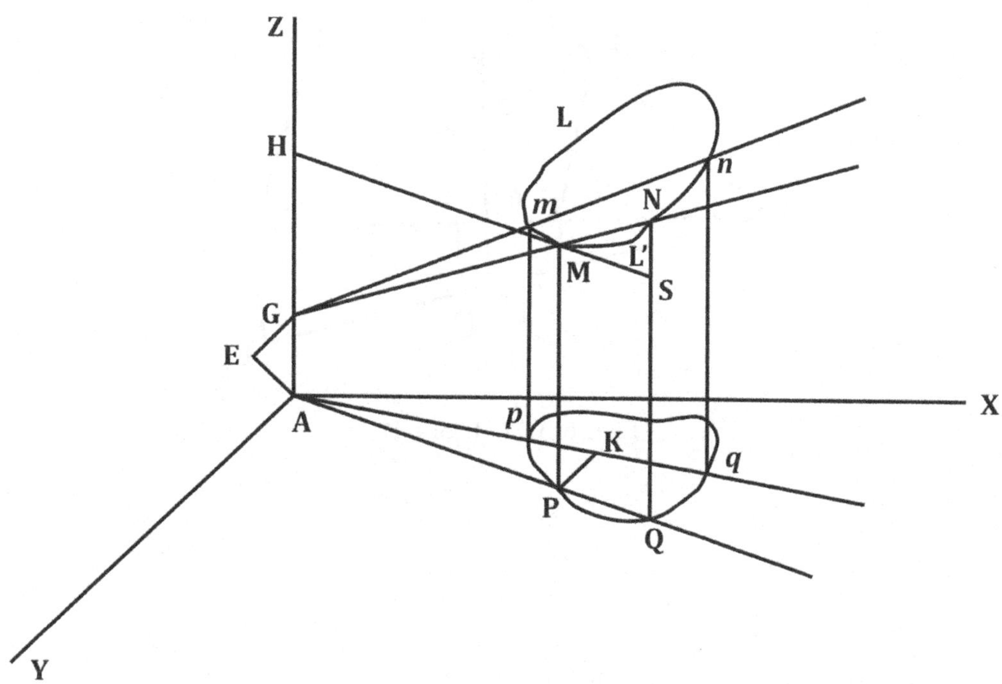

Figure A.15: Figure 15.

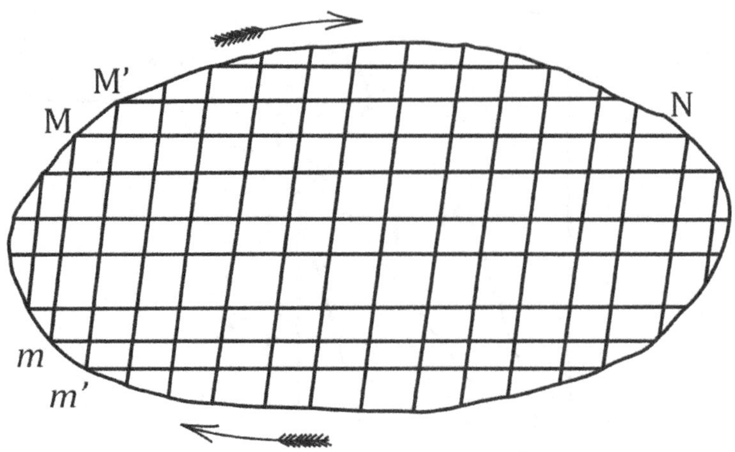

Figure A.16: Figure 16.

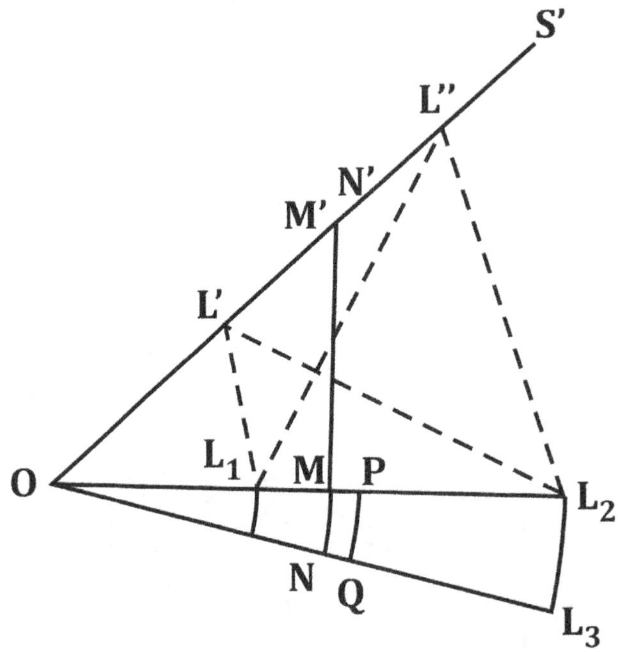

Figure A.17: Figure 17.

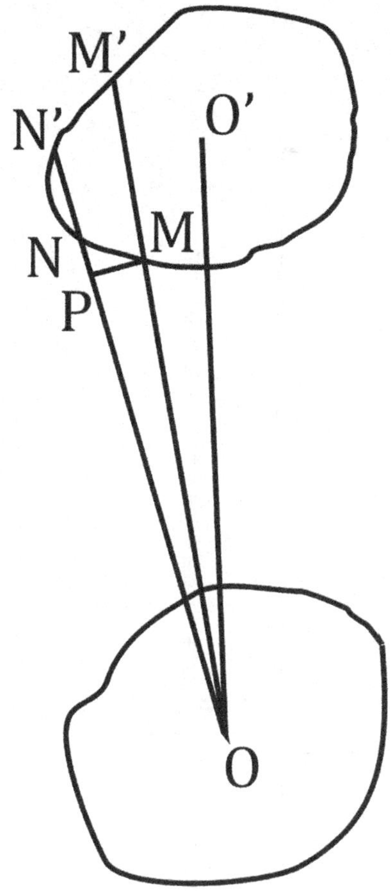

Figure A.18: Figure 18.

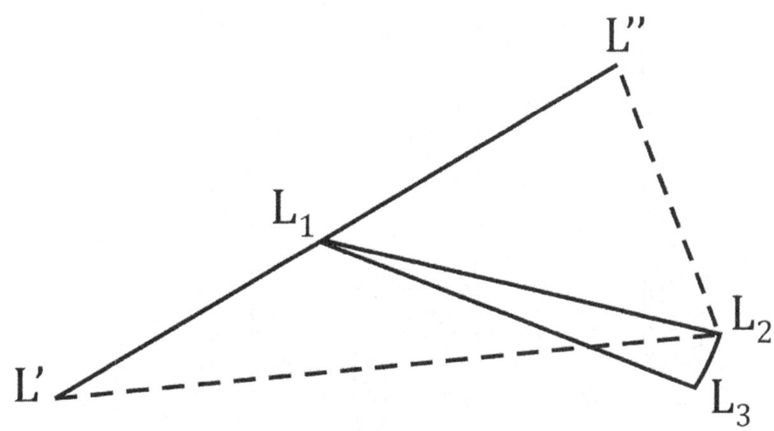

Figure A.19: Figure 19.

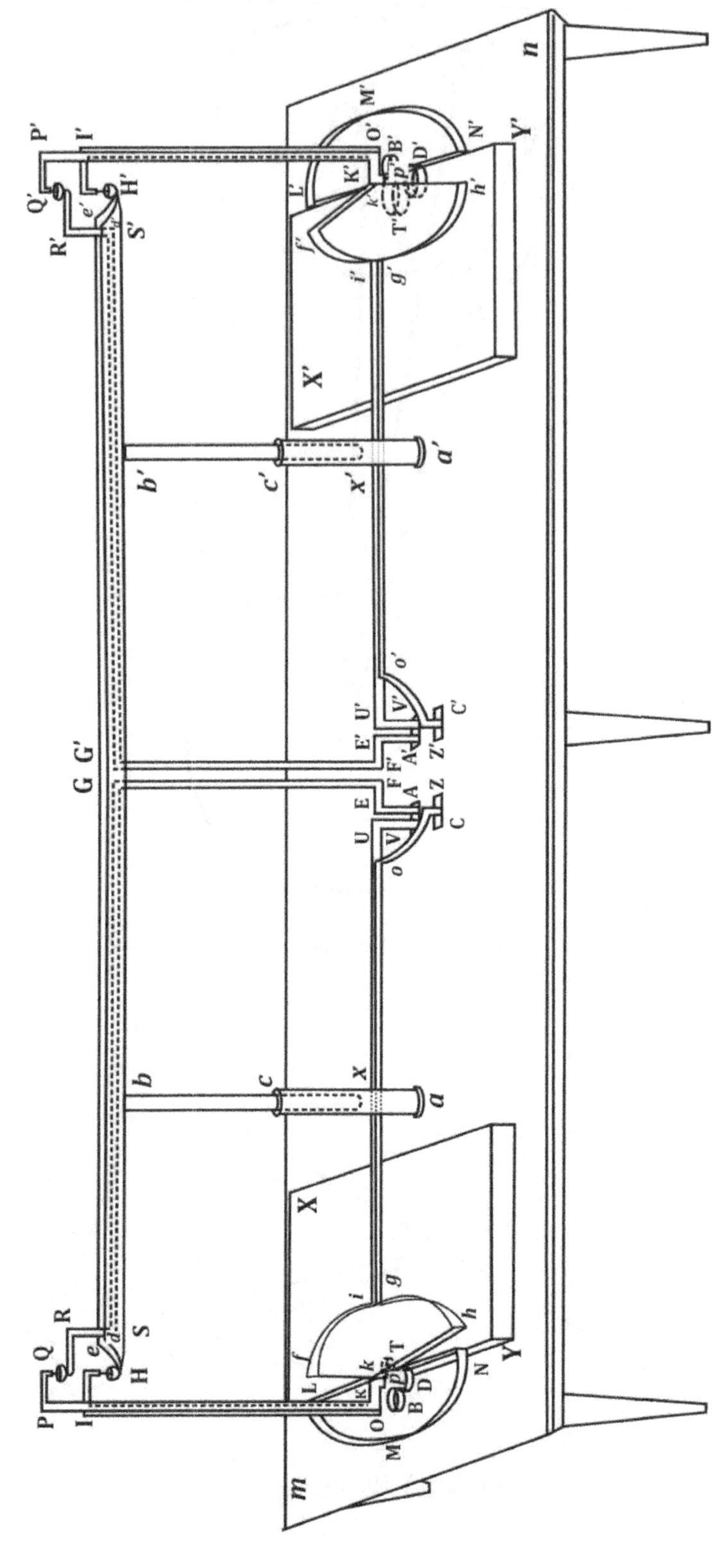

Figure A.20: Figure 20.

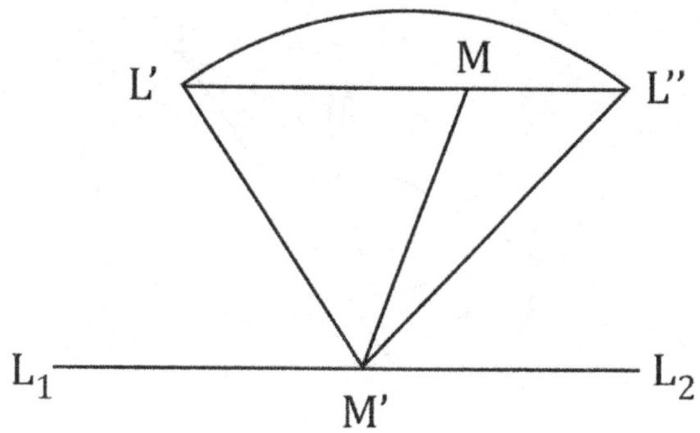

Figure A.21: Figure 21.

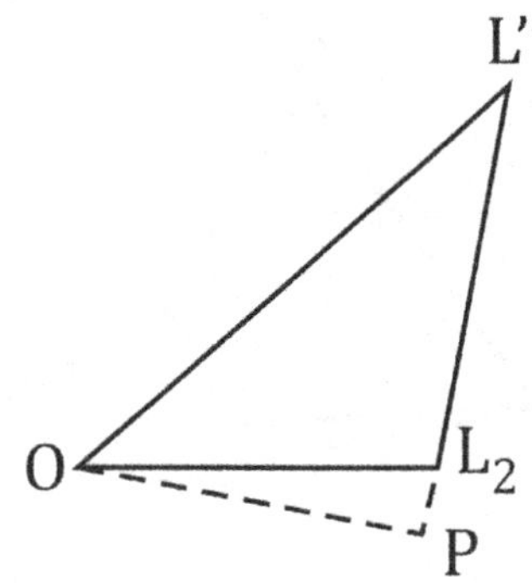

Figure A.22: Figure 22.

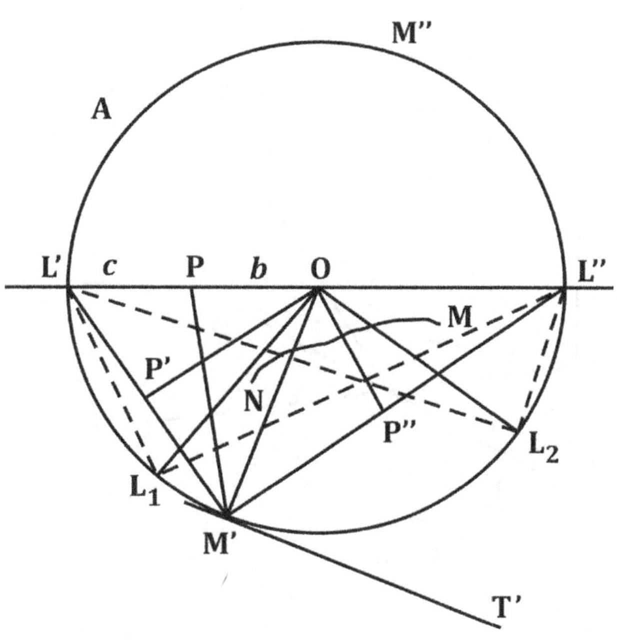

Figure A.23: Figure 23.

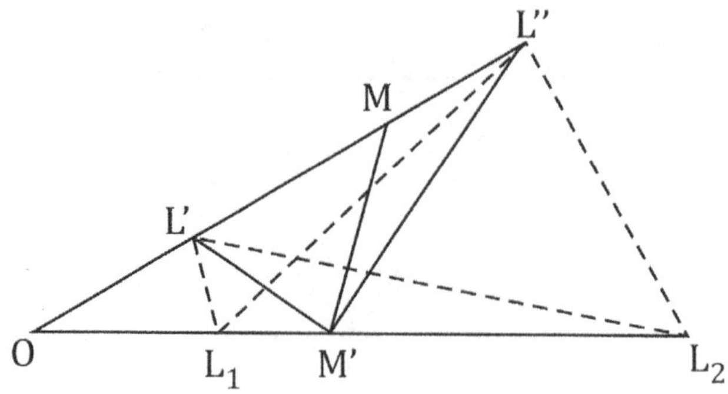

Figure A.24: Figure 24.

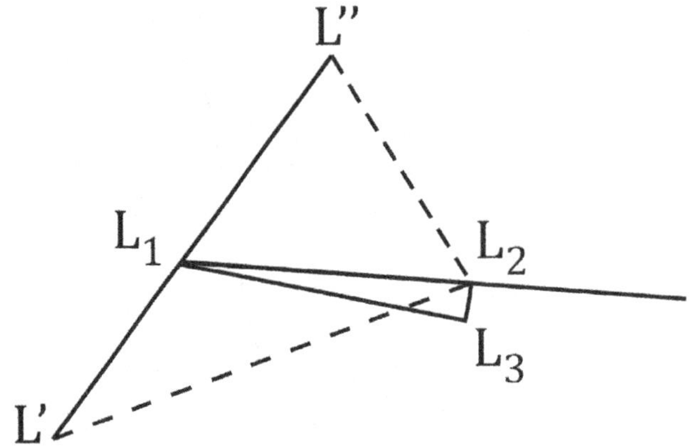

Figure A.25: Figure 25.

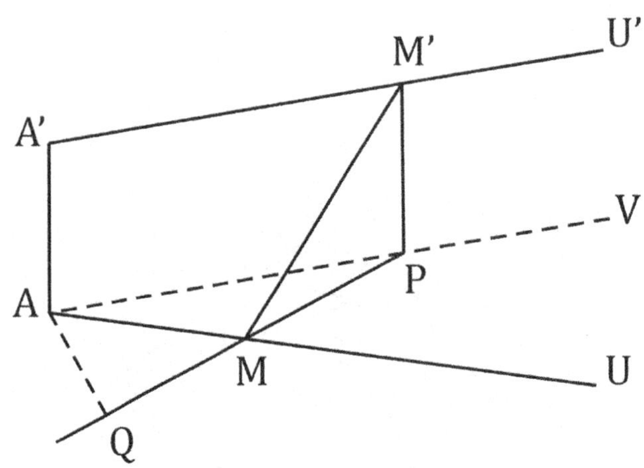

Figure A.26: Figure 26.

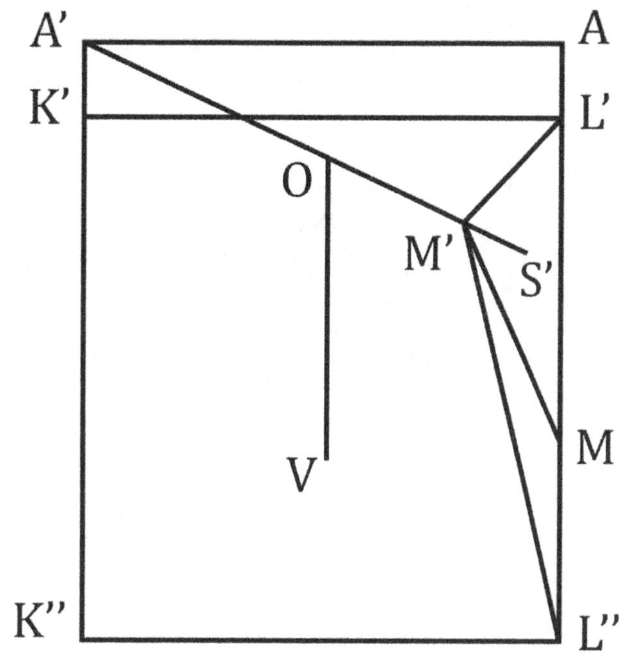

Figure A.27: Figure 27.

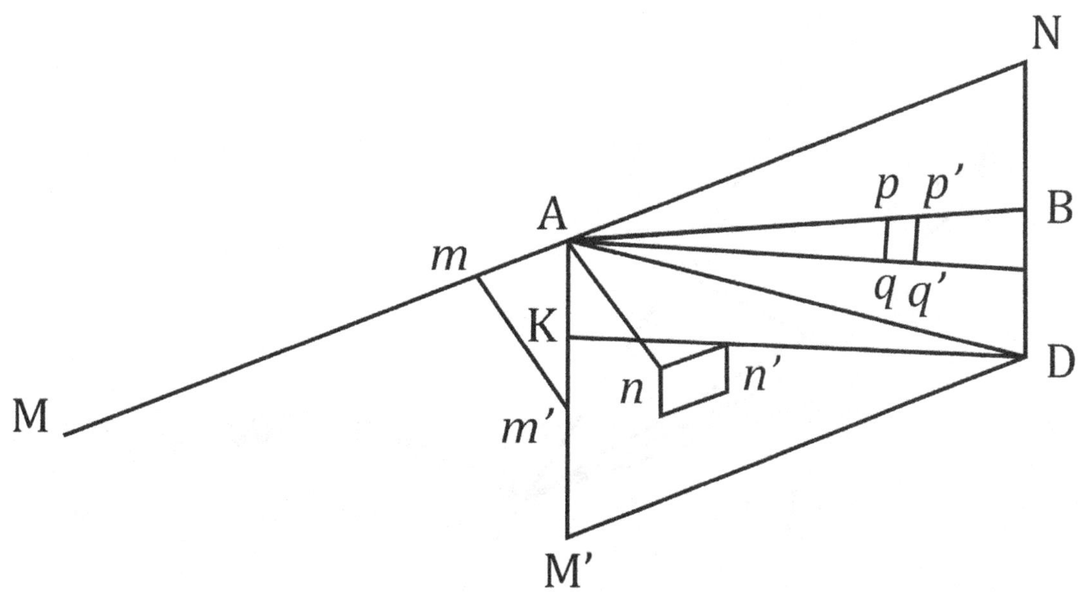

Figure A.28: Figure 28.

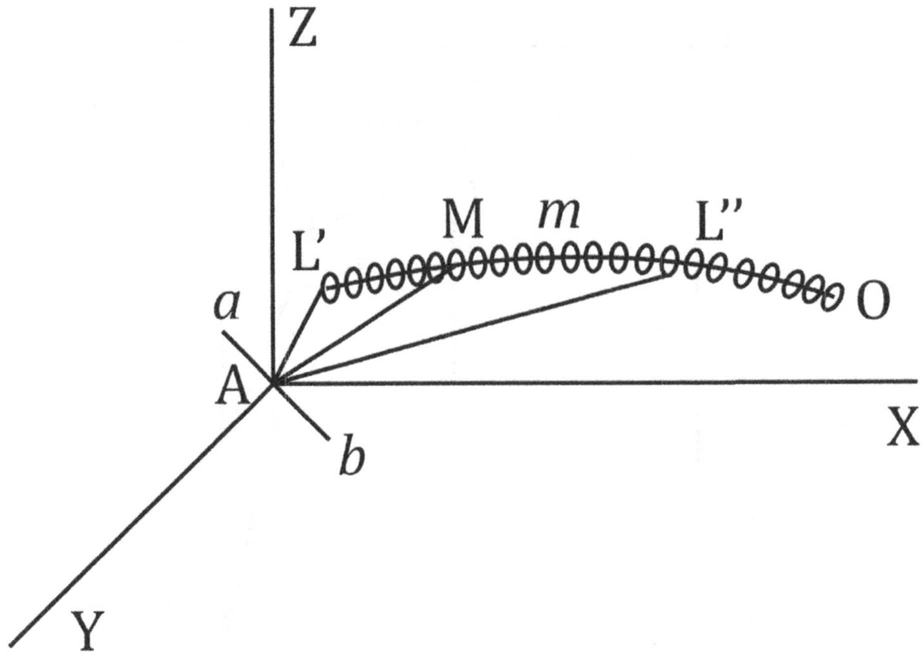

Figure A.29: Figure 29.

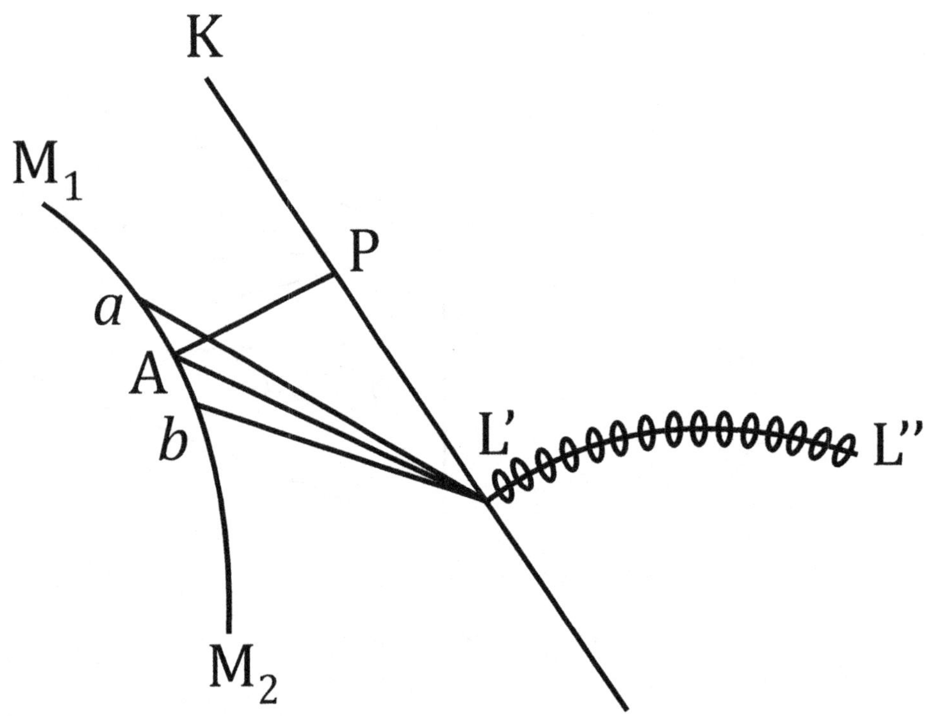

Figure A.30: Figure 30.

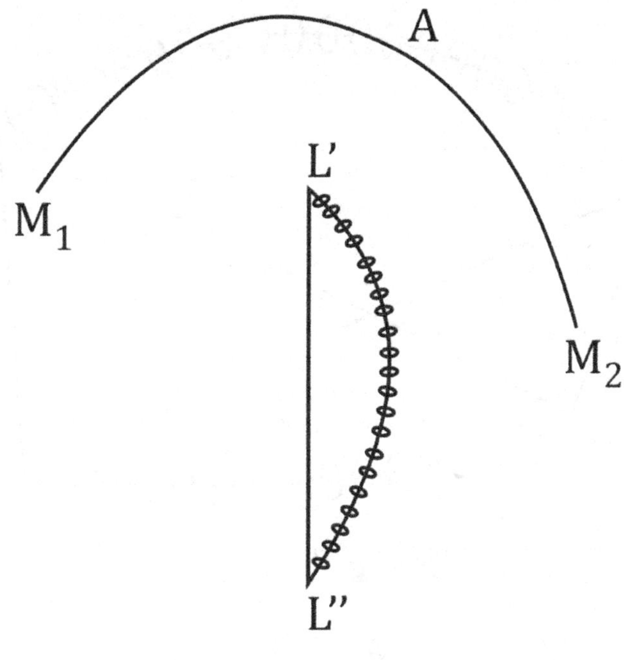

Figure A.31: Figure 31.

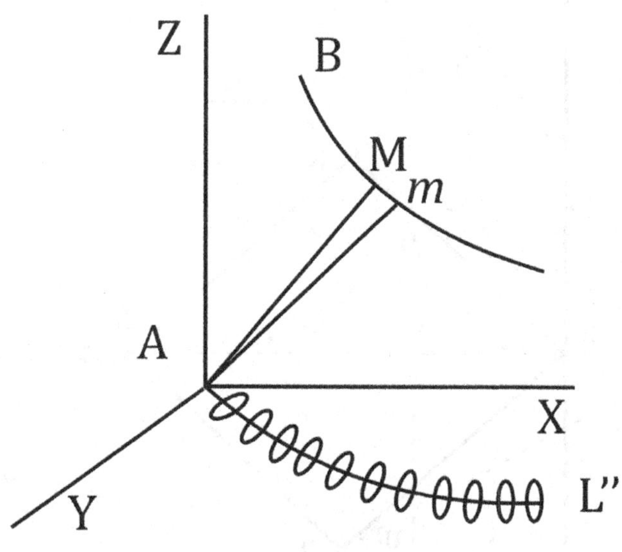

Figure A.32: Figure 32.

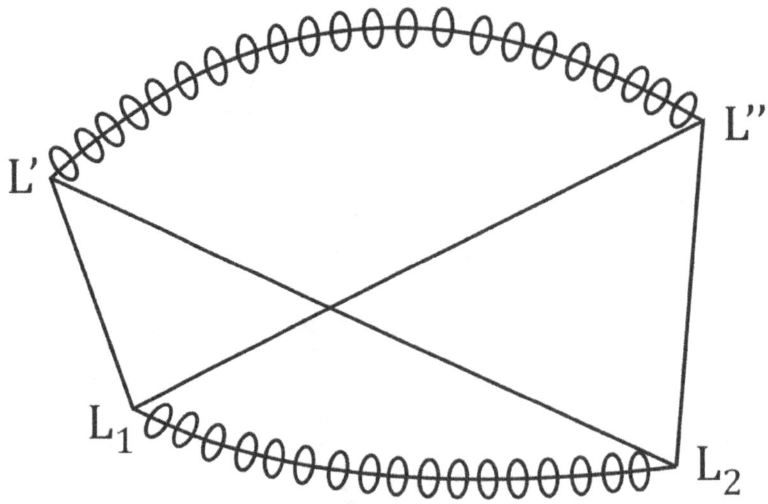

Figure A.33: Figure 33.

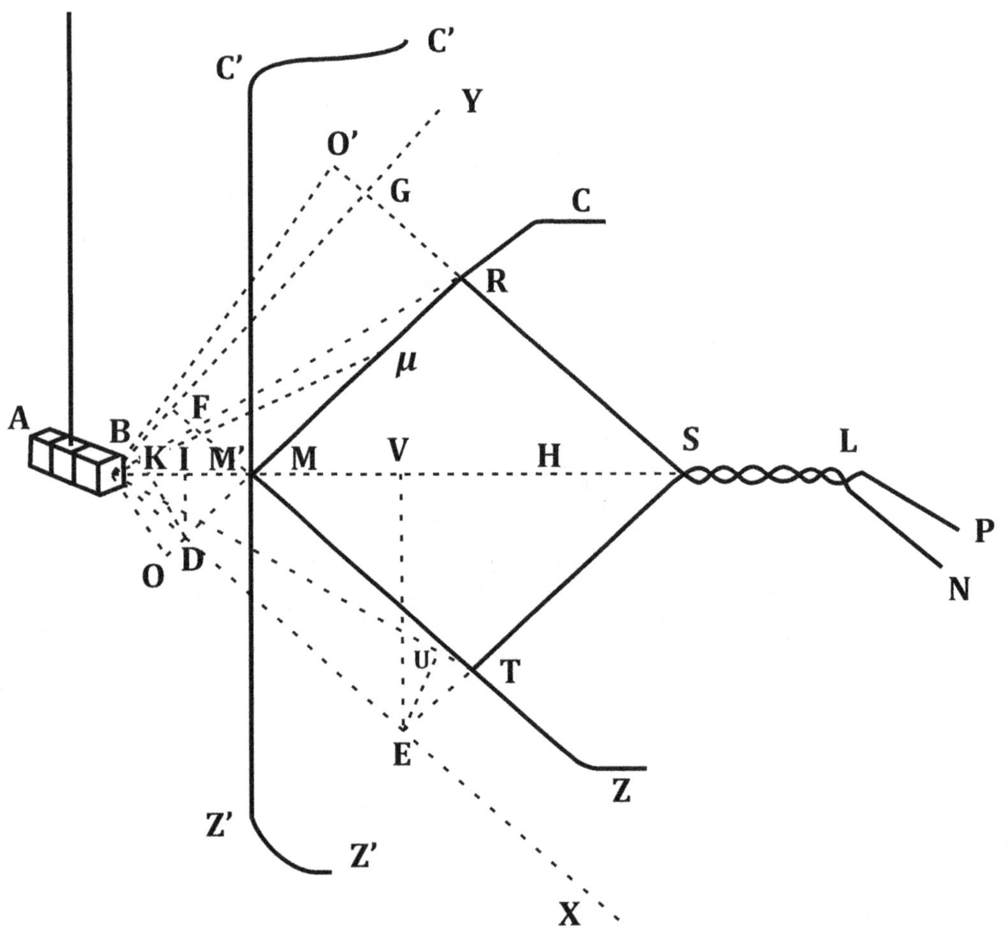

Figure A.34: Figure 34.

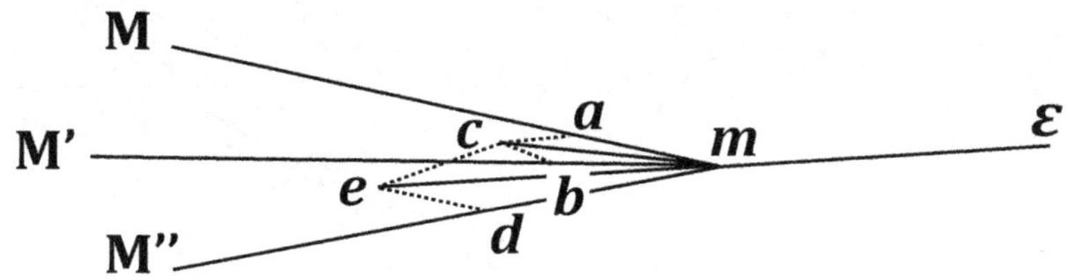

Figure A.35: Figure 35.

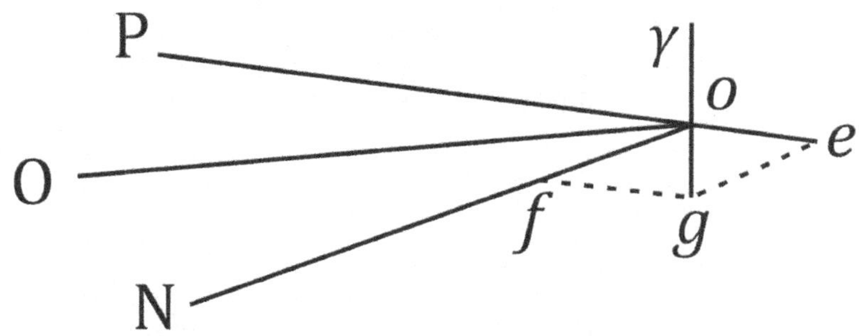

Figure A.36: Figure 36.

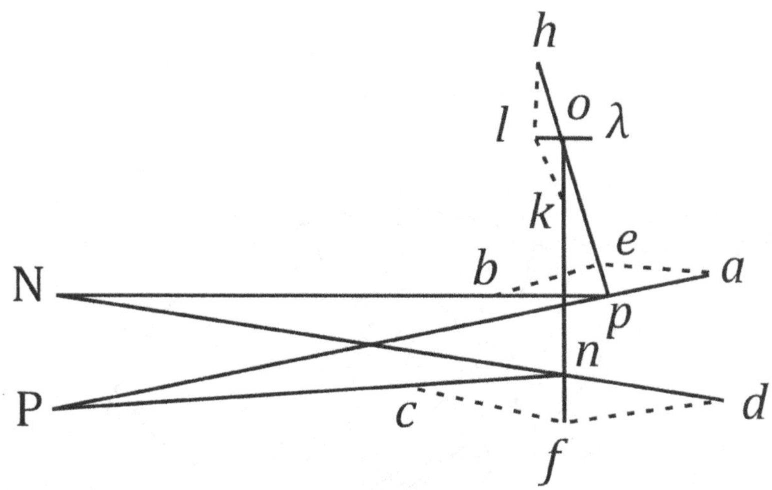

Figure A.37: Figure 37.

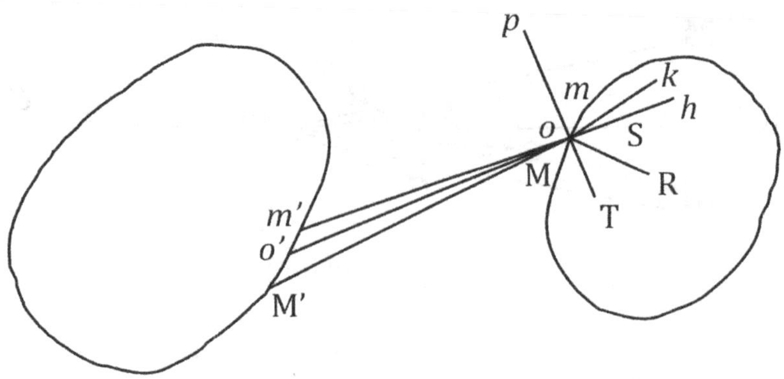

Figure A.38: Figure 38.

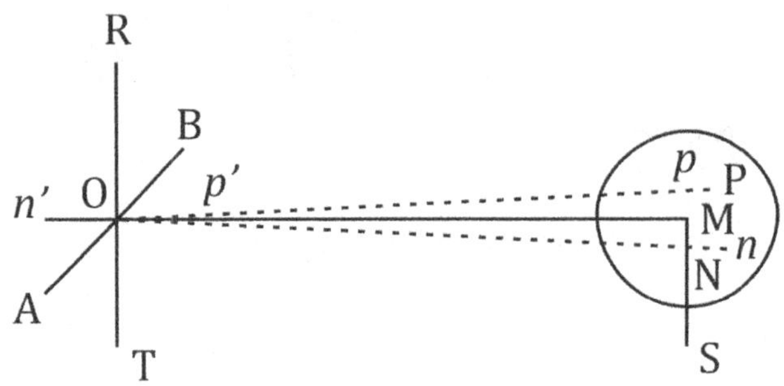

Figure A.39: Figure 39.

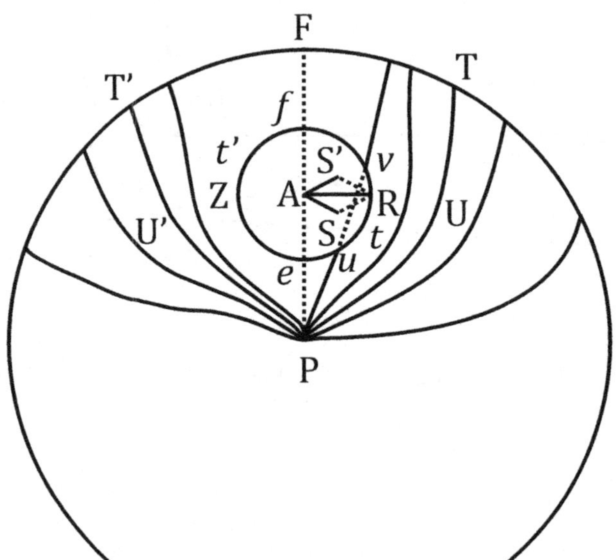

Figure A.40: Figure 40.

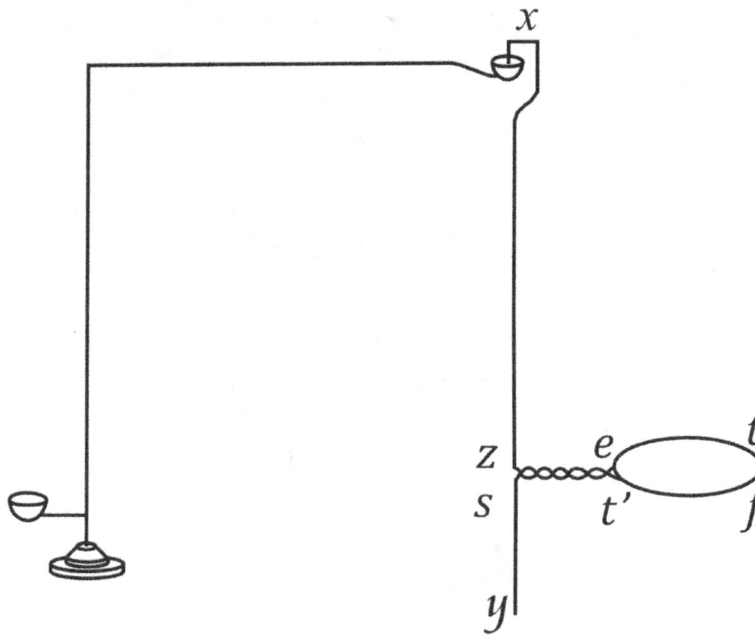

Figure A.41: Figure 41.

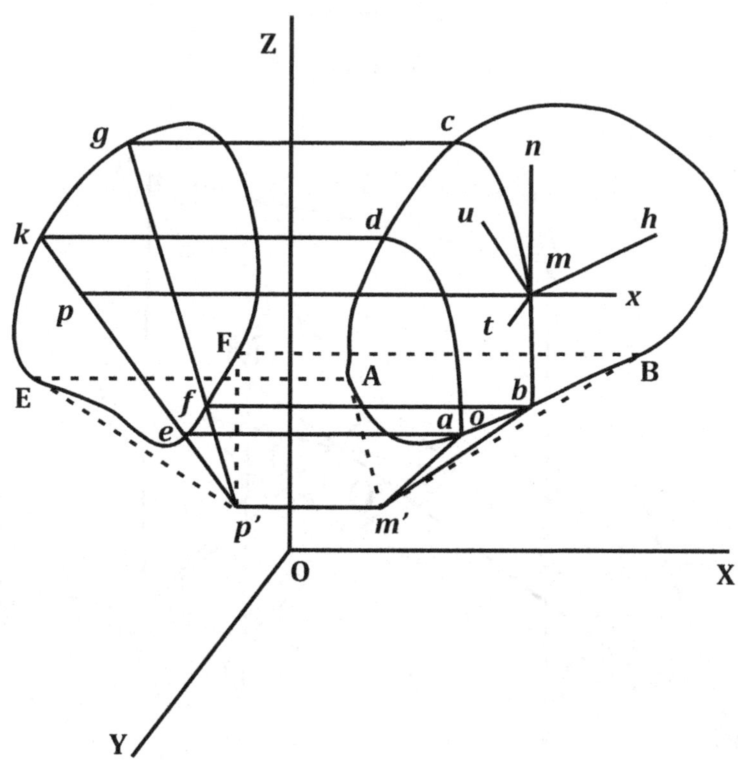

Figure A.42: Figure 42.

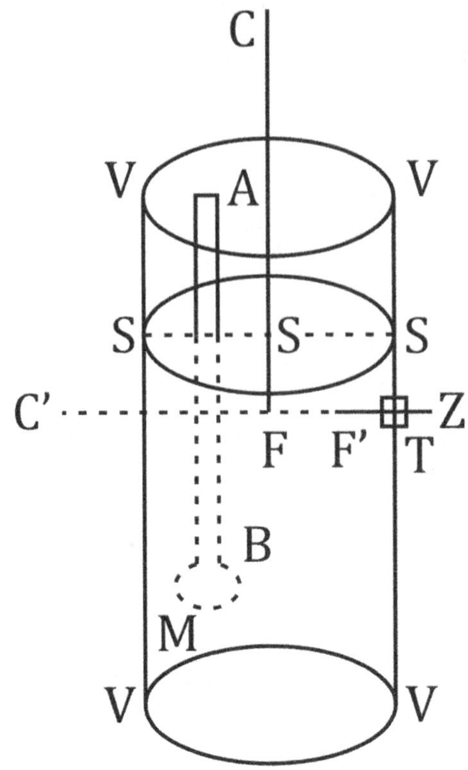

Figure A.43: Figure 43.

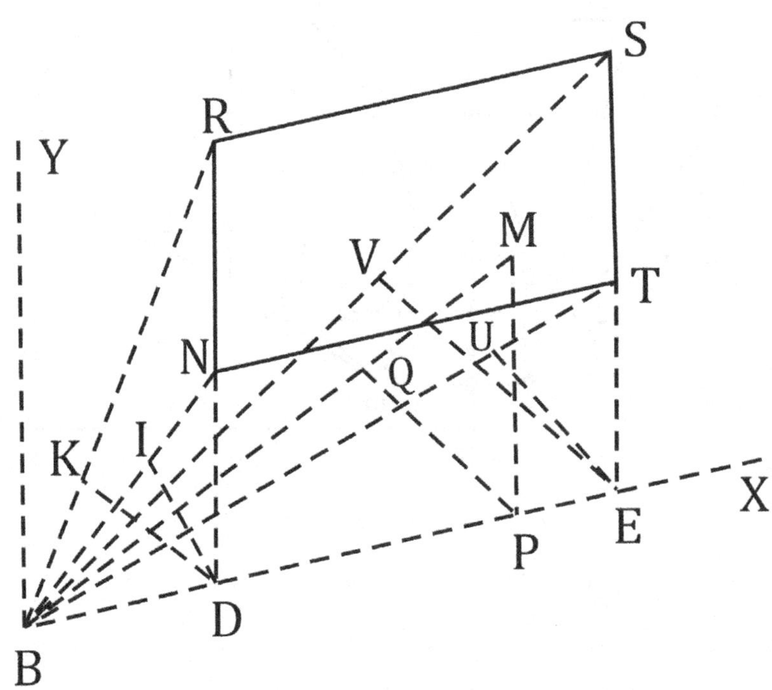

Figure A.44: Figure 44.

Bibliography

[Ampère, a] Ampère, A.-M. Analyse des mémoires lus par M. Ampère a l'Académie des Sciences, dans les séances des 18 et 25 septembre, des 9 et 30 octobre 1820. 20 pages. Reprint of Annales Générales des Sciences Physiques, Vol. 6, pp. 238-257 (1820).

[Ampère, b] Ampère, A.-M. Archives de l'Académie des Sciences de Paris.

[Ampère, c] Ampère, A.-M. Lettre de André-Marie Ampère à Gerrit Moll. Letter L1019 of 1821. Available at: `www.ampere.cnrs.fr`.

[Ampère, d] Ampère, A.-M. Lettre de André-Marie Ampère à Jean-Jacques Ampère. Letter L590 of 19 and 25 September 1820. Available at: `www.ampere.cnrs.fr`.

[Ampère, e] Ampère, A.-M. Lettre de André-Marie Ampère à Michael Faraday. Letter L622 of 10 July 1822. Available at: `www.ampere.cnrs.fr`.

[Ampère, 1814] Ampère, A.-M. (1814). Lettre de M. Ampère à M. le comte Berthollet sur la détermination des proportions dans lesquelles les corps se combinent d'après le nombre et la disposition respective des molécules dont les parties intégrantes sont composées. *Annales de Chimie et de Physique*, 90:43–86.

[Ampère, 1815] Ampère, A.-M. (1815). Letter from M. Ampère to Count Berthollet, on the determination of the proportions in which bodies are combined, according to the respective number and arrangement of the molecules of which their integrant molecules are composed. *The Philosophical Magazine and Journal*, 45:41–43, 109–116, 188–193 and 344–349.

[Ampère, 1816a] Ampère, A.-M. (1816a). Essai d'une classification naturelle pour les corps simples. *Annales de Chimie et de Physique*, 1:295–308 and 373–394.

[Ampère, 1816b] Ampère, A.-M. (1816b). Essay towards a natural classification of simple bodies. *The Philosophical Magazine and Journal*, 47:438–446.

[Ampère, 1820a] Ampère, A.-M. (1820a). Analyse des mémoires lus par M. Ampère a l'Académie des Sciences, dans les séances des 18 et 25 septembre, des 9 et 30 octobre 1820. *Annales Générales des Sciences Physiques*, 6:238–257.

[Ampère, 1820b] Ampère, A.-M. (1820b). Exposition du moyen par lequel il est facile de s'assurer directement, et par des expériences précises, de l'exactitude de la loi des attractions et répulsions des courans électriques, suivie de quelques observations sur cette loi. Mémoire lu le 26 décembre 1820. *Annales de Mines*, 5:553–558.

[Ampère, 1820c] Ampère, A.-M. (1820c). Mémoire présenté à l'Académie royale des Sciences, le 2 octobre 1820, où se trouve compris le résumé de ce qui avait été lu à la même Académie les 18 et 25 septembre 1820, sur les effets des courans électriques. *Annales de Chimie et de Physique*, 15:59–76.

[Ampère, 1820d] Ampère, A.-M. (1820d). Mémoire sur l'expression analytique des attractions et répulsions des courans électriques. Lu le 4 décembre 1820. *Annales de Mines*, 5:546–553.

[Ampère, 1820e] Ampère, A.-M. (1820e). Note sur un Mémoire lu à l'Académie royale des Sciences, dans la séance du 4 décembre 1820. *Journal de Physique, de Chimie, d'Histoire Naturelle et des Arts*, 91:226–230.

[Ampère, 1820f] Ampère, A.-M. (1820f). Suite du Mémoire sur l'Action mutuelle entre deux courans électriques, entre un courant électrique et un aimant ou le globe terrestre, et entre deux aimants. *Annales de Chimie et de Physique*, 15:170–218.

[Ampère, 1821a] Ampère, A.-M. (1821a). Lettre de M. Ampère a M. Erman, Secrétaire perpétuel de l'Académie royale. *Journal de Physique, de Chimie, d'Histoire Naturelle et des Arts*, 92:304–309.

[Ampère, 1821b] Ampère, A.-M. (1821b). Note sur deux Mémoires lus par M. Ampère à l'Académie royale des Sciences, le premier dans la séance du 26 décembre 1820; le second dans les séances des 8 et 15 janvier 1821. *Journal de Physique, de Chimie, d'Histoire Naturelle et des Arts*, 92:160–165.

[Ampère, 1821c] Ampère, A.-M. (1821c). Note sur un appareil à l'aide duquel on peut vérifier toutes les propriétés des conducteurs de l'électricité voltaïque, découvertes par M. Ampère. *Annales de Chimie et de Physique*, 18:88–106.

[Ampère, 1821d] Ampère, A.-M. (1821d). Réponse de M. Ampère a la lettre de M. Van Beck, sur une nouvelle expérience électro-magnétique. *Journal de Physique, de Chimie, d'Histoire Naturelle et des Arts*, 93:447–467.

[Ampère, 1821e] Ampère, A.-M. (1821e). Suite de la Note sur un Appareil à l'aide duquel on peut vérifier toutes les propriétés des conducteurs de l'électricité voltaïque, découvertes par M. Ampère. *Annales de Chimie et de Physique*, 18:313–333.

[Ampère, 1822s] Ampère, A.-M. (1821-1822s). Nouvelles Expériences électro-magnétiques. *Mémoires de l'Académie royale des Sciences de l'Institut de France*, 5:44–47. Histoire de l'Académie. Analyse des travaux de l'Académie royale des sciences pendant l'année 1821, partie mathématique; par M. le chevalier Delambre, secrétaire perpétuel. Despite this date, the volume was published only in 1826.

[Ampère, 1822a] Ampère, A.-M. (1822a). Addition au Mémoire précédent. In Ampère, A.-M., editor, *Recueil d'Observations Électro-dynamiques*, pages 286–292. Crochard, Paris. Despite this date, the volume of the Recueil was only published in 1823.

[Ampère, 1822b] Ampère, A.-M. (1822b). Additions au Mémoire précédent — Extrait d'un Mémoire présenté à l'Académie royale des Sciences, dans la séance du 16 septembre 1822. In Ampère, A.-M., editor, *Recueil d'Observations Électro-dynamiques*, pages 319–324. Crochard, Paris. Despite this date, the volume of the Recueil was only published in 1823.

[Ampère, 1822c] Ampère, A.-M. (1822c). De l'Action qu'exerce la Terre sur les conducteurs voltaïques. In Ampère, A.-M., editor, *Recueil d'Observations Électro-dynamiques*, pages 259–261. Crochard, Paris. Despite this date, the volume of the Recueil was only published in 1823.

[Ampère, 1822d] Ampère, A.-M. (1822d). Expériences relatives à de nouveaux phénomènes électro-dynamiques. *Annales de Chimie et de Physique*, 20:60–74.

[Ampère, 1822e] Ampère, A.-M. (1822e). Expériences relatives aux nouveaux phénomènes électro-dynamiques que j'ai obtenus au mois de décembre 1821. In Ampère, A.-M., editor, *Recueil d'Observations Électro-dynamiques*, pages 237–250. Crochard, Paris. Despite this date, the volume of the Recueil was only published in 1823.

[Ampère, 1822f] Ampère, A.-M. (1822f). Exposé méthodique des phénomènes électro-dynamiques et des lois de ces phénomènes. *Journal de Physique, de Chimie, d'Histoire Naturelle et des Arts*, 95:248–257.

[Ampère, 1822g] Ampère, A.-M. (1822g). Exposé méthodique des phénomènes électro-dynamiques et des lois de ces phénomènes. *Bulletin de la Société Philomatique*, November:177–183.

[Ampère, 1822h] Ampère, A.-M. (1822h). Exposé méthodique des phénomènes électrodynamiques et des lois de ces phénomènes. In Ampère, A.-M., editor, *Recueil d'Observations Électro-dynamiques*, pages 325–344. Crochard, Paris. Despite this date, the volume of the Recueil was only published in 1823.

[Ampère, 1822i] Ampère, A.-M. (1822i). Exposé sommaire des nouvelles Expériences électro-magnétiques faites par différens Physiciens, depuis le mois de mars 1821, lu dans la séance publique de l'Académie royale des Sciences, le 8 avril 1822. *Jounal de Physique*, 94:61–66.

[Ampère, 1822j] Ampère, A.-M. (1822j). Exposé sommaire des nouvelles Expériences électro-magnétiques faites par différens Physiciens, depuis le mois de mars 1821, lu dans la séance publique de l'Académie royale des Sciences, le 8 avril 1822. In Ampère, A.-M., editor, *Recueil d'Observations Électro-dynamiques*, pages 199–206. Crochard, Paris. Despite this date, the volume of the Recueil was only published in 1823.

[Ampère, 1822k] Ampère, A.-M. (1822k). Extrait d'un Mémoire lu à l'Académie royale des Sciences, dans la séance du 16 septembre 1822. *Bulletin des Sciences de la Société Philomatique de Paris*, (October):145–147.

[Ampère, 1822l] Ampère, A.-M. (1822l). Extrait d'une Lettre de M. Ampère à M. Faraday (Paris, 18 avril 1823). In Ampère, A.-M., editor, *Recueil d'Observations Électro-dynamiques*, pages 365–378. Crochard, Paris. Despite this date, the volume of the Recueil was only published in 1823. This letter does not appear in the Table of Contents at the end of the volume.

[Ampère, 1822m] Ampère, A.-M. (1822m). Extrait d'une Lettre de Mr. Ampère au Prof. De La Rive sur des expériences électro-magnétiques. In Ampère, A.-M., editor, *Recueil d'Observations Électro-dynamiques*, pages 252–258. Crochard, Paris. Despite this date, the volume of the Recueil was only published in 1823.

[Ampère, 1822n] Ampère, A.-M. (1822n). Lettre de Mr. Ampère a Mr. Erman, secrétaire de l'Académie Royale de Berlin. In Ampère, A.-M., editor, *Recueil d'Observations Électro-dynamiques*, pages 113–120. Crochard, Paris. Despite this date, the volume of the Recueil was only published in 1823.

[Ampère, 1822o] Ampère, A.-M. (1822o). Mémoire sur la Détermination de la formule qui représente l'action mutuelle de deux portions infiniment petites de conducteurs voltaïques. Lu à l'Académie royale des Sciences, dans la séance du 10 juin 1822. *Annales de Chimie et de Physique*, 20:398–421.

[Ampère, 1822p] Ampère, A.-M. (1822p). Notes sur cet exposé des nouvelles Expériences relatives aux Phénomènes produits par l'action électrodynamique, faites depuis le mois de mars 1821. In Ampère, A.-M., editor, *Recueil d'Observations Électro-dynamiques*, pages 207–236. Crochard, Paris. Despite this date, the volume of the Recueil was only published in 1823.

[Ampère, 1822q] Ampère, A.-M. (1822q). Notice sur les Expériences électro-magnétiques de MM. Ampère et Arago, lue à la séance publique de l'Académie royale des Sciences de Paris, le 2 avril 1821. In Ampère, A.-M., editor, *Recueil d'Observations Électro-dynamiques*, pages 109–112. Crochard, Paris. Reprinted from *Bibliothèque universelle*, Vol. 17, pp. 16-20 (1821). Despite this date, the volume of the Recueil was only published in 1823.

[Ampère, 1822r] Ampère, A.-M. (1822r). Notice sur les nouvelles expériences electro-magnétiques faites par différens physiciens, depuis le mois de mars 1821, lue dans la séance publique de l'académie royale des sciences, le 8 avril 1822. *Journal de Physique, de Chimie, d'Histoire Naturelle et des Arts*, 94:61–66.

[Ampère, 1822t] Ampère, A.-M. (1822t). Nouvelles expériences électro-magnétiques de MM. Faraday, Ampère, H. Davy, et De La Rive. *Bulletin des Sciences de la Société Philomatique de Paris*, 9:21–23.

[Ampère, 1822u] Ampère, A.-M. (1822u). Observation additionnelle par M. Ampère. In Ampère, A.-M., editor, *Recueil d'Observations Électro-dynamiques*, pages 354–364. Crochard, Paris. Despite this date, the volume of the Recueil was only published in 1823.

[Ampère, 1822v] Ampère, A.-M. (1822v). Premier Mémoire De l'Action exercée sur un courant électrique, par un autre courant, le globe terrestre ou un aimant. In Ampère, A.-M., editor, *Recueil d'Observations Électro-dynamiques*, pages 3–68. Crochard, Paris. Despite this date, the volume of the Recueil was only published in 1823.

[Ampère, 1822w] Ampère, A.-M. (1822w). *Recueil d'observations électro-dynamiques, contenant divers mémoires, notices, extraits de lettres ou d'ouvrages périodiques sur les sciences, relatifs à l'action mutuelle de deux courans électriques, à celle qui existe entre un courant électrique et un aimant ou le globe terrestre, et à celle de deux aimans l'un sur l'autre.* Crochard, Paris. Despite this date, the volume was only published in 1823, as there is on page 345 an extract made by Savary of a work he presented to the Academy of Sciences in 1823.

[Ampère, 1822x] Ampère, A.-M. (1822x). Réponse de M. Ampère à la Lettre de M. Van Beck, sur une nouvelle Expérience électro-magnétique. In Ampère, A.-M., editor, *Recueil d'Observations Électro-dynamiques*, pages 169–198. Crochard, Paris. Despite this date, the volume of the Recueil was only published in 1823.

[Ampère, 1822y] Ampère, A.-M. (1822y). Second mémoire sur la Détermination de la formule que représente l'action mutuelle de deux portions infiniment petites de conducteurs voltaïques. In Ampère, A.-M., editor, *Recueil d'Observations Électro-dynamiques*, pages 293–318. Crochard, Paris. Despite this date, the volume of the Recueil was only published in 1823.

[Ampère, 1822z] Ampère, A.-M. (1822z). Untitled note appended to Auguste de La Rive (1822): "Sur l'action qu'exerce le globe terrestre sur une portion mobile du circuit voltaïque," Annales de Chimie et de Physique, 21: 24-48. *Annales de Chimie et de Physique*, 21:48–53.

[Ampère, 1823a] Ampère, A.-M. (1823a). *Exposé Méthodique des Phénomènes Électro-dynamiques, et des Lois de Ces Phénomènes.* Bachelier, Paris.

[Ampère, 1823b] Ampère, A.-M. (1823b). Extrait d'une Lettre de M. Ampère à M. Faraday. *Annales de Chimie et de Physique*, 22:389–400.

[Ampère, 1823c] Ampère, A.-M. (1823c). Mémoire sur la théorie mathématique des phénomènes électro-dynamiques uniquement déduite de l'expérience, dans lequel se trouvent réunis les Mémoires que M. Ampère a communiqués à l'Académie royale des Sciences, dans les séances des 4 et 26 décembre 1820, 10 juin 1822, 22 décembre 1823, 12 septembre et 21 novembre 1825. *Mémoires de l'Académie Royale des Sciences de l'Institut de France*, 6:175–387. Despite this date, the work was only published in 1827.

[Ampère, 1823d] Ampère, A.-M. (1823d). New Electro-Magnetical Experiments. *The Quarterly Journal of Science, Literature, and the Arts*, 14:441–442.

[Ampère, 1824a] Ampère, A.-M. (1824a). Description d'un Appareil électro-dynamique. *Annales de Chimie et de Physique*, 26:390–411.

[Ampère, 1824b] Ampère, A.-M. (1824b). *Description d'un Appareil Électro-dynamique, construit par M. Ampère.* Crochard et Bachelier, Paris.

[Ampère, 1824c] Ampère, A.-M. (1824c). Extrait d'un Mémoire sur les Phénomènes électro-dynamiques. Communiqué à l'Academie royale des Sciences le 22 décembre 1823. *Annales de Chimie et de Physique*, 26:134–162.

[Ampère, 1824d] Ampère, A.-M. (1824d). Note sur une expérience relative à la nature du courant électrique faite para MM. Ampère et Becquerel. *Annales de Chimie et de Physique*, 27:29–31. Cette note a été lue par M. Ampère, dans la séance de l'Académie du 12 avril 1824.

[Ampère, 1824e] Ampère, A.-M. (1824e). *Précis de la Théorie des Phénomènes Électro-dynamiques, pour servir de supplément à son* Recueil d'Observations électro-dynamiques *et au* Manuel d'Électricité dynamique *de M. Demonferrand.* Crochard et Bachelier, Paris.

[Ampère, 1824f] Ampère, A.-M. (1824f). Suite de l'Extrait, fait par M. Ampère, de son Mémoire sur les Phénomènes électro-dynamiques. *Annales de Chimie et de Physique*, 26:246–258.

[Ampère, 1825a] Ampère, A.-M. (1825a). A letter from M. Ampere to M. Gerhardi on various electro-dynamic phenomena. *The Philosophical Magazine and Journal*, 66:368–373.

[Ampère, 1825b] Ampère, A.-M. (1825b). Lettre de M. Ampère à M. Gerhardi sur divers phénomènes électro-dynamiques. *Annales de Chimie et de Physique*, 29:373–381.

[Ampère, 1825c] Ampère, A.-M. (1825c). Memoir on a new electro-dynamic experiment, on its application to the formula representing the mutual action of the two elements of voltaic conductors, and on some new results deduced from that formula. *The Philosophical Magazine and Journal*, 66:373–387.

[Ampère, 1825d] Ampère, A.-M. (1825d). Mémoire sur une nouvelle Expérience électro-dynamique, sur son application à la formule qui répresente l'action mutuelle de deux élémens de conducteurs voltaïques, et sur de nouvelles conséquences déduites de cette formule. *Annales de Chimie et de Physique*, 29:381–404. Lu à l'Académie royale des Sciences, dans la séance du 12 septembre 1825.

[Ampère, 1825e] Ampère, A.-M. (1825e). *Mémoire sur une nouvelle expérience électro-dynamique, sur son application a la formule qui représente l'action mutuelle de deux élémens de deux conducteurs voltaïques, et sur de nouvelles conséquences déduites de cette formule; suivi d'une lettre a M. Gherardi, relative a l'explication de quelques phénomènes électro-dynamiques.* Crochard et Bachelier, Paris.

[Ampère, 1825f] Ampère, A.-M. (1825f). On the nature of the electric current. *The Quarterly Journal of Science, Literature, and the Arts*, 18:381–382.

[Ampère, 1825g] Ampère, A.-M. (1825g). Suite du Mémoire de M. Ampère sur une nouvelle Expérience électro-dynamique, sur son application à la formule qui représente l'action mutuelle de deux élémens de conducteurs voltaïques, et sur de nouvelles conséquences déduites de cette formule. *Annales de Chimie et de Physique*, 30:29–41.

[Ampère, 1826a] Ampère, A.-M. (1826a). *Description d'un Appareil Électro-dynamique, construit par M. Ampère*. Crochard et Bachelier, Paris. 2nd édition.

[Ampère, 1826b] Ampère, A.-M. (1826b). Extrait d'un Mémoire sur l'action exercée par un circuit électro-dynamique, formant une courbe plane dont les dimensions sont considerées comme infiniment petites; sur la manière d'y ramener celle d'un circuit fermé, quelles qu'en soient la forme et le grandeur; sur deux nouveaux instruments destinés à des expériences propres à rendre plus directe et à vérifier la détermination de la valeur de l'action mutuelle de deux éléments de conducteurs; sur l'identité des forces produites par des circuits infiniment petits, et par des particules d'aimant; enfin sur un nouveau théorème relatif à l'action de ces particules, lu à l'Académie royale des sciences dans sa séance du 21 novembre 1825. *Correspondence Mathématique et Physique royaume des Pays-Bas*, 2:35–47.

[Ampère, 1826c] Ampère, A.-M. (1826c). Note sur quelques Phénomènes électro-magnétiques. *Annales de Chimie et de Physique*, 32:432–443.

[Ampère, 1826d] Ampère, A.-M. (1826d). Note sur une nouvelle expérience électro-dynamique de M. Ampère, qui constate l'action d'un disque métallique en mouvement sur une portion de conducteur voltaïque pliée en hélice ou en spirale. *Nouveau Bulletin des Sciences de la Société Philomatique de Paris*, page 134.

[Ampère, 1826e] Ampère, A.-M. (1826e). Sequel of the memoir of M. Ampere on a new electro-dynamic experiment, on its application to the formula representing the mutual action of the two elements of voltaic conductors, and on new results deduced from that formula. *The Philosophical Magazine and Journal*, 67:37–45.

[Ampère, 1826f] Ampère, A.-M. (1826f). *Théorie des Phénomènes Électro-dynamiques, Uniquement Déduite de l'Expérience*. Méquignon-Marvis, Paris.

[Ampère, 1827a] Ampère, A.-M. (1827a). Mémoire sur l'action mutuelle d'un conducteur voltaïque et d'un aimant. *Nouveaux Mémoires de l'Académie Royale des Sciences et Belles-Lettres de Bruxelles*, 4:1–70. Présenté a la Séance du 28 octobre 1826.

[Ampère, 1827b] Ampère, A.-M. (1827b). On the action of a moving metallic disc, upon a portion of a voltaic conductor. *The Quarterly Journal of Science, Literature, and the Arts*, 23:228–229.

[Ampère, 1827c] Ampère, A.-M. (1827c). Supplément. Lettre a M. le Docteur Gherardi. *Nouveaux Mémoires de l'Académie Royale des Sciences et Belles-Lettres de Bruxelles*, 4:71–88.

[Ampère, 1828] Ampère, A.-M. (1828). Note sur l'Action mutuelle d'un Aimant et d'un Conducteur voltaïque. *Annales de Chimie et de Physique*, 37:113–139.

[Ampère, 1829] Ampère, A.-M. (1829). Démonstration élémentaire du principle de la gravitation universelle. *Annales de Mathématiques Pures et Appliquées*, 20:89–96.

[Ampère, 1832] Ampère, A.-M. (1832). Note de M. Ampère sur une expérience de M. Hippolyte Pixii, relative au courant produit par la rotation d'un aimant, à l'aide d'un appareil imaginé par M. Hippolyte Pixii. *Annales de Chimie et de Physique*, 51:76–79. Communiquée à l'Académie dans sa Séance du 29 octobre 1832.

[Ampère, 1833] Ampère, A.-M. (1833). M. Ampère's communication to the Academy of Sciences upon an experiment of M. Pixii relative to a current produced by the rotation of a magnet with an improved apparatus, Oct. 29, 1832. *The American Journal of Science and Arts*, 24:144–145. Translated from the Annales de Chim. et de Phys., Sept., 1832; by Oliver P. Hubbard, Assistant in the Chemical Department in Yale College.

[Ampère, 1834] Ampère, A.-M. (1834). *Essai sur la Philosophie des sciences*. Bachelier, Paris.

[Ampère, 1835a] Ampère, A.-M. (1835a). Note by M. Ampère on heat and light considered as the results of vibratory motion. *The Philosophical Magazine and Journal*, 7:342–349.

[Ampère, 1835b] Ampère, A.-M. (1835b). Note de M. Ampère sur la chaleur et la lumière considérées comme résultant de mouvemens vibratoires. *Annales de Chimie et de Physique*, 58:433–444.

[Ampère, 1843] Ampère, A.-M. (1843). *Essai sur la Philosophie des sciences,* second partie. Bachelier, Paris.

[Ampère, 1883] Ampère, A.-M. (1883). *Théorie Mathématique des Phénomènes Électro-dynamiques Uniquement Déduite de l'Expérience.* Hermann, Paris.

[Ampère, 1885a] Ampère, A.-M. (1885a). Analyse des Mémoires lus a l'Académie les 11 et 26 décembre 1820 et les 8 et 15 janvier 1821. In Joubert, J., editor, *Collection de Mémoires relatifs a la Physique,* Vol. II: *Mémoires sur l'Électrodynamique*, pages 136–140. Gauthier-Villars, Paris.

[Ampère, 1885b] Ampère, A.-M. (1885b). De l'action exercée sur un courant électrique par un autre courant, le globe terrestre ou un aimant. In Joubert, J., editor, *Collection de Mémoires relatifs a la Physique,* Vol. II: *Mémoires sur l'Électrodynamique*, pages 7–53. Gauthier-Villars, Paris.

[Ampère, 1885c] Ampère, A.-M. (1885c). De l'action qu'exerce la Terre sur les conducteurs voltaïques. In Joubert, J., editor, *Collection de Mémoires relatifs a la Physique,* Vol. II: *Mémoires sur l'Électrodynamique*, pages 305–307. Gauthier-Villars, Paris.

[Ampère, 1885d] Ampère, A.-M. (1885d). Expériences relatives aux nouveaux phénomènes électro-dynamiques obtenus au mois de décembre 1821. In Joubert, J., editor, *Collection de Mémoires relatifs a la Physique,* Vol. II: *Mémoires sur l'Électrodynamique*, pages 192–204. Gauthier-Villars, Paris.

[Ampère, 1885e] Ampère, A.-M. (1885e). Exposé sommaire des nouvelles expériences électromagnétiques faites par différens physiciens, depuis le mois de mars 1821, lu dans la séance publique de l'Académie royale des Sciences, le 8 avril 1822. In Joubert, J., editor, *Collection de Mémoires relatifs a la Physique,* Vol. II: *Mémoires sur l'Électrodynamique*, pages 238–244. Gauthier-Villars, Paris.

[Ampère, 1885f] Ampère, A.-M. (1885f). Extrait d'un Mémoire sur les Phénomènes Électrodynamiques; présenté à l'Académie royale des Sciences (Institut de France), dans la séance du 22 décembre 1823. In Joubert, J., editor, *Collection de Mémoires relatifs a la Physique,* Vol. II: *Mémoires sur l'Électrodynamique*, pages 395–410. Gauthier-Villars, Paris.

[Ampère, 1885g] Ampère, A.-M. (1885g). Extrait d'une lettre adressée a M. Faraday (Paris, 10 juillet 1822). In Joubert, J., editor, *Collection de Mémoires relatifs a la Physique,* Vol. II: *Mémoires sur l'Électrodynamique*, pages 293–299. Gauthier-Villars, Paris.

[Ampère, 1885h] Ampère, A.-M. (1885h). Extrait d'une lettre de M. Ampère a M. Faraday (Paris, 18 avril 1823). In Joubert, J., editor, *Collection de Mémoires relatifs a la Physique,* Vol. II: *Mémoires sur l'Électrodynamique*, pages 384–394. Gauthier-Villars, Paris.

[Ampère, 1885i] Ampère, A.-M. (1885i). Extrait d'une Lettre de Mr. Ampère au Prof. De La Rive sur des éxperiences électro-magnétiques (Paris, le 12 juin 1822). In Joubert, J., editor, *Collection de Mémoires relatifs a la Physique,* Vol. II: *Mémoires sur l'Électrodynamique*, pages 153–157. Gauthier-Villars, Paris.

[Ampère, 1885j] Ampère, A.-M. (1885j). Mémoire sur l'expression mathématique des attractions et des répulsions des courants électriques. In Joubert, J., editor, *Collection de Mémoires relatifs a la Physique,* Vol. II: *Mémoires sur l'Électrodynamique*, pages 128–135. Gauthier-Villars, Paris. Mémoire lu à l'Académie des Sciences, le 4 décembre 1820.

[Ampère, 1885k] Ampère, A.-M. (1885k). Note lue a la séance du 24 juin 1822. In Joubert, J., editor, *Collection de Mémoires relatifs a la Physique,* Vol. II: *Mémoires sur l'Électrodynamique*, pages 290–292. Gauthier-Villars, Paris.

[Ampère, 1885l] Ampère, A.-M. (1885l). Note relative au mémoire de M. Savary. In Joubert, J., editor, *Collection de Mémoires relatifs a la Physique,* Vol. II: *Mémoires sur l'Électrodynamique*, pages 376–383. Gauthier-Villars, Paris.

[Ampère, 1885m] Ampère, A.-M. (1885m). Notes sur cet exposé des nouvelles expériences relatives aux phénomènes produits par l'action électrodynamiques, faites depuis le mois de mars 1821. In Joubert, J., editor, *Collection de Mémoires relatifs a la Physique,* Vol. II: *Mémoires sur l'Électrodynamique*, pages 244–269. Gauthier-Villars, Paris.

[Ampère, 1885n] Ampère, A.-M. (1885n). Notice sur quelques expériences nouvelles relatives à l'action mutuelle de deux portions de circuit voltaïque et a la production des courants électriques par influence, et sur les circonstances dans lesquelles l'action électrodynamique doit, d'après la théorie, produire dans un conducteur mobile autour d'un axe fixe un mouvement de rotation continu, ou donner a ce conducteur une direction fixe. In Joubert, J., editor, *Collection de Mémoires relatifs a la Physique,* Vol. II: *Mémoires sur l'Électrodynamique*, pages 329–337. Gauthier-Villars, Paris. Présenté à l'Academie royale des Sciences, le 16 septembre 1822.

[Ampère, 1885o] Ampère, A.-M. (1885o). Réponse a la lettre de M. van Beck, sur une nouvelle expérience électromagnétique. In Joubert, J., editor, *Collection de Mémoires relatifs a la Physique,* Vol. II: *Mémoires sur l'Électrodynamique*, pages 212–237. Gauthier-Villars, Paris. Letter of 1821.

[Ampère, 1885p] Ampère, A.-M. (1885p). Second mémoire sur la détermination de la formule qui représente l'action mutuelle de deux portions infiniment petites de conducteurs voltaïques, lu à l'Académie royale des Sciences, le 10 juin 1822. In Joubert, J., editor, *Collection de Mémoires relatifs a la Physique,* Vol. II: *Mémoires sur l'Électrodynamique*, pages 270–289. Gauthier-Villars, Paris.

[Ampère, 1887a] Ampère, A.-M. (1887a). Lettre a M. le d'Gherardi. Supplemént au mémoire sur l'action mutuelle d'un conducteur voltaïque et d'un aimant. In Joubert, J., editor, *Collection de Mémoires relatifs a la Physique,* Vol. III: *Mémoires sur l'Électrodynamique*, pages 275–288. Gauthier-Villars, Paris.

[Ampère, 1887b] Ampère, A.-M. (1887b). Lettre de M. Ampère a M. Gherardi, sur divers phénomènes électro-dynamiques. In Joubert, J., editor, *Collection de Mémoires relatifs a la Physique,* Vol. III: *Mémoires sur l'Électrodynamique*, pages 217–223. Gauthier-Villars, Paris.

[Ampère, 1887c] Ampère, A.-M. (1887c). Mémoire communiqué a l'Academie royale des Sciences dans sa Séance du 21 novembre 1825, faisant suite au Mémoire lu dans la Séance du 12 septembre. In Joubert, J., editor, *Collection de Mémoires relatifs a la Physique,* Vol. III: *Mémoires sur l'Électrodynamique*, pages 194–202. Gauthier-Villars, Paris.

[Ampère, 1887d] Ampère, A.-M. (1887d). Mémoire sur la Théorie Mathématique des Phénomènes Électrodynamiques, Uniquement Déduite de l'Expérience. In Joubert, J., editor, *Collection de Mémoires relatifs a la Physique,* Vol. III: *Mémoires sur l'Électrodynamique*, pages 1–193. Gauthier-Villars, Paris.

[Ampère, 1887e] Ampère, A.-M. (1887e). Mémoire sur l'action mutuelle d'un conducteur voltaïque et d'un aimant. In Joubert, J., editor, *Collection de Mémoires relatifs a la Physique,* Vol. III: *Mémoires sur l'Électrodynamique*, pages 224–274. Gauthier-Villars, Paris.

[Ampère, 1887f] Ampère, A.-M. (1887f). Précis d'un Mémoire lu a l'Académie royale des Sciences dans sa Séance du 21 novembre 1825. In Joubert, J., editor, *Collection de Mémoires relatifs a la Physique,* Vol. III: *Mémoires sur l'Électrodynamique*, pages 203–216. Gauthier-Villars, Paris.

[Ampère, 1921] Ampère, A.-M. (1921). *Mémoires sur Électromagnétisme et l'Électrodynamique.* Gauthier-Villars, Paris.

[Ampère, 1936] Ampère, A.-M. (1936). Lettre de M. Ampère à M. Faraday, 13 avril 1833. In Launay, L. d., editor, *Correspondance du Grand Ampère*, volume 2, pages 763–770. Gauthier-Villars, Paris.

[Ampère, 1958] Ampère, A.-M. (1958). *Théorie Mathématique des Phénomènes Électro-dynamiques Uniquement Déduite de l'Expérience.* Blanchard, Paris. Edited by E. Bauer.

[Ampère, 1964] Ampère, A.-M. (1964). *Extraits du mémoire de l'action exercée sur un courant électrique par un autre courant, le globe terrestre ou un aimant.* In L. Poudensan, *Ampère, un Savant dans le Texte*, pages 26–50. Gauthiers-Villars, Paris.

[Ampère, 1965a] Ampère, A.-M. (1965a). The effects of electric currents. In R. A. R. Tricker, *Early Electrodynamics — The First Law of Circulation*, pages 140–154, New York. Pergamon. Partial English translation by O. M. Blunn of Ampère's works published in *Annales de Chimie et de Physique*, Vol. 15, pp. 59-76 and 170-218 (1820).

[Ampère, 1965b] Ampère, A.-M. (1965b). On the Mathematical Theory of Electrodynamic Phenomena, Experimentally Deduced. In R. A. R. Tricker, *Early Electrodynamics — The First Law of Circulation*, pages 155–200, New York. Pergamon. Partial English translation by O. M. Blunn of Ampère's work "Mémoire sur la théorie mathématique des phénomènes électro-dynamiques uniquement déduite de l'expérience", *Mémoires de l'Académie royale des Sciences de l'Institut de France*, Vol. 6, pp. 175-387 (1823), issued 1827.

[Ampère, 1982] Ampère, A.-M. (1982). *Notice sur la Vie et les Travaux d'A.M. Ampère.* In C. Blondel, *A.-M. Ampère et la Création de l'Électrodynamique*, Annexe I, pages 171–173. Bibliothèque Nationale, Paris.

[Ampère, 1990] Ampère, A.-M. (1990). *Théorie Mathématique des Phénomènes Électro-dynamiques Uniquement Déduite de l'Expérience.* Éditions Jacques Gabay, Sceaux. Réimpression du Mémoire fondamental d'André-Marie Ampère paru en 1827 dans les *Mémoires de l'Académie Royale des Sciences de l'Institut de France*, année 1823, Tome VI, pp. 175-388.

[Ampère, 1993] Ampère, A.-M. (1993). Letter 654 of Ampère to Faraday, 13 April 1833. In James, F. A. J. L., editor, *The Correspondence of Michael Faraday*, volume 2. IEE, London.

[Ampère, 2012] Ampère, A.-M. (2012). Mathematical Theory of Electrodynamic Phenomena, Uniquely Derived from Experiments. Translated by M. D. Godfrey. Available at: https://archive.org/details/AmpereTheorieEn and https://sites.google.com/site/michaeldgodfrey/physics-information-and-communication.

[Ampère and Babinet, 1822a] Ampère, A.-M. and Babinet, J. (1822a). *Exposé des Nouvelles Découvertes sur l'Électricité et le Magnétisme de MM. Oersted, Arago, Ampère, H. Davy, Biot, Erman, Schweiger, de La Rive, etc.* Méquignon-Marvis, Paris.

[Ampère and Babinet, 1822b] Ampère, A.-M. and Babinet, J. (1822b). *Exposé des Nouvelles Découvertes sur l'Électricité et le Magnétisme de MM. Oersted, Arago, Ampère, H. Davy, Biot, Erman, Schweiger, de La Rive, etc.* Paris. Supplément à la traduction française de la 5e édition du *Systeme de Chimie* de T. Thomson, pp. 166-256.

[Ampère and Dulong, 1824a] Ampère, A.-M. and Dulong, P. L. (1824a). Abstract of the report on M. Rouseau's memoir respecting a new method of measuring the power of bodies to conduct electricity. *Annals of Philosophy*, 8:39–41.

[Ampère and Dulong, 1824b] Ampère, A.-M. and Dulong, P. L. (1824b). Rapport sur un Mémoire de M. Rousseau relatif à un nouveau moyen de mesurer la conductibilité des corps pour l'électricité. *Annales de Chimie et de Physique*, 25:373–379.

[Anderton, 2007] Anderton, R. J. (2007). Einstein, ether and unified field. *The General Science Journal.* Available at: www.gsjournal.net/Science-Journals/Essays/View/408.

[Anonymous, 1823] Anonymous (1823). Nouveaux Résultats obtenus par M. Savary sur les phénomènes électro-dynamiques. *Annales de Chimie et de Physique*, 23:413–415.

[Arago, 1854a] Arago, F. (1854a). Notices Biographiques, Ampère. In Barral, J.-A., editor, *Oeuvres Complètes*, pages 1–116. Gide et J. Baudry, Paris. Vol. 2. Notice lue par extraits à l'Académie des Sciences le 21 août 1839.

[Arago, 1824] Arago, F. J. D. (1824). *Annales de Chimie et de Physique*, 27:363.

[Arago, 1825] Arago, F. J. D. (1825). *Annales de Chimie et de Physique*, 28:325–326.

[Arago, 1826] Arago, F. J. D. (1826). Note concernant les Phénomènes magnétiques auxquels le mouvement donne naissance. *Annales de Chimie et de Physique*, 32:213–223.

[Arago, 1854b] Arago, F. J. D. (1854b). Du magnétisme de rotation. In Barral, J.-A., editor, *Oeuvres Complètes de François Arago*, volume 4, pages 424–448. Gide et J. Baudry, Paris.

[Archibald, 1989] Archibald, T. (1989). Energy and the mathematization of electrodynamics in Germany, 1845-1875. *Archives Internationales d'Histoire des Sciences*, 39:276–308.

[Assis, 1991] Assis, A. K. T. (1991). Wilhelm Eduard Weber (1804-1891) — Sua vida e sua obra. *Revista da Sociedade Brasileira de História da Ciência*, 5:53–59.

[Assis, 1992] Assis, A. K. T. (1992). *Curso de Eletrodinâmica de Weber.* Setor de Publicações do Instituto de Física da Universidade Estadual de Campinas — UNICAMP, Campinas. Notas de Física IFGW Número 5. Available at www.ifi.unicamp.br/~assis and www.bibliotecadigital.unicamp.br/document/?down=60362.

[Assis, 1994] Assis, A. K. T. (1994). *Weber's Electrodynamics.* Kluwer Academic Publishers, Dordrecht. ISBN: 0792331370.

[Assis, 1998] Assis, A. K. T. (1998). *Mecânica Relacional.* Editora do Centro de Lógica, Epistemologia e História da Ciência da UNICAMP/FAPESP, Campinas. ISBN: 8586497010. Available at: www.ifi.unicamp.br/~assis.

[Assis, 1999a] Assis, A. K. T. (1999a). *Relational Mechanics.* Apeiron, Montreal. ISBN: 0968368921. Available at: www.ifi.unicamp.br/~assis.

[Assis, 1999b] Assis, A. K. T. (1999b). *Uma Nova Física.* Editora Perspectiva, São Paulo. ISBN: 8527301997.

[Assis, 2004] Assis, A. K. T. (2004). The principle of physical proportions. *Annales de la Fondation Louis de Broglie*, 29:149–171.

[Assis, 2008] Assis, A. K. T. (2008). *Arquimedes, o Centro de Gravidade e a Lei da Alavanca.* Apeiron, Montreal. ISBN: 9780973291179. Available at: www.ifi.unicamp.br/~assis.

[Assis, 2010a] Assis, A. K. T. (2010a). *Archimedes, the Center of Gravity, and the First Law of Mechanics: The Law of the Lever.* Apeiron, Montreal, 2nd edition. ISBN: 9780986492648. Available at: www.ifi.unicamp.br/~assis.

[Assis, 2010b] Assis, A. K. T. (2010b). *The Experimental and Historical Foundations of Electricity.* Apeiron, Montreal. ISBN: 9780986492631. Available at: www.ifi.unicamp.br/~assis.

[Assis, 2010c] Assis, A. K. T. (2010c). *Os Fundamentos Experimentais e Históricos da Eletricidade.* Apeiron, Montreal. ISBN: 9780986492617. Available at: www.ifi.unicamp.br/~assis.

[Assis, 2010d] Assis, A. K. T. (2010d). Wilhelm Weber's works translated into English. *21st Century Science & Technology*, Vol. 22, No. 4:67–69.

[Assis, 2011a] Assis, A. K. T. (2011a). *Arquimedes, o Centro de Gravidade e a Lei da Alavanca.* Editora Livraria da Física, São Paulo. ISBN: 9788578611057.

[Assis, 2011b] Assis, A. K. T. (2011b). *Os Fundamentos Experimentais e Históricos da Eletricidade.* Editora Livraria da Física, São Paulo. ISBN: 9788578610975.

[Assis, 2013] Assis, A. K. T. (2013). *Mecânica Relacional e Implementação do Princípio de Mach com a Força de Weber Gravitacional.* Apeiron, Montreal. ISBN: 9780986492693. Available at: www.ifi.unicamp.br/~assis.

[Assis, 2014] Assis, A. K. T. (2014). *Relational Mechanics and Implementation of Mach's Principle with Weber's Gravitational Force.* Apeiron, Montreal. ISBN: 978-0-9920456-3-0. Available at: www.ifi.unicamp.br/~assis.

[Assis, 2015a] Assis, A. K. T. (2015a). Das prinzip der physikalischen Größenvershältnisse. Translated (2015) from the English by M. Pohl. Original paper: A. K. T. Assis, "The principle of physical proportions," Annales de la Fondation Louis de Broglie, Volume 29, pp. 149-171 (2004). Available at: www.ifi.unicamp.br/~assis.

[Assis, 2015b] Assis, A. K. T. (2015b). *Eletrodinâmica de Weber: Teoria, Aplicações e Exercícios.* Editora da Unicamp, Campinas, 2nd edition. e-ISBN: 978-85-268-1240-6.

[Assis, 2015c] Assis, A. K. T. (2015c). *The Experimental and Historical Foundations of Electricity.* Apeiron, Montreal. Book in Russian translated from the English version by A. Baraov. ISBN: 978-0-9920456-9-2.

[Assis and Bueno, 1995] Assis, A. K. T. and Bueno, M. (1995). Longitudinal forces in Weber's electrodynamics. *International Journal of Modern Physics B*, 9:3689–3696.

[Assis and Bueno, 2000] Assis, A. K. T. and Bueno, M. (2000). Bootstrap effect in classical electrodynamics. *Revista Facultad de Ingenieria de la Universidad de Tarapaca (Chile)*, 7:49–55.

[Assis and Bueno, 1996] Assis, A. K. T. and Bueno, M. A. (1996). Equivalence between Ampère and Grassmann's forces. *IEEE Transactions on Magnetics*, 32:431–436.

[Assis and Chaib, 2006] Assis, A. K. T. and Chaib, J. P. M. C. (2006). Nota sobre o magnetismo da pilha de Volta — tradução comentada do primeiro artigo de Biot e Savart sobre eletromagnetismo. *Cadernos de História e Filosofia da Ciência*, 16:303–309.

[Assis and Chaib, 2012] Assis, A. K. T. and Chaib, J. P. M. C. (2012). Ampère's motor: Its history and the controversies surrounding its working mechanism. *American Journal of Physics*, 80:990–995. DOI: 10.1119/1.4746698.

[Assis and Chaib, 2011] Assis, A. K. T. and Chaib, J. P. M. d. C. (2011). *Eletrodinâmica de Ampère: Análise do Significado e da Evolução da Força de Ampère, Juntamente com a Tradução Comentada de Sua Principal Obra sobre Eletrodinâmica.* Editora da Unicamp, Campinas. ISBN: 9788526809383.

[Assis et al., 2000] Assis, A. K. T., Fukai, J., and Carvalho, H. B. (2000). Weberian induction. *Physics Letters A*, 268:274–278.

[Assis and Guala-Valverde, 2000] Assis, A. K. T. and Guala-Valverde, J. (2000). Mass in relational mechanics. *Apeiron*, 7:131–132.

[Assis and Guala-Valverde, 2002] Assis, A. K. T. and Guala-Valverde, J. (2002). On absolute and relative motions in physics. *Journal of New Energy*, 6:8–12.

[Assis and Guala-Valverde, 2003] Assis, A. K. T. and Guala-Valverde, J. (2003). Frequency in relational mechanics. *Annales de la Fondation Louis de Broglie*, 28:83–97.

[Assis and Hernandes, 2007] Assis, A. K. T. and Hernandes, J. A. (2007). *The Electric Force of a Current: Weber and the Surface Charges of Resistive Conductors Carrying Steady Currents.* Apeiron, Montreal. ISBN: 9780973291155. Available at: `www.ifi.unicamp.br/~assis`.

[Assis and Hernandes, 2009] Assis, A. K. T. and Hernandes, J. A. (2009). *A Força Elétrica de uma Corrente: Weber e as Cargas Superficiais de Condutores Resistivos com Correntes Constantes*, volume 73 of *Coleção Acadêmica*. Edusp and Edufal, São Paulo and Maceió. ISBNs: 9788531411236 and 9788571774315.

[Assis and Hernandes, 2013] Assis, A. K. T. and Hernandes, J. A. (2013). *Elektrischer Strom und Oberflächenladungen: was Wilhelm Weber schon vor mehr als 150 Jahre wußte.* Apeiron, Montreal. German translation by H. Härtel. ISBN: 9780992045609. Available at: `www.ifi.unicamp.br/~assis`.

[Assis and Peixoto, 1992] Assis, A. K. T. and Peixoto, F. M. (1992). On the velocity in the Lorentz force law. *The Physics Teacher*, 30:480–483.

[Assis et al., 2002] Assis, A. K. T., Reich, K., and Wiederkehr, K. H. (2002). Gauss and Weber's creation of the absolute system of units in physics. *21st Century Science & Technology*, Vol. 15, No. 3:40–48.

[Assis et al., 2004] Assis, A. K. T., Reich, K., and Wiederkehr, K. H. (2004). On the electromagnetic and electrostatic units of current and the meaning of the absolute system of units — For the 200th anniversary of Wilhelm Weber's birth. *Sudhoffs Archiv*, 88:10–31.

[Assis et al., 2009] Assis, A. K. T., Ribeiro, J. E. A., and Vannucci, A. (2009). The field concepts of Faraday and Maxwell. In Cattani, M. S. D., Crispino, L. C. B., Gomes, M. O. C., and Santoro, A. F. S., editors, *Trends in Physics: Festscrift in Homage to Prof. José Maria Filardo Bassalo*, pages 31–38. Editora Livraria da Física, São Paulo.

[Assis et al., 2007] Assis, A. K. T., Souza Filho, M. P., Caluzi, J. J., and Chaib, J. P. M. C. (2007). From electromagnetism to electrodynamics: Ampère's demonstration of the interaction between current carrying wires. In Costa, M. F., Dorrio, B. V., and Reis, R., editors, *Proceedings of the 4th International Conference on Hands-on Science*, pages 9–16. University of Azores, Ponta Delgada.

[Assis and Thober, 1994] Assis, A. K. T. and Thober, D. S. (1994). Unipolar induction and Weber's electrodynamics. In Barone, M. and Selleri, F., editors, *Frontiers of Fundamental Physics*, pages 409–414, New York. Plenum Press.

[Assis and Torres Silva, 2000] Assis, A. K. T. and Torres Silva, H. (2000). Comparison between Weber's electrodynamics and classical electrodynamics. *Pramana Journal of Physics*, 55:393–404.

[Assis and Wiederkehr, 2003] Assis, A. K. T. and Wiederkehr, K. H. (2003). Weber quoting Maxwell. *Mitteilungen der Gauss-Gesellschaft*, 40:53–74.

[Assis et al., 2011] Assis, A. K. T., Wiederkehr, K. H., and Wolfschmidt, G. (2011). *Weber's Planetary Model of the Atom*, volume 19 of *Nuncius Hamburgensis — Beiträge zur Geschichte der Naturwissenschaften*. Tredition Science, Hamburg. Edited by G. Wolfschmidt. ISBN: 9783842402416.

[Assis et al., 2014] Assis, A. K. T., Wiederkehr, K. H., and Wolfschmidt, G. (2014). *O Modelo Planetário de Weber para o Átomo*. Apeiron, Montreal. ISBN: 9780992045654. Available at: `www.ifi.unicamp.br/~assis`.

[Assis and Zylbersztajn, 2001] Assis, A. K. T. and Zylbersztajn, A. (2001). The influence of Ernst Mach in the teaching of mechanics. *Science and Education*, 10:137–144.

[Atherton, 1989] Atherton, W. A. (1989). Gauss and Weber: An unlikely partnership. *Electronics and Wireless World*, 95:521–522.

[Bednarek, 2002] Bednarek, S. (2002). Unipolar motors and their application to the demonstration of magnetic field properties. *American Journal of Physics*, 70:455–461.

[Benseghir, 1989] Benseghir, A. (1989). Formation des concepts d'électrocinétique: un point de vue historique. *Bulletin de l'Union des Physiciens*, 713:451–465. Congrés S. F. P. de Strasbourg - Juillet 1987.

[Benseghir and Blondel, 2007] Benseghir, A. and Blondel, C. (2007). Histoire des sciences et didactique de la physique. Available at: `www.ampere.cnrs.fr/parcourspedagogique/agora/recherchesdidactiques`.

[Benseghir and Closset, 1993] Benseghir, A. and Closset, J.-L. (1993). Prégnance de l'explication électrostatique dans la construction du concept de circuit électrique: points de vue historique et didactique. *Didaskalia*, 2:31–47.

[Benseghir and Closset, 2001] Benseghir, A. and Closset, J.-L. (2001). From electrostatics to electrodynamics: historical and present difficulties. In Viennot, L., editor, *Reasoning in Physics: The Part of Common Sense*, chapter 11, pages 173–189. Springer, Berlin.

[Bertrand, 1874] Bertrand, J. (1874). Démonstration des théorèmes relatifs aux actions électrodynamiques. *Journal de Physique Théorique et Appliquée*, 3:297–306 and 335–343.

[Biermann, 1971] Biermann, K. R. (1971). Der Brief Alexander von Humboldts an Wilhelm Weber von Ende 1831 — ein bedeutendes Dokument zur Geschichte der Erforschung des Geomagnetismus. *Monatsberichte der Deutschen Akademie der Wissenschaften zu Berlin (DDR)*, 13, Heft 3:234–242.

[Biot, 1821a] Biot, J. B. (1821a). *Précis élémentaire de Physique expérimentale*, volume II. Deterville, Paris, second edition.

[Biot, 1821b] Biot, J. B. (1821b). Sur l'Aimantation imprimée aux métaux par l'électricité en mouvement: lu à la séance publique de l'Académie des sciences, le 2 avril 1821. *Journal des Savants*, Avril:221–235.

[Biot, 1824] Biot, J. B. (1824). *Précis élémentaire de Physique expérimentale*, volume II. Deterville, Paris, third edition.

[Biot, 1885a] Biot, J. B. (1885a). Précis élémentaire de Physique expérimentale, seconde édition (Paris: Deterville, 1821), Vol. II, p. 123. In Joubert, J., editor, *Collection de Mémoires relatifs a la Physique — Tome II: Mémoires sur l'Électrodynamique*, pages 116–117, Paris. Gauthier-Villars.

[Biot, 1885b] Biot, J. B. (1885b). Précis élémentaire de Physique expérimentale, troisième édition (Paris: Deterville, 1824), Vol. II, pp. 704-723. In Joubert, J., editor, *Collection de Mémoires relatifs a la Physique — Tome II: Mémoires sur l'Électrodynamique*, pages 80–127, Paris. Gauthier-Villars.

[Biot and Savart, 1820] Biot, J. B. and Savart, F. (1820). Note sur le magnétisme de la pile de Volta. *Annales de Chimie et de Physique*, 15:222–223.

[Biot and Savart, 1824] Biot, J. B. and Savart, F. (1824). Sur l'aimantation imposée aux métaux par l'électricité en mouvement. In J. B. Biot, *Précis Élémentaire de Physique Expérimentale,* vol. II, troisième édition, pages 704–774. Gauthier-Villars, Paris.

[Biot and Savart, 1885] Biot, J. B. and Savart, F. (1885). Sur l'aimantation imprimée aux métaux par l'électricitè en mouvement, in: *Précis élémentaire de Physique expérimentale*, troisième édition (Paris: Deterville, 1824), Vol. II, pp. 704-774. In Joubert, J., editor, *Collection de Mémoires relatifs a la Physique — Tome II: Mémoires sur l'Électrodynamique*, pages 80–127, Paris. Gauthier-Villars.

[Biot and Savart, 1965a] Biot, J. B. and Savart, F. (1965a). Magnetization of metals by electricity in motion. In R. A. R. Tricker, *Early Electrodynamics — The First Law of Circulation*, pages 119–139, New York. Pergamon. Extracts from Biot's *Précis elémentaire de Physique expérimentale*, Vol. II, 3rd edition, pp. 707-723 and 740-746, 1824 (Translated by O. M. Blunn).

[Biot and Savart, 1965b] Biot, J. B. and Savart, F. (1965b). Note on the magnetism of Volta's battery. In R. A. R. Tricker, *Early Electrodynamics — The First Law of Circulation*, pages 118–119, New York. Pergamon. Translated by O. M. Blunn.

[Blondel, 1978] Blondel, C. (1978). Sur les premières recherches de formule électrodynamique par Ampère (octobre 1820). *Revue d'Histoire des Sciences*, 31:53–65.

[Blondel, 1982] Blondel, C. (1982). *A.-M. Ampère et la Création de l'Électrodynamique (1820-1827).* Bibliothèque Nationale, Paris.

[Blondel, 1989] Blondel, C. (1989). Vision physique "éthérienne", mathématisation "laplacienne": l'électrodynamique d'Ampère. *Revue d'Histoire des Sciences*, 42:123–137.

[Blondel, 2004] Blondel, C. (2004). Ampère, le "Newton de l'électricité". *Bulletin de la Sabix*, 37:56–64.

[Blondel, 2005] Blondel, C. (2005). Ampère et l'Histoire de l'Électricité. Site of the CNRS coordinated by Christine Blondel. Available at: `www.ampere.cnrs.fr`.

[Blondel and Wolff, a] Blondel, C. and Wolff, B. Ampère jette les bases de l'électrodynamique (septembre 1820-janvier 1821). Available at: `www.ampere.cnrs.fr/parcourspedagogique`.

[Blondel and Wolff, b] Blondel, C. and Wolff, B. Ampère lays the foundations of electrodynamics (september 1820-january 1821). Available at: `www.ampere.cnrs.fr/parcourspedagogique`. Translated by A. Butrica.

[Blondel and Wolff, c] Blondel, C. and Wolff, B. In search of a newtonian law of electrodynamics (1820-1826). Available at: `www.ampere.cnrs.fr/parcourspedagogique`.

[Blondel and Wolff, d] Blondel, C. and Wolff, B. A la recherche d'une loi newtonienne pour l'électrodynamique (1820-1826). Available at: `www.ampere.cnrs.fr/parcourspedagogique`.

[Blondel and Wolff, e] Blondel, C. and Wolff, B. Que dit l'article ELECTRICITE de l'Encyclopédie? Available at: `www.ampere.cnrs.fr/parcourspedagogique`.

[Blondel and Wolff, f] Blondel, C. and Wolff, B. Teinturiers et tubes de verre: Gray et Dufay. Available at: `www.ampere.cnrs.fr/parcourspedagogique`.

[Blondel and Wolff, 2006] Blondel, C. and Wolff, B. (2006). Video: La pile de Volta en court-circuit. Available at: `www.ampere.cnrs.fr/parcourspedagogique`.

[Boss et al., 2012] Boss, S. L. B., Assis, A. K. T., and Caluzi, J. J. (2012). *Stephen Gray e a Descoberta dos Condutores e Isolantes: Tradução Comentada de Seus Artigos sobre Eletricidade e Reprodução de Seus Principais Experimentos*. Editora Cultura Acadêmica da Unesp, São Paulo. Available at: `www.culturaacademica.com.br/catalogo-detalhe.asp?ctl_id=354`.

[Bouty, 1883] Bouty, E. (1883). R. Felici — Note sur une expérience d'Ampère, p. 243-249. *Journal de Physique Théorique et Appliquée*, 2:527–529.

[Bueno and Assis, 1997a] Bueno, M. and Assis, A. K. T. (1997a). Equivalence between the formulas for inductance calculation. *Canadian Journal of Physics*, 75:357–362.

[Bueno and Assis, 1997b] Bueno, M. and Assis, A. K. T. (1997b). Proof of the identity between Ampère and Grassmann's forces. *Physica Scripta*, 56:554–559.

[Bueno and Assis, 1997c] Bueno, M. and Assis, A. K. T. (1997c). Self-inductance of solenoids, bi-dimensional rings and coaxial cables. *Helvetica Physica Acta*, 70:813–821.

[Bueno and Assis, 2001] Bueno, M. and Assis, A. K. T. (2001). *Inductance and Force Calculations in Electrical Circuits*. Nova Science Publishers, Huntington, New York. ISBN: 1560729171.

[Bueno and Assis, 2015] Bueno, M. and Assis, A. K. T. (2015). *Cálculo de Indutância e de Força em Circuitos Elétricos*. Apeiron, Montreal, 2nd edition. ISBN: 978-1-987980-01-1. Available at: `www.ifi.unicamp.br/~assis`.

[Bueno and Assis, 1995] Bueno, M. A. and Assis, A. K. T. (1995). A new method for inductance calculations. *Journal of Physics D*, 28:1802–1806.

[Bueno and Assis, 1998] Bueno, M. d. A. and Assis, A. K. T. (1998). Deriving force from inductance. *IEEE Transactions on Magnetics*, 34:317–319.

[Bunchaft, 2001] Bunchaft, F. (2001). Velocidades de desligamento nas interações de tipo weberiano e na força gravitacional relativística de Schrödinger. In Pessoa Jr., O., editor, *Fundamentos da Física*, volume 2, pages 73–80. Editora Livraria da Física, São Paulo.

[Bunchaft and Carneiro, 1997] Bunchaft, F. and Carneiro, S. (1997). Weber-like interactions and energy conservation. *Foundations of Physics Letters*, 10:393–401.

[Caneva, 1980] Caneva, K. L. (1980). Ampère, the etherians, and the Oersted connexion. *The British Journal for the History of Science*, 13:121–138.

[Chaib and Assis, 2007a] Chaib, J. P. M. C. and Assis, A. K. T. (2007a). Ampère e a origem do magnetismo terrestre. In *et al.*, P. W. G., editor, *Anais do I Simpósio de Pesquisa em Ensino e História de Ciências da Terra e do III Simpósio Nacional sobre Ensino de Geologia no Brasil*, pages 315–320. Instituto de Geociências da Unicamp, Campinas.

[Chaib and Assis, 2007b] Chaib, J. P. M. C. and Assis, A. K. T. (2007b). Distorção da obra eletromagnética de Ampère nos livros didáticos. *Revista Brasileira de Ensino de Física*, 29:65–70.

[Chaib and Assis, 2007c] Chaib, J. P. M. C. and Assis, A. K. T. (2007c). Experiência de Oersted em sala de aula. *Revista Brasileira de Ensino de Física*, 29:41–51.

[Chaib and Assis, 2009a] Chaib, J. P. M. C. and Assis, A. K. T. (2009a). A concepção dos elementos de corrente e da intensidade de corrente em Ampère. In Beltran, M. H. R., Saito, F., Santos, R. N. d., and Wuo, W., editors, *História da Ciência e Ensino: tendências e construção de interfaces*, pages 30–34. Editora Livraria da Física, São Paulo.

[Chaib, 2009] Chaib, J. P. M. d. C. (2009). *Análise do Significado e da Evolução do Conceito de Força de Ampère, juntamente com a Tradução Comentada de sua Principal Obra sobre Eletrodinâmica*. PhD thesis, University of Campinas — UNICAMP, Campinas, Brazil. Supervisor: A. K. T. Assis. Available at `webbif.ifi.unicamp.br/teses` and at `www.ifi.unicamp.br/~assis`.

[Chaib and Assis, 2007d] Chaib, J. P. M. d. C. and Assis, A. K. T. (2007d). Sobre os efeitos das correntes elétricas — Tradução comentada da primeira obra de Ampère sobre eletrodinâmica. *Revista da Sociedade Brasileira de História da Ciência*, 5:85–102.

[Chaib and Assis, 2009b] Chaib, J. P. M. d. C. and Assis, A. K. T. (2009b). Sobre os efeitos das correntes elétricas (segunda parte) — Tradução da primeira obra de Ampère sobre eletrodinâmica. *Revista Brasileira de História da Ciência*, 2:118–145.

[Chaib and Assis, 2013] Chaib, J. P. M. d. C. and Assis, A. K. T. (2013). Motor de Ampère: elementos para um ensino crítico de física. In Silva, C. C. and Prestes, M. E. B., editors, *Aprendendo Ciência e sobre Sua Natureza: abordagens históricas e filosóficas*, pages 55–70. Tipographia Editora Expressa, São Carlos.

[Cheuvreux, 1869] Cheuvreux, H. (1869). *Journal et Correspondance de André-Marie Ampère (de 1793 à 1804)*. Claye, Paris.

[Cheuvreux, 1873] Cheuvreux, H. (1873). *The Story of His Love being the Journal and Early Correspondence of André-Marie Ampère with His Family Circle during the First Republic, 1793-1804*. Richard Bentley and Son, London.

[Chiaverina, 2004] Chiaverina, C. (2004). The simplest motor? *The Physics Teacher*, 42:553.

[Clausius, 1877] Clausius, R. (1877). Ueber die Ableitung eines neuen elektrodynamischen Grundgesetzes. *Journal für die reine und angewandte Mathematik*, 82:85–130.

[Coulomb, 1935a] Coulomb, A. (1935a). First memoir on electricity and magnetism. In Magie, W. F., editor, *A Source Book in Physics*, pages 408–413, New York. McGraw-Hill. Original publication in French in 1785.

[Coulomb, 1935b] Coulomb, A. (1935b). Second memoir on electricity and magnetism. In Magie, W. F., editor, *A Source Book in Physics*, pages 413–420, New York. McGraw-Hill. Original publication in French in 1785.

[Coulomb, 1785a] Coulomb, C. A. (1785a). Premier mémoire sur l'électricité et le magnétisme: Construction et usage d'une balance électrique, fondée sur la propriété qu'ont les fils de métal, d'avoir une force de réaction de torsion proportionelle à l'angle de torsion. Détermination expérimentale de la loi suivant laquelle les élémens des corps électrisés du même genre d'électricité, se repoussent mutuellement. *Mémoires de l'Académie royale des Sciences de l'Institut de France*, 88:569–577. Published in 1788.

[Coulomb, 1785b] Coulomb, C. A. (1785b). Second mémoire sur l'électricité et le magnétisme, où l'on détermine, suivant quelles loix de fluide magnétique, ainsi que le fluide électrique, agissent, soit par répulsion, soit par attraction. *Mémoires de l'Académie royale des Sciences de l'Institut de France*, 88:578–611. Published in 1788.

[Crowe, 1985] Crowe, M. J. (1985). *A History of Vector Analysis — The Evolution of the Idea of a Vectorial System*. Dover, New York.

[Curé, 2009] Curé, J. C. (2009). *Einstein on Trial or Metaphysical Principles of Natural Philosophy*. Shirley Ramsey, Caracas, 3rd edition.

[Darrigol, 2000] Darrigol, O. (2000). *Electrodynamics from Ampère to Einstein*. Oxford University Press, Oxford.

[Davy, 1821] Davy, H. (1821). On the magnetic phenomena produced by electricity. *Philosophical Transactions*, 111:7–19.

[Davy, 1823] Davy, H. (1823). On a new phenomenon of electro-magnetism. *Philosophical Transactions*, 113:153–159.

[Davy, 1824] Davy, H. (1824). Sur un nouveau phénomène électro-magnétique. *Annales de Chimie et de Physique*, 25:64–71.

[Davy, 1885] Davy, H. (1885). Sur un nouveau phénomène électromagnétique. In Joubert, J., editor, *Collection de Mémoires relatifs a la Physique,* Vol. II: *Mémoires sur l'Électrodynamique*, pages 207–211. Gauthier-Villars, Paris.

[De la Borne, 1821] De la Borne (1821). Note sur l'aimantation par l'électricité. *Annales de Chimie et de Physique*, 16:193–195.

[de la Rive, 1822a] de la Rive, A. (1822a). De l'Action qu'exerce le globe terrestre sur une portion mobile de circuit voltaïque. *Bibliotheque Universelle des Sciences, Belles-Lettres, et Arts*, 21:29–48. Memoire lu à la Société de Physique et d'Histoire Naturelle de Genève le 4 Septembre 1822.

[de la Rive, 1822b] de la Rive, A. (1822b). Mémoire sur l'Action qu'exerce le globe terrestre sur une portion mobile de circuit voltaïque. *Annales de Chimie et de Physique*, 21:24–48.

[de la Rive, 1822c] de la Rive, A. (1822c). Mémoire sur l'Action qu'exerce le globe terrestre sur une portion mobile du circuit voltaïque; par M. de La Rive fils. In Ampère, A.-M., editor, *Recueil d'Observations Électro-dynamiques*, pages 262–286. Crochard, Paris. Despite this date, the volume of the Recueil was only published in 1823.

[de la Rive, 1885] de la Rive, A. (1885). Mémoire sur l'Action qu'exerce le globe terrestre sur une portion mobile du circuit voltaïque; Par M. de La Rive fils. In Joubert, J., editor, *Collection de Mémoires relatifs a la Physique,* Vol. II: *Mémoires sur l'Électrodynamique*, pages 308–328. Gauthier-Villars, Paris.

[Demonferrand, 1823] Demonferrand, J. B. F. (1823). *Manuel d'Électricité dynamique, ou Traité sur l'action mutuelle des conducteurs électriques et des aimans, et sur la nouvelle théorie du magnétisme.* Bachelier, Paris.

[Drury, 1997] Drury, D. M. (1997). The electromagnetic momentum of a charged particle in Weber's theory. *Galilean Electrodynamics*, 8:75–78.

[Du Fay,] Du Fay, C. F. d. C. A letter from Mons. Du Fay, F. R. S. and of the Royal Academy of Sciences at Paris, to His Grace Charles Duke of Richmond and Lenox, concerning electricity. Translated from the French by T. S. M D. *Philosophical Transactions*, 38:258–266, 1733–4.

[Du Fay, 1733] Du Fay, C. F. d. C. (1733). Quatriéme mémoire sur l'électricité. De l'attraction et répulsion des corps électriques. *Mémoires de l'Académie Royale des Sciences*, pages 457–476.

[Duhem, 1886] Duhem, P. (1886). Sur la loi d'Ampère. *Journal de Physique Théorique et Appliquée*, 5:26–29.

[Duhem, 1892] Duhem, P. (1892). *Leçons sur l'Électricité et le Magnétisme*, volume III. Gauthier-Villars, Paris.

[Editor-News, 1914] Editor-News (1914). Experimental demonstration of an Ampere molecular current in a nearly perfect conductor. *Nature*, 93:481.

[Einstein, 1952] Einstein, A. (1952). On the electrodynamics of moving bodies. In A. Einstein, H. A. Lorentz, H. Weyl and H. Minkowski*, The Principle of Relativity*, pages 35–65, New York. Dover.

[Einstein, 1978] Einstein, A. (1978). Sobre a electrodinâmica dos corpos em movimento. In A. Einstein, H. Lorentz, H. Weyl and H. Minkowski*, O Princípio da Relatividade*, Lisboa, 2nd edition. Fundação Calouste Gulbenkian. Pages 47-86. Translated by M. J. Saraiva.

[Einstein and Infeld, 1938] Einstein, A. and Infeld, L. (1938). *The Evolution of Physics.* Cambridge University Press, London.

[Ettingshausen, 1879] Ettingshausen (1879). Sur les expériences fondamentales de l'électrodynamique. *Journal de Physique Théorique et Appliquée*, 8:70–71.

[Ettingshausen, 1878a] Ettingshausen, A. v. (1878a). Ueber Ampère's elektrodynamische Fundamentalversuche. *Sitzungsberichte der Mathematisch-Naturwissenschaftlichen Classe der Kaiserlichen Akademie der Wissenschaften in Wien*, 15:12–13.

[Ettingshausen, 1878b] Ettingshausen, A. v. (1878b). Ueber Ampère's elektrodynamische Fundamentalversuche. *Sitzungsberichte der Mathematisch-Naturwissenschaftlichen Classe der Kaiserlichen Akademie der Wissenschaften in Wien*, 77:109–134.

[Faraday, 1821a] Faraday, M. (1821a). Historical sketch of electro-magnetism. *Annals of Philosophy*, 2:195–200.

[Faraday, 1821b] Faraday, M. (1821b). Historical sketch of electro-magnetism. *Annals of Philosophy*, 2:274–290.

[Faraday, 1821c] Faraday, M. (1821c). Sur les Mouvemens électro-magnétiques et la théorie du magnétisme. *Annales de Chimie et de Physique*, 18:337–370.

[Faraday, 1822a] Faraday, M. (1822a). Description of an electro-magnetical apparatus for the exhibition of rotatory motion. *The Quarterly Journal of Science, Literature, and the Arts*, 12:283–285.

[Faraday, 1822b] Faraday, M. (1822b). Historical sketch of electro-magnetism. *Annals of Philosophy*, 3:107–121.

[Faraday, 1822c] Faraday, M. (1822c). Mémoire sur les Mouvemens électro-magnétiques et la théorie du magnétisme. In Ampère, A.-M., editor, *Recueil d'Observations Électro-dynamiques*, pages 125–158. Crochard, Paris. Despite this date, the volume of the Recueil was only published in 1823.

[Faraday, 1822d] Faraday, M. (1822d). New electro-magnetic apparatus. *The Quarterly Journal of Science, Literature, and the Arts*, 12:186–187.

[Faraday, 1822e] Faraday, M. (1822e). On some new electro-magnetical motions, and on the theory of magnetism. *The Quarterly Journal of Science, Literature, and the Arts*, 12:74–96.

[Faraday, 1885] Faraday, M. (1885). Mémoire sur les mouvements électromagnétiques et la théorie du magnétisme. In Joubert, J., editor, *Collection de Mémoires relatifs a la Physique*, Vol. II: *Mémoires sur l'Électrodynamique*, pages 158–183. Gauthier-Villars, Paris.

[Faraday, 1952a] Faraday, M. (1952a). Description of an electro-magnetical apparatus for the exhibition of rotary motion. In Hutchins, R. M., editor, *Great Books of the Western World*, Vol. 45: *Lavoisier, Fourier, Faraday*, pages 807–809. Encyclopaedia Britannica, Chicago. Reprint of the Quarterly Journal of Science, Vol. 12, pp. 283-285 (1822).

[Faraday, 1952b] Faraday, M. (1952b). Electro-magnetic rotation apparatus. In Hutchins, R. M., editor, *Great Books of the Western World*, Vol. 45: *Lavoisier, Fourier, Faraday*, page 807. Encyclopaedia Britannica, Chicago. Reprint of the Quarterly Journal of Science, Vol. 12, pp. 186-187 (1822).

[Faraday, 1952c] Faraday, M. (1952c). *Experimental Researches in Electricity*, volume 45, pp. 257-866 of *Great Books of the Western World*. Encyclopaedia Britannica, Chicago.

[Faraday, 1952d] Faraday, M. (1952d). On some new electro-magnetical motions and on the theory of magnetism. In Hutchins, R. M., editor, *Great Books of the Western World*, Vol. 45: *Lavoisier, Fourier, Faraday*, pages 795–807. Encyclopaedia Britannica, Chicago. Reprint of the Quarterly Journal of Science, Vol. 12, pp. 74-96 (1822).

[Farrar, 1826] Farrar, J. (1826). *Elements of Electricity, Magnetism, and Electro-magnetism, embracing the Late Discoveries and Improvements, digested into the Form of a Treatise; being the Second Part of a Course of Natural Philosophy*. Cambridge University Press, Cambridge. This volume, with the exception of the notes, is selected from Biot's *Précis Elémentaire de Physique*, third edition, printed at Paris in 1824, and translated with such alterations as were found necessary in order to adapt it to the English reader.

[Featonby, 2006] Featonby, D. (2006). Inspiring experiments exploit strong attraction of magnets. *Physics Education*, 41:292–295.

[Felici, 1882] Felici, R. (1882). Nota ad una esperienza dell'Ampère. *Nuovo Cimento*, 11:243–249. This paper is signed only as "R." But it is known that the author was R. Felici, one of the Editors of the *Nuovo Cimento*.

[Feynman et al., 1964] Feynman, R. P., Leighton, R. B., and Sands, M. (1964). *The Feynman Lectures on Physics*. Addison-Wesley, Reading. Volume 2: Mainly Electromagnetism and Matter.

[Fourier, 1952] Fourier, J. B. J. (1952). Analytical Theory of Heat. In *Great Books of the Western World, Vol. 45*, pages 161–251, Chicago. Encyclopaedia Britannica.

[Franksen, 1981] Franksen, O. I. (1981). *H. C. Ørsted — A Man of the Two Cultures.* Strandbergs Forlag, Birkerød.

[Fresnel, 1885a] Fresnel, A. (1885a). Comparaison de la supposition des courants autour de l'axe avec celle des courants autour chaque molécule. In Joubert, J., editor, *Collection de Mémoires relatifs a la Physique,* Vol. II: *Mémoires sur l'Électrodynamique*, pages 141–143. Gauthier-Villars, Paris.

[Fresnel, 1885b] Fresnel, A. (1885b). Deuxième note sur l'hipothèse des courants particulaires. In Joubert, J., editor, *Collection de Mémoires relatifs a la Physique,* Vol. II: *Mémoires sur l'Électrodynamique*, pages 144–147. Gauthier-Villars, Paris.

[Fukai, 2003] Fukai, J. (2003). *A Promenade Along Electrodynamics.* Vales Lake Publishing, Pueblo West.

[Fukai and Assis, 2003] Fukai, J. and Assis, A. K. T. (2003). Testing Mach's principle in electrodynamics. *Canadian Journal of Physics*, 81:1239–1242.

[Fukai and Kinzer, 1997] Fukai, J. and Kinzer, E. T. (1997). Compatibility of Weber's force with Maxwell's equations. *Galilean Electrodynamics*, 8:53–55.

[Galeczki and Marquardt, 1995] Galeczki, G. and Marquardt, P. (1995). Über die longitudinalen Ampèreschen Kräfte und Webers Elektrodynamik. *Fusion*, 16:15–17.

[Gardelli, 2004] Gardelli, D. (2004). Concepções de interação física: Subsídios para uma abordagem histórica do assunto no ensino médio. Master's thesis, Instituto de Física, Universidade de São Paulo, Brasil.

[Gillmor, 1971a] Gillmor, C. S. (1971a). *Coulomb and the Evolution of Physics and Engineering in Eighteenth-Century France.* Princeton University Press, Princeton.

[Gillmor, 1971b] Gillmor, C. S. (1971b). Coulomb, Charles Augustin. In Gillispie, C. C., editor, *Dictionary of Scientific Biography,* Vol. 3, pages 439–447. Charles Scribner's Sons, New York.

[Gluckman, 1999] Gluckman, A. G. (1999). On W. E. Weber's "Grundprincip der Elektrodynamik" and related earlier and contemporary studies. *Physics Essays*, 12:682–698.

[Graneau, 1990a] Graneau, P. (1990a). Far-action versus contact action. *Speculations in Science and Technology*, 13:191–201.

[Graneau, 1990b] Graneau, P. (1990b). Has the mystery of inertia been solved? In Bartocci, U. and Wesley, J. P., editors, *Proceedings of the Conference on Foundations of Mathematics and Physics*, pages 129–136, Blumberg, Germany. Benjamin Wesley Publisher.

[Graneau, 1990c] Graneau, P. (1990c). Interconnecting action-at-a-distance. *Physics Essays*, 3:340–343.

[Graneau, 1990d] Graneau, P. (1990d). The riddle of inertia. *Electronics and Wireless World*, 96:60–62.

[Graneau, 1990e] Graneau, P. (1990e). Some cosmological consequences of Mach's principle. *Hadronic Journal Supplement*, 5:335–349.

[Graneau, 1994] Graneau, P. (1994). *Ampere-Neumann Electrodynamics of Metals.* Hadronic Press, Palm Harbor, 2nd edition.

[Graneau and Graneau, 1993] Graneau, P. and Graneau, N. (1993). *Newton Versus Einstein — How Matter Interacts with Matter.* Carlton Press, New York.

[Graneau and Graneau, 1996] Graneau, P. and Graneau, N. (1996). *Newtonian Electrodynamics.* World Scientific, Singapore.

[Grassmann, 1845] Grassmann, H. (1845). Neue Theorie der Electrodynamik. *Annalen der Physik*, 64:1–18.

[Grassmann, 1877] Grassmann, H. (1877). Zur Elektrodynamik. *Journal für die reine und angewandte Mathematik*, 83:57–64.

[Grassmann, 1965] Grassmann, H. (1965). A new theory of electrodynamics. In R. A. R. Tricker, *Early Electrodynamics — The First Law of Circulation*, pages 201–214, New York. Pergamon.

[Grattan-Guinness, 1990a] Grattan-Guinness, I. (1990a). *Convolutions in French Mathematics, 1800-1840*, volume 2. Birkhäuser, Basel.

[Grattan-Guinness, 1990b] Grattan-Guinness, I. (1990b). *Convolutions in French Mathematics, 1800-1840*, volume 1. Birkhäuser, Basel.

[Grattan-Guinness, 1991] Grattan-Guinness, I. (1991). Lines of mathematical thought in the electrodynamics of Ampère. *Physis*, 28:115–129.

[Grattan-Guinness, 2000] Grattan-Guinness, I. (2000). Daniel Bernoulli and the varieties of mechanics in the 18th century. *Nieuw Archief voor Wiskunde*, 3:242–249.

[Gray,] Gray, S. Several experiments concerning electricity. *Philosophical Transactions*, 37:18–44, 1731–2.

[Griffiths, 1989] Griffiths, D. J. (1989). *Introduction to Electrodynamics*. Prentice Hall, Englewood Cliffs, 2nd edition.

[Gross, 2009] Gross, A. G. (2009). The continuity of scientific discovery and its communication: the example of Michael Faraday. *Journal of Biomedical Discovery and Collaboration*, 4:3–18.

[Guala-Valverde, 1997] Guala-Valverde, J. (1997). Centrifugal potential energy in a machian universe. *Galilean Electrodynamics*, 8:100.

[Guala-Valverde, 1998] Guala-Valverde, J. (1998). Gravitational mutual energy in Weber-type mechanics. *Physics Essays*, 11:164–165.

[Guala-Valverde, 1999a] Guala-Valverde, J. (1999a). *Inercia y Gravitacion*. Fundacion Julio Palacios, Neuquen, Argentina. In collaboration with J. Tramaglia and R. Rapacioli. Available at: `www.educ.ar/sitios/educar/recursos/ver?id=90380`.

[Guala-Valverde, 1999b] Guala-Valverde, J. (1999b). Inertial mass in Mach-Weber-Assis theory. *Apeiron*, 6:202–204.

[Guala-Valverde, 2001] Guala-Valverde, J. (2001). A new theorem in relational mechanics. *Apeiron*, 8:132–138.

[Guala-Valverde, 2002] Guala-Valverde, J. (2002). Why homopolar devices cannot be additive? *Spacetime & Substance*, 3:186–187.

[Guala-Valverde, 2004] Guala-Valverde, J. (2004). Distant matter in physics. *Annales de la Fondation Louis de Broglie*, 29:635–640.

[Guala-Valverde et al., 2005] Guala-Valverde, J., Achilles, R., and Blas, R. (2005). Inertial mass: a changing entity? — Weber vs. Einstein, Weber plus Einstein or none? *Apeiron*, 12:351–352.

[Guala-Valverde and Mazzoni, 2001] Guala-Valverde, J. and Mazzoni, P. (2001). The unipolar dynamotor: a genuine relational engine. *Apeiron*, 8:41–52.

[Guala-Valverde et al., 2002] Guala-Valverde, J., Mazzoni, P., and Achilles, R. (2002). The homopolar motor: A true relativistic engine. *American Journal of Physics*, 70:1052–1055.

[Haertel, 2012a] Haertel, H. (2012a). Spannung und Oberflächenladungen: Was Wilhelm Weber schon vor mehr als 150 Jahren wusste. *Praxis der Naturwissenschaften - Physik in der Schule*, 5:25–31.

[Haertel, 2012b] Haertel, H. (2012b). Tensão e cargas superficiais — o que Wilhelm Weber já sabia há 150 anos. *Caderno Brasileiro de Ensino de Física*, 29:1015–1029. Translated by A. K. T. Assis. DOI: 10.5007/2175-7941.2012v29n3p1015.

[Harman, 1982] Harman, P. M. (1982). *Energy, Force, and Matter — The Conceptual Development of Nineteenth-Century Physics*. Cambridge University Press, Cambridge.

[Heald and Marion, 1995] Heald, M. A. and Marion, J. B. (1995). *Classical Electromagnetic Radiation*. Saunders, Fort Worth, 3rd edition.

[Hecht, 1996] Hecht, L. (1996). The significance of the 1845 Gauss-Weber correspondence. *21st Century Science & Technology*, 9(3):22–34.

[Hecht, 1997] Hecht, L. (1997). Wilhelm Weber: giant of 19th century physics. *21st Century Science & Technology*, 10(2):53–54.

[Heering, 1992] Heering, P. (1992). On Coulomb's inverse square law. *American Journal of Physics*, 60:988–994.

[Heilbron, 1981] Heilbron, J. L. (1981). The electrical field before Faraday. In Cantor, G. N. and Hodge, M. J. S., editors, *Conceptions of Ether: Studies in the History of Ether Theories 1740-1900*, pages 187–213. Cambridge University Press, Cambridge.

[Heilbron, 1982] Heilbron, J. L. (1982). *Elements of Early Modern Physics.* University of California Press, Berkeley.

[Heilbron, 1999] Heilbron, J. L. (1999). *Electricity in the 17th and 18th Centuries — A Study in Early Modern Physics.* Dover, New York.

[Hofmann, 1982] Hofmann, J. R. (1982). *The Great Turning Point in André-Marie Ampère's Research in Electrodynamics: A Truly "Crucial" Experiment.* PhD thesis, Graduate Faculty of Arts and Sciences, University of Pittsburgh, Pittsburgh.

[Hofmann, 1987] Hofmann, J. R. (1987). Ampère's invention of equilibrium apparatus: a response to experimental anomaly. *The British Journal for the History of Science*, 20:309–341.

[Hofmann, 1996] Hofmann, J. R. (1996). *André-Marie Ampère, Enlightenment and Electrodynamics.* Cambridge University Press, Cambridge.

[Hottecke, 2000] Hottecke, D. (2000). How and what can we learn from replicating historical experiments? A case study. *Science & Education*, 9:343–362.

[Humboldt and Biot, 1804] Humboldt, A. v. and Biot, J. B. (1804). Sur les variations du magnétisme terrestre a différentes latitudes. *Journal de Physique, de Chimie, d'Histoire Naturelle et des Arts*, 59:429–450.

[Huray, 2010] Huray, P. G. (2010). *Maxwell's Equations.* Wiley: IEEE, Hoboken.

[James, 1993] James, F. A. J. L. (1993). *The Correspondence of Michael Faraday*, volume 2. IEE, London.

[Jeans, 1941] Jeans, J. (1941). *The Mathematical Theory of Electricity and Magnetism.* Cambridge University Press, Cambridge. Reprint of the 5th edition of 1925.

[Jefimenko, 1973] Jefimenko, O. D. (1973). *Electrostatic Motors: Their History, Types, and Principles of Operation.* Electret Scientific, Star City.

[Jones, 1870a] Jones, B. (1870a). *Faraday's Life and Letters*, volume 1. Longmans, Green, and Co., London, 2nd edition.

[Jones, 1870b] Jones, B. (1870b). *Faraday's Life and Letters*, volume 2. Longmans, Green, and Co., London, 2nd edition.

[Joubert, 1885] Joubert, J. (1885). *Collection de Mémoires relatifs a la Physique*, volume 2: *Mémoires sur l'Électrodynamique.* Gauthiers-Villars, Paris.

[Jungnickel and McCormmach, 1986] Jungnickel, C. and McCormmach, R. (1986). *Intellectual Mastery of Nature — Theoretical Physics from Ohm to Einstein*, volume 1-2. University of Chicago Press, Chicago.

[Kagan, 2005] Kagan, D. (2005). Upright homopolar motor. *The Physics Teacher*, 43:68.

[Kapoor, 1981] Kapoor, S. C. (1981). Dumas, Jean-Baptiste-André. In Gillispie, C. C., editor, *Dictionary of Scientific Biography,* Vol. 4, pages 242–248. Charles Scribner's Sons, New York.

[Kastler, 1977] Kastler, A. (1977). Ampère et les lois de l'électrodynamique. *Revue d'Histoire des Sciences*, 30:143–157.

[Kinzer and Fukai, 1996] Kinzer, E. T. and Fukai, J. (1996). Weber's force and Maxwell's equations. *Foundations of Physics Letters*, 9:457–461.

[Kirchner, 1957] Kirchner, F. (1957). Determination of the velocity of light from electromagnetic measurements according to W. Weber and R. Kohlrausch. *American Journal of Physics*, 25:623–629.

[Koestler, 1973] Koestler, A. (1973). *The Sleepwalkers: a history of man's changing vision of the universe.* Penguin, Middlesix.

[Kohlrausch and Weber, 1857] Kohlrausch, R. and Weber, W. (1857). Elektrodynamische Maassbestimmungen insbesondere Zurückführung der Stromintensitäts-Messungen auf mechanisches Maass. *Abhandlungen der Königl. Sächs. Gesellschaft der Wissenschaften, mathematisch-physische Klasse*, 3:221–290. Reprinted in Wilhelm Weber's *Werke*, Vol. 3, H. Weber (ed.), (Springer, Berlin, 1893), pp. 609-676.

[Krapas and Silva, 2005] Krapas, S. and Silva, M. C. d. (2005). Forças que atuam a distância: representações em livros didáticos do ensino médio e a história da ciência. In Moreira, M. A., Sahelices, C. C., and Villagrá, J. M., editors, *Actas do II Encuentro Iberoamericano sobre Investigación Básica em Educación em Ciências*, pages 351–366. Servicios de Publicaciones de la Universidad de Burgos, Burgos.

[Krapas and Silva, 2008] Krapas, S. and Silva, M. C. d. (2008). O conceito de campo: polissemia nos manuais, significados na física do passado e da atualidade. *Ciência & Educação*, 14:15–33.

[Lagrange, 1788] Lagrange, J. L. (1788). *Méchanique Analitique.* Veuve Desaint, Paris.

[Lagrange, 1811] Lagrange, J. L. (1811). *Méchanique Analitique*, volume 1. Courcier, Paris.

[Lagrange, 1997] Lagrange, J. L. (1997). *Analytical Mechanics*, volume 191 of *Boston Studies in the Philosophy of Science.* Kluwer, Dordrecht. Translated and edited by A. Boissonnade and V. N. Vagliente. Translated from the *Mécanique analytique, novelle édition* of 1811.

[Larson, 1982] Larson, D. B. (1982). *The Neglected Facts of Science.* North Pacific Publishers, Portland. Available at: `www.reciprocalsystem.com/nfs/index.htm`.

[Laumont, 1820] Laumont, G. d. (1820). Note sur les expériences électro-magnétiques de MM. Oersted, Ampère et Arago, relatives à l'identité de l'aimant avec l'électricité. *Annales de Mines*, 5:535–546.

[Laumont, 1822] Laumont, G. d. (1822). Additions au mémoire précédent — note sur les expériences électromagnétiques de MM. OErsted, Ampère, Arago et Biot. In Ampère, A.-M., editor, *Recueil d'Observations Électro-dynamiques*, pages 69–80. Crochard, Paris. Despite this date, the volume of the Recueil was only published in 1823.

[Launay, 1925] Launay, L. d. (1925). *Le Grand Ampère d'aprés des Documents Inédits.* Librarie Académique Perrin, Paris.

[Launay (ed.), 1936a] Launay (ed.), L. d. (1936a). *Correspondance du Grand Ampère*, volume 2. Gauthier-Villars, Paris.

[Launay (ed.), 1936b] Launay (ed.), L. d. (1936b). *Correspondance du Grand Ampère*, volume 1. Gauthier-Villars, Paris.

[Launay (ed.), 1943] Launay (ed.), L. d. (1943). *Correspondance du Grand Ampère*, volume 3. Gauthier-Villars, Paris.

[Lauridsen and Abrahamsen, 1998] Lauridsen, E. K. and Abrahamsen, N. (1998). The history of astatic magnet systems and suspensions. *Centaurus*, 40:135–169.

[Littré, 1877] Littré, E. (1872-1877). *Dictionnaire de la Langue Française.* Available at: `francois.gannaz.free.fr/Littre/accueil.php`.

[Lorentz, 1895] Lorentz, H. A. (1895). *Versuch einer Theorie der Electrischen und Optischen Erscheinungen in Bewegten Körpern.* E. J. Brill, Leiden. Abschmitt I (Die Grundgleichungen für ein System in den Aether eingelagerter Ionen), §12 (Der zweite Theil der auf die ponderable Materie wirkenden Kraft), pp. 21-22.

[Luhr, 2000] Luhr, J. (2000). Die Entstehung eines Demonstrationsexperiments: Zur Geschichte der Ampère'schen Stromwaage. In Heering, P., Rieß, F., and Sichau, C., editors, *Im Labor der Physikgeschichte — Zur Untersuchung historischer Experimentalpraxis*, pages 135–156. BIS der Carl von Ossietzky, Oldenburg.

[Mach, 1960] Mach, E. (1960). *The Science of Mechanics — A Critical and Historical Account of Its Development*. Open Court, La Salle, sixth edition. Translated by J. McCormack.

[Magnaghi and Assis, 2008] Magnaghi, C. P. and Assis, A. K. T. (2008). Sobre a eletricidade excitada pelo simples contato entre substâncias condutoras de tipos diferentes — Uma tradução comentada do artigo de Volta de 1800 descrevendo sua invenção da pilha elétrica. *Caderno Brasileiro de Ensino de Física*, 25:118–140.

[Marazzini and Tucci, 1983] Marazzini, P. and Tucci, P. (1983). The physical meaning of Faraday's law from an historical point of view. *European Journal of Physics*, 4:170–179.

[Marinov, 1990] Marinov, S. (1990). *The Thorny Way of Truth*, volume VIII. International Publishers East-West, Graz.

[Martins, 1986] Martins, R. d. A. (1986). Ørsted e a descoberta do eletromagnetismo. *Cadernos de História e Filosofia da Ciência*, 10:89–114.

[Maxwell, 1865] Maxwell, J. C. (1865). A dynamical theory of the electromagnetic field. *Philosophical Transactions*, 155:459–512. Doi: 10.1098/rstl.1865.0008.

[Maxwell, 1954] Maxwell, J. C. (1954). *A Treatise on Electricity and Magnetism*. Dover, New York.

[Maxwell, 1965] Maxwell, J. C. (1965). A dynamical theory of the electromagnetic field. In Niven, W. D., editor, *The Scientific Papers of James Clerk Maxwell*, pages 526–597 (vol. 1), New York. Dover. Article originally published in 1864.

[Mendelson, 2006] Mendelson, K. S. (2006). The story of c. *American Journal of Physics*, 74:995–997. DOI: 10.1119/1.2238887.

[Meyer, 1972] Meyer, H. W. (1972). *A History of Electricity and Magnetism*. Burndy Library, Norwalk, Connecticut.

[Monteiro et al., 2010] Monteiro, M. A. A., Germano, J. S. E., Monteiro, I. C. d. C., and Gaspar, A. (2010). As atividades de demonstração e a teoria de Vigotski: um motor elétrico de fácil construção e de baixo custo. *Caderno Brasileiro de Ensino de Física*, 27:371–384.

[Munoz, 2007] Munoz, A. M. (2007). Motor homopolar. *Revista Eureka sobre Enseñanza y Divulgación de las Ciencias*, 4:352–354.

[Neild, 1996] Neild, G. H. (1996). William Cruickshank (FRS-1802): Clinical chemist. *Nephrology Dialysis Transplantation*, 11:1885–1889.

[Nersessian, 1989] Nersessian, N. J. (1989). Faraday's field concept. In Gooding, D. and James, F. A. J. L., editors, *Faraday Rediscovered*, pages 175–187. Macmillan Press, Basingstoke.

[Newton, 1934] Newton, I. (1934). *Mathematical Principles of Natural Philosophy*. University of California Press, Berkeley. Cajori edition.

[Newton, 1952a] Newton, I. (1952a). *Mathematical Principles of Natural Philosophy*, volume 34, pp. 1-372 of *Great Books of the Western World*. Encyclopaedia Britannica, Chicago. Translated by A. Motte and revised by F. Cajori.

[Newton, 1952b] Newton, I. (1952b). *Optics*, volume 34, pp. 377-544 of *Great Books of the Western World*. Encyclopaedia Britannica, Chicago.

[Newton, 1979] Newton, I. (1979). *Opticks*. Dover, New York.

[Newton, 1990] Newton, I. (1990). *Principia — Princípios Matemáticos de Filosofia Natural.* Nova Stella/Edusp, São Paulo. Livro I: O Movimento dos Corpos. Portuguese translation by T. Ricci, L. G. Brunet, S. T. Gehring and M. H. C. Célia.

[Newton, 1996] Newton, I. (1996). *Óptica.* Edusp, São Paulo. Portuguese translation, introduction and notes by A. K. T. Assis. ISBN: 8531403405.

[Newton, 1999] Newton, I. (1999). *The Principia: Mathematical Principles of Natural Philosophy.* University of California Press, Berkeley. A new translation by I. B. Cohen and A. Whitman, assisted by J. Budenz.

[Newton, 2008] Newton, I. (2008). *Principia — Princípios Matemáticos de Filosofia Natural.* Edusp, São Paulo. Livro II: O Movimento dos Corpos (em Meios com Resistência). Livro III: O Sistema do Mundo (Tratado Matematicamente). Portuguese translation by A. K. T. Assis. ISBN: 9788531410895.

[Newton, 2010] Newton, I. (2010). *Principia — Princípios Matemáticos de Filosofia Natural.* Folha de São Paulo, São Paulo. Livro III: O Sistema do Mundo (Tratado Matematicamente). ISBN: 9788563270306. Coleção Folha de São Paulo: Livros que Mudaram o Mundo, Volume 9. Portuguese translation by A. K. T. Assis.

[OErsted et al., 1822] OErsted, Ampère, Arago, and Biot (1822). Note sur les expériences électro-magnétiques de MM. OErsted, Ampère, Arago et Biot. In Ampère, A.-M., editor, *Recueil d'Observations Électro-dynamiques*, pages 69–92. Crochard, Paris. Despite this date, the volume of the Recueil was only published in 1823.

[Oersted, 1814] Oersted, H. C. (1814). Des forces électriques considerérées comme des forces chimiques. *Journal de Physique, de Chimie, d'Histoire Naturelle et des Arts*, 78:338–374.

[Oersted, 1820] Oersted, H. C. (1820). Experiments on the effect of a current of electricity on the magnetic needle. *Annals of Philosophy*, 16:273–277. Translated from a printed account drawn up in Latin by the author and transmitted by him to the Editor of the *Annals of Philosophy.*

[Oersted, 1965] Oersted, H. C. (1965). Experiments on the effect of a current of electricity on the magnetic needle. In R. A. R. Tricker, *Early Electrodynamics — The First Law of Circulation*, pages 113–117, New York. Pergamon. Translation from Thomson's *Annals of Philosophy*, October 1820. Translated from a printed account drawn up in Latin by the author and transmitted by him to the Editor of the *Annals of Philosophy.*

[Onnes, 1967] Onnes, H. K. (1967). *Nobel Lectures, Physics 1901-1921*, pages 306–336. Elsevier, Amsterdam. Nobel Lecture, December 11, 1913, Investigations into the properties of substances at low temperatures, which have led, amongst other things, to the preparation of liquid helium.

[O'Rahilly, 1965] O'Rahilly, A. (1965). *Electromagnetic Theory — A Critical Examination of Fundamentals.* Dover, New York.

[Ørsted, 1986] Ørsted, H. C. (1986). Experiências sobre o efeito do conflito elétrico sobre a agulha magnética. *Cadernos de História e Filosofia da Ciência*, 10:115–122. Translated by R. d. A. Martins.

[Ørsted, 1998a] Ørsted, H. C. (1998a). Experiments on the effect of a current of electricity on the magnetic needle. In Jelved, K., Jackson, A. D., and Knudsen, O., editors, *Selected Scientific Works of Hans Christian Ørsted*, pages 417–420. Princeton University Press, Princeton. Reprinted from Thomson's *Annals of Philosophy*, Vol. 16, pp. 273-276 (1820).

[Ørsted, 1998b] Ørsted, H. C. (1998b). Experiments on the effect of the electric conflict on the magnetic needle. In Jelved, K., Jackson, A. D., and Knudsen, O., editors, *Selected Scientific Works of Hans Christian Ørsted*, pages 413–416. Princeton University Press, Princeton. Article originally written in Latin in 1820.

[Ørsted, 1998c] Ørsted, H. C. (1998c). New electro-magnetic experiments. In Jelved, K., Jackson, A. D., and Knudsen, O., editors, *Selected Scientific Works of Hans Christian Ørsted*, pages 421–424. Princeton University Press, Princeton. Paper originally published in German in 1820.

[Ørsted, 1998d] Ørsted, H. C. (1998d). A new electromagnetic experiment disproving Ampère's theory. In Jelved, K., Jackson, A. D., and Knudsen, O., editors, *Selected Scientific Works of Hans Christian Ørsted*, pages 539–541. Princeton University Press, Princeton. Reprinted from *Videnskabernes Selskabs Oversigter*, 1829-1830, pp. 22-26. Reports of the meetings held on April 16, 1830 and April 30, 1830, respectively. Originally published in Danish.

[Ørsted, 1998e] Ørsted, H. C. (1998e). Note on the discovery of electromagnetism. In Jelved, K., Jackson, A. D., and Knudsen, O., editors, *Selected Scientific Works of Hans Christian Ørsted*, pages 425–429. Princeton University Press, Princeton. Reprinted from *Videnskabernes Selskabs Oversigter*, 1820-1821, pp. 12-21. Based primarily on the reports of the meetings held on January 5 and April 6, 1821. Originally published in Danish.

[Ørsted, 1998f] Ørsted, H. C. (1998f). Observations on electro-magnetism. In Jelved, K., Jackson, A. D., and Knudsen, O., editors, *Selected Scientific Works of Hans Christian Ørsted*, pages 430–445. Princeton University Press, Princeton. Reprinted from Thomson's *Annals of Philosophy*, Vol. 2, pp. 321-337 (1821).

[Ørsted, 1998g] Ørsted, H. C. (1998g). Thermo-electricity. In Jelved, K., Jackson, A. D., and Knudsen, O., editors, *Selected Scientific Works of Hans Christian Ørsted*, pages 542–580. Princeton University Press, Princeton. Reprinted from *The Edinburgh Encyclopaedia*, conducted by David Brewster, LL.D., Vol. 18, pp. 573-589. Edinburgh 1830.

[Pais, 1982] Pais, A. (1982). *'Subtle is the Lord...' — The Science and the Life of Albert Einstein.* Oxford University Press, Oxford.

[Pais, 1986] Pais, A. (1986). *Inward Bound — Of Matter and Forces in the Physical World.* Clarendon Press, Oxford.

[Phipps Jr., 1990a] Phipps Jr., T. E. (1990a). Toward modernization of Weber's force law. *Physics Essays*, 3:414–420.

[Phipps Jr., 1990b] Phipps Jr., T. E. (1990b). Weber-type laws of action-at-a-distance in modern physics. *Apeiron*, 8:8–14.

[Phipps Jr., 1992] Phipps Jr., T. E. (1992). Derivation of a modernized Weber force law. *Physics Essays*, 5:425–428.

[Phipps, Jr., 1996] Phipps, Jr., T. E. (1996). Clock rates in a machian universe. *Toth-Maatian Review*, 13:5910–5917.

[Phipps Jr., 2013] Phipps Jr., T. E. (2013). *Old Physics for New: a worldview alternative to Einstein's relativity theory.* Apeiron, Montreal, 2nd edition.

[Pilet, 1981] Pilet, P. E. (1981). Prevost, Jean-Louis. In Gillispie, C. C., editor, *Dictionary of Scientific Biography,* Vol. 11, pages 132–133. Charles Scribner's Sons, New York.

[Poisson, 1822a] Poisson, S. D. (1821-1822a). Mémoire sur la théorie du magnétisme. Lu à l'Académie royale des Sciences le 2 Février 1824. *Mémoires de l'Académie royale des Sciences de l'Institut de France*, 5:247–338. Histoire de l'Académie. Analyse des travaux de l'Académie royale des sciences pendant l'année 1821, partie mathématique; par M. le chevalier Delambre, secrétaire perpétuel. Despite this date, the volume was published only in 1826.

[Poisson, 1822b] Poisson, S. D. (1821-1822b). Second Mémoire sur la théorie du magnétisme. Lu à l'Académie royale des Sciences le 27 Décembre 1824. *Mémoires de l'Académie royale des Sciences de l'Institut de France*, 5:488–533. Histoire de l'Académie. Analyse des travaux de l'Académie royale des sciences pendant l'année 1821, partie mathématique; par M. le chevalier Delambre, secrétaire perpétuel. Despite this date, the volume was published only in 1826.

[Poisson, 1823] Poisson, S. D. (1823). Mémoire sur la théorie du magnétisme en mouvement. *Mémoires de l'Académie Royale des Sciences de l'Institut de France*, 6:441–570. Despite this date, the work was only published in 1827.

[Poisson, 1826] Poisson, S. D. (1826). Mémoire sur la Théorie du Magnétisme en mouvement. *Journal de Chimie et de Physique*, 32:225–240.

[Poisson, 1833] Poisson, S. D. (1833). *Traité de Mécanique*, volume 2. Bachelier, Paris, 2nd edition.

[Poisson, 1842] Poisson, S. D. (1842). *A Treatise of Mechanics*, volume 2. Longman, London. Translated by H. H. Harte.

[Potier, 1884] Potier, A. (1884). *Collection de Mémoires relatifs a la Physique*, volume 1: *Mémoires de Coulomb*. Gauthiers-Villars, Paris.

[Poudensan, 1964] Poudensan, L. (1964). *Ampère, un Savant dans le Texte*. Gauthier-Villars, Paris.

[Purcell, 1980] Purcell, E. M. (1980). *Eletricidade e Magnetismo*. Edgard Blücher, São Paulo. Curso de Física de Berkeley, Volume 2. Translated by W. Wajntal, A. d. Oliveira, E. Cavallari, R. O. Zangari and J. Talpe.

[Ragusa, 1992] Ragusa, S. (1992). Gravitation with a modified Weber force. *Foundations of Physics Letters*, 5:585–589.

[Reich, 1977] Reich, K. (1977). *Carl Friedrich Gauß — 1777/1977*. Heinz Moos Verlag, München.

[Reitz et al., 1982] Reitz, J. R., Milford, F. J., and Christy, R. W. (1982). *Fundamentos da Teoria Eletromagnética*. Editora Campus, Rio de Janeiro, 3rd edition. Translated by R. B. Sander and C. Duarte.

[Reynard, 1870] Reynard, F. (1870). Nouvelle théorie des actions électrodynamiques. *Annales de Chimie et de Physique*, 19:272–328.

[Ribeiro, 2008] Ribeiro, J. E. A. (2008). Sobre a Força de Lorentz, os Conceitos de Campo e a "Essência" do Eletromagnetismo Clássico. Master's thesis, Instituto de Física, Universidade de São Paulo, Brasil.

[Ribeiro et al., 2008] Ribeiro, J. E. A., Vannucci, A., and Assis, A. K. T. (2008). The multiple definitions of 'field' in the context of Electromagnetism. In Batista, M. G. P., editor, *Proceedings of the VI Taller Internacional "ENFIQUI 2008" — La Enseñanza de la Física y la Química*, pages 1–4, Matanzas, Cuba. Universidad Pedagógica "Juan Marinello".

[Robinson, 1962] Robinson, M. (1962). A history of the electric wind. *American Journal of Physics*, 30:366–372.

[Roller and Roller, 1957] Roller, D. and Roller, D. H. D. (1957). The Development of the Concept of Electric Charge. In Conant, J. B., editor, *Harvard Case Studies in Experimental Science*, chapter 8, pages 541–639. Harvard University Press, Cambridge.

[Rosenfeld, 1957] Rosenfeld, L. (1957). The velocity of light and the evolution of electrodynamics. *Il Nuovo Cimento*, Supplement to vol. 4:1630–1669.

[Rosenfeld, 1981] Rosenfeld, L. (1981). Kirchhoff, Gustav Robert. In Gillispie, C. C., editor, *Dictionary of Scientific Biography,* Vol. 7, pages 379–383, New York. Charles Scribner's Sons.

[Rosmorduc, 1977] Rosmorduc, J. (1977). Ampère et l'optique: une intervention dans le débat sur la transversalité de la vibration lumineuse. *Revue d'Histoire des Sciences et Leurs Applications*, 30:159–167.

[Savary, 1822] Savary, F. (1822). Extrait fait par M. Savary du Mémoire qu'il a lu à l'Académie royale des Sciences, le 3 février 1823. In Ampère, A.-M., editor, *Recueil d'Observations Électro-dynamiques*, pages 345–354. Crochard, Paris. Despite this date, the volume of the Recueil was only published in 1823.

[Savary, 1823a] Savary, F. (1823a). Extrait d'un Mémoire lu à l'Académie des Sciences, le 3 février 1823. *Annales de Chimie et de Physique*, 22:91–100.

[Savary, 1823b] Savary, F. (1823b). Mémoire sur l'application du calcul aux phènoménes électro-dynamiques. *Journal de Physique, de Chimie, d'Histoire Naturelle et des Arts*, 96:1–25.

[Savary, 1823c] Savary, F. (1823c). *Mémoire sur l'Application du Calcul aux Phènoménes Électro-dynamiques*. Bachelier, Paris.

[Savary, 1885a] Savary, F. (1885a). Addition au Mémoire précédent. In Joubert, J., editor, *Collection de Mémoires relatifs a la Physique,* Vol. II: *Mémoires sur l'Électrodynamique*, pages 370–375. Gauthier-Villars, Paris. Ce complément du Mémoire de Savary a été présenté à l'Académie le 28 juillet 1823 (*Journal de Physique*, t. XCVI; 1823).

[Savary, 1885b] Savary, F. (1885b). Mémoire sur l'Application du Calcul aux Phénomènes électrodynamiques. In Joubert, J., editor, *Collection de Mémoires relatifs a la Physique,* Vol. II: *Mémoires sur l'Électrodynamique*, pages 338–370. Gauthier-Villars, Paris. Mémoire lu à l'Académie des Sciences, le 3 février 1823. Reprinted from Journal de Physique, de Chimie, d'Histoire Naturelle et des Arts, Vol. 96, pp. 1-25 (1823).

[Savary and Ampère, 1821] Savary, F. and Ampère, A.-M. (1821). Notes relatives au Mémoire de M. Faraday. *Annales de Chimie et de Physique*, 18:370–379.

[Savary and Ampère, 1822] Savary, F. and Ampère, A.-M. (1822). Notes relatives au Mémoire de M. Faraday. In Ampère, A.-M., editor, *Recueil d'Observations Électro-dynamiques*, pages 158–167. Crochard, Paris. Despite this date, the volume of the Recueil was only published in 1823.

[Savary and Ampère, 1885] Savary, F. and Ampère, A.-M. (1885). Notes relatives au Mémoire de M. Faraday. In Joubert, J., editor, *Collection de Mémoires relatifs a la Physique,* Vol. II: *Mémoires sur l'Électrodynamique*, pages 184–191. Gauthier-Villars, Paris.

[Schlichting and Ucke, 2004a] Schlichting, H. J. and Ucke, C. (2004a). Der einfachste Elektromotor der Welt. *Physik in Unserer Zeit*, 35:272–273.

[Schlichting and Ucke, 2004b] Schlichting, H. J. and Ucke, C. (2004b). A fast, high-tech, low cost electric motor construction. This article is a modified version which was published originally in German in the journal: *Physik in unserer Zeit* **35** (2004), 272-273. English translation by J. Williams. Available at: `users.physik.tu-muenchen.de/cucke/english.htm`.

[Schlichting and Ucke, 2004c] Schlichting, H. J. and Ucke, C. (2004c). Un motor eléctrico de construcción sencilla, bajo coste y alta tecnología. Este artículo es una versión modificada del original publicado en alemán en la revista *Physik in unserer Zeit* **35** (2004), 272-273. Available at: `users.physik.tu-muenchen.de/cucke/english.htm`.

[Schreier, 1993] Schreier, W. (1993). Die Stellung der Gebrüder Weber in der Geschichte der Medizin, Naturwissenschaften und Technik. In Berg, G., Eisenberg, W., Hergert, W., and Schreier, W., editors, *II. Weber — Symposium Die Gebrüder Weber — Wegbereiter interdisziplinärer Forschung in Halle und Leipzig am 16. Oktober und 18. November 1993*, pages 1–9. Fachbereich Physik der Martin-Luther-Universität Halle-Wittenberg und Wilhelm-Weber-Gesellschaft e. V.

[Schweigger, 1821] Schweigger, J. S. C. (1821). Sur l'electro-magnétisme. *Bibliotheque Universelle des Sciences, Belles-Lettres, et Arts*, 16:197–200.

[Silva and Krapas, 2007] Silva, M. C. d. and Krapas, S. (2007). Controvérsia ação a distância/ação mediada: abordagens didáticas para o ensino das interações físicas. *Revista Brasileira de Ensino de Física*, 29:471–479.

[Silva and Laburú, 2009] Silva, O. H. M. d. and Laburú, C. E. (2009). Motor elétrico de Faraday: uma montagem para museus e laboratórios didáticos. *Caderno Brasileiro de Ensino de Física*, 26:478–491.

[Snelders, 1975] Snelders, H. A. M. (1975). The reception in the Netherlands of the discoveries of electromagnetism and electrodynamics (1820-1822). *Annals of Science*, 32:39–54.

[Souza Filho et al., 2007] Souza Filho, M. P., Chaib, J. P. M. C., Caluzi, J. J., and Assis, A. K. T. (2007). Demonstração didática da interação entre correntes elétricas. *Revista Brasileira de Ensino de Física*, 29:605–612.

[Steinle, 2000] Steinle, F. (2000). Experiment, Instrument und Begriffsbildung: Ampère, das Galvanometer und der Stromkreis. In Meinel, C., editor, *Instrument - Experiment*, pages 98–108. GNT-Verlag, Berlin/Diepholz.

[Steinle, 2003] Steinle, F. (2003). The practice of studying practice: Analyzing research records of Ampère and Faraday. In Holmes, F. L., Renn, J., and Rheinberger, H.-J., editors, *Reworking the Bench: Laboratory Notebooks in the History of Science*, pages 93–117. Kluwer, Dordrecht.

[Steinle, 2005] Steinle, F. (2005). *Explorative Experimente. Ampère, Faraday und die Ursprünge der Elektrodynamik.* Franz Steiner Verlag, Stuttgart.

[Stewart, 2007] Stewart, S. M. (2007). Some simple demonstration experiments involving homopolar motors. *Revista Brasileira de Ensino de Física*, 29:275–281.

[Sugimoto and Kawada, 2006] Sugimoto, N. and Kawada, H. (2006). The homopolar motor and its evolution. *The Physics Teacher*, 44:313.

[Symon, 1971] Symon, K. R. (1971). *Mechanics.* Addison-Wesley, Reading, third edition.

[Tajmar, 2013] Tajmar, M. (2013). Propellantless propulsion with negative matter generated by electric charges. *AIAA Joint Propulsion Conference*, AIAA-2013-3913.

[Tajmar, 2015] Tajmar, M. (2015). Derivation of the Planck and fine-structure constant from Assis's gravity model. *Journal of Advanced Physics*, 4:219–221.

[Tajmar and Assis, 2015a] Tajmar, M. and Assis, A. K. T. (2015a). Particles with negative mass: production, properties and applications for nuclear fusion and self-acceleration. *Journal of Advanced Physics*, 4:77–82. DOI: dx.doi.org/10.1166/jap.2015.1159.

[Tajmar and Assis, 2015b] Tajmar, M. and Assis, A. K. T. (2015b). Teilchen mit negativer Masse: Produktion, Eigenschaften und Anwendungen für die Kernfusion und die Selbst-Beschleunigung. Translated (2015) from the English by M. Pohl. Original paper: M. Tajmar and A. K. T. Assis, "Particles with negative mass: production, properties and applications for nuclear fusion and self-acceleration," Journal of Advanced Physics, Vol. 4, pp. 77-82 (2015). Available at: www.ifi.unicamp.br/~assis.

[Taton, 1981] Taton, R. (1981). Pouillet, Claude-Servais-Mathias. In Gillispie, C. C., editor, *Dictionary of Scientific Biography,* Vol. 11, pages 110–111. Charles Scribner's Sons, New York.

[Tennenbaum, 1996] Tennenbaum, J. (1996). An introduction to 'The significance of the 1845 Gauss-Weber correspondence'. *21st Century Science & Technology*, 9(3):2–5.

[Tennenbaum, 1997] Tennenbaum, J. (1997). Wilhelm Weber's 'Aphorisms' — On the hypotheses of temporal ordering. *21st Century Science & Technology*, 10(2):50–51.

[Thomson, 1885] Thomson, J. J. (1885). Report on electrical theories. *Report of the British Association for the Advancement of Science (B. A. Report)*, pages 97–155.

[Thomson, 1921] Thomson, J. J. (1921). *Elements of the Mathematical Theory of Electricity and Magnetism.* Cambridge University Press, Cambridge, 5th edition.

[Tombe, 2012a] Tombe, F. D. (2012a). Maxwell's original equations. *The General Science Journal.* Available at: www.gsjournal.net/Science-Journals/Essays/View/3889.

[Tombe, 2012b] Tombe, F. D. (2012b). The significance of Maxwell's equations. *The General Science Journal.* Available at: www.gsjournal.net/Science-Journals/Essays/View/4258.

[Torres-Silva et al., 2015] Torres-Silva, H., López-Bonilla, J., López-Vázquez, R., and Rivera-Rebolledo, J. (2015). Weber's electrodynamics for the hydrogen atom. *Indonesian Journal of Applied Physics*, 5:39–47.

[Tricker, 1962] Tricker, R. A. R. (1962). Ampère as a contemporary physicist. *Contemporary Physics*, 3:453–468.

[Tricker, 1965] Tricker, R. A. R. (1965). *Early Electrodynamics — The First Law of Circulation.* Pergamon, Oxford.

[Varney and Fisher, 1980] Varney, R. N. and Fisher, L. H. (1980). Electromotive force: Volta's forgotten concept. *American Journal of Physics*, 48:405–408.

[Verdet, 1868] Verdet, E. (1868). *Cours de Physique*, Tome Premier. In Fernet, E., editor, *Oeuvres d'É. Verdet,* Tome II. Victor Masson et Fils, Paris.

[Verdet, 1872] Verdet, E. (1872). Conférences de Physique faites a l'École Normale, Première Partie. In Gernez, D., editor, *Oeuvres d'É. Verdet,* Tome IV. G. Masson, Paris.

[Volta, 1800a] Volta, A. (1800a). On the electricity excited by the mere contact of conducting substances of different kinds. *Philosophical Transactions*, 90:403–431. Letter in French from A. Volta to J. Banks dated March 20, 1800. It was read before the Royal Society in June 26, 1800.

[Volta, 1800b] Volta, A. (1800b). On the electricity excited by the mere contact of conducting substances of different kinds. *Philosophical Magazine*, 7:289–311.

[Volta, 1964] Volta, A. (1964). On the electricity excited by the mere contact of conducting substances of different kinds. In B. Dibner, *Alessandro Volta and the Electric Battery,* pages 111–131. Franklin Watts, New York. Translated from the author's paper published in French in the Philosophical Transactions for 1800, Part 2.

[Warkulwiz, 2013] Warkulwiz, V. P. (2013). *Universe without Space and Time.* Albertus Magnus Apostolate for Religion and Science, Bensalem.

[Weber and Weber, 1893] Weber, E. H. and Weber, W. (1893). *Wilhelm Weber's Werke,* E. Riecke (ed.), volume 5, *Wellenlehre auf Experimente gegründet oder über die Wellen tropfbarer Flüssigkeiten mit Anwendung auf die Schall- und Lichtwellen.* Springer, Berlin. Originally published in 1825.

[Weber, 1846] Weber, W. (1846). Elektrodynamische Maassbestimmungen — Über ein allgemeines Grundgesetz der elektrischen Wirkung. *Abhandlungen bei Begründung der Königl. Sächs. Gesellschaft der Wissenschaften am Tage der zweihundertjährigen Geburtstagfeier Leibnizen's herausgegeben von der Fürstl. Jablonowskischen Gesellschaft (Leipzig)*, pages 211–378. Reprinted in Wilhelm Weber's *Werke*, Vol. 3, H. Weber (ed.), (Springer, Berlin, 1893), pp. 25-214.

[Weber, 1848] Weber, W. (1848). Elektrodynamische Maassbestimmungen. *Annalen der Physik und Chemie*, 73:193–240. Reprinted in Wilhelm Weber's *Werke*, Vol. 3, H. Weber (ed.), (Springer, Berlin, 1893), pp. 215-254.

[Weber, 1852a] Weber, W. (1852a). Elektrodynamische Maassbestimmungen insbesondere über Diamagnetismus. *Abhandlungen der Königl. Sächs. Gesellschaft der Wissenschaften, mathematisch-physische Klasse*, 1:485–577. Reprinted in Wilhelm Weber's *Werke*, Vol. 3, H. Weber (ed.), (Springer, Berlin, 1893), pp. 473-554.

[Weber, 1852b] Weber, W. (1852b). Elektrodynamische Maassbestimmungen insbesondere Widerstandsmessungen. *Abhandlungen der Königl. Sächs. Gesellschaft der Wissenschaften, mathematisch-physische Klasse*, 1:199–381. Reprinted in Wilhelm Weber's *Werke*, Vol. 3, H. Weber (ed.), (Springer, Berlin, 1893), pp. 301-471.

[Weber, 1864] Weber, W. (1864). Elektrodynamische Maassbestimmungen insbesondere über elektrische Schwingungen. *Abhandlungen der Königl. Sächs. Geselschaft der Wissenschaften, mathematisch-physische Klasse*, 6:571–716. Reprinted in Wilhelm Weber's *Werke*, Vol. 4, H. Weber (ed.), (Springer, Berlin, 1894), pp. 105-241.

[Weber, 1871] Weber, W. (1871). Elektrodynamische Maassbestimmungen insbesondere über das Princip der Erhaltung der Energie. *Abhandlungen der Königl. Sächs. Gesellschaft der Wissenschaften, mathematisch-physische Klasse (Leipzig)*, 10:1–61. Reprinted in Wilhelm Weber's *Werke*, Vol. 4, H. Weber (ed.), (Springer, Berlin, 1894), pp. 247-299.

[Weber, 1872] Weber, W. (1872). Electrodynamic measurements — Sixth memoir, relating specially to the principle of the conservation of energy. *Philosophical Magazine*, 43:1–20 and 119–149.

[Weber, 1878] Weber, W. (1878). Elektrodynamische Maassbestimmungen insbesondere über die Energie der Wechselwirkung. *Abhandlungen der Königl. Sächs. Gesellschaft der Wissenschaften, mathematisch-physische Klasse, (Leipzig)*, 11:641–696. Reprinted in Wilhelm Weber's *Werke*, Vol. 4, H. Weber (ed.), (Springer, Berlin, 1894), pp. 361-412.

[Weber, 1892a] Weber, W. (1892a). *Wilhelm Weber's Werke,* E. Riecke (ed.), volume 2, *Magnetismus.* Springer, Berlin.

[Weber, 1892b] Weber, W. (1892b). *Wilhelm Weber's Werke,* W. Voigt, (ed.), volume 1, *Akustik, Mechanik, Optik und Wärmelehre.* Springer, Berlin.

[Weber, 1893] Weber, W. (1893). *Wilhelm Weber's Werke,* H. Weber (ed.), volume 3, *Galvanismus und Elektrodynamik,* first part. Springer, Berlin.

[Weber, 1894a] Weber, W. (1894a). Elektrodynamische Maassbestimmungen insbesondere über den Zusammenhang des elektrischen Grundgesetzes mit dem Gravitationsgesetze. In Weber, H., editor, Wilhelm Weber's *Werke,* Vol. 4, pages 479–525, Berlin. Springer.

[Weber, 1894b] Weber, W. (1894b). *Wilhelm Weber's Werke,* H. Weber, (ed.), volume 4, *Galvanismus und Elektrodynamik,* second part. Springer, Berlin.

[Weber, 1966] Weber, W. (1966). On the measurement of electro-dynamic forces. In Taylor, R., editor, *Scientific Memoirs,* Vol. 5, pages 489–529, New York. Johnson Reprint Corporation.

[Weber, 2007] Weber, W. (2007). Determinations of electrodynamic measure: concerning a universal law of electrical action, 21st Century Science & Technology, posted March 2007, translated by S. P. Johnson, edited by L. Hecht and A. K. T. Assis. Available at: `www.21stcenturysciencetech.com/translation.html`.

[Weber, 2008] Weber, W. (2008). Determinations of electrodynamic measure: particularly in respect to the connection of the fundamental laws of electricity with the law of gravitation, 21st Century Science & Technology, posted November 2008, translated by G. Gregory, edited by L. Hecht and A. K. T. Assis. Available at: `www.21stcenturysciencetech.com/translation.html`.

[Weber and Kohlrausch, 2008] Weber, W. and Kohlrausch, R. (2008). Sobre a quantidade de eletricidade que flui através da seção reta do circuito em correntes galvânicas. *Revista Brasileira de História da Ciência,* 1:94–102. Translated by A. K. T. Assis.

[Weber and Weber, 1894] Weber, W. and Weber, E. (1894). *Wilhelm Weber's Werke,* F. Merkel and O. Fischer (editors), volume 6, *Mechanik der menschlichen Gehwerkzeuge. Eine anatomisch-physiologische Untersuchung.* Springer, Berlin. Originally published in 1836.

[Wesley, 1987] Wesley, J. P. (1987). Weber electrodynamics extended to include radiation. *Speculations in Science and Technology,* 10:47–61.

[Wesley, 1990a] Wesley, J. P. (1990a). Evidence for Weber-Wesley electrodynamics. In Bartocci, U. and Wesley, J. P., editors, *Proceedings of the Conference on Foundations of Mathematics and Physics,* pages 289–343, Blumberg. Benjamin Wesley Publisher.

[Wesley, 1990b] Wesley, J. P. (1990b). Weber electrodynamics, Part I. General theory, steady current effects. *Foundations of Physics Letters,* 3:443–469.

[Wesley, 1990c] Wesley, J. P. (1990c). Weber electrodynamics, Part II. Unipolar induction, Z-antenna. *Foundations of Physics Letters,* 3:471–490.

[Wesley, 1990d] Wesley, J. P. (1990d). Weber electrodynamics, Part III. Mechanics, gravitation. *Foundations of Physics Letters,* 3:581–605.

[Wesley, 1991] Wesley, J. P. (1991). *Selected Topics in Advanced Fundamental Physics.* Benjamin Wesley Publisher, Blumberg.

[Wesley, 2002] Wesley, J. P. (2002). *Selected Topics in Scientific Physics.* Benjamin Wesley Publisher, Blumberg.

[Whipple, 1934] Whipple, R. S. (1934). The evolution of the galvanometer. *Journal of Scientific Instruments,* 11:37–43.

[Whittaker, 1973] Whittaker, E. T. (1973). *A History of the Theories of Aether and Electricity,* volume 1: *The Classical Theories.* Humanities Press, New York.

[Wiederkehr, 1960] Wiederkehr, K. H. (1960). Wilhelm Webers Stellung in der Entwicklung der Elektrizitätslehre. Dissertation, Hamburg.

[Wiederkehr, 1967] Wiederkehr, K. H. (1967). *Wilhelm Eduard Weber — Erforscher der Wellenbewegung und der Elektrizität (1804-1891)*, volume 32 of *Grosse Naturforscher*, H. Degen (ed.). Wissenschaftliche Verlagsgesellschaft, Stuttgart.

[Wiederkehr, 1993a] Wiederkehr, K. H. (1993a). Das Experiment von Wilhelm Weber und Rudolf Kohlrausch 1855 und Maxwells elektromagnetische Lichttheorie. In Schröder, W., editor, *The Earth and the Universe (A Festschrift in honour of Hans-Jürgen Treder)*, pages 452–463. Interdivisional Commission on History of the International Association of Geomagnetism and Aeronomy, Bremen-Rönnebeck.

[Wiederkehr, 1993b] Wiederkehr, K. H. (1993b). W. Weber und die Entwicklung der Elektrodynamik. In Berg, G., Eisenberg, W., Hergert, W., and Schreier, W., editors, *II. Weber — Symposium Die Gebrüder Weber — Wegbereiter interdisziplinärer Forshcung in Halle und Leipzig am 16. Oktober und 18. November 1993*, pages 39–54. Fachbereich Physik der Martin-Luther-Universität Halle-Wittenberg und Wilhelm-Weber-Gesellschaft e. V.

[Wiederkehr, 1994] Wiederkehr, K. H. (1994). Wilhelm Weber und Maxwells elektromagnetische Lichttheorie. *Gesnerus, Part. 3/4*, 51:256–267.

[Wiederkehr, 2004] Wiederkehr, K. H. (2004). Ein bisher unveröffentlichter Brief von Rudolf Kohlrausch an Andreas v. Ettingshausen von 1854, das Kohlrausch-Weber-Experiment von 1854/55 und die Lichtgeschwindigkeit in Wilhelm Webers Elektrodynamik. *NTM International Journal of History and Ethics of Natural Sciences, Technology and Medicine*, 12:129–145.

[Williams, 1981] Williams, L. P. (1981). Ampère, André-Marie. In Gillispie, C. C., editor, *Dictionary of Scientific Biography,* Vol. 1, pages 139–147, New York. Charles Scribner's Sons.

[Williams, 1983] Williams, L. P. (1983). What were Ampère's earliest discoveries in electrodynamics? *Isis*, 74:492–508.

[Williams, 1989a] Williams, L. P. (1989a). André-Marie Ampère. *Scientific American*, January:72–79.

[Williams, 1989b] Williams, L. P. (1989b). Faraday and Ampère: a critical dialogue. In Gooding, D. and James, F. A. J. L., editors, *Faraday Rediscovered*, chapter 5, pages 83–104. Macmillan Press, Hants.

[Wilson, 1760] Wilson, B. (1759-1760). Farther experiments in electricity. *Philosophical Transactions*, 51:896–906.

[Wise, 1981] Wise, M. N. (1981). German concepts of force, energy, and the electromagnetic ether: 1845–1880. In Cantor, G. N. and Hodge, M. J. S., editors, *Conceptions of Ether — Studies in the History of Ether Theories 1740–1900*, pages 269–307, Cambridge. Cambridge University Press.

[Wolff and Blondel, 2005] Wolff, B. and Blondel, C. (2005). Video: L'expérience d'Oersted. Available at: `www.ampere.cnrs.fr/parcourspedagogique`.

[Wong, 2009a] Wong, H. K. (2009a). Levitated homopolar motor. *The Physics Teacher*, 47:124.

[Wong, 2009b] Wong, H. K. (2009b). Motional mechanisms of homopolar motors & rollers. *The Physics Teacher*, 47:463–465.

[Woodruff, 1968] Woodruff, A. E. (1968). The contributions of Hermann von Helmholtz to electrodynamics. *Isis*, 59:300–311.

[Woodruff, 1981] Woodruff, A. E. (1981). Weber, Wilhelm Eduard. In Gillispie, C. C., editor, *Dictionary of Scientific Biography,* Vol. 14, pages 203–209, New York. Charles Scribner's Sons.

[Zylbersztajn, 1994] Zylbersztajn, A. (1994). Newton's absolute space, Mach's principle and the possible reality of fictitious forces. *European Journal of Physics*, 15:1–8.

[Zylbersztajn and Assis, 1999] Zylbersztajn, A. and Assis, A. K. T. (1999). Sobre a possível realidade das forças fictícias: uma visão relacional da mecânica. *Acta Scientiarum*, 21:817–822.

Made in the USA
Monee, IL
11 September 2026

62770613R00299